Springer Textbooks in Earth Sciences, Geography and Environment

The Springer Textbooks series publishes a broad portfolio of textbooks on Earth Sciences, Geography and Environmental Science. Springer textbooks provide comprehensive introductions as well as in-depth knowledge for advanced studies. A clear, reader-friendly layout and features such as end-of-chapter summaries, work examples, exercises, and glossaries help the reader to access the subject. Springer textbooks are essential for students, researchers and applied scientists.

More information about this series at http://www.springer.com/series/15201

Edoardo Martinetto · Emanuel Tschopp
Robert A. Gastaldo

Editors

Nature Through Time

Virtual Field Trips Through the Nature of the Past

 Springer

Editors
Edoardo Martinetto
Department of Earth Sciences
University of Turin
Turin, Italy

Emanuel Tschopp
Division of Paleontology
American Museum of Natural History
New York, NY, USA

Robert A. Gastaldo
Department of Geology
Colby College
Waterville, ME, USA

ISSN 2510-1307   ISSN 2510-1315   (electronic)
Springer Textbooks in Earth Sciences, Geography and Environment
ISBN 978-3-030-35060-4   ISBN 978-3-030-35058-1   (eBook)
https://doi.org/10.1007/978-3-030-35058-1

This Springer imprint is published by the registered company Springer Nature Switzerland AG
The registered company address is: Gewerbestrasse 11, 6330 Cham, Switzerland

Foreword

There have been countless books and articles written on the many aspects of paleontology and Earth's history—all with attempts to document and explain the changing biota and geography of our planet. However, we tend to write with the terms/jargon familiar to our specific fields, typically with a rather "dry" and impersonal "scientific" style targeting an audience of our peers. These colleagues already know the topic is interesting and usually do not need much convincing. Making these subjects appealing and interesting to more diverse audiences can be a challenge.

There are several paleobiological themed books that have been written by one or more authors in an engaging style targeting broader and sometimes younger audiences. The books of Stephen J. Gould and Kirk Johnson come to mind. Less frequently are books written by teams of scientists that have such broad appeal.

This book represents the culmination of a considerable effort by over 100 scientists from around the world to provide a virtual tour of the Earth's past biota and its response to changes in continental configurations and climate. Written like a tour guide to the past, these authors start with relatively recent Earth's history (Quaternary) and continue backward in time, moving from familiar environments to progressively older and more bizarre landscapes and seascapes of the past. Each chapter/unit is written by a different group of scientists whose expertise on these topics is internationally recognized. Edoardo Martinetto, Emanuel Tschopp, and Robert A. Gastaldo did a great job of convincing busy paleontologists to devote quality time to this project and, aided with multiple rounds of editing to provide a cohesive style, aiming to make for fun, informative, and digestible reading for individuals of varied backgrounds. Each unit includes its own summary and conclusions and ends with a set of study questions that are helpful to assess whether the reader learned the main points from the text. Even the colorful chapter titles should draw you in, for example: "The last three millions of unequal spring thaws," "When and why nature gained angiosperms," "Postcards from the Mesozoic: Forest landscapes with giant flowering trees, enigmatic seed ferns, and other naked-seed plants," "How to Live with dinosaurs: Ecosystems across the Mesozoic," "The coal farms of the late Paleozoic," and "Back to the beginnings: The Silurian-Devonian as a time of major innovation in plants and their communities."

The editors indicate that this work is written for an international audience of early undergraduate students. Hence, they have targeted readers who have gained some previous exposure to the geological and/or biological sciences. The glossary of technical terminology, and numerous accompanying images, will be useful to those new to the topics discussed in the book. Hopefully those students who are more advanced in their training and experiences will also find this book interesting for catching up on topics outside of their own areas of specialization. Enjoy….

Steven R. Manchester
President of the International Organisation of Palaeobotany
Florida Museum of Natural History
University of Florida
Gainesville, FL, USA

Book Summary

This book, and its extensive supplementary material, may be seen as a travel guide of a personal, emotional visit to understand not only the types of plants and animals that lived on the land and in the oceans but also the reasons why and how these biological systems changed over time. Combining both of these aspects with state-of-the-art scientific instruction is a powerful teaching strategy exploited in this international course. Although much could be written about each selected topic, the main focus of the work is on past terrestrial environments and the various steps that nature, on its own, has taken over its historical walk. However, we did not discard a few excursions into the marine realm. The main features of natural systems and their transformation are illustrated by telling stories of key sites in specific geological context. Therefore, highly specialized details, such as geochemical and paleobiological issues, are kept to a minimum. We have adopted a strategy of providing a larger overview and how these case studies may enlighten us about how our planet operates, both then and now. The stories are told by a collection of leading authorities from all over the world who describe their favorite and insightful localities in a virtual tour, reconstructing deep-time natural systems. Many of the most famous fossil localities are well known. Our aim is to highlight mainly lesser known and less conventional topics, using up-to-date and scientifically accurate information. We include artistic reconstructions of the highest fidelity, which will be attractive for the reader, simulating a walk through nature of deep time, around the world. With this, we would like to thank all the contributors to this book. It was a pleasure to see the various parts coming together. Even more so, it is our pleasure to present this project to our readers, now. We hope you enjoy!

Preface

Humanity has established itself across our planet. We inhabit the high latitude arctic regions to the warm and humid tropics, and live at altitudes ranging from the highest mountain peaks to the coasts. All of this has occurred over a geologically instant in time. Each culture experiences a different view of the plants and animals that surround it in space and time. Most cultures incorporate ideas about the natural world based primarily on the influence nature has had on their development. Historically, many cultures have kept records on what aspects of its biosphere have remained the same and what aspects have changed over time. It is not unusual for today's youth to hear stories from their elders about which plants and animals once were commonly seen and which ones no longer are common or have been lost from the region. The same story has been repeated for millennia, and we are certain that the children of several historical figures were told similar tales. Nevertheless, such cultural memories are records of a brief moment in geologic time influenced by our activities on today's ecosystems. Our communities organized, or modified, landscapes by building cities and associated agricultural, navigational, and geomorphic modifications, as well as a host of other activities in the recent past. Our planet, however, has a much longer biological history. During deep time, many different floras covered the landscape which was inhabited by different faunas over hundreds of millions of years. Our oceans were inhabited by invertebrates and vertebrates, both familiar yet strange. The fossil record preserved in these paloeenvironments shows that our planet has undergone profound biological changes, which have influenced many of the abiotic processes that operate across the planet. These deep-time floras and faunas are highlighted here as part of an innovative learning program designed for an international audience. A combination of experts, mostly from various research fields in the discipline of paleontology, has allowed for the development of a highly varied perspective on deep-time planet Earth.

Organization of the Book

This book is designed to provide the reader with deep-time insights about Earth's biosphere based on a variety of unique fossil localities scattered over most of the Earth's landmasses and over the last approximately 400 million years. These case studies, or postcards of a virtual journey, are intended to compartmentalize several key aspects of the how and why the world around us today differs from the past. These insights have been assembled and integrated based on the expertise of 96 authors, each of whom is a specialist in his or her own right. Throughout the book, and in many chapters, we have taken an unorthodox approach in how we explore our planet's history. Many, if not all, similar book projects have organized their texts as a "march through time" beginning with the earliest evidence of life and ending with the world around us. This "forward" approach is common in the geosciences and evolutionary biology. Our approach begins with the world we know and continues back in time to the same planet on which we exist, but under both analog and non-analog conditions that have been responsible for its modification and change over hundreds of millions of years. This approach allows us to identify puzzling observations in specific ecosystems, such as the highly divergent body plans of today's major vertebrate groups. By visiting various ecosystems across the planet back in time, we will find increasingly more intermediate organisms that can explain observations made earlier. The decision to take this unorthodox approach followed considerable discussion with many colleagues about pedagogy and the anticipated outcomes. We think that such an exploratory approach to paleontology highlights the wonders of scientific research in our field, in general, and will inspire students as it does ourselves. However, the reader cannot expect this book to be a complete guide to the main features of natural systems and their transformation through time.

Box P1: Subdivision of Geological Time

The International Commission on Stratigraphy (ICS) periodically publishes an updated version of the ICS International Chronostratigraphic Chart (http://www.stratigraphy.org/index.php/ics-chart-timescale) whose primary objective is to precisely define global chronostratigraphic units (systems, series, and stages). These are the global standard time units (periods, epochs, and ages) of the geologic time scale, fundamental for expressing the history of the Earth. People who are not familiar with geology often ask why so many odd names are used to define time units; would a simple number, in millions of years, not be better? The answer is no, because it is difficult to precisely date many stratigraphic sections, and a global reference of dates is fixed in a single stratigraphic section (the Global Standard Stratotype Point and Section) to which all the other sections have to be correlated by means of paleontological content, magnetic stratigraphy, radiometric dating*, etc. Some time units are formally subdivided into an Early, Middle, and Late part, and the use of capital letters indicates that these are formally recognized in the most updated ICS chart or in particular subcharts; For example, Early Pleistocene (not early!) in the global chronostratigraphical correlation table for the last 2.7 million years (http://www.stratigraphy.org/upload/QuaternaryChart1.JPG).

It is not possible to present a comprehensive and exhaustive view of our biosphere and the changes it has experienced over the past 400 million years of Earth's history in about as many pages. Four hundred million years is a long time, a very long time (Box P1). As such, we have selected topics of broad paleoenvironmental interest and developed 15 educational modules that allow for virtual visits to gain insights into our rich geological heritage. These modules are called units, and are composed of a book chapter, illustrated with a limited number of images, but accompanied by image-rich supplementary material organized as a natural science university lecture at the beginning of the twenty-first century (i.e., slide show). Although we follow a "backward in time" approach with the organization of the chapters, and within most units, there are some specific topics where a traditional forward approach still continues to be necessary to understand the conditions that are responsible for changes in the biosphere. There are definitive and demonstrable reasons why landscapes and the biota they support have changed over time, which, we know, is due, in large part, to changes in landmass distribution and climate. These changes imply that a set of conditions or circumstances existed before a definite event, and those conditions were transformed after its occurrence. Thus, some critical intervals that are the topic of single units are described in a forward approach (e.g., units 3, 4, and 10). Below, we present an outline of each chapter's organization.

The editors wanted to avoid a "factory" approach to each chapter, such that they do not follow a single, constant, and predictable formula in the text. Hence, each group of authors was given the freedom to organize their chapter in the way they felt most appropriate for their material. Generally, the chapters begin with an introduction to the topic that informs the reader

---

* The asterisk designates terms explained in the Glossary.

about the questions to be addressed, illustrated, and answered in the unit. The introduction is where the reader begins a new episode of a virtual journey around the world and back in time. The subsequent sections provide some kind of "field notes," through which we gain an understanding of deep-time ecosystems and the plant-and-animal associations. In several chapters, they are framed as "time-travel postcards" from planet Earth. These postcards illustrate localities and facts through time windows, from younger to older, in an attempt to provide objective data and reduce interpretations in the confines of these case studies. Of course, interpretations of observations and data are the most exciting and engaging topics, and these are neither omitted nor overlooked. Rather, interpretations usually are delayed until the final sections of a chapter where they are synthesized in a bigger picture. In fact, each chapter ends in a series of conclusions, about how or why the deep-time scenarios are of relevance for the evolution of our current world. In those chapters providing postcards, the authors have used them to tell interesting stories about the time interval's paleogeography, environmental change, biogeography, biological evolution, and extinction events that have played one or more roles in reshaping our planet.

Although we attempted to use a language that was easily understandable for an international audience of early undergraduate students, it was not always easy to avoid technical terms. In fact, given the breadth of this book, which covers numerous concepts in geology, evolutionary biology, paleobotany, and paleozoology, it is not surprising that some of these may be common knowledge for students (and researchers!) of some fields, but not of others. The words and expressions are explained in more detail in the glossary and are indicated by an asterisk at first occurrence when they appear in the text. More complex concepts that need an explanation, but would interrupt the flow of the text, are described in boxes throughout the chapters. Three of them (Boxes P1, P2, and P3) are of importance in several units, and are presented here, together with general information on climate and extinction/extirpation* of organisms.

Climate

Among the definitive reasons why landscapes and the biosphere they support have changed over time, climate plays a consistent and central role. In the early part of the twentieth century, Wladimir Köppen (Box P2) understood how temperature and precipitation shape climate of

Box P2: Köppen Climate Types

After half a century of studies, Köppen (1936) classified the Earth's climate into five basic types (Fig. P1): A (tropical pluvial), B (dry), C (temperate pluvial), D (boreal), and E (snow). Each Köppen climate type is further subdivided by seasonal precipitation and temperature and designated by a system of notations. For example, the designation Cfa would indicate a temperate climate with a hot but not dry summer (Fig. P1). Over time, several modifications have been proposed to this classification, but the world map of climates presented by Köppen (1936) is still very similar to those shown by recent publications (e.g., Peel et al. 2007).

any region, and how plant distribution and growth forms reflect these thermal and hydrologic conditions.

Today, there are two variants of this classification scheme: the modified Köppen–Geiger system (e.g., Peel et al. 2007) and the Köppen–Trewartha modification (Trewartha and Horn 1980). The terms used to describe particular climate types are different in the two variants. For example, the first variant recognizes a broad temperate zone, whereas the second distinguishes

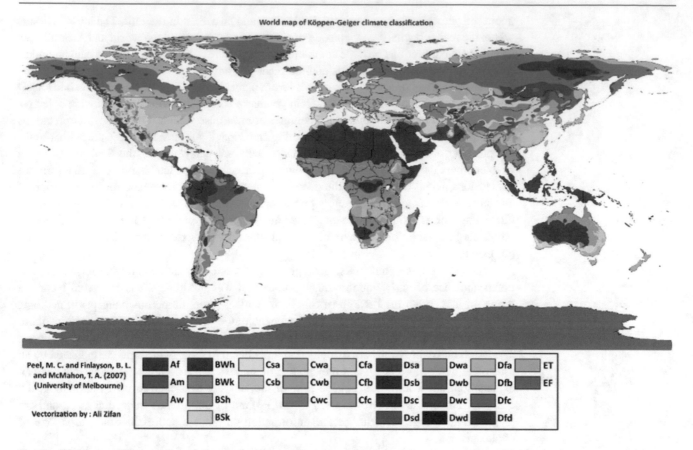

World map of Köppen-Geiger climate classification

Peel, M. C. and Finlayson, B. L.
and McMahon, T. A. (2007)
(University of Melbourne)

Vectorization by : Ali Zifan

Af	BWh	Csa	Cwa	Cfa	Dsa	Dwa	Dfa	ET
Am	BWk	Csb	Cwb	Cfb	Dsb	Dwb	Dfb	EF
Aw	BSh		Cwc	Cfc	Dsc	Dwc	Dfc	
	BSk				Dsd	Dwd	Dfd	

Fig. P1 World map of Köppen–Geiger climate classification. In recent modifications, the names of Köppen's (1936) basic types are often altered as follows: tropical = A in the legend, arid = B, temperate = C, cold or continental or boreal = D, polar = E. From Peel et al. (2007)

a subtropical and narrow temperate zone within the broad temperate zone of the Köppen–Geiger system. Actually, there is no broadly used climate classification scheme that allows us to apply the concept of warm temperate in opposition to subtropical. Subtropical can be opposed to cool temperate, boreal, dry (arid), and tropical. It is somewhat confusing that different authors actually use the definition warm temperate for what is defined a subtropical climate by others. Even more confusingly, the terms temperate and subtropical often are applied to biomes and vegetation types and impart a different meaning with respect to climate classification (see Martinetto et al., Chap. 2).

Extraterrestrial factors are responsible for changes in climate over various times scales of the Phanerozoic and, most likely, influenced climate in the Precambrian too. A number of climate cycles are controlled in response to how our planet orbits the sun. Long-term effects occurred then, and now, on scales of 10s of thousands to 100s of thousands of years, but also on shorter time scales. In combination, three orbital parameters are responsible for the flux

Box P3: Milanković Orbital Parameters

Milanković's model explains how variations in our planet's position and orientation, relative to the sun, alter global climate. One extraterrestrial factor, termed "eccentricity," influences long-term global climate and reflects the overall geometry of the planet's orbit around the sun. As the term eccentricity implies, it is related to how elliptical the planet's path is around our star. The more elliptical the shape of the orbit, the greater the variation in how much sunlight reaches the planet during the orbital year, which, in turn, influences climate extremes. When ellipticity is extreme, the amount of sun's energy reaching the planet also varies most between the nearest and farthest positions, maximizing the magnitude of seasonal climate change. When the planet's orbit is more concentric, or evenly distributed, sunlight reaching the planet is more constant throughout the year. Hence, seasonal changes in climate are less extreme under these conditions. Variation in seasonal climate is also influenced by two other factors.

Two additional extraterrestrial factors that impact Earth's climate are known as precession and variation in axial obliquity; both are more rapid processes than eccentricity. We know that our planet's spin is wobbly rather than perfectly rotating along a central axis fixed at two points. This wobble is known as precession. One wobble cycle takes somewhere between 19,000 and 24,000 years to complete, and is influenced by the sun and moon's gravitational effect on tidal forces. The combination of the precession with whether Earth is nearer or farther from the sun can affect the severity of the seasons in one hemisphere compared to the other. Over the course of a few 10s of thousands of years, the tilt also changes moving toward or away from the sun, a process termed variation in axial obliquity. The planet's axial tilt varies from 21.5° to 24.5° over time frames of approximately 41,000 years. As one pole becomes oriented more towards the sun, that hemisphere receives more sunlight and heat (irradiation) than its counterpart. A greater tilt results in greater seasonal variation in the amount of sunlight received by that hemisphere, which results in more extreme variation in seasonal temperatures, whereas the opposite pole will experience less extreme variation in seasonal temperatures. At or near the equator, overall irradiation is lessened. As the planet's axial tilt approaches 24.5° this orientation promotes more severe seasonality (colder winters and warmer summers), whereas climate becomes less seasonal as our planet's tilt approaches 21.5° (warmer winters and cooler summers). In combination, in- and out-of-phase Milanković orbital factors influence short (10,000 year) to long (100,000–400,000 year) climate cycles on Earth [U1205]. These, in turn, promote the formation and advance of ice sheets when the (paleo)geographic position of continents are at high latitudes near the poles (Fig. 12.1).

Recently, in a geologic sense, our planet has experienced oscillations in icehouse-to-hothouse conditions over the past 34 million years, beginning in the Oligocene (Zachos et al. 2001). The advances and retreats in polar glaciers of the late Cenozoic (see Martinetto et al., Chap. 1) are not unlike what the planet experienced during the Late Paleozoic Ice Age (see Gastaldo et al., Chap. 12).

from icehouse∗-to-hothouse∗ climates that were first identified and mathematically described by Milanković (Box P3).

Extinction and Extirpation

Large-scale changes in climate on a global scale, or climate change over smaller areas on a regional scale, alter the physicochemical conditions under which plants and animals can live at various points in time. Animals depend on ecosystems of primary producers. When the conditions under which these primary producers are altered or modified, and "the old" are replaced by "the new," there is a cascading effect felt throughout the ecosystem. Earth has witnessed the devastating effects of changing climate, on very short (Cretaceous-Paleogene), moderate (end-Permian crisis), or longer time scales (end Triassic). Some of these events have resulted in the whole-scale extinction of life forms; others have "pushed" plants and animals to expand, alter, or contract their biogeographic ranges. In many instances, the term "extinction" has been used to refer to both conditions. We follow the conclusions drawn by Smith-Patten et al. (2015)

Fig. P2 Milanković orbital parameters are modulated by three extraterrestrial factors: Eccentricity = elliptical orbit around the sun varies with periods of 100,000 and 430,000 years. Obliquity = planetary tilt varies 22.1°–24.5° over 41,000 years. Precession = Earth's slow wobble changes orientation of rotational axis over 23,000 years. Earth's climate is affected by these factors, individually and in combination, over time. In combination, they impact the location of solar energy around the Earth and seasonality and the contrasts between the seasons. In-phase and out-of-phase periodicities amplify or dampen these effects

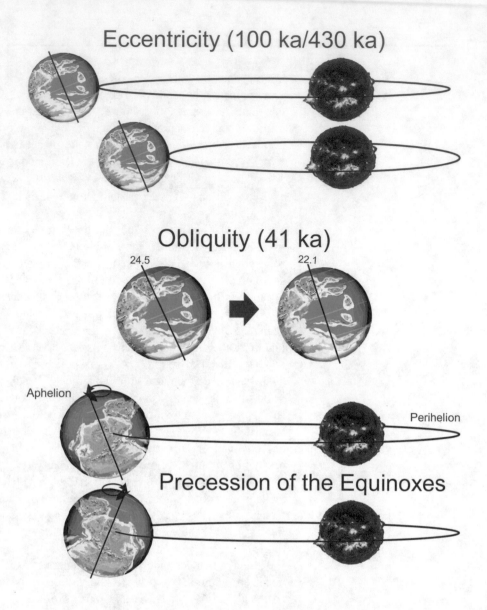

Fig. P3 Curves of Milanković orbital cycles: precession, obliquity, and eccentricity curves. In combination, they impact the location of solar energy around the Earth and seasonality and the contrasts between the seasons. In-phase and out-of-phase periodicities amplify or dampen these effects. These, in turn, promote the formation and advance of ice sheets when the (paleo)geographic position of continents is at high latitudes near the poles

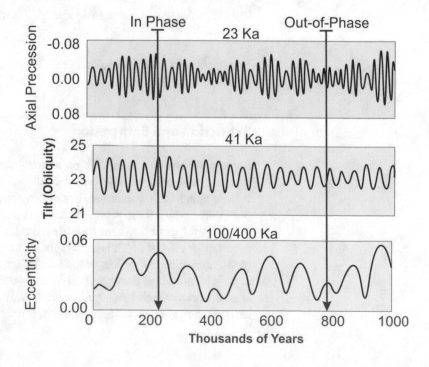

when using the following two terms in the book. Extinction is the irreversible death of a taxon, whereas extirpation is the loss of a taxon in a specific area. Often in the literature, extirpation is incorrectly described as a "local extinction," a "regional extinction," or a "population extinction." If plants are capable of establishing themselves, via spores* or seeds*, in a new territory, or if animals migrate from one region to another, leaving no trace or evidence of their existence in their former area, these organisms have not undergone an extinction. They just have found themselves a more suitable place to live.

Turin, Italy   Edoardo Martinetto
New York, NY, USA   Emanuel Tschopp
Waterville, ME, USA   Robert A. Gastaldo

References

Köppen W (1936) Das geographische System der Klimate. Gerbrüder Bornträger, Berlin

Peel MC, BL Finlayson, McMahon TA (2007) Hydrology and earth system sciences: updated world map of the Köppen-Geiger climate classification (supplement) map in PDF, University of Melbourne. https://commons.wikimedia.org/w/index.php?curid=47086879. Accessed Jan 2019

Smith-Patten BD, Bridge ES, Crawford PHC, Hough DH, Kelly JF, Patten MA (2015). Is extinction forever? Public Underst Sci 24(4):481–495.

Trewartha GT, Horn LH (1980) An introduction to climate. McGraw-Hill, New York

Zachos J, Pagani M, Sloan L, Thomas E, Billups K (2001) Trends, rhythms, and aberrations in global climate 65 Ma to present. Science 292:686–693

Electronic supplementary material
A slide presentation and an explanation of each slide's content is freely available to everyone upon request via email to one of the editors: edoardo.martinetto@unito.it, ragastal@colby.edu, tschopp.e@gmail.com

Contents

The Last Three Millions of Unequal Spring Thaws

Edoardo Martinetto, Adele Bertini, Sudarshan Bhandari,
Angela A. Bruch, Eugenio Cerilli, Marco Cherin,
Judith H. Field, Ivan Gabrielyan, Franco Gianotti,
Andrea K. Kern, Frank Kienast, Emily L. Lindsey,
Arata Momohara, Cesare Ravazzi,
and Elizabeth R. Thomas

Abstract

Evidence from various climate proxies provides us with increasingly reliable proof that only in the past 10 millennia were natural systems more or less as we see them at the present (without considering human impact). Prior to 10,000 years ago, natural systems repeatedly changed under the influence of an unstable climate. This is particularly true over the last one million years. During these times, terrestrial environments were populated by a diver-

sity of large animals that did not survive either the last dramatic climate change or the increasing power of humans. The volume of continental ice covering the land and its impact on the planet's physiography* and vegetation have varied consistently. We can try to imagine extreme conditions: the very cold springtimes of the full glacials*, and the warm springtimes of the rapid deglaciation phases, with enormous volumes of water feeding terrifying rivers. Most of this story is frozen in the ice cover of Greenland and Antarctica, the deep layers of which have been reached by human coring activities only over the past half century. Shorter cores have been drilled in high-altitude ice caps (e.g., in the Andes) that provide insight into other parts of the planet. The interpretation of the signals locked into the

Electronic supplementary material A slide presentation and an explanation of each slide's content is freely available to everyone upon request via email to one of the editors: edoardo.martinetto@unito.it, ragastal@colby.edu, tschopp.c@gmail.com

*The asterisk designates terms explained in the Glossary.

E. Martinetto (✉) · F. Gianotti
Department of Earth Sciences, University of Turin, Turin, Italy
e-mail: edoardo.martinetto@unito.it; franco.gianotti@unito.it

A. Bertini
Department of Earth Sciences, University of Florence, Florence, Italy
e-mail: adele.bertini@unifi.it

S. Bhandari
Paleo Labo Co. Ltd., Toda, Saitama, Japan

A. A. Bruch
ROCEEH Research Centre, Senckenberg Research Institute and Natural History Museum, Frankfurt/M, Germany
e-mail: angela.bruch@senckenberg.de

E. Cerilli
Soprintendenza Speciale Archeologia Belle Arti e Paesaggio di Roma, Rome, Italy

M. Cherin
Department of Physics and Geology, University of Perugia, Perugia, Italy
e-mail: marco.cherin@unipg.it

J. H. Field
School of Biological, Earth and Environmental Sciences, The University of New South Wales, Sydney, NSW, Australia
e-mail: judith.field@unsw.edu.au

I. Gabrielyan
Department of Higher Plant Systematic and Geography, Institute of Botany after A. Takhtajan of NAS RA, Yerevan, Armenia

A. K. Kern
Institute of Geoscience, University of São Paulo, São Paulo, Brazil

F. Kienast
Senckenberg Research Institute and Natural History Museum, Research Station of Quaternary Palaeontology, Weimar, Germany
e-mail: fkienast@senckenberg.de

E. L. Lindsey
La Brea Tar Pits and Museum, Los Angeles, CA, USA
e-mail: elindsey@tarpits.org

A. Momohara
Graduate School of Horticulture, Chiba University, Matsudo, Chiba, Japan
e-mail: arata@faculty.chiba-u.jp

C. Ravazzi
Consiglio Nazionale delle Ricerche - Istituto per la Dinamica dei Processi Ambientali, Laboratorio di Palinologia e Paleoecologia, Milan, Italy
e-mail: cesare.ravazzi@idpa.cnr.it

E. R. Thomas
British Antarctic Survey (BAS), Cambridge, UK
e-mail: lith@bas.ac.uk

© Springer Nature Switzerland AG 2020
E. Martinetto et al. (eds.), *Nature through Time*, Springer Textbooks in Earth Sciences, Geography and Environment,
https://doi.org/10.1007/978-3-030-35058-1_1

ice cores led to the reconstruction of climatic curves covering approximately the past 800 millennia. In addition, long sediment cores have been recovered from thousands of lakes across the globe and yielded data useful to estimate climatic trends based on pollen* records. In the past one to three million years, the continents and oceans were in roughly their present-day locations. Environmental factors, including tectonics (mountain uplift or closure of ocean gateways), interacted with the overall long-term oscillation in atmospheric carbon-dioxide concentration, which, in turn, influenced vegetation cover and ecosystem composition. Well-established glacial-interglacial* cycles impacted biotic dispersal* events at mid-to-high latitudes and determined the geographical restriction and expansion of tropical and subtropical (warm-temperate) biomes around the globe. This book chapter constitutes an imaginary field trip, presenting the reader with exemplary records of environments, plants, large mammals, and hominins impacted by cooling and warming phases, glaciations, changes in rainfall patterns, and sea level culminating in the world of today.

1.1   Introduction

Seasons, as perceived on Earth at high latitudes or at mountain elevations at mid-latitudes, may provide the best insight into nature's changes through time. Every year, spring thaw reconnects the natural elements: the sun rises higher and higher in the sky, the atmosphere gets warmer, and solid water, stored under- and aboveground during winter, starts to melt triggering plant growth and animal activities[1]. However, during a short human lifetime, one commonly sees that every spring is unlike another. Some seasons are warmer, some are drier, some melt an abundance of snow, and others have little to no snow to melt. We can imagine a never-ending sequence of springtimes, more or less repeating the same themes. But, natural records tell us it's not exactly like that! "Normal" spring thaws occur today in pleasant landscapes, with lakes, woodlands, and rivers, nice places enjoying a mild summer. Geological records tell us a different story about the past. At times, these records may show clear signs of the presence of thousands of meters of ice. The records tell of large glaciers that built up, moved on, stopped, retreated, and then began accumulating, again, with the cycle repeated numerous times. Is it hard to image what a terrible spring thaw may have been experienced at the mouth of a large ice-filled valley during a phase of strong warming? Probably, yes.

Of course, there are no spring thaws close to the poles and in the tropics, but even in these areas, some signals of climate change, which dramatically affected the mid-to-high latitudes, have been locked into the geological record (see Chap. 12). In those intertropical areas where ice and frost never

arrived, such as central Africa, central South America, northern Australia, and lowland New Guinea, fossils can tell us stories of drastic biotic changes. Our first imaginary field trip allows us to observe deep-time records of plants and animals (including humans), and the environments in which they lived focused on several sites scattered throughout most of the present continents. We will follow the pathways traced by scientists who were able to detect the effects of changes that occurred in the Earth's natural systems during the last three million years. Selected examples (Fig. 1.1) of the recent past will be examined in areas affected by extensive glaciation, including the European Alps, in regions less affected by ice cover, such as southern Caucasus, the Himalaya foothills and Siberia, and in non-glaciated intertropical (central Africa, etc.) and extratropical areas (California, northern Mediterranean, and Japan). Before starting our survey, we need to become acquainted with at least one indispensable tool: the names given to subdivisions of geological time over the last three million years (Box 1.1).

We will meet all of the names explained in Box 1.1 during our imminent virtual visit to several sites around the world

---

Box 1.1: Subdivisions of the Last Three Million Years—Labels and Names

The reason to extend our journey back over the last three million years is to witness the various climatic phases with different paleotemperatures experienced by our planet. In this time interval, most of the globally recognized climatic phases, either warm (interglacial), cold (glacial), or intermediate (stadial* or interstadial*), are given defining names and/or labels. The analysis of stable-oxygen isotopes extracted from calcite* skeletons of marine planktonic foraminifera, entombed in bottom sediments of the ocean, has permitted the development of a uniform label for all the globally recognized climatic phases. In fact, more than 100 marine isotope stages (MIS) have been numbered, and, by convention, odd numbered isotope stages (e.g., MIS 1) correspond to warm intervals, and even numbers (e.g., MIS 2) to cold intervals (Fig. 1.2). Just a few of the most recent climatic phases have been given proper names (e.g., Eemian, a northwest European regional stage the name of which is based on a particular locality in the Netherlands). In another case, the name of the climate phase is based on plants, precisely *Dryas octopetala*, a creeping member of the rose family growing in cold, treeless habitats. In fact, at the end of the nineteenth century, clay deposits commonly containing *Dryas* fossil leaves were discovered at low altitude in Denmark (where the plant does not grow today) and became known as *Dryas* clays (Birks 2008). Further research demonstrated that the deposits pre-

---

[1]The U-codes are referred to the slides provided as Electronic supplementary material.

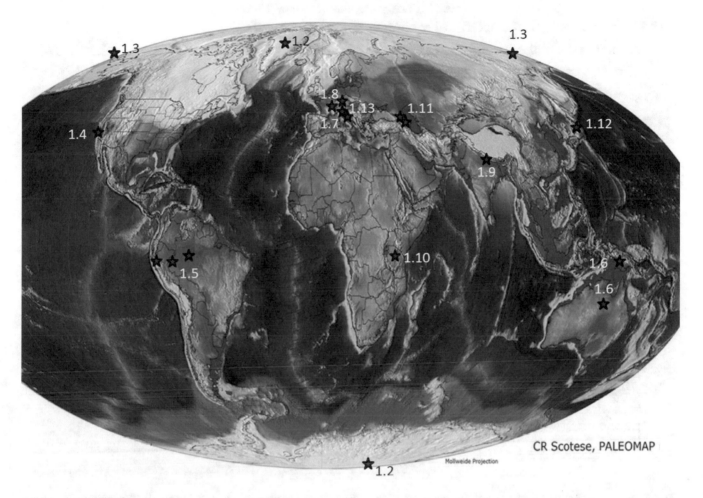

Fig. 1.1 Paleogeographic map of the last glacial maximum (c. 21,000 years before present) from Scotese (2014). The numbers refer to the sub-chapters in which each site is treated (courtesy of the PALEOMAP project)

serving the leaf fossils belonged to two "*Dryas*" cold time intervals before the warm Holocene, still commonly named today as the Younger Dryas (11.7–12.9 ka, i.e., thousands of years ago) and Oldest Dryas (c. 15–17 ka). The Dryas intervals are part of a longer cold phase termed the last glacial period or last ice age (c. 115–11.7 ka). This time witnessed a peak of maximum glacial expansion [U0102] across our planet, which is known as the last glacial maximum (LGM). The globally recognized time span of the LGM lasted from about 21,000 to approximately 27,000 years ago, but actually the LGM has no formal stratigraphic status in geological time. This is because each ice sheet in a different part of the planet reached its maximum extent at different times, which, in turn, is the reason why ice cores from different parts of the globe record a slightly different story. According to some authors (Lambeck et al. 2014), the global glaciation maximum can be identified by sea-level minima and, in this way, the global LGM can be dated to 21,000–29,000 years ago. Let's see now which formal names are given to the geological time of this chapter's focus.

The geological time (Box P1) between today and three million years ago mainly falls in the Quaternary Period (0–2.58 Ma, i.e., millions of years ago). This period is subdivided into two epochs: Holocene (0–11.7 ka) and Pleistocene (from 11.7 ka to 2.58 Ma). The Upper (11.7–126 ka) and Middle (126–781 ka) Pleistocene are indicated as stages in the Global Chronostratigraphic Chart 2018/08 (Box P1) and do not have further subdivisions. In contrast, the Early Pleistocene is further subdivided (Head and Gibbard 2015) into the Calabrian (0.78–1.8 Ma) and Gelasian Stage (1.8–2.58 Ma). The oldest portion of the last three million years falls in the Piacenzian (2.58–3.60 Ma), the age of the Pliocene Epoch that closes the Neogene Period, the last of all the names that we will need to know.

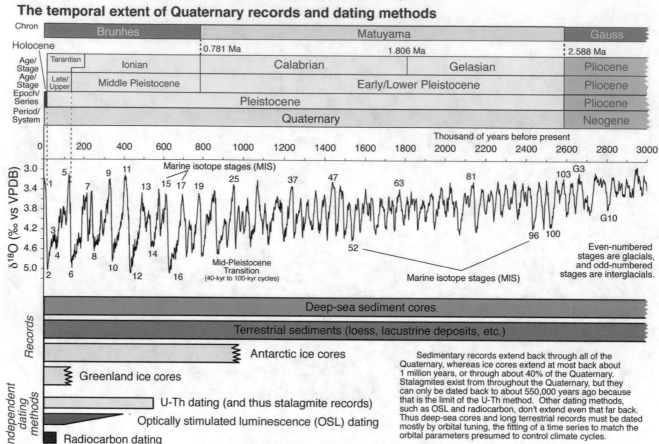

Fig. 1.2 Subdivisions of the Quaternary and temporal extent of Quaternary records and dating methods. Reproduced with permission from Bruce Railsback's Fundamentals of Quaternary Research website, http://railsback.org/FQS/FQS.html

[U0103]. As in many chapters in this book, we will travel backwards in time and the journey will allow us to understand how different deep-time states are from the present, especially when only a small geographical area of Earth's surface is covered in ice. Let's start directly at the poles to look for signals of how natural systems changed from a few years ago to many 1000 years before present.

1.2 Antarctica and Greenland, Ice Cores

Elizabeth R. Thomas

The polar ice caps [U0104] contain a valuable archive of past climate. Ice cores are cylinders of ice, extracted using an ice-core drill of various diameters [U0105]. Manually operated shallow ice-core drills are successful at reaching depths of up to 20 m, whereas cores obtained from deeper depths require electromechanical or thermal drilling technologies (Fig. 1.3). Manual and mechanical drills operate with a rotating drill barrel with cutters at the head. As the barrel rotates, an outer circle is cut away until the inner barrel is filled with ice. Thermal drills operate with a heating ring that melts the edges of the core barrel, filling the central barrel with undisturbed ice. These techniques are applied in both valley and continental glaciers, unlocking the yearly records buried therein.

In central Greenland and Antarctica, where surface melt is negligible, snow is buried in distinct annual layers. The snow that falls contains information about the climate at the time the snow fell, providing a climate history or "time capsule" trapped in the ice layers. The seasonal deposition on the ice cap of sea salt or dust, along with the seasonal cycle in the stable-isotope record (Box 1.2), provides a method for counting annual layers. This approach is much the same as counting tree rings. At high snow accumulation sites, this method can provide accurate dating back several 1000 years. The NGRIP (North Greenland Ice Core Project) ice core from summit Greenland includes some 60,000 years of climate history (Svensson et al. 2008), whereas the WAIS (West Antarctic Ice Sheet) Divide ice core from Antarctica provides climate data from the past 31,000 years (Sigl et al. 2016). A relationship exists between the thickness of any annual layer and its depth in the ice. Snow compaction

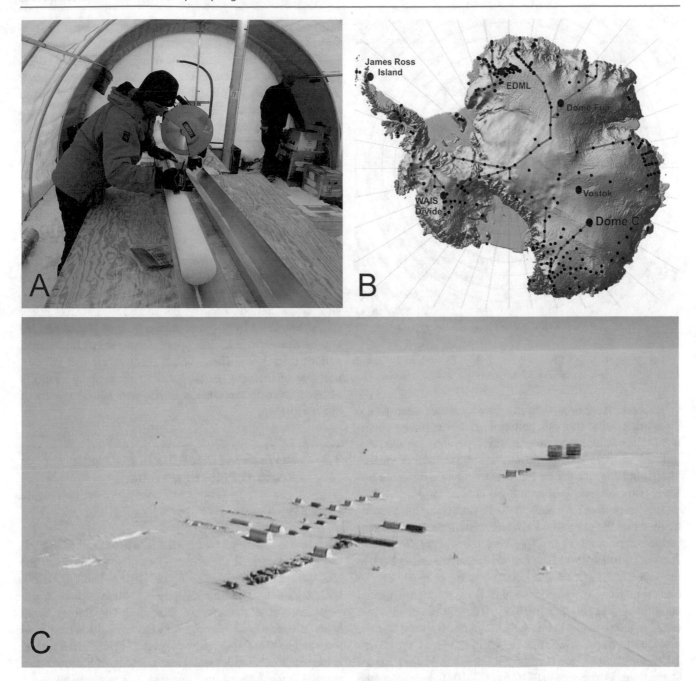

Fig. 1.3 Ice coring in Antarctica. (**a**) Liz Thomas processing an ice core in Antarctica. (**b**) Map of Antarctica with position of all ice cores including Dome C site (75.10′S 123.35′E). (**c**) Image of ice-core drilling site Dome C. Ice coring in Antarctica implies working in some of the most hostile environments on the planet. The monotony of long travels over featureless snow is only interrupted by blizzards (Images courtesy of Liz Thomas (**a**), British Antarctic Survey (**c**))

occurs as a function of depth, producing thinner layers towards the bottom of the ice sheet. As such, annual layer counting is not possible for the deep ice cores or at low accumulation sites, such as the Antarctic Plateau, where the thicknesses of annual layers are less than the sampling resolution. At these sites, numerical flow models are required to produce a depth-age profile. Both the annual layer and modeled age scales benefit from the use of dating unique events locked in the ice, known as tie points. For example, volcanic eruptions provide an excellent method for fixing an ice-core age scale to a historically dated event. The chemical fallout from a large stratospheric eruption can be measured in ice cores from both hemispheres, allowing ice-core age scales to be linked to a standard volcanic chronology. The ash, dust, and sea salt locked in the ice reflect global and hemispherical wind patterns.

The stable-water isotope ratio of snow in the high latitudes provides a method of reconstructing past surface temperature. An atom of any chemical element can exist with different numbers of neutrons, creating distinct isotopes—atoms with different "weights." Oxygen has three stable isotopes, ^{16}O, ^{17}O, and ^{18}O, while hydrogen has two, ^{1}H and ^{2}H. The number on the left shows how many neutrons are in a specific isotope. Thus, water and snow can exist in both a "heavy" ($^{2}H_{2}^{18}O$) and a "light" ($^{1}H_{2}^{16}O$) form in the natural environment. As we will see in other chapters of this book, stable isotopes of various elements and molecules can inform us about different conditions in the environment at the time the element or molecule formed. In the polar regions, colder air results in snow that is isotopically "light," while snow crystallized in a warm atmosphere will contain more "heavy" isotopes of water [U0108]. Thus, the stable-water isotope ratio has been successfully used as a paleo-thermometer.

Ice cores have been used to reconstruct past atmospheric circulation, wind strength, pollution, and even sea ice conditions in the adjacent oceans. However, perhaps the most significant ice-core reconstructions are those of past surface temperature, which can be reconstructed based on stable isotopes (Box 1.2) and greenhouse gas concentration.

Ice cores from Greenland and Antarctica have documented that the climate system has existed in many different stable states over time. The Vostok ice core, drilled in Antarctica in 1985, was the first to retrieve ice dating back to the last glacial period (150,000 years ago), capturing the transition from extreme cold glacial to warm interglacial conditions (Jouzel et al. 1987). These millennial scale changes are related to variations in the Earth's orbit and the amount of solar radiation that the Earth receives (Box P3). In 2004, the Dome C ice core of the European Project for Ice Coring in Antarctica (EPICA) [U0104] extended the Antarctic climate record back more than 740,000 years [U0106] capturing the past eight glacial and interglacial cycles (Fig. 1.4a, EPICA Community Members 2004).

The first deep ice cores in Greenland were drilled in the 1990s, with the parallel GRIP (Greenland Ice Core Project) and GISP-2 (Greenland Ice Sheet Project) cores capturing about 100,000 years of Northern Hemisphere climate variability. The NGRIP [U0107] and NEEM (North Greenland Eemian Ice Drilling) ice cores, completed in 2003 and 2010, respectively, extended the Greenland record back to the previous interglacial about 128,000 years ago. The Greenland ice cores revealed that climate could change very rapidly. Abrupt jumps in temperature, known as Dansgaard-Oeschger events,

occurred throughout the last glacial period (between 115,000 and 14,000 years ago) (Dansgaard et al. 1993), with temperature changes of more than 15 °C (centigrade degrees) occurring in less than a decade (Fig. 1.4b; Steffensen et al. 2008; Thomas et al. 2009). Abrupt climate change is also observed during the Holocene. For example, an abrupt cooling event was captured in the Greenland ice cores 8200 years ago, when temperatures dropped by as much as 6 °C (Thomas et al. 2007). It is possible to understand the relationship between these trends in climate directly from what is trapped in the ice.

Air bubbles trapped in ice cores provide a record of past trace-gas concentrations. Ice cores have been instrumental in demonstrating that the concentration of greenhouse gases, such as carbon dioxide (CO_2) and methane (CH_4), in our atmosphere today is outside of the range of natural variability over the recent past. The Vostok and Dome C records were the first to reveal the natural variability in these gases, with concentrations ranging between 150 and 200 ppmv (parts per million by volume) during the cold glacial periods and between 250 and 280 ppmv during warm interglacial periods (Fig. 1.4b). These values demonstrate that the current rate of increase (c. 20 ppmv in the past 10 years) is unprecedented in the context of the past 800,000 years of climate history.

1.3 Permafrost Deposits in the Arctic Region of Northern Siberia

Frank Kienast

We now travel through North Siberia, a broad Eurasian territory to the south of the Arctic Ocean, extending from the Ural Mountains eastwards to the Bering Strait [U0109]. Different tundra* landscapes, passing southwards to taiga* vegetation, now characterize this area. The Bering Strait hosts extensive wetland tundra in the coastal lowlands and more or less closed taiga forests inland. This modern setting is the result of a fundamental restructuring of northern ecosystems at the dawn of the Holocene. It was a time when the decay of ice sheets determined the amount of global (eustatic*) sea-level rise across the northern Siberian shelves. Concurrently, increasing precipitation and thawing of ice-rich permafrost (thermokarst processes) resulted in a rapid loss of xeric* grasslands, which coincided with the demise of the large animals that lived here during the Pleistocene (see Chap. 3). This radical restructuring seems to be unique in the Quaternary and differs from the situation at the beginning of the last interglacial (Eemian).

Quaternary glaciation in Siberia was limited to mountains and relatively small ice sheets in northwestern Siberia (Astakhov et al. 2016), centered on the Kara Shelf, and reached a maximum geographical extent during the Middle

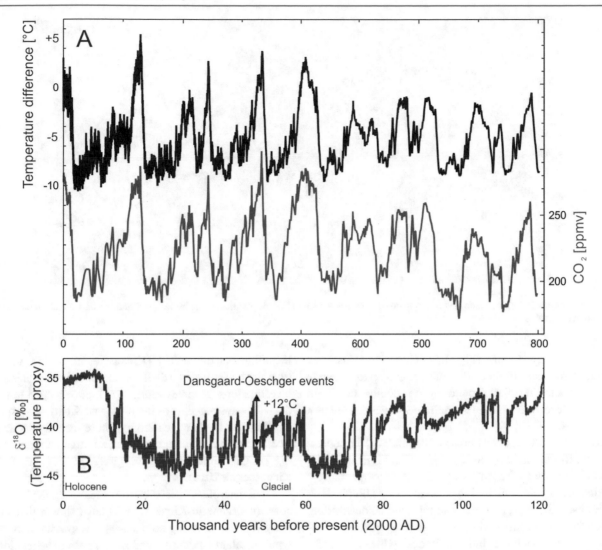

Fig. 1.4 Climate changes detected in ice cores. (**a**) Dome C temperature change from present day (derived from stable-water isotopes; EPICA community members 2004) and concentration of carbon dioxide (ppmv) from a composite of ice-core records (Buizert et al. 2015) spanning the past c. 800,000 years. (**b**) Stable-water isotope record from NGRIP showing the abrupt climate change during the last glacial and Holocene (Svensson et al. 2008)

Pleistocene (Möller et al. 2015). At this time, even the central Arctic Ocean was covered by an ice shelf up to 1000 m thick (Jakobsson et al. 2016). Eastern Siberia, including its vast Arctic shelves, was not covered by large Quaternary ice sheets due to limited precipitation. The landscape was continuously populated by plants and animals adapted to cold-and-dry conditions (Kienast 2013). Whereas the formation of immense ice sheets in the higher latitudes of North America, Greenland, and Europe regionally extirpated terrestrial life southwards to about 50° N, most of Siberia was affected rather indirectly by the ice ages. Tremendous volumes of water were locked away in the northern ice sheets resulting in a eustatic sea-level fall of up to 120 m during the LGM. This, in turn, resulted in the emergence of the shallow shelf areas bordering northeast Siberia and in the formation of an extensive landmass, Beringia. The landmass eventually connected Eurasia and North America. This connection provided a highway for animals, including humans, to extend their ranges following the expansion of plant communities. The drop in temperature triggered the formation of continuous permafrost, several 100 m thick in parts of Beringia, evidenced by the presence of fine-grained deposits with characteristic ice wedges (the Yedoma ice complex; Schirrmeister et al. 2013). The coastal plains expanded and shifted up to 800 km in a seaward direction, in response to the drop in sea level, which affected regional climate. In eastern Siberia, colder winters occurred in response to an increase in the seasonal temperature gradient across the land surface. Due to thermal contraction of the frozen ground during winter, frost cracks in characteristic polygonal patterns formed. These were filled [U0110], again, in springtime, when melt water descended into the cracks, immediately

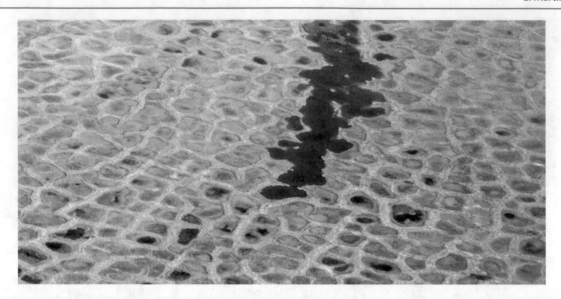

Fig. 1.5 Ice-wedge polygons in the east Siberian coastal lowlands. Due to low evaporation/transpiration, wetland tundra vegetation prevails today (photo courtesy of A Schneider)

freezing to form ice veins (Fig. 1.5). These ice veins grew eventually, over tens of thousands of years, to up to 6 m wide wedges penetrating the ground up to depths of 50 m (Fig. 1.6)! Records locked away in these wedges can tell us a lot of stories.

Sediments and organic remains accumulated in depressions that formed in ice-wedge polygons and remained preserved in a quasi-frozen state. Syngenetic permafrost deposits, accessible at natural exposures such as bluffs along riverbanks, sea coasts, or at thaw slumps, are an outstanding archive for the reconstruction of Quaternary paleoenvironments in non-glaciated high latitudes [U0111–U0112]. Biological remains range from tiny seeds [U0113] (Ashastina et al. 2018) to complete carcasses of such iconic megaherbivores as the woolly mammoth [U0114] (Fisher et al. 2012), woolly rhinoceros (Boeskorov et al. 2011), steppe bison (Boeskorov et al. 2016), and wild horse (Boeskorov et al. 2018). Consistent remains of all of these organisms, including organic parts, are conserved in excellent condition (Fig. 1.7). The animals were primarily grazers and, as such, they were intrinsically tied to grassland vegetation—the mammoth steppe [U0115] (Guthrie 1990).

The character of Beringian cold-stage vegetation was initially contentious because pollen-and-spore (palynological) data recovered from lake deposits reflected a polar desert or Arctic fell-field vegetation, neither of which are able to support large populations of herbivores (Cwynar and Ritchie 1980). But, pollen data are of limited value in Arctic settings owing to low taxonomic resolution and because these ecosystems do not have any modern analogs (Birks and Birks 2000). Only in the past 20 years of Russian and German collaborative research, systematic paleontological studies, including paleobotanical (Kienast et al. 2005), vertebrate (mammalian), insect (Sher et al. 2005), and microfossil investigations (Andreev et al. 2011), have helped to unravel the paleoenvironments of the northeast Siberian coastal lowlands during the late Quaternary. More recently, locations in interior Yakutia have been studied, adding to our understanding of Quaternary paleoecology (Ashastina et al. 2018; Kienast et al. 2018).

Plant macrofossil records from sites along the Laptev Sea coast (overview in Kienast 2013) have shown that the primary vegetation during cold stages was grassland of various types, similar to modern meadow steppes* (a vegetation type classified as *Festucetalia lenensis*), tundra steppes* (classified as *Carici rupestris-Kobresietea*), and productive meadows* (Reinecke et al. 2017). This set of xeric plant communities, supplemented by aquatic and semi-emergent pioneer vegetation, did not fundamentally change through any Pleistocene cold stage. It appears that this vegetation remained stable during interstadials, where more stable climatic conditions are recorded in the soil development (pedogenesis*; Zanina et al. 2011). In inland locations away from the coast, steppe vegetation was dominant during cold intervals and cold-adapted tundra-steppe species were either scarce or absent in these areas, indicating a latitudinal temperature gradient (Ashastina et al. 2018; Kienast et al. 2018). Woody plants are scanty in all records, everywhere, but it is likely that a few trees maintained an isolated and restricted distribution. This condition is based on the presence of scattered larch-and-birch remains in cold-stage deposits of Batagay (Ashastina et al. 2018) and Central Yakutia (Kienast et al. 2018). Another clue is provided by the rapid expansion of trees, along with forest-dwelling invertebrates, at the

Fig. 1.6 Ice-complex deposits exposed along the coast of Oyogos Yar at the Dmitri Laptev Strait (**a**) and at the worldwide largest thaw slump near Batagay in the Yana Highlands, inland Yakutia, Russia (**b**). The headwall is c. 70 m high (trees on upper exposure are c. 7–8 m tall). The upper c. 40 m of exposure consists of the Yedoma ice complex (photo F Kienast)

beginning of the Holocene in NE Siberia. It is fortunate that the Siberian deposits allow us to extend our observations on the distribution of trees back to the last interglacial (c. 120 ka).

The last interglacial is represented in organic-rich deposits, in places accessible at the base of the ice complex in eastern Siberia (Fig. 1.7a). Past biological communities are well documented by a rich fossil record of invertebrates and plant macrofossils along the coast of the Dmitry Laptev Strait and in inland Yakutia (Kienast et al. 2008, 2011; Ashastina et al. 2018). The coastal records come from sites that are only about 80 km apart, but the fossils (Fig. 1.7d)

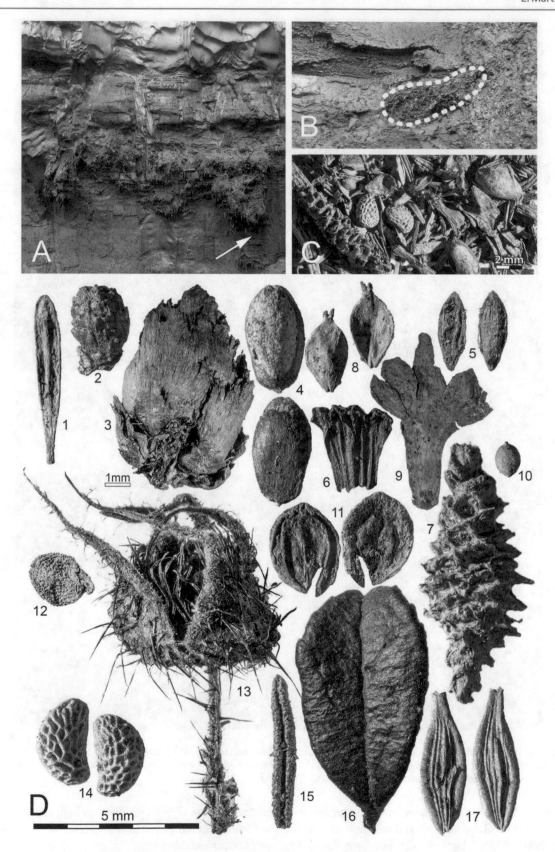

Fig. 1.7 Organic accumulations with macroscopic plant remains preserved in permafrost and accessible at the Batagay mega-thaw slump, in the Yana Highlands (Ashastina et al. 2018). (**a**) Fill of a former depression, below the base of the Yedoma ice complex, deposited during the last interglacial about 125 ka. White arrow points to position of sampling for paleobotanical-paleoecological studies. (**b**) Ground squirrel nest

reflect a very different vegetation. Fossils from the southern site indicate the presence of an open forest tundra with few *Larix gmelinii* (larch) and *Alnus incana* (alder) and abundant alder and birch shrubs (*Alnus alnobetula* subsp. *fruticosa, Betula fruticosa, B. divaricata, B. nana*), interspersed with grassland patches. In contrast, an open variant of subarctic shrub tundra, also with shrubby birch and alder, and grassland patches, but without trees, existed at the northern site. Grassland species in both localities are diagnostic mostly for tundra-steppe environments (*Kobresietea* communities) and, to a lesser extent, for meadow steppes. Such grassland vegetation indicates a drier climate than today, which may be odd due to its current close proximity to the seacoast. However, a more northern position of the coastline during the last interglacial must be assumed because of intense tectonic activity and regional subsidence in the Laptev Shelf since then. The open forest tundra detected at the mainland site proves that the northern tree line during the last interglacial was situated almost 300 km north of its current position. This is confirmed by a recent vegetational reconstruction for interior Yakutia: a wooded, open coniferous forest, similar to modern larch taiga but with an admixture of birch trees. That vegetation is not present at the site today. Similar to the more northern coastal locations, grassland species indicate a rather open vegetation. These interpretations are supported by the presence of light-demanding shrubs, including wild rose and raspberry, co-occurring with shrub alder and shrub birch. A more openly vegetated landscape might be the result of the impact of wildfires, evidenced by abundant charcoal remains, or, alternatively, of large herbivores [U0116]. In fact, the abundance of *Urtica dioica* (stinging nettle), a nitrophilous* pioneer plant, indicates eutrophication* and disturbance of the plant cover by large animals—a clue for the possible survival of megafauna* in this area during the last interglacial.

In summary, the paleoecological data indicate that grassland vegetation and meadow steppes, in particular, were persistent in Yakutia throughout the Late Pleistocene. This vegetation prevailed during the cold stages and was more restricted in its distribution during warm stages. The persistence of steppes and tundra steppes in the north indicates climatic continuity in this region. These conditions may be a consequence of a continental climate with scarce

rainfall, possibly associated with a greater landmass due to a different tectonic setting during warm phases prior to the Holocene. The successive sea-level drop during the LGM, down to at least 120 m lower than today's base level (shoreline*), connected Eurasia to America. Theoretically, a marathon-prone terrestrial herd of animals could have moved along the Pacific coast, without getting its feet wet, from the Philippines to Patagonia, passing through California. Evidence for such a scenario exists in North America.

1.4 Rancho La Brea

Emily Lindsey

When driving through Los Angeles, California [U0117], now the third-largest city in North America, crossing the city just south of Hollywood, the smell of asphalt in the air is not because they have just repaved some street. It comes from several open-asphalt seeps (Fig. 1.8) that have existed there for more than 50,000 years. During all that time, and continuing today, these seeps have acted as episodic traps for a variety of organisms [U0119]. When fossils were first discovered here, at the end of the nineteenth century, Rancho La Brea was a large Mexican land grant. Since then, millions of well-preserved vertebrate [U0120], invertebrate, and plant fossils have been recovered from excavations at the site [U0118]. The quantity, variety, and remarkable preservation of the remains, which comprise more than 600 species of plants and animals (Fig. 1.9a, b), have enabled scientists to study how species and ecosystems responded to late Quaternary climatic changes, human impacts, and extinctions over the past 50 millennia. For instance, there is evidence for size variation in large carnivores across climatic transitions. Dire wolves (O'Keefe et al. 2014) were larger during the LGM and smaller after its end, while saber-toothed cats (Meachen et al. 2014) were smaller during the LGM and larger afterwards. Climate-induced size variation also has been inferred for Rancho La Brea owls (Husband 1924). Evidence from trees demonstrates that carbon starvation, physiological adaptive limitations, and reduced productivity of C3 plants occurred during the LGM (Ward et al.

Fig. 1.7 (continued) in Yedoma ice complex deposits stuffed with plants as winter storage for hibernation, dated to the onset of the LGM about 30 ka. (**c**) Detail of picked plant remains from bulk samples taken from last interglacial deposits, sample position illustrated in A. Needles and seeds of *Larix gmelinii*, pyrenes of *Rubus idaeus*, catkin rachis and bracts of *Alnus alnobetula* subsp. *fruticosa* and of *Betula pendula,* and seeds of *Urtica dioica* are visible. (**d**) Macrofossils of vascular plants characteristic of the last interglacial vegetation at Batagay. (1) *Larix gmelinii* needle, (2) *Larix gmelinii* fascicle*, (3)

Larix gmelinii cone (for 1–3, note the smaller scale bar), (4) *Larix gmelinii* seed, (5) *Puccinellia* sp. caryopsis, (6) *Alnus alnobetula* subsp. *fruticosa* catkin bract, (7) *Alnus alnobetula* subsp. *fruticosa* catkin rachis, (8) *Betula* sp. nutlet, (9) *Betula pendula* catkin bract, (10) *Urtica dioica* seed, (11) *Corispermum crassifolium* seed, (12) *Stellaria jacutica* seed, (13) *Rubus idaeus* inflorescence (note the smaller scale), (14) *Rubus idaeus* pyrene, (15) *Ledum palustre* leaf, (16) *Vaccinium vitis-idaea* leaf, (17) *Sonchus arvensis* cypsela. Images courtesy of K Ashastina (**b, c**)

Fig. 1.8 Excavations at Rancho La Brea in the nineteenth century

2005; Gerhart et al. 2012). Other studies have highlighted the importance of dietary flexibility as a key to survival in extant large mammals and birds (Fox-Dobbs et al. 2006; DeSantis and Haupt 2014). More recently, microfossil studies (e.g., Holden et al. 2015, 2017) have sought to reconstruct late-Quaternary southern California ecosystems [U0121] to understand fine-scale climatic and ecological changes in this temperate region. Paleoenvironmental studies have also provided very fresh and interesting results further south, in the American tropics [U0122].

1.5 Vegetation Change in South American Tropics

Andrea K. Kern

Exploring the South American tropics today (Fig. 1.10), we find large communities of organisms that contribute to high rates of decay, and, consequently, there are few sites across the vast area in which organic matter accumulates. A fast decay rate is, of course, favorable to the complex

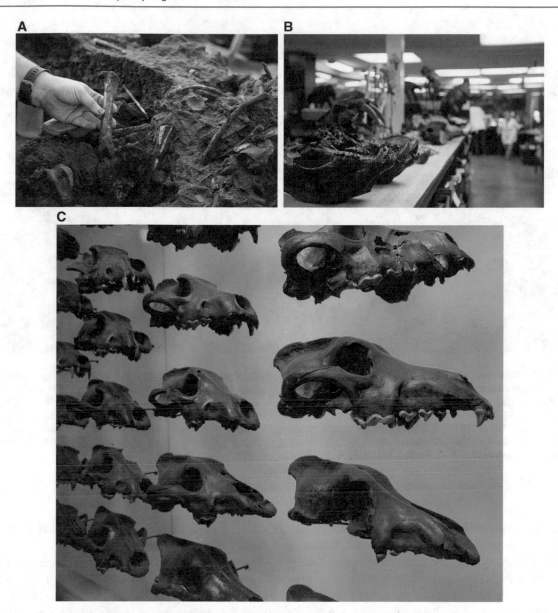

Fig. 1.9 Pleistocene fossils from Rancho La Brea and their storage. (**a**) Mammal bones from a fossil-bearing deposit preserved at the museum. (**b**) Storage room at the La Brea Tar Pits and Museum. (**c**) Skulls of dire wolves

carbon cycle that is necessary to maintain the high diversity and productivity of the tropics. However, it also means that the potential for fossilization is low, very low. Therefore, the number of records covering various glacial-interglacial cycles remains thinly distributed over large distances and is mainly limited to fossil pollen (Fig. 1.11). Nonetheless, the last few decades have seen an increasing number of Holocene and Late Pleistocene climate-and-vegetation reconstructions. Even though few in number, these records contribute to an ongoing conversation on the effect of glacial cooling on the tropical biomes. And, the conversation focuses on questions on its intensity, duration, and geographical extension of dry climates.

Temperature reconstructions from the LGM report an approximately 5 °C cooling in the lowlands (Colinveaux et al. 1996; van der Hammen and Hooghiemstra 2000; Bush et al. 2004) and a remarkable 8–12 °C cooling in the Andes (Thompson et al. 1995). These changes are linked to rainfall patterns that were more regionally complex, as moisture availability is mainly dependent on Atlantic sea surface temperatures (source of Amazonian moisture) and the influence of a shifting intertropical convergence zone∗ (Baker and Fritz 2015 and references therein). Similar to the reconstructions from the ice cores (see Sect. 1.2), based on records in the Andes (Knüsel et al. 2003), it is possible to reconstruct paleoclimate in the tropics using stable-oxygen isotopic ratios locked in the speleothems∗ of caves. These geochemi-

Fig. 1.10 South American vegetation of today. (**a**) Southeastern Amazonia: vegetation around Lake Amendoim in Serra do Sul dos Carajás showing rainforest and altitudinal savanna. (**b**) Andes: altitudinal puna grassland vegetation around Lake Lagunillas (Peru) west of Lake Titicaca at an elevation of 4100 m. (**c**) Andes: dry vegetation of the Altiplano above 4000 m in Peru dominated by high-elevation tree *Polylepis* sp. (Rosaceae). (**d**) Lowland Amazonia: tropical broad-leaved evergreen rainforest vegetation in the Adolpho Ducke Botanical Garden in Manaus (Brazil). (**e**) Lowland Amazonia: riparian forest vegetation in a tributary of southern Rio Tapajós (Brazil) during dry season with low river levels. (**f**) Lowland Amazonia: flooded forest in Rio Negro (Brazil) close to Manaus during wet season and high river level. Images courtesy of JTF Guimaraes (**a**), SF Fritz (**b, c**), AK Kern (**d, e, f**)

cal records come from drill cores and show trends in rainfall variation across the western and eastern Amazon, forming a wet-dry dipole pattern during the last 250,000 years (Cheng et al. 2013). In addition, strong oscillations in rainfall in eastern Amazonia were less intense than rainfall in western Amazonia and, thus, may have been a factor responsible for the higher biodiversity in this region. Climatic variation across tropical South America also was influenced by orbital cycles (Box P3) as well as shorter variations on decadal, centennial, and millennial scale patterns (e.g., Baker and Fritz 2015 and references therein). These long- and short-term climate cycles impacted lake levels (e.g., Bush et al. 2004), expansion and contraction of Andean glaciers (Thompson et al. 1995), and alternations in the river-dominated landscapes of the lowlands. The effect of climate variation on vegetation has been apparent across Amazonia and the Andes and is important to understand the evolution of its currently exceptional biodiversity.

Now let's start a journey to South America to discover how fossil plants reveal the extent to which natural systems of a recent past were different from today (Fig. 1.10). Our first stops are in the forests of western Amazonian lowlands, southeastern Amazonia, and the Andes.

1.5.1　Western Amazonian Lowlands

The longest available records of vegetational dynamics in the Amazonian lowlands are found on the Hill of the Six Lakes in northwestern Brazil (Amazonas state). These records cover the environmental history of the last 170,000 years. Located on a structural high at an elevation of about 300 m, the Hill has a hot and fully humid climate (after Köppen 1918), supporting a tropical evergreen rainforest and forested swamp vegetation dominated by the tree *Mauritia flexuosa* (moriche palm) (Bush et al. 2004). Palynological and geochemical data from lake cores (Lake Pata, Dragão, and Verde) covering the last 170,000 years consistently show the presence of a tropical rainforest in the abundance of pollen from different woody angiosperms, most probably trees (Anacardiaceae [cashew or sumac family], *Alchornea*, *Cassia*, *Copaifera*, *Ilex* [holly], Melastomataceae/Combretaceae, Myrtaceae [myrtle family], and Moraceae/Urticaceae [mulberry/nettle family]). During glacial stages, a strong temperature decrease is indicated by the presence of plant taxa known, today, mainly from higher elevations (Colinveaux et al. 1996; D'Apolito et al. 2013). These higher-altitude plants include the gymnosperm tree

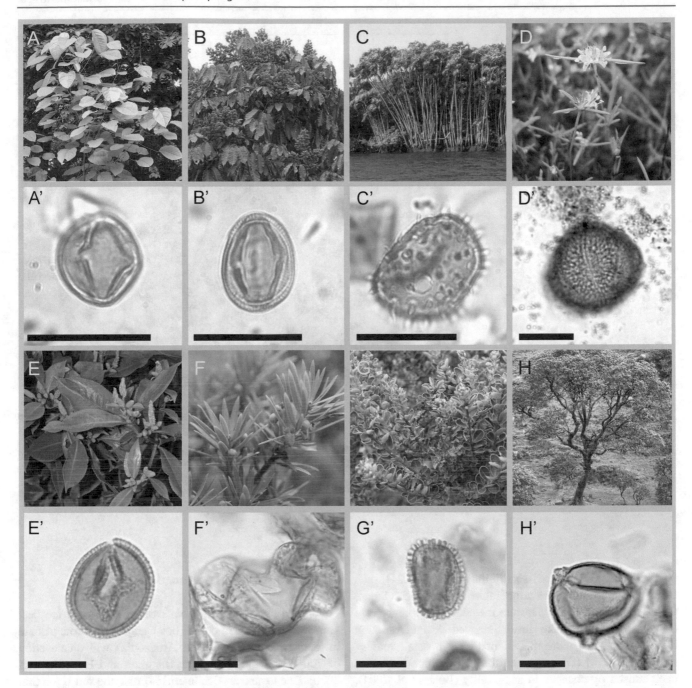

Fig. 1.11 Examples of common Pleistocene pollen (**a'–j'**) in the different vegetation zones across Amazonia and the Andes. All images using light microscopy of pollen recovered from fluvial terraces around Tefé (western Amazon, Brazil) except image (**e'**), which originates from a marine core in proximity to the Brazilian continental margin. Growth forms of parental plants are shown (**a–j**). (**a'**) *Alchornea* sp. (Euphorbiaceae)—25 μm scale bar. (**b'**) *Schefflera* sp. (Araliaceae)—25 μm scale bar. (**c'**) *Mauritia* sp. (palms, Arecaceae)—25 μm scale bar. (**d'**) *Borreria* sp. (Rubiaceae)—20 μm scale bar. (**e'**) *Hedyosmum* sp. (Chloranthaceae)—20 μm scale bar. (**f'**) *Podocarpus* sp. (Podocarpaceae)—20 μm scale bar. (**g'**) *Ilex* sp. (Aquifoliaceae)—20 μm scale bar. (**h'**) *Polylepis* sp. (Rosaceae)—20 μm scale bar. (**a**) *Alchornea* sp. (**b**) *Schefflera* sp. (**c**) *Mauritia* sp. (**d**) *Borreria* sp. (**e**) *Hedyosmum* sp. (**f**) *Podocarpus* sp. (**g**) *Ilex* sp. (**h**) *Polylepis* sp. (Images courtesy of JTF Guimaeres (**a, b, d**), H Hooghiemstra (**c, g, h**), DL Nickrent, M Costea, JF Barcelona, PB Pelser, & K Nixon 2006 onwards PhytoImages http://www.phytoimages.siu.edu (**e**), TK Akabane (**f**))

Podocarpus (yellowwood, plum pine) and several woody angiosperms (*Humiria*, *Weinmannia*, *Hedyosmum*), some of which could be shrubs (Ericaceae, *Myrsine*). Although *Ilex* and *Podocarpus* have species common in the lowland for-ests, their diversity and occurrence predominately in cooler phases indicate an elevational range shift down the Andes into the tropical forest (occasionally representing c. 1000 m difference). Considering modern vegetation zones (Fig. 1.12)

Fig. 1.12 Schematic drawing of the altitudinal belts characterizing climate and vegetation in South America following Stadel (1991) and references therein

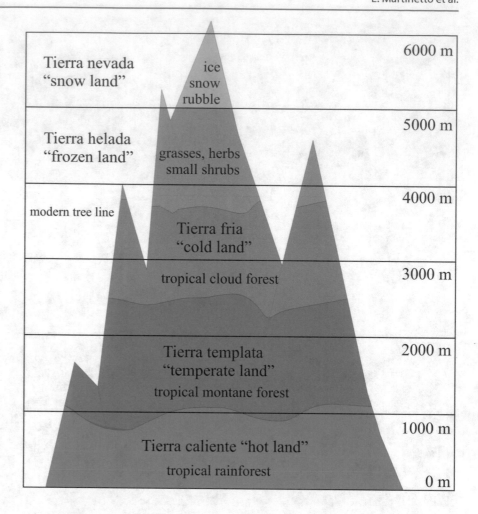

[U0123] and climatic gradients in the Andes [U0124], such shifts would account for a temperature drop of 4–5 °C (e.g., Colinveaux et al. 1996). Despite the constant presence of a tropical rainforest, lake levels showed severe variation through time, and there is evidence for a possible hiatus∗ during the LGM (D'Apolito et al. 2013), which may have been caused by changes in seasonality (Bush et al. 2004). Other western Amazonian lowland records support the appearance of upland taxa (van der Hammen and Hooghiemstra 2000) and additional high-elevation elements, such as *Alnus* (alder) along the Madeira River during the LGM in southwestern Brazil (Cohen et al. 2014) and *Drimys* (mountain pepper) on the slope and lowlands surrounding the Andes (e.g., Bush et al. 2004). Vegetational shifts from tropical rainforest to tropical savanna, however, did occur during the LGM, at least in southwestern Brazil (van der Hammen and Hooghiemstra 2000) and northwestern Bolivia (Burbridge et al. 2004).

1.5.2 Southeastern Amazonia

Pollen and spore records from eastern Amazonia are best known from the Serra do Sul dos Carajás, a narrow plateau in southern Pará (southeastern Amazonia), with an elevation of 700–900 m (e.g., van der Hammen and Hooghiemstra 2000). Due to an iron-rich lateritic∗ crust, the soil is too thin to support plants of the tropical rainforest. The vegetation, here, is dominated by higher-altitude savanna established around lakes and occasional *Mauritia flexuosa* swamps, which grow in a hot tropical climate with a distinct dry season (Aw after Köppen 1918). Forest patches of *Ficus* (fig), Melastomataceae, Anacardiaceae, and *Alchornea* are still part of the present landscape. Yet, the main components of the modern vegetation include Poaceae (grasses), Cyperaceae (sedges), Myrtaceae, and Asteraceae (daisy family), together with *Borreria*, *Byrsonima*, and Mimosoideae. And, tropical vegetation is still present on mountain slopes and in lower elevations (Fabaceae [legumes], *Alchornea*, Euphorbiaceae, Melastomataceae, Anacardiaceae, Moraceae, and Meliaceae). The longest palynological record in this region (Hermanowski et al. 2012) documents a single phase of forested vegetation on the plateau for more than 50,000 years, when lake level rose and fire was infrequent. Cold phases are indicated by the presence of cold-adapted plants, such as *Myrsine, Ilex, Hedyosmum, Styrax* (snowbell), *Weinmannia*, and *Podocarpus*, which also grew in this region more than 50,000 years ago and during the interval between 30,000 and 25,000 years ago. During the LGM, a period of severe dry conditions resulted in the disappearance of tropical vegetation from slope environments. Climate variations continued during the Holocene, before modern-like conditions were established beginning approximately 3400 years ago (Hermanowski et al. 2012).

1.5.3 Andes

Andean paleoenvironmental reconstructions are best known from the Bogotá high plain (Colombia, c. 2550 m) and the area around Lake Titicaca (Peru/Bolivia border, c. 3810 m). Combined, these records allow for an understanding of tropical mountain vegetation in both the Northern and Southern Hemispheres. The Bogotá high plain (or Sabana de Bogotá) is an intramontane basin* in the Eastern Cordillera, which was partially covered by a paleolake. Its sedimentary record includes lake (lacustrine) and river (fluvial) deposits dating back to more than three million years ago. Vegetation changes reconstructed using pollen records [U0126] appear to parallel the global temperature change and, thus, changes in pCO$_2$, whereas they appear less correlated to changes in precipitation patterns (Hanselman et al. 2011). Glacial-interglacial variations triggered shifts in the position of the upper tree line (Hooghiemstra and Ran 1994), which expanded from a minimum elevation of about 2000 m during the LGM to a maximum elevation of approximately 3500 m during MIS 5e. Evolutionary adaption of endemic* taxa (e.g., *Borreria*), as well as the arrival from North America of alder and oak (*Alnus, Quercus*), contributed to biodiversity differences found in each glacial cycle (Hooghiemstra and Ran 1994). Glacial stages commonly caused an expansion of alpine tundra vegetation, known as páramo. Today, this vegetation type is distributed at elevations above 3200–3400 m and comprised of Poaceae and herbaceous plants (e.g., *Valeriana*, Caryophyllaceae, *Geranium, Plantago*). Andean forest was present during warm stages, when *Podocarpus, Weinmannia*, and *Myrica*, together with *Alnus* and *Quercus*, formed forests around the paleolake. At these times, other plants including members of Melastomataceae, Moraceae/Urticaceae, *Acalypha*, and *Alchornea* populated warmer periods (Hooghiemstra and Ran 1994; Bogotá-A et al. 2011).

Such vegetational trends differ from the Lake Titicaca records.

Lake Titicaca is located on the Andean high plateau, the so-called Altiplano*, a region of semi-humid climate. During glacial stages, extremely low pollen concentrations in the lake sediments are interpreted to reflect a barren landscape and poorly developed vegetation. Wind-born (transported) pollen grains [U0125] dominate the assemblage (e.g., *Podocarpus*). In contrast, sediments recording transitions to interglacials preserve pollen of the high-elevation plant *Polylepis*, established in the area before the occurrence of grass pollen, and an increased charcoal concentration indicative of dry vegetation and wildfire. High proportions of Amaranthaceae pollen are used to identify extreme dry events during MIS 9 and MIS 5e (Hanselman et al. 2011). The water level in Lake Titicaca varied in response to changes in rainfall patterns. Lake level dropped almost 200 m, which is comparable to the drying of lakes at the Salar de Uyuni, southwest Bolivia (Baker et al. 2001), and phases of glacier retreat in Peru (Thompson et al. 1995). Hence, our South American trip can be concluded by writing in our notebook: "Quaternary climate change left obvious signs in the fossil record of past vegetation in both mountain and tropical lowland settings." Fossils also show that a diversified animal community populated South American Quaternary paleoenvironments and had to cope with both climate change and the arrival of humans (Barnosky and Lindsey 2010), often without success. In fact, several species disappeared during the Quaternary, especially components of the megafauna, such as the giant ground sloth *Eremotherium* [U0127]. Unfortunately, we don't have the time to study these enigmatic South American animals in detail, but we will focus more on the megafauna during our following stop in Oceania.

1.6 Sahul (Oceania) and Its Extinct Megafauna

Judith H. Field

Traveling back just a few 1000 years, and to where Australia now is positioned [U0131], we would find a landmass quite a bit larger than as we know it today. This Pleistocene landmass comprised Australia, Tasmania, and New Guinea and is called Sahul. The two straits now present between these modern landmasses were exposed following global sea-level drop during cold stages. Climate-related factors can be invoked to help explain one of the most impressive recent changes in Sahul's natural systems: the extinction of a giant and diverse Pleistocene megafauna. Questions over whether climate change, human activities, or a combination of both are responsible for its demise have generated a particularly

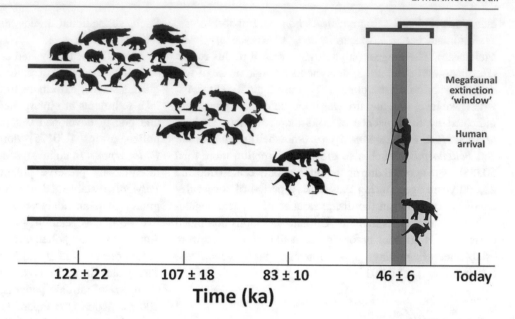

Fig. 1.13 Megafaunal extinction times in Sahul (after Wroe et al. 2013)

robust debate (Wroe et al. 2013). Sahul and its modern-day "remnants" (Australia, New Guinea, and Tasmania) have yielded a patchy fossil record, and many sites are undated. The suite of extinct Pleistocene megafauna from this region consists of around 90 large vertebrate taxa identified from a range of preservational contexts. The largest component of the megafauna was the 2.8 ton browsing marsupial *Diprotodon optatum*, whereas the approximately 100 to 130 kg marsupial lion, *Thylacoleo carnifex,* is considered the world's most specialized mammalian carnivore. *Varanus priscus* is the largest true (monitor) lizard that ever existed, attaining a length of nearly 5 m, similar in size to the modern Komodo dragon from Indonesia. Although some researchers have argued that all megafauna disappeared shortly after human arrival, three sites indicate that a number of species survived well beyond this time (i.e., after 50–55 ka). Seton rock-shelter contains evidence of a single instance of mega-fauna persisting after the LGM. And, two archeological sites–Cuddie Springs in southeastern Australia and Nombe rock-shelter in the New Guinea highlands–have a demon-strated co-occurrence of extinct megafauna and people (Fig. 1.13; Wroe et al. 2013). We provide here short post-cards about the last two sites.

1.6.1 Nombe Rock-Shelter

Nombe rock-shelter (Mountain 1991; Sutton et al. 2010) has yielded a record dated from just over 25,000 years ago to the recent past. The complex stratigraphic relationships in this site have led some researchers to speculate over its integrity (e.g., Johnson 2006). However, re-dating of the sequence has

confirmed the original proposed chronology (Denham and Mountain 2016). Archeological remains are first encoun-tered around 25,500 years ago, and bones from a number of extinct and extant species have been identified in these sedi-ments. It is unclear what the relationship may have been between megafaunal species and the archeological record, as there is no evidence for an association other than a co-occurrence in the same strata. The now extinct species iden-tified include *Protemnodon nombe, Protemnodon tumbuna* (similar to a very large wallaby), *Dendrolagus noibano* (a tree kangaroo), and an unidentified diprotodontid (wombat-like marsupial).

1.6.2 Cuddie Springs

Cuddie Springs [U0132] is an open site in southeastern Australia (Fig. 1.14) where the fossil deposits are preserved in a treeless pan on the floor of an ancient ephemeral* lake (Field and Dodson 1999). Investigations over nearly two decades revealed a stratified deposit of human occupation and fossil megafauna in the lead-up to the LGM, when conditions were more arid, and perhaps cooler, than today. The site has yielded the first, and only, clear evidence of an unequivocal associa-tion of megafauna with humans in Sahul. Two distinct strati-graphic intervals (SU6A and underlying SU6B) have been shown to contain a human-megafauna association (bones and stone artifacts), and these can be correlated to the hydrology of the lake. When people first arrived, about 40,000 years ago, the claypan on the lake floor was a waterhole/swamp (Field et al. 2002). At least seven species of megafauna (Fig. 1.15) were identified in SU6B, where the bone assemblage is mostly

Fig. 1.14 Excavations at Cuddie Springs, Australia

Fig. 1.15 Example of extinct megafauna from Sahul, these particular species are found in the lowest beds (c. 35–40 ka) at Cuddie Springs. (**a**) *Diprotodon optatum*, a giant browsing marsupial estimated to c. 2500 kg in weight. (**b**) *Genyornis newtoni*, a large flightless bird weighing up to 275 kg. (**c**) *Sthenurus* sp. a short-faced giant wallaby. Illustrations by Anne Musser, courtesy of the Australian Museum, Sydney

made up of the giant flightless bird *Genyornis newtoni* (Fillios et al. 2010). Flaked stone artifacts are scattered throughout this level. After the lake entered a period of extended drought (SU6A), there is an increase in stone artifacts, bones of extant species, and charcoal. The resource base broadened to include a range of plant foods as evidenced by the first appearance of grinding stones (Fullagar and Field 1997; Fullagar et al. 2008). Megafauna [U0133] is just one facet of a range of food resources exploited during this period, as environmental conditions deteriorated and grasslands developed in the lead-up to the LGM (30–19 ka; Lambeck and Chappell 2002). By around 28,000 years ago, megafaunal bones are no longer found in the fossil record. Throughout the record at Cuddie Springs, there is no evidence for any specialized targeting of megafauna. It appears that the waterhole provided a resource focus for both humans and animals.

In conclusion, during interglacials, the conditions were warmer and wetter in many parts of Sahul. During the last glacial period (c. 11.7–30 ka), vegetation changes occurred even in the ever-wet tropics (Williams et al. 2009; Morley 2000). But, the effects of climate change were more apparent in semiarid and temperate zones and extreme in the periglacial* Tasmanian environments (Hesse et al. 2004). In Australia, for example, the Late Pleistocene cold intervals promoted aridity, expanding from the center to cover nearly 70% of the continent. In the southern regions, conditions were much colder and vegetation low and sparse. In New Guinea, with a central mountain belt extending the length of the island, glaciers expanded considerably during the LGM (Prentice et al. 2005). Even if the analysis of the impact of such dramatic environmental changes on the megafauna is still open to future research, studies carried out during the last decades have contributed significantly to our knowledge on Sahul's extinctions. We can now shift our attention to a place where the extinct megafauna is rather well known for more than the past two centuries. That location is "old" Europe.

1.7 A Window on the Extinct Interglacial Megafauna of Europe

Eugenio Cerilli

Hundreds of fossil sites in Europe, studied for more than a century, provide extraordinary information on the history of megafaunal extinctions in the last 300,000 years. During the LGM the same large mammal species (e.g., mammoth and woolly rhino) that we met in Siberia (Sect. 1.3) extended their range to southern Europe and into the Iberian Peninsula. This megafauna differed during older interglacial intervals and representative species can be encountered by visiting a single site where there is a mass occurrence of fossil mammal remains in southern Europe [U0135]. The site of La Polledrara di Cecanibbio lies about 22 km northwest of an ancient and historically important town, Rome. Here, impressive remains of large extinct mammals, including the elephant *Palaeoloxodon antiquus*, the rhino *Stephanorhinus* cf. *hemitoechus,* and the European water buffalo *Bubalus murrensis*, are preserved. They are associated with the wild ancestors of domesticated cattle (the European auroch *Bos primigenius*) and horse (*Equus ferus*). Extant species are represented by a deer (*Cervus elaphus*) along with several smaller-sized animals. The site was excavated between 1985 and 2013, exposing an area of about 1200 m² in an ancient riverbed, cut into a bank of compact volcaniclastic deposits (Castorina et al. 2015; Santucci et al. 2016). In 2000, about 900 m² of the site was covered with a structure, and the site opened as a museum to visitors (Fig. 1.16a). The site is characterized by fluvial∗ deposits (Fig. 1.16b) overlain by marsh-and-wetland sediments deposited during a phase of sea-level highstand∗ (Anzidei et al. 2012). The initial sedimentation dates to around 325,000 years ago (Pereira et al. 2017) and, therefore, developed at the beginning of MIS 9, a warm interglacial interval. The fauna consists of over 20,000 skeletal remains and is dominated by large mammals, with *Bos primigenius* (auroch) and *Palaeoloxodon antiquus* (straight-tusk elephant) the most abundant, followed by *Cervus elaphus* (red deer). Few, additional remains demonstrate the presence of *Sus scrofa* (wild boar), *Stephanorhinus* cf. *hemitoechus* (rhinoceros), *Bubalus murrensis* (water buffalo), *Equus ferus* (horse), *Canis lupus* (gray wolf), *Vulpes vulpes* (red fox), *Meles meles* (badger), *Felis silvestris* (wildcat), *Macaca sylvanus* (Barbary macaque), *Lepus* sp. (hare), murids (*Apodemus sylvaticus, Microtus (Iberomys)* cf. *breccensis*), and arvicolids (*Pliomys* cf. *episcopalis, Arvicola* sp.). The herpetofauna (reptiles) and the avifauna (birds) also are well represented, and the abundance of anseriformes (geese, ducks) is in agreement with the interpretation of a wetland area. Such reconstructed conditions can explain the large amount of skeletal elements deposited in this area.

The area is reconstructed as a waterlogged mire (Fig. 1.16c) in which the accumulation of fine-grained clastic∗ material formed mud traps. These swampy conditions resulted in the death of three elephants. And, surrounding the skeleton of one elephant (Fig. 1.16d) are about 600 stone artifacts. Their characteristics and the taphonomic∗ analysis on the elephant bones document human activity related to the exploitation of the carcass by scavenging (Santucci et al. 2016). Indeed, the site acted as a source of food and raw material for the hunter-gatherer bands of hominins (*Homo heidelbergensis*) that frequented the floodplain adjacent to the river. Humans and their activities are documented by hundreds of artifacts including flint pebbles and flakes (denticulates, notches, sidescrapers and end scrapers, multiple-use tools), tools made out of elephant bone, and several bones intentionally fractured for marrow extraction. However, human skeletal elements are only represented by a deciduous second maxillary molar of a juvenile aged between 5 and 10 years old. It would be interesting to know the vegetational and climatic conditions under which this child lived.

Based on the faunal data, the landscape at La Polledrara was likely characterized by a dense tree cover interspersed with open spaces, with both plants and animals living in a moderately humid temperate climate (Palombo et al. 2005). To date, no pollen or macrofloral evidence recovered from the site confirms this hypothesis. Regionally, though, there is a better palynological record. Vegetation in the area around Rome is well documented by a long pollen record dated to the end of MIS 9. This comes from the volcanic lake sediments of Valle di Castiglione (Bertini et al. 2014) and indicates a phase when temperate forests prevailed. Pollen analysis of several sites in central Italy shows that such phases fell within short interglacial intervals, which broke the monotony of the steppes (with *Artemisia* [ragweed] and chenopods) and grasslands dominating the long-lasting glacial intervals. Such shifts in vegetation are the primary evidence for climate change during the Late-Middle Pleistocene of central Italy, where glaciers rarely advanced and formed large end moraine systems∗. The last end moraine systems provide the best evidence of late Quaternary climate cooling at mid-latitudes, as we will see below.

Fig. 1.16 Extinct megafauna of southern Europe. Aspects of the Middle Pleistocene megafaunal site La Polledrara di Cecanibbio, near Rome. (**a**) View of a portion of the museum area with elephant tusk and bone accumulation. (**b**) Ancient riverbed with mammal skeletal remains oriented by flowing water. (**c**) Overview of the museum area with a proposed reconstruction of the waterlogged paleoenvironment and its fauna. (**d**) One of the elephants trapped in the mud. All photos courtesy of Soprintendenza Speciale Archeologia Belle Arti e Paesaggio (SSABAP) of Rome

1.8   The Tale of Mid-Latitude End Moraine and Lake Systems

Franco Gianotti and Cesare Ravazzi

When traveling through mountains around the globe, especially at middle latitudes, hills often can be observed where the valleys open up onto the plains. These hills are termed end moraines if the materials to build them were transported by ancient glaciers down to their terminus located at the lowest elevation. The glacially sculpted morphology of the valleys and the distribution of end moraine systems at the valley outlets are the main evidence for the advance and retreat of the glaciers during the Quaternary in the mid-latitude mountain ranges. End moraine systems form wide and long-term amphitheaters of lateral and frontal deposits, along with hanging plains (kame terraces*) made of glacigenic* sediments (till*, glaciolacustrine*, and glaciofluvial* deposits). These geomorphologic features were built by the Pleistocene piedmont-glacier lobes which flowed out of their trunk valleys onto the plains. Several large end moraine systems are preserved in the eastern side of the Andes, in New Zealand,

and around the Eurasian mountains (Hughes et al. 2016). These systems also are known in the Rocky Mountains (Ehlers et al. 2011), where glaciers could spread out over the piedmont plains.

These hilly moraine complexes are dated from the end of the Calabrian to the final part of the Late Pleistocene (from 21–18 ka to before 781 ka, e.g., Granger and Muzikar 2001). The ages of these deposits indicate that they formed in a discontinuous, but cyclic, way during the acmes of the 100,000-year glacial-interglacial cycles that characterized the third part of the Quaternary. These eccentricity-driven cycles are evidenced from the Early-Middle Pleistocene transition (EMPT) at about 900,000 years ago (see Head and Gibbard 2015). The history of the 41,000-year glacial-interglacial obliquity cycles (about forty minor glaciations from 2.6–0.9 Ma) has been lost in the glacial valleys due to subsequent erosion. But, these can be partially reconstructed from the lacustrine*, floodplain, and alluvial* records preserved downstream or in areas lateral to these catchments. Let's now head to an iconic set of end moraine systems of the mid-latitudes, surrounding the European Alps [U0136], to see which signals of Quaternary climate changes are locked in these archives.

1.8.1 The Puzzle of Glacial-Forced Diversity of Terrestrial Ecosystems in the Alpine Late Cenozoic

Glaciation deeply affected the recent geological and paleo-biological history of the Alps [U0137], inspiring mountaineering students at the onset of the nineteenth century. Indeed, detailed observations on glacial deposits and on the plants and animals preserved in these sedimentary successions resulted in the classical theory of multifold glaciation established by Penck et al. (1894). Their monumental treatise not only examined the effects of multiple glaciations on the Alpine relief but also recognized the value that the floral and faunal assemblages had in characterizing phases of substantial glacier retreat from the Alpine forelands and the valley floors. They coined the term interglacial and promoted the concept of using biological proxies to reconstruct changes in climate over time. This idea must be recognized for its significance, because now climate could be reconstructed without direct evidence of glacier retreat or contraction. Once the complexity of global climate history was clarified using late Cenozoic marine-and-polar records, the Alpine record became a tangled puzzle. Several glacial advances, together with ecological factors including elevation, isolation of soil-bearing areas, and evolutionary biodiversity, forced changes in terrestrial mountain ecosystems. All these variables acted in consort over both time and space, with effects ranging from the large to small scale (i.e., endemism). Using a multi-proxy approach helps to disentangle the puzzle.

A good example of a multi-proxy approach was applied to the study of the Ivrea morainic amphitheater (IMA), situated on the southern side of the Alps. That is our next stop [U0138]. The IMA is a 505 km² wide end-moraine system built by the Dora Baltea glacier that carved the Aosta Valley. This glacier formed in a 3400 km² wide catchment area comprising some of the highest peaks in Europe (e.g., Monte Bianco). The IMA is considered to be one of the most notable amphitheaters of Alpine glaciation (Penck et al. 1894). Its impressive size is due to the Serra d'Ivrea [U0139], which is the largest and most regular moraine found in the Alps (Fig. 1.17). Interestingly, today, Ivrea is a quite small town enjoying a mild temperate climate, but it is estimated that the area was covered, only 25,000 years ago, by ice that attained a thickness of about 500 m. The sedimentary records left in the IMA are very remarkable.

Moraines at the top of the IMA are weathered to various degrees and can be distinguished into three different units (pedogroups*) based on the color of the soils' surface (various yellowish reds of 2.5YR, 5YR, and 7.5YR according to the Munsell* soil color chart). Moreover, paleosols* and organic-rich lake-to-marsh deposits occur in the glacigenic successions, interpreted as interglacial or interstadial deposits. These crop out along stream incisions cutting the outer and inner edges of the IMA or are accessible via drill cores recovered in deep boreholes. An integration of these data indicates at least nine stratigraphic units representing glacigenic successions that correspond to at least nine main glacial expansions (Fig. 1.18a; Gianotti et al. 2008, 2015). Hence, it is possible to evaluate a potentially complete succession that can be correlated to the sequence of the main Quaternary glaciations as recognized using marine oxygen-isotope stratigraphy (Fig. 1.18b; Lisiecki and Raymo 2005). The IMA record ranges from MIS 2 (14–29 ka) to MIS 20–22 (>790 ka) (Box 1.1). Few, if any, plant megafossils are preserved in these sediments. In contrast, the palynological record is more complete. This microfossil record is used to establish trends in climate varying from interglacials with a warm climate, interstadials with a cool-temperate climate, and possibly glacials with a cold climate. The fossil-plant record, thus, better constrains the local sedimentary succession when correlated with the global paleoclimate curve established using the MIS record (Fig. 1.18b). For instance, the oldest pollen assemblage from a lacustrine body comes from an interstadial bracketed by the two distinct glacial peaks of the MIS 18 (712–761 ka). It shows the presence of a coniferous forest adapted to a relatively cold climate, with sedges (Cyperaceae) and ferns (Filicales) reflecting a humid landscape. The pollen record can be used to identify temporal differences in the vegetation.

In a more central, younger part of the amphitheater, pollen spectra from a 1 m thick, organic-rich lacustrine muddy (gyttja*) layer interbedded* between tills are dominated by *Abies* (fir). This conifer coexisted with *Corylus* (hazelnut),

Fig. 1.17 A characteristic end moraine system of the Alps, the Ivrea morainic amphitheater. (**a**) Serra d'Ivrea lateral moraine and part of the internal plain of the amphitheater (view from the Masino Castle). (**b**) Oblique satellite view of the IMA and the Aosta Valley with Monte Bianco, the highest peak in western Europe. (**c**) Map of a part of northwestern Italy showing the location of the IMA and RAMA amphitheaters in the Alps (Images courtesy of GoogleEarth (**b**))

deciduous *Quercus,* and *Ulmus* (elm), indicating the presence of dense fir-dominated forest at a regional scale growing under temperate climatic conditions (Gianotti and Pini 2011). This layer possibly correlates with the MIS 11c interstadial of the northeast Italian Azzano Decimo core (Pini et al. 2009). If so, the glacial deposit beneath the gyttja equates to MIS 12 (c. 420–480 ka) and the one above the gyttja equates to MIS 10 (c. 340–370 ka) (Fig. 1.18b). The realization that these deposits originated from the advance and retreat of mountain glaciers led to early ideas about how they formed.

Sir Charles Lyell proposed his iceberg theory (Lyell 1833) in which he believed that glacial sediments were deposited by floating ice in a marine environment. This idea spread in Italy as it was supported by the occurrence of dropstones∗, as well as reworked fossil marine mollusks and zooplankton (foraminifera) in stratified silty sand of IMA outcrops. However, it is now clear that fossils in the IMA come from fossiliferous sandstones plucked and abraded by the movement of the Dora Baltea glacier across Pliocene marine bedrock. These erratics∗, then, were deposited by the glacier as lacustrine sediments (MIS 18 glacial) in sub-marginal till, forming the frontal moraines (MIS 6 glacial), and in ice-marginal glaciolacustrine deposits, forming kame terraces (MIS 2, LGM). The climatic events triggered by major Quaternary glacial phases are well recorded in Early to Middle Pleistocene lake sediments.

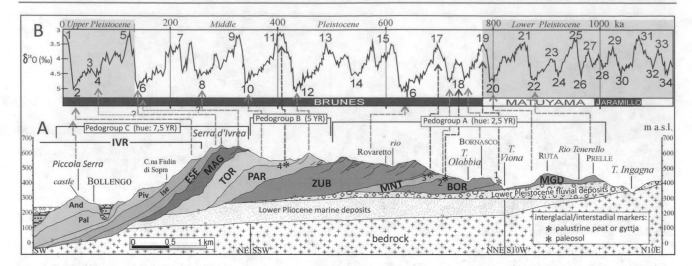

Fig. 1.18 Phases of climate change recorded in the Ivrea morainic amphitheater. (**a**) Geological cross-section of the IMA across the north-eastern sector. Glacial sedimentary cycles (synthems*) are indicated with black acronyms. Interstadial or interglacial markers are shown with asterisks. (1) Bosa paleosol. (2) Sorgente Solfurea peat layer. (3) Comunità paleosol. (4) Gianetto gyttja layer. (**b**) Correlation of the IMA sequence with the global stable-isotope curve from Lisieki and Raymo (2005), which is a synthesis of 57 benthic marine oxygen-isotope records (modified from Gianotti et al. 2015)

Fig. 1.19 The glacial cycles as recorded in the stacked marine-oxygen isotopic record for the last 1500 ky of the Quaternary period (from Lisiecki and Raymo 2005). An isotopic threshold is used to separate the most pronounced glacial phases (in blue) from interglacials and intermediate climate intervals (in red). The climatic events, triggered by major Quaternary glacial phases, are recorded by ecosystem changes in two lake sediments of the European Alps: Leffe and Piànico-Sèllere

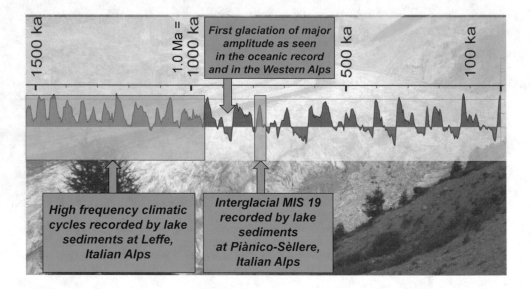

1.8.2 Lacustrine Interglacial Archives of the European Alps

Cesare Ravazzi

Without any need to divert our trip from the southern slope of the Alps [U0140], we stay in Italy and introduce two paleo-lakes, known as Piànico-Sèllere and Leffe, where parts of the puzzle of Alpine paleobiological history are settled [U0150]. Here, fine sediments blanketed lake bottoms and mires, where very organic-rich deposits preserve plant remains. These successions offer well-ordered stratigraphic archives enhanced by virtually continuous sedimentation. Coupling these two archives provides a reference for how the "preglacial" Pleistocene (Calabrian) forest biodiversity responded to major glacial cycles in the Alpine realm (Fig. 1.19).

A spectacular archive of nearly 21,000 annual varves* (Fig. 1.20b; Ravazzi et al. 2014) is preserved in the paleolake archive of the Piànico-Sèllere Basin [U0141], close to the present Iseo Lake [U01042, U0143] sitting at an elevation of 300 m (Fig. 1.20a). This unique record spans the complete MIS 19 interglacial (760–790 ka), as well as the glacial termination of MIS 20 (c. 790 ka), and part of the next glaciation, MIS 18 (760–720 ka; Fig. 1.21a; Ravazzi et al. 2014). Glacigenic deposits occur both at the base and at the top of the succession, making the Piànico-Sèllere varves [U0144] an interglacial deposit in classical terms. Visiting this key-site (Fig. 1.21b) has been a unique opportunity for students

Fig. 1.20 Location and stratigraphic record of the Piànico-Sèllere paleolake. (**a**) Panoramic view of the Iseo Lake, southern slope of the European Alps, Italy. The Piànico-Sèllere paleolake is located to the right (asterisk). (**b**) Detail of the sedimentary record of the Piànico-Sèllere paleolake, showing the typical laminated* sedimentary structures formed at the undisturbed deep lake bottom. Whitish calcite laminations are annual varves and preserve microfossils, pollen, abundant leaf compressions* of deciduous and evergreen woody plants, and vertebrate remains. Each winter season is marked by a very thin (0.1 mm), dark organic layer. In combination with the white laminations, annual resolution of the deposits allows for the recognition of temperate climatic phases. The absolute chronology of the varves set was obtained by paleomagnetic analyses coupled with radiometric dating of a volcanic ash layer, the thickness of which is indicated by the finger width. The upper section is marked by contrasting brownish laminae overlain by blue clay. These darkening layers reflect the increasing detrital sediment supply as the temperate phase ends after the MIS 19 interglacial and the onset of a stadial phase (from Ravazzi 2013). (**c**) Entire skeleton of a deer with antlers preserved in the varves of Piànico-Sèllere, assigned to MIS 19, earliest Middle Pleistocene (Photo courtesy of Giovanni Cattaneo (**c**))

for more than a century. Each yearly layer (lamina) is made of biochemical calcite (lime), in which biological proxies, including fossil pollen (Rossi 2004) and diatoms, are preserved. The fact that the sediment is lime ($CaCO_3$) allows for the development of geochemical climate proxies using stable carbon-and-oxygen isotopes locked away in the mineral (Mangili et al. 2007). Several annual layers also preserve leaves and fruits (Martinetto 2009), wood, mollusks (Esu

Fig. 1.21 Stratigraphical and paleontological record of the early Middle Pleistocene Piànico-Sèllere succession. (**a**) Piànico-Sèllere composite section showing the position of a tephra layer and a fossil deer. (**b**) Outcrop called "main" section. (**c**) Leaf compression of *Acer* gr. *cappadocicum* (Images courtesy of Sabrina Rossi (**a**), E Martinetto (**b, c**))

and Gianolla 2009), fish, and complete mammal skeletons, including a male deer with antlers (Fig. 1.20c; Govoni et al. 2006). The vegetation growing on that landscape, though, is not what we normally see in Italy, today.

The macroflora of the Pianico-Sèllere interglacial interval is remarkably rich in so-called Pontic species. These plants currently grow in the Pontic Mountains of Turkey and in the Caucasus and have relic distributions of species in wet spots of southern Europe. Two plants, *Rhododendron ponticum* var. *sebinense* and *Acer* gr. *cappadocicum* (maple) (Fig. 1.21c), are probably the most striking examples of this community. *Rhododendron ponticum* is a characteristic species of other early Middle Pleistocene macrofloral assemblages in the Alps (i.e., the Hötting breccia near Innsbruck; Denk 2006; Sanders and Spötl 2014). Several other species are closely related, or identical, to relics living in the Alps today (e.g., *Acer opulifolium*). On the other hand, strong pollen-producing conifers and members of the Juglandaceae (walnut family), which mark the "preglacial" Early Pleistocene in the nearby Leffe record (*Tsuga* [hemlock], *Cedrus* [cedar], *Carya* [hickory], *Pterocarya* [wingnuts], *Juglans* sect. *Cardiocaryon* [white walnut or butternut]), are completely absent. This latter group likely was regionally lost (extirpated) from the Italian Alps in response to the first major glacial event(s), in the final phases of the Early

Pleistocene. In turn, some tree species that are widespread in temperate deciduous forests of Europe, today, are missing (*Acer pseudoplatanus*, *Fagus sylvatica* [beech]). Their absence indicates that these plants had yet to extend their biogeographic range into the Circum-Alpine region by the beginning of Middle Pleistocene.

The Leffe archive, only 5 km distant from the Pianico-Sèllere paleolake, preserves a "preglacial" Early Pleistocene paleoecosystem in the Alps. Major glacial advances on the Italian side of the Alps began either at MIS 20 or MIS 22 (810–795 ka or 880–850 ka, respectively; Muttoni et al. 2007). There is a record of climate cycles back to the Calabrian Stage (1.1–1.8 Ma), but the amplitude in the range of temperature during this time was lower. These climate dynamics are captured by the pollen record from the long lake-to-swamp (paludal*) succession preserved at Leffe, currently at an elevation of 450 m in the Italian Alps. It is hard to image, at present, that the Leffe paleolake was not surrounded by high mountains. Those mountains were uplifted beginning one million years ago. At that time, an estimate of the area's paleoelevation was only about 200 m. High-frequency climate cycles affected forest ecosystems covering most of the alpine relief (Fig. 1.22). Cyclical climates resulted in a change from mixed coniferous-hardwood forests (*Tsuga, Abies, Cedrus, Picea* [spruce], *Fagus* dominated) to temperate

Fig. 1.22 Reconstruction of the Early Pleistocene paleoenvironment around the Leffe paleolake, located on the southern slope of the European Alps, Italy. The view is from the west and shows forest belts during one of the several warm-temperate phases cyclically punctuating the Alpine Early Pleistocene between 1.6 and 1 Ma. A littoral* swamp forest is represented by *Glyptostroboxylon*-rich brown-coal seams interbedded with lake carbonates. This swamp facies also included

floating mats that, most probably, trapped elephants and preserved their skeletons. Terrestrial forest communities vary from mixed coniferous-hardwood forests to warm-temperate mesophytic deciduous forest. The south-facing sunny slope (left in the view) supported cedar forests (*Cedrus* sp.), as inferred by high pollen accumulation rates transported down by air and rivers to the lake surface (Drawing by Paola Rota)

deciduous forest (*Carya, Pterocarya, Juglans, Quercus, Acer*
sect. *Saccharina, Aesculus* [buckeye], *Rhododendron,
Staphylea* [bladdernut]) [U0149] to open dry forests (*Celtis-
Eucommia-Quercus*) (Ravazzi and Rossignol Strick 1995).
Most of the tree species identified at Leffe are now extinct,
although living members of closely related plants still grow in
temperate regions of southeastern Asia and North America
(see Chap. 2). On the other hand, typical species common in
middle to late Quaternary temperate stages are missing
(*Carpinus betulus* [hornbeam], *Fagus sylvatica, Acer pseudoplatanus*). This is also true for cold boreal species (e.g., *Pinus
cembra* [pine]) whose evolutionary history continues to be
debated. Wetlands included woody plants extirpated from
other parts of Europe and include species of Cupressaceae,
Liquidambar (sweetgum), *Nyssa* (tupelo) (Ravazzi and van
der Burgh 1994; Ravazzi 1995), and a gallery of vines (*Vitis*
[grape], *Parthenocissus*). Despite the proximity of the Leffe
Lake to spillways of mountain glaciers, there is no sedimentary evidence of glacial-sourced rivers or of treeless phases in
this long Calabrian lake record of the preglacial Early
Pleistocene of continental western Europe (see, e.g., West
1980, for an overview). What about the history of other high
mountain terrains?

1.9 Himalaya Foothills, Nepal

Sundarshan Bhandari and Arata Momohara

Having been impressed by the records of biotic and climatic
change around the Alps, we move now eastward in Eurasia
[U0151], to the foothills of the highest mountain range in the
world: the Himalayas (Fig. 1.23). We visit the Kathmandu
Valley [U0152], located at an altitude of about 1200 m and
hosting a large intermontane sedimentary basin. The valley is
framed by mountains with elevations of 2400–2800 m and
warm-temperate (=subtropical) evergreen broad-leaved forest
covering the valley floor and the lower slopes. Woody plants
including *Schima wallichii* (tea family), *Castanopsis* (chin-
quapin), *Eurya* (Pentaphylacaceae), and several species of
oak (*Quercus incana, Q. lanuginosa*, and *Q. semecarpifolia*)
grow up to elevations of 2300 m (Malla et al. 1976). The
Kathmandu Basin has never been glaciated and is filled by a
fluviolacustrine and fluvial to fluvio-deltaic sedimentary succession of Plio-Pleistocene to Holocene age, the result of
damming in the southern part of the valley (Sakai et al. 2006).
The basin covers an area of 400 km^2 and sediments fill it to a
depth of more than 500 m in the center (Fujii and Sakai 2002).

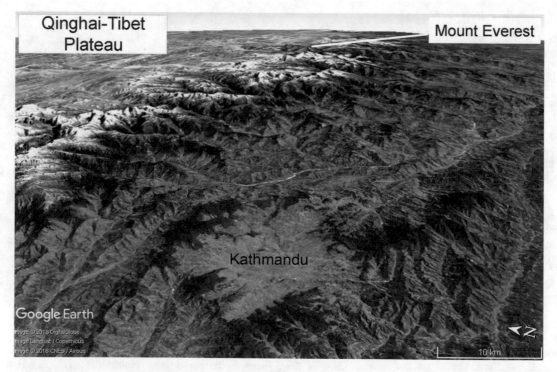

Fig. 1.23 Kathmandu Valley panorama showing also the position of Mt. Everest, Earth's highest peak above sea level (8848 m). Image courtesy
of GoogleEarth

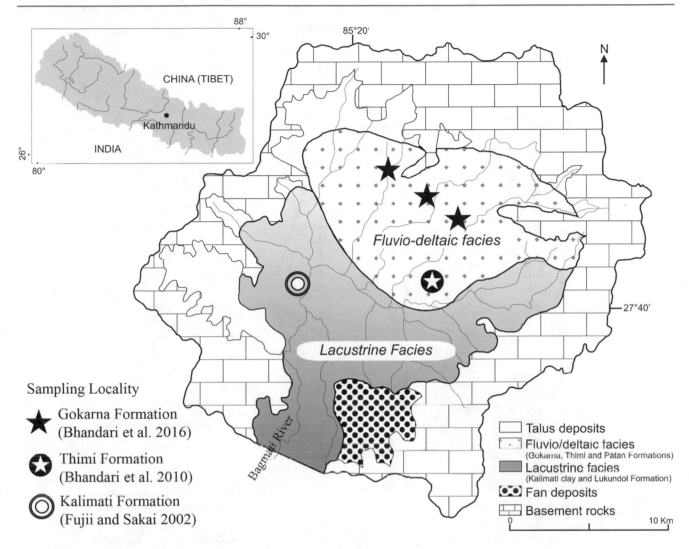

Fig. 1.24 Geological map of the Kathmandu Valley showing pollen and plant-macrofossil localities (redrawn after Sakai et al. 2001)

Sediments outcropping in the valley contain well-preserved pollen and plant macrofossils (e.g., Nakagawa et al. 1996; Paudayal and Ferguson 2004; Bhandari et al. 2016).

The sedimentary record, and the plant fossils preserved therein, is time transgressive from the south to north in the basin. That is, older deposits occur to the south and younger deposits occur in the north and central part of the basin [U0153]. This character developed in response to rapid uplift of the Mahabharat Mountain Range in the south and a northward tilting of the Kathmandu Basin, leading to the gradual northward shift of river and lake sedimentation. The basal Lukundol Formation∗, in an orientation following the tilt of the basin, is overlain by the Kalimati, Gokarna, Thimi, and Patan formations (Fig. 1.24). Studies of macrofossil assemblages in stratigraphic context, carried out in the Gokarna and Thimi formations (Fig. 1.25), have clarified the history of the extant vegetation and traced environmental change in the area since the Late Pleistocene.

The available sedimentary record allows us to imagine a field trip, backwards in time, in the Kathmandu area. Our first stop can be fixed between 20,000 and 30,000 years ago, when we encounter conditions that differ from today. The plant fossils (*Abies, Pinus wallichiana, Tsuga, Picea, Taxus* [yew], and *Quercus* subgen. *Lepidobalanus*) found in the Thimi Formation grew under the influence of a cold climate during the second half of the final glacial period of the Late Pleistocene (Bhandari et al. 2010; Ooi 2001). Stepping further back in time, we can observe different landscapes of the Gokarna Formation, which is our second stop.

Plant macrofossils (Fig. 1.26) from the Gokarna Formation reveal that the upper part of the formation (dated to c. 49 ka) is dominated by such cool-temperate taxa as *Abies, Pinus, Picea smithiana, Tsuga dumosa* (hemlock), *Taxus wallichiana* (yew), *Quercus* subgen. *Lepidobalanus,* and *Betula.* In contrast, the middle and lower parts (Besigaon, Mulpani, and Dhapasi members, dated from c. 50–53 ka) are

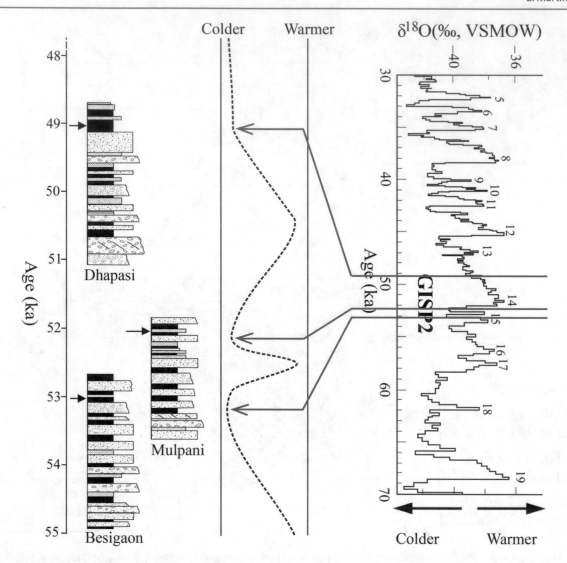

Fig. 1.25 Correlation based on ^{14}C ages of the Late Pleistocene succession in the Himalayan Foothills with the ice core of the GISP-2. The climatic trend reconstructed on the basis of plant assemblages of the Kathmandu Basin is in agreement with that reconstructed in Greenland's ice core on the basis of the oxygen-isotope curve (after Bhandari et al. 2016)

characterized by the dominance of taxa requiring subtropical growing conditions, similar to today. These subtropical plants include *Eurya*, *Ficus*, *Morus*, *Zizyphus* (buckhorn family), *Stephania* (Menispermaceae), *Quercus* subgen. *Cyclobalanopsis*, *Pyracantha* (firethorn), and *Carpinus*. The diversity of plant taxa and occurrence of moisture-loving plants indicate the presence of sufficient rainfall in this area for their growth and reproduction, possibly the result of a very active Indian monsoon system (Bhandari et al. 2016). The climatic information derived from the fossils is fairly local or regional. But, the temperature drop detected from older to younger assemblages can be correlated with the overall global cooling trend of the GISP-2 oxygen-isotope curve of the LGM (Fig. 1.25). We are not able to add further stops to our trip from 53,000 back to about 600,000 years

ago here, because there is a time gap in the plant-fossil record, and no information about vegetation exists.

The next stop hosts the lacustrine deposits of the Kalimati Formation, about 0.6 million years old, that provide a palynological record revealing both the vegetation and prevailing climate (Fujii and Sakai 2002). Subtropical, evergreen broad-leaved oak forests were present throughout the depositional interval without any major shift in vegetation and, thus, overall climate. There is another time gap in the terrace deposit located between the Kalimati and Lukundol formations, which means that we have no information about the local vegetation from about 0.6 to 0.8 million years ago. Deeper in time, the oldest lacustrine Lukundol Formation (0.8–1.1 Ma; Goddu et al. 2007) can be subdivided into three pollen zones. Palynomorphs preserved in the uppermost (and therefore

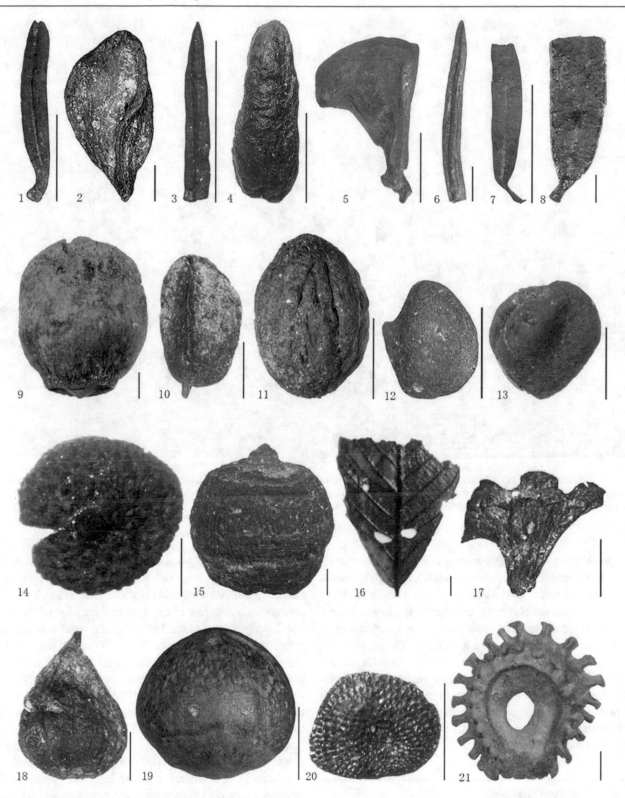

Fig. 1.26 Macrofossils from the Quaternary sediments of the Kathmandu Valley. (1) *Abies* sp. (leaf), (2) *Abies* sp. (seed), (3) *Picea smithiana* (leaf), (4) *Pinus wallichiana* (female cone), (5) *Pinus roxburghii* (female cone scale), (6) *Pinus* sp. (needle), (7) *Tsuga dumosa* (leaf), (8) *Taxus wallichiana* (leaf), (9) *Taxus wallichiana* (seed), (10) *Pyracantha* sp. (endocarp), (11) *Zizyphus* sp. (endocarp), (12) *Ficus* sp. (endocarp), (13) *Morus* sp. (endocarp), (14) *Quercus* subgen. *Lepidobalanus* (cupule∗), (15) *Quercus* subgen. *Cyclobalanopsis* (young fruit), (16) *Quercus* sp. (leaf), (17) *Betula* sp. (fruiting scale of infructescence), (18) *Carpinus* sp. (nutlet), (19) *Corylus ferox* (nut), (20) *Eurya* sp. (seed), (21) *Stephania* sp. (endocarp). Scale of 1, 3, 5, 7, 11, 14–16, 19 = 5 mm; scale of 2, 6, 8–10, 12, 13, 15, 17, 18, 20, 21 = 1 mm. Scale of 5 = 50 mm

Fig. 1.27 The eastern part of Olduvai Gorge, where more than 50 m of sediment accumulated and most of the stratigraphic units (Olduvai Beds) are exposed. The dark rocks at the base of the gorge are the base-ment basalts dated about two million years ago. The well-distinguishable rust-colored levels are Bed III (Photo courtesy of M Cherin)

youngest) Zone-III indicate subtropical to cool-temperate conditions (Bhandari and Paudayal 2007). A slightly cooler, temperate climate was detected in the underlying Zone-II. Finally, the pollen assemblage of the lowest Zone-I indicates a vegetation that grew under a warmer, definitely subtropical climate, again, similar to today. Hence, although we envision the plant cover across the foothills of the Himalayas from a twenty-first-century viewpoint, this landscape has not remained static over time. Rather, changes in the prevailing climate over the past million years or more have dramatically altered the vegetational composition able to grow in the region.

1.10   Olduvai Gorge

Marco Cherin

We move now to the intertropical belt to observe the impact of late Cenozoic-Quaternary climate change on natural systems that are far away from the poles. We travel to Africa [U0155] and visit Olduvai Gorge [U0156], an approximately 40 km long dry-river valley located in the Ngorongoro Conservation Area in northern Tanzania (Fig. 1.27). The conservation area lies at the southeastern edge of the extensive Serengeti ecosystem. The name of the site is an incorrect transliteration of the Maasai word "Oldupai," used by local communities to indicate the East African wild sisal (*Sansevieria ehrenbergii*), which grows abundantly in the area. The sediments exposed in Olduvai Gorge all lie above basement basalts and are divided into seven beds, collectively referred to as the Olduvai Beds [U0157] (Hay 1976). Lower beds (Beds I and II; c. 2.0–1.3 Ma) are relatively richer in fossils (Fig. 1.28) [U0159] and stone tools and were deposited in and around a paleolake (Lake Olduvai). The paleolake varied in size, sometimes being small and highly saline and, at other times, larger with fresher water. The upper beds (c. 1.3 Ma–10 ka) mostly represent deposition in fluvial environments and include numerous paleosols. The entire succession contains several tuff layers, each corresponding to eruptive episodes of one of the numerous volcanoes present in the area. These "marker tuffs" were

Fig. 1.28 Replicas of some of the most significant fossil mammals found in the oldest deposits (Beds I–II) of Olduvai Gorge, exposed in the local museum (Photo courtesy of M Cherin)

dated with radiometric∗ methods, thus allowing for a high-resolution chronological sequence which increases the importance of the Olduvai Gorge stratigraphy.

Olduvai Gorge is recognized as one of the most important paleontological and archeological sites in the world and has hosted international groups of researchers for more than a century (Fig. 1.29) [U0158]. The abundance of fossils and stone tools found in the stratigraphic succession allows for the reconstruction of the evolution of east African ecosystems over the last two million years. This record also provides insight into some crucial phases of early human evolution. Much of its fame is due to the research activities of Mary and Louis Leakey, who have been working at the site for more than 30 years. Among their most important findings are Olduvai Hominid (OH) 5 and OH 7 representing, respectively, the holotypes∗ of *Paranthropus boisei* (Leakey 1959) and *Homo habilis* (Leakey et al. 1964). These were the first ancient hominin fossils discovered in east Africa (Blumenschine et al. 2012). We can decipher the climates in which our ancestors lived by analyzing the chemical properties of the ancient soils.

Stable-oxygen-and-carbon isotopic analyses carried out on fossil soils, both from Olduvai and the Turkana Basin

(Kenya), indicate the presence of open savanna grasslands around one million years ago and woodland around three million years ago (de Menocal 2004). But, generally, global climate changes produced weaker effects in these intertropical ecosystems when compared to vegetation in the mid-latitudes. As the plants that sprawled over the landscape were different, so were the mammals using them as a resource. The modern African faunas started to appear on these landscapes at the end of the Early Pleistocene (c. 1 Ma), with only a few species of the African megafauna experiencing extinction during the Late Pleistocene (Turner and Antón 2007). Conversely, most of the fossils found in the Early Pleistocene deposits of Olduvai and eastern Africa, in general, belong to extinct mammal species, often phylogenetically related to the ones now living on the continent.

1.11   Southern Caucasus

Angela A. Bruch and Ivan Gabrielyan

The previous section documents places that have seen the appearance and development of our direct ancestors of the

Fig. 1.29 Members of the Italian-Tanzanian THOR (Tanzania Human Origins Research) project during surface survey activities at Olduvai Gorge (Photo courtesy of M Cherin)

genus *Homo*. As far as we know, currently, the first species of *Homo* appeared at the beginning of the Pleistocene in eastern Africa and started to spread into Eurasia from at least two million years ago. One of the most impressive examples of direct paleoanthropological evidence of hominins in Eurasia is the fossil site of Dmanisi in the Republic of Georgia, southern Caucasus [U0161]. Here, fossil remains of *Homo* have been uncovered in layers dated about 1.8 million years ago (Ferring et al. 2011). What vegetational landscape did our ancestors occupy?

As we've now seen elsewhere, we understand the nature of these Early Pleistocene environments using the plant fossils preserved in the sediments. Detailed analyses, covering a long time span, are used to reconstruct the region by integrating data from Georgia with those of well-dated sites in the Vorotan Basin in Armenia. These records point to a diverse and dynamic vegetational pattern in which early men lived, thus providing a broader scenario for the phase of early human occupation detected at Dmanisi. Especially in the southern Caucasus, with its variable topographic relief, an understanding of the altitudinal and spatial differentiation of vegetational units and their response to climate change is crucial, as these have played roles in human evolution. A

quantitative assessment of fossil-plant assemblages allows insights into the climate. By comparing short-term vegetational changes over different climatic events and in different regions, it is possible to understand the mechanisms of climatic influence on local vegetation. Global climatic signals can be translated into the local setting, and it is possible to interpret the general climate even during phases without a meaningful plant-fossil record. Thanks to the integration of all results obtained from Armenia and Georgia, we can lead the reader to a walk through the spatial and temporal distribution of Early Pleistocene forests and mosaic landscapes in the southern Caucasus, animated by variations in climate.

1.11.1 Vorotan Basin, Armenia

The southern part of Armenia is a paradise for plant macrofossil studies and provides insight into changing southern Caucasus environments during the Calabrian. Extensive outcrops of the Sisian Formation (Fig. 1.30) in the Vorotan River Basin expose lake sediments that cover a stratigraphic age from about 1.3 to 0.9 million years ago [U0162]. These deposits include the Jaramillo subchron (Kirscher et al.

Fig. 1.30 Early Pleistocene plant fossil-bearing lacustrine succession cropping out at Shamb in the Vorotan Basin, Armenia (Photo courtesy of I Gabryelian)

2014) that allows us to recognize time using the magnetic signal locked in the rocks. The precise age control, based on magnetostratigraphic and radiometric dating, constrains a high-resolution pollen analysis of one section across the Matuyama/Jaramillo reversal, spanning 1.12–1.035 million years ago (MIS 33-MIS 30). The mean resolution per sample is of approximately 250 years (Scharrer 2013). The data indicate a clear vegetational response on orbitally forced climatic change. Open vegetation occupied the area during the less pronounced MIS 33/34 cycles, whereas broad-leaved deciduous forests expanded during the cooling phase of the very warm and humid MIS 31. Subsequently, needle-leaved (coniferous) forests occupied the region during the long, cool, and humid MIS 30. In general, the preservation of numerous macrofloral assemblages (Fig. 1.31), along with those from other late Early Pleistocene sections, are constrained to warm and humid climatic phases. Plant species compositions [U0163] are very similar to forests occurring today along the southern coasts of the Black Sea and Caspian Sea. During warmer and more humid periods of the Early Pleistocene, these communities must have expanded their biogeographic range considerably (Gabrielyan and Kovar-Eder 2011).

Generally, the vegetation in this part of Armenia oscillated between warm and humid phases and cooler, drier phases. During the former, forests and mosaic landscapes expanded and these contrast with the cooler phases where, similar to today, open landscapes prevail with *Artemisia* steppes (ragweed-dominated) and remnant forests growing on northern slopes and along the rivers (Ollivier et al. 2010; Joannin et al. 2010). In contrast, western Georgian lowland data document permanent forest cover throughout the Early Pleistocene (Shatilova et al. 2002; Kirscher et al. 2017).

1.11.2 Dmanisi (Georgia)

The well-known Early Pleistocene locality of Dmanisi is one of the richest paleoanthropological sites of the world [U0164] and yielded the earliest fossil bones of the genus *Homo* outside Africa (Lordkipanidze et al. 2013). The site (Fig. 1.32a), which also contains a rich assemblage of stone tools, is located in southern Georgia, on a promontory at the confluence of two rivers. Based on radiometric dating of the underlying basalt and magnetostratigraphic analysis of the fossil-bearing sediments, the age of the Dmanisi Paleolithic site can be constrained between about 1.85 and 1.76 million years ago (Ferring et al. 2011). Although stone tools occur throughout the entire section, human and nonhuman mammal bones are restricted to one level just on top of the paleomagnetic reversal marking the Olduvai to Matuyama transition (1.77 Ma; Ferring et al. 2011).

The human fossils unearthed are extraordinary proof of the high morphological variability of *Homo erectus*, which is somehow surprising for a single hominin species (Fig. 1.32c). In fact, so-called Skull 5 is the most complete early human skull, to date [U0165]. This fossil combines features that are attributed to two different species (*H. erectus* and *H. habilis*), thus raising fundamental questions on the human species concept (Lordkipanidze et al. 2013; Rightmire et al. 2017). Yet, fossils of our relatives constitute but a very small part of the paleontological record.

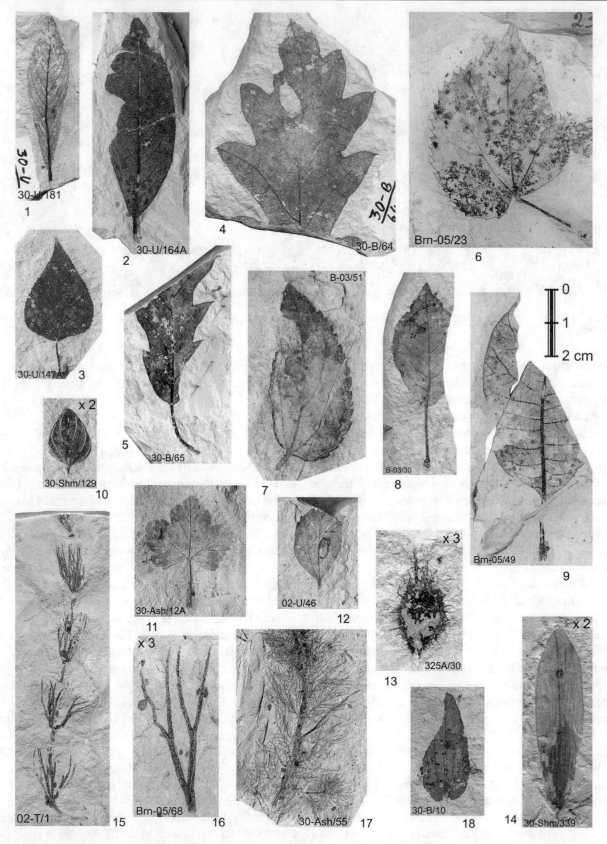

Fig. 1.31 Fossil-plant remains (mostly leaves) from the Vorotan Basin, Early Pleistocene of Armenia. (1) *Salix* cf. *alba*. (2) *Salix* cf. *pseudomedemii*. (3) *Populus* cf. *nigra*. (4, 5) *Quercus* cf. *cerris*. (6) *Tilia* cf. *begoniifolia*. (7) *Celtis* cf. *caucasica*. (8) *Prunus* cf. *padus*. (9) *Cotinus* cf. *coggygria*. (10) *Thymus* cf. *kotschyanus*. (11) *Ribes* cf. *orientale*. (12) *Spiraea* cf. *crenata*. (13) *Calligonum* cf. *polygonoides*. (14) *Fraxinus* cf. *oxycarpa*. (15, 16) *Ceratophyllum* cf. *demersum*. (17) *Myriophyllum* cf. *spicatum*. (18) *Potamogeton* cf. *perfoliatus*

Fig. 1.32 The Dmanisi fossil site in Georgia. (**a**) Excavation site. (**b**) Landscape reconstruction. (**c**) One of the famous hominid skulls yielded from this site (Images courtesy of Elen Hakobyan (**a**), Hessisches Landesmuseum Darmstadt (**b**), AA Bruch (**c**))

The overwhelmingly rich and well-preserved fossil record comprises a great variety of animals from different habitats (Blain et al. 2014). Forest or riverine (riparian*) dwellers are represented by several species of deer hunted by felids. Open grassland was inhabited by ostrich, horse, and several genera of antelope, as well as the high-speed hunter cheetah. Rhino and southern elephant inhabited the transitional area of the forest steppe, whereas mountainous-rocky regions are evidenced by an extinct form of mountain sheep, the pika, and, to date, the oldest evidence of ibex (wild mountain goat) (Vekua 1995). There is no flux of African animals detected in the Dmanisi fauna, and the

human dispersal out of Africa may have occurred earlier, as suggested by older archeological finds from China now dated to 2.1 million years ago (Zhu et al. 2018). The climate at the time of entombment and preservation of the animals, based on amphibian and reptile data (Blain et al. 2014), is estimated as a mean annual temperature (MAT) of 13.1 °C, which was 3.1 °C higher than today. Estimates of mean annual precipitation (MAP) around 630 mm are slightly lower than the current level (700 mm), with more humid winters accompanied by a drier season during summer months. These conditions favored the preservation of animal remains over those of plants.

Although infrequently preserved, fossil-plant remains generally confirm the mosaic features of the landscape with areas of forests, steppe, and riparian vegetation (Fig. 1.32b). Most importantly, phytolith∗ assemblages occur throughout the succession encompassing the entire time of early human occupation. These silica-rich plant remains (see Chap. 3) indicate the development from more humid, more forested landscapes to drier, more open landscapes during this time (Messager et al. 2010). Plant-fossil deposits in the western Georgian lowland, dating between 1.8 and 1.0 million years ago, document a permanent forest cover throughout the Early Pleistocene (Shatilova et al. 2002; Kirscher et al. 2017), marking a difference from the less-forested neighboring regions. This area, which today experiences a higher MAP (>700 mm) than Dmanisi, has been more constantly wet during that time. For this reason, western Georgia is considered a Pleistocene refuge for plants that do not tolerate excessive drought and cold. However, during the Early-Middle Pleistocene transition, some of these plants were extirpated here, such as *Sequoia* (redwood) [U0165]. In synthesis, fossil-plant studies paint a picture of a diverse and dynamic vegetation in which early humans lived in the southern Caucasus.

1.12 Central Japan

Arata Momohara

We now will visit a landmass at the eastern margin of the Eurasian continent [U0166], the Japanese archipelago, which is influenced strongly by humid oceanic air masses. The past natural systems of Japan can be reconstructed on the basis of abundant paleontological data. Quaternary sedimentary basins are well developed and stratigraphic age control has been refined by using widespread tephra∗ beds correlated to paleomagnetic events. In some instances, biostratigraphic correlation using calcareous nannoplankton has been applied. These methods make it possible to trace changes in the spatial and temporal distribution of island plants in the context of glacial-interglacial climate change in a high latitude setting (Momohara 2016). Plant macrofossil assemblages during the LGM, scattered across central Japan, demonstrate the presence, distribution, and composition of refugia for humid temperate plants. Two particular sites provide outstanding information about the inequality of climate in different Early and Middle Pleistocene intervals.

1.12.1 Last Glacial Maximum

Most of the lowland and mountain zones in central and western Japan [U0128] during the LGM (c. 24.4 and 20.1 ka), in

MIS 2, were covered by forest comprised of pinaceous conifers, including *Picea, Abies, Tsuga, Pinus* subgen. *Haploxylon,* and *Larix* (larch). Cold-tolerant broad-leaved trees, such as *Betula* and *Alnus,* also were part of the landscape (Tsukada 1985). The distribution of common and/or dominant trees under the present humid warm- and cool-temperate condition, including *Cryptomeria, Fagus, Quercus, Pterocarya, Carpinus, Zelkova,* and *Acer,* was limited to small populations in the lowlands south of 38° N latitude. Several coeval pollen assemblages in western Japan exhibit the dominance of deciduous broad-leaved forest taxa with the occurrence of temperate conifers preserved as macrofossils. Using pollen and fruit-and-seed data from sediments in the Hanamuro River site (Momohara et al. 2016), Ibaraki Prefecture, central Japan, it has been possible to reconstruct the distribution patterns of a highly diverse vegetation (Fig. 1.33) [U0129]. The occurrence of thermophilous plants, including *Selaginella remotifolia* (club moss), indicates that the region experienced a paleotemperature similar to that of the present cool-temperate zone [U0130], which is covered by deciduous broad-leaved trees. Pinaceous subarctic conifers expanded their biogeographic range to warmer temperature conditions than where they grow today. The probable reason for their expansion is that moderate precipitation, lower than what the region experiences at the present, confined temperate broad-leaved trees to the humid valley bottoms. This physiologic constraint hampered their competition with subarctic conifers on the better-drained slopes.

1.12.2 Middle Pleistocene to Pliocene

Two areas of central Japan preserve excellent records of the organization of ecosystems in the Middle and Early Pleistocene. Before we travel back in time to visit these ecosystems in which the records are preserved, we have to look at the geology of the two sites today: central Kinki District and the South Niigata Prefecture. In central Kinki, thick deposits composed mainly of fluvial sediments accumulated beginning in the Pliocene (c. 3.5 Ma), with sedimentation continuing to the present (Yoshikawa and Mitamura 1999; Satoguchi and Nagahashi 2012). Marine sediments in the stratigraphy correlate with interglacial stages since MIS 37 (c. 1.25 Ma) and are interbedded with the fluvial deposits (Yoshikawa and Mitamura 1999). Plant macrofossils from the site were first studied intensively by Miki (1938, 1948), and their stratigraphic positions were later correlated in detail in several sections (e.g., Huzita 1954; Itihara 1960). In the South Niigata Prefecture, fluvial and marine sediments were deposited in a coastal basin along the Sea of Japan, which, now, crop out in adjacent hilly areas. The deposits accumulated between 0.7 and 2.6

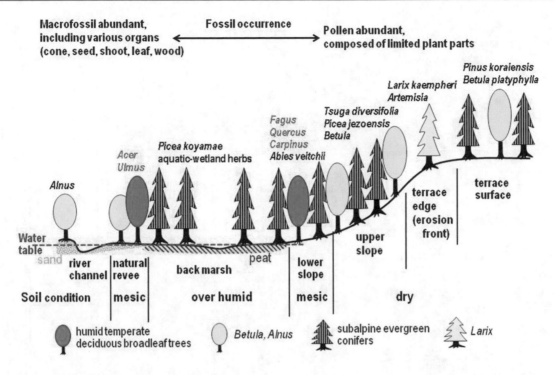

Fig. 1.33 Paleovegetation distribution in the middle part of the Kanto Plain, central Japan, in the LGM based on occurrence of pollen and plant macrofossils (modified from Momohara 2017)

million years ago and are known as the Uonuma Group which attains a thickness of more than 1500 m (Kazaoka 1988). Abundant tephra beds are distributed throughout the succession (Satoguchi and Nagahashi 2012) and provide excellent chronostratigraphic control for the plant macrofossil assemblages (Momohara 2016; Momohara et al. 2017). As in other parts of the world, marine sediments deposited during interglacial transgressions are correlated, in part, to MIS (Urabe et al. 1995). Our virtual postcards will touch progressively older sediments of both the central Kinki and South Niigata areas.

Our next stop brings us to the late Middle Pleistocene, between 0.12 and 0.4 million years ago. It is cold: we are in the middle of a glacial phase. Subalpine plants dominate the assemblages and there is no evidence of any Neogene relict such as *Metasequoia* (dawn redwood). We know that these relicts are absent based on data from late Middle Pleistocene (MIS 11) pollen profiles collected from a sediment core in Lake Biwa, Shiga Prefecture. These profiles exhibit clear cyclic patterns reflecting sedimentation during glacial and interglacial climates. Deciduous broad-leaved tree pollen (*Fagus*, deciduous *Quercus*, occasionally evergreen *Quercus*) dominates during interglacial peaks, but is replaced by temperate conifer pollen (*Cryptomeria* and *Sciadopitys*) and dominant pinaceous conifer pollen (*Picea*, *Abies*, *Tsuga*, *Pinus* subgen. *Haploxylon*) during glacial peaks (Miyoshi et al. 1999). Macrofossil assemblages consisting of cool-temperate and subalpine plants, indicating glacial stages,

also occur in deeper time, at the end and at the beginning of the Calabrian (Momohara 2016).

Fluctuations in paleotemperature reflecting glacial-interglacial cycles are evident in river and lake deposits of our second stop: South Niigata. Changes in plant macrofossil assemblages [U0167] during the MIS 53 to 63 interval indicate considerable changes of vegetation and climate, with a MAT shift from 6.6 °C to 11.0 °C (Momohara et al. 2017). This temperature falls within the present geographical range for cool-temperate, deciduous broad-leaved forests in Japan. It should be noted that MIS 60 was the coldest of the glacial stages, with temperatures nearly equivalent to those during the LGM. Major plant extirpations [U0168] in and around MIS 60 are not recorded in the sedimentary succession and indicate there was little vegetational response to the temperature decline. This response is attributed to a floral recovery after the end of the glacial stage. Plants residing in glacial refugia in the southern basins expanded their biogeographic range. A distinct floral change occurs in and around 1.3 million years ago. *Metasequoia* was extirpated earlier here than in the central Kinki District, and *Stewartia monadelpha* (Theaceae) and *Chamaecyparis obtusa* (false cypress), which are endemic to the present-day Pacific Ocean-side mountains, also disappeared [U0168]. The change may have been in response to active tectonics and uplift of mountain terranes in central Japan, which hindered plant dispersal from southern refuges and promoted a differentiation of local microclimates (Momohara 2018). However, we need to

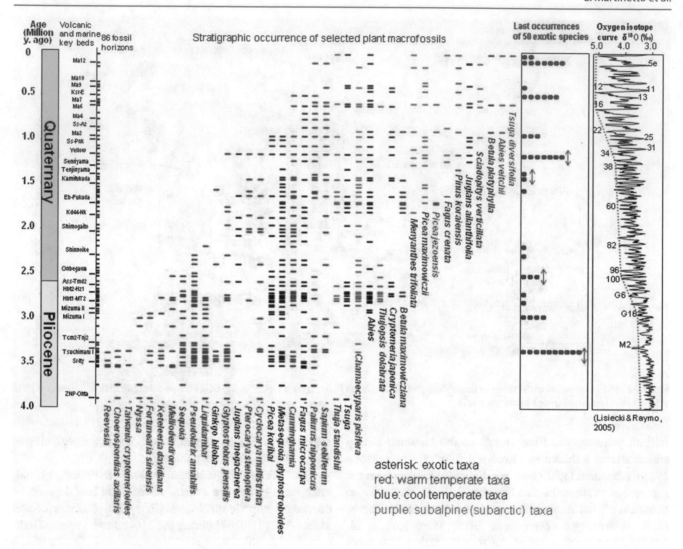

Fig. 1.34 Stratigraphic change in the macrofossil flora from 0 to 4 Ma in the Kinki District (modified from Momohara 2016). Number of dots and arrows in the last occurrences indicate number of species and range of estimated age, respectively

move again to central Kinki, and travel deeper in time, to learn more about Neogene relict plants.

In central Kinki, fossil assemblages preserved between 2.4 and 2.8 million years ago are rich in cool-temperate elements that are found in modern forests of Japan. However, there are a number of plant-fossil assemblages preserved in fluvial sediments that are dominated by remains of *Metasequoia*, an indicator for interglacial climate conditions in the country. Evidently, this tree had been a dominant component of wetland forests during Early Pleistocene warm-temperate intervals, but was extirpated from Japan later in the epoch. By going further back in time, we would encounter a higher and higher number of plants that do not currently grow in the country (Fig. 1.34). As explained elsewhere in this volume (Chap. 2), the more constant subtropical climate of the Pliocene (between 2.8 and 3.5 Ma) favored the occurrence of a high number of thermophilous* plants, which are now endemic to central and southwestern China.

1.13 The Intermontane Basins of Northern Apennines (Italy)

Adele Bertini

To find another place where, similar to Japan, we can discover rich records of the terrestrial biota dating from one to three million years ago, we need to travel back to Europe [U0169] and, specifically, to the Italian mountain range of the northern Apennines (NA) [U0170]. Here several Neogene-Quaternary basins formed during the Tertiary in the context of the Africa-Eurasia collisional belt. Among them are two intermontane basins, filled with continental (fluvial and lacustrine) deposits, that are best known for their extensive fossil fauna and plant remains. These are the Upper Valdarno Basin (UV; e.g., Bertini 2013), located about 40 km southeast of Florence (Tuscany), and the Tiberino Basin (TB; Martinetto et al. 2014), east of Perugia (Umbria)

[U0170]. Their sedimentary successions provide, as a whole, the best history of floristic and vegetational changes in central Italy over approximately 3.3 million years, spanning the Pliocene-Pleistocene boundary. Faunal remains are also present and exceptionally abundant at the Pietrafitta locality of the TB (Martinetto et al. 2014). We will now virtually open older and older time windows to look at this region's outstanding fossil biota and witness vegetation/climate dynamics over several thousands of years. Figure 1.35 illustrates the main paleoenvironmental changes for the UV, based primarily on fossil-pollen assemblages.

1.13.1 Interval 1.5–2.6 Ma

Our first window opens on the time slice from 1.5 to 1.7 million years ago facing the Pietrafitta locality in the Tiberino Basin. We focus our attention on the animals. What an interesting fauna of extinct species are found: large southern elephants (*Mammuthus meridionalis*), rhinos (*Stephanorhinus*), relatives of deer (*Pseudodama farnetensis, Praemegaceros obscurus*), bovids, bears (*Ursus etruscus*), monkeys (*Macaca sylvanus*), beavers (*Castor fiber*), arvicolids (muroid rodents), swans and other birds, snakes, turtles, frogs, and fishes. Thinking that these animals lived together in central Italy during the Early Pleistocene isn't fantasy. Fossil evidence of these vertebrate remains is exceptionally preserved in a lignite seam excavated at this locality (Martinetto et al. 2014). But, what vegetation supported such a diverse Early Pleistocene megafauna? To answer that question, we must move north about 80 km to UV sites.

Fossil-pollen assemblages in the UV [U0174] open another time window back to 1.8–2.0 million years ago (Fig. 1.35; Bertini 2013). Here, the pollen record represents a repetitive development of open vegetation (steppe), dry open woodland, deciduous forest, mixed forest, and conifer forest (often dominated by *Picea* and/or *Abies*). These changes in vegetation reflect glacial-interglacial cycles induced by climate change. Climates shifted from dry and cold to warmer but still dry, then to warm and humid, and finally to cool and continued humid (e.g., Bertini 2010, 2013). The pattern is not restricted to the 1.8- to 2.0-million-year interval in these rocks, but is more widespread in time.

Curiously, if we could look out of our central Italian virtual window for a much longer interval of time, from 1.5 to 2.5 million years ago, we would observe varying abundances of the same trees, the same shrubs and vines, and the same herbs appearing, disappearing, and reappearing. We would see the same cyclical changes of vegetation and climate time after time after time. And, we would need to go back just a little more than 2.5 million years to be part of a remarkable regional event. Between 2.5 and 2.6 million years ago, the UV alluvial plain was covered by dunes of white sand trans-ported by rivers and wind (fluvio-eolian) (Fig. 1.35) [U0174]. The vegetation growing during this dry interval was treeless and dominated by herbaceous species adapted to an open environment. The pollen assemblages record a major pulse of the mostly herbaceous (or shrubby) plant *Artemisia*, which is frequent in today's steppe* vegetation. This *Artemisia* pulse is considered to represent the oldest event in which a glacial phase was so intense that forests were extirpated in the area and replaced by open vegetation (Bertini et al. 2010). But, this phase of eolian sand dunes was short lived.

1.13.2 Interval 2.7–3.2 Ma

Forest plants actually dominate all landscapes before about 2.8 million years ago, but different players occupy the area at different times. Around 2.7 million years ago, altitudinal coniferous trees (especially *Picea* and *Fagus*) grew in the UV, suggesting that cool-temperate climate conditions prevailed. Slightly deeper in geological time, from 2.8 to 3.2 million years ago (during the Piacenzian), the UV was a lake surrounded by luxuriant swamps, which were rich in plants requiring a subtropical climate (e.g., *Glyptostrobus* [Chinese swamp cypress] and *Taxodium* [bald cypress] along with *Nyssa* [Tupelo gum], *Carya, Sequoia* [redwood], *Symplocos* [sweet leaf]; Figs. 1.36 and 1.37). The large mammals living in these swamps comprised mastodons, tapirs, bears, and bovids (Rook et al. 2013), all of which were well adapted to a forest habitat. Plants like *Engelhardia* (Juglandaceae), which grew outside the swamps and possibly in well-drained forest soils, are evidence for analogous climatic conditions. This is not a surprise because the time interval between 2.8 and 3.2 million years ago is the globally recognized mid-Piacenzian warm period. It is during this time interval when peat accumulated in these forests. It was also a perfect interval for the growth of the large trees of *Glyptostrobus*. These trees formed the spectacular fossil forest at Dunarobba, with the same type of flora and paleoenvironmental conditions as the UV subtropical assemblages. However, it remains unclear if these sites were truly contemporaneous, as well-established dates for Dunarobba are still lacking (Box 1.3).

1.13.3 Environmental Changes from 3.3 to 1.5 Ma in the Central Mediterranean Area

A reconsideration of what we have seen through our stepped time windows facing the intermontane depressions of central Italy suggests that this area, located between 42–43° N and 11–12° E, was affected by the first apparent changes in temperature and precipitation during the maximum expansion of Arctic ice about 2.6 million years ago. Together with a new,

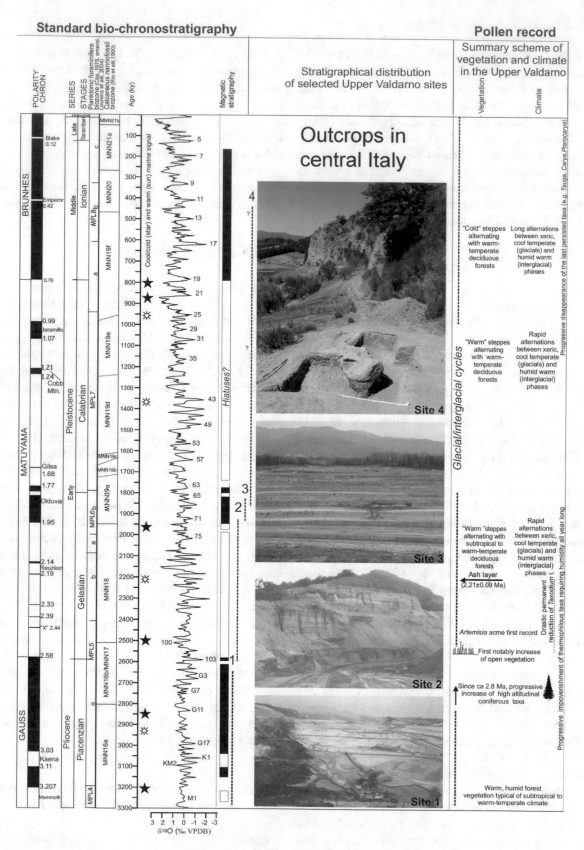

Fig. 1.35 Summary of vegetation and climate change in the Upper Valdarno Basin from 1.0 to 3.3 Ma. Panoramic photos and stratigraphic locations of selected sites (modified from Bertini 2013). Site 1: Panoramic view of the Piacenzian lacustrine cycle at Santa Barbara. Site 2: Panoramic view on the 35 m thick dominant eolian deposits at Rena Bianca (Gelasian). Site 3: Panoramic view of the Poggio Rosso succession in the SOLAVA clay pit; the Pratomagno ridge is visible in the background. An impressive accumulation of fossil bones, largely preserved in anatomical connection and dated at 1.87 Ma, occurs mid-way in the succession. A progressive increase of aridity during a glacial stage had severe consequences on the faunal communities, inducing the migration of open plain dwellers and the trapping of other residents around a few residual shrinking water bodies. Hyenas acted both as regulator of game populations debilitated by drought and major bone accumulators (Bertini et al. 2010). Site 4: Late Early to Middle Pleistocene deposits (fluvial, palustrine, alluvial fan) from the northern side of the Arno River. In the foreground, the latest discovered skull of *Mammuthus meridionalis* (photo courtesy Paolo Mazza)

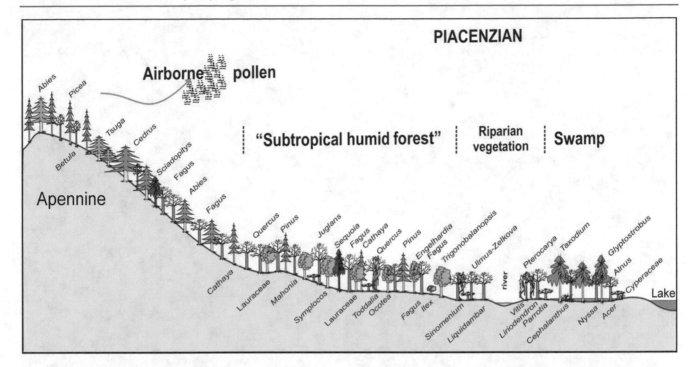

Fig. 1.36 Reconstruction of a Piacenzian vegetational transect across central Italy, modified from Bertini and Martinetto (2011). The proportion of evergreen trees is based on the micro- and macroflora assemblages of the Upper Valdarno Basin (Bertini and Roiron 1997; Bertini 2013)

Box 1.3: The Fossil Forest of Dunarobba

The famous and impressive Dunarobba fossil forest (Fig. 1.38), located in the TB, about 80 km south of the UV, preserves similar thermophilous plants [U0175] (Martinetto et al. 2014). It would be extremely temping to place the growth of this forest at about the same time as the UV subtropical swamps (Piacenzian) using the rich assemblage of fossil plants preserved in both localities. Unfortunately, the dating of the Dunarobba locality is not well constrained and it has often been assigned to the Gelasian (2.6–1.8 Ma). The forest features several large trunks, up to 7 m long, preserved in an upright position [U0176]. These grew in a swamp dominated by the conifer *Glyptostrobus europaeus* (see Chap. 2). Other woody plants include the angiosperms *Alnus*, *Cephalanthus* (buttonbush), and *Cornus* (dogwood), accompanied by herbaceous monocotyledons (*Carex*, *Cladium*, *Schoenoplectus* [sedges]). The occurrence of *Cryptomeria*, *Eurya*, and *Sinomenium* (now restricted to a single species in China), plants that today grow in the modern mixed mesophytic forests* of East Asia, suggest the presence of similar forests in well-drained parts of the Dunarobba paleoenvironment.

developing topography resulting from the rise of the Apennines, the landscape formerly dominated by subtropical trees turned to coniferous/beech forest and, subsequently, to steppe vegetation. As local elevations changed, both altitudinal arboreal taxa and herbs occupied the region, creating new complex patterns of biological competition in the communities, already cyclically changing under the effects of climate. The case studies presented as postcards, together with high-resolution datasets from other central Mediterranean sites, provide a sound understanding of several biological and climatic changes in this broader area.

The climate cycles detected in central Italy are quite similar to those described, even more recently, from other sections of the central Mediterranean area. These include key marine and continental succession in southern Italy, such as Semaforo-Vrica (Suc et al. 2010) and Montalbano Jonico (Bertini et al. 2015), and from the Balkan peninsula, such as Tenaghi Philippon (Pross et al. 2015) and Lake Ohrid (Sadori et al. 2016). Conversely, climate cycles in central Italy differ from those documented in northern Italian sites, where the glacial periods are marked by the development of coniferous forest (e.g., Bertini 2001, 2010). The expansion of this forest type is linked to a decline of temperature but not to a drastic decrease in humidity.

In summary, the pollen studies in the central Mediterranean area have informed us about five aspects of its recent geo-

Fig. 1.37 Microphotographs of selected pollen types from the Upper Valdarno and corresponding modern plants. Fossil pollen: (**a**) *Cathaya,* (**b**) *Nyssa,* (**c**) *Carya,* (**d**) *Symplocos,* (**e**) cf. *Craigia,* (**f**) *Tsuga.* Modern plants: (**g**) *Cathaya argyrophylla,* (**h**) *Symplocos cochinchinensis,* (**i**) *Craigia yunnanensis,* (**j**) *Nyssa sylvatica,* (**k**) *Carya* (Images courtesy of H Nimsch (**g**), P Pelser (**h**), Kunming Institute of Botany (**i**), Pinelands Nurseries (**j**), SR Manchester (**k**))

Fig. 1.38 The Fossil Forest of Dunarobba at the time of its discovery (AD 1987, photo by Chiara Benigni)

logic history. First, we have learned how climate and vegetation varied over the past three million years and, secondly, that subtropical ecosystems were extirpated, close to 1.4–1.2 million years ago. The flora we see around us, today, did not appear in a single burst of colonization and expansion. Rather, our third insight is that progressive changes in climate controlled progressive changes in plant composition resulting in its modern aspect. We have documented the history of biogeographic range expansion, contraction, and systematic extinction of plants in response to regional and global events in this region. And, lastly, we learned that refugial areas have played a significant role, over time, in the conservation of Pleistocene plants.

1.14   Synthesis of Three Million Years of Changes in Natural Systems

Our virtual field trip through the last three million years allowed us to reach areas across the globe, each of which was affected in a number of different ways by the late Cenozoic ice age. These areas extend from the European Alps, where the natural conditions of a recent past were very different from the present ones, to temperate regions less affected by

ice cover, including California, northern Mediterranean, southern Caucasus, the Himalaya foothills, and Japan. One thing that can be said about these vegetated landscapes is that they all were different than those of today. This is also true of the natural systems in tropical areas, including South America, Tanzania, Australia, and New Guinea. The vegetation that grew here, before 11,000 years ago, was not exactly what we find there at the present. In fact, vegetation, biomes, and environments across the planet were impacted from the range expansions and contractions of plant communities and the migration and disappearance of animals over various time scales. Almost everywhere we find records of past environments documenting the presence of plants and animals that do not correspond to our modern world. All these trends are attributed to climate change.

In the course of the twentieth century, marine sediments have recorded climate signals and climate proxies that have been revealed by the results of deep drilling programs, including the International Ocean Discovery Program (IODP). This nearly continuous record of climate and time has allowed us to better understand and interpret the discontinuous geological and paleontological records of the continental deposits formed during the last three million years. The paleoclimatic record obtained from ocean floor sediments extends deeper in time (Lisiecki and Raymo 2005), even if the resolution in deep time is at the scale of thousands of years. Since the late twentieth and beginning of the twenty-first centuries, this marine record has been augmented by the availability of long polar ice-core data that provide a very detailed record of the climatic oscillations over the past 800,000 years. The resolution of time in these ice cores may be on the order of decades or even years (Thomas et al. 2009).

Ice cores provide profound insights into past natural conditions, allowing for the reconstruction of deep-time atmospheric circulation patterns, wind strength, sea surface temperature, sea ice conditions, and greenhouse gas concentrations. Millennial scale climate changes are known to be influenced by variations in the Earth's orbit (Box P3) and the amount of solar radiation that the Earth receives across the polar to equatorial gradient. The continuous and detailed records unlocked from ocean floor sediment and ice cores have been correlated to the sedimentary sequences of the terrestrial realm.

The impact of climatic oscillations on continental vegetation at different latitudes has been demonstrated by studies of long pollen records. These records originate mostly from lake sediments, some of which (e.g., Pross et al. 2015; Sadori et al. 2016) compete with those from ice cores in their detail and completeness of the climatic trends over time. Of course, each region has its own vegetational history. The authors of this chapter have selected only a few localities where we have been able to reconstruct an almost continuous history of vegetation, in some cases with the accompanying fauna, for more than two million years.

Paleobotanical data from the Neogene document that vegetation and community stability has never been the rule, and changes have occurred over the last three million years in consort with climate variability. We recognized that fossil-plant assemblages are small "windows" into deep time and, combined, represent short and/or discontinuous records of past terrestrial systems, even if correlation to the continuous record of the marine isotope stages is often possible (e.g., Sects. 1.7 and 1.9). Greater insight into the dynamic history of plant communities is provided by fossil-rich lacustrine and fluvial sedimentary archives. Here, the incorporation of a multidisciplinary approach results in a more comprehensive reconstruction of paleoenvironments.

Over the last three million years, terrestrial ecosystems have experienced an impressive reorganization in two steps. The first step occurred around 2.7 million years before present. Here, the onset of the Northern Hemisphere glaciation (Shackleton et al. 1984) resulted in a significant spread of open biomes, including steppe and tundra. Such biomes are detected for the first time at mid-latitudes, in the previously densely forested Europe (e.g., Italy, see Sect. 1.12). The warm-temperate (= subtropical) forests that covered large parts of Eurasia prior to 2.7 million years ago began to con-

tract in area during the interval from 3.3 to 2.4 million years ago. As a result, several tens of woody genera were extirpated from Europe (Eiserhardt et al. 2015; Martinetto et al. 2017) and many warm-temperate plants endemic to present central and southwestern China were extirpated from Japan (Fig. 1.34). Later, from 2.7 to 0.9 million years ago, landscapes in southern Europe, Caucasus, Himalaya, Japan, and central Africa experienced cyclical changes. Animal and plant species went extinct and other species were extirpated over small or larger regional areas. In general, though, the fossil record indicates an absence of extreme climatic shifts or large biological turnover. Gradually, the biosphere acquired characteristics of the modern condition as elements of older floras (Martinetto 2015; Momohara 2016) did not survive into the late Early Pleistocene. At that point in time, all the Middle Pleistocene floras (e.g., Italy, starting from the 0.78-million-year-old Pianico flora, Japan, and probably everywhere) preserve plant fossils of species currently living in these geographies.

The second phase of plant extinctions and extirpations occurred during the EMPT (see Head and Gibbard 2015). This event was previously known as the mid-Pleistocene transition or mid-Pleistocene revolution (Berger and Jansen 1994). The interval from 1.4 to 0.4 million years ago was a time characterized by a change in the pattern of glacio-eustatic oscillations. Glacial advances and retreats were of lower amplitude and higher frequency before about 0.9 million years ago and higher amplitude combined with lower frequency afterwards. Based on paleoclimatic data, temperature oscillations were less intense in tropical areas (e.g., central Africa) than those detected at the mid-latitudes (de Menocal 2004). And, most of the biodiversity was concentrated and conserved in the tropics during the last three million years. Those organisms inhabiting the broad extratropical territories, where resource availability varied over time, had to cope with climate change. High-elevation geographies, like the Qinghai-Tibet plateau, may have been the laboratories in which taxa evolved adaptations to life in a cold climate and were favored when the climate was still mild. When the cold pulses spread over the continents, these preadapted organisms, such as the woolly rhinos (Deng et al. 2011), were able to conquer large areas. The largest of the present landmasses, Eurasia, was so deeply affected by climate change that forest biomes, once widespread across the region (see Chap. 2), were restricted in a few small refugia to the west (Magri et al. 2017) and in limited areas to the east (Momohara 2018). Indeed, central Japan is one refugium as is the Kathmandu Basin at the foot of the Himalaya. Here, despite the persistence of forests, the fossil-plant record indicates an overall climatic change during the deposition of the Gokarna Formation. The foothills of the Himalaya record at least three climate warm-and-cold phase cycles (with three cold peaks), between about 53,000 and 49,000 years, even in this non-glaciated area. These repeated changes in climate occurred over the last one million years, and the most intense pulses determined the extinction, extirpation, and dispersal of plants and animals, resulting in the formation of the modern biotopes.

We have seen that the fossil sites we visited almost invariably provide records of extinct animals and of extirpated plants (e.g., *Acer* gr. *cappadocicum* at Pianico). Physiographic and climatic barriers must have played a major role in determining the biotic composition of different ecological zones. The biotic interchanges between Africa and Europe, for example, have always been rather limited. Dispersal routes between these continents may have been via today's desert regions, during wetter and less arid phases, and through the Caucasian mountain ranges. Here, Pliocene-Quaternary volcanic and tectonic events resulted in a complex geomorphology, favoring a mosaic of vegetation types. However, the arrival of *Homo* from Africa was not impeded by these landscapes, as evidenced by the 1.8-million-year-old skeletal remains from the Georgian site of Dmanisi. The unexpected discovery of these hominid fossils indicates that human expansion into new territories occurred after the crucial phases of their early evolution in Africa (see Sect. 1.10).

The rare occurrence of early humans often is associated with abundant remains of other animals, mostly belonging to extinct species, which may have been a favored food resource. The extinction of these animals, all over the last three million years, appears to be concentrated during several events related to intervals of climate change. It is known that a small percentage of megafauna went extinct in central Africa during the Late Pleistocene and early Holocene (Sandom et al. 2014), whereas many large animals did not survive beyond 5000 years before present in North America, South America, Eurasia, and Australia.

The so-called American megafaunal extinction interval (c. 13 ka; Koch and Barnosky 2006) is well documented at the Rancho La Brea site. Here, the rich vertebrate fossil record brackets a time interval approximately coincident with the onset of the Younger Dryas cold phase (12.9–11.7 ka; Kennett and West 2008). The impact of climate change affected mammalian predators at La Brea (Meachen et al. 2014; O'Keefe et al. 2014). Between 50,000 and 11,000 years ago, climate shifted from a distinct cooling phase to a rapid warming, resulting in a change in skull size of both dire wolves and saber-toothed cats. After the end of the last ice age (c. 11 ka), glaciers receded, climate warmed, and dire wolves in this region became smaller and more graceful, adapting to take smaller prey. In contrast, saber-toothed cat populations became larger and adapted to taking larger prey after the end of the ice age. Both mammals are now extinct, the causes of whether climate, anthropogenic activity, or both are still debated.

Dire wolves and saber-toothed cats are not the only extinct megafauna. The American megafaunal extinction also affected giant beavers, giant sloths, and the last mastodons. Victims of the megafaunal extinction in Eurasia include mammoths, woolly rhinos, and cave bears. Across the Sahul paleocontinent, giant wombats (*Diprotodon optatum*) and marsupial lions (*Thylacoleo carnifex*) are iconic extinct animals, but, in reality, around 90 large vertebrate taxa disappeared from Sahul sometime during the Pleistocene. The majority of these losses (54 taxa) clearly took place within the last 400,000 years.

Explanations for megafaunal extinction have centered on climatic change, even if Sahul shows a notable contrast with the dramatic biotic change patterns in the temperate biota of the Northern Hemisphere. The Australian biota persisted through climatic oscillations over multiple glacial cycles, which has led to its high biodiversity and endemism in many localities. Nevertheless, a progressive increase of seasonal dryness and aridity probably influenced the profound ecosystem collapse and reduction in food sources (Miller et al. 2005). Both factors, combined, could explain megafaunal extinction. Human involvement in the disappearance of some species remains possible, but unproven.

In Sahul, the loss of most species occurred before the arrival of humans (c. 50–45 ka), and some authors believe that human activities may have just compounded an extinction process well under way (Wroe et al. 2013). Other workers believe that megafaunal extinctions were linked to human occupation and climate change played little, if any, role in its demise (Sandom et al. 2014). This last view is supported by the low rate of Late Pleistocene mammal extinctions in central Africa, where the megafauna has the longest history of coexistence and coevolution with humans. Fossil assemblages from sites like Olduvai Gorge prove that Early and Middle Pleistocene extinctions have been a consistent feature in central Africa. Hence, the evolution of new megafaunal species on the continent may have been less affected by long-term human hunting pressures. Sandom et al. (2014) suggest that the most severe Late Pleistocene extinctions occurred to a region (e.g., Australia) after the arrival of modern humans. Yet, the history of Africa, Australia, and other continents is very different, and generalizations about the factors influencing the destiny of the megafauna may need to be re-evaluated. A good compromise between these two views would be to consider each case of extinction separately, because both climate and human activities may have affected this process to different degrees and at different times in different geographical regions.

1.15   Conclusions

The last three million years has been a time of change in response to a series of glacial-interglacial cycles, with a periodicity of 100,000 years in the last million years and

41,000 years from one to three million years ago, as a consequence of extraterrestrial factors (Box P3). Ice sheets and mountain glaciers covered the high latitudes, both in the northern and southern continents. During the most intense glaciations, an enormous volume of water, originating from the oceans as condensate, was stored in continental ice sheets. Thus, global sea level dropped to at least 120 m lower than today's base level, resulting in dramatic changes to the continent's geography. Eurasia was periodically connected to America through the broad Beringia land bridge. Australia, New Guinea, and Tasmania formed the extensive Sahul paleocontinent. Changes in climate strongly affected the high and mid-latitudes, but also had effects in the tropics. It is well documented in South America that plants occupying higher elevations extended their range down slope to lower elevations during glacial intervals. Forest-plant communities varied more and more in their species composition and were very different from today's forests back to three million years ago. However, all the plant genera we find as part of Pleistocene and latest Pliocene forests are still living today, and only some species experienced extinction during the Quaternary. Conversely, many genera of large animals (megafauna) met their demise either due to climate change or to human hunting (or both). Africa is the continent where extinctions have been more limited, such that most of its Pleistocene megafauna can still be seen in nature parks today. One member of that megafauna, the primate genus *Homo,* also had its birthplace in Africa beginning about two million years ago. Unlike other Pleistocene megafauna, though, this animal successfully expanded its range beginning around 50,000 years ago to inhabit all of the planet's continents and altering Earth history in the process.

For Deeper Learning

- Dmanisi paleoanthropological site: Gabunia et al. (2000); Lordkipanidze et al. (2007, 2013); Ferring et al. (2011)
- Kathmandu Basin evolution: Yoshida and Igarashi (1984); Yoshida and Gautam (1988); Sakai et al. (2001, 2006)
- Kathmandu Basin pollen and plant macrofossils: Yoshida and Igarashi (1984); Igarashi et al. (1988); Nakagawa et al. (1996); Fujii and Sakai (2001, 2002); Ooi (2001); Paudayal and Ferguson (2004); Paudayal (2005, 2006, 2011); Bhandari and Paudayal (2007); Bhandari et al. (2009, 2010, 2011a, 2011b, 2016)
- Lake levels and climate cycles in South America: Bush et al. (2004); Hanselman et al. (2011); Hermanowski et al. (2012)
- Start of major glacial advances on the Italian side of the Alps c. 850 ka ago: Muttoni et al. (2003, 2007); Ravazzi et al. (2014)

Questions

1. What characteristics define the Quaternary interval as either icehouse or hothouse?
2. What ecological dynamics separate the concepts of extirpation and extinction?
3. During which part of the Quaternary did the mammoth steppe exist, and across which latitudinal (or hemispherical) gradient is it restricted?
4. How do the Quaternary fossil plants of Japan compare with those of the Caucasus?
5. What physical conditions may have played a role in the preservation of extensive mammal bone assemblages at Quaternary sites?
6. Explain how data from pollen studies provide insights into the response of plants to climate change.
7. Just 1.5 million years ago, the forests of southern Europe comprised plants that today grow only in China. What factors may have controlled these relationships?
8. Why is the fossil forest at Dunarobba, Italy, considered a unique fossil site?
9. Although roughly contemporaneous in time, why do the Rancho La Brea and the Cuddie Springs sites preserve different animals?
10. The Late Pleistocene Siberian flora grew at high northern latitudes not covered by ice. What rainfall features permitted an ice-free environment when there was a thick ice cover at lower latitude (e.g., Sweden)?
11. Explain the major changes in global vegetational patterns recorded in the Pliocene to Middle Pleistocene sediment record. What impact(s) did these have on the trajectory of plants into the Holocene?

Acknowledgments The authors are thankful to E Hakobyan for the help in the preparation of text and images, to R Sardella for useful information about vertebrate fossil assemblages of Europe, to C Scotese for the permission to use his paleogeographic map, and to TK Akabane, K Ashastina, JF Barcelona, C Benigni, M Costea, H Hooghiemstra, JTF Guimaeres, DL Nickrent, P Mazza, K Nixon, and PB Pelser for the photographs provided. Special thanks to Soprintendenza Speciale Archeologia Belle Arti e Paesaggio (SSABAP) of Rome for allowing us to use images of the Polledrara di Cecanibbio site without charges.

References

Andreev AA, Schirrmeister L, Tarasov PE, Ganopolski A, Brovkin V, Siegert C, Wetterich S, Hubberten H-W (2011) Vegetation and climate history in the Laptev Sea region (Arctic Siberia) during Late Quaternary inferred from pollen records. Quat Sci Rev 30:2182–2199

Anzidei AP, Bulgarelli GM, Catalano P, Cerilli E, Gallotti R, Lemorini C, Milli S, Palombo MR, Pantano W, Santucci E (2012) Ongoing research at the late Middle Pleistocene site of La Polledrara di

Cecanibbio (central Italy), with emphasis on human-elephant relationships. Quat Int 255:171–187

Ashastina K, Kuzmina S, Rudaya N, Troeva E, Schoch WH, Römermann C, Reinecke J, Otte V, Savvinov G, Wesche K, Kienast F (2018) Woodlands and steppes: Pleistocene vegetation in Yakutia's most continental part recorded in the Batagay permafrost sequence. Quat Sci Rev 196:38–61

Astakhov V, Shkatova V, Zastrozhnov A, Chuyko M (2016) Glaciomorphological map of the Russian federation. Quat Int 420:4–14

Baker PA, Fritz SC (2015) Nature and causes of Quaternary climate variation of tropical South America. Quat Sci Rev 124:31–47

Baker PA, Rigsby CA, Seltzer GO, Fritz SC, Lowenstein TK, Bacher NP, Veliz C (2001) Tropical climate changes at millennial and orbital timescales in the Bolivian Altiplano. Nature 409:698–701

Barnosky AD, Lindsey EL (2010) Timing of Quaternary megafaunal extinction in South America in relation to human arrival and climate change. Quat Int 217(1–2):10–29

Berger WH, Jansen E (1994) Mid-Pleistocene climate shift: the Nansen connection. In: Johannessen OM et al (eds) The polar oceans and their role in shaping the global environment, AGU geophysical monograph, vol 85, pp 295–311

Bertini A (2001) Pliocene climatic cycles and altitudinal forest development from 2.7 Ma in the Northern Apennines (Italy): evidence from the pollen record of the Stirone section (~5.1 to ~2.2 Ma). Géobios 34(3):253–265

Bertini A (2010) Pliocene to Pleistocene palynoflora and vegetation in Italy: state of the art. Quat Int 225:5–24

Bertini A (2013) Climate and vegetation in the Upper Valdarno basin (central Italy) as a response to northern hemisphere insolation forcing and regional tectonics in the late Pliocene-early Pleistocene. Ital J Geosci (Boll Soc Geol Ital) 132(1):137–148

Bertini A, Magi M, Mazza P, Fauquette S (2010) Impact of short-term climatic events on latest Pliocene land settings and communities in Central Italy (Upper Valdarno basin). Quat Int 225:92–105

Bertini A, Magri D, Martinetto E, Sadori L, Vassio E (2014) The Pleistocene flora of central Italy. In: Kustatscher E, Roghi G, Bertini A, Miola A (eds) Palaeobotany of Italy, 2nd edn. Naturmuseum Sudtirol, Bolzano, p 3080338

Bertini A, Roiron P (1997) Evolution de la végétation et du climat pendant le Pliocène moyen, en Italie centrale: apport de la palynologie et de la macroflore à l'étude du bassin du Valdarno supérieur (coupe de Santa Barbara). Comptes rendus de l'Académie des sciences. Série 2. Sci Terre Planets 324(9):763–771

Bertini A, Toti F, Marino M, Ciaranfi N (2015) Vegetation and climate across the Early–Middle Pleistocene transition at Montalbano Jonico, southern Italy. Quat Int 383:74–88

Bhandari S, Momohara A, Paudayal KN (2009) Late Pleistocene plant macro-fossils from the Gokarna Formation of the Kathmandu Valley, Central Nepal. Bull Dep Geol 12:75–88

Bhandari S, Momohara A, Uhl D, Paudayal KN (2016) Paleoclimatic significance of the late Quaternary plant macrofossils from the Gokarna Formation, Kathmandu Valley, Nepal. Rev Palaeobot Palynol 228:98–112

Bhandari S, Paudayal KN (2007) Palynostratigraphy and palaeoclimatic interpretation of the Plio-Pleistocene Lukundol Formation from the Kathmandu Valley, Nepal. J Nepal Geol Soc 35:1–10

Bhandari S, Paudayal KN, Momohara A (2010) Late Pleistocene plant macrofossils from the Thimi Formation (Madhyapur Thimi section) of the Kathmandu Valley, Central Nepal. J Nepal Geol Soc 40:31–48

Bhandari S, Paudayal KN, Momohara A (2011a) Late Quaternary plant macrofossils assemblages from the Besigaon section of the Gokarna Formation, Kathmandu Valley, Central Nepal. J Nepal Geol Soc 42:1–12

Bhandari S, Paudayal KN, Momohara A (2011b) Climate change on the basis of plant macrofossil assemblages from the late Quaternary sediments from the Mulpani section of the Gokarna Formation, Kathmandu Valley, Nepal. J Stratigr Assoc Nepal 7:47–58

Birks HH (2008) The Late-Quaternary history of arctic and alpine plants. Plant Ecol Divers 1:135–146

Birks HH, Birks HJB (2000) Future uses of pollen analysis must include plant macrofossils. J Biogeogr 27:31–35

Blain HA, Agustí J, Lordkipanidze D, Rook L, Delfino M (2014) Paleoclimatic and paleoenvironmental context of the early Pleistocene hominins from Dmanisi (Georgia, Lesser Caucasus) inferred from the herpetofaunal assemblage. Quat Sci Rev 105:136–150

Blumenschine RJ, Stanistreet IG, Masao FT (2012) Olduvai Gorge and the Olduvai landscape paleoanthropology project. J Hum Evol 63:247–250

Boeskorov GG, Lazarev PA, Sher AV, Davydov SP, Bakulina NT, Shchelchkova MV, Binladen J, Willerslev E, Buigues B, Tikhonov AN (2011) Woolly rhino discovery in the lower Kolyma River. Quat Sci Rev 30:2262–2272

Boeskorov GG, Potapova OR, Protopopov AV, Plotnikov VV, Agenbroad LD, Kirikov KS, Pavlov IS, Shchelchkova MV, Belolyubskii IN, Tomshin MD, Kowalczyk R, Davydov SP, Kolesov SD, Tikhonov AN, van der Plicht J (2016) The Yukagir Bison: the exterior morphology of a complete frozen mummy of the extinct steppe bison, *Bison priscus* from the early Holocene of northern Yakutia, Russia. Quat Int 406:94–110

Boeskorov GG, Potapova OR, Protopopov AV, Plotnikov VV, Maschenko EN, Shchelchkova MV, Petrova EA, Kowalczyk R, van der Plicht J, Tikhonov AN (2018) A study of a frozen mummy of a wild horse from the Holocene of Yakutia, East Siberia, Russia. Mamm Res 63:307–314

Bogotá-A RG, Groot MHM, Hooghiemstra H, Lourens LJ, van der Linden M, Berrio JC (2011) Rapid climate change from north Andean Lake Fúquene pollen records driven by obliquity: implications for a basin-wide biostratigraphic zonation for the last 284 ka. Quat Sci Rev 30:3321–3337

Buizert C, Adrian B, Ahn J, Albert M, Alley RB, Baggenstos D, Bauska TK, Bay RC, Bencivengo BB, Bentley CR, Brook EJ, Chellman NJ, Clow GD, Cole-Dai J, Conway H, Cravens E, Cuffey KM, Dunbar NW, Edwards JS, Fegyveresi JM, Ferris DG, Fitzpatrick JJ, Fudge TJ, Gibson CJ, Gkinis V, Goetz JJ, Gregory S, Hargreaves GM, Iverson N, Johnson JA, Jones TR, Kalk ML, Kippenhan MJ, Koffman BG, Kreutz K, Kuhl TW, Lebar DA, Lee JE, Marcott SA, Markle BR, Maselli OJ, McConnell JR, McGwire KC, Mitchell LE, Mortensen NB, Neff PD, Nishiizumi K, Nunn RM, Orsi AJ, Pasteris DR, Pedro JB, Pettit EC, Buford Price P, Priscu JC, Rhodes RH, Rosen JL, Schauer AJ, Schoenemann SW, Sendelbach PJ, Severinghaus JP, Shturmakov AJ, Sigl M, Slawny KR, Souney JM, Sowers TA, Spencer MK, Steig EJ, Taylor KC, Twickler MS, Vaughn BH, Voigt DE, Waddington ED, Welten KC, Wendricks AW, White JWC, Winstrup M, Wong GJ, Woodruff TE (2015) Precise interpolar phasing of abrupt climate change during the last ice age. Nature 520:661

Burbridge RE, Mayle FE, Killenn T (2004) Fifty-thousand-year vegetation and climate history of Noel Kempff Mercado National Park, Bolivian Amazon. Quat Res 61:215–230

Bush BM, De Oliveira PE, Colinvaux PA, Miller MC, Moreno JE (2004) Amazonian paleoecological histories: one hill, three watersheds. Palaeogeogr Palaeoclimatol Palaeoecol 214:359–393

Castorina F, Masi U, Milli S, Anzidei AP, Bulgarelli GM (2015) Geochemical and Sr-Nd isotopic characterization of Middle Pleistocene sediments from the paleontological site of La Polledrara di Cecanibbio (Sabatini Volcanic District, central Italy). Quat Int 357:253–263

Cheng H, Sinha A, Cruz FW, Wang X, Edwards RL, d'Horta FM, Ribas CC, Vuille M, Stott LD, Auler AS (2013) Climate change patterns in Amazonia and biodiversity. Nat Commun 4:1411

Cohen MLC, Rossetti DF, Pessenda LCR, Friaes YS, Oliveira PR (2014) Late Pleistocene glacial forest of Humaitá - Western Amazonia. Palaeogeogr Palaeoclimatol Palaeoecol 415:37–47

Colinveaux PA, DeOlivera PE, Moreno JE, Miller MC, Bush MB (1996) A long pollen record from lowland Amazonia: forest cooling in glacial times. Science 274:85–88

Cwynar LC, Ritchie JC (1980) Arctic steppe-tundra: a Yukon perspective. Science 208:1375–1377

D'Apolito C, Absy ML, Latrubesse E (2013) The hill of six lakes revisited: new data and re-evaluation of a key Pleistocene Amazon site. Quat Sci Rev 76:140–155

Dansgaard W, Johnsen SJ, Clausen HB, Dahl-Jensen D, Gundestrup NS, Hammer CU, Hvidberg CS, Steffensen JP, Sveinbjörnsdottir AE, Jouzel J, Bond G (1993) Evidence for general instability of past climate from a 250-kyr ice-core record. Nature 364:218–220

Deng T, Wang X, Fortelius M, Li Q, Wang Y, Tseng ZJ, Takeuchi GT, Saylor JE, Säilä LK, Xie G (2011) Out of Tibet: Pliocene woolly rhino suggests high-plateau origin of ice age megaherbivores. Science 333(6047):1285–1288

Denham T, Mountain M-J (2016) Resolving some chronological problems at Nombe rock shelter in the highlands of Papua New Guinea. Archaeol Ocean 51(Suppl. 1):73–83

Denk T (2006) *Rhododendron ponticum* L. var. *sebinense* (Sordelli) Sordelli in the Late Pleistocene flora of Hötting, Northern Calcareous Alps: witness of a climate warmer than today? Veroffenlichungen des Tiroler Landesmuseums Ferdinandeum 86:43–66

DeSantis LR, Haupt RJ (2014) Cougars' key to survival through the Late Pleistocene extinction: insights from dental microwear texture analysis. Biol Lett 10(4):20140203

Ehlers J, Gibbard PL, Hughes PD (eds) (2011) Quaternary glaciations - extent and chronology. A closer look. Developments in quaternary science, vol 15. Elsevier, Amsterdam, pp 2–1108

Eiserhardt WL, Borchsenius F, Plum CM, Ordonez A, Svenning JC (2015) Climate-driven extinctions shape the phylogenetic structure of temperate tree floras. Ecol Lett 18(3):263–272

EPICA Community Members (2004) Eight glacial cycles from an Antarctic ice core. Nature 429:623–628

Esu D, Gianolla D (2009) The malacological record from the Middle Pleistocene Pianico-Sellere Basin (Bergamo, northern Italy). Quat Int 204:3–10

Ferring R, Oms O, Agusti J, Berna F, Nioradze M, Shelia T, Tappe M, Vekua A, Zhvania D, Lordkipanidze D (2011) Earliest human occupations at Dmanisi (Georgian Caucasus) dated to 1.85-1.78 Ma. Proc Natl Acad Sci U S A 108:10432–10436

Field J, Dodson J (1999) Late Pleistocene megafauna and human occupation at Cuddie Springs, southeastern Australia. Proc Prehist Soc 65:275–301

Field J, Dodson J, Prosser I (2002) A Late Pleistocene vegetation history from the Australian semi-arid zone. Quat Sci Rev 21(8–9):1005–1019

Fillios M, Field J, Charles B (2010) Investigating human and megafauna co-occurrence in Australian prehistory: mode and causality in fossil accumulations at Cuddie Springs. Quat Int 211:123–143

Fisher DC, Tikhonov AN, Kosintsev PA, Rountrey AN, Buigues B, van der Plicht J (2012) Anatomy, death, and preservation of a woolly mammoth (*Mammuthus primigenius*) calf, Yamal Peninsula, northwest Siberia. Quat Int 255:94–105

Fox-Dobbs K, Stidham TA, Bowen GJ, Emslie SD, Koch PL (2006) Dietary controls on extinction versus survival among avian megafauna in the late Pleistocene. Geology 34(8):685–688

Fujii R, Sakai H (2001) Palynological study of the drilled sediments from the Kathmandu Basin and its palaeoclimate and sedimentological significance. J Nepal Geol Soc 25(Sp. Issue):53–61

Fujii R, Sakai H (2002) Paleoclimatic changes during the last 2.5 myr recorded in the Kathmandu Basin, Central Nepal Himalayas. J Asian Earth Sci 20:255–266

Fullagar R, Field J (1997) Pleistocene seed grinding implements from the Australian arid zone. Antiquity 71:300–307

Fullagar R, Field J, Kealhofer L (2008) Grinding stones and seeds of change: starch and phytoliths as evidence of plant food processing. In: Rowan YM, Ebeling JR (eds) New approaches to old stones: recent studies of ground stone artifacts. Equinox Publishing P/L, London, pp 159–172

Gabrielyan I, Kovar-Eder J (2011) The genus *Acer* from Lower/Middle Pleistocene diatomites of the Sisian Formation, Syunik region, South Armenia. Rev Palaeobot Palynol 165:111–134

Gabunia L, Vekua A, Lordkipanidze D, Swisher CC, Reid Ferring R, Justus A, Nioradze M et al (2000) Earliest Pleistocene hominid cranial remains from Dmanisi, Republic of Georgia: taxonomy, geological setting, and age. Science 288(5468):1019–1025

Gerhart LM, Harris JM, Nippert JB, Sandquist DR, Ward JK (2012) Glacial trees from the La Brea tar pits show physiological constraints of low CO2. New Phytol 194(1):63–69

Gianotti F, Forno MG, Ivy-Ochs S, Kubik PW (2008) New chronological and stratigraphical data on the Ivrea Amphitheatre (Piedmont, NW Italy). Quat Int 190:123–135

Gianotti F, Forno MG, Ivy-Ochs S, Monegato G, Pini R, Ravazzi C (2015) Stratigraphy of the Ivrea Morainic amphitheatre (NW Italy). An updated synthesis. Alp Mediterr Quat 28(1):29–58

Gianotti F, Pini R (2011) Stratigraphical subdivision of the Middle Pleistocene glacigenic sequence of the Ivrea amphitheatre (Piedmont, NW Italy). Abstracts XVIII INQUA Congress Bern 2011, 21-27 July. Quat Int 279–280:165

Goddu SR, Appel E, Gautam P, Oches EA, Wehland F (2007) The lacustrine section at Lukundol, Kathmandu basin, Nepal: dating and magnetic fabric aspects. J Asian Earth Sci 30:73–81

Govoni L, Paganoni A, Sala B (2006) The mammal fauna of the Piànico-Sèllere Basin. In: Donegana M, Ravazzi C (eds) INQUA-SEQS conference quaternary stratigraphy and evolution of the alpine region in the European and global framework. Field trip guide. INQUA-SEQS, Milano, pp 63–71

Granger DE, Muzikar PF (2001) Dating sediment burial with in situ produced cosmogenic nuclides: theory, techniques, and limitations. Earth Planet Sci Lett 188:269–281

Guthrie RD (1990) Frozen fauna of the mammoth steppe - the story of blue babe. University of Chicago Press, Chicago, 323 pp

Hanselman JA, Bush MB, GoslingWD CA, Knox C, Baker PA, Fritz SC (2011) A 370,000-year record of vegetation and fire history around Lake Titicaca (Bolivia/Peru). Palaeogeogr Palaeoclimatol Palaeoecol 305:201–214

Hay R (1976) Geology of the Olduvai Gorge: a study of sedimentation in a semiarid basin. University of California Press, Berkeley-Los Angeles-London, 203 pp

Head MJ, Gibbard PL (2015) Formal subdivision of the quaternary system/period: past, present, and future. Quat Int 383:4–35

Hermanowski B, Lima da Costa M, Carvalho AT, Behling H (2012) Palaeoenvironmental dynamics and underlying climatic changes in southeast Amazonia (Serra Sul dos Carajas, Brazil) during the late Pleistocene and Holocene. Palaeogeogr Palaeoclimatol Palaeoecol 365–366:227–246

Hesse PP, Magee JW, van der Kaars S (2004) Late Quaternary climates of the Australian arid zone: A review. Quat Int 118–119:87–102

Holden AR, Erwin DM, Schick KN, Gross J (2015) Late Pleistocene galls from the La Brea Tar Pits and their implications for cynipine wasp and native plant distribution in southern California. Quat Res 84(3):358–367

Holden AR, Southon JR, Will K, Kirby ME, Aalbu RL, Markey MJ (2017) A 50,000 year insect record from Rancho La Brea, Southern California: insights into past climate and fossil deposition. Quat Sci Rev 168:123–136

Hooghiemstra H, Ran ETH (1994) Late and middle Pleistocene climatic change and forest development in Colombia: pollen record

Funza-2 (2–158 m core interval). Palaeogeogr Palaeoclimatol Palaeoecol 109:211–246

Hughes ALC, Gyllencreutz R, Lohne ØS, Mangerud J, Svendsen JI (2016) The last Eurasian ice sheets – a chronological database and time-slice reconstruction, DATED-1. Boreas 45:1–45

Husband RA (1924) Variability in *Bubo virginianus* from Rancho La Brea. Condor 26(6):220–225

Huzita K (1954) Stratigraphic significance of the plant remains contained in the late Cenozoic formations in central Kinki, Japan. J Inst Polytech Osaka City Univ Ser G 2:75–88

Igarashi Y, Yoshida M, Tabata H (1988) History of vegetation and climate in the Kathmandu Valley. Proc Ind Nat Sci Acad 54A(4):550–563

Itihara M (1960) Some problems of the quaternary sedimentaries in the Osaka and Akasi areas, Japan. J Inst Polytech Osaka City Univ Ser G 5:13–30

Jakobsson M, Nilsson J, Anderson L, Backman J, Björk G, Cronin TM, Kirchner N, Koshurnikov A, Mayer L, Noormets R, O'Regan M, Stranne C, Ananiev R, Barrientos Macho N, Cherniykh D, Coxall H, Eriksson B, Flodén T, Gemery L, Gustafsson Ö, Jerram K, Johansson C, Khortov A, Mohammad R, Semiletov I (2016) Evidence for an ice shelf covering the central Arctic Ocean during the penultimate glaciation. Nat Commun 7:10365

Joannin S, Cornée JJ, Münch P, Fornari M, Vasiliev I, Krijgsman W, Nahapetyan S, Gabrielyan I, Ollivier V, Roiron P, Chataigner C (2010) Early Pleistocene climate cycles in continental deposits of the Lesser Caucasus of Armenia inferred from palynology, magnetostratigraphy, and 40Ar/39Ar dating. Earth Planet Sci Lett 291(1–4):149–158

Johnson C (2006) Australia's mammal extinctions: A 50 000 year history. Cambridge University Press, Cambridge

Jouzel J, Lorius C, Petit JR, Genthon C, Barkov NI, Kotlyakov VM, Petrov VM (1987) Vostok ice core: a continuous isotope temperature record over the last climatic cycle (160,000 years). Nature 329:403–408

Kazaoka O (1988) Stratigraphy and sedimentary facies of the Uonuma Group in the Higashikubiki Hills, Niigata Prefecture, Central Japan. Earth Sci (Chikyu Kagaku) 42:61–83. (in Japanese with English Abstract)

Kennett JP, West A (2008) Biostratigraphic evidence supports Paleoindian population disruption at c.12.9 ka. Proc Natl Acad Sci USA 105:E110

Kienast F (2013) Plant macrofossil records – Arctic Eurasia. In: Elias SA, Mock C (eds) Encyclopedia of quaternary science, vol 3, 2nd edn. Elsevier, Amsterdam, pp 733–745

Kienast F, Rudaya N, Maschenko EN, Potapova O, & Protopopov AV (2018) The habitat of the Megin Mammoth: vegetation, environments and climate in Central Yakutia during the late Weichselian. In: 5th European Conference on Permafrost (EUCOP), 23 June to 1 July 2018, Chamonix-Mont Blanc, France

Kienast F, Schirrmeister L, Siegert C, Tarasov P (2005) Palaeobotanical evidence for warm summers in the East Siberian Arctic during the last cold stage. Quat Res 63:283–300

Kienast F, Tarasov P, Schirrmeister L, Grosse G, Andreev AA (2008) Continental climate in the East Siberian Arctic during the last interglacial: implications from palaeobotanical records. Glob Planet Chang 60:535–562

Kienast F, Wetterich S, Kuzmina S, Schirrmeister L, Andreev AA, Tarasov P, Nazarova L, Kossler A, Frolova L, Kunitsky VV (2011) Paleontological records indicate the occurrence of open woodlands in a dry inland climate at the present-day Arctic coast in western Beringia during the Last Interglacial. Quat Sci Rev 30:2134–2159

Kirscher U, Gabrielyan I, Scharrer S, Bruch AA, Kuiper K, Bachtadse V (2014) High resolution magnetostratigraphy and radiometric dating of early Pleistocene lake sediments from Southern Armenia. Quat Int 328–329:31–44

Kirscher U, Oms O, Bruch AA, Shatilova I, Chochishvili G, Bachtadse V (2017) The Calabrian in the Western Transcaucasian basin (Georgia): Paleomagnetic constraints from the Gurian regional stage. Quat Sci Rev 160:96–107

Knüsel S, Ginot P, Schotterer U, Schwikowski M, Gäggeler HW, Francou B, Taupin JD (2003) Dating of two nearby ice cores from the Illimani, Bolivia. J Geophys Res Atmos 108:D6

Koch PL, Barnosky AD (2006) Late Quaternary extinctions: state of the debate. Annu Rev Ecol Evol Syst 37:215–250

Köppen W (1918) Klassifikation der Klimate nach Temperatur, Niederschlag und Jahreslauf. Petermanns Geogr Mitt 64:193–203

Lambeck K, Chappell J (2002) Sea level change through the last glacial cycle. Science 292:679–686

Lambeck K, Rouby H, Purcell A, Sun Y, Sambridge M (2014) Sea level and global ice volumes from the last glacial maximum to the holocene. Proc Natl Acad Sci 111:15296–15303

Leakey LSB (1959) A new fossil skull from Olduvai. Nature 184:491–493

Leakey LSB, Tobias PV, Napier JR (1964) A new species of the genus Homo from Olduvai Gorge. Nature 202:7–9

Lisiecki LE, Raymo ME (2005) A Pliocene-Pleistocene stack of 57 globally distributed benthic $\delta^{18}O$ records. Paleoceanography 20:1003–1020

Lordkipanidze D, Jashashvili T, Vekua A, Ponce de León MS, Zollikofer CPE, Rightmire GP, Pontzer H et al (2007) Postcranial evidence from early Homo from Dmanisi, Georgia. Nature 449(7160):305

Lordkipanidze D, Ponce de León MS, Margvelashvili A, Rak Y, Rightmire GP, Vekua A, Zollikofer CPE (2013) A complete skull from Dmanisi, Georgia, and the evolutionary biology of early Homo. Science 342(6156):326–331

Lyell C (1833) Principles of geology, vol 3. John Murray, London

Magri D, Di Rita F, Aranbarri J, Fletcher W, González-Sampériz P (2017) Quaternary disappearance of tree taxa from Southern Europe: timing and trends. Quat Sci Rev 163:23–55

Malla SB, Shrestha AB, Rajbhandary SB, Shrestha TB, Adhikari PM, Adhikari SR (1976) Flora of Langtang and cross section vegetation survey (central zone). Bull Dept Med Plants 6:269

Mangili C, Brauer A, Plessen B, Moscariello A (2007) Centennial-scale oscillations in oxygen and carbon isotopes of endogenic calcite from a 15,500 varve year record of the piànico interglacial. Quat Sci Rev 26:1725–1735

Martinetto E (2009) Palaeoenvironmental significance of plant macrofossils from the Piànico Formation, Middle Pleistocene of Lombardy, North Italy. Quat Int 204:20–30

Martinetto E (2015) Monographing the Pliocene and early Pleistocene carpofloras of Italy: methodological challenges and current progress. Palaeontogr Abt B 293:57–99

Martinetto E, Bertini A, Basilici G, Baldanza A, Bizzarri R, Cherin M, Gentili S, Pontini MR (2014) The plant record of the Dunarobba and Pietrafitta sites in the Plio-Pleistocene palaeoenvironmental context of central Italy. Alp Mediterr Quat 27(1):29–72

Martinetto E, Momohara A, Bizzarri R, Baldanza A, Delfino M, Esu D, Sardella R (2017) Late persistence and deterministic extinction of "humid thermophilous plant taxa of East Asian affinity" (HUTEA) in southern Europe. Palaeogeogr Palaeoclimatol Palaeoecol 467:211–231

Meachen JA, O'Keefe FR, Sadleir RW (2014) Evolution in the sabretooth cat, *Smilodon fatalis*, in response to Pleistocene climate change. J Evol Biol 27:714–723

de Menocal P (2004) African climate change and faunal evolution during the Pliocene-Pleistocene. Earth Planet Sci Lett 220:3–24

Messager E, Lordkipanidze D, Kvavadze E, Ferring CR, Voinchet P (2010) Palaeoenvironmental reconstruction of Dmanisi site (Georgia) based on palaeobotanical data. Quat Int 223–224:20–27

Miki S (1938) On the change of flora of Japan since the upper Pliocene and the floral composition at the present. Jpn J Bot 9:213–251

Miki S (1948) Floral remains in Kinki and adjacent districts since the Pliocene with description of 8 new species. Mineral Geol (Kobutsu to Chishitsu) 2:105–144. (in Japanese with English abstract)

Miller G, Fogel M, Magee J, Gagan M, Clarke S, Johnson B (2005) Ecosystem collapse in Pleistocene Australia and a human role in megafaunal extinction. Science 309:287–290

Miyoshi N, Fujiki T, Morita Y (1999) Palynology of a 250-m core from Lake Biwa: a 430,000-year record of glacial-interglacial vegetation change in Japan. Rev Palaeobot Palynol 104:267–283

Möller P, Alexanderson H, Funder S, Hjort C (2015) The Taimyr Peninsula and the Severnaya Zemlya archipelago, Arctic Russia: a synthesis of glacial history and palaeo-environmental change during the Last Glacial cycle (MIS 5e–2). Quat Sci Rev 107:149–181

Momohara A (2016) Stages of major floral change in Japan based on macrofossil evidence and their connection to climate and geomorphological changes since the Pliocene. Quat Int 397:92–105

Momohara A (2017) Plio-Pleistocene geomorphological development of the Japanese Islands and floral and vegetation changes. Quat Res (Tokyo) 56:251–264. (in Japanese with English abstract)

Momohara A (2018) Influence of mountain formation on floral diversification in Japan, based on macrofossil evidence. In: Hoorn C, Perrigo A, Antonelli A (eds) Mountains, climates and biodiversity. Wiley, Hoboken, pp 459–473

Momohara A, Ueki T, Saito T (2017) Vegetation and climate histories between MIS 63 and 53 in the early Pleistocene in central Japan based on plant macrofossil evidences. Quat Int 455:149–165

Momohara A, Yoshida A, Kudo Y, Nishiuchi R, Okitsu S (2016) Paleovegetation and climatic conditions in a refugium of temperate plants in central Japan in the Last Glacial Maximum. Quat Int 425:38–48

Morley RJ (2000) Origin and evolution of tropical rain forests. Wiley, London, 362 pp

Mountain M-J (1991) Highland New Guinea hunter-gatherers: the evidence of Nombe Rockshelter, Simbu with emphasis on the Pleistocene. Unpublished PhD thesis, ANU

Muttoni G, Carcano C, Garzanti E, Ghielmi M, Piccin A, Pini R, Rogledi S, Sciunnach D (2003) Onset of major Pleistocene glaciations in the Alps. Geology 31(11):989–992

Muttoni G, Ravazzi C, Breda M, Pini R, Laj C, Kissel C, Mazaud A, Garzanti E (2007) Magnetostratigraphy of the Leffe lacustrine succession (Southern Alps, Italy): evidence for an intensification of glacial activity in the Alps at Marine Isotope Stage 22 (0.87 Ma). Quat Res 67:161–173

Nakagawa T, Yasuda Y, Tabata H (1996) Pollen morphology of Himalayan *Pinus* and *Quercus* and its importance in palynological studies in Himalayan area. Rev Palaeobot Palynol 91:317–329

O'Keefe FR, Binder WJ, Frost SR, Sadlier RW, Van Valkenburgh B (2014) Cranial morphometrics of the dire wolf, *Canis dirus*, at Rancho La Brea: temporal variability and its links to nutrient stress and climate. Palaeontol Electron 17(1):24

Ollivier V, Gabrielyan I, Gasparyan B, Chataigner C, Joannin S, Cornée JJ, Guillou H, Scaillet S, Munch P, Krijgsman W (2010) Quaternary volcano-lacustrine patterns and palaeobotanical data in southern Armenia. Quat Int 223/224:312–326

Ooi N (2001) Last Glacial plant macrofossils discovered in the Kathmandu valley, Nepal. Jpn J Hist Bot 10:1

Palombo MR, Filippi ML, Iacumin P, Longinelli A, Barbieri M, Maras A (2005) Coupling tooth microwear and stable isotope analyses for palaeodiet reconstruction: the case study of Late Middle Pleistocene *Elephas (Palaeoloxodon) antiquus* teeth from Central Italy (Rome area). Quat Int 126–128:153–170

Paudayal KN (2005) Late Pleistocene pollen assemblages from the Thimi Formation, Kathmandu Valley, Nepal. Island Arc 14(4):328–337

Paudayal KN (2006) Late Pleistocene pollen assemblages from the Gokarna Formation, Kathmandu Valley. J Nepal Geol Soc 33:33–38

Paudayal KN (2011) High resolution palynostratigraphy and climate from the late Quaternary Besigaon section belonging to Gokarna Formation in the Kathmandu Valley. J Stratigr Assoc Nepal 7:33–38

Paudayal KN, Ferguson DK (2004) Pleistocene palynology of Nepal. Quat Int 117:69–79

Penck A, Brückner E, du Pasquier L (1894) Le Système glaciaire des Alpes. Guide Congr Géol Int Zurich Bull Soc Sc Nat Neuchatel 22:86

Pereira A, Nomade S, Faulguères C, Bahain JJ, Tombret O, Garcia T, Voichet P, Bulgarelli GM, Anzidei AP (2017) 40Ar/39Ar and ESR/U-series data for the La Polledrara di Cecanibbio archaeological site (Lazio, Italy). J Archaeol Sci Rep 15:20–29

Pini R, Ravazzi C, Donegana M (2009) Pollen stratigraphy, vegetation and climate history of the last 215 ka in the Azzano Decimo core (plain of Friuli, north-eastern Italy). Quat Sci Rev 28:1268–1290

Prentice ML, Hope GS, Maryunani K, Peterson JA (2005) An evaluation of snowline data across New Guinea during the last major glaciation and area-based glacier snowlines in the Mt. Jaya region of Papua Indonesia during the Last Glacial Maximum. Quat Int 138:93–117

Pross J, Koutsodendris A, Christanis K, Fischer T, Fletcher WJ, Hardiman M, Kalaitzidis S, Knipping M, Kotthoff U, Milner AM, Müller UC, Schmiedl G, Siavalas G, Tzedakis PC, Wulf S (2015) The 1.35-Ma-long terrestrial climate archive of Tenaghi Philippon, northeastern Greece: evolution, exploration and perspectives for future research. Newsl Stratigr 48:253–276

Ravazzi C (1995) Paleobotany of the biogenic unit of the Leffe Formation (early Pleistocene, N-Italy): brief report on the status of the art. Il Quat Ital J Quat Sci 8(2):435–442

Ravazzi C (2013) Un lago di 800 mila anni fa a Sovere. Guida alla scoperta di un calendario di 50 mila anni: piante, animali e ceneri vulcaniche nel bacino di Pianico-Sèllere. CNR-IDPA, Milano, 96 pp

Ravazzi C, Badino F, Pinti D, Scardia G (2014) The stratigraphic setting of the Pianico-Sèllere Basin (Early/Middle Pleistocene, Italian Alps). An updated framework. 2 mid-conference field trip "the quaternary of the Italian Alps – vegetation, palaeoenvironment and climate". In: 9th European palaeobotany palynology conference. CLEUP, Padova, pp 4–12

Ravazzi C, Rossignol Strick M (1995) Vegetation change in a climatic cycle of early Pleistocene age in the Leffe basin (Northern Italy). Palaeogeogr Palaeoclimatol Palaeoecol 117:105–122

Ravazzi C, Van Der Burgh J (1994) Coniferous woods in the early Pleistocene brown coals of the Leffe Basin (Lombardy, Italy). Riv Ital Paleontol Stratigr 100(4):597–620

Reinecke J, Troeva E, Wesche K (2017) Extrazonal steppes and other temperate grasslands of northern Siberia-Phytosociological classification and ecological characterization. Phytocoenologia 47:167–196

Rightmire GP, Ponce de León MS, Lordkipanidze D, Margvelashvili A, Zollikofer CPE (2017) Skull 5 from Dmanisi: Descriptive anatomy, comparative studies, and evolutionary significance. J Hum Evol 104:50–79

Rook L, Croitor R, Delfino M, Ferretti MP, Gallai G, Pavia M (2013) The Upper Valdarno Plio-Pleistocene vertebrate record: an historical overview, with notes on palaeobiology and stratigraphic significance of some important taxa. Ital J Geosci (Boll Soc Geol Ital) 132:104–125

Rossi S (2004) Analisi pollinica della sequenza lacustre di Piànico-Sèllere (Italia). Tesi di Dottorato in Cotutela in Scienze Naturalistiche e Ambientali, Univ. St. Milano e Univ. d'Aix-Marseille III

Sadori L, Koutsodendris A, Panagiotopoulos K, Masi A, Bertini A, Combourieu-Nebout N, Francke A, Kouli K, Joannin S, Mercuri AM, Peyron O, Torri P, Wagner B, Zanchetta G, Sinopoli G, Donders TH (2016) Pollen-based paleoenvironmental and paleoclimatic change at Lake Ohrid (south-eastern Europe) during the past 500 ka. Biogeosciences 13:1423–1437

Sakai H, Fujii R, Kuwahara Y, Uprety BN, Shrestha SD (2001) Core drilling of the basin fill sediments in the Kathmandu Valley for paleoclimatic study: preliminary results. J Nepal Geol Soc 25(Sp. Issue):9–18

Sakai H, Yahagi W, Fujii R, Hayashi T, Upreti BN (2006) Pleistocene uplift of the Himalayan frontal ranges recorded in the Kathmandu and Siwalik basins. Palaeogeogr Palaeoclimatol Palaeoecol 241:16–27

Sanders D, Spötl C (2014) The Hötting breccia – a Pleistocene key site near Innsbruck. In: Kerschner H, Krainwer K, Spötl C (eds) From the foreland to the Central Alps. DEUQUA Excursions, Tyrol, pp 82–94

Sandom C, Faurby S, Sandel B, Svenning JC (2014) Global late Quaternary megafauna extinctions linked to humans, not climate change. Proc R Soc B Biol Sci 281(1787):20133254

Santucci E, Marano F, Cerilli E, Fiore I, Lemorini C, Palombo MR, Anzidei AP, Bulgarelli GM (2016) *Palaeoloxodon* exploitation in the late Middle Pleistocene site of Polledrara di Cecanibbio (Rome, Italy). Quat Int 406:169–182

Satoguchi Y, Nagahashi Y (2012) Tephrostratigraphy of the Pliocene to Middle Pleistocene series in Honshu and Kyushu Islands, Japan. Island Arc 21:149–169

Scharrer S (2013) Frühpleistozäne Vegetatiosentwicklung im Südlichen Kaukasus -Pollenanalytische Untersuchungen an Seesedimenten im Vorotan-Becken (Armenien). PhD thesis submitted to Frankfurt University. http://publikationen.ub.uni-frankfurt.de/frontdoor/index/index/docId/30108

Schirrmeister L, Froese D, Tumskoy V, Grosse G, Wetterich S (2013) Yedoma: late Pleistocene ice-rich syngenetic permafrost of Beringia. In: Elias SA (ed) The encyclopedia of quaternary science, vol 3. Elsevier, Amsterdam, pp 542–552

Scotese CR (2014) The PALEOMAP project PaleoAtlas for ArcGIS, version 2, volume 1, Cenozoic plate tectonic, paleogeographic, and paleoclimatic reconstructions, maps 1–15. PALEOMAP Project, Evanston

Shackleton NJ, Backman J, Zimmerman H, Kent DV, Hall MA, Roberts DG, Schnitker D, Baldauf JG, Desprairies A, Homrighausen R, Huddlestun P, Keene JB, Kaltenback AJ, Krumsiek KAO, Morton AC, Murray JW, Westbergsmith J (1984) Oxygen isotope calibration of the onset of ice-rafting and history of glaciation in the North Atlantic region. Nature 307:620–623

Shatilova I, Rukhadze L, Mchedlishvili N, Makharadze N (2002) The main stages of the development of vegetation and climate of Western Georgia during the Gurian (Eopleistocene) time by the pollen records. Bull Georgian Acad Sci 166(3):624–627

Sher AV, Kuzmina SA, Kuznetsova TV, Sulerzhitsky LD (2005) New insights into the Weichselian environment and climate of the East Siberian Arctic, derived from fossil insects, plants, and mammals. Quat Sci Rev 24:533–569

Sigl M, Fudge TJ, Winstrup M, Cole-Dai J, Ferris D, McConnell JR, Taylor KC, Welten KC, Woodruff TE, Adolphi F, Bisaux M, Brook EJ, Buizert C, Caffee MW, Dunbar NW, Edwards R, Geng L, Iverson M, Koffman B, Layman L, Maselli OJ, McGwire K, Muscheler R, Nishiizumi K, Pasteris DR, Rhodes RH, Sowers TA (2016) The WAIS divide deep ice core WD2014 chronology – part 2: annual-layer counting (0–31 ka BP). Clim Past 12:769–786

Stadel C (1991) Altitudinal belts in the tropical Andes: their ecology and human utilization. Yearb Conf Lat Am Geogr 17/18:1–370

Steffensen JP, Andersen KK, Bigler M, Clausen HB, Dahl-Jensen D, Fischer H, Goto-Azuma K, Hansson M, Johnsen SJ, Jouzel J, Masson-Delmotte V, Popp T, Rasmussen SO, Röthlisberger R, Ruth U, Stauffer B, Siggaard Andersen M-L, Sveinbjörnsdóttir AE, Svensson A, White JW (2008) High resolution Greenland ice core data show abrupt climate change happens in few years. Science 321(5889):680–684

Suc JP, Combourieu Nebout N, Seret G, Popescu SA, Klotz S, Gautier F, Clauzon G, Westgate J, Insinga D, Sandhu AS (2010) The Crotone series: a synthesis and new data. Quat Int 219:121–133

Sutton A, Mountain M-J, Aplin K, Denham T (2010) Archaeozoological records for the highlands of New Guinea: A review of current evidence. Aust Archaeol 69:41–58

Svensson A, Andersen KK, Bigler M, Clausen HB, Dahl-Jensen D, Davies SM, Johnsen SJ, Muscheler R, Parrenin F, Rasmussen SO, Röthlisberger R, Seierstad I, Steffensen JP, Vinther BM (2008) A 60 000 year Greenland stratigraphic ice core chronology. Clim Past 4:47–57

Thomas ER, Wolff EW, Mulvaney R, Johnsen SJ, Steffensen JP, Arrowsmith C (2009) Anatomy of a Dansgaard-Oeschger warming transition: high-resolution analysis of the North Greenland ice core project ice core. J Geophys Res-Atmos 114:D08102

Thomas ER, Wolff EW, Mulvaney R, Steffensen JP, Johnsen SJ, Arrowsmith C, White JWC, Vaughn B, Popp T (2007) The 8.2 ka event from Greenland ice cores. Quat Sci Rev 26:70–81

Thompson LG, Mosley-Thompson E, Davis ME, Lin PN, Henderson KA, Cole-Dai J, Bolzan JF, Liu K (1995) Late Glacial stage and Holocene tropical ice core records from Huascaran, Peru. Science 269:46–50

Tsukada M (1985) Map of vegetation during the last glacial maximum. Quat Res 23:369–381

Turner A, Antón M (2007) Evolving Eden. An illustrated guide to the evolution of the African large-mammal Fauna. Columbia University Press, New York

Urabe A, Tateishi M, Kazaoka O (1995) Depositional cycle of marine beds and relative sea-level changes of the Plio-Pleistocene Uonuma Group, Niigata, central Japan. Mem Geol Soc Jpn 45:140–153. (in Japanese with English Abstract)

Van der Hammen T, Hooghiemstra H (2000) Neogene and quaternary history of vegetation, climate, and plant diversity in Amazonia. Quat Sci Rev 19:725–742

Vekua A (1995) Die Wirbeltierfauna des Villafranchium von Dmanisi und ihre biostratigraphische Bedeutung. Jahrb Römisch-Ger Zentralmus Mainz 42(1):77–180

Ward JK, Harris JM, Cerling TE, Wiedenhoeft A, Lott MJ, Dearing MD, Coltrain JB, Ehleringer JR (2005) Carbon starvation in glacial trees recovered from the La Brea tar pits, southern California. Proc Natl Acad Sci 102(3):690–694

West RG (1980) The preglacial Pleistocene of Norfolk and Suffolk coasts. Cambridge University Press, New York

Williams M, Cook E, van der Kaars S, Barrows T, Shulmeister J, Kershaw P (2009) Glacial and deglacial climatic patterns in Australia and surrounding regions from 35 000 to 10 000 years ago reconstructed from terrestrial and near-shore proxy data. Quat Sci Rev 28(23–24):2398–2419

Wroe S, Field J, Archer M, Grayson DK, Price GJ, Louys J, Faith JT, Webb GE, Davidson I, Mooney SD (2013) Climate change frames debate over the extinction of megafauna in Sahul (Pleistocene Australia-New Guinea). Proc Natl Acad Sci 110(22):8777–8781

Yoshida M, Gautam P (1988) Magnetostratigraphy of Plio-Pleistocene lacustrine deposits in the Kathmandu Valley, central Nepal. Proc Ind Nat Sci Acad 54A(30):410–417

Yoshida M, Igarashi Y (1984) Neogene to quaternary lacustrine sediments in the Kathmandu Valley. Nepal J Nepal Geol Soc 4:73–100

Yoshikawa S, Mitamura M (1999) Quaternary stratigraphy of the Osaka Plain, central Japan and its correlation with oxygen isotope record from deep sea cores. J Geol Soc Jpn 105:332–340. (in Japanese with English abstract)

Zanina OG, Gubin SV, Kuzmina SA, Maximovich SV, Lopatina DA (2011) Late-Pleistocene (MIS 3-2) palaeoenvironments as recorded by sediments, palaeosols, and ground-squirrel nests at Duvanny Yar, Kolyma lowland, northeast Siberia. Quat Sci Rev 30:2107–2123

Zhu Z, Dennell R, Huang W, Wu Y, Qiu S, Yang S, Rao Z, Hou Y, Xie J, Han J, Ouyang T (2018) Hominin occupation of the Chinese Loess Plateau since about 2.1 million years ago. Nature 559:7715

Triumph and Fall of the Wet, Warm, and Never-More-Diverse Temperate Forests (Oligocene-Pliocene)

2

Edoardo Martinetto, Nareerat Boonchai,
Friðger Grímsson, Paul Joseph Grote, Gregory Jordan,
Marianna Kováčová, Lutz Kunzmann, Zlatko Kvaček,
Christopher Yusheng Liu, Arata Momohara,
Yong-Jiang Huang, Luis Palazzesi, Mike Pole,
and Ulrich Salzmann

Abstract

Large areas of Earth's continents were covered by temperate forests before the dramatic increase of the human population in the past two millennia. Prior to human expansion, temperate forests were more extensive in the Neogene (23–2.6 Ma) when climate at the middle latitudes was slightly warmer and more equable than at the present. These temperate forests exhibited a high diversity of plant taxa, higher than today in several geographical areas. Such high diversity in the past can be explained by two reasons. First, angiosperms originated in the Cretaceous and underwent an important phylogenetic diversification during and shortly after that period. These new plant lineages easily dispersed between North America and Eurasia, and bio-

Electronic supplementary material A slide presentation and an explanation of each slide's content is freely available to everyone upon request via email to one of the editors: edoardo.martinetto@unito.it, ragastal@colby.edu, tschopp.e@gmail.com

*The asterisk designates terms explained in the Glossary.

E. Martinetto (✉)
Department of Earth Sciences, University of Turin, Turin, Italy
e-mail: edoardo.martinetto@unito.it

N. Boonchai
Palaeontological Research and Education Centre,
Mahasarakham University, Maha Sarakham, Thailand

F. Grímsson
Department of Botany and Biodiversity Research, University
of Vienna, Vienna, Austria
e-mail: fridgeir.grimsson@univie.ac.at

P. J. Grote
The Northeastern Research Institute of Petrified Wood and Mineral
Resources, (Nakhon Ratchasima Rajabhat University),
Nakhon Ratchasima, Thailand

G. Jordan
School of Biological Sciences, University of Tasmania,
Hobart, TAS, Australia
e-mail: greg.jordan@utas.edu.au

M. Kováčová
Department of Geology and Palaeontology,
Comenius University in Bratislava, Bratislava, Slovak Republic
e-mail: marianna.kovacova@uniba.sk

L. Kunzmann
Section Palaeobotany, Senckenberg Natural History Collections
Dresden, Dresden, Germany
e-mail: Lutz.Kunzmann@senckenberg.de

Z. Kvaček
Institute of Geology and Palaeontology, Faculty of Science,
Charles University, Prague, Czech Republic
e-mail: kvacek@natur.cuni.cz

C. Y. Liu
Department of Earth & Environmental Sciences, University
of Missouri, Kansas City, Missouri, USA
e-mail: Y.Liu@umkc.edu

A. Momohara
Graduate School of Horticulture, Chiba University, Chiba, Japan
e-mail: arata@faculty.chiba-u.jp

Y.-J. Huang
Key Laboratory for Plant Diversity of East Asia, Kunming Institute
of Botany, Chinese Academy of Sciences, Kunming, China
e-mail: huangyongjiang@mail.kib.ac.cn

L. Palazzesi
Department of Paleontology, Museo Argentino de Ciencias
Naturales, Buenos Aires, Argentina
e-mail: lpalazzesi@macn.gov.ar

M. Pole
Queensland Herbarium, Brisbane Botanic Gardens,
Toowong, QLD, Australia

U. Salzmann
Department of Geography and Environmental Sciences,
Northumbria University, Newcastle upon Tyne, UK
e-mail: Ulrich.Salzmann@northumbria.ac.uk

E. Martinetto et al. (eds.), *Nature through Time*, Springer Textbooks in Earth Sciences, Geography and Environment,
https://doi.org/10.1007/978-3-030-35058-1_2

geographic range expansions continued across other continents. Second, since the Eocene/Oligocene transition (c. 34 Ma), several members of tropical/subtropical lineages adapted to cooler conditions and entered the warmer temperate realm. An equable climate with abundant precipitation in widespread areas provided a suitable habitat for moisture-requiring woody plants. The higher floristic diversity in the Neogene compared to the present is best illustrated by European fossil plants and, to a lesser extent, by those in North America. The area covered by temperate forests in South America decreased consistently after the late Miocene, and the dominant woody plants of the Neogene remained only in the westernmost regions. A floristic impoverishment is not clearly documented in Australia, where there was a much higher diversity of conifers in the Oligocene-Miocene than today. Beginning some 6 million years ago, several global intervals of colder and/or drier climate reduced the habitat of those taxa that required nonfreezing temperatures and moisture, finally resulting in a large mass extirpation/extinction of thermophilous plants in western Eurasia. This turnover occurred primarily between 3.5 and 1.0 million years ago. The trend was different in eastern Eurasia where extirpation/extinction has been rather limited. In conclusion, the mid-latitudes of all the continents witnessed a triumph of the extension and diversity of temperate forests from about 34 to 3 million years ago (Oligocene-Pliocene) and, in many temperate places, these grew under wetter and warmer conditions than today.

2.1 Introduction—A Visit to Temperate Forests Back in Time

Edoardo Martinetto and Ulrich Salzmann

This chapter is designed to illustrate how the natural world around us has changed over more than 30 million years. To understand the context of this biological change, it is important to know the subdivisions of the Paleogene and Neogene Periods. The Paleogene (c. 66–23 Ma) is subdivided into the Paleocene (c. 66–56 Ma), the Eocene (c. 56–34 Ma), and the Oligocene (c. 34–23 Ma) Epochs, whereas the Neogene (c. 23–2.6 Ma) includes the Miocene (c. 23–5.3 Ma) and Pliocene (5.3–2.6 Ma) Epochs. Here, we concentrate on ecosystems from the Oligocene to the Pliocene. Some of the best examples of how natural systems changed beginning about 34 million years ago to the present are provided by one of the more familiar vegetation types found in the temperate ecoregion* (Olson et al. 2001): the temperate forest (de Gouvenain and Silander 2016). But, does the term "temperate" have an unmistakable definition? The answer to that question is "absolutely not." It is applied differently not only in ecoregion versus climate classification but also in two largely used climate classification systems. The Köppen-Geiger system (Köppen 1936) recognizes a broad temperate zone, whereas the Köppen-Trewartha modification (Trewartha and Horn 1980) distinguishes a subtropical and narrow temperate zone within the broad temperate zone of the former (Box P2). However, the Köppen-Geiger system provides an idea of the climatic parameters under which the temperate-forest ecoregion grows. The MAT is below 18 °C, but the average temperature of the coldest month is above −3 °C, only slightly below freezing temperature, and precipitation is higher than in dry climates (the precise boundary parameters defining dry climates are ambiguous; see Trewartha and Horn 1980). The temperate-forest ecoregion differs climatically from the boreal and tropical forest ecoregions, whereas the dry ecoregions differ even more. These dry areas are not covered by forest, but are typified by grassland/savanna, shrubland, or desert plants. The climatic differences identified by some workers have been used to suggest using the adjective "temperate" in some categories of the current biome* classifications, for example, "temperate deciduous forest" (Salzmann et al. 2008). But, let's now see where to find temperate forests on Earth.

Today, many of us live in regions where a temperate forest would grow if humans had not disturbed the landscape. These regions include East Asia, Europe, and North America, but some areas are either included, or not, in the maps of temperate forests produced by different authors (compare Olson et al. 2001; Silander 2001; Fischer et al. 2013; de Gouvenain and Silander 2016). One of the main disputed regions is southern China, a very important area that provides partial analogs for a widespread type of mid-Cenozoic forest. This region is alternatively treated as temperate or subtropical. We follow the latitudinal vegetation zonation by Osawa (1993) for East Asia in this chapter, which includes south China in the subtropical belt, but considers this merely a part of the temperate vegetation zone. Hence, areas of south China with a MAT below 18 °C (temperate areas, according to the Köppen-Geiger system) are covered by subtropical forest, but these are considered part of the temperate vegetation belt. The forests of East Asia occupy, from south to north, the subtropical, cool temperate, and cold temperate zones. The distinction of such zones is justified by the restricted distribution of several tree taxa in the formerly continuous forest cover along the eastern coast of Eurasia, from southern China to the Russian Far East. But, actually, China has very few forest patches due to extensive deforestation.

We know that the resources of the land and timber were needed for several purposes as humanity spread across the globe and established countries over the past few centuries or millennia. To fuel our own expansion and migration, we cut down most of the trees, altering large areas that, once, were forested, leaving small isolated patches, even if rarely unaltered. However, before human intervention began, temperate forests already had changed over the course of deep geological time.

2.1.1 Past Diversity of Temperate Forests

The most apparent discrepancy between temperate forests of today and those in deep time is the taxonomic diversity of trees. If we walk through a standing forest in different geographical areas (e.g., Europe and New Zealand) to count the number of tree genera, and then examine fossil-rich plant localities in the same area and repeat the counting exercise, we would be surprised by the higher number of tree genera among the fossil assemblages. Of course, this difference is meaningful only if both the modern and the fossil-tree record represent temperate forests, because tree diversity is usually higher in tropical versus temperate forests. The map published by the Food and Agriculture Organization of the United Nations (FAO 2006) documents the number of tree genera in each country [U0203]. The map shows that small countries that lie completely in the temperate-forest area (e.g., Germany) generally have a lower diversity than countries of comparable size in the tropical ecoregions (e.g., Ecuador). The modern diversity of these genera and species is also not the same in the temperate forests of different parts of the globe. For example, the Chinese temperate forests are the most diverse on Earth (de Gouvenain and Silander 2016). This fact is mainly due to the reduction and loss of taxa in specific areas, extirpation (Smith-Patten et al. 2015), or to the irreversible death and loss of taxa through extinction. These two conditions are not always easy to distinguish on the basis of fossil-plant data. Hence, the recognition of any decrease in diversity over time often needs to be ascribed to events of "extirpation/extinction" (see Preface).

Another factor that affects how we view ancient temperate-forest diversity is our ability to obtain precise data about tree diversity in these forests through time. In addition to the several complications implicit in assessing the taxonomic diversity of fossil floras (e.g., Signor 1994; Collinson and Hooker 2003; Wilf and Johnson 2004), professional papers on plant-fossil assemblages rarely summarize the number of genera or species present in any locality. However, fossil floras that can be assigned to temperate-forest paleobiomes* are well documented in the paleobotanical literature (e.g., Wolfe 1977; McIver and Basinger 1999; Herman et al. 2009). These publications estimate a maximum number of 45 (McIver and Basinger 1999) or 47 (Wolfe 1977) genera for Paleogene temperate-forest trees. The same estimates for the Neogene temperate floras are higher values (Table 2.1), with a maximum of 61 for a single locality in the Pliocene of northern Italy (Martinetto et al. 2018) and 121 for the collective Neogene floras of this area (Bertini and Martinetto 2011). Clues pointing to a high Neogene versus Paleogene diversity are also provided by Mai (1995, p. 285) for Europe, where 62 genera are reported in the Paleogene and 84 in the Neogene. Furthermore, Manchester et al. (2009, p. 12) provided a synthesis of the past distribution of eastern Asian endemic seed-plant genera (mostly temperate) in the Northern Hemisphere. They show the occurrence of 20 genera in the Paleogene and 28 in the Neogene. However, additional research will confirm when the peak in temperate tree generic diversity occurred. Regardless of when in time peak diversity existed, there are other documented changes in temperate forests with respect to their biogeographic expansion and geographic location.

Table 2.1 Comparison of the number of canopy genera (trees + climbers) between the mid-Cenozoic (34–2 Ma) and the present

Selected temperate-forest sample area	North central Italy	Japan (Honshu Island)	Australia and New Zealand)	North central Thailand	Iceland	Antarctica
Temperate forest today	Dominant	Dominant	Limited	Very limited	Absent	Absent
Number of canopy genera in the modern flora	44	102	68	c. 360 (tropical!), only 4 temperate	3 (subarctic)	0
Number of canopy genera in the "Neogene" fossil flora 34–2 Ma (tf = temperate forest)	121 (tf)	135 (tf)	33 (mostly tf)	31 temperate (and c. 100 tropical)	30 (tf)	2
Number of canopy genera present both in the fossil (34–2 Ma) and the modern flora	40	98	27	17 temperate (and c. 100 tropical)	2	0
Number of canopy genera present in the fossil (34–2 Ma) but not in the modern flora	81	37	6	14 (temperate)	28	2
Number of canopy genera present in the modern but not in the fossil (34–2 Ma) flora	6	4	41	>200 (tropical!)	1 (*Sorbus*)	0

The results, when converted in percentage, indicate that 64% of the genera that grew in Italy during the mid-Cenozoic were later extirpated. Generic extirpation appears to be more moderate in Japan (27%). In contrast, the flora of Australia comprises significantly more genera today than in the mid-Cenozoic. The data for Thailand are difficult to analyze because it is a tropical area today, with influences from the temperate zone in the past. However, an analysis restricted to temperate plants indicates that there was a major loss of generic diversity since mid-Cenozoic. Source of data: Italy from Bertini and Martinetto 2011), Japan from Momohara (2016), Iceland from Denk et al. (2011), Australia and New Zealand from Kooyman et al. (2014), Thailand from unpublished data collected by PJ Grote, Antarctica from Rees-Owen et al. (2018)

2.1.2 Past Distribution of Temperate Forests

We use the compilation by Olson et al. (2001) to present a global map of the potential expanse of the temperate-forest ecoregion today (Fig. 2.1a). This map will allow us to start our journey into its past history. Paleogeographic maps of vegetational distribution at a specific point in geologic time help us to understand the changing forest landscape during the Neogene. Unfortunately, it is not possible to create pre-

cise vegetational maps for many geological stages, because reconstructing the spatial distribution of forests is often hampered by the scarcity of available paleoecological records. One example of a high-resolution map is that of Salzmann et al. (2008). These authors introduced a global data-model hybrid-vegetation reconstruction for the late Pliocene that compared and synthesized paleobotanical data with the outputs of the BIOME4 computer model (a coupled biogeography and biogeochemistry model). This model is controlled

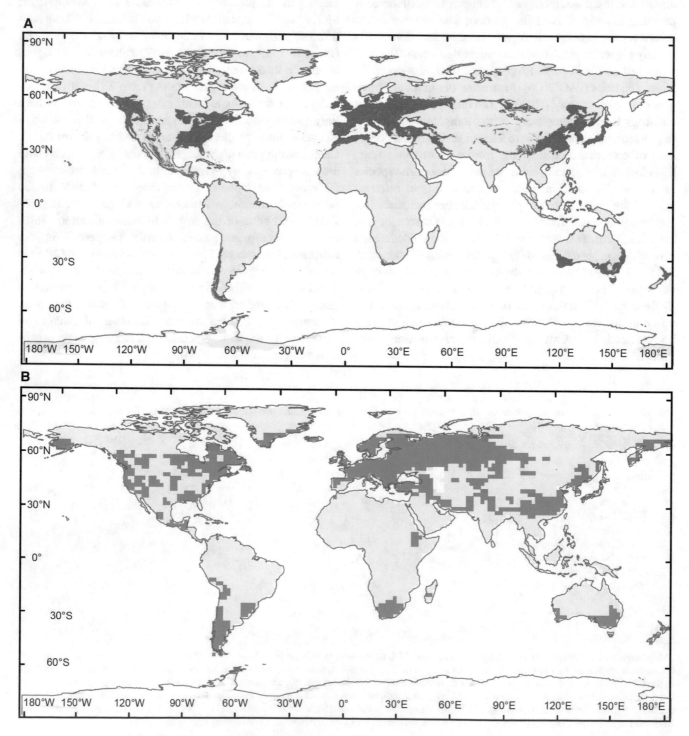

Fig. 2.1 Modern and Pliocene distribution of temperate-forest ecoregions, in green. (**a**) Modern distribution after Olson et al. (2001); the stars correspond to the main localities visited in this chapter. (**b**) Reconstructed late Pliocene temperate-forest distribution (after Salzmann et al. 2008, 2013)

by climatology derived from a Hadley Center* general circulation model (GCM). Model results show that temperate-forest cover was greater during the warmer late Pliocene (Fig. 2.1b) than at the present. Warm temperate (subtropical) broad-leaved forest became the dominant Pliocene vegetation in most parts of Europe and Southeast Asia. These forests consisted of a combination of floral elements for which we have no modern analog. This is because, currently, their relatives are scattered in different areas under cool temperate, subtropical, and even tropical climates (e.g., the conifer tree *Glyptostrobus*, having today relict distribution in tropical southern China, Vietnam, and Laos). In northeastern North America and Norway, late Pliocene temperate forests extended northward to 60° N and 70° N, respectively. The expansion of temperate forest to these higher latitudes occurred in response to a globally warmer climate and was accompanied by a northward shift of evergreen boreal taiga forests. As a consequence of taiga expansion, areas of tundra were reduced. Both data-model vegetation reconstructions and paleontological proxy data indicate that the extension of temperate forests could have been even broader during the early-middle Miocene (23–12 Ma), when they probably formed a continuous belt from east to west across both Eurasia and North America (Leopold et al. 2007; Utescher et al. 2011; Pound et al. 2012).

The history of temperate forests continues to be written as new results and insights are published. These studies provide volumes of data that require a monographic treatise to describe. However, the purpose of the current chapter is to introduce the reader to the topic by means of virtual field trips to selected key sites around the world. At each site we will examine the plant-fossil record including the characteristics and taxonomic diversity of temperate-forest communities over the time span from 3 to 34 million years ago.

2.2   Western Eurasia

Marianna Kováčová, Lutz Kunzmann, Zlatko Kvaček and Edoardo Martinetto

We start our field trip on the largest continuous landmass of today's world, Eurasia. This territory is located completely in the Northern Hemisphere and conventionally subdivided in the "political" continents of Europe and Asia (Fig. 2.1a). The only natural feature that supports this geographical separation is a broad, arid, and poorly populated belt stretching across Eurasia from the Ural Mountains (north to south), east to the Caspian Sea. This physiographic belt divides a mostly humid and densely populated area to the west, mainly coinciding with Europe, and a heterogeneous and locally even more densely populated territory to the east, which is the main part of Asia. Temperate forests represent the potential vegetation that, without domesticated animal production and human disturbance, would cover most of Europe today. This vegetation type also covers the eastern part of Asia, but is largely absent in its center, and is limited geographically in the western part (southern Caspian area, Turkey).

The present diversity of tree genera in European forests is very low in comparison to the forests of Asia and eastern North America. This was not the case in the past. Several plant-fossil assemblages [U0205, U0206] in Europe provide evidence of many genera that do not grow in this area today (Mai 1995). Traveling to five places in ancient Europe demonstrates this point.

2.2.1   Western Alps

The highest peaks of Europe lie in the western Alps, in an area where the borders of Italy, France, and Switzerland meet (see Gianotti and Ravazzi, in Martinetto et al., Chap 1). These areas were already positioned at higher elevations about 30 million years ago and a broad marine inlet existed at the southern foothills of these rising mountains (northwestern Italy). Before the Quaternary ice ages began, this basin was filled by Pliocene sediments ranging in age from 5 to 2.6 million years. Abundant plant remains are preserved in the sedimentary succession (Martinetto 2015). These assemblages demonstrate that the lower alpine slopes hosted a flora (Fig. 2.2) that was very different from the one we can observe today. This is especially true of the flora that flourished, here, between 4 and 3 million years ago where there is a notable abundance of taxa belonging to several tropical to subtropical lineages that were adapted to temperate conditions [U0207]. Interestingly, these lineages (i.e., *Cinnamomum* [cinnamon] and *Ocotea* [Lauraceae], *Eurya* [Pentaphylaceae], *Ilex* [holly], *Litsea* [may chang], and *Symplocos* [Ericales]) coexisted with typical cool temperate elements, such as *Carpinus* (hornbeam), *Fagus* (beech), and *Liriodendron* (tulip tree) (Bertini and Martinetto 2011). All these woody plants are known currently from several warm temperate forests of China, and many are not found in Europe at the present. The Pliocene fossil floras from northwestern Italy, despite the well-known incompleteness of the fossil record, prove that 44 different angiosperm and conifer tree genera populated the region, whereas only 25 genera now grow in this area. Integrated analyses of pollen, leaves, and fruits/seeds (Bertini and Martinetto 2011) formed the basis for a reconstruction of a lush "subtropical humid forest*" (Fig. 2.3), which has the best present analogs in a few sites located in Southeast Asia. This forest community covered the lower slopes of the Alps, facing a narrow coastal belt in the early Pliocene (Martinetto et al. 2018). The coastal plain gradually advanced toward the sea in the late Pliocene, and swamps developed in which the waterlogged environment

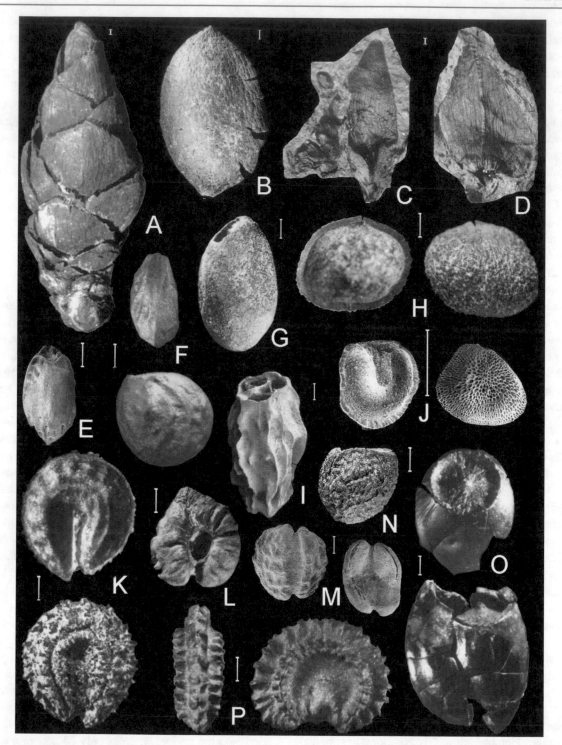

Fig. 2.2 Fossil cones, fruits, and seeds from the early Pliocene of northwestern Italy belonging to plant genera that were later extirpated from Europe and are mostly distributed in East Asia today. (**a**) *Cathaya vanderburghii*, cone; (**b**) *Cephalotaxus* cf. *rhenana*, seed; (**c**) *Pseudolarix schmidtgenii*, seed close to a cone scale in abaxial* view; (**d**) *Pseudolarix schmidtgenii*, cone scale in adaxial* view; (**e**) *Alangium deutschmannii*, endocarp; (**f**) *Meliosma canavesana*, the same endocarp from two sides; (**g**) *Turpinia ettingshausenii*, endo- carp; (**h**) *Mallotus maii*, seed, internal and external view; (**i**) *Symplocos schereri*, endocarp; (**j**) *Eurya stigmosa*, seeds, internal and external view; (**k**) *Cyclea palatinati-bavariae*, endocarp, internal and external view; (**l**) *Tetrastigma chandlerae*, seed; (**m**) *Ehretia europaea*, endocarp from both sides; (**n**) *Zanthoxylum* cf. *ailanthiforme*, seed; (**o**) *Sargentodoxa gossmannii*, seeds; (**p**) *Sinomenium cantalense*, endocarp. Scale bars 1 mm

Fig. 2.3 Reconstruction of the early Pliocene lush "subtropical humid forest," rich in evergreen trees, at the southern foothills of the western Alps. The physiographic setting is based on data collected at the Ca' Viettone site (Martinetto et al. 2018). The sea is colored in dark blue and the the coastal lakes in light blue. The vegetation of the coastal plain (mainly swamp) is not shown. The hypothetical higher-elevation vegetation with deciduous angiosperms and conifers is reconstructed mainly on pollen data summarized by Bertini and Martinetto (2011)

provided suitable conditions for the burial and preservation of fossil forests. A few of these are in outcrop today due to the successive uplift of this territory (Martinetto et al. 2007). Evidence for this distribution and expansion of Pliocene temperate forests can be found in parts of the Italian Piedmont region. Here, a short 20 min trip by car would bring us from the wealth of the early Pliocene exotic fruit-and-seed assemblages (Fig. 2.2) at the Ca' Viettone site, which was close to the mountains (not far from Ivrea, see Gianotti and Ravazzi, in Martinetto et al., Chap. 1) (Fig. 2.4a), to the swamp of the Stura di Lanzo river, which was positioned further seaward on the coastal plain. This area was vegetated by familiar "European" plant genera, such as *Alnus*, though being dominated by the conifer *Glyptostrobus* (Fig. 2.4b). This Neogene member of the Cupressaceae was a close relative of the trees that, today, survive only in restricted areas of eastern Eurasia and are assigned to the biological species *Glyptostrobus pensilis* (Fig. 2.7a, b) [U0223]. The fossils of the European Neogene are placed into a single fossil species, the *Glyptostrobus europaeus* "whole plant" (Vassio et al. 2008), which was the main conifer of the contemporary Eurasian swamps [U0222].

2.2.2 German Brown Coals

Elements of Neogene temperate forests were common components of peat swamps (see Gastaldo et al., Chap. 12, for how they form) that, through burial and fossilization processes, have become today's brown coal deposits [U0208]. If

we were to enter one of the few brown coal mines still operating in Germany (these were more abundant in the twentieth century), we would see huge machines hauling large volumes of sand, mud, and coal and producing extensive outcrop exposures [U0210]. Sieving some of the river sands would provide us with mummified plant remains, among which fruits and seeds are indicative of the plant genera that once lived in these forests. We would not be surprised to find *Carpinus* (hornbeam), *Fagus* (beech), and *Quercus* (oak) as in today's flora, but we would also encounter abundant "strange" American elements, including *Carya* (hickory), *Magnolia*, and East Asian plants, such as *Cinnamomum*, *Mastixia* (Cornaceae) [U0211, U0212], *Sinomenium* (Menispermaceae), and *Rehderodendron* (Styracaceae; Mai 1995). Although at least 10 million years older, this fossil flora is very similar to those of the Pliocene of northwestern Italy, with the addition of *Mastixia* and related fossil genera.

2.2.3 Bílina

The brown coal mine at Bílina in the Czech Republic [U0214] (see also Chap. 11) is one in which the richest assemblages of beautifully preserved leaves of Neogene age can be found in Europe (Fig. 2.5). This site yields not only leaves with well-preserved venation and marginal features [U0213, U0215] but also their resistant outer cuticular* coatings. It is possible to detail cellular characters with small fragments of cuticle* that aid in the identification of

Fig. 2.4 Relevant fossil plant-bearing outcrops at the foothills of the western Alps, northwestern Italy. (**a**) Early Pliocene continental sediments at the Ca' Viettone site (Martinetto et al. 2018). (**b**) Detail of late Pliocene Stura di Lanzo Fossil Forest, just 15 km from the Ca' Viettone site (Martinetto et al. 2007)

Fig. 2.5 Fossil-leaf assemblage from the Miocene Bílina mine, Czech Republic. Most leaf impressions are alder, *Alnus julianiformis*, *Laria rueminiana* (center), and *Paliurus tiliifolius* (upper right). Scale bar 1 cm (photo Zlatko Kvaček)

Fig. 2.6 Large upright fossil trunks assigned to *Glyptostrobus* in the Bükkábrány Fossil Forest, Miocene of Hungary. Photo courtesy of Boglarka Erdei

the parent plant of any leaf. Again, the plants preserved in these buried forest remnants do not match the modern forest flora of the surrounding area. Rather, the fossils include palms (*Calamus*), "exotic" conifers (*Taxodium* [bald cypress], *Tetraclinis* [Sictus tree]) and broad-leaved trees (*Carya*, *Craigia* [Tiliaceae], *Engelhardia* [Juglandaceae], *Liquidambar* [sweet gum], and *Nyssa* [Tupelo gum]), and many relatives of the laurel family (now found mainly in subtropical America and Southeast Asia).

2.2.4 Paratethys

The sediments exposed at Bílina formed at the margin of a large basin that was located in eastern Eurasia during the Neogene, named Paratethys. The physical dimensions and salinity of this seaway varied over time, and fossil assemblages preserved in deposits surrounding it provide a clear picture of the regional vegetation. Miocene landscapes

around Paratethys have been reconstructed by means of paleobotanical studies from several exceptionally preserved sites [U0216] (e.g., Kvaček et al. 2006; Kováčová et al. 2011). Hence, it has been possible to characterize a few Miocene vegetational types [U0218] and reconstruct their dynamics, primarily using trends in pollen (Ivanov et al. 2011) [U0219]. In one instance, though, an entire forest of standing in situ trees (Fig. 2.6) was unearthed at the Bükkábrány Fossil Forest site in Hungary [U0220, U0221]. The huge upright conifer stumps (Erdei et al. 2009) represent a wetland forest dominated by *Glyptostrobus europaeus,* a fossil plant that we have already seen in northwestern Italy. The persistence of this tree in Europe into the Holocene, suggested only by the presence of fossil pollen (Biltekin et al. 2015), is debated by many paleo- and neobotanists. Additionally, these conifer-dominated paleoenvironments have yielded several vertebrate and invertebrate assemblages, especially in Italy and Spain, that have allowed the Neogene coastal

Fig. 2.7 Extant and fossil *Glyptostrobus*. (**a**) Stand of living *Glyptostrobus* trees in China. (**b**) Detail of terminal shoots with a cone. (**c**) Fossil shoots with cones of *Glyptostrobus europaeus* from a late Pliocene fossil forest of northwestern Italy (Stura di Lanzo: Martinetto et al. 2007). Photo by A Momohara (**a**), Mike Clayton (**b**), E Martinetto (**c**)

plain biota to be reconstructed [U0224]. Wandering through these densely forested temperate environments, we would have been able to spot more than 150 species including mastodons, tapirs, and rhinos, as well as smaller animals such as various kinds of reptiles and amphibians, rodents, bats, birds, fishes, and mollusks (see Colombero et al. 2017).

2.2.5 Mediterranean

Today, the landmasses facing the Mediterranean Sea, both to the west (Iberian Peninsula) and to the east (Greece, Turkey), have an intensely to moderately summer-dry Mediterranean climate. The most arid places are characterized by open vegetation. But, the potential vegetation is still classified as temperate forest in more humid and wetter areas (Olson et al. 2001), even if the plant species are quite different from those found in the temperate forests of the summer-wet western Eurasian climate. An imaginary visit to these same Mediterranean localities, between 9 and 13 million years ago, allows us to experience a more densely forested landscape. This landscape is reflected in fossil-plant assemblages where taxa are preserved that require a humid temperate climate (subtropical in the broad sense). These assemblages occur in the Iberian Peninsula (Barrón et al. 2010), southern France, central Italy, central Anatolia (Turkey), and the Greek islands Chios, Crete, Gavdos, and Samos (Denk et al. 2018). Such fossil records prove that broad areas of the Mediterranean were covered by temperate forests during intervals of the Neogene, which were wetter than today.

2.3 Eastern Eurasia

Arata Momohara and Yong-Jiang Huang

A visit back in time and back to the eastern part of Eurasia would reveal that temperate forests were persistent and located in a belt straddling the Himalaya Mountains to southwestern China and Japan. Several paleofloras document these assemblages in the Pliocene, Miocene, and Oligocene of China (Liu and Zheng 1995). Temperate megafossil floras can be traced to at least the early Oligocene in southwestern China (Linnemann et al. 2017), but modern species do not appear until the late Pliocene in the area (Huang et al. 2016). These southwestern China floras have experienced an overall southward biogeographic expansion into climate zones formerly occupied by warm temperate forests during the late Cenozoic. This range expansion was thought to be caused by the tectonic activity of the Tibetan Plateau in the north (Jacques et al. 2014), although an overall global cooling trend also may have played a role. Temperate elements of the oak family (Fagaceae), represented by *Quercus* (subgenus* *Cyclobalanopsis*), were common during the late Miocene, while those represented by mountain oaks (section* *Heterobalanus*) were dominant during the late Pliocene (Huang et al. 2016). The late Pliocene establishment of extensive and pure forests of mountain oaks in the northern part of southwestern China is one of the most profound events of temperate vegetation evolution in this area. Earlier, during the Oligocene and early Miocene, many cool temperate, subtropical, and tropical plants lived in present-day Thailand and Vietnam [U0226]. Paleobotanical studies in central Japan indicate the presence of wetland forests dominated by *Glyptostrobus* (see Sect. 2.2) and *Metasequoia* (dawn redwood) in the latest Pliocene (2.6 Ma) [U0225] (Yamakawa et al. 2017). These two genera continued to be major components of wetland forests in Japan until the late

Early Pleistocene, when intensifying climate change and tectonic movements diminished their habitats in lowland settings (Momohara 2016).

2.4 Iceland

Friðger Grímsson

Civilizations are accustomed to living on large continents, which makes a visit to a rather small and isolated island curious. However, Iceland [U0228] deserves this postcard stop because its sediments and sedimentary rocks preserve a rich trove of fossil plants documenting floras of various ages. And, more importantly, changes in Neogene vegetation identify the island as a key player of intercontinental dispersal* [U0229]. Nowadays, natural Icelandic forests are scarce and species diversity low, clearly bearing the mark of growth close to the Arctic Circle. The only woody taxa that grew to the size of trees when the first settlers arrived (around the year 874 AD) on the island were birches (*Betula pubescens*) and rowans (*Sorbus aucuparia*). On the other hand, Neogene plant fossils show that conditions had been favorable for woody tree taxa, and Iceland was covered with various mixed broad-leaved and conifer forests that flourished under a warm and humid climate for millions of years (Denk et al. 2011). Floras from the middle Miocene, the 15-million-year-old Selárdalur-Botn Formation and the 12-million-year-old Brjánslækur-Seljá Formation (e.g., Denk et al. 2005; Grímsson et al. 2007), comprise numerous woody angiosperms and conifers. The Brjánslækur-Seljá flora [U0239] has yielded 33 woody angiosperms and 10 gymnosperms (conifers and *Ephedra*) (Denk et al. 2011). Here, swamp, floodplain, and lake-margin forests were dominated by species of *Alnus* (alder), *Glyptostrobus*, *Populus* (poplar, aspen), *Pterocarya* (wingnut), and *Salix* (willow). Well-drained lowland and lake-margin forests were characterized by trees of *Acer* (maple), *Betula* (birch), *Carpinus*, *Carya*, *Fraxinus* (ash), *Liriodendron* (tulip tree), *Lonicera* (honeysuckle), *Magnolia*, *Platanus* (sycamore), *Pterocarya*, *Rhododendron* (Ericaceae), *Sassafras* (Lauraceae), *Tetracentron* (Trochodendraceae), *Ulmus* (elm), and *Viburnum* (Adoxaceae) [U0231]. Forests at higher elevation, for example, in the foothills and montane regions, were richer in conifers including *Abies* (fir), *Cathaya* (Pinaceae), *Cryptomeria* (Cupressaceae), *Picea* (spruce), *Sciadopitys* (Sciadopityaceae), *Sequoia* (redwood), and *Tsuga* (hemlock). The fossil records of Iceland indicate that plants were able to disperse between North America/Greenland and Europe via the island (the North Atlantic Land Bridge) well into the late Miocene (Grímsson and Denk 2005, 2007; Denk et al. 2010, 2011). The Icelandic Neogene plant assemblages also indicate that there were three major climate-related (cooling) vegetation shifts. The first cooling occurred between 12 and 10 million years ago, a second between 8 and 4 million years ago, and the last in the Pleistocene (Denk et al. 2013).

2.5 North America

Christopher Y. Liu

We now continue our travels across the Icelandic bridge from Eurasia to North America. This route compels us to cross at least three seaways, today (but not 10 Ma). We have seen in unit 1 that, just 12,000 years ago, Eurasia was connected to North America by Beringia, a land bridge that existed between Alaska and eastern Siberia. This connection also was present in deeper time (>3 Ma) when a continuous cover of temperate forests is reconstructed from Eurasia to North America (Tiffney and Manchester 2001). There are several outstanding North American plant-fossil sites that permit the reconstruction of Neogene temperate forests (summaries in Tiffney 1985; Graham 1999; Leopold et al. 2007). However, these sites are scattered across the continent, and large temporal and geographical gaps exist. Hence, the past geographical distribution and extent of the temperate forests cannot be reconstructed with precision. Salzmann et al.'s (2011) models of temperate forests on the continent propose that their biogeographical range, beginning 3 million years ago, was similar to today. Despite a fragmentary fossil record, an interesting fact is firmly established. The flora of these Neogene forests was more similar to the Eurasian flora than at the present (Tiffney and Manchester 2001; Leopold et al. 2007). The Gray fossil site in Tennessee in the United States best exemplifies this fact and is our next postcard stop (Fig. 2.8) [U0235].

Discovered in May 2000, this site contains dark gray to blackish sediments [U0236] that yield a high diversity of plant and animal fossils, enriching the ancient biota that was barely known in southeastern North America before its discovery. All fossil-plant genera are found in the contemporary floras of Eurasia, and two genera (i.e., *Sargentodoxa* and *Sinomenium*, see Fig. 2.2o, p) now only occur in eastern Eurasia. The abundance of vines (e.g., *Sinomenium*, *Vitis* [grape], and *Sargentodoxa*), along with dominant angiosperm and coniferous trees (*Quercus*, *Carya*, *Ulmus*, *Liquidambar*, and *Pinus*) and shrubs (*Staphylea* [bladdernut]), provides evidence of a well-developed, forested landscape [U0239] (Liu and Jacques 2010). In addition to paleobotanical evidence, vertebrate fossils provide insight into the region's paleoecology.

A diverse assemblage of vertebrate taxa is associated with the fossil plants. Their age has been dated to the transition interval between the Pliocene and Miocene (4.5–6.5 Ma) on the basis of the recognized chronologic distribution of mammals such as *Plionarctos* (a bear) and *Teleoceras* (a rhinoceros) [U0238]. Remains of fossil salamanders, turtles, tapirs, and other mammals indicate a unique paleoecological environment that developed in the southern Appalachians during the late Neogene [U0239]. Vegetation transitioned from C3-dominated forests to C4 grasslands (see Saarinen et al., Chap. 3), and global climate deterioration began in response to southern polar glaciation (Salzmann et al. 2011). The rare occurrence of equids (horses) along with a prolific number of tapirs supports the presence of an extensive forested landscape, which included localized ponds surrounded by trees. The Gray locality is considered to be a late Neogene forest refugium during the time when North American grasslands began to expand across the continent. Grasslands are, today, the most widespread vegetation type, which is also true in temperate South America, our next point of interest.

2.6 South America

Luis Palazzesi

We now travel to the southern part of South America, where extensive grasslands cover a part of Argentina [U0240]. Today, obligate-wet taxa grow in limited patches of temperate forest on the well-watered slopes in the westernmost regions of Patagonia (Andean region). These plants consist (Fig. 2.10) of the Nothofagaceae or southern beech (*Nothofagus pumilio*, *N. antarctica*, and *N. obliqua*), Podocarpaceae (yellow wood), Proteaceae (*Embothrium coccineum*), Winteraceae (*Drimys winteri*), and a variety of ferns. The predominant open vegetation of Patagonia is, in part, a result of a progressive decrease in MAP, beginning in the Miocene when an increase in arid-adapted taxa first occurs in the fossil record (Palazzesi and Barreda 2012; Palazzesi et al. 2014). There are thick marine sedimentary successions exposed along the Patagonian coast in which a rich fossil-pollen data set is preserved.

Pollen analysis showed that, prior to the appearance of arid-adapted plants, many taxa requiring wet growing conditions occupied the region. The timing of the major transition from an open vegetation to a typically tree-dominated landscape across non-Andean Patagonia is difficult to estimate, although it may have occurred during the Miocene [U0242] (Palazzesi et al. 2014). Diverse forests [U0241] occupied broad areas of southern South America until the Miocene and included cool temperate and subtropical lineages (e.g., Barreda and Palazzesi 2007). Interestingly, a significant number of taxa with a clear southern biogeographic connection also occur in the fossil record (Box 2.1). Some of these are *Eucalyptus*, *Gymnostoma* (Casuarinaceae), *Dacrycarpus*

Fig. 2.8 Aerial view of the excavations (left) and close-up of laminated layers (right) at the Gray Fossil Site, Tennessee, North America. Photo by Michael Zavada and Steven Wallace

Box 2.1: Impact of Paleogeography on the Modern Patagonian Flora

The modern Patagonian flora is a consequence of paleogeographic changes experienced by South America through the late Cenozoic. In fact, the ongoing physical breakup of Gondwana resulted in profound changes in climate that led to the glaciation of Antarctica and the separation of these landmasses. These developments ran in parallel across the "Southern Connection" (the combination of South America, Antarctica, and Oceania) [U0244]. Biogeographic exchange of plant groups among these continents continued through the Paleogene, even if, by Miocene time, the continents had sepa-rated and their plate trajectories followed different pathways. These paleogeographic and paleoclimatic shifts were important drivers influencing South America's flora, which includes several families with (at least) two different evolutionary histories. The first history involves those plants covering Patagonia well before the final isolation of South America from Antarctica (>34 Ma). This component decreased in abundance and diversity as a consequence of the Andean uplift, during the late Miocene. In contrast, the second history involves those plant groups that have increased in abundance and diversity just after South America became isolated from Antarctica (c. 25 Ma).

Fig. 2.9 The succession of lithologies at Puerto Madryn in Patagonia, where pollen trends indicate the impoverishment of the tree flora in the late Miocene, marking a retreat of temperate forests in South America. Photo by L Palazzesi

(Podocarpaceae), *Papuacedrus* (Cupressaceae), and *Orites* (Proteaceae) (Wilf et al. 2013). Usually, Neogene sediments (Puerto Madryn; Fig. 2.9) preserve a moderate plant diversity, although the proportion of wet-forest taxa is minimized, or close to zero. In contrast, Paleocene sedimentary rocks are rich in fossil pollen of plants belonging to the wet Patagonian forests (e.g., Podocarpaceae pollen) [U0243].

2.7 Oceania

Gregory Jordan and Mike Pole

The last place we visit is Oceania (better known as the Australian continent). Here, temperate forests now cover eastern Australia, New Zealand, and Tasmania. The other landmasses in the region are mostly young oceanic islands or larger continental fragments with poor fossil records (e.g., New Caledonia and Fiji). Therefore, the focus of this section is on Australia and New Zealand, where fossil assemblages are best preserved and the history of floral changes has been reconstructed (Box 2.2). Environmental change in Australia's recent past has been described by Field (in Martinetto et al.,

Box 2.2: A Brief History of Floral Change in Oceania

Oceania, like many other parts of the world, has seen a considerable change in its flora during the Oligocene to Quaternary. Such change is a consequence of evolution, extinction/extirpation, and varying degrees of immigration. Members of the ancient New Zealand forests exhibit both extirpation/extinction and evolutionary processes, as fire- and drought-tolerant groups (eucalypts, acacias, and non-rainforest Casuarinaceae) diminished in numbers and several rainforest groups (e.g., *Coprosma* [Rubiaceae], *Pittosporum* [cheesewood], *Pseudopanax* [Araliaceae]; Greenwood and Atkinson 1977; Perrie and Shepherd 2009) diversified during this time. In addition, alpine and open-habitat groups also experienced significant radiations (e.g., Asteraceae [daisy family], *Carmichaelia* [Fabaceae, legumes], *Hebe/Veronica*) (Wagstaff and Garnock-Jones 1998; Wagstaff et al. 1999; Wagstaff and Breitwieser 2004). In Australia, many temperate-forest species (especially conifers and rainforest species) went extinct beginning in the Miocene and

Fig. 2.10 Living taxa of the westernmost regions of Patagonia (Andean region) that require moist and humid growing conditions. (**a**) Nothofagaceae, *Nothofagus alpina*. Credit: Paco Garin. (**b**) Winteraceae, *Drimys winteri*. Credit: Thomas Stone. (**c**) Proteaceae, *Embothrium coccineum*, Torres del Paine National Park, Chile. Credit: Paweł Drozd on Wiki Drozdp. Licensed under the Creative Commons Attribution-Share Alike 3.0 Unported license

continued into the Pleistocene. Their loss is probably due to climate change, although the cooling was less drastic in this part of the planet than in the Northern Hemisphere. Other species, especially those tolerating fire and drought, increased in diversity during the Quaternary. The island's flora was supplemented by immigration of taxa, especially from the north. These immigrant species now dominate the parts of Australia closest to, and ecologically most similar to, Southeast Asia and Indonesia (Sniderman and Jordan 2011). However, southern Australia also contains some well-dispersed groups (i.e., Asteraceae) that arrived from other southern continents, including South America (see Sect. 2.6).

Chap. 1). However, Australia is not a good place for a virtual visit anytime between 2.6 and 10 million years ago. This is because the late Neogene deposits preserve a poor and scant plant macrofossil record. A more comprehensive view on the continent's past vegetation is found in older Neogene (10–23 Ma) localities. Here, plant fossils provide a somehow surprising postcard in which *Eucalyptus*, the most abundant and landscape-building tree in the living Australian vegetation, is surprisingly sparse (Hill et al. 2016). Instead, most fossil floras are dominated by trees characteristic of rainforests, especially *Nothofagus*, and members of the conifers and Lauraceae. It is not known when the highest diversity of temperate trees occurred in Australia, and there is no evidence that Neogene diversity was higher than today. The evidence we do have indicates that rainforest groups contracted dramatically after the Pliocene until this biome became restricted to the small patches they now occupy. Rainforest conifers, in particular, declined from the exceptional diversity found in the Oligocene. In contrast, the diversity of rainforest trees was greater than at present until as recent as the Early Pleistocene in Tasmania, and diversity in the early Neogene was even higher. Dry climate vegetation expanded through this time and, presumably, the number of species was greater. It is in New Zealand where a higher diversity of plants occurred in the Neogene when compared to the temperate forests of the island, today.

The plants that formed New Zealand's Neogene temperate forests are well documented in a very rich local paleoflora found in Miocene (17–18 Ma) sediments from St. Bathans' Blue Lake (Fig. 2.11a) [U0246]. Collections made at this locality include many leaves that are preserved with minimal alteration of the organic tissues (Fig. 2.11b). These include 13 conifers, 144 flowering plant types, and two cycad-like taxa. Hence, the fossils [U0247] that, likely, originated from an area of just a few hectares surrounding the depositional site record the past presence of more conifer types that now grow in all of New Zealand, where treeless vegetation is widespread [U0248]. When compared with the archipelago's Neogene flora, the modern vegetation is of much lower diversity [U0249].

2.8   Cradles and Routes for Plant Diversity

Edoardo Martinetto

We have witnessed during our global tour that temperate forests were comprised of a high diversity of woody plants, especially from 34 to about 5 or 3 million years ago. We have provided only a partial accounting of arboreal genera over this time because a complete inventory is beyond the scope of this chapter. Neogene diversity is greater than pre-Oligocene landscapes and greater than the inventories of living tree gen-

era in the same areas at the present. Iceland, hosting three tree genera, today, is the best example. These diversity trends indicate that temperate forests were never more diverse than in the Oligocene to Pliocene. Do we have an explanation for this triumph of large autotrophic* terrestrial organisms?

There are several ideas about how, and why, temperate forests flourished around the world in the not-so-distant past. One explanation is climate. The expansion of this forest type may have been promoted by wetter and warmer conditions in many places during that time interval. A wetter seasonality, in conjunction with evolution, certainly played a key role. In fact, recent phylogenetic analyses on forest-tree lineages (e.g., the order Fagales: Larson-Johnson 2016) indicate that these plants experienced increasing diversification beginning 60 million years ago, an evolutionary trend that lasted for 30 million years. Many genera that now live in temperate climates were widespread at mid-latitudes in the Eocene and existed as minor elements in tropical floras (Leopold et al. 2007). When Oligocene or Miocene global cooling reduced tropical taxa at these low latitudes, temperate groups outcompeted and replaced them, rising to dominance. However, the origination of new taxa probably occurred in rather restricted and localized areas, which requires dispersal as another fundamental process to explain the plant diversity we find in the fossil record [U0232] (Donoghue 2008). For example, Denk and Grimm (2009) provide a good, well-documented case study of the evolution and dispersal history of the genus *Fagus* [U0233, U0234].

Fagus originated in North America in the Paleocene, about 60 million years ago, and rapidly spread to Greenland and eastern Eurasia. By the Oligocene (34 Ma), the taxon reached western Eurasia. Then, a complex interchange of taxa between Eurasia and North America enhanced the diversity of the genus, and this is particularly apparent in East Asia. The routes over which beech extended its biogeographic range were also available to other taxa. In the early Cenozoic, plants could easily disperse between northwestern North America and NE Eurasia via Bering archipelago and bridge (currently the Bering Strait) [U0234]. Dispersal was also possible between Greenland/North America and Europe (Scandinavia/Great Britain). This dispersal route was facilitated by the North Atlantic Land Bridge (Greenland-Scotland Transverse Ridge) throughout Eocene to Miocene times. Paleobotanical evidence indicates that dispersal occurred via the Greenland-Scotland Transverse Ridge until about 10 million years ago (late Miocene) and via the Greenland-Iceland part of the mid-oceanic ridge until about 6 million years ago (latest Miocene; Denk et al. 2011). There is a different story in the Southern Hemisphere.

The temperate floras of the Southern Hemisphere were strongly modified by range expansion of plants growing in the Northern Hemisphere. Range expansions began in the late Paleocene and continued into the latest Neogene, mainly

A

B

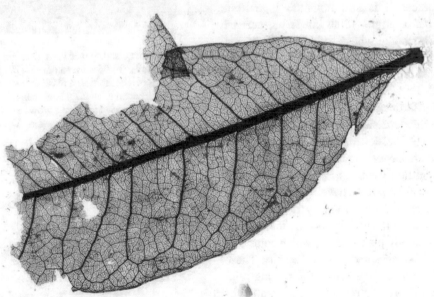

Fig. 2.11 The Miocene sediments (c. 17–18 Ma) cropping out at St. Bathans' Blue Lake. (**a**) General view of the outcrop with prevailing white sand in which the richest local paleoflora of New Zealand's Neogene has been collected. (**b**) A "mummified" leaf showing excellent preservation of resistant organic material at Blue Lake. Photo by M Pole

from 30 to 3 million years ago. South America received many plant taxa from the neotropics and now has relatively few arborescent* genera in common with Australasia. In contrast, New Zealand and southern Australia shared the same Australasian sources, and, hence, both exhibit a strong floristic relationship having 80% of vascular genera in common. This percentage may have been even greater in the middle Cenozoic, when Australian genera such as *Eucalyptus, Casuarina,* and *Acacia* were also present in New Zealand (Markgraf et al. 1995).

2.9   The Late Cenozoic Extirpation/ Extinction of Plants

Yong-Jiang Huang, Gregory Jordan, Edoardo Martinetto and Arata Momohara

As we will see in the following chapters (in particular Delfino et al., Chap. 10), mass extinctions have altered Earth's biosphere at least five times during the Phanerozoic. However, biodiversity of plants and animals also has experienced smaller and more localized extinctions over time in different parts of the globe. For example, Eiserhardt et al. (2015) have shown that an uneven relationship [U0250] exists between phylogenetic diversification* and plant extirpation/extinction in different parts of the Northern Hemisphere during the Neogene. Western Eurasia is the only region where the woody plant-fossil record from 2 to 34 million years ago dramatically differs from the composition of the modern flora. Studies of fruits and seeds document that at least 100 woody genera that are now absent in the native flora of Western Europe were part of the vegetation there only 4–3 million years ago. All of these plants either were extirpated from the region or are now extinct. *Actinidia* (Fig. 2.12) [U0251] represents a good example of a genus that was extirpated from western Eurasia, whereas it is still common in eastern Eurasia. This genus includes the edible kiwi fruit, derived from wild plants in China, and whose fruits were improved by cultivation in New Zealand. Once cultivated, the plants were reintroduced to Europe and into North America (Ferguson et al. 1990). Fossil seeds from central (Mai 1995) and southern Europe (Martinetto 2015) prove that *Actinidia* grew there during the Neogene, and a few population centers persisted until the Early Pleistocene. Ultimately, environmental change affected these refugia and the genus was lost in the area about 1.8 million years ago. Many stories, more or less similar to that of *Actinidia*, have a single possible explanation. Research carried out in the twentieth century (Box 2.3) showed that western Eurasia was the theater of a massive extirpation of woody plant genera, which coincided with a real extinction of species and loss of diversity. These species are thought to have differed from their living relatives in East Asia and/or North America. Their extirpation or extinction has been attributed, by some, to the rise of the southern European mountain chains, oriented in an east-to-west direction. However, recent paleobotanical data demonstrate that many of these extirpated/extinct woody taxa grew to the south of such mountain chains, along the shores of the Mediterranean, during the Pliocene and Early Pleistocene (Magri et al. 2017; Martinetto et al. 2017). Hence, the increasingly dry conditions around this sea, and on continental North Africa, may have played a significant role in pre-

Fig. 2.12 Extinct and extant kiwi (*Actinidia*). (**a**) Extant *Actinidia arguta* with fruits. (**b**) Fossil seed of *Actinidia faveolata* recovered from the sediments of a late Pliocene fossil forest in northwestern Italy (Stura di Lanzo: Martinetto et al. 2007), scale bar 1 mm. The genus *Actinidia* includes the edible kiwi fruit and grows wild only in eastern Eurasia. The genus was also present in western Eurasia, evidenced by fossil seeds dating from c. 20 to 1.8 Ma

Box 2.3: The Lost European Flora as a Tear in a Political "Curtain"

The lost European flora, including several tens of extirpated plant genera, was first uncovered in the nineteenth century. But, our understanding of Neogene forest diversity and the paleoecological relationships began during the second half of the twentieth century, when Europe was split into an eastern and western block by the "Iron Curtain*." Both communication and travel across this political boundary were very difficult at that time, and scientific research was one of the few possible ways to establish an east-to-west link between people. Research generated a positive feedback loop in the paleobotanical community, especially with the intensity of eastern scientific activity (Fig. 2.13), resulting in an impressive number of fossil collections and the resultant publication of extensive data sets. Many Neogene temperate-forest species were established or revised by a small number of researchers including Erwin Knobloch and Zlatko Kvaček in Czechoslovakia (now the Czech Republic) and Dieter H. Mai and Harald Walther in Eastern Germany (now united Germany) [U0209].

venting the southward extension of their range distribution. Recent research also suggests that there was neither a single, late Cenozoic extirpation or extinction event affecting most taxa. Rather, the flora experienced a sequence of extirpation/extinction events, affecting different areas at different times (Martinetto et al. 2017). This pattern resulted in the conservation of the most sensitive taxa in warm and wet relict niches to the south. A very similar pattern has been demonstrated for Japanese floras, where a smaller number of woody plants experienced extinction or extirpation.

Most genera extirpated from western Eurasia survived in eastern Eurasia (Fig. 2.14). Their survival may be the result of this region not having experienced as dramatic environmental change as other places and, possibly, to a Miocene trans-Eurasian biogeographic distribution. Many woody plant genera are known to have had a holoarctic∗ distribution in the Miocene, and several are preserved in fossil deposits of both eastern Asia and North America (e.g., *Carya, Hamamelis* [witch hazel], *Liriodendron, Meliosma, Nyssa, Sassafras, Symplocos*). A few genera persisted solely in North America (*Leitneria* [corkwood], *Sequoia, Taxodium* [bald cypress]). Several other taxa, which better tolerated seasonally dry conditions, were able to survive in western Eurasia or northern Africa (e.g., *Coriaria, Liquidambar, Myrtus* [myrtle], *Ocotea, Styrax* [snowbell], *Tetraclinis, Zelkova*) and continue to be part of today's flora in these regions. It is in East Asia where the woody plant diversity of the Northern Hemisphere is best preserved. Despite this, some East Asian regions, such as Japan, have experienced relatively severe extirpation/extinction of woody species as a consequence of their isolation and absence of any physical land bridges. The species losses in Japan occurred chiefly after the Pliocene (Fig. 2.15), when global temperature declined rapidly followed by frequent transitions between glacial∗ and interglacial stages (Momohara 2016). Indeed, Japan may represent one of the key regions in East Asia that has undergone significant species loss. Most of the taxa extirpated from Japan (and, in fact, from Europe) have occupied refugia in mainland Asia, mainly in China and the Indo-Chinese Peninsula. For example, many early Cenozoic relict taxa that once lived in other regions of the Northern Hemisphere have an obvious relict concentration in moun-

Fig. 2.13 Influential European paleobotanists who were very active during the second half of the twentieth century, in the eastern side of the "Iron Curtain." From left to right: Harald Walther, Zlatko Kvaček, Erwin Knobloch, Dieter Hans Mai. These scientists collected an impressive amount of material and amassed a significant data set about Neogene fossil plants, which has improved our knowledge about the floristic diversity of these temperate forests

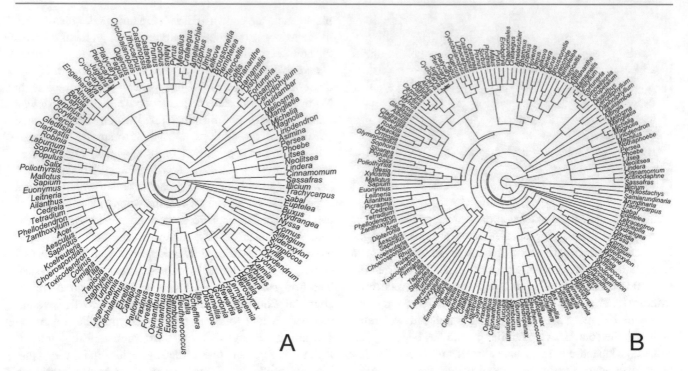

Fig. 2.14 Cladograms published by Eiserhardt et al. (2015) showing the extirpated temperate-forest genera in black versus those still growing in the area in red. Note the dominance of black in Europe (**a**), indicating that most genera that grew there in the mid-Cenozoic were extirpated in the late Cenozoic. Conversely, the dominant red color for East Asia (**b**) indicates that most of the mid-Cenozoic genera persisted there since the Neogene

tainous areas of central China (Huang et al. 2015). However, there still are a few exceptions.

Recent discoveries indicate that some temperate genera, such as *Cedrus* (cedar), occurred in southwestern China during the Pliocene, but its modern members live in the areas around the Mediterranean and Himalayas (Su et al. 2013). *Cedrus* is one of a few examples, together with *Sequoia*, of late Cenozoic plant extirpations in China. The loss of both genera is thought to be in response to the strengthening of seasonal drought associated with the East Asian monsoon (Su et al. 2013; Zhang et al. 2015). According to Markgraf et al. (1995), both the timing and nature of the decline in Cenozoic forest diversity occur in parallel in both hemispheres. For instance, these authors argued that different systems of mountainous terrains surrounded by oceans, including Japan, New Zealand, and southern South America, experienced a near-identical reduction of woody species diversity at the end of the Cenozoic. The reduction of woody species is detected also in cooler rainforests of southern Australia (especially in Tasmania; Jordan 1997) but not in the warmer Australian rainforests. Here, the range extension of plants previously growing in Asia/Malaysia into warm Australian rainforests may have balanced or outnumbered those taxa that were lost by either extinction or extirpation (Sniderman et al. 2013).

Today, a very high plant diversity, much higher than in other mid-latitude floristic regions∗, is found in areas with a Mediterranean climate of the Australian Southwest Floristic Region. Here, sclerophyll∗ taxa evolved in response to long-term climate stability in southwest Australia during the late Cenozoic (Sniderman and Jordan 2011). This condition promoted an increasing species richness by encouraging genetic divergence between populations and discouraging plant dispersal. Given the spectacular recent sclerophyll diversity, any loss during the late Cenozoic would hardly seem possible. The situation is different in southeastern Australia, where the present-day diversity is relatively low. A recent study (Sniderman et al. 2013) showed that a very high-diversity sclerophyll flora existed under high-rainfall, summer-wet climates in the Early Pleistocene. The regional loss of sclerophyll diversity occurred at the same time as diversity loss in the cooler rainforests. This phenomenon cannot be explained by substitution of species of one ecological type by another but, rather, a general extirpation due to climate change. Both ecological types suffered subsequent extirpation/extinction, and gymnosperms have been more prone to these losses than angiosperms.

Fig. 2.15 Exotic taxa extirpated from the central Kinki District (Japan) since the late Pliocene (total 50 species), and numerical age of each last occurrence. Scale bars 5 mm

2.10 The East Asian Refuge of Eurasian Plant Diversity

Edoardo Martinetto, Nareerat Boonchai, Paul J. Grote, Christopher Y. Liu and Arata Momohara

The persistence of a continuous moist climate belt, extending north to south along the eastern coast of Asia, over the last 20 million years (Morley 2001; Guo et al. 2008; Quan et al. 2014), was probably the main factor permitting the survival of several Tertiary woody plant genera only here. Studies on paleofloras of southeastern Asia are advancing quickly, and there are clues that a long-term climate regime was accompanied by vegetational stability in this area. Recently, the first Oligocene mummified fossil flora was found in East Asia [U0253] (Quan et al. 2016), occurring in the Nanning Basin, Guangxi, southern China. More than 500 stumps, 1000 fruits and seeds [U0254], and abundant leaf impressions were recovered. Fruits of Fagaceae and Theaceae are dominant, and the living equivalents of the fossil plants

grow, today, in Guangxi. However, oscillations in climate are documented throughout the interval from 50 to 30 million years ago (Quan et al. 2012), and it is demonstrated that the geographical position of the tropical/temperate transition has moved [U0252].

Current research shows that the Paleogene tropical/temperate transitional zone was positioned more to the south than at present. Data from Thailand, based on pollen records (Songtham et al. 2003, 2005), indicate that the climate in northern Thailand was temperate during the late Oligocene or early Miocene. During the middle Miocene, fossil-plant assemblages are indicative of tropical conditions, the same as today. However, macrofossil assemblages from the late Oligocene or early Miocene indicate a mix of tropical (*Mangifera* [mango], *Semecarpus* [varnish tree]) (Sawangchote et al. 2009, 2010) and temperate plants (*Sciadopitys, Sequoia*), demonstrating that the area was part of the transitional zone between climate regimes.

The tropical/temperate boundary in the Neogene (Fig. 2.16) occurred somewhere close to the modern border between India and China (Morley 1998; Mehrotra et al.

Fig. 2.16 Map of the modern temperate/tropical transition in Southeast Asia and latitudinal vegetation zonation after Osawa (1993). The subtropical belt is considered part of the temperate zone. To the right, four modern examples of East Asian forests: (**a**) Tropical monsoon forest in Xishuangbanna, southwestern China (*Dipterocarpus, Lithocarpus*). (**b**) Subtropical evergreen forest in Okinawa, southern Japan (*Castanopsis, Cyclobalanopsis*). (**c**) Cool temperate deciduous broad-leaved forest in Mt. Odaigahara, central Japan (*Fagus*). (**d**) Mixed coniferous and deciduous broad-leaved forest in Hokkaido, Japan (*Picea, Abies, Quercus*)

2005; Jacques et al. 2015). In fact, a more or less uniform tropical moist climate persisted throughout peninsular India. This condition is confirmed by the wide distribution of the tropical rainforest family Dipterocarpaceae in the fossil record along with other genera with similar climatic requirements (Mehrotra et al. 2005). Moist tropical forests were present in Indochina at least from the early Miocene (Morley 2000), and Dipterocarpaceae have been prominent in Thailand since then (Morley 2000). The middle Miocene fossil record, summarized by Jacques et al. (2015), indicates a strong floristic link with a continuous tropical forest zone across India and between Southeast Asia and southern China. During most of the Neogene, large areas of China were covered by temperate forests, whereas temperate fossil forms in India have only been recorded in a single Plio-Pleistocene locality (Mehrotra et al. 2005). The dispersal route of a large number of temperate elements to India is thought to have originated from Yunnan via Tibet, and was favored by the development of amenable climatic conditions to the south of the Himalayans in the late Cenozoic. During the Pleistocene, giant legume trees with possible affinity to *Koompassia* (Kempas or Tualang; Fig. 2.17) [U0256, U0257], one of the tallest tropical trees in the southeastern Asian rainforest, first appeared in northwestern Thailand [U0255] (Philippe et al.

2013). An episode of wetter mid-Pleistocene climate, here, is indicated by the fact that these trees are restricted, today, to Thailand's southernmost provinces, Malaysia, Indonesia, and Papua New Guinea. Even during the coldest Pleistocene intervals, temperate forests remained in southeastern Asia, probably in the numerous mountain ranges that offered refugia (Fig. 2.18). When Plio-Pleistocene climate deterioration promoted the extirpation of Paleogene-Neogene plant genera from parts of East Asia, amenable temperatures accompanied by weak seasonal drought continued in central China permitting the growth of these genera in relict habitats [U0259] (Huang et al. 2015).

2.11 Conclusions

The diversification and dispersal of angiosperm and gymnosperm lineages beginning in the Cretaceous-Paleogene determined the coexistence of a large diversity of plant genera in the temperate forests on Earth. Since the Eocene/Oligocene transition (34 Ma), plant diversity increased as members of tropical/subtropical lineages adapted to cooler conditions and extended their ranges into the warmer temperate forests (i.e., the subtropical forests). An equable climate

Fig. 2.17 Petrified tree trunks with possible affinity to *Koompassia* from the Middle Pleistocene of Thailand, and living *Koompassia*. (**a**) Petrified trunk No. 1, 69 m long. (**b**) Petrified trunk No. 7, 38.7 m long. (**c**) Recent tree of *Koompassia malaccensis*. Photos by N Boonchai (**a**, **b**), and courtesy of Niruth Tangsiri (**c**)

across large areas of the Northern Hemisphere, much warmer and wetter than today, occurred during the Miocene (23–12 Ma). These conditions provided a suitable habitat for moisture-requiring woody plants, and their adaptation and evolution controlled the triumph of temperate forests. In response to the presence of several land bridges between continents, many woody plant genera extended their biogeographic ranges from their places of origination to all northern temperate regions. North America, Greenland, Iceland, and Eurasia witnessed a uniform and never-more-diverse landscape. Since the end of the Miocene, and more consistently during the Pliocene and Pleistocene, global climatic deterioration in response to Antarctic glaciation triggered the demise of the temperate forest. The most thermophilous elements of the once, high-diversity forests were extirpated from the northern regions in response to several cold pulses. In western Eurasia, most of these elements were able to survive in southern Europe until the late Pliocene, and a few persisted into the Early Pleistocene. Subsequently, though, all these plants disappeared. The southern European moun-

tain ranges do not seem to have played any relevant role in stopping the southward "escape" of trees, which was more severely hampered by increasing dry conditions around the Mediterranean and northern Africa. Indeed, western Eurasia was the theater of a large extirpation/extinction of thermophilous plants in the Plio-Pleistocene, whereas extirpation/extinction was rather limited in eastern Eurasia. The case for temperate forests in the Southern Hemisphere is also different. Here, a uniform flora, sharply distinct from those of the Northern Hemisphere, was present by the early-middle Cenozoic. This situation was fostered because floral exchange between austral continents was rather easy due to the presence of a "Southern Connection" that existed until about 25 million years ago. Yet, there is a close parallel in the timing and nature of the late Cenozoic decline in forest diversity of the Northern (Europe, Japan) and Southern Hemisphere (New Zealand, South America). The only exception to this trend is recorded in the Australian rainforests, which were enriched in diversity during the Quaternary by the immigration of taxa from southeastern Asia.

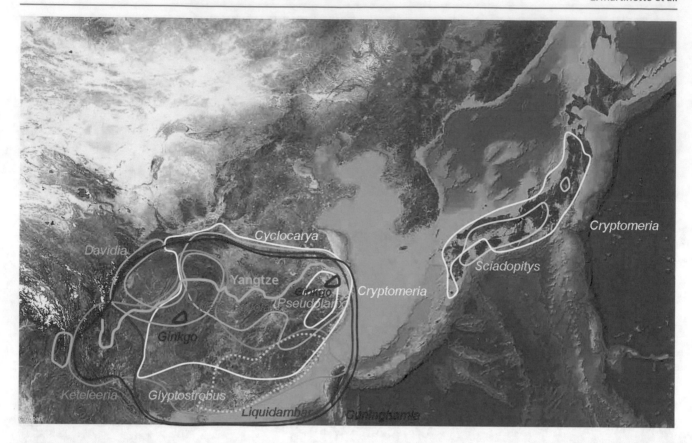

Fig. 2.18 East Asian refugia of several temperate plant genera that had a broader distribution during the mid-Cenozoic. Image by A Momohara, map courtesy of Google Maps

Questions

1. What is ambiguous when the word "temperate" is used with reference to climate and forests?
2. What ecological dynamics separate the concepts of extirpation and extinction?
3. How could beech trees (*Fagus*), with rather large fruits, expand their biogeographic range distribution from North America to Eurasia?
4. What biological processes are needed to explain higher plant diversity in temperate forests?
5. How do the Neogene fossil plants of Europe compare with the living ones of China?
6. Using the information available in units 1 and 2, how would you describe the environmental changes experienced at the southern foothills of the western Alps (e.g., near Ivrea)?
7. Do you think t hat the fossil forest of Bükkábrány had high species diversity, and, regardless if yes or no, why?
8. Although roughly contemporaneous in time, the Blue Lake (New Zealand) and Bílina (Czech Republic) sites preserved different plants. Why?
9. The Miocene flora of Iceland grew at a high northern latitude, but represented a temperate forest. What temperature and rainfall conditions permitted the growth of this biome?
10. What are the major changes in plant macrofossil assemblages from the Pliocene to Pleistocene record in Japan?
11. The southern European mountain chains do not seem to have played a major role in plant extirpation/extinction. Why?

Acknowledgments The authors are thankful to Guido Grimm and Thomas Denk for constructive discussions and data contributions. Boglarka Erdei kindly provided the photographs of the Bükkábrány Fossil Forest. Mike Zavada and Steven Wallace are thanked for providing images of the Gray Fossil Site. Suggestions received by Steven R. Manchester consistently improved the text. Financial support was provided by the University of Turin.

References

Barreda V, Palazzesi L (2007) Patagonian vegetation turnovers during the Paleogene-early Neogene: origin of arid-adapted floras. Bot Rev 73:31–50

Barrón E, Rivas-Carballo R, Postigo-Mijarra JM, Alcalde-Olivares C, Vieira M, Castro L, Pais J, Valle-Hernández M (2010) The Cenozoic vegetation of the Iberian Peninsula: a synthesis. Rev Palaeobot Palynol 162(3):382–402

Bertini A, Martinetto E (2011) Reconstruction of vegetation transects for the Messinian/Piacenzian of Italy by means of comparative analysis of pollen, leaf and carpological records. Palaeogeogr Palaeoclimatol Palaeoecol 304:230–246

Biltekin D, Popescu S-M, Suc J-P, Quézel P, Yavuz N, Jiménez-Moreno G, Safra A, Çağatay MN (2015) Anatolia: a plant refuge area during the last 23 million years according to pollen records from Turkey. Rev Palaeobot Palynol 215:1–22

Collinson ME, Hooker JJ (2003) Paleogene vegetation of Eurasia: framework for mammalian faunas. In: Reumer JWF, Wessels W (eds) Distribution and migration of Tertiary mammals in Eurasia. A volume in honour of Hans de Bruijn, vol 10. Deinsea, Rotterdam, pp 41–83

Colombero S, Alba DM, D'Amico C, Delfino M, Esu D, Giuntelli P, Pavia G (2017) Late Messinian mollusks and vertebrates from Moncucco Torinese, north-western Italy. Paleoecological and paleoclimatological implications. Palaeontol Electron 20:1–66

de Gouvenain RC, Silander JA (2016) Temperate forests. In: Reference Module in Life Sciences. Elsevier, Amsterdam

Denk T, Grimm GW (2009) The biogeographic history of beech trees. Rev Palaeobot Palynol 158(1):83–100

Denk T, Grímsson F, Kvaček Z (2005) The Miocene floras of Iceland and their significance for late Cainozoic North Atlantic biogeography. Bot J Linn Soc 149:369–417

Denk T, Grímsson F, Zetter R (2010) Episodic migration of oaks to Iceland: evidence for a North Atlantic "land bridge" in the latest Miocene. Am J Bot 97:276–287

Denk T, Grímsson F, Zetter R, Símonarson LA (2011) Late Cainozoic floras of Iceland. 15 million years of vegetation and climate history in the northern North Atlantic. Springer, Dordrecht, p 854

Denk T, Grimm GW, Grímsson F, Zetter R (2013) Evidence from "Köppen signatures" of fossil plant assemblages for effective heat transport of gulf stream to subarctic North Atlantic during Miocene cooling. Biogeosciences 10:7927–7942

Denk T, Zohner CM, Grimm GW, Renner SS (2018) Plant fossils reveal major biomes occupied by the late Miocene Old-World Pikermian fauna. Nat Ecol Evol 2(12):1864

Donoghue MJ (2008) A phylogenetic perspective on the distribution of plant diversity. PNAS 105(Supplement 1):11549–11555

Eiserhardt WL, Borchsenius F, Plum CM, Ordonez A, Svenning JC (2015) Climate-driven extinctions shape the phylogenetic structure of temperate tree floras. Ecol Lett 18(3):263–272

Erdei B, Dolezych M, Hably L (2009) The buried Miocene forest at Bükkábrány (Hungary). Rev Palaeobot Palynol 155(1–2):69–79

FAO 2006. Number of native forest tree species. http://www.fao.org/forest-resources-assessment/past-assessments/fra-2005/maps-and-figures/en/

Ferguson AR, Seal AG, Davison RM (1990) Cultivar improvement, genetics and breeding of kiwifruit. Acta Hortic 282:335–334

Fischer A, Marshall P, Camp A (2013) Disturbances in deciduous temperate forest ecosystems of the northern hemisphere: their effects on both recent and future forest development. Biodivers Conserv 22(9):1863–1893

Graham A (1999) Late Cretaceous and Cenozoic history of north American vegetation (North of Mexico). Oxford University Press, New York. 350 pp

Greenwood RM, Atkinson IAE (1977) Evolution of the divaricating plants in New Zealand in relation to moa browsing. Proc N. Z. Ecol Soc 24:21–33

Grímsson F, Denk T (2005) *Fagus* from the Miocene of Iceland: systematics and biogeographical considerations. Rev Palaeobot Palynol 134:27–54

Grímsson F, Denk T (2007) Floristic turnover in Iceland from 15 to 6 Ma - extracting biogeographical signals from fossil floras assemblages. J Biogeogr 34:1490–1504

Grímsson F, Denk T, Símonarson LA (2007) Middle Miocene floras of Iceland – the early colonization of an island? Rev Palaeobot Palynol 144:181–219

Guo ZT, Sun B, Zhang ZS, Peng SZ, Xiao GQ, Ge JY, Hao QZ, Qiao YS, Liang MY, Liu JF, Yin QZ, Wei JJ (2008) A major reorganization of Asian climate by the early Miocene. Clim Past 4(3):153–174

Herman AB, Akhmetiev MA, Kodrul TM, Moiseeva MG, Yakovleva AI (2009) Flora development in northeastern Asia and northern Alaska during the Cretaceous–Paleogene transitional epoch. Stratigr Geol Correl 17(1):79–97

Hill RS, Beer YK, Hill KE, Maciunas E, Tarran MA, Wainman CC (2016) Evolution of the eucalypts - an interpretation from the macrofossil record. Aust J Bot 64(8):600–608

Huang Y, Jacques FMB, Su T, Ferguson DK, Tang H, Chen W, Zhou Z (2015) Distribution of Cenozoic plant relicts in China explained by drought in dry season. Sci Rep 5:14212

Huang Y, Jia L, Wang Q, Mosbrugger V, Utescher T, Su T, Zhou Z (2016) Cenozoic plant diversity of Yunnan: a review. Plant Divers 38:271–282

Ivanov D, Utescher T, Mosbrugger V, Syabryaj S, Djordjevic-Milutinovic D, Molchanoff S (2011) Miocene vegetation and climate dynamics in Eastern and Central Paratethys (Southeastern Europe). Paleogeogr Palaeoclimatol Palaeoecol 304(3–4):262–275

Jacques FMB, Su T, Spicer RA, Xing Y-W, Huang Y-J, Zhou Z-K (2014) Late Miocene southwestern Chinese floristic diversity shaped by the southeastern uplift of the Tibetan Plateau. Palaeogeogr Palaeoclimatol Palaeoecol 411:208–215

Jacques FMB, Shi G, Su T, Zhou Z (2015) A tropical forest of the middle Miocene of Fujian (SE China) reveals Sino-Indian biogeographic affinities. Rev Palaeobot Palynol 216:76–91

Jordan GJ (1997) Evidence of Pleistocene plant extinction and diversity from Regatta Point, western Tasmania, Australia. Bot J Linn Soc 123(1):45–71

Kooyman RM, Wilf P, Barreda VD, Carpenter RJ, Jordan GJ, Kale Sniderman JM, Allen A, Brodribb TJ, Crayn D, Feild TS et al (2014) Paleo-Antarctic rainforest into the modern old world tropics: the rich past and threatened future of the "southern wet forest survivors". Am J Bot 101(12):2121–2135

Köppen W (1936) Das geographische system der Klimate. Gerbrüder Bonträger, Berlin

Kováčová M, Doláková N, Kováč M (2011) Miocene vegetation pattern and climate change in the northwestern Central Paratethys domain (Czech and Slovak Republic). Geol Carpath 62(3):251–266

Kvaček Z, Kovar-Eder J, Kováč M, Doláková N, Jechorek H, Parashiv V, Kováčová M, Sliva L (2006) Evolution of landscape and vegetation in the Central Paratethys area during the Miocene. Geol Carpath 57(4):295–310

Larson-Johnson K (2016) Phylogenetic investigation of the complex evolutionary history of dispersal mode and diversification rates across living and fossil Fagales. New Phytol 209(1):418–435

Leopold EB, Reinink-Smith L, Liu G (2007) Overview of Alaskan Tertiary Floras-building on the work of Jack Wolfe. Cour Forschungsinst Senck 258:129–138

Linnemann U, Su T, Kunzmann L, Spicer RA, Ding W-N, Spicer TEV, Zieger J, Hofmann M, Moraweck K, Gärtner A, Gerdes A, Marko L, Zhang S-T, Li S-F, Tang H, Huang J, Mulch A, Mosbrugger V,

Zhou Z-K (2017) New U-Pb dates show a Paleogene origin for the modern Asian biodiversity hot spots. Geology 46(1):3–6

Liu YSC, Jacques FMB (2010) *Sinomenium macrocarpum* sp. nov. (Menispermaceae) from the Miocene-Pliocene transition of Gray, northeast Tennessee, USA. Rev Palaeobot Palynol 159:112–122

Liu YSC, Zheng YH (1995) Neogene Floras. In: Li X et al (eds) Fossil Floras of China through the geological ages. Guangdong Science and Technology Press, Guangzhou, pp 506–551

Magri D, Di Rita F, Aranbarri J, Fletcher W, González-Sampériz P (2017) Quaternary disappearance of tree taxa from Southern Europe: timing and trends. Quat Sci Rev 163:23–55

Mai DH (1995) Tertiäre Vegetationsgeschichte Europas. Methoden und Ergebnisse. Verlag G. Fischer, Jena. 691 pp

Manchester SR, Chen Z, Lu A, Uemura K (2009) Eastern Asian endemic seed plant genera and their paleogeographic history throughout the Northern Hemisphere. J Syst Evol 47:1–42

Markgraf V, McGlone M, Hope G (1995) Neogene paleoenvironmental and paleoclimatic change in southern temperate ecosystems–a southern perspective. Trends Ecol Evol 10(4):143–147

Martinetto E (2015) Monographing the Pliocene and early Pleistocene carpofloras of Italy: methodological challenges and current progress. Palaeontogr Abt B 293:57–99

Martinetto E, Scardia G, Varrone D (2007) Magnetobiostratigraphy of the Stura di Lanzo fossil forest succession. Riv Ital Paleontol Stratigr 113(1):109–125

Martinetto E, Momohara A, Bizzarri R, Baldanza A, Delfino M, Esu D, Sardella R (2017) Late persistence and deterministic extinction of "humid thermophilous plant taxa of East Asian affinity" (HUTEA) in southern Europe. Palaeogeogr Palaeoclimatol Palaeoecol 467:211–231

Martinetto E, Tema E, Irace A, Violanti D, Ciuto M, Zanella E (2018) High-diversity European palaeoflora favoured by early Pliocene warmth: new chronological constraints from the Ca' Viettone section, NW Italy. Palaeogeogr Palaeoclimatol Palaeoecol 496:248–267

McIver EE, Basinger JF (1999) Early Tertiary floral evolution in the Canadian high Arctic. Ann Mo Bot Gard 86(2):523–545

Mehrotra RC, Liu XQ, Li CS, Wang YF, Chauhan MS (2005) Comparison of the Tertiary flora of southwest China and northeast India and its significance in the antiquity of the modern Himalayan flora. Rev Palaeobot Palynol 135(3):145–163

Momohara A (2016) Stages of major floral change in Japan based on macrofossil evidence and their connection to climate and geomorphological changes since the Pliocene. Quat Int 397:92–105

Morley RJ (1998) Palynological evidence for Tertiary plant dispersals in the SE Asian region in relation to plate tectonics and climate. In: Biogeography and geological evolution of SE Asia. Backhuys, Leiden, pp 211–234

Morley RJ (2000) Origin and evolution of tropical rain forests. Wiley, London. 362 pp

Morley RJ (2001) Why are there so many primitive angiosperms in the rain forests of Asia-Australasia. In: Faunal and floral migrations and evolution in SE Asia-Australasia, vol 1. Balkema, Lisse, pp 185–199

Olson DM, Dinerstein E, Wikramanayake ED, Burgess ND, Powell GV, Underwood EC, D'amico JA, Itoua I, Strand HE, Morrison JC, Loucks CJ, Allnutt TF, Ricketts TH, Kura Y, Lamoreux V, Wettengel WW, Hedao P, Kassen KR (2001) Terrestrial ecoregions of the world: a new map of life on earth. Bioscience 51(11):933–938

Osawa M (1993) Latitudinal pattern of mountain vegetation zonation in southern and eastern Asia. J Veg Sci 4:13–18

Palazzesi L, Barreda V (2012) Fossil pollen records reveal a late rise of open-habitat ecosystems in Patagonia. Nat Commun 3:1294

Palazzesi L, Barreda VD, Cuitiño JI, Guler MV, Tellería MC, Santos RV (2014) Fossil pollen records indicate that Patagonian desertification was not solely a consequence of Andean uplift. Nat Commun 5:3558

Perrie LR, Shepherd LD (2009) Reconstructing the species phylogeny of *Pseudopanax* (Araliaceae), a genus of hybridising trees. Mol Phylogenet Evol 52:774–783. https://doi.org/10.1016/j.ympev.2009.05.030

Philippe M, Boonchai N, Ferguson DK, Jia H, Songtham W (2013) Giant trees from the middle Pleistocene of northern Thailand. Quat Sci Rev 65:1–4

Pound MJ, Haywood AM, Salzmann U, Riding JB (2012) Global vegetation dynamics and latitudinal temperature gradients during the Mid to Late Miocene (15.97–5.33 Ma). Earth Sci Rev 112(1–2):1–22

Quan C, Liu YSC, Utescher T (2012) Paleogene temperature gradient, seasonal variation and climate evolution of northeast China. Palaeogeogr Palaeoclimatol Palaeoecol 313:150–161

Quan C, Liu Z, Utescher T, Jin J, Shu J, Li Y, Liu YSC (2014) Revisiting the Paleogene climate pattern of East Asia: a synthetic review. Earth-Sci Rev 139:213–230

Quan C, Fu Q, Shi G, Liu Y, Li L, Liu X, Jin J (2016) First Oligocene mummified plant Lagerstaetten at the low latitudes of East Asia. Sci China Earth Ssci 59:445–448

Rees-Owen RL, Gill FL, Newton RJ, Ivanović RF, Francis JE, Riding JB, Vane CH, dos Santos RAL (2018) The last forests on Antarctica: reconstructing flora and temperature from the Neogene Sirius group, Transantarctic Mountains. Org Geochem 118:4–14

Salzmann U, Haywood AM, Lunt DJ, Valdes PJ, Hill DJ (2008) A new global biome reconstruction and data-model comparison for the middle Pliocene. Glob Ecol Biogeogr 17:432–447

Salzmann U, Williams M, Haywood AM, Johnson AL, Kender S, Zalasiewicz J (2011) Climate and environment of a Pliocene warm world. Palaeogeogr Palaeoclimatol Palaeoecol 309(1):1–8

Salzmann U, Dolan AM, Haywood AM, Chan W-L, Hill DJ, Abe-Ouchi A, Otto-Bliesner B, Bragg F, Chandler MA, Contoux C, Dowsett HJ, Jost A, Kamae Y, Lohmann Lunt DJ, Pickering SJ, Pound MJ, Ramstein G, Rosen-bloom NA, Sohl L, Stepanek C, Ueda H, Zhang Z (2013) Challenges in reconstructing terrestrial warming of the Pliocene revealed by data-model discord. Nat Clim Chang 3:969–974

Sawangchote P, Grote PJ, Dilcher DL (2009) Tertiary leaf fossils of *Mangifera* (Anacardiaceae) from Li Basin, Thailand as examples of the utility of leaf marginal venation characters. Am J Bot 96:2048–2061

Sawangchote P, Grote PJ, Dilcher DL (2010) Tertiary leaf fossils of *Semecarpus* (Anacardiaceae) from Li Basin, northern Thailand. Thai Forest Bulletin (Botany) 38:8–22

Signor (1994) Biodiversity in Geological Time. Am Zool 34:23–32

Silander JA (2001) Temperate forests. In: Encyclopedia of biodiversity. Princeton University, New Jersey, pp 607–626

Smith-Patten BD, Bridge ES, Crawford PHC, Hough DH, Kelly JF, Patten MA (2015) Is extinction forever? Public Underst Sci 24(4):481–495

Sniderman JMK, Jordan GJ (2011) Extent and timing of floristic exchange between Australian and Asian rain forests. J Biogeogr 38:1445–1455

Sniderman JK, Jordan GJ, Cowling RM (2013) Fossil evidence for a hyperdiverse sclerophyll flora under a non–Mediterranean-type climate. Proc Natl Acad Sci 110(9):3423–3428

Songtham W, Ratanasthien B, Mildenhall DC, Singharajwarapan S, Kandharosa W (2003) Oligocene-Miocene climatic changes in northern Thailand resulting from extrusion tectonics of southeast Asian landmass. ScienceAsia 29:221–233

Songtham W, Ratanasthien B, Watanasak M, Mildenhall DC, Singharajwarapan S, Kandharosa W (2005) Tertiary basin evolution in northern Thailand: a palynological point of view. Nat Hist Bull Siam Soc 53(1):17–32

Su T, Liu Y-S, Jacques FMB, Huang Y-J, Xing Y-W, Zhou Z-K (2013) The intensification of the east Asian winter monsoon contributed to the disappearance of *Cedrus* (Pinaceae) in southwestern China. Quat Res 80:316–325

Tiffney BH (1985) Perspectives on the origin of the floristic similarity between eastern Asia and eastern North America. J Arnold Arbor 66:73–94

Tiffney BH, Manchester SR (2001) The use of geological and paleontological evidence in evaluating plant phylogeographic hypotheses in the northern hemisphere Tertiary. Int J Plant Sci 162:S3–S17

Trewartha GT, Horn LH (1980) An introduction to climate. McGraw-Hill, New York

Utescher T, Bruch AA, Micheels A, Mosbrugger V, Popova S (2011) Cenozoic climate gradients in Eurasia -a palaeo-perspective on future climate change? Palaeogeogr Palaeoclimatol Palaeoecol 304:351–358

Vassio E, Martinetto E, Dolezych M, Van der Burgh J (2008) Wood anatomy of the *Glyptostrobus europaeus* "whole-plant" from a Pliocene fossil forest of Italy. Rev Palaeobot Palynol 151:81–89

Wagstaff SJ, Breitwieser I (2004) Phylogeny and classification of *Brachyglottis* (Senecioneae: Asteraceae): an example of a rapid species radiation in New Zealand. Syst Bot 29(4):1003–1010

Wagstaff SJ, Garnock-Jones PJ (1998) Evolution and biogeography of the *Hebe* complex (Scrophulariaceae) inferred from ITS sequences. N Z J Bot 36:425–437

Wagstaff SJ, Heenan PB, Sanderson MJ (1999) Classification, origins, and patterns of diversification in New Zealand Carmichaelinae (Fabaceae). Am J Bot 86:1346–1356

Wilf P, Johnson KR (2004) Land plant extinction at the end of the cretaceous: a quantitative analysis of the North Dakota Megafloral record. Paleobiology 30(3):347–368

Wilf P, Cúneo NR, Escapa IH, Pol D, Woodburne MO (2013) Splendid and seldom isolated: the paleobiogeography of Patagonia. Annu Rev Earth Planet Sci 41:561–603

Wolfe JA (1977) Paleogene Floras from the Gulf of Alaska region. In: Professional Paper of the United States Geological Survey, vol 997. United States Government Publishing Office, Washington, D.C, pp 1–108

Yamakawa C, Momohara A, Saito T, Nunotani T (2017) Composition and paleoenvironment of wetland forests dominated by *Glyptostrobus* and *Metasequoia* in the latest Pliocene (2.6 Ma) in central Japan. Palaeogeogr Palaeoclimatol Palaeoecol 467:191–210

Zhang J-W, D'Rozario A, Adams JM, Li Y, Liang X-Q, Jacques FM, Zhou Z-K (2015) *Sequoia maguanensis*, a new Miocene relative of the coast redwood, *Sequoia sempervirens*, from China: implications for paleogeography and paleoclimate. Am J Bot 102(1):103–118

Aridity, Cooling, Open Vegetation, and the Evolution of Plants and Animals During the Cenozoic

3

Juha Saarinen, Dimitra Mantzouka, and Jakub Sakala

Abstract

The development of grassland ecosystems across most continents was a multistage process involving the appearance of open-habitat grasses in the Paleogene, the mid-late Cenozoic spread of C3 grass-dominated habitats, and, finally, the Late Neogene expansion of C4 grasses at tropical and subtropical latitudes. In addition, the timing of these evolutionary and ecological events varied across continents and between regions. The middle Miocene witnessed a climate optimum at a global scale, but, soon thereafter, beginning about 14 million years ago, a global cooling trend commenced that was accompanied by environmental change. This drop in global temperature resulted in the spread of more arid and seasonally dry conditions that, in turn, resulted in the expansion of increasingly open woodlands. Changes in resources also led to modifications in the distribution and abundance of mammalian herbivores, favoring forms with hypsodont teeth. In Europe, where humid climates generally persisted throughout the Neogene, the Iberian Peninsula was the first region to experience increasingly seasonal aridity in the early Miocene. Open environments under an arid climate remained confined to Southern Europe and only expanded into Northern Italy in the latest Miocene. The spread of aridity to this region is associated with the final stage of the Messinian Salinity Crisis. In Eastern Asia, generally humid conditions existed before the late middle Miocene but were replaced by a midlatitude arid belt in the late middle–late Miocene. The East Asian summer monsoon, an atmospheric circulation pattern of the present, only influenced the region later, with the beginning of eolian red clay deposition, sometime between seven and eight million years ago. A similar pattern is recorded elsewhere. Biomes changed from tropical forest to steppes across broad areas of southern South America during the early–middle Miocene. Hence, a general global pattern of vegetational turnover is witnessed in the Neogene. More open woodlands and/or grasslands appeared over all continental landmasses in response to changes from rather humid conditions to more seasonally dry, semiarid, and arid climates at various times during the Miocene. In this chapter, we will explore these vegetational changes to understand if they were accompanied by changes in faunal composition, reshaping the world's biosphere.

Electronic supplementary material A slide presentation and an explanation of each slide's content is freely available to everyone upon request via email to one of the editors: edoardo.martinetto@unito.it, ragastal@colby.edu, tschopp.e@gmail.com

*The asterisk designates terms explained in the Glossary.

J. Saarinen (✉)
University of Helsinki, Helsinki, Finland
e-mail: juha.saarinen@helsinki.fi

D. Mantzouka
National and Kapodistrian University of Athens, Faculty of Geology and Geoenvironment, Athens, Greece

J. Sakala
Charles University in Prague, Institute of Geology and Palaeontology, Prague 2, Czech Republic
e-mail: rade@natur.cuni.cz

3.1 Introduction

The spread of open landscapes (Box 3.1) and grasslands is one of the most characteristic phenomena of the Cenozoic Era (c. 66–0 Ma). Their presence across the globe expanded during the Neogene and Quaternary (c. 23–0 Ma). It was not a simple, continuous process, but a complex one with strong temporal and spatial variations and flux. The geological evidence documenting the change is locked away in ancient soils (paleosols), not consistently exposed at the surface due to recent cover, and the paleontological proxies that provide crucial evidence of changes in plant-and-animal turnover are not often apparent in the field (e.g., microscopic phytoliths and remains of vertebrates of small body size and stature). Hence, this chapter is organized differently from others in

Box 3.1: Modern Open Habitats and Grasses

Today, open, semi arid, and arid environments range from savannas and woodland-grassland mosaics to shrublands, grasslands, and almost barren deserts (Fig. 3.1). These biomes evolved largely during the Cenozoic, following a global cooling and drying trend, and increased seasonality of the climate (Zachos et al. 2001). These drastic climatic changes affected the evolution of plant-and-animal communities. Most importantly, with the appearance of grasslands, animals adapted to life in an open habitat, and many mammals evolved a grazing lifestyle. Grasses are the most important group of plants in open habitats today. They contribute around 25% of the global terrestrial primary production (Sage 2004) and are one of the most diverse plant groups, occupying forest, wetland, and various other environments. Grasses are most abundant in open savanna and grassland because they tolerate seasonal drought, frequent fires, and heavy grazing pressure better than trees and many non-grass herbs. Reasons why these plants are successful are their fast regenerative capabilities and adaptations to water-stressed conditions. However, the most basal* grasses, today, are minor elements of the understory vegetation of tropical rainforests, reflecting the humble origins of the group. Grasses evolved several different photosynthetic pathways to adapt to a changing planet. More than 50% of currently living grass species utilize the C4 photosynthesis pathway and are mostly adapted to relatively arid, warm, and open environments (Sage 2004). In contrast, equally specialized cold-adapted, C3-photosynthesizing grasses (a group called Pooideae) are significant components of temperate and cold environments (e.g., Edwards and Smith 2010). All C4 grasses belong to a specialized clade* named PACMAD (Panicoideae-Arundinoideae-Chloridoideae-Micrairoideae-Aristidoideae-Danthonioideae), in which this photosynthetic pathway evolved in several tribes separately, especially during the middle Miocene (Vicentini et al. 2008). The other major clade of grasses, the so-called BEP (Bambusoideae-Ehrhartoideae-Pooideae), have remained C3 photosynthesizing throughout their evolutionary history. C3 grasses include bamboos, rice, and several temperate and cold-adapted grasses. How did this diversity evolve?

There are many different disciplinary approaches used to outline the history of grasses. Phylogenetic analyses based on morphological and molecular characteristics have provided valuable information on their origins and diversification. Data derived from fossil pollen, macrofossils, phytoliths (microscopic silica inclusions), and the ancient paleosols, in which these fossils are preserved, are direct evidence of their presence. Analyses of stable carbon isotopes from paleosol carbon and carbonate have added another dimension in a multidisciplinary approach to understand these

Fig. 3.1 Examples of natural open environments from East Africa. Upper left: Arid semi-desert and shrubland from Tsavo East National Park, Kenya. Upper right: Grassland from Nairobi National Park, Kenya. Lower left: Wooded grassland savanna from Tsavo East National Park, Kenya. Lower right: Mesic woodland-grassland mosaic from Aberdare National Park, Kenya. Photos by J Saarinen

plants. Combined, all these data provide a means to reconstruct the evolutionary history, dynamics, and spatial-and-temporal distribution of open-environment vegetation and grasses during the Cenozoic. And, we now know that the rise to dominance of grasses in open environments occurred globally during the Neogene (e.g., Strömberg 2011). Interestingly, details about grasslands also are available from the animals that consumed them, fossilized herbivores.

this volume. The reader is invited to follow the evidence that scientists use to reconstruct the history of open environments, instead of being accompanied in a virtual visit of outstanding sites and ecosystems back in time. We will examine the history of the main player, the grasses, which have a fossil record beginning in the Cretaceous, and how this resource influenced a turnover in mammal faunas we now consider characteristic of planet Earth. Adaptations of herbivorous mammals to these novel, open habitats, in turn, were followed by the evolution of defensive mechanisms in plants.

One evolutionary feature of many plant groups, especially abundant in grasses, and believed to have evolved as a defense mechanism against herbivory is the phytolith. Phytoliths are composed of silica, a mineral component in soils that is taken up by the plant during growth. Chemical processes in the plant recrystalize the element in specialized cells or cell walls, and, because of their chemical nature, these features are refractory and resistant to decomposition. Phytolith shapes and properties are unique to different species, and once the grass dies and decays, the phytoliths remain as a soil component and can be retrieved. The presence of phytoliths in the living plant is especially efficient protection against herbivorous insects and other small invertebrates. But, their presence also accelerates tooth wear in herbivorous mammals, creating selective pressure for functional durability of dentition.

To compensate for the increased tooth wear, grazing mammals have evolved teeth with an increased tooth-crown height, so-called hypsodont teeth (Fig 3.2a; Fortelius 1985; Janis and Fortelius 1988). Other adaptations evolved for feeding on grasses, one of which is the morphology of the occlusal surface of molar crowns. These adaptive features are the focus of ecomorphological analyses of herbivorous mammals. Their results have provided valuable evidence that has largely confirmed, but also greatly increased, our understanding of the spread of open landscapes (Box 3.2). There are three basic evolutionary stages commonly found in teeth of large herbivorous mammals. Bunodont* molars, where cusps are separate, evolved for efficient crushing of fruits, seeds, and other plant parts. In contrast, lophodont and selenodont molars, where cusps are fused or elongated into cutting ridges, are efficient for cutting of leaves and other tough plant parts. Lastly, complex plagiolophodont, tubular, and lamellar molars evolved for increasingly efficient shearing of tough and abrasive plants (Fig. 3.2b). Consequently, it has been thought that the evolution of hypsodont dentition in herbivorous mammals is an adaptation to grazing on the abrasive grasses. However, several more recent studies have shown that the earliest hypsodont mammals, several exclusively South American ungulate lineages, appeared before the development and expansion of any significant grass-dominated habitat on the continent (e.g., Fortelius et al. 2002; Eronen et al. 2010a; Madden 2015). Thus, this tooth configuration did not evolve, necessarily, as a response to a diet that included grasses. Rather, this tooth configuration evolved in response to the general presence of abrasives that increase tooth-wear rates, which can also be airborne dust. Indeed, the surprisingly early evolution of hypsodont dentition in the Eocene of these South American forms indicates, perhaps, that the accumulation of highly abrasive volcanic ash could have played a significant role (e.g., Strömberg et al. 2013). Hypsodonty* and other quantitative traits of mammalian teeth and skeletons can be used for reconstructing past environments when averaged across fossil-mammal communities (Box 3.2). These approaches are called ecometrics, and we will next have a closer look at those.

Box 3.2: Fossil Mammals Reveal Vegetation and Climate
Mammal ecometrics are known to correlate with climatic variables (temperature, precipitation), primary productivity, and general habitat openness, rather than with the abundance and distribution of grasses specifically. Analysis of the mean hypsodonty score in herbivorous mammal communities has been shown to reflect rainfall patterns globally, providing a useful tool for reconstructing the spread of arid, open environments during the Neogene (the last 23 My; Fortelius et al. 2002; Eronen et al. 2010a). Teeth are assigned a value of 1 (brachydont*, low-crowned), 2 (mesodont*, medium-high-crowned), or 3 (hypsodont, high-crowned) (Fig. 3.2a–c), and these are averaged for the fossil assemblage. Mean hypsodonty and another tooth feature, mean longitudinal loph count* (the number of elongated ridges between cusps) of molar teeth in ungulate communities, serve as proxies for primary productivity. This is because hypsodonty is correlated with precipitation and longitudinal loph count is correlated with temperature (Liu et al. 2012). Details of vegetational structure also can be obtained from other dental features, as well as from postcranial ecometrics

of herbivorous mammals (Žliobaitė et al. 2016, 2018). For example, the structural fortification of cusps in ungulate molars reflects grass-dominated feeding in relatively closed or wetland habitats (Fig. 3.2h). Ecometric analyses of the postcranial skeletons can provide information about general habitat openness. Adaptations, such as increased cursoriality* in many groups of ungulates (as for instance, bovids), reflect a more efficient locomotion in open environments and can be recognized based on the proportions of the limb bones (e.g., Kappelman 1991; Kovarovic and Andrews 2007). The analysis of body-size distribution in mammal communities (cenogram analysis*; Legendre 1986) reflects habitat openness and aridity, as medium-sized species tend to be rare in arid, open environments. Body-size distribution in fossil-mammal assemblages, especially when there is a gap in such medium-sized (1–10 kg) species, reflects habitat openness in modern environments. This is true even if body size is a complex characteristic and its connection with climate and environmental variables depends on taxon-specific ecological strategies, community structures, population densities, and, above all, resource availability (e.g., McNab 2010; Saarinen 2014). It is the availability of food resources, which are the constraints of diets, that controls herbivore populations. Hence, it is possible to decipher mammalian diets from the analysis of teeth.

Several methods for analyzing diets of fossil herbivorous mammals have been developed, and they provide valuable additional information about the response of the mammalian herbivores to their environments. These range from the microscopic to the mesoscopic and include geochemistry of the tooth, itself. Microscopic wear marks on teeth, such as scratches and pits, are easily assessed on the worn facets of molars. Their characteristics and variety reflect the proportion of grass versus leaf matter that an animal ingests while feeding (Rensberger 1978; Walker et al. 1978). The benefit of microwear analysis is that it is applicable to virtually all kinds of teeth. And, in theory, microwear analysis should give consistent results for mammals with very different tooth morphologies. However, microwear analysis only reveals the last few meals of the animal instead of any long-term average dietary signal (e.g., Rivals et al. 2010). Furthermore, factors other than diet, such as mineral particles ingested with food, may obscure the dietary signal provided by these data (Rivals et al. 2010). In mesowear analysis (Fortelius and Solounias 2000), the relief and shape of worn cusps and/or lophs are analyzed. This proxy reflects the long-term average amount of grass in an animal's diet. Grasses abrade the enamel during mastication resulting in progressively lower relief and a blunter shape of these molar characters in proportion to an increasing grass dietary component. Evidence of a herbivore's diet is locked away in tooth enamel, and the analysis of stable carbon isotopes, from either dental enamel or bone, allows for an understanding of what an animal ate (Lee-Thorp and van der Merwe 1987). From stable isotopes alone, it is not possible to discriminate exactly what was an animal's diet. Rather, this technique can only be used to compare the proportions of C3 and C4 plant matter ingested as part of the animal's diet. Nevertheless, it is a useful metric because values indicating a high proportion of C4 plants are indicative of tropical and midlatitude environments where C4 grasses dominate (Cerling et al. 1997). Now that we learned a bit about the methods, let us travel back to the Cretaceous to find some of the earliest grasses in the history of Earth.

3.2 Cretaceous: The Origin of Grasses (Poaceae)

Grasses originated during the Cretaceous but remained a relatively minor component of the global flora for tens of millions of years until the mid-Cenozoic. The earliest macrofossils of grass-like monocots, leaf-and-spikelet macrofossils of *Programinitis burmitis*, are preserved in Early Cretaceous Burmese amber (Poinar 2004). They are identified as possible early bambusoid (bamboo) forms in tropical forest habitats. Pollen grains associated with these early grass-like plant fossils are assigned to the genus *Monoporites* and, when found isolated without macrofossils, used to infer the presence of early grass relatives (Poales). This pollen grain is more abundant in the Late Cretaceous (Jacobs et al. 1999) where other microfossils of grasses are known, too.

The earliest undisputed fossil evidence of true grasses (Poaceae) are phytoliths discovered in coprolites of titanosaur sauropods (a group of the gigantic long-necked dinosaurs; see Chaps. 7 and 8) from the Late Cretaceous of India (Prasad et al. 2005). Therefore, to discover the earliest undisputed fossil evidence of true grasses, we now travel to India to examine these fossil feces. The phytolith record from titanosaur coprolites shows a remarkable early diversity of grasses. Representatives include not only basal grasses (Pueliodeae) but also both major lineages of derived grasses, the BEP and the PACMAD (Prasad et al. 2011). Interestingly, the groups of grasses that are generally believed to be the most primitive forms, the Anomochlooideae and Pharoideae, are absent. These grasses, today, are restricted to tropical rainforests, an environment in which grasses are rare. The abundance of grass phytoliths in these Late Cretaceous

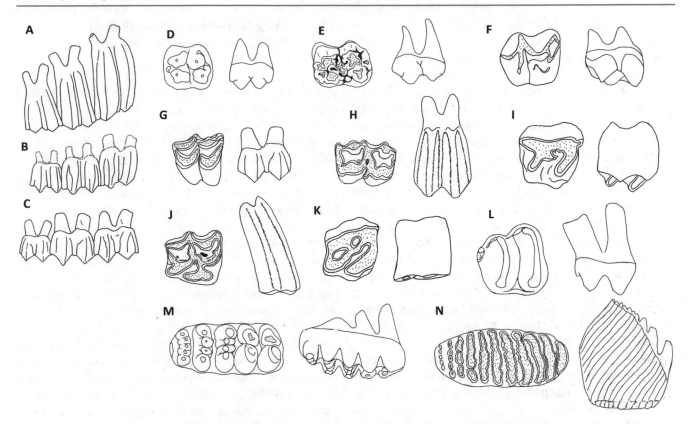

Fig. 3.2 Molar morphology of herbivorous mammals shows adaptations to feeding on different kinds of vegetation and reflects vegetation and climate. (**a–c**) Variation in tooth crown height. (**a**) Hypsodont (high-crowned) molars are an adaptation to increased tooth wear rates. This dentition evolved in many lineages of herbivorous mammals in open, dry environments. (**b**) Mesodont molars are intermediate between brachydont and hypsodont and reflect incipient adaptation to living in open, dusty environments and/or feeding on abrasive grasses. (**c**) Brachydont (low-crowned) molars are the primitive condition in mammals and are retained in herbivorous mammals living in forests and/or feeding on low-abrasive vegetation. (D-N) Examples of molar crown types of herbivorous mammals. (**d**) Bunodont, brachydont molar (white-lipped peccary, *Tayassu pecari*), a primitive condition in herbivorous mammals where cusps are separate. (**e**) Bunodont, brachydont molar with fortifying invaginated enamel for the processing of a wider range of harder or more abrasive (plant) foods (bearded pig, *Sus barbatus*). (**f**) Buno-lophodont, brachydont molar (chalicotheriid perissodactyl *Anisodon grande*). This is an intermediate form between the bunodont and lophodont architectures, where some cusps are separated and others are fused into lophs (into a cutting, W-shaped outer loph in this case). This occurs in species that feed relatively unselectively on fruit and leaves in forests. (**g**) Selenodont, brachydont molar (moose, *Alces alces*), typical of ruminants and camels, where the cusps are elongated into crescent-shaped crests. (**h**) Selenodont, hypsodont molar (aurochs, *Bos primigenius*), where buccal cusps are fortified by thickened enamel, and cement fills cavities in the tooth. This tooth morphology is typical of fresh grass grazers in relatively wooded environments. (**i**) Lophodont, mesodont molar (black rhinoceros, *Diceros bicornis*), where an ectoloph (an outer, cheek-side cutting ridge) is present, and the anterior and posterior cusp pairs are fused into transverse lophs. This is typical of some browsing groups including rhinoceroses, early browsing horses, and some hyraxes. (**j**) Plagiolophodont, hypsodont molar (wild horse, *Equus ferus*), where the lophs are folded and fused, and dental cement fills the gaps between the lophs, creating a uniform occlusal surface with complex enamel edges. This is typical of derived horses and other specialized grazing ungulates. (**k**) Another example of plagiolophodont, hypsodont molar (white rhinoceros, *Ceratotherium simum*). (**l**) Bilophodont, brachydont molar (deinotheriid proboscidean *Deinotherium bozasi*), typical for browsers feeding in forests and woodlands, where the anterior and posterior cusp pairs are fused into transverse lophs. (**m**) Bunodont, brachydont molar with multiplied transverse cusp rows (lophids). This is the typical tooth morphology of gomphotheriid proboscideans, the ancestral group for modern elephants, which was versatile and enabled gomphotheres to utilize several types of plant foods ranging from fruits to blades of grass. (**n**) Lamellar, hypsodont molar (Asian elephant, *Elephas maximus*), found in elephants and some rodents, where transverse lophs have increased in number and are bound together by dental cement. The multiplied lophs together with hypsodonty make a durable tooth that is efficient for shearing tough and abrasive plants such as grasses. The images are not in the same scale. Drawings **a–c**: from Fortelius et al. (2002), with permission; drawings **d–n**: J Saarinen

Indian coprolites indicates that they were a part of the diverse plant diet of titanosaurs (Prasad et al. 2005). Intense herbivory by vertebrates and invertebrates on early grasses is undoubtedly connected with the evolution of phytoliths as a structural defense mechanism. But, phytoliths are no match for wildfire, evidence of which is found as charcoal often associated with phytoliths in paleosols. It is suggested that forest fires facilitated the spread of angiosperms during the Cretaceous and might have enabled early grasses to enter forest habitats. Their appearance in forests may also have influenced early mammalian evolution (see also Williamson and Brusatte in Chap. 7).

Late Cretaceous mammals of India, in addition to titanosaurs, may have relied on grasses as part of their diets. Burrowing gondwanatherian mammals, which were the earliest mammals with hypsodont cheek teeth, could have evolved high-crowned molars as a means to feed on grasses (e.g., Prasad et al. 2005). Gondwanatheres probably would have resembled stocky, burrowing rodents, such as ground hogs or marmots, in appearance and ecology (Williamson and Brusatte in Chap. 7). However, the evolution of hypsodonty in many Cenozoic mammal groups has been shown to reflect overall environmental openness and aridity rather than an adaption to grazing specifically (Fortelius et al. 2002; Madden 2015). Indeed, the paleoenvironmental conditions in India are interpreted as having been comparatively open and arid (Samant and Mohabey 2009). Such landscapes promoted early grass diversification in a setting where the accumulation of airborne dust and volcanic ash on plants also could have been a factor in the evolution of hypsodonty in this and other groups (e.g., Strömberg et al. 2013).

The latest Cretaceous (Maastrichtian) rocks of India document extensive basaltic activity in the western part of the continent, coincident with a changing coastal zone. The emplacement of the Deccan Traps, a large igneous province, and associated volcanism is thought to have played a role in the evolution of hypsodonty. The latest Cretaceous events also are associated with the most recent of the big five mass extinctions in geological history, the so-called K/Pg event at the end of the Cretaceous (Schoene et al. 2019). This mass extinction greatly affected many animal-and-plant groups, many of which went extinct as a consequence of this crisis. We will hear more about this event in other chapters. Here, we will concentrate on how this event influenced open environments and their inhabitants in the Paleogene.

3.3 Paleogene: The Early Evolution of Open Environments and Open-Habitat Grasses

The Paleogene period (c. 66–23 Ma) is the time from the end-Cretaceous mass extinction of large, nonavian dinosaurs and many other groups of animals and plants, until the beginning of the Neogene, when modern vegetation zones were first established (see Chaps. 1 and 2). Continental landscapes of the Paleogene are characterized by the early spread of open environments in parts of the globe where climate cooled and increasing seasonality led to drying. Under these conditions, open-adapted grasses diversified and became an important vegetational element in several parts of the world.

3.3.1 Paleocene–Eocene (66–34 Ma): The Warm, Humid, and Forest-Dominated World

Global temperatures increased during the Paleocene after a dramatic, but very short-term, cooling event associated with the end-Cretaceous mass extinction. Temperatures peaked about 55 to 52 million years ago (Zachos et al. 2008), as discussed in more detail in the following Chap. 4. A warm and humid climate prevailed across nearly all paleoenvironments in which extensive forests grew (Strömberg 2011). During the early Eocene thermal maximum, our journey would involve hiking through subtropical deciduous forests near the poles, as evidenced by the fossil communities from the high arctic Greenland, Spitzbergen, Kamchatka, and Ellesmere Island. Here, fossil remains of thermophilous plants are common, including the trees *Metasequoia* (dawn redwood), *Nyssa* (tupelo), *Magnolia*, and *Ginkgo* (Collinson and Hooker 2003). However, as a time traveler, you would probably have wanted to avoid the alligators, which were also common in these northern, high-latitude environments during the early Eocene (Erlebe and Greenwood 2012). Landscapes in Europe, Southern Asia, and much of North America largely resembled tropical rainforests, although elements of these forests include plant taxa that are non-analogs with modern tropical rainforests (e.g., Jacobs et al. 1999). Mangroves* and marshlands fringing costal areas were typical across the world, as indicated by paleobotanical evidence from the London Clay, UK, and many Eocene sites in Africa (Collinson and Hooker 2003; Jacobs et al. 2010). During the late Eocene, incipient climatic cooling led to changes in vegetation. Paratropical/subtropical* evergreen forests were replaced by increasingly mixed deciduous/evergreen subtropical/temperate forests (see Preface) in the higher latitudes. But, forests and woodlands remained the dominant landscape overall across the globe (Collinson and Hooker 2003) with little to virtually no evidence for grass-dominated environments (Jacobs et al. 1999; Strömberg 2011). One possible exception may be found in some middle Eocene African paleoenvironments (see below), but let's first explore the distribution of grasses among the forests of the Northern Hemisphere.

Macrofossils of grasses have been discovered from the early Eocene deposits of the London Clay, UK (Jacobs et al. 1999), North America (Crepet and Feldman 1991), and from the middle-late Eocene Baltic amber, Estonia (Sadowski et al. 2016). These rare records of grass or grass-related plants do not indicate the presence of any substantial open grassy habitat, though. Their remains are quite rare and could represent either wetland grasses or, in some cases, other plant taxa mimicking their morphology, such as sedges. West

African and Egyptian middle Eocene sites include some of the earliest paleobotanical records where grasses were abundant, associated with mangrove vegetation and palms (Jacobs et al. 2010). This grass- and palm-dominated assemblage probably indicates the presence of wetlands rather than arid open environments. The middle Eocene locality of Mahenge, Tanzania, preserves some of the earliest well-founded paleobotanical evidence of a more arid woodland setting. The site is interpreted as having been similar to extant miombo woodlands (currently covering much of Central and Southern Africa). However, evidence for the presence of grasses is lacking in the Mahenge assemblage of Tanzania (Jacobs and Herendeen 2004), indicating complex evolutionary patterns of open habitats and grasses. A similar story can be observed in Asia.

The Paleocene and early Eocene deposits of central Eastern Asia host what is probably the oldest evidence of arid and at least partly open, Cenozoic habitats. A trip to these ecosystems reveals the beginning of a drying trend in the interiors of continents. Here, several fossil floras preserve significant proportions of the arid-adapted shrubs *Nitraria* (nitre or dillon bush) and *Ephedra* (Jacobs et al. 1999). Although these plants are, today, definitive evidence for a dry landscape, grasses are rare in those assemblages. Here, the arid-adapted shrubs sometimes are found to co-occur with forest or woodland taxa. Hence, these paleoenvironments were likely semiarid shrubland-woodland mosaics; they were not open grasslands, which is also reflected in the ecometrics of accompanying fossil mammal communities.

The ecomorphology of mammals preserved in association with Paleocene and Eocene forest-and-woodland paleoenvironments reflects their adaption to these landscapes. The animals generally lack hypsodont cheek dentition, as this character had yet to evolve in large herbivorous mammals. Yet, several South American notoungulates (endemic hoofed mammals that originated separately from other continents) are notable exceptions (see below). The evolutionary trends in herbivorous mammal teeth, especially the increase in the number of ridges (lophedness*) of molar teeth (Fig. 3.2), reflect increasing specialization to leaf browsing diets (e.g., Jernvall et al. 1996). A notable shift in ungulate dentition occurs during the Eocene with an increasing diversity of tooth-morphology types. The more complex tooth morphology implies the evolution of an increasingly browsing diet (comprising dicotyledonous leaves and branches) in herbivorous mammals. These changes in dentition parallel the change in habitats from tropical to subtropical and temperate forests (see Chap. 2) at high and middle latitudes (e.g., Blondel 2001). Paleodietary analyses based on tooth wear also indicate an almost exclusive browsing diet for large Eocene herbivorous mammals, at least in Eurasia (e.g., Blondel 2001) and North America (e.g., Mihlbachler and Solounias 2006; Boardman and Secord 2013). To date, there

is no conclusive evidence of any grazing-adapted, or even highly specialized, open-habitat mammal nor the adoption of a grazing diet among mammals during either the Paleocene or Eocene. Yet, hypsodonty did appear.

The evolution of hypsodont dentition is found first in some endemic South American ungulates (notoungulates). In some areas, including Patagonia, this happened much earlier than on any other continent. The earliest evidence for increasing hypsodonty occurs as early as the middle Eocene (c. 40 Ma) or possibly even earlier (Ortiz-Jaureguizar and Cladera 2006). This increase in tooth crown height originally was interpreted to indicate that aridification and the spread of open grassland environments occurred in South America earlier than in any other part of the world. Additional evidence in support of early Patagonian grasslands was thought to include the presence of extensive *Coprinisphaera* trace fossil assemblages in the paleosols of the Gran Barranca Member of the Sarmiento Formation. These trace fossils are considered to be a modern analog of dung-beetle brood balls and interpreted to indicate the presence of large grazing mammals in an open, grass-dominated environment (Bellosi et al. 2010). Furthermore, rare phytoliths discovered from the middle Eocene of Patagonia have been used to support the interpretation (Zucol et al. 2010). But, science is an evolving paradigm. Interpretations change as new data are acquired and older data are reexamined. Recent reanalysis of the phytolith assemblage from Gran Barranca indicates that it is dominated by phytoliths of forest-inhabiting palms and woody dicotyledons, and few are characteristic of grasslands (Strömberg et al. 2013). The true grass phytoliths include bambusoid, open-habitat PACMAD, and some open-habitat pooid morphotypes, but comprise only about 6% of the assemblage (Strömberg et al. 2013). We now know that dung beetles also frequently occur in more closed habitats. Thus, although some opening across the landscape may have occurred, a trip to the middle Eocene of Patagonia would have kept the traveler enveloped in wooded environments in which peculiar endemic mammals thrived (Fig. 3.3), at least until the Oligocene.

Fossil assemblages recovered from Oligocene landscapes record a slight increase in the proportion of grasses (see Sect. 3.3.2). Environments in the Sarmiento Formation include woodland and savanna-like settings, but grass-dominated areas remained rare (Strömberg et al. 2013). Thus, the evolution of hypsodont dentition in Eocene notoungulates does not seem to correspond with a substantial development of grass-dominated environments or any aridification trend. Rather, the appearance of this trait may, again, be an adaption to the ingestion of the volcanic ash from soils in the Andean volcanic province across the Sarmiento landscape. The earliest notoungulates to evolve hypsodont dentitions were relatively small-sized forms (interatheriids and archaeohyracids), which would have been feeding close to the

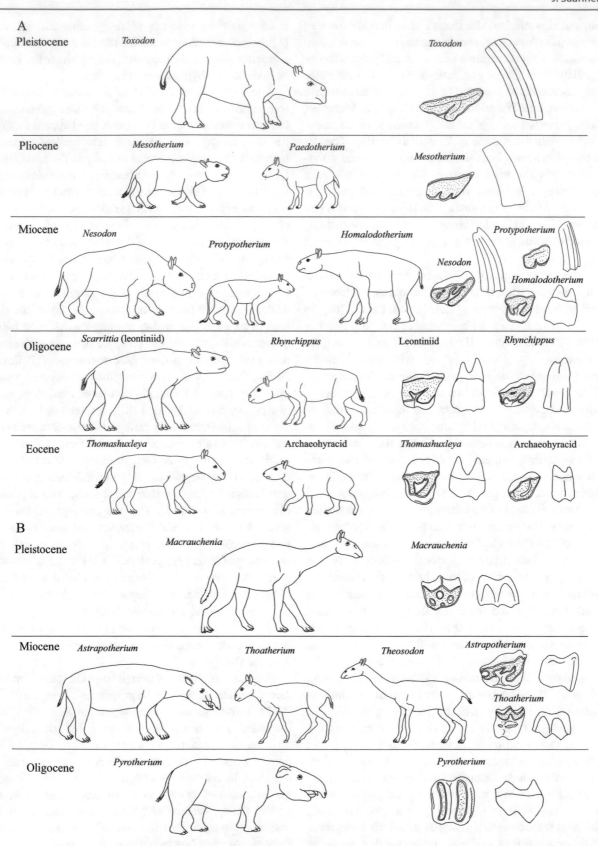

Fig. 3.3 Life reconstruction outlines (left) and molar teeth (right) of South American endemic ungulates (hoofed mammals). (**a**) Notoungulates were house cat to hippopotamus-sized, relatively robustly built hoofed mammals, many of which adapted to feeding on abrasive plants and evolved extremely high-crowned dentitions early in their life history (exceptions were the Eocene to Miocene leontiniids and homalodotheres that retained low-crowned dentitions and browsing diets). Eocene: *Thomashuxleyia* (Isotemnidae), an archeohyracid;

ground. The distance above the soil surface that any animal feeds might also explain why not all endemic South American ungulates evolved hypsodont teeth. For example, the high-browsing macraucheniid litopterns (extinct three-toed ungulate), as well as some groups of large notoungulates, such as leontiniids and probably high-browsing homalodotheres (see Fig. 3.3), retained relatively low-crowned dentitions throughout their evolutionary history (e.g., Croft 2016).

3.3.2 Oligocene (34–23 Ma): The Time of Global Cooling, Initial Spread of Open Habitats, and the Origin of C4-Photosynthesizing Grasses

Dramatic cooling occurred during the Eocene–Oligocene transition (c. 34 Ma), initiating significant changes in environments and vegetation across the globe (Zachos et al. 2001). This cooling event was driven largely by tectonic and orogenic processes, including the opening of the Drake's Passage between South America and Antarctica, which led to the formation of the cold circum-Antarctic ocean current (Livermore et al. 2007) and the uplift and aridification of the Tibetan Plateau (Dupont-Nivet et al. 2007), respectively. Initially, global cooling was accompanied by a drop in atmospheric pCO_2 at the Eocene–Oligocene boundary (Christin et al. 2008). Despite the climatic changes, relatively humid conditions and forest/woodlands still largely dominated continental environments during the Oligocene. The impact of these climatic changes was particularly strong on the evolution of grasses.

The early Oligocene witnessed diversification and expansion of all grass clades. The C3-photosynthesizing cool-habitat pooid grasses began evolving from generally warm-adapted and plesiomorphic ("primitive") grass ancestors (Sandve and Fjellheim 2010). About the same time, some of the earliest C4-photosynthesizing grasses emerged in Western Europe (Urban et al. 2010). The advent of the C4 photosynthetic pathway is an innovation that evolved convergently somewhere between 12 and 20 times in various subclades* of PACMAD grasses (Vicentini et al. 2008; Bouchenak-Khelladi et al. 2014). This physiological pathway is most efficient under low atmospheric CO_2 pressures

and warm, and especially semiarid and arid, climatic conditions. Moreover, C4-photosynthesizing grasses regenerate fast and are, thus, resistant against strong herbivory pressures and wildfire (Fox and Koch 2003; Keeley and Rundel 2005). It is interesting to note that the first C4 grasses coincide with the drastic cooling during the early Oligocene when relatively humid conditions prevailed. It is plausible to think that decreasing atmospheric pCO_2 triggered the evolution of C4 photosynthesis in some grasses and other plant groups (e.g., chenopods, asters, sedges, and spurges, among others). But, their rise to ecological dominance required the appearance of additional climatic and environmental perturbations (Christin et al. 2008; Strömberg and McInerney 2011). Indeed, these plants remained as relatively rare elements in global communities until their ecological expansion in middle and low latitudes in the late Miocene, about eight million years ago. Evidence in support of this comes from stable-carbon isotopic data (e.g., Cerling et al. 1997). This late Miocene expansion corresponds primarily with a significant increase in aridity (e.g., Strömberg and McInerney 2011), but these environments were occupied by grasslands different from those of the present.

Arid shrublands and early grassy woodland habitats spread in parts of the world during the Oligocene. The earliest colonizers of these settings were C3 grasses. This is because the first C4 grasses had only just evolved and environmental conditions for their rise and expansion had yet to be realized. Shrublands and grassy woodland habitats appeared first in Eastern-Central Asia and the increasingly arid western inland areas of North America (Jacobs et al. 1999; Strömberg 2011). Below, we will travel to the various continents where Oligocene sediments are exposed and observe if we can spot at least some evidence for the appearance of C4 grasses. We begin our trip in the late Eocene and early Oligocene of Europe, followed by a whirlwind tour through Eastern Asia, North America, Africa, South America, and Australia.

3.3.2.1 Iberia and the Balkans
Our first stops are in the Iberian Peninsula and the Balkan region where indicators for the presence of increasingly arid-adapted sclerophyllous forest/woodland vegetation are preserved. The appearance of this vegetation type marks a

Fig. 3.3 (continued) Oligocene, from left to right: *Scarrittia* (Leontiniidae) and *Rhynchippus* (Notohippidae); Miocene, from left to right: *Nesodon* (Toxodontidae), *Protypotherium* (Interatheriidae), and *Homalodotherium* (Homalodotheriidae); Pliocene, from left to right: *Mesotherium* (Mesotheriidae) and *Paedotherium* (Hegetotheriidae); Pleistocene: *Toxodon* (Toxodontidae). (**b**) Pyrotheres, astrapotheres and litopterns. Pyrotheres included the large (hippopotamus-sized) browser *Pyrotherium* with low-crowned, bilophodont teeth, from the Oligocene. Astrapotheres were somewhat similar to pyrotheres and experienced their greatest diversification during the Miocene, when they included many

tapir-to-hippo-sized browsers, such as *Astrapotherium* (first animal on the left for Miocene). In contrast to the notoungulates, the litopterns (Litopterna) mostly retained relatively low-crowned dentitions and probably largely browse-dominated diets throughout their evolutionary history. In the Miocene, they included several relatively small, somewhat deer-like or camel/llama-like forms, some of which show convergence in limbs and skulls with browsing horses (such as *Thoatherium*). The litopterns persisted until the Late Pleistocene megafaunal* extinction in South America. The last litoptern was the large (roughly camel-sized), long-necked *Macrauchenia*. Drawings by J Saarinen

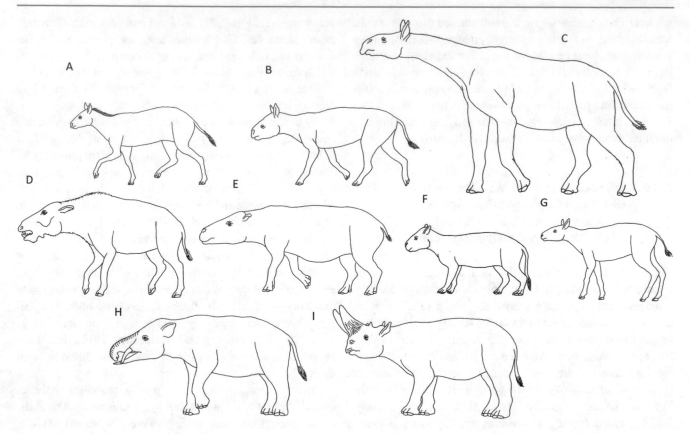

Fig. 3.4 Oligocene ungulates. (**a**) *Mesohippus*, a muntjac-sized, early three-toed horse from North America, with a duiker-like body shape, small skull, and brachydont, lophodont molars adapted for a browsing diet. (**b**) *Hyracodon*, a sheep-sized early relative of rhinoceroses adapted to running in the open, arid woodland and shrubland habitats of the North American western continental interior. (**c**) *Indricotherium* (*Paraceratherium*), the largest terrestrial mammal and a giant relative of rhinoceroses with long limbs and neck, that grew to a body mass estimated around 11 tons. They were browsers whose large size helped them survive seasonal shortages of food and water, and migrate efficiently in search of resources, in the arid, open woodlands and shrublands of Central Asia. (**d**) Entelodonts were large, pig-like animals with a massive skull and relatively long limbs, adapted to running in relatively open habitats. (**e**) *Anthracotherium magnum*. Anthracotheres were early relatives of hippos, with probably largely similar semi-aquatic lifestyles. They had brachydont molars and mostly browsing diets, and they were common in wetland environments, especially in Europe. (**f**) Oreodonts (Merycoidodontidae), such as *Merycoidodon*,

were among the most common and abundant North American ungulates in the Oligocene. They were rabbit-to-tapir-sized and possessed short, robust skulls in which mostly low-crowned, selenodont teeth developed for browse-dominated diets. These mammals are possibly distantly related to camels. (**g**) *Bachitherium*, an early ruminant from the Oligocene of Europe. The spread of early ruminants from Asia to Europe in the Oligocene probably was related to environmental changes such as increasingly seasonal climate. (**h**) *Palaeomastodon* was a member of the elephantoid proboscideans, the group that later gave rise to true elephants. Mature animals were roughly rhino-sized browsers living in the marshy forests of North and East Africa. (**i**) *Arsinoitherium* is the best-known representative of embrithopods, an early African group of herbivorous mammals closely related to elephants and their early relatives. These animals shared their habitats with early proboscideans such as *Palaeomastodon*, and their peculiar high-lophed molars may have been an adaptation to feeding on tougher plants in more open marshy habitats. The images are not in the same scale. Drawings by J Saarinen

floristic turnover from the humid woody forests to a landscape controlled by the presence of a seasonally dry climate (Collinson and Hooker 2003). Some of the earliest stable carbon-isotope evidence for the presence, but not dominance, of C4 grasses is found in the Oligocene of Spain (Urban et al. 2010). The rest of Europe remained mostly covered in subtropical and warm temperate forest elements. It appears that an important tectonic event, the closure of the Turgai Strait, had an effect on these regional floristic turnovers.

The Turgai Strait was a seaway that separated Europe and Asia during the Eocene, stretching from the present Caspian Sea northward to the Paleoarctic. The interchange of marine

waters between these areas remained open until the late Oligocene. Its closure at that time resulted in a remarkable turnover in the European vertebrate fauna. During this "Grande Coupure," many mammal groups of Asian origin, most notably ruminant artiodactyls, migrated to Europe (Fig. 3.4g). In general, artiodactyls (even-toed ungulates) became more diverse than perissodactyls (odd-toed ungulates), and several of their characteristic features indicate a trend toward browse-dominated diets. These new immigrants possessed both brachydont and fully selenodont dentitions, as well as an evolutionary innovation of foregut fermentation in ruminants. Browsing on leaves in increasingly seasonal environments was a derived feeding strategy compared to the less generalized frugivorous (fruit)-folivorous (leaf) diets of Eocene artiodactyls (Blondel 2001). Despite the evidence for the presence of grasses (even rare C4 grasses) in the European Oligocene, there is no conclusive evidence for any specialized grazing-adapted ungulate presently known from the times. Some early moschid ruminants (e.g., musk deer) may have been mixed feeders rather than pure browsers (Novello et al. 2010). But, the scarcity of grasses in the Oligocene forest-dominated environments did not favor the evolution of pure grazers among herbivorous mammals. Similar environments are documented elsewhere, each with an increasing seasonally dry component.

3.3.2.2 Pakistan, Mongolia, and Eastern Asia

Traveling eastward through Asia, we encounter a diversity of Oligocene landscapes. Forests dominated the area where Pakistan is located, today, in which a minor grass component is preserved (De Franceschi et al. 2008). These environments are interpreted based on both fossil pollen and macrofossils. Heading further north to Mongolia in central East Asia, the Oligocene Hsanda Gol Formation conserves a relatively arid and open shrubland/woodland. Here, evidence for grasses also is scarce. Hence, it is likely no grass-dominated habitats existed in this region (e.g., Métais et al. 2003; Prothero 2013). Nonetheless, a turnover in the East Asian mammal faunas did occur at about the same time as the one in Europe. Here, the brachydont, perissodactyl-dominated faunas of the Eocene were replaced by Oligocene assemblages dominated by hypsodont lagomorphs (rabbits) and rodents (Wang et al. 2007). Moreover, the rare and small-sized, earliest hypsodont, bovid-like ruminant, *Paleohypsodontus,* appeared in the late Oligocene of China and Pakistan. These mammals evolved an adaptation to feeding on abrasive plants in arid, open habitats (Métais et al. 2003), indicating that these habitats were expanding at that time. Many particularly large-sized ungulates evolved in Asia during the Oligocene. These include the gigantic indricotheriid perissodactyls, which stood up to 5 meters at their shoulders (Fortelius and Kappelman 1993; Prothero 2013) and were high browsers at a canopy level, and the giant, probably scavenging/omnivorous entelodont suiforms (pig-related artiodactyls), reaching 1.3 m at the shoulder, with wart-like projections from a skull that attained a length of 65 cm (Wang 1992). The large size of the ungulates, on average, in the Oligocene East Asian open environments may have been a response to a seasonally harsh, prevailing climate (e.g., Saarinen et al. 2014).

3.3.2.3 North America

Our diversion to the continental interior of North America in the Oligocene brings us to a similar setting to that of central East Asia. Here, open and increasingly dry shrubland/woodland habitats expanded in the continental interior.

Paleobotanical and paleopedological* (ancient soil structure) evidence indicates that some habitats were dominated by C3-dominated grasses making up to 13% of the vegetation (Retallack 2007; Strömberg 2011). Open-habitat grasses probably occupied the understory of forests and woodlands during the early Oligocene, and became more abundant in the increasingly open late Oligocene and early Miocene savanna-like environments (Strömberg 2011). The change in landscape is reflected in the large herbivorous mammal communities with several species adapted to increasingly open, probably shrubland habitats. These mammals include the rhinocerotoid perissodactyl *Hyracodon* (Fig. 3.4b), early camels such as *Poebrotherium*, and several oreodonts (merycoidodont artiodactyls, possibly distantly related to camel), such as *Sespia*. But, to date, there is no indication that any taxon evolved a specialized grazing diet. Dietary analyses of horses and oreodonts (Fig. 3.4a, f) indicate purely browse-based diets in the early Oligocene and a slightly more mixed-feeding diet in some species during the late Oligocene (Mihlbachler and Solounias 2006; Mihlbachler et al. 2011). Moreover, ungulates retained essentially similar browse-dominated dietary signals throughout the stratigraphic succession of the White River Formation, from the late Eocene to the early Oligocene (Boardman and Secord 2013).

3.3.2.4 Africa

Fossil evidence for a continental transition from forests to grassland across the African continent is limited in space and time. Paleobotanical remains indicate the presence of forest, swamp, and woodland environments across the continent during the Oligocene (Jacobs et al. 1999). In fact, there is even less evidence for open environments or the presence of grasses in the Oligocene than in the Eocene records from the continent (Jacobs et al. 2010). The rich mammal faunas from the early Oligocene of Fayum, Egypt, and the late Oligocene of Chilga, Ethiopia, also are indicative of animals adapted to forest and wetland paleoenvironments. These assemblages are dominated by forms possessing brachydont, bunodont, and lophodont teeth. Large herbivores include afrotherians (hyracoids [hyraxes], embrithopods [rhino-like elephantines], and early proboscideans [elephant relatives]), anthracotheres (hippopotamus-like ungulates) (Fig. 3.4e), and early primates (e.g., Werdelin and Sanders 2010).

3.3.2.5 Patagonia

We return for a second visit to South America, now during the Oligocene, in which the landscapes continue to be dominated by forest/woodland elements. We begin to find a greater proportion of arid-adapted shrubland and herbaceous vegetation in Patagonia no earlier than the late Oligocene (Strömberg 2011). And, as we have seen elsewhere, this change is accompanied by a higher diversity of hypsodont mammals, in this case rodents, xenarthrans and notoungulates, that has been used to interpret the presence of increased habitat openness (Ortiz-Jaureguizar and Cladera 2006). Dietary analyses of tooth micro- and mesowear of late Oligocene herbivorous mammals, including mylodontid sloths and notoungulates, seem to indicate the acquisition of grazing diets in several taxa. However, similar wear also is indicative of mastication in the presence of volcanic ash (Billet et al. 2009). Indeed, as discussed above (Sect. 3.3.1), there is no conclusive evidence for extensive Patagonian grassland habitats during the entire Oligocene (Strömberg et al. 2013).

3.3.2.6 Australia

The Oligocene rocks of Australia are one place where there is evidence of grasses in woodland or open forest environments. As early as the early–middle Eocene, records of grass pollen are reported from the continental interior where the vegetation was otherwise dominated by rainforest and dry forest taxa (Martin 2006). Most of what is known about the mammalian fauna at this time comes from a single locality on the continent, Riversleigh. Based on the ecomorphology of the fossil marsupial-mammal community preserved, here, the paleoenvironment in this area has been reconstructed as a relatively open forest (Travouillon and Legendre 2009).

3.4   Early Neogene (23–11 Ma): Early Expansion of C3-Photosynthesizing Grasses, Grasslands, and the Evolution of the Earliest Specialized Mammal Communities

After the drastic Oligocene cooling, global temperatures increased throughout the early (23–16 Ma) and middle Miocene (16–11 Ma), terminating in the middle Miocene thermal maximum (Zachos et al. 2001). The early Miocene saw the development of the first definitive C3 grass-dominated open environments in parts of the world, although forest and woodland environments were still clearly dominant globally (Strömberg 2011; Strömberg and McInerney 2011). Indeed, paleoenvironments during the early Miocene across Eurasia remained predominantly warm, humid, and forested. There is, mostly, relatively little difference between these landscapes and those of the late Paleogene, as shown by paleobotanical and vertebrate paleontological evidence (Fortelius et al. 2002; Kovar-Eder 2003; Eronen and Rössner 2007). Similar to our travels in the Oligocene, we will journey to the various continents to check how far the development of grasslands has advanced in the Miocene. We will start our journey in Europe.

3.4.1   Iberia and the Near East

The early Miocene of Europe was dominated by humid, warm temperate forest and wetland environments, where abundant subtropical elements coexisted, much like the Oligocene (Fortelius et al. 2002; Kovar-Eder 2003). However, as we saw above, the Iberian Peninsula already hosted arid, open environments with vegetation consisting of microphyllous* trees and shrubs (Kovar-Eder 2003) and abundant grasses. These grasses included C4-photosynthesizing members of the PACMAD clade (Urban et al. 2010). This mix of vegetation is seen in the fossil records of plant micro- and macrofossils and in the preservation of their stable carbon-isotope signatures. Interestingly, C4 grasses were more abundant in this area during the early Miocene than at present. On average, 15% of the Iberian vegetation consisted of grasses, with a quarter to more than a half of them being C4-photosynthesizing taxa (Urban et al. 2010). Further evidence in support of an open and arid peninsula is the appearance of the *Hispanotherium* fauna. This assemblage of mammals includes some of the earliest hypsodont ungulates in Europe as, for instance, the rhinoceros *Hispanotherium matritense* (Fortelius et al. 2002). Moreover, the abundance and utilization of grasses as a predominant food resource are reflected in the dental records of these mammals. A definitive mixed-feeding diet with a significant component of grass is indicated by dental mesowear analysis of some ungulates, such as early deer (DeMiguel et al. 2008), which lived alongside *Hispanotherium*. A similar fauna including ungulates also occurs further east, in Central Asia.

Traveling eastward through Eurasia, and a bit forward in time, we can observe an increasing number of open, rather dry landscapes. In the area where Turkey is now situated, arid environments, similar to the ones we just visited in Iberia, prevailed in the early Miocene (Fortelius et al. 2003). In contrast, most other regions of Europe remained forested. Phytoliths and sedimentological records from Turkey indicate early aridification, as well as the presence of grassy patches and dry woodland during the early and middle Miocene (Bestland 1990; Akgün et al. 2007). From the middle Miocene onward, continental environments became increasingly seasonally dry, arid, and open over larger areas. These physical changes resulted in a climatic division of the region (Fig. 3.5). That climatic subdivision separated Eurasia into a humid, forested "West" (~Europe) and the more arid/open "East" (eastern Mediterranean to China). Mean hypsodonty values in the ungulate communities that inhabited these eastern Mediterranean environments also point to increasingly dry conditions (Fortelius et al. 2002, 2006). Yet, pollen (Barrón et al. 2006), phytoliths (Pinilla and Bustillo 1997), and ungulate ecometrics (Fortelius et al. 2002, 2006) indicate that more humid and

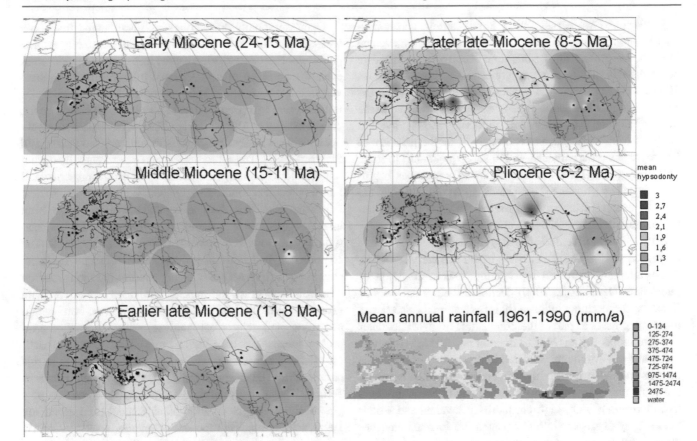

Fig. 3.5 Color-interpolated maps showing mean hypsodonty in fossil herbivorous mammal communities from the early Miocene to the Pliocene. Mean hypsodonty correlates with precipitation today (Eronen et al. 2010a). Blue color indicates communities with a low hypsodonty index (high precipitation, wooded environments), whereas green, yellow, and red colors represent progressively higher hypsodonty indices (dryer, more open environments). Reproduced from Fortelius et al. (2002) with permission

forested conditions existed in some parts of Europe during the middle Miocene than even during the early Miocene (e.g., Iberian Peninsula).

3.4.2   Eurasia

Although hypsodonty data from middle Miocene mammal communities indicates an increase in aridity in many areas, it does not yet suggest the development of extremely dry or open conditions anywhere in Eurasia. Early hypsodont bovids and rhinoceroses occur in faunas that were dominated by brachydont ruminants. One such locality is Tung Gur, China (Fortelius et al. 2003). Here, a diverse fauna of large mammals inhabited the opening woodland habitats, and we encounter C4 grasses while hiking through these wooded pastures. Their presence is recorded in the rocks of the southwestern Chinese Loess Plateau beginning in the middle Miocene (Dong et al. 2018) and is probably related to a particularly warm climate and the onset of the monsoon. In response to this regional climate change, the hypsodont hip-

parionine horses, found exclusively in North America until that time, began their migration to Eurasia at the beginning of the late Miocene. In contrast, deer-like brachydont, browsing anchitherine horses were present in Eurasia during the early and middle Miocene (Kaiser 2009; Eronen et al. 2010b).

3.4.3   North America

Earlier, we walked through Oligocene North America and witnessed the development of open shrubland and woodland environments (Sect. 3.3.2), whereupon these landscapes spread further during the early and middle Miocene. Focusing on the Midwest of the continent, we can explore how some of the earliest open, savanna-like environments developed. Evidence supporting the opening of environments comes from paleobotany, fossil-mammal paleontology, paleopedology, and stable-isotope geochemistry. After the initial spread of open-habitat grasses in the understory of Oligocene forests and woodlands, C3 grass-dominated open habitats spread during the early Miocene (Retallack 1997; Janis et al. 2000;

Strömberg and McInerney 2011). Stable-isotope geochemistry from soil carbonates indicates that a minority (on average 20%) of the grasses during this time occupying savanna-like environments were C4-photosynthesizing (Fox and Koch 2003). This observation is supported by phytoliths, but the earliest unequivocal fossils of C4 grasses are preserved in the middle Miocene (Strömberg and McInerney 2011). Phytolith analyses from Nebraska and Kansas indicate the presence of heterogenous environments, with a subsequent shift to more uniformly open and C4 grass-dominated landscapes beginning in the late Miocene. These landscapes continued into the early Pliocene (Strömberg and McInerney 2011). The vertebrate fossil record parallels the change in vegetation.

Hypsodonty increased in early equine and hipparionine horses, as well as some camelids, in the late early Miocene. These adaptations reflect the opening and aridification of landscapes (Janis et al. 2000). Detailed dietary analyses of horses and oreodonts indicate that these animals also began ingesting grasses as part of their diet at this time (Mihlbachler and Solounias 2006; Mihlbachler et al. 2011). However, woodland environments persisted in many parts of the continent as indicated by the high diversity of middle Miocene browsing ungulates. Such diversity could only be maintained in an extremely rich, high-productivity savanna and woodland setting (Janis et al. 2002). Extensive forested environments dominated in Africa during the early and middle Miocene.

3.4.4 Africa

Most of Africa was forested during the early Miocene with some localized areas partially open and, at least seasonally, dry. The earliest occurrences of such open grassland near the newly opened land connection between Eurasia and Africa are known from Saudi Arabia. Pollen assemblages, there, are composed of a high percentage of grasses (68%) and a low proportion of other nonwoody plants (26%) (Whybrow and McClure 1981; Jacobs et al. 2010). Yet, fossils from localities in Libya (Grossman et al. 2014), the Niger delta, and Cameroon provide little evidence of any open, grassy vegetation in Northern and Western Africa during that time. Only later, during the middle Miocene, is there evidence of an increasing component of grasses in Western Africa (Morley and Richards 1993; Jacobs et al. 2010). On the other side of the continent in East Africa, plant macrofossils, pollen, and soil structures indicate the presence of extensive tropical rainforest, similar to those found, today, in the Guineo-Congolian lowlands. Evidence for any expansive open environment or the presence of grasses, here, is scarce, with some notable exceptions, where we will travel next.

Open, seasonally dry settings are documented from both fossil plants and vertebrates, invertebrate assemblages, and paleosols. Paleobotanical evidence in Bukwa, Uganda, indicates the presence of dry, seasonal miombo-like woodland vegetation (Pickford 2002) in the earliest Miocene, about 22 million years ago (Bamford 2011). Fossil mollusks from the coeval locality of Meswa Bridge, Kenya, indicate a relatively open and seasonally dry forest or woodland, too (Tassy and Pickford 1983). These biomes expand geographically during the middle Miocene when partly open grassland-woodland mosaics became increasingly common in East Africa. Paleosols from Maboko and Nyakach, Kenya (c. 15 Ma), indicate the presence of dambo-like, seasonally dry, and seasonally waterlogged grasslands (Retallack et al. 2002). The composition and body-size distribution of fossil mammal assemblages from Maboko further support the interpretation of a woodland-grassland mosaic, similar to modern African savannas (Mihlbachler et al. 1998). Based on dental mesowear and microwear analyses, some gomphotheriid proboscideans (elephant-like) from Maboko show the earliest evidence of grass-dominated mixed-feeding diets in this part of Africa (Saarinen, unpublished data). Pollen from Fort Ternan, Kenya (c. 14 Ma), also indicate the presence of a grass-dominated paleoenvironment (Dugas and Retallack 1993), with a major component of open-adapted PACMAD taxa (Chloridoideae and Panicaeae). Today, all PACMAD taxa utilize C4 photosynthesis. Yet, stable carbon isotopic values obtained from these grassland paleosols indicate the absence of biomass that originated from plants using any C4 photosynthetic pathway. Thus, these Fort Ternan data indicate a landscape predominantly inhabited by a C3-photosynthesizing plant community and suggest that the PACMAD grasses here had not yet evolved a C4 photosynthetic pathway (Cerling et al. 1991). At the same time, C4 grasses become increasingly abundant in South America.

3.4.5 South America

The Miocene record of South American phytoliths, pollen, and stable carbon-isotope trends indicates an increasing grass component on a continent still dominated by forest and woodland vegetation (Kleinert and Strecker 2001; Strömberg et al. 2013). Evidence exists for the presence of some C4 grasses in the early Miocene of Patagonia, and pollen from Venezuela indicate that more open, grass-dominated habitats occurred all across the continent (Lorente 1986). Nonetheless, tooth-wear analyses of ungulates (Townsend and Croft 2008) and xenarthrans (anteaters, tree sloths, and armadillos; Saarinen and Karme 2017) from the rich early Miocene fossil assemblage of Santa Cruz, Argentina, indicate browse-based diets for all the species. Similar changes are observed in Australia.

3.4.6 Australia

There is a dearth of paleobotanical evidence from Australia from which we can reconstruct Neogene vegetation. What evidence exists indicates that forests and woodlands dominated these paleoenvironments, which became increasingly open in the Miocene when compared to the Oligocene. Sites such as Riversleigh have yielded pollen assemblages that contain 4–11% grasses, although these forms may have originated from wetland grasses rather than those adapted to semiarid and arid climates (Kershaw et al. 1994; McPhail and Hill 2002). By the middle Miocene, some species of kangaroo and wombat evolved adaptations to living in more open landscapes. These adaptions include an increase in hypsodonty (Dawson and Dawson 2006), as well as structural modifications of the hindlimb. The shape of the heal bone (calcaneus*), in particular, indicates that some Miocene kangaroos had the ability to move increasingly efficiently in open habitats by hopping, although they were still not as derived as the modern kangaroos in that respect (Janis et al. 2016).

3.5 The Late Neogene and Quaternary (11–0 Ma) Spread of Grasslands and the Rise of C4-Photosynthesizing Grasses to Ecological Dominance in Low and Midlatitudes

Global climate cooled drastically after the middle Miocene climatic optimum and launched the development of extensive drying and increasingly seasonal climate in the midlatitudes (Zachos et al. 2001). This, in turn, triggered one of the most significant environmental changes during the Cenozoic, the late Miocene expansion of grass-dominated environments that were populated, in large part, by C4 grasses (e.g., Cerling et al. 1997; Strömberg 2011). The spread of C4 grass-dominated environments occurred across all continents and is recorded by a significant change in stable carbon-isotope ratios in both soil carbonates and in the dental enamel of grazing ungulates and other herbivorous mammals (Cerling et al. 1997). Why C4 grasses expanded and became ecologically dominant in many low and midlatitude open environments continues to be debated. It is suggested that a drop in atmospheric pCO_2 could have given a physiological advantage to C4 grasses (Cerling et al. 1997). The mechanism behind the Late Miocene rise of C4 grasslands has remained poorly understood. It has been suggested that the climatic drying would have been a major factor driving this in low-latitude ecosystems (Strömberg and McInerney 2011). However, more recently, Pollisar et al. (2019) reported a synchronous increase in C4 grass-dominated environments across Africa, which is not connected to any significant change in climatic factors, suggesting instead that the main factor driving the spread of C4 grasses was a further drop in pCO_2 launched by the drastic cooling of high latitude environments. Increased seasonality and adaptation to the variation between dry and wet seasons may have driven both a second diversification event of C4 grasses and their rise to ecological dominance during the late Miocene (Osborne 2008; Vicentini et al. 2008). Molecular phylogeny indicates that grasses were originally warm-adapted and, in fact, the cold adapted C3 grasses of the subfamily Pooideae are specialized forms, and they became an important element in high-latitude open environments (Edwards and Smith 2010). The spread of C4 grasses is more likely due to changes in atmospheric pCO_2 and perhaps mid-latitude drying and opening of habitats than, solely, a change in temperature (Edwards and Smith 2010). Moreover, an increase in the frequency of fires, the result of increased seasonality and the high risk of fire during dry seasons (Keeley and Rundel 2005), or an increase in the intensity of herbivory by the diverse and abundant middle and late Miocene large mammals (Fox and Koch 2003) could have favored the fast-regenerating C4 grasses.

The climatic cooling of the late Miocene was followed in the Pliocene (5–2.6 Ma) by more stable and even slightly warmer conditions, until a more drastic drop in the global temperatures led to the Pleistocene ice age climate (Herbert et al. 2016). The Pleistocene is characterized by strong climatic oscillations and recurrent glacial advances and retreats in the Northern Hemisphere (see Chap. 1). These oscillations were driven by cyclic changes in the orbit and axis orientation of the Earth (Walker and Lowe 2007; see also Box P3). Open, grass-dominated environments continued to spread in many mid-latitudinal parts of the world resulting in the development of completely treeless steppes and prairies during the Pliocene and, especially, the Pleistocene (e.g., Ding and Yang 2000; Fortelius et al. 2002, 2006; Strömberg 2011). Aridification in many midlatitudinal areas continued into the Holocene (11 ka—present), which is seen as the desertification of the Sahara (Claussen et al. 1999), Sahel (Gonzalez 2001), and Central Asia (Jin et al. 2012), partly due to human impact including overgrazing. Let's now travel again through the continents to observe directly what changed these landmasses experienced during the late Neogene.

3.5.1 Eurasia

The late Miocene landscape of Eurasia remained largely dominated by forests and woodlands, but there were significant changes in the composition of plant-and-animal communities. This is especially true in the eastern Mediterranean and in Asia, where the landscape became increasingly arid and open. But, let's first return to the Iberian Peninsula where a drastic environmental change occurred during the late Miocene. This significant change mostly affected the mammal communities of Western Europe and is an event known as the "Vallesian crisis" that occurred during the early late Miocene (the mid-Vallesian, c. 10 Ma). Specialized forest-adapted mammals, such as tapirs and tragulids (mouse deer), declined in diversity and were replaced by more open woodland-adapted taxa, including early bovids (Agustí and Moya-Sola 1990). A significant migration event is marked by the arrival of the hypsodont, grazing-adapted hipparionine horses in Europe approximately 10.5 million years ago. However, the first hipparionines to appear in Europe (*Hippotherium primigenium*) were forest-adapted and exhibited predominantly browsing and mixed-feeding strategies rather than grazing (e.g., Kaiser 2004). The Old World Savanna Paleobiome (OWSP) originated and spread in Eurasia and Africa during the late Miocene and gradually disappeared during the Pliocene. Remnants of this paleobiome gave rise to the modern South Asian and African savanna faunas (Kaya et al. 2018).

By traveling to the eastern Mediterranean during the latest Miocene, we will be able to explore savanna-like environments that supported diverse herbivorous mammal faunas. These eastern Mediterranean habitats had a seasonal climate and included a significant component of grasses. Nevertheless, they were not open grasslands. Both paleobotanical evidence and the structure, paleoecology, and ecometrics of the mammal faunas indicate that these environments existed as semi-open woodland savannas with grasses, probably analogous to extant Indian woodland savannas rather than true open grasslands (Solounias et al. 2010). In contrast to earlier mammal faunas, the latest Miocene "*Hipparion* faunas" of Eurasia were populated by numerous mixed-feeding taxa along with some large mammals that were specialized grazers (Fig. 3.6). These included hipparionine horses (e.g., *Hipparion*, *Cremohipparion*; Bernor et al. 2014), bovids (e.g., *Tragoportax*, *Gazella*; Passey et al. 2007; Solounias et al. 2013), giraffes (e.g., *Samotherium*; Solounias et al. 2013), and gomphotheriid proboscideans (*Choerolophodon pentelici*; Konidaris et al. 2016). The end of the Miocene witnessed the partial desiccation of the Mediterranean due to the closure of the basin by tectonic events. The area underwent a period of aridification and salt deposition. This event, known as the Messinian Salinity Crisis, had an effect on the environments and migration pathways around the Mediterranean (Rouchy and Caruso 2006), whose spa-

---

Box 3.3: Eastern Mediterranean Fossil Plants and Climate Gradients

Although climate evolution and spatial gradients have been intensely studied for the midlatitudes of the late Miocene of western Eurasia (e.g., Bruch et al. 2011; Ivanov et al. 2012), comparatively few comprehensive studies are available for the lower latitudes of the eastern Mediterranean. The degree of seasonality and how humid or dry the climate was during the late Miocene in that area are still debated. It is assumed that the southeastern Mediterranean realm experienced about the same temperatures in the late Miocene as today (Utescher et al. 2011). Therefore, the area would have experienced no drastic cooling during the later Neogene (e.g., Kovar-Eder 2003). This interpretation is supported by the recently recovered Gavdos megaflora from southernmost Greece (Mantzouka et al. 2015) and a number of other coeval localities from Crete. The Cretian sites include Vrysses (latest Tortonian to Messinian, c. 7.5–6.0 Ma), Makrilia (late Tortonian, c. 8.6–7.7 Ma), and Pitsidia (early Tortonian, c. 10.5 Ma). An almost coeval (Messinian, c. 7–6 Ma) rich macroflora is available from Vegora, Macedonia, Northern Greece.

Climate data derived from the Gavdos flora and the other Greek localities show no signs of a distinct climate aridification trend nor changes toward a summer-dry ("etesian wind pattern") regime during the Messinian crisis. However, the possibility exists that the plant records may be biased and preserve plants that lived during the humid part of the Tortonian climate cycles when preservation potential was high. Tortonian temperatures estimated from four southern Greek megafloras are comparable to those of the present day or even slightly lower. Similar climatic conditions were also detected at higher latitude (Utescher et al. 2011), and this indicates that the late Neogene south-to-north climatic gradient across Europe was lower than the present one.

---

tial climatic gradients during the late Miocene are still poorly known (Box 3.3).

Travelling eastward we notice that open and dry environments are more widely spread than before, with abundant hypsodont and mesodont bovids and horses. These grazers/mixed feeders are the dominant component in most of the Mediterranean, Central Asian, and Chinese localities of the time. Eastern and Central Europe remained forested under humid conditions during the latter part of the epoch (Fig. 3.5).

Fig. 3.6 Tentative reconstructions of a middle Miocene forest from Central Europe (**a**) and a late Miocene (Pikermian) woodland-savanna from the eastern Mediterranean (**a**), with examples of characteristic ungulate species. Mammals figured in (**a**): 1. *Anchitherium aurelianense*, a browsing, deer-like horse with brachydont dentition. 2. *Anisodon grande*, a high-browsing chalicothere (clawed perissodactyl) distantly related to tapirs and rhinoceroses. 3. *Dicrocerus elegans*, a small, brachydont deer with simple antlers. 4. *Hoploaceratherium tetradactylum*, a browsing, hornless (aceratherine) rhinoceros. 5. *Dorcatherium* sp., a tragulid (mouse deer). 6. *Conohyus* sp., an omnivorous tetraconodontine pig. Mammals figured in (**b**): 1. *Ancylotherium pentelicum*, a large browsing chalicothere with brachydont dentition. 2. *Ceratotherium neumayri*, a mixed-feeding rhinoceros with mesodont dentition and an early relative of African black and white rhinoceroses. 3. *Microstonyx major*, a large, omnivorous pig. 4. *Tragoportax amalthea*, a brachydont-mesodont, mixed-feeding bovid (antelope). 5. *Palaeoryx pallasi*, a mesodont, mixed-feeding bovid (antelope). 6. *Gazella* sp., a brachydont-mesodont, mixed-feeding bovid (antelope). 7. *Samotherium major*, a mesodont, mixed-feeding giraffe. 8. *Hipparion* sp., a hypsodont, mixed-feeding or grazing horse. Drawings by J Saarinen

In contrast, western and eastern Mediterranean regions and Central Asia were relatively dry (Fortelius et al. 2006; Eronen et al. 2010a).

Cyclical changes in the Pleistocene climate caused dramatic periodic changes in biome distributions and environments in Eurasia. There was a general shift during the Early Pleistocene from forests to more open savanna-like landscapes, especially in Southern Europe (Bertini in Chap. 1). Similar changes occurred as far north as England, where semi-open, boreal grassland-heathland environments prevailed during the cool glacial stages and woodland environments during the warmer interglacials (e.g., West 1980). Later during the Pleistocene, forest successions were characteristic of the warm interglacial stages across Europe and northern Eurasia. In contrast, especially during the Late Pleistocene glacial stages, continental ice sheets and mountain ranges blocked moisture from reaching the interior of Eurasia. The drastic Late Pleistocene glaciation promoted long-term, cold-and-dry, high-pressure climates that resulted in the periodic spread of "mammoth steppe" vegetation over the continent (Guthrie 2001). Modern levels of hypsodonty were reached in the ungulate faunas during the Pliocene (Fortelius et al. 2002, 2006), but diets remained variable during the Pleistocene, reflecting the prevailing vegetation structure. Tooth-wear-based dietary analyses have shown that the composition of food resources available to ungulate communities varied in the Pleistocene of Europe according to habitat openness and, particularly, the presence of grasses in the environments (Saarinen et al. 2016; Saarinen and Lister 2016).

3.5.2 China

Revisiting the Chinese Loess Plateau during the late Miocene shows that climatic changes affected the extent of open steppe-like environments. Here, C4 grasses were present and formed a part of herbivore diets (Passey et al. 2009). However, the boundary between the southern forest-dominated landscapes and the northern steppe-desert environments shifted geographically over time. In fact, the forest-steppe transitional zone shifted northward between seven and four million years ago as a consequence of a humid climatic phase, which followed a more arid period characterized by the expansion of C4 grasses in the northern parts of the Loess Plateau (Passey et al. 2009). Stable carbon-isotope evidence from Southern Asia also indicates the expansion and spread of C4 grasses in Pakistan during the late Miocene. By about six million years ago, at the time of the Messinian crisis in Europe, pure C4 grasslands were established in Eastern Asia (Quade and Cerling 1995). Interestingly, mean hypsodonty values declined during the latest Miocene in some parts of the Loess Plateau when compared to assemblages of the earlier late Miocene, but the

variance in the values increased. This trend probably is due to the intensification of the monsoon climate, which began about eight million years ago, and led to a strong contrast between more humid and forested areas and more arid, open areas in China. With a greater variance in climate across the continent, mammal populations responded. Mean hypsodonty in ungulate communities became spatially more variable, despite a generally increasing trend through time (Fortelius et al. 2002, 2006).

The geographic extent of open habitats increased in Central and Eastern Asia significantly during the Pliocene (5–2.6 Ma) (Fortelius et al. 2006; Ding and Yang 2000), while most of Europe remained humid and forested (Kovar-Eder 2003). The development of cold-and-arid, open environments in the Tibetan Plateau triggered the evolution of many cold-adapted mammal species, such as the woolly rhinoceros. These taxa became abundant in the open "mammoth steppe" environments during glacial events of the Pleistocene (2.6 Ma–11 ka) in northern Eurasia (Deng et al. 2011).

3.5.3 North America

Back in North America, we can witness a significant increase in the geographic extent of open habitats occupied by grasses, including abundant C4 photosynthesizing ones. This biome replaced the wooded, high-productivity savanna environments of the middle Miocene in the dry continental interior. The expansion of these grasslands is recorded in paleosols of the late Miocene and Pliocene (c. 8–2 Ma; Strömberg 2011; Strömberg and McInerney 2011) and is interpreted to represent increasing seasonality and aridity. Beginning about eight million years ago, the continental interior of North America witnessed the development of the first truly open, C4 grass-dominated grasslands. Evidence in support of this interpretation is based on grass phytoliths and soil carbonates (Strömberg and McInerney 2011). The late Miocene and Pliocene spread of grasslands also is reflected in the diets of ungulates. For example, the earliest horses with specialized grazing diets (rather than mixed feeding) emerged in North America during this time (Mihlbachler et al. 2011). The subsequent climatic cooling during the Pliocene and Pleistocene greatly affected these biomes.

The glacial-interglacial cyclicity of Pleistocene climate had drastic effects on North American environments. Compared to Eurasia, a much greater area of the continent was covered by ice sheets during the zenith of glacial stages (Williams et al. 1998; Guthrie 2001). Environmental differences were pronounced between the heavily forested eastern part of North America and the dry, grassland-dominated western half of the continent (Williams et al. 1998). Southeastern North America even retained subtropical forests and wetland swamps, populated by thermophilous trees

(*Taxodium*, *Nyssa*) and mammals (*Tapirus*) that now inhabit jungles of South America, Central America, and Southeast Asia (see Liu in Chap. 2). The continental interior and western areas were dominated by open prairie environments (Williams et al. 1998), inhabited by open-adapted, grazing, or mixed-feeding mammals. The fauna included bison (*Bison* sp.; Rivals et al. 2007), horses (*Equus*; Mihlbachler et al. 2011), and the Columbian mammoth (*Mammuthus columbi*; Rivals et al. 2012). North of the great Pleistocene ice sheets, the mammal fauna was similar to the Siberian mammoth steppes, whereas south of the ice terminus were forests and grassland habitats (Williams et al. 1998; Guthrie 2001). In the Alaskan mammoth steppe, grazing ungulates were clearly dominant, and even taxa that are usually browse-dominated feeders, such as deer, were feeding primarily on grass (Rivals et al. 2010). Grasslands also expanded during this time in East Africa.

3.5.4 Africa

Forests in the late Miocene of Eastern Africa were increasingly replaced by C4 grass-dominated, arid communities. Stable carbon-isotope values from soil carbonates reveal the drastic increase in C4 grasses during this time (Cerling et al. 1997). The same signal is recorded in the isotopic signatures of the dental enamel of many herbivorous mammals, including early elephants from Lothagam, West Turkana (Kenya). The geochemistry of elephant teeth indicates a shift to diets based largely on C4 grasses (Cerling et al. 1999; Uno et al. 2011). However, this shift to increasingly open-and-arid environments, and the dominance of C4 grasses, proceeded in a mosaic-like manner. Data indicate their first occurrence in the most arid areas, such as the Lake Turkana region, and elsewhere, thereafter. Paleoenvironmental proxies, including plant fossils (Jacobs et al. 2010), mammal faunas (Roche et al. 2013), ecometrics (Fortelius et al. 2016), and paleoecology indicate the presence of woodland-grassland mosaics in East Africa during the late Miocene. Local differences are known throughout the Pliocene and especially during the Pleistocene. For example, the Lake Turkana region was at the center of soil acidification and the spread of open habitats during the Pliocene (Fortelius et al. 2016). In contrast, paleobotanical and mammal evidence indicate more wooded habitats in other areas, including Laetoli, Tanzania, and Hadar, Ethiopia (e.g., Bonnefille et al. 2004; Bamford 2011). Thus, during the Pliocene, the spread of C4 grasses happened in the context of grassy woodlands rather than in the context of open grassland landscapes. Further south, stable carbon-isotope evidence from Langebaanweg, South Africa, shows no evidence of C4 grasses at least until the early Pliocene, four million years ago (Franz-Odendaal

et al. 2002). C4-photosynthesizing grasses of the family Andropogoneae, which today dominate in open grasslands with heavy grazing and frequent burning, probably became increasingly abundant during the late Miocene (Retallack 1992). Arid, open C4 grasslands became widespread in East Africa during the Pleistocene, as indicated by stable carbon-isotope analyses of soils from sites such as Olduvai, Tanzania (Plummer et al. 1999), Lake Turkana region (Levin et al. 2011), Kanjera, and Olorgesailie in Kenya (Plummer et al. 1999). The aridification over large parts of Africa continued into the Late Pleistocene and especially the Holocene (11 ka–present) (Claussen et al. 1999; Gonzalez 2001). Other Southern Hemisphere continents show a similar trend.

3.5.5 South America and Australia

The development of open grassland environments in South America and Australia during the late Miocene largely parallels what happened in the other continents. In South America, stable carbon-isotope analyses from calcic paleosols indicate that the first C4 grass-dominated habitats emerged in the late Miocene (c. 7 Ma). They appear first in the lower latitudes (Kleinert and Strecker 2001) where unequivocal C4 grass-grazing ungulates in South America appear (MacFadden et al. 1996). In Australia, sclerophyllous forests with ground cover grasses are interpreted as the main components of late Miocene landscapes in the northern part of the continent. Pollen and phytolith records indicate that the first truly open grassland environments, with abundant PACMAD grasses and desert shrublands, emerged somewhat later. These landscapes are recognized during the Pliocene and Pleistocene (starting c. 4 Ma and peaking c. 2.5 Ma) in northern and southern Australia (Kershaw et al. 1994; McPhail and Hill 2002). Kangaroos and wombats further adapted to open landscapes in northern Australia during the late Miocene and spread across the continent during the Pliocene (Dawson and Dawson 2006).

3.6 Conclusions

Grasses originated during the Late Cretaceous and diversified into the two major derived clades (BEP and PACMAD) at that time, perhaps in more open environments of the Gondwana continent (possibly Africa or the Indian subcontinent) and facilitated by wildfire. Titanosaur sauropods fed on these grasses in India. Grasses survived the end-Cretaceous mass-extinction event, but remained rare and sparsely distributed, throughout the Paleogene. Paratropical and subtropical forest-and-wetland environments dominated Paleocene and Eocene landscapes, globally, although there is some evidence of abundant (possibly wetland) grasses in middle Eocene deposits of Africa. During the Eocene, at least some Central Asian and East African areas were covered in seasonally dry woodland and woodland/shrubland environments. Here, though, there is no indication of grasslands. Contrary to earlier interpretations, extensive grasslands also seem to be absent in South America, where paleosols with extensive dung-beetle brood balls are not associated with any abundance of grass phytoliths (Strömberg et al. 2013). The early hypsodonty displayed by endemic South American ungulates from this period more likely reflects an adaption to a high volume of volcanic ash than to ingestion of refractory silica phytoliths that form in grasses.

The first C4 grasses originated at the Eocene–Oligocene transition, probably in response to the drastic decline in atmospheric pCO_2. But, these forms were not dominant in any Paleogene environment. Increasingly open woodlands and shrublands, with grasses constituting a minor element of the vegetation, emerged in the continental interior of North America, Central Asia, Spain, and locally in South America, during the Oligocene. Subsequently in the early and middle Miocene, increasingly grass-rich and open savanna-like environments, with a minor C4 grass component, emerged in the continental interior of North America, South America, and Spain. Open-adapted, hypsodont ungulates with increasingly mixed-feeding diets coevolved in these early grassy savannas. At about the same time, C4 grasses expanded their biogeographic range into Asia and East Africa. This expansion might also reflect the fact that C4-photosynthesizing grasses emerged in many lineages of PACMAD grasses. But, this diversification and convergent evolution of the same pathway in various groups did not lead to their ecological dominance.

Open, grass-dominated environments spread globally during the late Miocene. Their expansion followed the cooling, midlatitudinal drying trend and increasing seasonality of global climate. The first C4 grass-dominated grasslands and savannas appear during the late Miocene in the continental interior of North America and in parts of South America, East Africa, and Southern Asia. The rise of C4 grasses into an ecologically dominant plant group in low-latitude open environments probably happened primarily in response to a drop in pCO_2 associated with high-latitude cooling, and possibly mid-latitude drying and increased wildfire regime. In addition, C4 grasses grow fast, which gives them an advantage in recovering from herbivory. The likely intensified periodic herbivory due to increased seasonality could, therefore, have affected C3 grasses more than C4 grasses. During the Pliocene, C4 grasses also became abundant in Central Asia and Australia. These environmental changes are reflected in increasingly open-adapted and hypsodont-dominated ungulate faunas, of which many were grazers (such as bovids, horses, and elephants).

Beginning in the Pliocene and continuing until the present, open dry grasslands spread across parts of the globe including East Africa, Southern Asia and China, and the continental interior of North America. In contrast, other areas including Europe remained humid and forested. Almost completely treeless grassland habitats, now characteristic of East African grassland savannas, North American prairies, and Central Asian steppes, are probably largely a Pliocene–Holocene phenomenon. We have little evidence for their existence prior to a few million years ago. During the Pleistocene glaciation events, vast open, grassy "mammoth steppe" habitats spread into northern Eurasia and Alaska. Desertification proceeded in large geographic areas during the Holocene, in response to both changes in climate and anthropogenic activities.

In sum, the evolution and development of today's grasslands can be traced in detail since the origin of the earliest grasses in the Cretaceous. Their rise to dominance on all continents is relatively well understood and was largely controlled by climate. Expanding grasslands also had a strong impact on the evolution of mammals. These animals adapted to the novel, major food source through time, evolving high-crowned teeth that can resist the increased wear rate caused by abrasive materials such as grass phytoliths and dust on dentition during their lifespans.

For Deeper Learning

- Browsing diets for Eocene large herbivorous mammals, North America: Mihlbachler and Solounias (2006); Mihlbachler et al. (2011); Boardman and Secord (2013); Métais et al. (2009)
- Climate evolution and spatial gradients, Miocene of western Eurasia: Mosbrugger et al. (2005); Bruch et al. (2006, 2011); Utescher et al. (2007); Ivanov et al. (2012)
- Miocene of North America: Janis et al. (2004); Jardine et al. (2012)
- Australia: Martin (2006)
- Desertification and human impact, including overgrazing: Claussen et al. (1999); Gonzalez (2001); Jin et al. (2012)
- Drastic cooling and drop in atmospheric pCO_2 at the Eocene-Oligocene boundary: Zachos et al. (2001); Christin et al. (2008); Vicentini et al. (2008)
- Effect of volcanic ash in the tooth-wear signal: Shockey and Anaya (2001); Billet et al. (2009); Croft and Weinstein (2008)
- Geographic extent of open habitats during Neogene: Fortelius et al. (2002, 2003, 2006); Ding and Yang (2000), Eronen et al. (2010a); Kovar-Eder (2003); Eronen and Rössner (2007)
- Hipparionine horses: Solounias et al. (2010, 2013); Bernor et al. (2014)

- Miocene bovids: Bibi (2007)
- Isotopic signatures recorded in teeth: Cerling (1992), Cerling et al. (1997, 1999); Uno et al. (2011).
- Late Cretaceous diversity of grass phytoliths from India: McPhail and Hill (2002); Piperno and Sues (2005); Prasad et al. (2005); Samant and Mohabey (2009); Edwards and Smith (2010)
- Levels of hypsodonty: Fortelius et al. (2002, 2003, 2006)
- Mesowear analysis: Fortelius and Solounias (2000); Eronen et al. (2014); Kaiser (2009); Kaiser et al. (2013), Rivals et al. (2007, 2010); Mihlbachler et al. (2011); Kaiser et al. (2013); Saarinen et al. (2016); Saarinen and Lister (2016); Saarinen and Karme (2017)
- Treeless steppe and prairie, Pliocene and Pleistocene: Ding and Yang (2000); Fortelius et al. (2002, 2003, 2006); Strömberg (2011), Strömberg and McInerney (2011)

Questions

1. What metrics taken from mammal teeth can be used to infer which environmental factors? Why don't these metrics always work?
2. Provide an example of an area and time when mammal teeth wrongly indicate open grasslands.
3. What are phytoliths and how can these be used to document the evolutionary history of grasslands?
4. What physiological features differentiate PACMAD grasses from BEP grasses?
5. What evidence exists to establish the first appearance of grasses in the fossil record, and when, in time, does it occur?
6. How can the morphologies of mammalian teeth provide evidence for the presence of grasslands in deep time?
7. Which conditions favored the evolution of open grasslands?
8. Compare and contrast the evolutionary history and timing of grassland expansion in the Oligocene of North America and Africa, Iberia/Balkans and Pakistan/Mongolia/Eastern Asia, and South America and Australia.
9. Compare and contrast the evolutionary history and timing of grassland expansion in the Early Neogene of Iberia/Near East and Eurasia, North America and Africa, and South America and Australia.
10. Compare and contrast the evolutionary history and timing of grassland expansion in the late Neogene–Quaternary of North America and Africa, Eurasia and China, and South America and Australia.

Acknowledgments Juha Saarinen was working as a post-doctoral researcher at the University of Helsinki, in the frame of two projects funded by the Academy of Finland: "ECHOES" and "Behavioural and morphological adaptation to environmental change: the example of African Neogene to Quaternary Proboscidea" (post-doctoral research project).

References

Agustí J, Moya-Sola S (1990) Mammal extinctions in the Vallesian (Upper Miocene). In: Kauffman EG (ed) Extinction events in earth history, vol IV. Springer-Verlag, Berlin, pp 425–432

Akgün F, Kayseri MS, Auiraz MS (2007) Paleoclimatic evolution and vegetational changes during the Late Oligocene-Miocene period in Western and Central Anatolia (Turkey). Palaeogeogr Palaeoclimatol Palaeoecol 253:56–90

Bamford MK (2011) Fossil leaves, fruits and seeds. In: Harrison T (ed) Paleontology and geology of Laetoli: human evolution in context – volume 1: geology, geochronology, paleoecology and paleoenvironment. Springer, Dordrecht, pp 235–252

Barrón E, Lassaletta L, Alcalde-Olivares C (2006) Changes in the early Miocene palynoflora and vegetation in the east of the Rubielos de Mora Basin (SE Iberian Ranges, Spain). Neues Jahrb Geol Palaontol Abh 242:171–204

Bellosi ES, Laza JH, Sánchez V, Genise JF (2010) Ichnofacies analysis of the Sarmiento Formation (middle Eocene – early Miocene) at Gran Barranca, central Patagonia. In: Madden RH, Carlini AA, Vucetich MG, Kay RF (eds) The paleontology of Gran Barranca: evolution and environmental change through the middle Cenozoic of Patagonia. Cambridge University Press, New York, pp 302–312

Bernor RL, Semprebon G, Damuth J (2014) Maragheh ungulate mesowear: interpreting paleodiet and paleoecology from a diverse fauna with restricted sample. Ann Zool Fenn 51:201–208

Bestland EA (1990) Sedimentology and paleopedology of Miocene alluvial deposits at the Pasalar hominoid site, western Turkey. J Hum Evol 19:363–377

Bibi F (2007) Dietary niche partitioning among fossil bovids in late Miocene C3 habitats: consilience of functional morphology and stable isotope analysis. Palaeogeogr Palaeoclimatol Palaeoecol 253:529–538

Billet G, Blondel C, de Muizon C (2009) Dental microwear analysis of notoungulates (Mammalia) from Salla (Late Oligocene, Bolivia) and discussion on their precocious hypsodonty. Palaeogeogr Palaeoclimatol Palaeoecol 274:114–124

Blondel C (2001) The Eocene-Oligocene ungulates from Western Europe and their environment. Palaeogeogr Palaeoclimatol Palaeoecol 168:125–139

Boardman GS, Secord R (2013) Stable isotope paleoecology of White River ungulates during the Eocene–Oligocene climate transition in northwestern Nebraska. Palaeogeogr Palaeoclimatol Palaeoecol 375:38–49

Bonnefille R, Potts R, Chalie F, Jolly D, Peyron O (2004) High-resolution vegetation and climate change associated with Pliocene *Australopithecus afarensis*. PNAS 101:12125–12129

Bouchenak-Khelladi Y, Slingsby JA, Verboom GA, Bond WJ (2014) Diversification of C4 grasses (Poaceae) does not coincide with their ecological dominance. Am J Bot 101:300–307

Bruch AA, Utescher T, Mosbrugger V, Gabrielyan I, Ivanov DA (2006) Late Miocene climate in the circum—Alpine realm–a quantitative analysis of terrestrial palaeofloras. Palaeogeogr Palaeoclimatol Palaeoecol 238:270–280

Bruch AA, Utescher T, Mosbrugger V, NECLIME Members (2011) Precipitation patterns in the Miocene of Central Europe and the development of continentality. Palaeogeogr Palaeoclimatol Palaeoecol 304:202–211

Cerling TE (1992) Development of grasslands and savannas in East Africa during the Neogene. Palaeogeogr Palaeoclimatol Palaeoecol 97:241–247

Cerling TE, Quade J, Ambrose SH, Sikes NE (1991) Fossil soils, grasses, and carbon isotopes from Fort Ternan, Kenya: grassland or woodland? J Hum Evol 21:295–306

Cerling TE, Harris JM, MacFadden BJ, Leakey MG, Quade J, Eisenmann V, Ehleringer JH (1997) Global change through the Miocene/Pliocene boundary. Nature 389:153–158

Cerling TE, Harris JM, Leakey MG (1999) Browsing and grazing in elephants: the isotope record of modern and fossil proboscideans. Oecologia 120:364–374

Christin P-A, Besnard G, Samaritani E, Duvall MR, Hodkinson TR, Savolainen V, Salamin N (2008) Oligocene CO2 decline promoted C4 photosynthesis in grasses. Curr Biol 18:37–43

Claussen M, Kubatzki C, Brovkin V, Ganopolski A (1999) Simulation of an abrupt change in Saharan vegetation in the mid-Holocene. Geophys Res Lett 26:2037–2040

Collinson ME, Hooker JJ (2003) Paleogene vegetation of Eurasia: framework for mammalian faunas. Deinsea 10:41–83

Crepet WL, Feldman GD (1991) The earliest remains of grasses in the fossil record. Am J Bot 78:1010–1014

Croft DA (2016) Horned armadillos and rafting monkeys - the fascinating fossil mammals of South America. Indiana University Press, Bloomington

Croft DA, Weinstein D (2008) The first application of mesowear method to endemic South American ungulates (Notoungulata). Palaeogeogr Palaeoclimatol Palaeoecol 269:103–114

Dawson TJ, Dawson L (2006) Evolution of arid Australia and consequences for vertebrates. In: Merrick JR, Archer M, Hickey GM, Lee MSY (eds) Evolution and biogeography of Australasian vertebrates. AUSCIPUB, Oatlands, pp 51–70

De Franceschi D, Hoorn C, Antoine PO, Cheema IU, Flynn LJ, Lindsay EH, Marivaux L, Metais G, Rajpar R, Welcomme J-L (2008) Floral data from the mid-Cenozoic of Central Pakistan. Rev Palaeobot Palynol 150:115–129

DeMiguel D, Fortelius M, Azanza B, Morales J (2008) Ancestral feeding state of ruminants reconsidered: earliest grazing adaptation claims a mixed condition for Cervidae. BMC Evol Biol 8:1–13

Deng T, Wang X, Fortelius M, Li Q, Wang Y, Tseng ZJ, Takeuchi GT, Sylor JE, Säilä LK, Xie G (2011) Out of Tibet: Pliocene woolly rhino suggests high-plateau origin of ice age megaherbivores. Science 333:1285–1288

Ding ZL, Yang SL (2000) C3/C4 vegetation evolution over the past 7.0 Myr in the Chinese loess plateau: evidence from pedogenic carbonate d13 C. Paleogeography, Paleoclimatology, Paleoecology 160:291–299

Dong J, Liu Z, An Z, Liu W, Zhou W, Qiang X, Lu F (2018) Mid-Miocene C4 expansion on the Chinese loess plateau under an enhanced Asian summer monsoon. J Asian Earth Sci 158:153–159

Dugas DP, Retallack GJ (1993) Middle Miocene fossil grasses from fort Ternan, Kenya. J Paleontol 67:113–128

Dupont-Nivet G, Krijgsman W, Lagreis C, Abels HA, Dai S, Fang X (2007) Tibetan plateau aridification linked to global cooling at the Eocene-Oligocene transition. Nature 445:635–638

Edwards EJ, Smith SA (2010) Phylogenetic analyses reveal the shady history of C4 grasses. PNAS 107:2532–2537

Erlebe JJ, Greenwood DR (2012) Life at the top of the greenhouse Eocene world - a review of the Eocene flora and vertebrate fauna from Canada's high Arctic. Geol Soc Am Bull 124:3–23

Eronen JT, Rössner GE (2007) Wetland paradise lost: Miocene community dynamics in large herbivorous mammals from the German Molasse Basin. Evol Ecol Res 9:471–494

Eronen JT, Puolamaki K, Liu L, Lintulaakso K, Damuth J et al (2010a) Precipitation and large herbivorous mammals II: application to fossil data. Evol Ecol Res 12:235–248

Eronen JT, Evans AR, Fortelius M, Jernvall J (2010b) The impact of regional climate to the evolution of mammals: a case study using fossil horses. Evolution 64:398–408

Eronen JT, Kaakinen A, Liu L-P, Passey BH, Tang H, Zhang Z-Q (2014) Here be dragons: Mesowear and tooth enamel isotopes of the classic Chinese "Hipparion" faunas from Baode, Shanxi Province, China. Ann Zool Fenn 51:227–244

Fortelius M (1985) Ungulate cheek teeth: developmental. functional and evolutionary interrelations Acta Zoologica Fennica 180:1–76

Fortelius M, Kappelman J (1993) The largest land mammal ever imagined. Zool J Linnean Soc 107:85–101

Fortelius M, Solounias N (2000) Functional characterization of ungulate molars using the abrasion-attrition wear gradient: a new method for reconstructing paleodiets. Am Mus Novit 3301:1–35

Fortelius M, Eronen JT, Jernvall J, Liu L, Pushkina D, Rinne J, Tesakov A, Vislobokovo I, Zhang Z, Zhou L (2002) Fossil mammals resolve regional patterns of Eurasian climate change during 20 million years. Evol Ecol Res 4:1005–1016

Fortelius M, Eronen J, Liu L, Pushkina D, Tesakov A, Vislobokova I, Zhang Z (2003) Continental-scale hypsodonty patterns, climatic paleobiogeography and dispersal of Eurasian Neogene large mammal herbivores. Deinsea 10:1–11

Fortelius M, Eronen J, Liu L, Pushkina D, Tesakov A, Vislobokova I, Zhang Z (2006) Late Miocene and Pliocene large land mammals and climatic changes in Eurasia. Palaeogeogr Palaeoclimatol Palaeoecol 238:219–227

Fortelius M, Zliobaité I, Kaya F, Bibi F, Bobe R, Leakey L, Leakey M, Patterson D, Rannikko J, Werdelin L (2016) An ecometric analysis of the fossil mammal record of the Turkana Basin. Philos Trans R Soc B 371:1–13

Fox DL, Koch PL (2003) Tertiary history of C4 biomass in the Great Plains, USA. Geology 31:809–812

Franz-Odendaal TA, Lee-Thorp JA, Chinsamy A (2002) New evidence for the lack of C4 grassland expansions during the early Pliocene at Langebaanweg, South Africa. Paleobiology 28:378–388

Gonzalez P (2001) Desertification and a shift of forest species in the West African Sahel. Clim Res 17:217–228

Grossman A, Liutkus-Pierce C, Kyongo B, M'Kirera F (2014) New fauna from Loperot contributes to the understanding of Early Miocene catarrhine communities. Int J Primatol 35:1253–1274

Guthrie DR (2001) Origin and causes of the mammoth steppe: a story of cloud cover, woolly mammal tooth pits, buckles, and inside-out Beringia. Quat Sci Rev 20:549–574

Herbert TD, Lawrence KT, Tzanova A, Cleaveland-Patterson L, Caballero-Gill R, Kelly CS (2016) Late Miocene global cooling and the rise of modern ecosystems. Nat Geosci 9:843–847

Ivanov D, Utecher T, Ashraf AR, Mosbrugger V, Bozukov V, Djorgova N, Slavomirova E (2012) Late Miocene palaeoclimate and ecosystem dynamics in southwestern Bulgaria – a study based on pollen data from the Gotse-Delchev Basin. Turk J Earth Sci 21:187–211

Jacobs BF, Herendeen PS (2004) Eocene dry climate and woodland vegetation in tropical Africa reconstructed from fossil leaves from northern Tanzania. Palaeogeogr Palaeoclimatol Palaeoecol 213:115–123

Jacobs BF, Kingston JD, Jacobs LL (1999) The origin of grass-dominated ecosystems. Ann Mo Bot Gard 86:590–643

Jacobs BF, Pan AD, Scotese CR (2010) A review of the Cenozoic vegetation history of Africa. In: Werdelin L, Sanders WJ (eds) Cenozoic mammals of Africa. University of California Press, Berkeley, pp 57–72

Janis CM, Fortelius M (1988) On the means whereby mammals achieve increased functional durability of their dentitions, with special reference to limiting factors. Biol Rev 63:197–230

Janis CM, Damuth J, Theodor JM (2000) Miocene ungulates and terrestrial primary productivity: where have all the browsers gone? PNAS 97:237–261

Janis CM, Damuth J, Theodor JM (2002) The origins and evolution of the North American grassland biome: the story from the hoofed mammals. Palaeogeogr Palaeoclimatol Palaeoecol 177:183–198

Janis CM, Damuth J, Theodor JM (2004) The species richness of Miocene browsers, and implications for habitat type and primary productivity in the North American grassland biome. Palaeogeogr Palaeoclimatol Palaeoecol 207:371–398

Janis CM, Damuth J, Travouillon KJ, Figueirido B, Hand SJ, Archer M (2016) Palaeoecology of Oligo-Miocene macropodoids determined from craniodental and calcaneal data. Mem Mus Vic 74:209–232

Jardine PE, Janis CM, Sahney S, Benton MJ (2012) Grit not grass: concordant patterns of early origin of hypsodonty in Great Plains ungulates and Glires. Palaeogeogr Palaeoclimatol Palaeoecol 365-366:1–10

Jernvall J, Hunter J, Fortelius M (1996) Molar tooth diversity, disparity, and ecology in Cenozoic ungulate radiations. Science 274:1489–1492

Jin L, Chen F, Morrill C, Otto-Bliesner BL, Rosenbloom N (2012) Causes of early Holocene desertification in arid Central Asia. Clim Dyn 38:1577–1591

Kaiser TM (2004) The dietary regimes of two contemporaneous populations of *Hippotherium primigenium* (Perissodactyla, Equidae) from the Vallesian (Upper Miocene) of Southern Germany. Palaeogeogr Palaeoclimatol Palaeoecol 198:381–402

Kaiser TM (2009) *Anchitherium aurelianense* (Equidae, Mammalia): a brachydont "dirty browser" in the community of herbivorous large mammals from Sandelzhausen (Miocene, Germany). Paläontol Z 83:131–140

Kaiser TM, Müller DWH, Fortelius M, Schulz E, Codron D, Clauss M (2013) Hypsodonty and tooth facet development in relation to diet and habitat in herbivorous ungulates: implications for understanding tooth wear. Mammal Rev 43:34–46

Kappelman J (1991) The paleoenvironment of *Kenyapithecus* in Fort Ternan. J Hum Evol 20:95–129

Kaya F, Bibi F, Žliobaitė I, Eronen JT, Hui T, Fortelius M (2018) The rise and fall of the Old World savannah fauna and the origins of the African savannah biome. Nat Ecol Evol 2:241–246

Keeley JE, Rundel PW (2005) Fire and the Miocene expansion of C4 grasslands. Ecol Lett 8:683–690

Kershaw AP, Martin HA, McEwen Mason JC (1994) The Neogene: a time of transition. In: Hill RS (ed) History of the Australian vegetation: cretaceous to recent. Cambridge University Press, Cambridge, pp 299–327

Kleinert K, Strecker MR (2001) Climate change in response to orographic barrier uplift: paleosol and stable isotope evidence from the late Neogene Santa Maria basin, northwestern Argentina. Geol Soc Am Bull 113:728–742

Konidaris GE, Koufos GD, Kostopoulos DS, % Merceron G (2016) Taxonomy, biostratigraphy and paleoecology of *Choerolophodon* (Proboscidea, Mammalia) in the Miocene of SE Europe – SW Asia: implications for phylogeny and biogeography. J Syst Palaeontol 14:1–27.

Kovar-Eder J (2003) Vegetation dynamics in Europe during the Neogene. Deinsea 10:373–392

Kovarovic K, Andrews P (2007) Bovid postcranial ecomorphological survey of the Laetoli paleoenvironment. J Hum Evol 52:663–680

Lee-Thorp J, van der Merwe NJ (1987) Carbon isotope analysis of fossil bone apatite. S Afr J Sci 83:712–715

Legendre S (1986) Analysis of mammalian communities from the late Eocene and Oligocene of Southern France. Palaeovertebrata 16:191–212

Levin NE, Brown FH, Behrensmeyer AK, Bobe R, Cerling TE (2011) Paleosol carbonates from the Omo Group: isotopic records of local and regional environmental change in East Africa. Palaeogeogr Palaeoclimatol Palaeoecol 307:75–89

Liu L, Puolamäki K, Eronen JT, Mirzaie Ataabadi M, Hernesniemi E, Fortelius M (2012) Dental functional traits of mammals resolve productivity in terrestrial ecosystems past and present. Proc R Soc B 279:2793–2799

Livermore R, Hillenbrand C-D, Meredith M, Eagles G (2007) Drake Passage and Cenozoic climate: an open and shut case? Geochem Geophys Geosyst 8. https://doi.org/10.1029/2005GC001224

Lorente MA (1986) Palynology and palynofacies of the Upper Tertiary in Venezuela. Dissertationes Botanicae 99, Stuttgart, Germany

MacFadden BJ, Cerling TE, Prado J (1996) Cenozoic terrestrial ecosystem evolution in Argentina: evidence from carbon isotopes of fossil mammal teeth. PALAIOS 11:319–327

Madden RH (2015) Hypsodonty in mammals. Cambridge University Press, Cambridge

Mantzouka D, Kvaček Z, Teodoris V, Utescher T, Tsaparas N, Karakitsios V (2015) A new late Miocene (Tortonian) flora from Gavdos Island in southernmost Greece evaluated in the context of vegetation and climate in the Eastern Mediterranean. N Jb Geol Paläont (Abh) 275:47–81

Martin HA (2006) Cenozoic climatic change and the development of the arid vegetation in Australia. J Arid Environ 66:533–563

McNab BK (2010) Geographic and temporal correlations of mammalian size reconsidered: a resource rule. Oecologia 164:13–23

McPhail MK, Hill RS (2002) Paleobotany of the Poaceae. Flora of Aust 43:37–70

Métais G, Antoine P-O, Marivaux L, Welcomme J-L, Ducrocq S (2003) New artiodactyl ruminant mammal from the late Oligocene of Pakistan. Acta Palaeontol Pol 48:375–382

Métais G, Welcomme J-L, Ducrocq S (2009) New lophiomerycid ruminants from the Oligocene of the Bugti Hills (Balochistan, Pakistan). J Vertebr Paleontol 29:231–241

Mihlbachler MC, Solounias M (2006) Coevolution of tooth crown height and diet in oreodonts (Merycoidodontidae, Artiodactyla) examined with phylogenetically independent contrasts. J Mamm Evol 13:11–36

Mihlbachler MC, McCrossin ML, Benefit BR (1998) Body size distribution and habitat structure of Maboko Island, middle Miocene, Kenya. SVP meeting 1998 Abstract

Mihlbachler MC, Rivals F, Solounias N, Semprebon GM (2011) Dietary change and evolution of horses in North America. Science 331:1178–1181

Morley RJ, Richards K (1993) Gramineae cuticle: a key indicator of late Cenozoic climatic change in the Niger Delta. Rev Palaeobot Palynol 77:119–127

Mosbrugger V, Utescher T, Dilcher DL (2005) Cenozoic continental climatic evolution of Central Europe. PNAS 102:14964–14969

Novello A, Blondel C, Brunet M (2010) Feeding behavior and ecology of the Late Oligocene Moschidae (Mammalia, Ruminantia) from La Milloque (France): evidence from dental microwear analysis. C R Palevol 9:471–478

Ortiz-Jaureguizar E, Cladera GA (2006) Paleoenvironmental evolution of southern South America during the Cenozoic. J Arid Environ 66:498–532

Osborne CP (2008) Atmosphere, ecology and evolution: what drove the Miocene expansion of C4 grasslands? J Ecol 96:35–45

Passey BH, Eronen JT, Fortelius M (2007) Paleodiets and paleoenvironments of late Miocene gazelles from North China: evidence from stable carbon isotopes. Vertebrata PalAsiatica 45:118–127

Passey BH, Ayliffe LK, Kaakinen A, Zhang Z, Eronen JT, Zhu Y, Zhou L, Cerling TE, Fortelius M (2009) Strengthened East Asian summer monsoons during a period of high-latitude warmth? Isotopic evidence from Mio-Pliocene fossil mammals and soil carbonates from northern China. Earth Planet Sci Lett 277:443–452

Pickford M (2002) Early Miocene grassland ecosystem at Bukwa, Mount Elgon, Uganda. C R Palevol 1:213–219

Pinilla A, Bustillo A (1997) Silícofitolitos en secuencias arcillosas con silcretas Mioceno Medio, Madrid. In: Pinilla A, Juan-Tresserras J, Machado MJ (eds) The state of the art of phytoliths in soils and plants, pp 255–265

Piperno DR, Sues H-D (2005) Dinos dined on grass. Science 310:1126–1128

Plummer T, Bishop L, Ditchfield P, Hicks J (1999) Research on Late Pliocene Oldowan sites at Kanjera South, Kenya. J Hum Evol 36:151–170

Poinar GO (2004) *Programinis burmitis* gen. et sp. nov., and *P. laminatus* sp. nov., Early Cretaceous grass-like monocots in Burmese amber. Aust Syst Bot 17:497–504

Pollisar PJ, Rose C, Uno KT, Phelps SR, deMenocal P (2019) Synchronous rise of African C4 ecosystems 10 million years ago in the absence of aridification. Nat Geosci 12:657–660

Prasad V, Strömberg CAE, Alimohammadian H, Sahni A (2005) Dinosaur coprolites and the early evolution of grasses and grazers. Science 310:1177–1180

Prasad V, Strömberg CAE, Leaché AD, Samant B, Patnaik R, Tang L, Mohabey DM, Ge S, Sahni A (2011) Late Cretaceous origin of the rice tribe provides evidence for early diversification in Poaceae. Nat Commun 2:480. https://doi.org/10.1038/ncomms1482.

Prothero DR (2013) Rhinoceros giants: the paleobiology of indricotheres. Indiana University Press, Bloomington

Quade J, Cerling TE (1995) Expansion of C4 grasses in the late Miocene of Northern Pakistan: evidence from stable isotopes in paleosols. Palaeogeogr Palaeoclimatol Palaeoecol 115:91–116

Rensberger JM (1978) Scanning electron microscopy of wear and occlusal events in some small herbivores. In: Butler PM, Joysey KA (eds) Development, function and evolution of teeth. Academic Press, New York, pp 415–438

Retallack GJ (1992) Middle Miocene fossil plants from Fort Ternan (Kenya) and evolution of African grasslands. Paleobiology 18:383–400

Retallack GJ (1997) Neogene expansion of the North American prairie. PALAIOS 12:380–390

Retallack GJ (2007) Cenozoic paleoclimate on land in North America. J Geol 115:271–294

Retallack GJ, Wynn JG, Benefit BR, McCrossin ML (2002) Paleosols and paleoenvironments of the middle Miocene, Maboko Formation, Kenya. J Hum Evol 42:659–703

Rivals F, Solounias N, Mihlbachler MC (2007) Evidence for geographic variation in the diets of late Pleistocene and early Holocene *Bison* in North America, and differences from the diets of recent *Bison*. Quat Res 68:338–346

Rivals F, Mihlbachler MC, Solounias N, Mol D, Semprebon GM, de Vos J, Kalthoff DC (2010) Paleoecology of the Mammoth Steppe fauna from the late Pleistocene of the North Sea and Alaska: separating species preferences from geographic influence in paleoecological dental wear analysis. Palaeogeogr Palaeoclimatol Palaeoecol 286:42–54

Rivals F, Semprebon G, Lister A (2012) An examination of dietary diversity patterns in Pleistocene proboscideans (*Mammuthus*, *Palaeoloxodon*, and *Mammut*) from Europe and North America as revealed by dental microwear. Quat Int 255:188–195

Roche D, Ségalen L, Senut B, Pickford M (2013) Stable isotope analyses of tooth enamel carbonate of large herbivores from the Tugen Hills deposits: Paleoenvironmental context of the earliest Kenyan hominids. Earth Planet Sci Lett 381:39–51

Rouchy JM, Caruso A (2006) The Messinian salinity crisis in the Mediterranean basin: a reassessment of the data and an integrated scenario. Sediment Geol 188-189:35–67

Saarinen J (2014) Ecometrics of large herbivorous land mammals in relation to climatic and environmental changes during the Pleistocene. Unigrafia, Helsinki, 42pp

Saarinen J, Karme A (2017) Tooth wear and diets of extant and fossil xenarthrans (Mammalia, Xenarthra) – applying a new mesowear approach. Palaeogeogr Palaeoclimatol Palaeoecol 476:42–54

Saarinen J, Lister AM (2016) Dental mesowear reflects local vegetation and niche separation in Pleistocene proboscideans from Britain. J Quat Sci 31:799–808

Saarinen J, Boyer AG, Brown JH, Costa DB, Ernest SKM, Evans AR, Fortelius M, Gittleman JL, Hamilton MJ, Harding LE, Lintulaakso K, Lyons SK, Okie JG, Sibly RM, Stephens PR, Theodor J, Uhen MD, Smith FA (2014) Patterns of body size evolution in Cenozoic land mammals: intrinsic biological processes and extrinsic forcing. Proc R Soc B Biol Sci 281:20132049. https://doi.org/10.1098/rspb.2013.2049

Saarinen J, Eronen J, Fortelius M, Seppä H, Lister AM (2016) Patterns of diet and body mass of large ungulates from the Pleistocene of Western Europe, and their relation to vegetation. Paleontol Electron 19.3.32A:1–58; paleo-electronic.org/content/2016/1567-pleistocene-mammal-ecometrics

Sadowski E-M, Schmidt AR, Rudall PJ, Simpson DA, Gröhn C, Wunderlich J, Seyfullah LJ (2016) Graminids from Eocene Baltic Amber. Rev Palaeobot Palynol 233:161–168

Sage RF (2004) The evolution of C4 photosynthesis. New Phytol 161:341–370

Samant B, Mohabey DM (2009) Palynoflora from Deccan volcano-sedimentary sequence (Cretaceous-Paleogene transition) of central India: implications for spatio-temporal correlation. J Biosci 34:811–823

Sandve SR, Fjellheim S (2010) Did gene family expansions during the Eocene–Oligocene boundary climate cooling play a role in Pooideae adaptation to cool climates? Mol Ecol 19:2075–2088

Schoene B, Eddy MP, Samperton KM, Keller CB, Keller G, Adatte T, Khadri SFR (2019) U-Pb constraints on pulsed eruption of the Deccan Traps across the end-Cretaceous mass extinction. Science 363:862–866. https://doi.org/10.1126/science.aau2422

Shockey BJ, Anaya F (2001) Grazing in a new Late Oligocene mylodontid sloth and a mylodontid radiation as a component of the Eocene-Oligocene faunal turnover and the early spread of grasslands/savannas in South America. J Mamm Evol 18:101–115

Solounias N, Rivals F, Semprebon GM (2010) Dietary interpretation and paleoecology of herbivores from Pikermi and Samos (late Miocene of Greece). Paleobiology 36:113–136

Solounias N, Semprebon GM, Mihlbachler MC, Rivals F (2013) Paleodietary comparisons of ungulates between the late Miocene of China, and Pikermi and Samos in Greece. In: Wang X, Flynn LJ, Fortelius M (eds) Fossil mammals of Asia: neogene biostratigraphy and chronology. Columbia University Press, New York, pp 676–692

Strömberg CAE (2011) Evolution of grasses and grassland ecosystems. Annu Rev Earth Planet Sci 39:517–544

Strömberg CAE, McInerney FA (2011) The Neogene transition from C3 to C4 grasslands in North America: assemblage analysis of fossil phytoliths. Paleobiology 37:50–71

Strömberg CAE, Dunn RE, Madden RH, Kohn MJ, Carlini AA (2013) Decoupling the spread of grasslands from the evolution of grazer-type herbivores in South America. Nat Commun 4:1478

Tassy P, Pickford M (1983) Un noveau mastodonte zygolophodonte (Proboscidea, Mammalia) dans le Miocène inférieur d'Afrique Orientale: systématique et paléoenvironnement. Geobios 16:53–77

Townsend KEB, Croft DA (2008) Diets of notoungulates from the Santa Cruz Formation, Argentina: new evidence from enamel microwear. J Vertebr Paleontol 28:217–230

Travouillon KJ, Legendre S (2009) Using cenograms to investigate gaps in mammalian body mass distributions in Australian mammals. Palaeogeogr Palaeoclimatol Palaeoecol 272:69–84

Uno KT, Cerling TE, Harris JM, Kunimatsu Y, Leakey MG, Nakatsukasa M, Nakaya H (2011) Late Miocene to Pliocene carbon isotope record of differential diet change among east African herbivores. PNAS 108:6509–6514

Urban MA, Nelson DM, Jiménez-Moreno G, Châteauneuf J-J, Pearson A, Hu FS (2010) Isotopic evidence of C4 grasses in southwestern Europe during the early Oligocene–middle Miocene. Geology 38:1091–1094

Utescher T, Erdei B, Francois L, Mosbrugger V (2007) Tree diversity in the Miocene forests of Western Eurasia. Palaeogeogr Palaeoclimatol Palaeoecol 253:226–250

Utescher T, Bruch AA, Micheels A, Mosbrugger V, Popova S (2011) Cenozoic climate gradients in Eurasia - a palaeo-perspective on future climate change? Palaeogeogr Palaeoclimatol Palaeoecol 304:351–358

Vicentini A, Barber JC, Aliscioni SS, Giussani LM, Kellogg EA (2008) The age of the grasses and clusters of origins of C4 photosynthesis. Glob Chang Biol 14:2963–2977

Walker M, Lowe J (2007) Quaternary science 2007: a 50-year retrospective. J Geol Soc 164:1073–1092

Walker A, Hoeck HN, Perez L (1978) Microwear of mammalian teeth as an indicator of diet. Science 201:908–910

Wang B (1992) The Chinese Oligocene: a preliminary review of mammalian localities and local faunas. In: Prothero DR, Berggren WA (eds) Eocene-Oligocene climatic and biotic evolution. Princeton University Press, Princeton, pp 529–547

Wang Y, Meng J, Ni X, Li C (2007) Major events in Paleogene mammal radiation in China. Geol J 42:415–430

Werdelin L, Sanders WJ (2010) Cenozoic mammals of Africa. University of California Press, Berkeley

West RG (1980) The pre-glacial Pleistocene of the Suffolk and Norfolk coasts. Cambridge University Press, Cambridge

Whybrow PJ, McClure HA (1981) Fossil mangrove roots and paleoenvironments of the Miocene of the eastern Arabian peninsula. Palaeogeogr Palaeoclimatol Palaeoecol 32:213–225

Williams M, Dunkerley D, De Deckker P, Kershaw P, Chappell J (1998) Quaternary environments. Arnold, London

Zachos JC, Pagani M, Sloan L, Thomas E, Billups K (2001) Trends, rhythms and aberrations in global climate 65 Ma to present. Science 292:686–693

Zachos JC, Dickens GR, Zeebe RE (2008) An early Cenozoic perspective on greenhouse warming and carbon-cycle dynamics. Nature 451:279–283

Žliobaitė I, Rinne J, Tóth AB, Mechenich M, Liu L, Behrensmeyer AK, Fortelius M (2016) Herbivore teeth predict climatic limits in Kenyan ecosystems. PNAS 45:12751–12756

Žliobaitė I, Tang H, Saarinen J, Fortelius M, Rinne J, Rannikko J (2018) Dental ecometrics of tropical Africa: linking vegetation types and communities of large plant-eating mammals. Evol Ecol Res 19:127–147

Zucol AF, Brea M, Bellosi E (2010) Phytolith studies in Gran Barranca (central Patagonia, Argentina): the middle-late Eocene. In: Madden RH, Carlini AA, Vucetich MG, Kay RF (eds) The paleontology of Gran Barranca: evolution and environmental change through the Middle Cenozoic of Patagonia. Cambridge University Press, Cambridge, pp 317–340

The Paleocene–Eocene Thermal Maximum: Plants as Paleothermometers, Rain Gauges, and Monitors

4

Melanie L. DeVore and Kathleen B. Pigg

Abstract

We travel back in time through this chapter and take a field trip to western North America during the Paleocene–Eocene Thermal Maximum (PETM), some 56 million years ago. Here, plant-and-animal fossils were discovered in the warmest interval of the last 500 million years, a condition that lasted only 200,000 years. We provide a brief review of what may have caused a massive influx of atmospheric carbon detected during the PETM. We contrast the PETM to similar ongoing thermal events that began during the Industrial Revolution and persist today. We discuss the tools that paleobotanists have devised to interpret climate from fossil leaf, pollen, and wood records, and present a brief overview of floral changes that occurred in western North America before, during, and right after this thermal maximum. Lastly, we explore how fossil data can be incorporated with ecological and systematic information into biogeographical models to predict how plants respond to climate change.

4.1 Introduction

The Paleocene–Eocene Thermal Maximum is among the warmest intervals in Earths's history (Fig. 4.1) [U0401, U0402]. This transition serves as the best analog we have for the modern temperature trends and the current global warming events that began with the Industrial Revolution and are culminating in today's greenhouse conditions. The input of carbon to the atmosphere played a central role in both of these thermal trends, and it is estimated that 2000 gigatons were generated during the PETM. The initial influx of carbon during the PETM is estimated to have taken place in less than 20,000 years, resulting in a global warming of 4–10° C (Wing and Currano 2013). This greenhouse interval is thought to have lasted 200,000 years before returning to pre-PETM conditions. In contrast, the Industrial Revolution, perhaps most poignantly illustrated in its contemporary novels of British author Charles Dickens, was fueled by the massive Paleozoic coal reserves mined in England and elsewhere. Since then, enough carbon emissions have been generated, with human influence [U0404], to raise the average global temperature by 1.5° C in less than only 300 years [U0403]. Estimates of annual CO_2 input into today's atmosphere are approximately 10 gigatons per year. In both instances, major influxes of carbon have resulted in changing Earth systems (Fig. 4.2).

To understand how the use of fossil fuels has changed global systems, and how we need to respond as a world community, it is necessary to evoke the spirits of the past, present, and future, as Charles Dickens did in *A Christmas Carol* (Dickens 1843). Approaching the issue in this manner will help us to acknowledge human-caused perturbation and address present environmental concerns. Paleobiologists have a uniquely broad perspective that allows us to look deep into the past to (1) document ecosystems millions of years old, (2) compare the past conditions with the present to help inform us about what is going on today, and (3) consider what could happen in the future and how to plan for it. Let us head back to the Paleocene (66–56 Ma) and see the impacts of a changing climate on ecosystems.

4.2 What Is the PETM?

What happens when the rate of carbon cycling drastically changes? It is no surprise that life, being carbon-based, is strongly influenced by the carbon cycle. One of the key

Electronic supplementary material A slide presentation and an explanation of each slide's content is freely available to everyone upon request via email to one of the editors: edoardo.martinetto@unito.it, ragastal@colby.edu, tschopp.e@gmail.com

*The asterisk designates terms explained in the Glossary.

M. L. DeVore (✉)
Georgia College, Milledgeville, GA, USA
e-mail: melanie.devore@gcsu.edu

K. B. Pigg
School of Life Sciences, Arizona State University,
Tempe, AZ, USA
e-mail: kpigg@asu.edu

E. Martinetto et al. (eds.), *Nature through Time*, Springer Textbooks in Earth Sciences, Geography and Environment,
https://doi.org/10.1007/978-3-030-35058-1_4

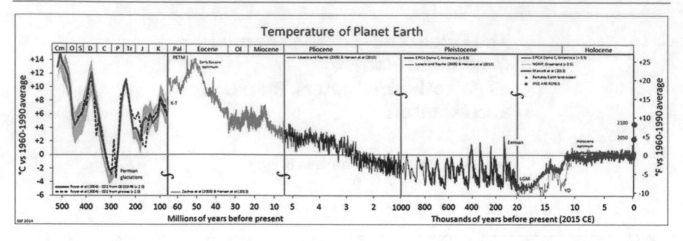

Fig. 4.1 Temperature of planet Earth over geological time. From: http://theconversation.com/sudden-global-warming-55m-years-ago-was-much-like-today-35505

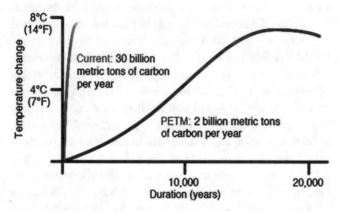

Fig. 4.2 Carbon output over time; the PETM and current conditions are compared

served in the skeletons of marine fossils is also indicative of an increase in the abundance of primary producers. This excursion from the normal range of carbon isotope values to one signaling a major shift in the fixing of carbon by autotrophic organisms is referred to as a carbon isotope excursion (CIE). The CIE that is recorded during the PETM was accompanied by a corresponding dissolution of calcium carbonate (lime), found in both the shells of invertebrates and limestone, itself, and was the result of acidification at levels sufficient to lower ocean pH drastically. At the same time, a global temperature increase of 4–10 °C has been identified by researchers using numerous climate proxies. But, what was the source of all the carbon?

There is still no consensus regarding the source or sources of carbon released into the atmosphere during the PETM. Many different authors have suggested various sources and reservoirs. These include release from (1) permafrost, (2) shallow deposits of coal and peat, (3) organic-rich marine sediment heated by volcanic intrusions, and/or (4) methyl hydrates (clathrates) from the ocean floor (World Ocean Review (WOR) 2010). More recently, volcanism associated with the North Atlantic Igneous Province has been proposed as a driver of the PETM (Gutjahr et al. 2017). Some workers propose that volcanism was more significant in its contribution of carbon than organic carbon reservoirs, although, the volcanic carbon would have been sequestered into organic carbon for Earth systems to stabilize.

The climate change of the PETM affected both marine and terrestrial ecosystems, with its greatest impacts occurring in the oceans. However, there is at least one continental region where numerous fossil floras captured the changes that occurred during this pivotal period of geologic time. That region is western North America. There are several underlying geologic reasons for the abundance of sedimentary basins in which plant remains accumulated and fossilized then. These geological processes initiated in the Cretaceous.

functions of ecosystems and global-level (Earth) systems is the cycling of nutrients, such as carbon, and the flow of energy. Every carbon atom travels through the cycle of transformation from one reservoir to another. This cycling is referred to as flux. You might imagine that one carbon atom, in one of the carbon dioxide molecules you exhaled today, was once part of a trilobite that scurried over the bottom sediments of a Cambrian sea (see Tinn et al., Chap. 14). That molecule would have moved through each of the major reservoirs (biosphere, hydrosphere, lithosphere, and atmosphere) multiple times.

Historically, the first evidence of a major change in the carbon cycle during the PETM came from 55 million-year-old sediment cores that showed an unusual carbon isotope signature (Zachos et al. 2007). A large volume of light carbon (^{12}C) was released into the atmosphere, absorbed, and incorporated into the bodies of autotrophs. Their death and incorporation into sediments and other carbon pools record this excursion. Because more ^{12}C is taken up by autotrophs relative to the heavier ^{13}C isotope, an increase in $\delta^{13}C$ pre-

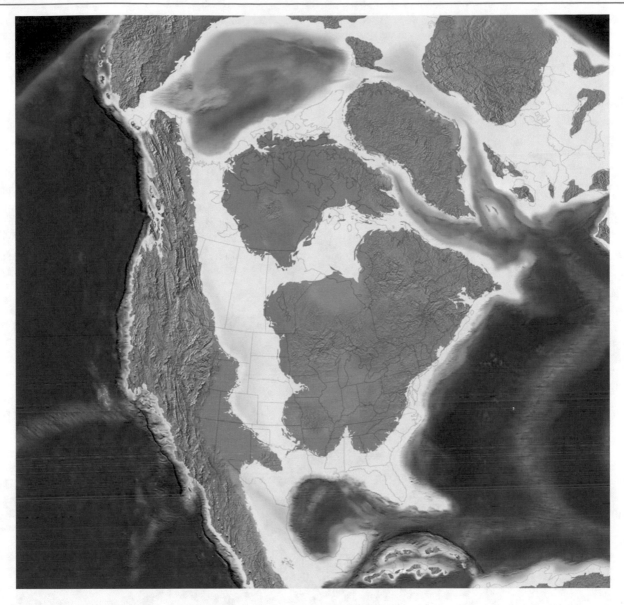

Fig. 4.3 Late Cretaceous intercontinental seaway. From https://i.redd.it/pnneetqg830z.jpg

During the Cretaceous, an intercontinental seaway, occupying a large foreland basin, dissected the middle of North America from the Gulf of Mexico to the Arctic Ocean (Fig. 4.3). From the end of the Cretaceous and throughout the Eocene, the basin was shortened physically as the seaway retreated and was subjected to deformation during the Laramide orogeny. That mountain-building event compressed the Earth's crust to form smaller basins and associated uplifted areas (Gingerich 1983; Hamilton 1988; Perry Jr et al. 1991; Flores et al. 1994; Johnson and Flores 1998). Some of the basins contain substantial fossil-plant remains from the late Paleocene through the early-middle Eocene age.

The geological processes that formed these basins and the deposition of plant materials into them occurred during the critical time period before, during, and after the PETM. Factors such as the topography, scale, and location of depositional environments (e.g., small vs large lakes; coastal versus interior upland) add to the character of the various floras. These variations allow us to visit the vegetation of a number of different areas (Box 4.1) much as we can visit lakes versus mountains, and deserts versus coastal areas, today.

Box 4.1: Early–Middle Eocene Floras of North America
Rich Paleocene and Eocene fossil leaf-compression* floras are preserved in some basins located from the Pacific Northwest to the western interior on both sides of the Paleocene–Eocene boundary. These include the

Williston Basin [U0421] of North Dakota, eastern Montana, and southern Saskatchewan; the Alberta Basin of Alberta, Canada [U0417]; the Green River, Powder River, Wind River, and Uinta and Bighorn Basins of Wyoming, Utah, and Colorado (Figs. 4.4 and 4.5) [U0416, U0426]. To the northwest in Washington State, low-lying coastal floras [U0429] extend into the Eocene where they can be contrasted with the early-middle Eocene floras of the upland Okanogan Highlands [U431] (Mustoe and Gannaway 1997; Mustoe et al. 2007; Pigg and DeVore 2010). Of the extensive Paleocene basins, the Williston Basin [U0421] is probably the most familiar because, currently, it is a major site of unconventional oil-and-gas exploration ("fracking") and development in North America.

The fossil floras of these basins have been studied both systematically and paleoecologically by a wide variety of scientists for well over a century (Brown 1962; Wolfe 1995; Pigg and DeVore 2010; Wing and Currano 2013; Manchester 2014). Early explorers included Lesquereux, Fountaine, Ward, and others in the 1860s as westward expansion across the continent began and regional mapping was done by the US Geological Survey. Studies in the mid-twentieth century (1930s–1960s) of Roland Brown, Ralph W. Chaney, and Harry D. MacGinitie provided a framework for understanding some of the most fossiliferous

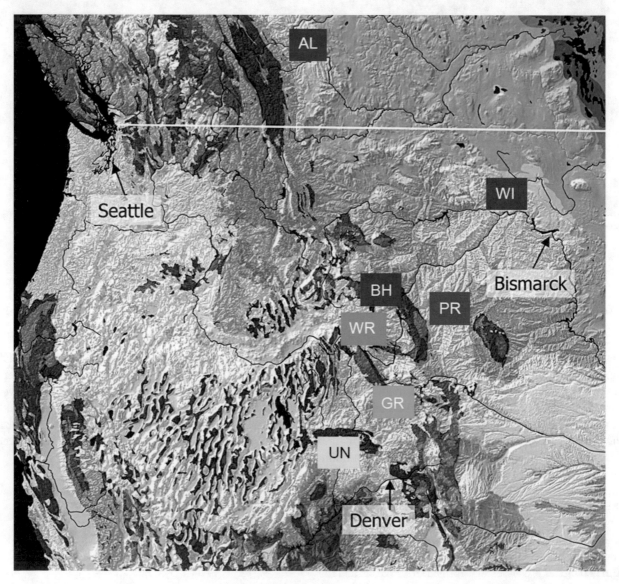

Fig. 4.4 Map of western North America showing localities of major depositional basins. Redrawn from Pigg and DeVore 2010

localities in western North America (Andrews 1980). Beginning in the 1960s to early 2000s, the prodigious works of Jack Wolfe and colleagues; Leo J. Hickey and his students Scott L. Wing, Kirk R. Johnson, David W. Taylor, and Walt Green; and other researchers refined techniques for documenting and analyzing paleoecological data [U0407], [U0409], [U0426], [U0410] (Spicer and Leopold 2006; Greenwood 2007; Wing et al. 2014). Hundreds of floras have been described, with some of the best studied occurring in several major late Paleocene–early Eocene basins (Wing et al. 1995; Pigg and DeVore 2010; Breedlovestrout et al. 2013; Wing and Currano 2013).

The western North American floras are in the right place and record the right time to capture this pivotal point and document this transition in Earth history. It is possible in some areas to document the transition in very high resolution. The best of these is the Bighorn Basin in Wyoming [U426] that has been studied extensively by Scott Wing and colleagues. We will visit these floras and show how they illustrate the PETM, below. But, first we need to consider some basic facts about plants and discuss some tools used to study them.

4.2.1   How to Use Leaves as Thermometers, Rain Gauges, and CO₂ Monitors

Now that we know where to study the fossil floras, we need the right tools to retrieve the climate information they store. Paleobotanists have devised ways to use the entire plant, as well as fossil leaves, wood, and pollen, to document and track both the climate and the biogeographical distributions of the plants during this interval. Focusing on the PETM, we will explain how plant fossils can be used as thermometers, rain gauges, and CO_2 monitors at discrete points in time. When combined, these proxies help us to understand past, present, and even future climate. Before that, we need to think about some basic truths about plants.

Why do plants and their individual structures, like leaves and wood, serve as proxies for temperatures, precipitation, and atmospheric concentrations of carbon dioxide? To address this question, we need to think about the basic role that plants play in the environment and their interrelationships with other biota. Plants and other autotrophs are primary producers, animals are consumers, and fungi and various other organisms acquire their nutrients through absorption. We, as consumers, are mainly concerned with acquiring nutrients. We do not operate like plants and are not used to thinking about the world in the same way. Plants both synthesize and metabolize their own food, as well as providing food for consumers.

Fig. 4.5 Localities studied by Roland W. Brown with addition of Canadian sites. From Manchester 2014

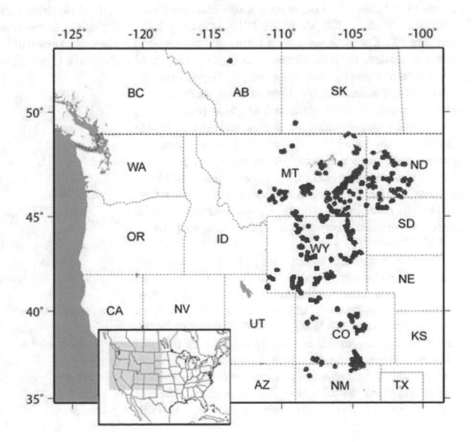

Unlike us, plants collect the basic building materials from the atmosphere (CO_2) and soil (as dissolved minerals) to build their own bodies. They harvest the energy they need to produce and metabolize nutrients by absorbing light energy. They convert this energy into a form that they, then, employ in building carbon-hydrogen bonds to create sugars. In turn, sugars are consumed or act as fundamental carbon skeletons for producing proteins, nucleic acids, and all the other molecules needed for life. Plants depend on water to transport dissolved minerals and nutrients up from the outermost root hairs in the soil to the uppermost gas-exchange structures (stomata) on their leaves. Evapotranspiration pulls the water through the plant body, and this process is driven by temperature and influenced by CO_2 concentrations in the atmosphere. Therefore, the temperature a plant is experiencing during growth, its sources of water from precipitation and soil moisture, and the amount of light present all impact the structure and function of the overall plant body and its integrated component parts, its roots, stems, and leaves. Consequently, plants reflect the history of the effects of light, water, and temperature with a surprising accuracy [U0405]. All we have to do is read the signals. It's time to become a plant whisperer.

People began reading the plants and noting how they reflect variations in temperature, precipitation, and CO_2 and nutrient availability, by noticing how annual variation in temperature and precipitation shapes the climates where plants live. Among the first to look at this issue systematically was Wladimir Köppen in the 1910s (Pidwirny 2006; see Box P2). Correlation between leaves and plant-growth forms in relation to climate became an important aspect of vegetation analysis by biogeographers, ecologists, and others interested in broader questions of plant distribution. In particular, a correlation between leaf shape (morphology) and climate was explored by Bailey and Sinnott (1916) [U0406]. They found that features, such as leaf size, leaf margin, and presence of drip tips, among others, were characteristic of particular biomes. Variations in leaf form (physiognomy) are largely influenced by rainfall and temperature, and can be predictive of climatic regime independent of the systematic group to which any plant can be assigned [U0408] (Wolfe 1979).

Over the past several decades, we realized that these same features could be recognized in fossil leaves. Paleobotanists have modified these techniques to use fossilized plant remains as thermometers, rain gauges, and CO_2 monitors (Wolfe 1990; Wilf 1997; Wilf et al. 1998; Spicer 2007, 2008; Greenwood et al. 2016). Several approaches are taken to quantify and analyze data from the plant-fossil record. These include using the entire plants as a proxy for climate [U0409] or looking more specifically at certain features of leaves [U0411], wood [U0412], and epidermal structure [U0412].

Below we consider these approaches, their value, and limitations.

4.2.1.1 Using the Entire Plant as Proxy

We need to understand not only plants but also what assumptions we make when we come to interpreting their fossil record. One of the major assumptions foundational to our study of the past is the concept of uniformitarianism. Uniformitarianism, often defined as "the present is the key to the past," was proposed by Hutton, the founder of geology, and used to decipher the regional stratigraphy near his hometown of Edinburgh (Hutton 1788). If we accept the premise that Earth processes in the past operated comparably to how they operate today, we can assume that ancient relatives of living plants occupied similar habitats and climates. This assumption, which applies most reliably to angiosperms, permits us to predict the climatic conditions of past environments. The approach, known as the nearest living relative (NLR) method [U0409], relies on the taxonomic identification of fossil and living plants (Greenwood et al. 2016) and is not without challenges (Box 4.2). The method identifies (1) a fossil plant with its closest living relative, (2) an assemblage of these fossil taxa with an extant community, and (3) the climate parameters associated with its ecological niche. A closely related approach developed by Mosbrugger and Utescher (1997) is called the "coexistence approach" (Fig. 4.6) [U0409] (Mosbrugger and Utescher 1997; Utescher et al. 2014; Grimm and Potts 2016). Both approaches rely on the ability to determine the environments where an assemblage of plants will have overlapping tolerances and, therefore, exist together [U0413] (Grimm and Potts 2016).

Box 4.2: Challenges of Nearest Living Relative (NLR) Method

One of the challenges of using NLR methods with fossil floras is that it is difficult to be certain of the level of taxonomic affiliation of the fossil plants to their modern relatives. The assumption that fossil plants would have the same range of physiological tolerances as their descendants is untested. And, in fact, this would be difficult to test in most instances. Use of NLR is further confounded by the tendency of large and diverse genera with broad geographic distributions to have developed similar morphological attributes to the different regions in which they grow. For example, we find pines (*Pinus*) growing in the Caribbean as well as covering boreal forests, and the different groups respond to their individual environments.

Plant taxonomists use evolutionary features in a lineage to classify and name plants. We know that different plants growing in a similar environment can

Fig. 4.6 The concept of the mutual climate range as used in the coexistence approach (Mosbrugger and Utescher 1997). From Grimm and Potts 2016

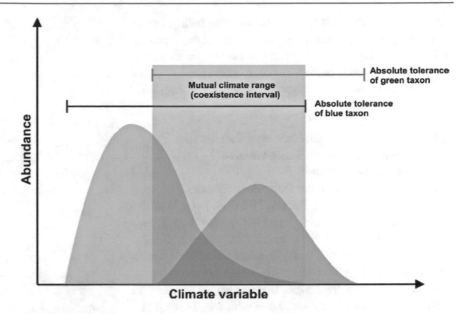

respond in similar ways, resulting in evolutionary convergence and unrelated plants having similar morphological features. NLR methods are promising because they enable paleobiologists (including palynologists; Graham 1999) to weigh in on climate analyses, but other methods can be applied to fossil plants as well and may be less problematic. These include comparative anatomy and morphology of leaves, wood anatomy, and pollen. All of these characters provide taxon-free thermometers and rain gauges for assessing paleoclimates.

4.2.1.2 Jack Wolfe and the Climate Leaf Analysis Multivariate Program (CLAMP)

The relationship identified by Bailey and Sinnott (1916) between certain features of angiosperm leaves with climate, or physiognomic characters, provides a means for using leaves as climatic indicators. Physiognomic characters are taxon-free [U0408]. That is, the same features are found in genera and species of very different angiosperm clades and lineages. These characters are the plant's developmental response to climate. Leaves demonstrate a wide range of phenotypic (morphological) plasticity that results from fine-tuning of developmental responses to climate variables. This assumption led Jack Wolfe to recognize that any well-preserved fossil leaf, even it could not be identified taxonomically, could be used to infer the climate under which that leaf grew. Wolfe acquired leaf assemblages from many different living plant communities around the world in which leaf physiognomy varied. Using these collections, he laid the groundwork for developing the first computerized program

to use leaf characters to estimate climate (Spicer and Leopold 2006; Greenwood 2007). The Climate Leaf Analysis Multivariate Program (CLAMP; Wolfe 1979) [U411] detects changes in the physiognomy of the leaves of woody dicots that correlate with changes in climate. CLAMP can be downloaded free and is currently curated by Robert Spicer and Jian Yang with the help of many colleagues (Yang et al. 2011). CLAMP requires investigators to score 31 different character states for, at least, 20 woody dicot morphotypes using a standardized template and scoresheet provided at the website.

CLAMP is an important tool for Earth scientists because it provides a means of obtaining terrestrial quantitative paleoclimatic data that can be correlated with marine-isotope temperature curves. For understanding the dynamics that operated during the PETM, CLAMP provides a quantitative proxy for temperature. The analysis of successive floras in stratigraphic context, then, could track changes in climate across the Paleocene–Eocene boundary [U0426]. The CLAMP database is not restricted to fossil floras around the PETM, though. The database includes Cretaceous floras as old as 100 million years, as well as modern sets of leaves. Hence, it is a particularly powerful and reliable tool for the younger Neogene (Wolfe 1995) and Quaternary assemblages. Yet, the program is not without constraints. Several difficulties are inherent in using the CLAMP model. First, it requires considerable time to score data points for each flora from which a temperature curve can be generated. Second, each researcher may code leaf morphotypes differently, which will influence the analytical results. And, lastly, errors in coding are easy to make, given fatigue.

Since Wolfe's work, Kirk Johnson and others have modified and codified the process of leaf morphotyping [U0410] (Spicer 2007). Despite its difficulties in application, the development of this tool led paleobotanists to build an entire scaffolding of leaf temperature data throughout the Cenozoic. It also set the stage to go back and consider which characters present in leaves may provide the crucial data needed to estimate paleoclimate and paleoprecipitation for any fossil angiosperm flora.

4.2.1.3 Leaf Margin Analysis and Calculation of Mean Annual Temperature and Precipitation

In addition to CLAMP, other techniques for calculating past climate factors have been devised [U0411]. Peter Wilf discovered, in the benchmark paper of Bailey and Sinnott (1916) [U0406] on leaf physiognomy and climate, that two morphological characters could be related to climate parameters (Wilf 1997; Wilf et al. 1998). Leaf margin (entire vs. toothed) strongly correlated with mean annual temperature (MAT), whereas leaf area correlated with mean annual precipitation (MAP). This insight led to a new approach that enabled researchers to focus on only two features, leaf margin and leaf size, for calculating confidence intervals associated with estimates of MAT or MAP. Using fewer leaf characters and obtaining better statistical confidence levels meant that this approach would be ideal for estimating climate change across the PETM in the western interior of North America. How the method works and how math is involved is explained in Box 4.3 (Temperature and Precipitation from Leaf Margin Analysis).

Box 4.3: Temperature and Precipitation from Leaf Margin Analysis

In leaf margin analysis (LMA), the mean annual temperature (MAT) is calculated using the following equation:

$$MAT(^{\circ}C) = (30.6 \times P) + 1.141 \qquad (4.1)$$

where P is the proportion of leaves present in the sample with entire margins and is expressed in the equation in decimal form. You may have noticed that this equation defines a line. The value 30.6 is the slope of the MAT vs. leaf margin regression line in Wolfe's (1979) modern dataset.

Here is an example of how the equation is applied to estimate the MAT for a set of fossil leaves with 30 different morphotypes, 10 with toothed margins and 20 with entire margins.

$$MAT(^{\circ}C) = (30.6 \times P) + 1.141$$

$$MAT(^{\circ}C) = (30.6 \times 0.6667) + 1.141$$

$$MAT(^{\circ}C) = 21.54$$

Mean annual precipitation is calculated from a set of fossil leaves by sorting morphotypes into categories based on Raukier-Webb size classification scheme. A template is available at the CLAMP website for obtaining the area values of leaves, which helps standardize area estimation.

$$MAP(mm) = 3.77(\%mesophylls* + \%macrophylls + \%megaphylls) + 47.0 \qquad (4.2)$$

A quick inspection of the three size classes used to estimate MAP indicates that the drier the climate, the higher proportion of woody dicots with small leaves in the flora. To refine the estimate of MAP more, additional data, from another source, the anatomical feature of woody dicots, would be helpful. Luckily for paleobotanists, there is a rich fossil record of anatomically preserved wood from which numerous characters also can be obtained.

4.2.2 Using Wood as a Means of Estimating Temperature and Precipitation

Like leaves, the development of wood is strongly influenced by climate, and the character of dicot woods, similar to extant plants, evolved by the Late Cretaceous [U0412]. Wood anatomists have carefully observed and reviewed wood anatomical characters in the fossil record (Wheeler and Baas 1991, 2019). Both fossil and extant woods can be studied with the online resource Inside Wood (InsideWood 2004-onwards; Wheeler 2011), also freely available. Using the characters identified by these workers takes the expertise of someone who can prepare the material and, then, carefully code and identify wood anatomical features. Once this exercise is completed, it is possible to extract the climate variables under which these trees grew in deep time. Wiemann et al. (1998, 1999) developed multivariate models to estimate and quantify (1) MAT, (2) MAP, (3) mean annual range in temperature, (4) cold month mean annual precipitation, and (5) length of the dry season. One more parameter, levels of atmospheric pCO_2 in terrestrial environments, would be invaluable for correlation with paleoclimate data obtained from leaf physiognomy and wood characters. These values could be calculated from fossil-and-extant leaf cuticles* that retain the number and distribution of stomata, a proxy known as stomatal index.

4.2.3 Stomatal Indices and CO$_2$

Proxy data using leaf stomata, the gas exchange structures normally located on the lower surface of a leaf, have been used for estimating pCO$_2$ concentration for more than 20 years [U0412] (Beerling and Kelly 1997; Doria et al. 2011). This is done using the relationship between the number and density of stomata per unit area (Poole and Kürschner 1999). Typically, fossil leaves with well-preserved cuticles or epidermal impressions are examined, and three main parameters are assessed: stomatal density, epidermal density, and stomatal index. Stomatal density (SD) is the ratio of the number of stomata per unit area, whereas epidermal density (ED) is the ratio of the number of epidermal cells per unit area. Generally, that unit area consists of a cm^2. The third parameter, stomatal index (SI), is the percentage of stomata per the total number of all cells within a given area of the leaf. The formula for SI follows:

$$SI - SD \times 100 / (SD + ED) \qquad (4.3)$$

Based on the inverse relationship between the SD and SI in relation to atmospheric CO$_2$ levels, the stomatal ratio was developed and refined to track pCO$_2$ (Chaloner and McElwain 1997; McElwain 1998). Very simply stated, when the partial pressure of CO$_2$ is high in the atmosphere, gas exchange is easier for leaves and the number of stomata per unit area is low. Conversely, when the pCO$_2$ is low, gas exchange is not as easy, and more stomata per unit area are needed for normal photosynthesis.

The use of stomatal indices to evaluate climate requires standardization to reconstruct pCO$_2$ levels, and ongoing research continues to refine our understanding of how different plant-growth habits respond (e.g., Daly and Gastaldo 2010). The value of this approach is that taxa with histories in deep time, for example, *Ginkgo biloba*, can be used to trace levels of CO$_2$ across key intervals (Beerling et al. 1998). If we assume that the relationship between SD and SI has not changed over time, complementary climate data can be derived from the same taxa used in other methodologies, such as the NLR principle, to track pCO$_2$ in deep time.

4.3 Tracking Floras Across the PETM

After the abrupt transitions of the Late Cretaceous K/Pg boundary event, the Paleocene floras changed gradually [U0414]. A study by Peppe (2010) of 44 localities in the early to middle Paleocene record in the Williston Basin, North Dakota, from 65.52 to about 58 million years ago, shows a slow increase in diversity. Peppe (2010) recognized three zones, or stratigraphic intervals, lasting for approximately one million years each, all in the Fort Union Formation. These are WB I (65.51–64 Ma), WB II (64–63 Ma), and WB III (c. 61–58.5 Ma). WB I and WB II are in the lower Ludlow Member and WB III in the overlying Tongue River Member of the formation. WB I is characterized by an average of 33 taxa, WB II by 36, and WB III by 24. Rarefaction calculations (a statistical technique to assess species richness from the results of sampling) are of similar magnitudes yielding results of 43, 42, and 28 taxa, respectively. These numbers contrast with an average of nearly 72 taxa in the Late Cretaceous, and indeed the least diverse Late Cretaceous site is more than twice the size of the most diverse early–middle Paleocene site in this formation. Similar patterns have been recognized for the Bighorn and Hanna Basins of Wyoming. One exception to this generality is the occurrence at Castle Rock, Colorado, of a tropical flora (Fig. 4.7) [U0415] (Johnson and Ellis 2002). The Castle Rock flora contains a diverse, thermophilic assemblage of over 100 taxa. Some very large leaves, many with drip tips similar to leaves borne by many modern tropical plants, are present. The authors suggest that this diversity may be the result of a microhabitat effect and not a widespread biome at the time.

The late Paleocene is well represented in western North America. Fossil-plant assemblages from this time are extensive, well known, and intensively studied. Leaf-compression floras are documented from the work of Roland Brown, Leo J Hickey, Steven R Manchester, Peter Crane, David Dilcher, and us, among others (Pigg and DeVore 2010; Zetter et al. 2011; Manchester 2014). These late Paleocene floras, with several exceptions, are generally categorized as "low diversity floras." For example, the Bison Basin flora was studied from ten adjacent quarries from the Fort Union Formation of Wyoming of mid-late Tiffanian age (59–61 Ma). This assemblage documents a mere 28 morphotypes with only 4 taxa comprising over 95% of the flora: *Metasequoia occidentalis* (Cupressaceae), *Corylites* sp. (Betulaceae), *Archeampelos acerifolia* (Cercidiphyllaceae?), and *Fortuna* cf. *marsiliodes*, a probable small aquatic herb (Gemmill and Johnson 1997). Rare reproductive structures include fruits of *Joffrea* (Cercidiphyllaceae), *Palaeocarpinus* (Betulaceae), and *Palaeocarya* (Juglandaceae). However, other sites preserve a more diverse flora.

Brown (1962) studied late Paleocene floras from over 450 localities in New Mexico, Colorado, Wyoming, Montana, North Dakota, and South Dakota [U0416]. Manchester (2014) revised and updated Brown's work, including references from the Paskapoo Formation of the Alberta Basin, Canada (Hoffman and Stockey 1999; Stockey et al. 2014), and additional floras of the Williston Basin (Golden Valley: Hickey 1977; Almont/Beicegel Creek: Crane et al. 1990 in North Dakota; Ravenscrag in Saskatchewan, McIver and Basinger 1993). Among the elements of Brown's taxa, with Manchester's revisions, are the following: bryophytes (Marchantiaceae and two mosses); lycopsids (*Selaginella*

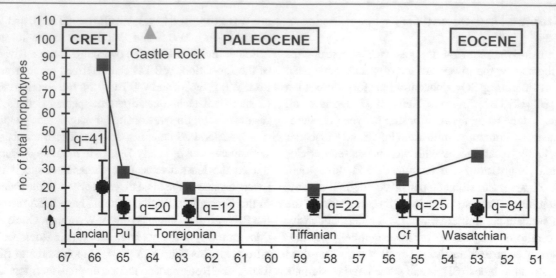

Fig. 4.7 Curve showing plant diversity of Castle Rock site (green triangle) in comparison to other Paleocene and Eocene localities. Redrawn from Johnson and Ellis 2002

Fig. 4.8 Joffre Bridge/ Munce's Hill floras. From Stockey et al. (2014)

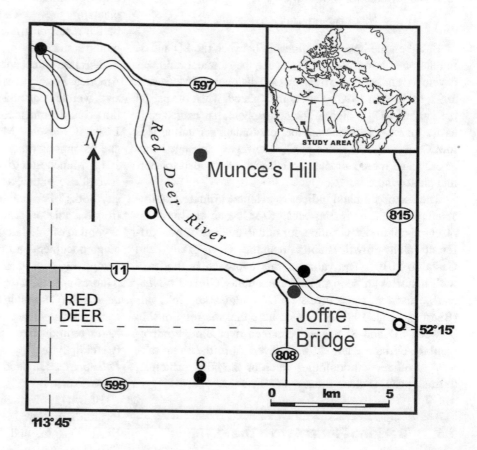

and *Isoetites*); *Equisetum*, ferns of *Woodwardia*, *Onoclea*, *Osmunda*, *Allantodiopsis*, and several additional forms; 2 cycads; *Ginkgo*; Araucariaceae?; *Amentotaxus*; Cupressaceae (5 genera); Nymphaeaceae (2, possibly 3 genera); Magnoliaceae; at least 9 Lauraceae; Chloranthaceae, a possible grass, possible Alismataceae; possibly 3 Zingiberales, Araceae (*Limnobiophyllum*); around a dozen types of palms; Sabiaceae; 3 genera in Nelumbonaceae; at least 7 types of Platanaceae; Trochodendraceae; Cannabaceae; legumes; Sapindaceae; Malvaceae; Juglandaceae; Rhamnaceae; Rutaceae?; Ulmaceae; Fagaceae; Betulaceae; Rosaceae; Vitaceae; Polygonaceae; Ericaceae; Cornaceae; Nyssaceae (3); Icacinaceae; Bignoniaceae; Elaeocarpaceae; and 22 forms of *Incertae sedis* (Brown 1962; Manchester 2014). One now can appreciate the diversity attained in landscapes of the late Paleocene.

Floras from the Joffre Bridge site near the town of Red Deer (Fig. 4.8), the nearby Munce's Hill locality, and several other sites in central Alberta, Canada [U0417], have yielded a suite of intriguing, well-preserved fossils. These assemblages provide us with details of both the plant's life history and the multiple organs needed to develop several whole-plant reconstructions. From Joffre Bridge we find aquatic ferns (*Azolla*) and monocots (*Limnobiophyllum*; Stockey et al. 1997). The floodplain plants *Joffrea speirsii* (Cercidiphyllaceae) and a sycamore have been reconstructed as whole plants based on leaves, inflorescences, fruits, pollen, and, most intriguingly, from floodplain deposits, small seedlings that demonstrate stages in early plant development [U0418] (Crane and Stockey 1985; Pigg and Stockey 1991). From nearby Munce's Hill [U0419], whole-plant reconstructions of the fern *Onoclea sensibilis* [U0420] were produced (Rothwell and Stockey 1991) as well as *Metasequoia foxii,* including details of development and additional taxa (Stockey et al. 2014).

Another quite informative flora of the late Paleocene (c. 57 Ma) is the Almont/Beicegel Creek flora of central and western North Dakota, respectively. The Almont flora occurs in Morton County, North Dakota, at the western edge of the prairie in shallow pits dug in agricultural land. In contrast, the Beicegel Creek localities crop out toward the tops of small fossiliferous buttes in the Little Badlands area of McKenzie County, western North Dakota (Fig. 4.9) [U0421]. This flora [U0422] contains 44 megafossil genera assigned to 28 families, with an additional number of unknowns, and also 12 families represented by spores and pollen (palynomorphs; Zetter et al. 2011). There are two reasons that make the Almont/Beicegel Creek flora so valued. The first is its diversity and large number of reproductive structures in comparison to a typical leaf-compression site. The second is the preservation of the plants, themselves. When the Almont site was first discovered in the late 1970s, it quickly became well known by collectors for its beautiful white *Ginkgo* leaves on a reddish brown matrix. Scientifically, it is a valuable resource because it bears a combination of well-preserved leaf surfaces along with fruits and seeds with anatomical structure (Crane et al. 1990; Pigg and DeVore 2010). With the discovery of the same flora at the Beicegel Creek site in the late 1990s, we found material that not only showed anatomical preservation in individual seeds but also the anatomy of stem, leaf, and reproductive remains. Some of these specimens could be prepared with cellulose acetate peels (Joy et al. 1956) to reveal details of leaf-and-fruit wall anatomy, including forms not seen at the original Almont site such as *Isoetes* and several filicalean ferns (Zetter et al. 2011).

By general inspection, the Almont/Beicegel flora preserves three plant associations, all occurring together at both localities (Crane et al. 1990; Pigg and DeVore 2010). The first group [U0423] contains plants of archaic* taxa typical of late Paleocene "depauperate" floras. These include *Ginkgo* and taxodiaceous conifers, dominated by *Metasequoia* (redwood), along with members of the Betulaceae (birches, alders, etc.), Trochodendraceae (two living genera in Southeast Asia), Cornales (dogwoods), and Platanaceae (plane trees). These taxa are very common at both sites. The

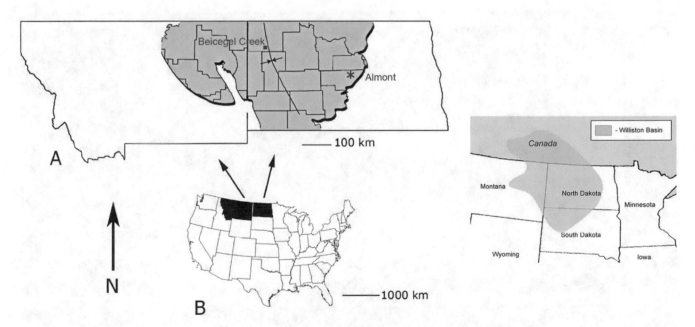

Fig. 4.9 Late Paleocene Almont and Beicegel Creek sites in the Williston Basin, North Dakota. (**a**) Continental United States showing Montana and North Dakota (in black). (**b**) Diagram of the Williston Basin in outline showing localities. (**c**) Extent of Williston Basin. (**a**) and (**b**) from Zetter et al. 2011; (**c**) from https://willistonbasinminerals.wordpress.com/williston-basin-overview/

second floristic group [U0424] includes rarer elements of tropical-subtropical forms that, although uncommon in the late Paleocene, will become very important elements in the warmer PETM and early to middle Eocene. Sabiaceae, Icacinaceae, Menispermaceae (moonseed family), Polygalaceae (milkwort family), and Zingiberales (banana/ginger families) occur here. Lastly, the third category [U0425] comprises early occurrences of plants that are common components of today's eastern deciduous temperate flora of North America. These plants include *Acer* (maple), *Aesculus* (buckeye), and members of the Ranunculaceae (buttercups), Juglandaceae (walnut family), Hamamelidaceae (witch-hazel), and Actinidiaceae (gooseberry family). The most recent review of this flora can be found in Zetter et al. (2011), with updates on the report of several new genera since that publication (*Ginkgo cranei*, *Paleoochna tiffneyi*; Zhou et al. 2012, Ickert-Bond et al. 2015). As a consequence of the continuing work on the Almont/Beicegel Creek flora, we now have a more complete picture of the affinities of these late Paleocene taxa. With this understanding of the late Paleocene, we now need to move to the best place to study floras of the PETM.

4.3.1 Floras that Traverse the Paleocene– Eocene Boundary

The most intensely, precisely, and accurately studied sites in western North America that straddle the Paleocene–Eocene boundary, and provide the highest-resolution data, are those in the Bighorn Basin (Figs. 4.10 and 4.11) [U0426] (Wing et al. 1995; Wing and Currano 2013). The floral database used to assess these assemblages has been developed by Scott Wing and associates over a period of several decades. Of all of the basins formed during the Laramide orogeny, the Bighorn Basin of northern Wyoming preserves the most complete and best-studied stratigraphic record of the PETM. That record begins in the late Paleocene and extends through the PETM into the early Eocene. This succession gives us the most detailed view of life before, during, and after the PETM. Thick sequences of sandstone, mudstone, lignite, and freshwater carbonate were deposited in the Bighorn Basin from the Paleocene through the early Eocene (Wing and Currano 2013). The basal unit, comprising the Fort Union Formation, is characterized by approximately 3000 m of Paleocene clastics* and lignites. It is overlain by the Willwood Formation (late Paleocene to early Eocene) that can be recognized by a distinctive set of oxidized mudstone (Paleocene to early Eocene). The Bighorn Basin assemblages were preserved in river channels, associated floodplains, and swamps (Wing 1984).

Conditions in the Bighorn Basin were wet and warm, during the late Paleocene, with MAT exceeding 15° C and MAP over 100 cm. The temperate rose quickly by 5° C during the PETM and then returned to previous late Paleocene conditions after approximately a 200,000-year excursion. Plant assemblages from more than 200 sites in the basin have been sampled and dicot leaves determined to a grade of morphotype, with only around 30% identifiable to any specific taxon (Wing and Currano 2013). Nonmetric multidimensional

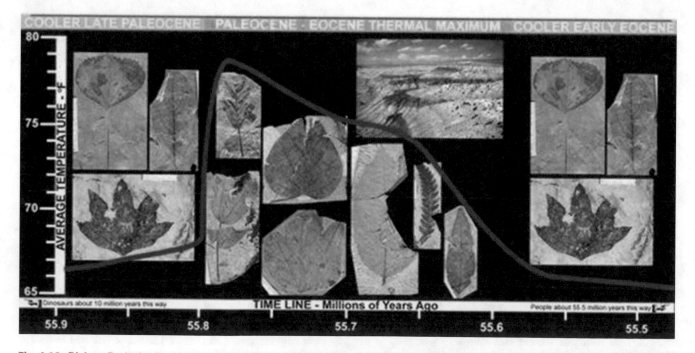

Fig. 4.10 Bighorn Basin timeline showing temperature and floristic changes from pre-PETM late Paleocene through PETM and post-PETM early Eocene. From https://naturalhistory2.si.edu/ete/ETE_People_Wing_ResearchThemes_Wyoming.html (Wing)

Fig. 4.11 Bighorn Basin
floras over time by
depositional environment.
From Wing and Currano 2013

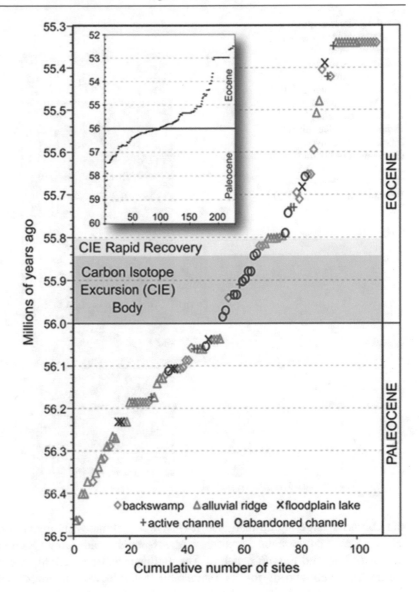

scaling (NMDS; an ordination technique used to identify gradients based on a distance or dissimilarity matrices) methods identified five types of leaf assemblages. The first are forms known only in the late Paleocene, whereas a second group consists of those plants only (or mostly) found during the PETM. The third plant group contains forms that occur throughout the late Paleocene, PETM, and into early Eocene, and the fourth group consists of only those plants occurring after the PETM. The last category identified forms present before and after, but not during the PETM. A general pattern emerges from the analysis.

Taxodiaceous conifers (*Metasequoia*, *Glyptostrobus*), *Corylites* (Betulaceae), *Averhoites*, *Phoebe*, and the fern *Allantoidiopsis* are found in the wetland and backswamp environments. *Ginkgo*, *Cercidiphyllum*, *Zizyphoides*, *Macginitiea* and other sycamores, and the monocot leaf *Zingiberopsis* are found on the floodplain in what is termed the alluvial ridge assemblages. The cornaleans (dogwoods)

disappear after the PETM and *Alnus* (alder) appears in the early Eocene. The major change that occurred during the PETM is the appearance of warm-tolerant taxa, especially legumes. Some of these plants persist in the Bighorn Basin only through the PETM. Some of the late Paleocene forms reappear in the overlying lower Eocene deposits, along with the introduction of new taxa. Our understanding of the landscape dynamics is not restricted, though, to the Bighorn Basin. Additional floras witnessing the PETM transition occur in the Pacific Northwest in coastal floras of the Chuckanut Formation (Fig. 4.12; Mustoe and Gannaway 1997; Mustoe et al. 2007; Breedlovestrout et al. 2013).

The Chuckanut Formation contains the lower Bellingham Bay Member, recently dated at around 57 million years ago (Breedlovestrout et al. 2013). It is overlain by the Governors Point and Slide Members, the latter of which is dated at 53.6 million years ago. Fossil plants are preserved in each unit. These three lower floras are characterized as subtropical

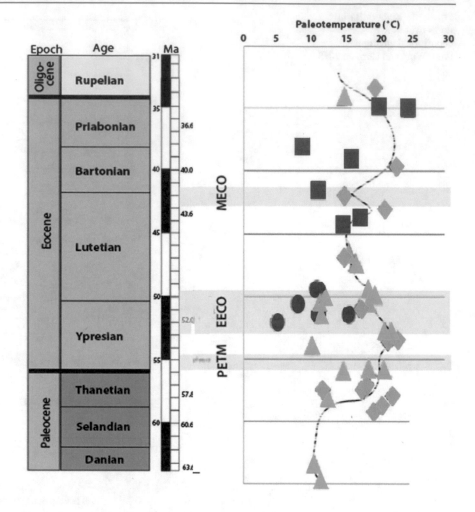

Fig. 4.12 Stratigraphic placement of localities before, during, and after the PETM. Mean annual temperature (MAT) of each locality abbreviation at left; at right MAT is summarized by locality; green triangles: Pacific Coast; blue squares: Inland West; yellow diamonds: Cordillarian Basins; red circles: Okanogan Highlands. Redrawn from Breedlovestrout et al. (2013)

and comprise large, entire-margined leaves, warm climate ferns, the taxodiaceous conifers *Metasequoia* and *Glyptostrobus,* and *Sabalites* palms. The Bellingham Bay Member is well known for its remarkable palm leaves (Fig. 4.13) [U0429]. These lower assemblages in the Chuckanut are contemporaneous with Bighorn Basin PETM floras. Plant assemblages change upsection. Above the Slide Member is the Padden Member, dated at about 49 to 48 million years ago. These rocks lack palms and the taxodiaceous conifers and, instead, preserve smaller, tooth-margined leaves and pine needles. Using the nearest living relative approach, these plants represent growth in a more temperate climate. Another thinner younger unit, the Maple Falls, contains rare palm leaves and appears to be a more microthermal assemblage. The Padden Member is contemporaneous with the early–middle Eocene upland floras of the Okanogan Highlands including the diverse Republic flora of northeastern Washington and related sites in British Columbia (Johnson 1996).

The Okanogan Highlands floras contain some of the earliest members of families that are important temperate genera today. These include a rich conifer assemblage, Betulaceae, Ulmaceae (elms), Rosaceae, Trochodendraceae,

Hamamelidaceae, and the temperate Sapindaceae *Acer* and *Aesculus*, to name a few [U0431-U0433] (Greenwood et al. 2016). It is particularly interesting to compare the upland, microthermal Okanogan Highlands floras, like Republic, with the more common megathermal floras of the early–middle Eocene [U0430]. The emergence of a temperate flora is clearly of systematic interest, because these assemblages record the first occurrence and earliest radiation of important temperate families like the Rosaceae [U0434]. In addition to systematics, however, these floras also document the distinctive morphological fingerprints of several processes characteristic of recent temperate forests (see Martinetto et al., Chap. 2). Presumably, these processes operated in a similar manner as today and were instrumental in the transition plants faced in entering a temperate, seasonal climate. We find evidence in the Republic flora of dormancy in the discovery of plants with distinctive leaf scar and terminal bud scale scars [U0435]. Dimorphic leaves, occurring on short-shoot and long-shoot systems, are another indication of seasonality that occurs in some plants that produce large, round leaves that overwinter and are present to photosynthesize upon bud break [U0436]. An additional process we can infer from morphology are leaves

Fig. 4.13 George Mustoe with palm frond from the Bellingham Bay Member, Chuckanut Formation, south of Bellingham, Washington. Courtesy of George Mustoe

of Rosaceae and Anacardiaceae with the intermediate leaf morphologies and disrupted venation patterns found in hybrids today [U0437]. Lastly, relatives of the modern onion show evidence of both sexual and asexual reproduction by fertile flowers and vegetatively produced bulbils on the same inflorescences [U0438]. All of these processes point to strategies important in the establishment of temperate biomes in the Northern Hemisphere.

4.4   Energy Flow and Its Impact on Herbivores

We have shown how techniques measuring the morphological features of leaves, such as CLAMP and LMA, and the anatomical structure of wood can provide temperature and meteorological data about past climates. These measurements provide the setting for which we can examine and understand how plants responded to changing climate

regimes. These techniques also allow us to understand how plants diversified along various evolutionary pathways. These pathways have resulted in their current biogeographic distribution and enduring presence in our world. We have shown how these types of data may also be of value in predicting how they will survive in the future. Now it is time to think about the broader consequences of how plants affect the rest of the biosphere.

One level of the biosphere that we often define by the processes operating therein is that of the ecosystem. These processes include, in particular, (1) energy flow and (2) nutrient cycling. So far, we have discussed elevated rates of increasing atmospheric carbon during the PETM and illustrated how stomatal indices can be used to track changes in atmospheric levels of CO_2. This is because stomata are the gateways of carbon dioxide movement into plants for photosynthesis. Photosynthesis is an essential component of both energy flow and nutrient cycling. By fixing carbon into the biosphere, this process is the ultimate source of nutrients in the carbon cycle.

By transforming light energy into carbon-hydrogen bond energy, photosynthesis is also the driver of energy flow through the metabolism of plants and the herbivores that consume them. As plants responded to a greenhouse world, it makes us question how the primary consumers of plants, predominately insects and mammals, changed as well.

Many fossil leaves show the effects of interactions with other biota. Such interactions include wood, leaf, and reproductive damage by insects that leave distinctive damage patterns. Entomologists use insect-mouth parts to classify insects, and different insect groups have evolved various patterns of reproductive (e.g., egg deposits, galls) and feeding behaviors. Hence, it is possible to identify which groups of insects and their young were feasting on foliage and using these resources for reproductive purposes. Conrad Labandeira is credited with much of the work documenting the relationship of damage patterns on plants throughout the geologic column in relation to their insect predators. Codification of these patterns and understanding of how they changed through time, in relation to major environmental events, have been further advanced through work by his colleagues and students (e.g., Peter Wilf, Kirk Johnson, Ellen Currano).

Studies of the diversity and types of insect-damage patterns occurring on leaves throughout the Bighorn Basin were correlated with leaf physiognomy (Currano et al. 2008). An association was found between the abundance and number of types of damage on angiosperm leaves with rising and falling temperatures. Focusing exclusively on PETM floras, the leaves of every species of plant were found to be extremely damaged and colonized by specialized insects. This increase in plant diversity and associated increases in productivity provided an ideal set of conditions for the diversity of thermophilic insects. These new insect groups expanded into new niches affiliated with feeding on angiosperm leaves [U0428] (Currano et al. 2008). This begs the question of how these circumstances influenced other herbivores, in particular, herbivorous mammals.

Herbivorous mammals in the Bighorn Basin became dwarfs during the PETM [U0427] (Gingerich 2003). It is known from the time of Bergmann (1847) that there is a correlation of body size with temperature among mammals. Larger bodied species inhabit cold regions in contrast to smaller bodied ones in warm climates. This phenomenon is known as Bergmann's rule. However, by itself, this correlation between body size and climate is not sufficient to infer how mammals responded to increasing temperatures during this time. Both Gingerich (2003) and subsequent investigators (D'Ambrosia et al. 2017) recognized a flaw in assuming that this phenomenon explained what they observed in mammalian faunas during past greenhouse climates. Bergmann's rule appears to apply to situations where there is warming in higher-latitude regions. Potentially, as was the case with insects, migration of animals into a newly created climate zone may be responsible for the pattern recorded in the fossil record. The movement into higher latitudes by lower latitude taxa could also explain the observed dwarfing of mammals in greenhouse faunas. Observations sometimes require, and operate in the realm of, multiple working hypotheses. This appears to be the case with whether or not dwarfism was a response to increasing temperatures during the PETM. Indeed, it is possible to suggest that body size decreases may also be a genuine evolutionary response to higher temperatures. A smaller body size during the PETM also may be associated with shorter reproductive cycles associated with climate change (e.g., drought, rainy seasons). It is beyond the scope of the current chapter to examine all of the potential drivers of mammalian dwarfism during the PETM. Details of current debates can be found in D'Ambrosia et al. (2017).

4.5 Connecting the Past, Present, and Future

So far in our virtual visit to the western interior of the United States, we have seen how some paleobotanists have tracked floras across the Paleocene–Eocene boundary and used them as recorders of terrestrial climate change before, during, and after the thermal event. Other paleobotanists have focused on the systematics and evolution of these plants and considered patterns of diversity in response to such major shifts in Earth's climate state. Together, both groups are aware of, and document, the position of plant lineages in space and time. Anyone interested in understanding the development of modern plant distributions is dependent on these data for constraining the history of any plant lineage. Hence, biogeographers are very interested in using these fossil records to interpret the climatic conditions under which ancestral lineages evolved and grew, the environmental parameters they needed to reproduce, and the conditions that allowed plants to maintain viable populations (Sanmartín 2012; Stigall 2012; Meseguer et al. 2013). Currently, there is a new kind of paleobotanical synthesis and that entails modeling (Fig. 4.14) [U0413].

Models are essential for providing a framework for understanding and integrating a network of factors influencing observations we make about nature. As new data are added, or as new scrutiny verifies or intimates processes that are not as we first thought, the model can be adjusted and advanced to serve as a representation of complex natural processes. Species distribution models (SDMs) are based on our current knowledge of the niches of modern taxa. These models are developed to project future vegetational distributions. The most significant underpinning of this modeling approach originates from Hutchinson (1957) who proposed the concept of fundamental and realized niches. There is a difference between these two paradigms.

The fundamental niche (Hutchinson 1957) is simply a set of biological and physical resources needed for a species to

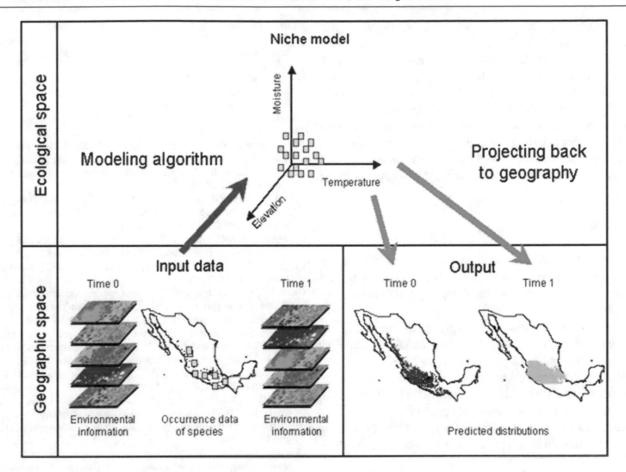

Fig. 4.14 Diagrammatic summary of the ecological niche modeling process. From Martinez-Meyer 2005

persist indefinitely. One can understand the fundamental niche of any organism by thinking about how it responds to the environment (Wiens et al. 2009). Is it possible for that species to shift from one subset of conditions, currently shaping its biogeographic distribution and role in a community (a realized niche), to another (realized) niche under new climate change conditions? In essence, it is possible for a species in its fundamental niche to evolve, or adapt, its lifestyle and life strategy to a new set of conditions, a realized niche. It is important to recognize that one or more of the following responses can happen when a species is perturbed by climate change. A species can shift geographic distribution, moving to another physical area in which the same, or similar, physical conditions exist. Alternatively, a species may adapt to the new climate reality and retain its biogeographic distribution. Also, the geographical distribution of a species may contract to a refugium, if available, a smaller area in which that organism's requirements for growth are found. If none of these responses happen, the least favorable response is that a species goes extinct. Hence, ecological niche models (ENM) permit the development of SDMs and predict how a species will respond to different climate scenarios. These approaches require expertise in bioinformatics

and software development (Fig. 4.11) [U0426]. Meseguer et al. (2013) and Nürk et al. (2015) provide examples that show how fossils can be incorporated in ENM.

4.6   Conclusions

The PETM is one of the most significant Earth historical events in recent geologic history and provides us with rare clues to how Earth systems reliant on angiosperms will respond to elevated levels of CO_2. In this chapter we have looked at how we can use plants to track and detect major changes in past climates [U0439, U0440]. We have looked at how the plants, themselves, and aspects of their communities have changed over the Paleocene–Eocene boundary in response to climate. And, finally, we mention how biogeographers, ecologists, and evolutionary biologists depend on fossil data to construct models to predict how plants changed their past distributions and how they may alter their biogeographies in the Anthropocene.

In the deep past, the change to a greenhouse world was much slower than what we are currently experiencing. Under a prolonged and protracted rise in global climate, plants could find ways to adapt, expand their biogeographic ranges, and function when they had time to evolve. Today, the rate of change in levels of CO_2 is considered to be rapid, and the associated rise in global temperature may be too quick for some plants to either adapt or evolve and survive. What will forests look at the end of the Anthropocene? The only thing we can say with certainly is those plants, with the greatest potential to evolve quickly, or those that have deep-seated genetic plasticity based on their past responses, will be the ones that will dominate the Earth's vegetation in the future.

Questions

1. What relationship between leaf shape and climate did Bailey and Sinnott recognize in their classic study, and how can it be applied to fossil angiosperm floras in deep time?
2. What are the challenges in applying the nearest living relative (NLR) approach to assess MAT and MAP to Cretaceous and Paleogene fossil floras?
3. How does leaf margin analysis (LMA) and CLAMP overcome some of the problems inherent in the NLR approach?
4. Why is wood anatomy a reliable paleoclimate proxy?
5. How does the concentration of atmospheric CO_2 affect the distribution of stomates* on the surface of a developing leaf?
6. What are the differences between stomatal density (SD), epidermal density (ED), and stomatal index (SI), and how can these be used to assess pCO_2 in the deep past?
7. What general trends in plant diversity are recorded in the fossil floras of the Williston and Big Horn Basins across the PETM?
8. How did increasing temperatures during the PETM affected both invertebrate and vertebrate communities in western North America?
9. Why and how can studies of the PETM and its associated changes in fauna and flora inform us about the present?

Acknowledgments The authors thank the editors of this volume for inviting our participation. We are also grateful to Deborah Freile, David R Greenwood, Steven R Manchester, George Mustoe, Selena Y Smith, Bruce H Tiffney, and Scott L Wing for sharing their ideas and contributions to our development of this chapter and to Robert A Gastaldo, Patricia G Gensel, and Jean-Phillipe Solves for editorial assistance [U0441].

References

Andrews HN (1980) The fossil hunters. In search of ancient plants. Cornell University Press, London

Bailey IW, Sinnott EW (1916) The climatic distribution of certain types of angiosperm leaves. Am J Bot 3:24–39

Beerling DJ, Kelly CK (1997) Stomatal density responses of temperate woodland plants over the past seven decades of CO_2 increase: a comparison of Salisbury (1927) with contemporary data. Am J Bot 84(11):1572–1583

Beerling DJ, McElwain JC, Osborne CP (1998) Stomatal responses of the 'living fossil' *Ginkgo biloba* L. to changes in atmospheric CO_2 concentrations. J Exp Bot 49(326):1603–1607

Bergmann C (1847) Über die Verhältnisse der Wärmeökonomie der Thiere zu ihrer Grösse. Göttiger Studien, Göttingen, p 3

Breedlovestrout RL, Evraets BJ, Parrish JT (2013) New Paleogene paleoclimate analysis of western Washington using physiognomic characteristics from fossil leaves. Palaeogeogr Palaeoclimatol Palaeoecol 392:22–40

Brown RW (1962) Paleocene floras of the Rocky Mountains and great Plains. United States Geological Survey Professional Paper 375. Department of the Interior, Washington, DC

Chaloner WG, McElwain J (1997) The fossil plant record and global climatic change. Rev Palaeobot Palynol 95(1–4):73–82

CLAMP ONLINE http://clamp.ibcas.ac.cn/CLAMP_Home.html. Accessed 1 Feb 2019

Crane PR, Stockey RA (1985) Growth and reproductive biology of *Joffrea speirsii* gen. et sp. nov., a *Cercidiphyllum*-like plant from the Late Paleocene of Alberta, Canada. Can J Bot 63:340–364

Crane PR, Manchester SR, Dilcher DL (1990) A preliminary survey of fossil leaves and well-preserved reproductive structures from the Sentinel Butte Formation (Paleocene) near Almont, North Dakota, Fieldiana, geology, new series no. 20 Publication 1418. Field Museum of Natural History, Chicago, pp 1–63

Currano ED, Wilf P, Wing SL, Labandeira CC, Lovelock EC, Royer DL (2008) Sharply increased insect herbivory during the Paleocene-Eocene thermal maximum. Proc. Natl. Acad. Sci U S A 105:1960–1964

Daly RG, Gastaldo RA (2010) The effect of orientation to growing-season sunlight on stomatal parameters of *Quercus rubra* around the Belgrade Lakes, Central Maine. PALAIOS 25:339–346

D'Ambrosia AR, Clyde WC, Fricke HC, Gingerich PD, Abels HA (2017) Repetitive mammalian dwarfing during ancient greenhouse warming events. Sci Adv 3:1–9

Dickens C (1843) A Christmas carol. Chapman and Hall, London

Doria G, Royer DL, Wolfe AP, Fox A, Westgate JA, Beerling DJ (2011) Declining atmospheric CO_2 during the late Middle Eocene climate transition. Am J Sci 311(1):63–75

Flores RM, Roberts SB, Perry WJ Jr (1994) Paleocene paleogeography of the Wind River, Bighorn, and Powder River basins, Wyoming. In: Flores RM, Mehring KT, Jones RW, Beck TL (eds) Organics and the rockies field guide. Public information circular 33. Wyoming Geological State Survey, Laramie, pp 1–16

Gemmill CEC, Johnson KR (1997) Paleoecology of a late Paleocene (Tiffanian) megaflora form the northern Great Divide Basin, Wyoming. PALAIOS 12(5):439–448

Gingerich PD (1983) Paleocene-Eocene faunal zones and a preliminary analysis of Laramide structural deformation in the Clark's Fork Basin, Wyoming. In: Wyoming Geological Association Guidebook of the 34th Annual Field Conference. Wyoming Geological Association, Casper, pp 185–195

Gingerich PD (2003). Mammalian responses to climate change at the Paleocene-Eocene boundary: Polecat Bench record in the northern Bighorn Basin, Wyoming. Special Papers-Geological Society of America, pp 463–478

Graham A (1999) Late Cretaceous and Cenozoic history of North American vegetation: north of Mexico. Oxford University Press, Oxford

Greenwood DR (2007) North American Eocene leaves and climates: from Wolfe and Dilcher to Burnham and Wilf. CFS Courier Forschungsinstitut Senckenberg 258:95–108

Greenwood DR, Pigg KB, Basinger JF, DeVore ML (2016) A review of paleobotanical studies of the early Eocene Okanagan (Okanogan)

Highlands floras of British Columbia, Canada and Washington, USA. Can J Earth Sci 53:548–564

Grimm GW, Potts AJ (2016) Fallacies and fantasies: the theoretical underpinnings of the coexistence approach for palaeoclimate reconstruction. Clim Past 12:611–622

Gutjahr M, Ridgwell A, Sexton PF, Anagnostou E, Pearson PN, Pälike H, Norris RD, Thomas E, Foster GL (2017) Very large release of mostly volcanic carbon during the Paleocene-Eocene Thermal Maximum. Nature 548(7669):583–577

Hamilton WB (1988) Laramide crustal shortening. Geological Society of America Memoir 171:27–38

Hickey LJ (1977) Stratigraphy and paleobotany of the Golden Valley Formation (Early Tertiary) of western North Dakota. Geological Society of America Memoir 150:1–183

Hoffman GL, Stockey RA (1999) Geological setting and paleobotany of the Joffre Bridge Roadcut fossil locality (Late Paleocene), Red Deer Valley, Alberta. Can J Earth Sci 36(12):2073–2084

Hutchinson MF (1957) Concluding remarks, Cold Spring Harbor Symposia on Quantitative Biology. In: Population studies: animal ecology and demography. Biological Laboratory, Cold Spring Harbor, pp 415–427

Hutton J (1788) Theory of the Earth: an Investigation of the Laws observable in the Composition, Dissolution, and Restoration of Land upon the Globe. In: Transactions of the Royal Society of Edinburgh, vol. I, Part II. Royal Society of Edinburgh, Edinburgh, pp 209–304. plates I and II

Ickert-Bond SM, Pigg KB, DeVore ML (2015) *Paleoochna tiffneyi* gen. et sp. nov. (Ochnaceae) from the late Paleocene Almont/Beicegel Creek flora, North Dakota, USA. Int J Plant Sci 176:892–900

InsideWood (2004-onwards) Published on the Internet. http://insidewood.lib.ncsu.edu/search. Accessed 1 Jan 2019

Johnson KR (1996) The role of the Republic flora in documenting floristic evolution of the Northern Hemisphere. Wash Geol 24.41–42

Johnson KR, Ellis B (2002) A tropical rainforest in Colorado 1.4 million years after the Cretaceous-Tertiary boundary. Science 28(296, 5577):2379–2383. https://doi.org/10.1126/science.107102

Johnson RC, Flores RM (1998) Developmental geology of coalbed methane from shallow to deep in Rocky Mountain basins and in Cook Inlet–Matanuska Basin, Alaska, U.S.A. and Canada. Int J Coal Geol 35(1–4):241–282. https://doi.org/10.1016/S0166-5162(97)00016-5

Joy KM, Willis AJ, Lacey WS (1956) A rapid cellulose peel technique in palaeobotany. Ann Bot 20:635–637

Manchester SR (2014) Revisions to Roland Brown's North American Paleocene flora. Acta Mus Nat Pragae, Ser. B Hist Nat 70(3–4):153–210. Praha. ISSN 1804–6479

Martinez-Meyer E (2005) Climate change and biodiversity: some considerations in forecasting shifts in species' potential distributions. Biodivers Inform 2, 42–55

McElwain JC (1998) Do fossil plants signal palaeoatmospheric carbon dioxide concentration in the geological past? Philos Trans R Soc Lond B 353:83–96

McIver EE, Basinger JF (1993) Flora of the Ravenscrag formation (Paleocene), southwestern Saskatchewan, Canada. Palaeontogr Can 10:1–167

Meseguer AS, Aldasoro JJ, Sanmartín I (2013) Bayesian inference of phylogeny, morphology and range evolution reveals a complex evolutionary history in St. John's wort (*Hypericum*). Mol Phylogenet Evol 67:379–403

Mosbrugger V, Utescher T (1997) The coexistence approach--a method for quantitative reconstructions of Tertiary terrestrial paleoclimate data using plant fossils. Palaeogeogr Palaeoclimatol 134(1–4):61–86

Mustoe GE, Gannaway WL (1997) Paleogeography and paleontology of the early Tertiary Chuckanut Formation, northwest Washington. Wash Geol 25:3–18

Mustoe GE, Dillhoff RM, Dillhoff TA (2007) Geology and paleontology of the early Tertiary Chuckanut Formation. In: Stelling P, Tucker DS (eds) Floods, faults, and fire: Geological field trips in Washington state and southwest British Columbia, vol 9. Geological Society of America, Boulder, pp 121–135

Nürk NM, Uribe-Convers S, Gehrke B, Tank DC, Blattner FR (2015) Oligocene niche shift, Miocene diversification – cold tolerance and accelerated speciation rates in the St. John's Worts (Hypericum, Hypericaceae). BMC Evol Biol 15:80. https://doi.org/10.1186/s12862-015-0359-4

Peppe DJ (2010) Megafloral change in the early and middle Paleocene in the Williston Basin, North Dakota, USA. Palaeogeogr Palaeoclimatol Palaeoecol 298(3–4):224–234

Perry WJ Jr, Weaver JN, Flores RM, Roberts SB, Nichols DJ (1991) Sequential Laramide deformation in Montana and Wyoming. In: Geological Society of America abstracts with Programs, vol 23(4). Geological Survey, Washington, DC, p 56

Pidwirny M (2006) Climate classification and climatic regions of the world. In: Fundamentals of physical geography, 2nd edn. Brown Publishers, Dubuque. Accessed 11 Feb 2019

Pigg KB, DeVore ML (2010) Floristic composition and transitions of late Paleocene to early Eocene floras in North America. In: Collinson ME, Pigg KB, DeVore ML (eds) Northern Hemisphere Paleogene floras and global change events. Bulletin of Geosciences, vol 85, pp 135–154

Pigg KB, Stockey RA (1991) Platanaceous plants from the Paleocene of Alberta, Canada. Rev Palaeobot Palynol 70:125–146

Poole I, Kürschner WM (1999) Stomatal density and index: the practice. In: Jones TP, Rowe NP (eds) Fossil plants and spores: modern techniques. Geological Society, London, pp 257–260

Rothwell GW, Stockey RA (1991) *Onoclea sensibilis* in the Paleocene of North America, a dramatic example of structural and ecological stasis. Rev Palaeobot Palynol 70(1–2):113–124

Sanmartín I (2012) Historical biogeography: evolution in time and space. Evol Educ Outreach 5:555–568

Spicer RA (2007) Recent and future developments of CLAMP: building on the legacy of Jack A. Wolfe. Courier Forschunginstitut Senckenberg 258:109–118

Spicer RA (2008) CLAMP. In: Gornitz V (ed) Encyclopedia of paleoclimatology and ancient environments. Springer, Dordrecht, pp 156–158

Spicer B, Leopold E (2006) Memorial to Jack Albert Wolfe (1936-2005). Geol Soc Am Mem 35:59–61

Stigall AL (2012) Using ecological niche modelling to evaluate niche stability in deep time. J Biogeogr 39:772–781

Stockey RA, Hoffman GL, Rothwell GW (1997) The fossil monocot *Limnobiophyllum scutatum*: resolving the phylogeny of Lemnaceae. Am J Bot 84:355–368

Stockey RA, Hoffman GL, Vavrek MJ (2014) Paleobotany and paleoecoloogy of the Munce's Hill fossil locality near Red Deer, Alberta, Canada. In: Stevens WD, Montiel OM, Raven PH (eds) Paleobotany and biogeography. A Festscrift for Alan Graham in his 80th year. Missouri Botanical Garden, St. Louis, pp 367–388

Utescher T, Bruch AA, Erdei B, François L, Ivanov D, Jacques FMB (2014) The coexistence approach--theoretical background and practical considerations of using plant fossils for climate quantification. Palaeogeogr Paleoclimatol Palaeoecol 410:58–73

Wheeler EA (2011) Inside wood--a web resource for hardwood anatomy. IAWA J 32(2):199–211

Wheeler EA, Baas P (1991) A survey of the fossil record for dicotyledonous wood and its significance for evolutionary and ecological wood anatomy. IAWA J 12:272–332

Wheeler EA, Baas P (2019) Wood evolution: Baileyan trends and Functional traits in the fossil record. IAWA J 2019:2–42

Wiemann MC, Wheeler EA, Manchester SR (1998) Dicotyledonous wood anatomical characters as predictors of climate. Palaeogeogr Palaeoclimatol Palaeoecol 139(1998):83–100

Wiemann MC, Manchester SR, Wheeler EA (1999) Paleotemperature estimation from dicotyledonous wood anatomical characters. PALAIOS 14:459–474

Wiens JA, Stralberg D, Jongsomjit D, Howell CA, Snyder MA (2009) Niches, models, and climate change: assessing the assumptions and uncertainties. Proc Natl Acad Sci 106(Supplement 2):19729–19736

Wilf P (1997) When are leaves good thermometers? A new case for leaf margin analysis. Paleobiology 23(3):373–390

Wilf P, Wing SL, Greenwood DR, Greenwood CL (1998) Using fossil leaves as paleoprecipitation indicators: an Eocene example. Geology 26(3):203–206

Wing SL (1984) Relation of paleovegetation to geometry and cyclicity of some fluvial carbonaceous deposits. J Sediment Petrol 54:52–66

Wing SL, Currano ED (2013) Plant response to a global greenhouse event 56 million years ago. Am J Bot 100(7):1234–1254

Wing SL, Alroy J, Hickey LJ (1995) Plant and mammal diversity in the Paleocene to early Eocene of the Bighorn Basin. Palaeogeogr Palaeoclimatol Palaeoecol 115(1–4):117–155. https://doi.org/10.1016/0031-0182(94)00109-L

Wing SL, Johnson KR, Peppe DJ, Green WA, Taylor DW (2014) The multi-stranded career of Leo J. Hickey. Bull Peabody Mus Nat Hist 55(2):69–78

Wolfe JA (1979) Temperature parameters of humid to mesic forests of eastern Asia and relation to forests of other regions of the Northern Hemisphere and Australasia. U S Geol Surv Prof Pap 1106:1–37

Wolfe JA (1990) Palaeobotanical evidence for a marked temperature increase following the Cretaceous/Tertiary boundary. Nature 343:153–156

Wolfe JA (1995) Paleoclimatic estimates from Tertiary leaf assemblages. Annu Rev Earth Planet Sci 23:119–142

World Ocean Review (WOR) (2010) 1. Living with the oceans. A report on the state of the world's oceans. https://worldoceanreview.com/en/wor-1/ocean-chemistry/climate-change-and-methane-hydrates. Assessed 15 Mar 2019

Yang J, Spicer RA, Spicer TEV, Li CS (2011) 'CLAMP Online': a new web-based palaeoclimate tool and its application to the terrestrial Paleocene and Neogene of North America. Palaeobiodivers Paleoenviron 91(3):163. https://doi.org/10.1007/s12549-011-0056-2. Accessed 15 Mar 2019

Zachos JC, Bohaty SM, John CM, McCarren H, Kelly DC, Nielsen T (2007) The Palaeocene-Eocene carbon isotope excursion: constraints from individual shell planktonic foraminifer records. Philos Trans R Soc 365:1829–1842

Zetter R, Farabee MJ, Pigg KB, Manchester SR, DeVore ML, Nowak MD (2011) Palynoflora of the late Paleocene silicified shale at Almont, North Dakota, USA. Palynology 35:1–33. https://doi.org/10.1080/01916122.2010.50164

Zhou Z, Quan C, Liu Y-S (2012) Tertiary *Ginkgo* ovulate organs with associated leaves from North Dakota, U.S.A., and their evolutionary significance. Int J Plant Sci 173(1):67–80. https://doi.org/10.1086/662651

When and Why Nature Gained Angiosperms

5

Jiří Kvaček, Clement Coiffard, Maria Gandolfo,
Alexei B. Herman, Julien Legrand, Mário Miguel Mendes,
Harufumi Nishida, Sun Ge, and Hongshan Wang

Abstract

Flowering plants, the angiosperms, are the most diverse group of plants on our planet. Today, they dominate most vegetation types, but their origin continues to remain a mystery. However, we continue to gain knowledge about their early evolution and history. It seems increasingly probable that their origin is associated with climatic and environmental changes in tropical areas and was coeval with the breakup of the supercontinent Gondwana. The first angiosperms appeared in the fossil record about 135 million years ago based on the occurrence of their rare pollen grains in fossil assemblages of North Gondwana and southwest Europe. Their evolution may be associated with climate perturbation and an overall change in wetland to mesophytic habitats, as this group is adapted to tolerate a seasonally dry climate. Soon after the first early angiosperms in the late Valanginian, higher angiosperms, the eudicots, are part of the fossil record of Africa. These initial flowering plants had small inconspicuous flowers and small fruits, and were most probably of small growth stature, likely herbs and shrubs. After angiosperms colonized mineral soils across the landscape, they expanded their habitats to aquatic environments and evolved strategies for their rapid dispersal in these settings. By the mid-Cretaceous (90–100 Ma), angiosperms conquered higher latitudes in both hemispheres and expanded into various tropical to warm temperate (= subtropical) environments. Chloranthoids, laurels, and plane trees experienced their heyday. In the Late Cretaceous, core* and higher eudicots evolved rapidly, and nearly all extant angiosperm families appeared by the end of the Cretaceous. Angiosperm clades developed a physiology capable of overcoming drought conditions by the Cenomanian. However, their expansion and colonization of mesophytic upland habitats only took place in the Late Cretaceous. Seasonally dry habitats, such as savannas, were inhabited by angiosperms in the Late Cretaceous, with the first evidence of graminoids.

Electronic supplementary material A slide presentation and an explanation of each slide's content is freely available to everyone upon request via email to one of the editors: edoardo.martinetto@unito.it, ragastal@colby.edu, tschopp.e@gmail.com

*The asterisk designates terms explained in the Glossary.

J. Kvaček (✉)
National Museum Prague, Prague, Czech Republic
e-mail: jiri.kvacek@nm.cz

C. Coiffard
Museum für Naturkunde, Berlin, Germany
e-mail: clcmcnt.coiffard@berlin.de

M. Gandolfo
LH Bailey Hortorium, Plant Biology Section, School of Integrative Plant Science, Cornel University, Ithaca, NY, USA
e-mail: mag4@cornell.edu

A. B. Herman
Geological Institute, Russian Academy of Sciences, Moscow, Russia
e-mail: alexeiherman@gmail.com

J. Legrand · H. Nishida
Department of Biological Sciences, Faculty of Science and Engineering, Chuo University, Tokyo, Japan
e-mail: legrand@bio.chuo-u.ac.jp; helecho@bio.chuo-u.ac.jp

M. M. Mendes
MARE - Marine and Environmental Sciences Centre, Faculty of Science and Technology, University of Coimbra, Coimbra, Portugal
e-mail: mmmendes@mail.telepac.pt

S. Ge
College of Paleontology, Shenyang Normal University, Shenyang, China
e-mail: sunge0817@163.com

H. Wang
Florida Museum of Natural History, Gainesville, FL, USA
e-mail: hwang@flmnh.ufl.edu

5.1 Introduction

The attractive colors, beautiful patterns, pleasing arrangements, and pleasant fragrances of flowers are all products of co-evolution. These features stand out and demand our attention when admiring flowering plants, either in a natural setting or in our home, because they have evolved to attract animal pollinators. Not all flowering plants require insect pollination. Several angiosperm lineages have followed different paths to manage their reproductive strategy, using either wind or water as the medium to carry pollen. Pollination is essential to transfer genetic material from one individual to another, facilitating cross-pollination, which is important in the evolutionary selection of inherited characters.

Today, angiosperms are the most common plants in the world, with nearly 300,000 species known. They occupy all environments from dry (semiarid) deserts, savannas, and cool temperate, subtropical, and tropical rainforests to coastal mangroves; their success has allowed them to adapt to even growing on the sea floor. Flowers, the most showy parts of their bodies, have been inspirations of beauty in human civilization for millennia. They are invaluable resources, providing us food and building materials. Even products such as Coca Cola® have their origin in flowering plants. However, many questions concerning their appearance and evolution remain. From where did the flowering plants originate? When did they appear for the first time? And, why are they so successful? We will attempt to answer these questions in an abbreviated manner in this chapter and visit several Cretaceous localities from which we have learned the answers.

5.1.1 Which Plant Groups May Be the Ancestors of Flowering Plants?

Over the past 100 years, or so, several hypotheses have emerged about which group or groups of plants might be the ancestors of angiosperms. The most common hypothesis focuses on ancestral stock(s) in one of three gymnosperm groups: Caytoniales, Bennettitales, or Erdmanithecales. It is increasingly probable that key evolutionary characters of angiosperms did not evolve together or simultaneously, but were part of a longer phylogenetic mosaic process that began in the gymnosperms.

The Mesozoic pteridosperms (seed-bearing plants with similarities in growth architecture to some ferns; see Gee et al., Chap. 6) are considered a good ancestral candidate for angiosperms (Fig. 5.1). *Caytonia*, belonging to an extinct order of seed plants known from the Late Triassic to latest Cretaceous, is the plant most frequently mentioned as ancestral

(Herendeen et al. 2017). These plants produced woody stems and branches with short shoots on which cupulate (enclosed) ovules developed a pollen-receiving apical tube (micropyle). Pteridosperm pollen typically possess bladders (sacci), which are characteristics of wind pollination. One of the major differences between *Caytonia* and angiosperms is in the access of pollen to a seed micropyle∗. In *Caytonia* [U0504], pollen is free to fall into the micropyle because its "proto-carpel∗" is not closed (Fig. 5.1). In contrast, angiosperm carpels∗ are closed, pollen always lands on a stigma, a specialized tissue of a carpel, and germination of the pollen grain produces a pollen tube that grows through the tissue to the egg (Herendeen et al. 2017). Other gymnosperm groups, besides the Caytoniales, are implicated as being ancestral to angiosperms.

The Erdtmanithecales, Bennettitales, and Gnetales are good candidates as angiosperm ancestors. The Erdtmanithecales are Cretaceous in age and only known from their reproductive structures and seeds associated with *Eucommiidites* pollen. The Bennettitales, a group with a cycad-like growth habit, developed a "flower-like" reproductive structure, and early workers thought this innovation was angiosperm-like. Gnetales, a small relict clade found living in dry environments, today, possess both homologous∗ (flowers, siphonogamy, lignin chemistry) and convergent (vessels, dicot-like leaves, an enveloped seed) characters with angiosperms (Doyle and Donoghue 1985). After some debate and recent phylogenetic analyses, these groups now are considered close relatives of angiosperms. The Gnetales, Bennettitales, and Erdtmanithecales are placed into a group called the anthophytes. Lately, a group of chlamydospermous∗ seeds related to these three taxa was described from the Early Cretaceous of Portugal and the USA [U0505] (Friis et al. 2009, 2013). Some of the seeds are quite complex structures, such as the seed envelope of *Tomcatia* described from the Barremian–Aptian of Portugal, that exhibits tepal-like projections. However, neither the pollen nor the vegetative parts of the plant are known.

It is widely accepted in the scientific community that more than 99% of extant angiosperm species can be assigned into one of three major angiosperm groups: (eu)magnoliids, eudicots, and monocots [U0501]. The (eu)magnoliids are a group of early diverging angiosperms distinct from the chlorantoids and the **ANA** group, which have only few extant species, and include **A**mborellales (today only found in New Caledonia), **A**ustrobaileyales (including star anise plant), Ceratophyllales (aquatic "hornworts"), Chloranthales (some of which are used in traditional medicines of Southeast Asia), and **N**ymphaeales ("water lilies"; APG 2016) (Fig. 5.2; Herendeen et al. 2017). Plants of these small groups are of particular interest because they represent relics of the early angiosperms.

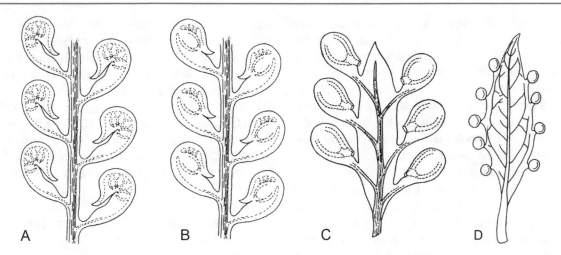

Fig. 5.1 Schematic model showing the derivation of an angiosperm carpel from a gymnosperm structure with ovule-bearing cupules (e.g., *Caytonia*). Each cupule contains several seeds (**a**). A hypothetical intermediate form shows only one ovule per cupule ripening to a seed (**b**). The following situation is that of an enlarged ovule-bearing axis (**c**), and in the last stage the planted axis (resembling a leaf) corresponds to the open angiosperm carpel (**d**). Modified from Friis et al. (2011)

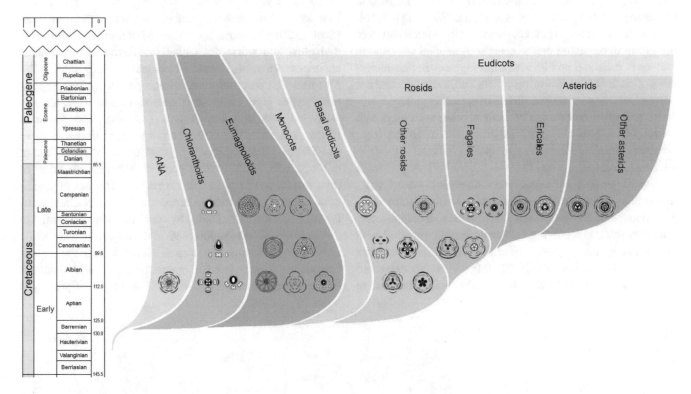

Fig. 5.2 Temporal ranges of angiosperm lineages, modified from Herendeen et al. (2017)

5.1.2 What Are the Key Characters of Flowering Plants?

Angiospermy, itself, is the visible character of flowering plants. Angiosperms develop their seeds in an entirely closed carpel, a plant tissue that envelops the ovule. After fertilization, a maturing seed forms. In contrast, partial closure of a tissue protecting the seed occurs among most gymnosperm groups. Gymnosperm seeds are partly closed either in cupules* (Caytoniales, ginkgos) or cones (Erdmanithecales, conifers) or embedded among leaf bases (Bennettitales). Thus, angiosperms are unique in their complete closure of a seed in a carpel and later in a mature fruit.

Showy flowers are the most conspicuous, and the most variable character, of flowering plants. Flowers of the most basal extant angiosperm *Amborella* are quite small, and this feature is common among flowers of Early Cretaceous age. As a consequence of co-evolution with insects during the Cretaceous, angiosperms developed conspicuous and flamboyant flowers along with odiferous volatiles as a means to attract insect pollinators. To date, it would appear that such volatiles (resinous oils) evolved first in early angiosperms, which is one reason why a number of the early divergent groups, such as *Illicium* (star anise), *Piper* (black and white peppercorn), and *Laurus* (bay leaf), are used as spices. Yet, there is a very low potential for the preservation of flowers and, even less, for their biochemical aromatics in the fossil record.

There are several characteristics that may be used to identify fossil angiosperms, but not all of these are definitive (Fig. 5.3). One key character is the architecture of their pollen grains. Angiosperm pollen is unique in its form, composition, and construction. The wall of the pollen grain consists of two layers: an internal intine and the external exine, which also is subdivided into inner and outer layers [U0503]. The (inner) endexine* lacks any evidence of layering (lamination), whereas the (outer) ektexine* consists of three laminae: the tectum, columellae, and foot layer. The construction of the angiosperm pollen wall is unique and used in palynology* as the most important character for the identification of the group in dispersed pollen spectra. A second angiosperm feature is their leaf architecture, although several characters are not exclusive in flowering plants and, therefore, are not unique. Leaves with several orders of typically pinnate*, palmate, reticulate, and parallel venation, often associated with angiosperms, also are found in gymnosperm groups (e.g., *Gnetum*). Another common angiosperm character is the conducting cell called a vessel in its wood. Vessels are present in nearly all groups, but some early diverging groups (e.g., *Amborella*, *Tetracentron*) have wood built only of tracheids*, a gymnosperm feature. Double fertilization, a complex fertilization process involving two male gametes joining with the female gametophyte*, on the other hand, is a character that is not detectable in fossils. Finding this feature preserved in a Mesozoic or Paleozoic fossil plant might help solve

some of the unknown. However, fossil plants do provide relevant information that can answer intriguing questions about the origin and diversification of angiosperms. We will now start our field trip back in time to relevant Cretaceous fossil-angiosperm localities around the world to try answering three fundamental questions about the evolution of the group: When did angiosperms first appear? Where did angiosperms appear for the first time? And, lastly, what was the most probable environment colonized by the first angiosperms? Let us start our field trip in Central Europe [U0506].

5.2 Angiosperms in the Cretaceous of Europe

Jiří Kvaček and Mário Miguel Mendes

The Cretaceous of Europe had a very different geography of its southern coast than what we recognize, currently. The Pyrenees, Alps, and Carpathian mountains did not exist (Scotese 2014), and Europe consisted of numerous islands in a shallow and warm sea called the Tethys. These islands experienced a strong, seasonal climate that promoted wildfires in which plentiful charcoal was produced. Two vegetational types grew on these islands similar to what we see, today, in the Madeira, Açores (Azores), and Canárias (Canary) Islands. Vegetation on the western slopes of these Cretaceous islands probably grew under a more humid microclimate, while eastern parts of islands were drier, or even semidesert in their climate regime. This set of conditions is found in the Iberian Peninsula. The Early Cretaceous floras of Spain are dominated by conifers, mainly assigned to the families Cheirolepidiaceae and Araucariaceae, while angiosperm remains are scarce. Nevertheless, the Lower Cretaceous strata from the Lusitanian Basin (western Portugal) have yielded rich floras including well-preserved

Fig. 5.3 Typical characters of angiosperms. (**a**) Schematic section of angiospermous ovule inside ovary. (**b**) Pollen wall. (**c**) Three main types of angiosperm leaves. Modified from Jud et al. (2015)

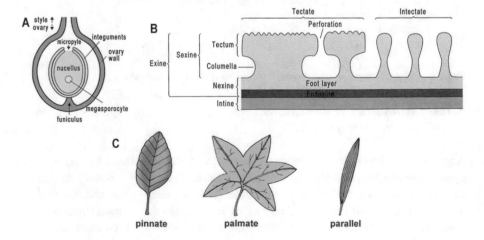

angiosperm seeds, fruits, flowers, and dispersed stamens (Fig. 5.4) (Box 5.1). These Portuguese floras are pivotal for understanding vegetational changes during the earliest phases of angiosperm radiation and diversification (e.g., Friis et al. 1999, 2006a, 2015a; Mendes et al. 2011; Mendes et al. 2014a, b; Mendes and Friis 2018). The same floras also include well-preserved and informative non-angiosperm plant groups, including a great diversity of conifer remains (e.g., Friis and Pedersen 2011; Mendes et al. 2011, 2014a, 2018; Mendes and Friis 2018). It is important to note that the Iberian Peninsula was nearly in the same geographic position as it is today, very close to Africa. It is very likely that the Iberian archipelago witnessed the first angiosperms to colonize Europe. These early forms appear in the Valanginian of Portugal, near Porto da Calada, sometime between 139 and 133 million years ago (Trincão 1990). Their presence is recorded by monosulcate and reticulate pollen grains (Fig. 5.5). More reliable angiosperm-pollen records [U0502] are better preserved and known from Hauterivian strata in England (Hughes 1994), sometime between 133 and 129.5 million years ago. Once established, angiosperms began to extend their biogeographic range (Box 5.2).

Box 5.1: When Did Angiosperms First Appear?

Fossils of flowering plants appeared about 135 million years ago in the late Valanginian as rare, single inaperturate and monocolpate pollen grains (Gübeli et al. 1984; Trevisan 1988; Brenner 1996). These grains make up a very small percentage of any pollen spectrum at that time. Younger rocks yield fossil evidence that flowering plants rose to dominance in different Cretaceous ecosystems across the planet at different times. For example, floras preserved in Hauterivian to Barremian (130 Ma) rocks are dominated by gymnosperms and not by flowering plants. All evidence, to date, indicates that angiosperms occupied patchy distributions in these forests. It is not until the Aptian that the greening of an angiosperm-dominated planet begins.

It took only a short time for the group to establish itself as an important component of the European vegetation. In the Barremian, 129.5 to 125 million years ago, a diversified angiosperm flora was well established in the peninsula. A diverse mesofossil flora was described by Friis et al. (2010b) from the late Barremian–early Aptian of Torres Vedras (Portugal), in the western Lusitanian Basin. This flora is very rich in angiosperm remains, dominated by non-eudicots, such as *Canrightia*, *Canrightiopsis* (Chloranthaceae; Fig. 5.5a, b), and *Mayoa* (Araceae) (Friis et al. 2006a). In

Box 5.2: Where Did Angiosperms First Appear?

As far as we can determine from the fossil record, flowering plants appeared for the first time in tropical North Gondwana. But, we cannot exclude an earlier occurrence of the group in southwestern Europe. There are three reports of late Valanginian pollen from Northern Gondwana. Rare angiosperm pollen grains are described from the Valanginian-Hauterivian of Morocco (Gübeli et al. 1984) and Israel (Brenner 1996) and from the Tuscan Nappe of Italy (Trevisan 1988), a region that originated near to the North Gondwana coast (e.g., Brogi and Giorgetti 2010). It is possible, though, that the Iberian Peninsula was the first land mass to witness the arrival of angiosperms in Europe. This is despite the fact that Valanginian pollen grains from Portugal (Trincão 1990), an array of islands close to North Gondwana at that time, are not unequivocally angiosperm in origin. If debate continues about the Portuguese fossils and these are determined not to be of angiosperm origin, the first unequivocal European occurrence of the group is in the Hauterivian of England (Hughes 1994). Since their first occurrence in the late Valanginian, angiosperms began to spread through latitudinal gradients to both poles (Fig. 5.6). By the Barremian, higher angiosperms (eudicots producing tricolpate pollen) appeared in Africa (Doyle et al. 1977; Doyle 1992) and spread across latitudes in the same manner as the basal angiosperms (Coiro et al. 2019).

addition, two different kinds of eudicot pollen grains have been recognized in the Torres Vedras flora. These plants are interpreted to have occupied the understory as shrubs in coniferous woodlands. Early Cretaceous woodlands of the Iberian archipelago were dominated by members of the cheirolepidiaceous (an extinct family with similar features to modern Cupressaceae), cupressaceous, and podocarpaceous conifers (Mendes et al. 2018).

Fossil occurrences demonstrate that angiosperms were also already well established in aquatic and mesophytic Early Cretaceous environments (Box 5.3). An important angiosperm, which may represent an early aquatic form, is reported from the early Barremian of northern Spain. The herbaceous *Montsechia* probably grew in the shallow water fringes of a freshwater lake (Gomez et al. 2015). Another presumed aquatic angiosperm is *Bevhalstia*, which grew in the Hauterivian to Barremian wetlands of England (Hill 1996). Once established, the diversification of the group quickly followed.

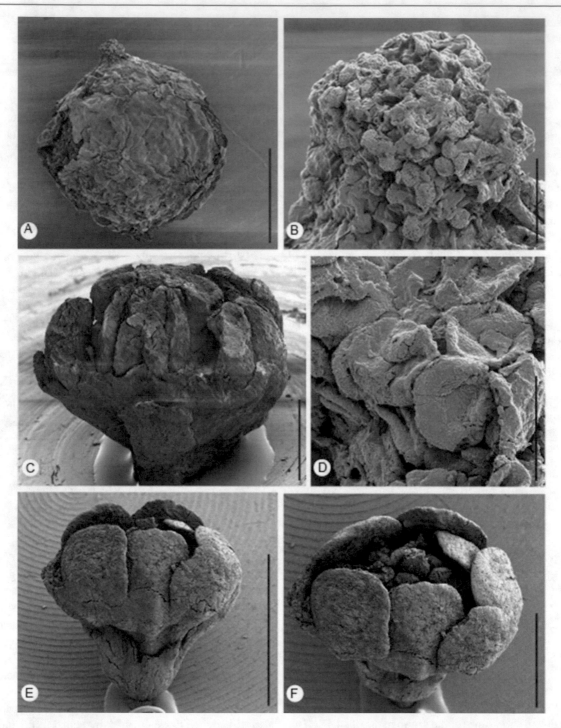

Fig. 5.4 SEM micrographs of selected angiosperm mesofossils from the Cretaceous of Portugal. (**a**) Fruit of *Canrightiopsis dinisii* from the Early Cretaceous of Juncal showing two or three stamen scars and a slightly raised, rounded stigmatic area. (**b**) Detail of fruit showing stigmatic area covered by amorphous substance with embedded pollen grains. (**c**) Lateral view of flower of *Kajanthus lusitanicus* from the Early Cretaceous of Juncal showing one narrow perianth part of the outer perianth whorls followed by the tightly adpressed parts of the inner perianth whorls enclosing the stamens and carpels. (**d**) Detail of *Kajanthus* showing a group of in situ pollen grains with circular shape, triaperturate organization, a perforate-punctate pollen wall, and verrucate aperture membrane. (**e** and **f**) Flower of *Saportanthus parvus* from the Early Cretaceous of Catefica showing rounded tepals surrounding several stamens. Scale bars = 1 mm (**e**), 500 μm (**a, f**), 250 μm (**c**), 50 μm (**b**), 20 μm (**d**)

Fig. 5.5 SEM micrograph of *Clavatipollenites* pollen grains from the Early Cretaceous of Catefica site, Lusitanian Basin, western Portugal. Scale bar 20 μm

Aptian floras, 125 to 113 million years ago, are rich in angiosperms with 140–150 taxa reported (Friis et al. 1999, 2010b) from Portugal. In addition to aquatic plants, such as the first representatives of water lilies (*Monetianthus*; Friis et al. 2001), angiosperm growth habits may have been small shrubs and trees. These growth strategies are interpreted for a variety of eudicots including *Teixeiraea* (von Balthazar et al. 2005), *Kajanthus* (Mendes et al. 2014b) (Fig. 5.5c, d), and buxalean flowers *Lusistemon/Lusicarpus* and *Valecarpus* (Pedersen et al. 2007). Angiosperms adapted and spread into other environmental and ecological niches, and their dispersion may have been favored by their small reproductive propagules, fruits, and seeds (Friis et al. 2010a). The exact growth strategies of these early forms, other than presumed herbs and small shrubs, are difficult to pinpoint because of a paucity of any angiosperm wood in Early Cretaceous strata. But, it wasn't very long before angiosperms appeared everywhere.

By the mid-Cretaceous, angiosperms had colonized probably all environments across Europe (Coiffard et al. 2006). In the Albian (113–100 Ma) and Cenomanian (100–94 Ma), flowering plants expanded their ecologies into river floodplain settings of various river architectures (Fig. 5.7). Vegetation grew on poorly developed (immature) soils adjacent to high-energy braided rivers, and these floras were dominated by ancestors of extant laurels similar to those occurring in North America at the same time. This vegetation consisted largely of only two lauralean angiosperms, *Mauldinia* and *Pragoladus*, growing with a similar habit as extant trees able to survive and regenerate in high-energy environments (Fig. 5.8). The inflorescences of *Mauldinia* were cone-like and, most likely, smelly, but possessed inconspicuous flowers. These were composed of bilobed partial inflorescences (lateral units) arranged in a helix along an elongated central axis (Eklund and Kvaček 1998). Terminal bracts of the partial inflorescences showed perforations interpreted to have contained volatile substances as a means to attract insects (Eklund

and Kvaček 1998). Floodplain soils in the areas between river channels were colonized by the ancestors of extant plane trees (*Platananthus*) with their characteristic globose inflorescences. Different angiosperms occupied riparian and floodplains of meandering river systems.

Meandering river systems provide a different suite of nutrient-rich soils than braided regimes. The greater availability of nutrients hosted a very diversified vegetation along the river banks (riparian elements) and across the floodplain, including numerous ferns, cycads, and conifers. But, here, the dominant vegetation type was the flowering plants. Large-leaved plane trees (*Ettingshausenia*, *Platananthus*) [U0508] and entire-margined lauroids (*Myrtoidea*, *Pandemophyllum*) were accompanied in these forests by the first angiosperm lianas (vines; e.g., *Hederago*). A number of tri-, penta-, and septa-foliate leaves (*Debeya*, *Araliaephyllum*; Fig. 5.9) of unknown systematic affinity, perhaps belonging to the Chloranthaceae, are also common. In these floodplains, with their prevailing seasonal climate in which dry periods were frequent, vegetation was not immune from wildfire.

Rivers surrounded by mesophytic vegetation were subject to regular fires as evidenced by the number of fossil assemblages in which charcoal is preserved. Periods of rain followed by periods of drought affected the vegetation. The herbaceous understory of flowering plants and ferns was affected particularly by burning. Here, flowering plants are characterized by leaves with reduced laminae*, and several forms display a cuticle having stomata surrounded by numerous papillae* (bumps). These plants, including the chloranthoid *Zlatkocarpus* (Fig. 5.10; Kvaček and Friis 2010), infructescences assigned to *Cathiaria* with its leaves referred to *Liriodendropsis*, and a number of unknown species (e.g., *Pecinovia*; Kvaček and Eklund 2003), have been interpreted as "upland" vegetation growing beside seed ferns (*Sagenopteris*) and bennettitaleans (*Zamites*). At the same time, several angiosperms evolved adaptations to salt tolerance and invaded the coastal zones.

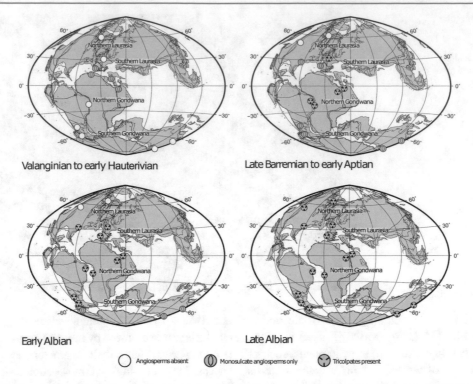

Valanginian to early Hauterivian

Late Barremian to early Aptian

Early Albian

Late Albian

○ Angiosperms absent   ◐ Monosulcate angiosperms only   ✲ Tricolpates present

Fig. 5.6 Distribution of angiosperms in the Lower Cretaceous based on pollen spectra. Modified from Coiro et al. (2019)

Box 5.3: What Is the Most Probable Environment of the First Angiosperms?

Angiosperms are thought to have evolved in a number of different continental environments. The earliest hypotheses believed the group originated as rainforest members of the magnoliids (Takhtajan 1969; Cronquist 1988). Other workers suggested that the earliest forms were fast-growing, drought-tolerant shrubs in open, subtropical, disturbed habitats (Stebbins 1974; Axelrod 1972). In contrast, another idea identified the basal angiosperms as the earliest herbaceous, pioneering weeds that tolerated disturbance (Taylor and Hickey 1992). Many workers argued that this hypothesis was consistent with the small size of angiosperm flowers and seeds and the rarity of angiosperm wood in the Early Cretaceous fossil record (Taylor and Hickey 1992; Friis et al. 1999, 2000). A similar hypothesis has been proposed by Feild et al. (2004) and is based on studies of ecological requirements of extant basal angiosperms. They proposed that wet, dark, and disturbed habitats were the entry point for the first angiosperms to establish the group across the landscape (Feild et al. 2004). However, Doyle (2012) argued that

Northern Gondwana, with its diversified Barremian pollen, was generally quite dry.

We must consider an additional aspect of the group, and that is its seed dormancy (Friis et al. 2015b). The ability of a minute seed, with its internal embryo, to remain dormant allowed early angiosperms to to wait for germination and seedling establishment when conditions were favorable. Recently, an additional hypothesis has been proposed by Lee et al. (2015) who interpret early angiosperms as plants of open habitats as well as the understory. They suggest that the earliest angiosperms possessed low transpiration rates and, therefore, were more suitable for growth beneath the canopy of larger trees. However, those angiosperms possessing small leaves, thereby mitigating their low transpiration cooling capacity, could also grow in open habitats. This latest hypothesis seems to be most inspiring, and we only can conclude that early angiosperms probably tolerated not only wet and dark but also seasonally dry, disturbed mesophytic upland environments. The rapid diversification and opportunistic* life strategy of flowering plants may have allowed the group to colonize many different habitats at the same time in the deep geologic past.

Fig. 5.7 The fossil site Pecínov Quarry, Czech Republic, exposes fresh and brackish water Cenomanian sediments of the Peruc-Korycany Formation, Cenomanian, Late Cretaceous

Salt-tolerant plants formed special communities in coastal areas and often occurred in monodominant stands similar to extant mangroves (Falcon-Lang et al. 2006). The dominant vegetation in this environment is characterized by the conifer *Frenelopsis* with its ovuliferous cones, assigned to *Alvinia* (Kvaček 2000), and the ginkgoalean plant *Nehvizdyella* with leaves named *Eretmophyllum* (Kvaček et al. 2005). The first angiosperm halophytes also are found here. The halophytic herb *Pseudoasterophyllites* (Fig. 5.11) [U0507] had needle-like leaves and very simple flowers comparable to flowers of *Ceratophyllum*. This plant has been shown to be an important evolutionary link between *Ceratophyllum* and the Chloranthaceae (Kvaček et al. 2016). Over time, angiosperms replaced gymnosperms in these, and other, settings.

Late Cretaceous floras of Western and Central Europe are dominated by angiosperms, and the proportion of gymnosperms in these assemblages was reduced to only one-fourth of the vegetation typically present in swampy and coastal estuarine∗ habitats. Angiosperms characterize all alluvial plain and continental interior (higher elevation) slope mesophytic vegetation. Alluvial plains were overgrown by lauroids and platanoids. Swamp vegetation was dominated by cypresses, with the first appearance of palms and pandans in

the Late Cretaceous (Kvaček and Herman 2004). Mesophytic vegetation, interpreted as having grown on slopes, was characterized by ancestors of the fagaleans (oaks and relatives) and juglandoids (hickories, walnuts, and relatives). Pollen from these groups is common and their association across parts of the planet at this time is recognized as the Normapolles complex (Friis et al. 2006b). In addition to producing a similar type of Normapolles pollen, this group possesses flowers composed of an ovary enclosed in the receptacle, with the stamens and other floral parts situated above (epigynous) and arranged in two or more symmetrically placed branches (dichasia). Many of those plants, such as *Budvaricarpus* (Heřmanová et al. 2011), *Caryanthus* (Friis 1983; Heřmanová et al. 2016), and *Zlivifructus* (Heřmanová et al. 2017), were probably part of the mesophytic vegetation. A higher diversity of flowering plants is preserved on islands of Late Cretaceous Europe. These plants, similar to those found on today's Mediterranean islands, were shrubs and small trees of the Ericales (heathers) (e.g., *Actinocalyx*, Friis 1985; *Paradinandra*, Schönenberger and Friis 2001; *Alericarpus* Knobloch and Mai 1986) and were common in the Late Cretaceous. The first eudicots in the lamiid group (the one including today's

Fig. 5.8 Reconstruction of braided river alluvial plain vegetation, *Mauldinia bohemica*, Cenomanian, Peruc Flora, by Jan Sovák

mint, basil, and rosemary) also appeared (*Scandianthus* Friis and Skarby 1982). Hence, the ancestors of nearly all extant angiosperm families were present in Europe by the Late Cretaceous. What about other regions of the Northern Hemisphere at higher paleolatitudes?

5.3 Cretaceous Flora of North Polar Regions

Alexei B Herman

Although we have evidence that angiosperms appeared in, what is now, Southern Europe early in the Cretaceous, these plants are missing in the coeval paleobotanical records of Northeast Asia, the Far East, Northern Alaska, the Yukon River Basin, and Northern and Western Canada. Early Cretaceous forests in these regions are typified by Mesozoic ferns, a seed-fern group (leptostrobaleans, often called czekanowskialeans), cycads, ginkgoes, and conifers, with rich plant-fossil assemblages reported even from high Arctic terranes (Herman and Spicer 2010). It was not until the late Albian (c. 105 Ma), which is regarded as a time of a global floristic turnover and rise to dominance of flowering plants, that angiosperms reached the Arctic regions of both Asia and North America

(Samylina 1968; Smiley 1969a, b). Interestingly, angiosperms soon outcompeted other gymnosperm and fern groups in these high latitudes, indicating that they were better adapted to such a harsh climate (Herman 2002, 2013). The first herbaceous angiosperms appeared in the Far East by the middle Albian and are preserved in coastal plain and lowland deposits of the Sea of Japan near Vladivostok (Golovneva et al. 2018). In Northeastern Asia and Northern Alaska, the first angiosperms appear by the middle-late Albian (c. 108 Ma). Initially, these plants were not major components of the vegetation and probably occupied disturbed habitats near river and stream channels. It was not until the latest Albian that angiosperms conquered territories at mid- and higher latitudes.

Angiosperm-dominated plant communities gradually invaded Asian continental interiors along river valleys and other disturbed environments. Their first occurrence in the region happened during the latest Albian and continued their advance into the Late Cretaceous. These communities often are associated with volcanic deposits (Fig. 5.12), where increased nutrients may have promoted their rapid growth and eventually replaced the pre-existing "mesophytic" fern-gymnosperm communities (Herman 2013). Based on paleoclimate proxies, the prevailing temperature and humidity of the mid-Cretaceous high Arctic were similar to a modern humid cool temperate, or even humid warm-temperate (=subtropical, see Box P2), region. The one main difference, though, was the presence of a pronounced sunlight seasonality regime typical of high-latitude areas. One locality where these conditions occur is in the angiosperm-rich, latest Albian to Cenomanian, Grebenka River basin (Chukotka) [U0509] (Spicer et al. 2002; Herman 2013). Here, the paleo-Grebenka River coastal floodplain had soils enriched in volcanic ash and lava. Plants similar to modern *Platanus* commonly grew on these floodplains, and their fossil leaves possess lobate laminae and globular infructescences borne on a common axis just like their modern relatives (Maslova and Herman 2006). In northern and western Siberia, platanoids were diverse, with several recognized genera and species (Golovneva and Nosova 2012), and they coexisted with cupressoid conifers in floodplain forests. Other angiosperms are found in these floras, but many cannot be assigned taxonomically to any modern counterpart (Fig. 5.13). One exception is the angiosperm group known as trochodendroids, characterized by leaves having three major veins.

By the Turonian, 94–90 million years ago, warm-temperate deciduous platanoid and coniferous forests were well established in the north polar regions. These plants grew in extreme conditions of winter darkness, lasting several months. In terms of a modern classification scheme, these are best described as a mixed coniferous and broad-leaved forests. However, there is no true modern analog because of the predominance of deciduous taxa, including deciduous

Fig. 5.9 *Araliaephyllum kowalewskianum* angiosperm leaf, Pecínov Quarry, Peruc-Korycany Formation, Cenomanian, Late Cretaceous, Czechia

conifers, and adaptations to an extreme polar light regime under seasonally warm temperatures. Platanoid-dominated vegetation existed in northern Siberia, Northeastern Asia, Northern Alaska, and the Yukon River Basin (Spicer and Herman 2010; Herman 2013; http://arcticfossils.nsii.org. cn/). Fossil wood anatomy exhibits well-constrained annual growth rings, a feature caused by a dramatic reduction of photosynthesis in response to prolonged darkness (for more info, see Herman et al. 2016). This light-regime seasonality was probably the major controlling factor in this extinct Arctic biome. Synchronous shedding of foliage (deciduousness) was the most common overwintering strategy of these polar plants, while year-round retention of leaves (evergreenness) was viable only for a small number of taxa. The propensity for nearly all trees to drop their leaves during winter months had an effect on the herbivores and their predators.

Deciduousness of the Arctic vegetation was important for plant-eating dinosaurs. Their herds probably behaved in a way similar to that of herbivorous artiodactyl herds (even-toed ungulates such as antelope, giraffe, and camel) of modern savannas and steppes. These herds travelled from south to north regularly every spring to graze on fresh near-polar plants (Herman et al. 2016), and regularly journeyed back south when platanoids and other plants were yellowing and shedding foliage. It appears that the large Arctic dinosaurs were driven to migrate 1200–1300 km southward when food and shelter in the high latitudes were compromised (Fiorillo et al. 2018). As they approached the lower latitudes, the availability of food, warmer temperatures, and a more consistent light regime provided conditions to survive over the winter months. Smaller forms may well have overwintered, in place, in burrows or under leaf litter, by lowering their metabolic rates sufficiently to reduce their need for food.

North polar forests still existed in Northeastern Asia and Northern Alaska as far north as 75–80° N (Herman et al. 2016) in the late Campanian (84–72 Ma) and Maastrichtian (72–66 Ma) (Herman et al. 2016). However, large-leaved platanoids were replaced by taxa more typical of Paleogene floras (Herman 2013). Such plants include relatives of modern hazelnut (genus *Corylites*), ferns (*Onoclea*), cupressoid conifers (*Parataxodium*, *Metasequoia*), and trochodendroids (only one living relative, today). It is interesting to note that these high-latitude floras show no evidence of a catastrophic event across the Cretaceous–Paleogene transition (Herman

Fig. 5.10 *Zlatkocarpus brnikensis* chloranthoid infructescence, Pecínov Quarry, Peruc-Korycany Formation, Cenomanian, Late Cretaceous, Czechia

Fig. 5.11 *Pseudoasterophyllites cretaceus*, a herbaceous halophytic plant associated with representatives of the Chloranthaceae and Ceratophyllaceae, Pecínov Quarry, Peruc-Korycany Formation, Cenomanian, Late Cretaceous, Czechia

2013). Instead, floral turnover was controlled, most likely, by long-term climate changes, evolution, and biogeographic range expansions and contractions. Hence, a pronounced floristic change at the end of the Cretaceous has not been detected in these Arctic regions. Moreover, the late Maastrichtian flora of Northeastern Asia was likely a source of many plant taxa preserved in the Paleocene floras of Alaska. These plants extended their biogeographic range into Northern Alaska from Northeastern Asia during the Paleocene warming and became prominent in the early Tertiary deciduous floras of the Arctic and adjacent regions.

5.4 When Did the First Flower Bloom in Japan?

Harufumi Nishida and Julien Legrand

Fossil evidence of the first appearance and later dispersal of angiosperms in Japan and East Asia is still limited. Recent studies are providing us insight into their history during the Cretaceous, when the Japanese archipelago was not emergent, as it is now, and was part of the eastern margin of the

Asian continent. The Japanese islands consist of two parts along the Median Tectonic Line before the opening of the Sea of Japan (c. 28 Ma). The Inner Zone was part of the eastern margin of the Asian continent, whereas the Outer Zone was composed of oceanic islands that had moved northward. This northward plate movement ultimately resulted in the collision with the Inner Zone during the Hauterivian, joining them together. Some authors claim that accretionary tectonics between the Asian continent and the Pacific Ocean floor resulted in the formation of Japan. Irrespective of the islands' geologic origins, the presence of two paleofloristic provinces, the Tetori province and the Ryoseki province, each comparable to the Inner and the Outer Zones, respectively, is inferred from both micro- and megafossil data.

The first Japanese angiosperms are in the late Barremian (126 Ma) pollen record as evidenced by the presence of the monosulcate grain, *Retimonocolpites*. This pollen occurs in

Fig. 5.12 Fossil site section in Grebenka River, Far East, Russia

the Nishihiro Formation of Wakayama in the Outer Zone (Fig. 5.14; Legrand et al. 2014). Barremian pollen grains reported from the Sorachi Formation of Hokkaido are highly diversified forms at that time, including trisulcates (Tanaka 2008). By contrast, the early Aptian *Clavatipollenites* from the Kitadani Formation of Fukui in the Inner Zone is the oldest angiosperm record there (Fig. 5.15). Unfortunately, no macrofloral remains are associated with any microfossil assemblage. Megafossils appear later in the late Albian, with a *Trimenia*-like seed *Stopesia* and *Icacinoxylon* wood, both from the Yezo Group of Hokkaido. Pollen and megafossil records increase dramatically following the Albian, suggesting a rapid angiosperm radiation in Japan during the remainder of the Cretaceous. Compared to the angiosperm diversification history from the Potomac Group of North America (Sect. 5.6), Japanese records occur earlier in both the appearance of the first pollen (late Barremian vs. early Aptian) and the first morphological radiation of these grains (late Aptian vs. early Albian) (Nishida and Legrand 2017). In contrast, Japanese records are in agreement with the known angiosperm inva-

sion and radiation history in Northeast Asia [U0510], and younger fossils have contributed to our understanding of angiosperm evolution. Among them are *Elsemaria* and *Protomonimia* (Fig. 5.16a, b).

5.5 Cretaceous Angiosperms in Asia

Jiří Kvaček and Sun Ge

In the Early Cretaceous, the southern coast of Asia was configured very differently than today. India was an island located in the middle of the Indian Ocean, and the Himalayas had yet to be formed. The appearance of angiosperms in the region first was reported from the Liaoning Lagerstätte* in China (125–121 Ma) and was once considered the oldest evidence of the group in the World (e.g., Sun et al. 1998).

China hosts a famous fossil biota at Jehol, also known as Rehe province, in which spectacular early birds and feathered dinosaurs occur in association with plants and other animals.

Fig. 5.13 Grabenka Late Cretaceous Flora.
(**a**) *Paraprotophyllum*, platanoid leaf, scale bar 30 mm;
(**b**) *Trochodendroides*, trochodendroid leaf, scale bar 10 mm; (**c**) *Corylites*, betuloid leaf, scale bar 10 mm. Far East, Russia

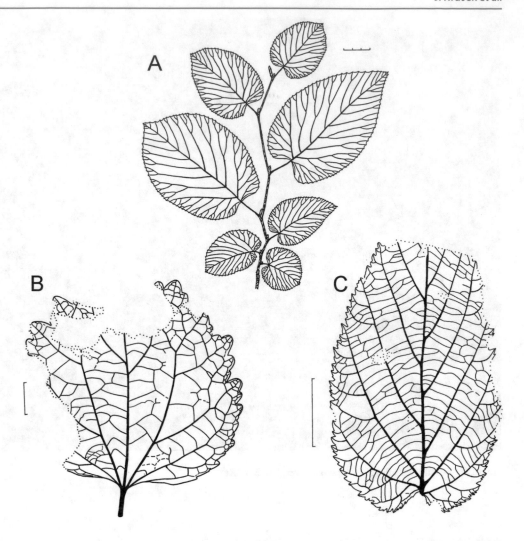

The sediments were deposited in large lakes and preserved a famous, small aquatic angiosperm: *Archaefructus* (Fig. 5.17a, b). *Archaefructus* is a nearly complete aquatic plant that has been the center of some controversy over the early origin of angiosperms. Both the age of the fossil and its paleobotanical interpretation have resulted in numerous debates. A Jurassic age originally was assigned to these sediments and later revised to an Early Aptian age (He et al. 2006). The plant, itself, consists of numerous very finely divided leaves on an erect axis with terminal reproductive structures [U0516]. The reproductive organ is interpreted variously by different authors. Sun et al. (1998) thought it represented a non-condensed flower with stamens and carpels distributed on a long axis. Later, Friis et al. (2003) interpreted the structure as an inflorescence consisting of numerous reduced flowers. The debate continues. *Archaefructus* grew submerged in a freshwater lake surrounded by higher elevations that included several active volcanoes. It is believed that only the inflorescences of the plant appeared on the lake surface, with the remainder of the plant body submerged. Vegetation that surrounded the lake also is preserved, and those plants are interpreted as coming from sparse forests and fern steppes, with shrubby gnetalean and bennettitalean gymnosperms. Although preserving a diverse fauna, there is no evidence for the presence of grasses in these Albian deposits.

Fossilized specimens of grasses are very rare in the stratigraphic record because of their growth habit and rate of decay following plant death. Yet, in the absence of megafossils, evidence of grasses is found in the form of phytoliths (see Saarinen et al., Chap. 3). The latest phytolith studies in China indicate that the first grasses grew in the Albian (Wu et al. 2018). Thereafter, records of mid-Cretaceous grasses are known from India and represent another important record of early graminoids (Prasad et al. 2011). However, each of these occurrences is solitary and unique, and an increasingly reliable graminoid record occurs only at the beginning of the Maastrichtian (Linder 1987). As we have seen on our journey to follow the evolution of grasses (Saarinen et al., Chap. 3), forested landscapes prevailed in many other parts of Asia.

Fig. 5.14   Fossil site at Kitadani, Fukui, bearing Early Cretaceous dinosaurs and plants, Japan

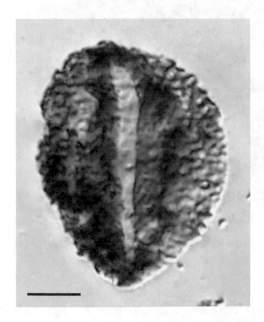

Fig. 5.15 *Retimonocolpites* pollen from the Nishihiro Formation, Early Cretaceous, Barremian, Japan. Laser image. Scale bar 5 μm

In Mongolia, para-tropical∗ coniferous and gymnospermous swamp forests of Aptian–Albian age have few angiosperms. The pollen record in these rocks is minimal (Nichols et al. 2006) and probably indicates a windblown origin from areas outside of the depositional site. Angiosperms are reported from the Albian of Kazakhstan (Shilin 1986). Here, Albian floras contain follicular (a unilocular fruit formed from a single carpel) fruiting structures including *Hyrcantha* (Vakhrameev and Krassilov 1979) and *Caspiocarpus* (Krassilov et al. 1983). The Late Cretaceous (typically Turonian) flora of Kazakhstan is rich in platanoids and lauroids (Shilin 1986; Frumin and Friis 1996), with one particular plant of probably buxaceous affinity, *Cathiaria*, and its associated leaves *Liriodendropsis* (Golovneva and Nosova 2012), which we have already encountered in the floodplains of Europe (Sect. 5.2). When angiosperms arrived in India, though, is another story.

The timing of angiosperm appearance in Southern Asia differs between India and Myanmar, once a part of the Indian paleoplate. There is no evidence, to date, of angiosperms in the Early Cretaceous of India (Drinnan and Crane 1989). Permineralized fossils of flowering plants appear in and around the time when the Deccan Traps were emplaced. The

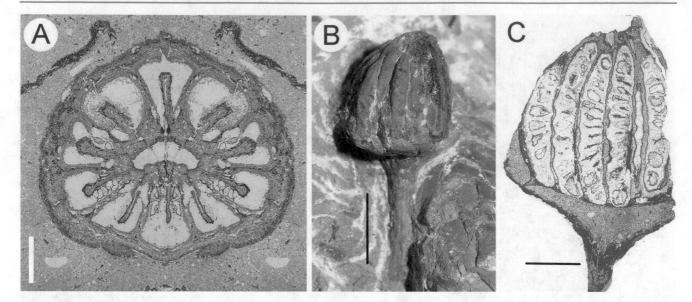

Fig. 5.16 Fruits from the Late Cretaceous of Japan. (a) Calcified fruit *Elsemaria kokubunii* from Hokkaido (Coniacian–Santonian), showing ten locules in cross section. Scale bar: 5 mm. (b) Permineralized fruit, *Protomonimia kasai-nakajhongii* in a calcified nodule from Hokkaido (Turonian). Scale bar: 10 mm. (c) Longitudinal section of *Protomonimia kasai-nakajhongii* showing multicarpelate fruiting structure. Scale bar: 5 mm

Fig. 5.17 Aquatic angiosperms. (**a**) *Archaefructus liaoningensis*; (**b**) *A. sinensis*, Yixian Formation, Hengdaozi Member, Aptian of China, scale bar = 15 mm

Maastrichtian Deccan chert (66 Ma) preserves exceptional angiosperm fruits and seeds [U0517]. From here *Indovitis*, the earliest member of the grape family, is described (Fig. 5.18; Manchester et al. 2013). A unique locality situated in northern Myanmar records earlier angiosperm fossils than in India. This locality produces insect and plant remains preserved in Kauri amber of mid-Cretaceous age (99 Ma). One excellent example is a pentamerous flower, *Tropidogyne*

Fig. 5.18 *Indovitis chitaleyae*. A virtual transverse section showing the ventral infolds of all five seeds from the Maastrichtian of Mahurzari, India, scale bar 1 mm, photo by Steven Manchester

(Cunoniaceae), which is very similar to its modern relative *Ceratopetalum* (Poinar and Chambers 2017). Let us now travel back to North America, this time to the midlatitudes, to search for angiosperms.

5.6   Cretaceous Angiosperms in Midlatitudes of North America

Hongshan Wang and Jiří Kvaček

The history of Cretaceous angiosperms in North America is well understood. Flowering plants first appeared in Barremian–Aptian sediments, slightly later than the first records of the group in Africa and Europe. Well-preserved angiosperm megafossils are documented from the Barremian–Cenomanian strata of a fluvial sedimentary series in the Potomac Group, a succession of deposits exposed along the banks of the Potomac River running through Washington, DC. These angiosperms are very similar to those described from coeval strata in Portugal. Twelve species occur from the middle to early late Albian megaflora at the Quantico locality, Virginia (Upchurch et al. 1994). At least seven angiosperm species show affinities to extant Laurales, Nelumbonaceae, and Platanaceae, and many are known for their largely irregular venation, for example, in the case of putative monocot leaves (Fig. 5.19; *Acaciaephyllum spatulatum*) (Doyle and Hickey 1976; Hickey and Doyle 1977; Doyle et al. 2008). A number of early angiosperm representatives also have very simple flowers (e.g., *Chloranthistemon crossmanensis*;

Fig. 5.19 *Acaciaephyllum spatulatum*. Putative monocot leaf, Potomac Group, Aptian, Dutch Cap locality, Virginia, USA. Scale bar 5 mm

Herendeen et al. 1993). These flowers are the same lineage reported from Portugal (*Canrightia*, *Canrightiopsis*) and, to some extent, are comparable to the extant relict family Chloranthaceae (Friis and Pedersen 2011; Friis et al. 2015a).

Apart from the basal angiosperms, the first eudicots also are reported from the Potomac Group (e.g., *Fairlingtonia*; Jud 2015), along with other eudicots assignable to the family Platanaceae. As seen elsewhere in this chapter, members of the plane-tree group have characteristic, globose, unisexual, sessile inflorescences attached on an axis. However, their leaves are not always similar to those of modern *Platanus*, but are pinnately compound as in the case of *Sapindopsis* (Fig. 5.20) [U0511] (Crane et al. 1993; Doyle and Endress 2010). If we envision ourselves along the banks of an ancient, Early Cretaceous Potomac River in the USA, we would not see a floodplain dominated by these trees. Rather, coniferous forests grew on these floodplains with the first platanoid trees flanking rivers, and fern prairies with thickets of coniferous woods were established on the uplands. The early angiosperms in these communities probably were shrubby

Fig. 5.20 *Sapindopsis magnifolia*, platanaceous leaf, Potomac Group, Aptian, White House Bluff, Virginia, USA. Scale bar 5 mm

Fig. 5.21 *Eoplatanus serrata*, platanaceous leaf, Dakota Formation, Albian–Cenomanian, Braun Ranch, Kansas, USA. Scale bar: 10 mm

and growing in the understory. Younger North American landscapes were different.

Mid-Cretaceous floras are common in both the western and eastern areas of the USA. The Dakota Formation, a fluvial to marginal marine sedimentary succession, was deposited along the eastern margin of the mid-Cretaceous (Albian–Cenomanian) Western Interior Seaway. These rocks are rich in plant fossils [U0512] commonly preserved in oxbow lakes (meander channel cutoffs) and in other floodplain deposits. Aquatic angiosperms with affinities to water lilies (*Aquatifolia* and *Brasenites*) and the Nymphaeales (Wang and Dilcher 2006) are common in these water bodies, as well as hornworts (*Donlesia*, Ceratophyllaceae; Dilcher and Wang 2009), which were surrounded by wooded forest. Swamp vegetation probably included *Archaeanthus* [U0513] and *Lesqueria* (Magnoliales; Crane and Dilcher 1984; Dilcher and Crane 1984), members of the Platanaceae (*Eoplatanus*; Fig. 5.21), and *Sapindopsis* (Wang et al. 2011; Wang and Dilcher 2018), as well as members of the Lauraceae (*Prisca* or *Mauldinia*, *Pandemophyllum*, and *Pabiania*). Flowering structures are preserved in addition to

eudicot leaves and wood. The Rose Creek flower (*Dakotanthus*) is the earliest known bisexual flower from Albian sediments, exposed near the border of Kansas and Nebraska (Manchester et al. 2018), along with members of the Combretaceae (*Dilcherocarpon*; Manchester and O'Leary 2010). *Dakotanthus* consists of five basally fused sepals, five free spatulate petals, a gynoecium with five styles, and two whorls of ten stamens (one whorl of five organized opposite the sepals and another whorl of five opposite the petals). Another early flower, *Mauldinia*, is a member of the laurel family, named after the Mauldin Mountain in Maryland, eastern USA. This fossil differs in being a compound inflorescence consisting of bilobed cladodia (a flattened, leaf-like stem) bearing sessile, trimerous flowers (Crane et al. 1993).

The mid-Cretaceous landscape of North America remained very similar to that of the Early Cretaceous, with an increasing number of angiosperm taxa replacing gymnosperms particularly in the floodplains. The reconstructed landscape at the Rose Creek locality envisions a river traversing floodplain forests rich in flowering plants. These plants have affinities to living plane trees, including *Eoplatanus* and *Platananthus*, and formed the canopy. Relatives of laurels, chloranthoids, and a number of unknown flowering plants grew in the understory. Angiosperm richness and diversity increased in these settings throughout the remainder of the Cretaceous.

By the Late Cretaceous, angiosperms had dominated nearly all plant assemblages in North America. The continent was divided into two floristic provinces by the Western Interior Seaway, which was a large inland sea that existed from the mid-Cretaceous to the early Paleogene and extended from north to south (Fig. 4.3). The eastern province was dominated by plants producing Normapolles pollen (see Sect. 5.2), which grew under a seasonally dry climate in open habitats (Graham 1999). We would encounter very dense woodlands comprised of ancestors of modern juglandoids and betuloids growing adjacent to gallery forests flanking rivers during the mid-Cretaceous in the region. During the Late Cretaceous, in addition to the established aquatic members of the water lilies (Nymphaeaceae, *Microvictoria*; Gandolfo et al. 2004) and laurels (Lauraceae, Calycanthaceae; Herendeen et al. 1994; Crepet et al. 2005), plane trees (Platanaceae; Pedersen et al. 1994) were common and higher eudicots were experiencing an evolutionary pulse of diversification (Crepet et al. 2018). Rapid evolution

is seen particularly in the Ericales (Nixon and Crepet 1993; Martinez et al. 2016), Malpighiales (Crepet and Nixon 1998), and plants of the Normapolles complex. In addition, the first palms appeared during the Santonian (86–84 Ma). Their characteristic foliage is found in the latest Cretaceous floras of North America and Europe. It is thought that palms and other monocots evolved probably in tropical areas and extended their range in response to changing climates across the continents. Reconstructions for the Late Cretaceous of southeastern North America, using fossil plant proxies (e.g., the absence of growth rings of petrified wood; Wheeler and Baas 1993), indicate a prevailing subhumid tropical climate. In the latest Cretaceous, though, climate became more seasonal with marked dry intervals recorded in the rock record (Graham 1999).

The second floristic province occurred on a narrow strip of emergent land in the western part of the USA and Canada. Here, during the Maastrichtian (72–66 Ma), a very different flora is characterized by *Aquilapollenites* pollen, the parental

Fig. 5.22 Outcrop of the Crato Formation, Early Cretaceous. Quarry SW of Nova Olinda, on the north-facing slope of the Chapada do Araripe table mountain, Brazil, photo by Karolin Moraweck

plant of which is still unknown. This pollen type dominates all the assemblages. Western North America is characterized by the presence of more deciduous species, including relatives of extant plane trees that formed low diversity forests with an understory of ferns and horsetails (Parish and Spicer 1988). Fossil floras in the Pacific Northwest and the Northern Rocky Mountain regions show a slightly different pattern. There, climate in the early Campanian became warmer, and the area witnessed angiosperm diversification involving taxa with modern affinities. We find the appearance and rise to prominence of many extant families and orders, including the Magnoliales, Laurales, Chloranthales, Nymphaeales, Menispermaceae, Trochdendrales, Plantanaceae, Fagales, and Arecales (palms; Crabtree 1987).

5.7 Cretaceous Angiosperms of Latin America

Maria A. Gandolfo

The Early Cretaceous paleogeography of South America also is very different from that of the present. It was part of the Gondwana supercontinent until the Barremian, when rifting began between South America and Africa, and separation brought marine waters to the southern Atlantic. It is in these rift basins on opposite sides of the proto-Atlantic that evidence for early angiosperms is found (Graham 2010). Records contain early angiosperm pollen of *Clavatipollenites* (Chloranthaceae) and *Tucanopollis* (an extinct plant associated with both *Ceratophyllum* and Chloranthaceae; Kvaček et al. 2016). These plants are interpreted to have grown under a rather seasonal climate.

Unlike the Western Interior Seaway of North America where swamp forests grew under a wet climate, South America was seasonally dry during the Early Cretaceous, before continental breakup and conifers and ferns dominated the vegetation. Flowering plants first appear in southern South America beginning in the late Barremian–early Aptian sediments of Patagonia, Argentina. These are small pollen grains (Archangelsky et al. 2009), which are used in conjunction with leaves, first found in Albian strata, to identify three major stages of angiosperm evolution in the region. The first angiosperm leaves are small and either entire-margined or dentate; but, their taxonomic affinity remains unknown (Cúneo and Gandolfo 2005; Passalia 2007). Younger assemblages are more diverse.

Fig. 5.23 *Jaguariba wiersemana*. Nymphaeacean angiosperm, Crato Formation, Early Cretaceous, Brazil. Scale bar 10 mm

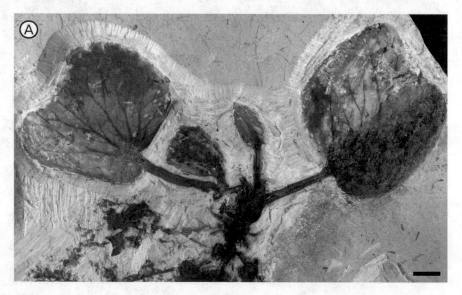

The most famous South American flora of Early Cretaceous age occurs in the Crato and Santana formations (Aptian, c. 115 Ma; Fig. 5.22), northeast Brazil, from where ancestors of water lilies (Mohr et al. 2008; Coiffard et al. 2013) and early flowering plants are described (Mohr and Eklund 2003). The Crato flora [U0514] is one of the few equatorial floras of this age in the world. Nearly 20 species of angiosperm leaves are described from here, and several examples preserve leaves that are organically connected to a parent plant (Fig. 5.23). In addition to angiosperms, other seed plants and a wide array of free-sporing plants are components of the landscape (Mohr et al. 2008). Gymnosperms include xerophilous conifers, rare cycads, and gnetophytes. This last group, the gnetophytes, is remarkably diverse with representatives of the Ephedraceae and Welwitschiaceae, the only two extant families on the planet, today. Looking down from the top of a Brazilian hill at the time, we would see a panoramic landscape of semidesert savanna with gallery forests flanking rivers.

The Crato angiosperms are remarkable in terms of their systematics and ecology. The flora is highly diverse, with the magnoliids, monocotyledons, and eudicots being the most common components (Mohr and Eklund 2003; Mohr et al. 2008). All major angiosperm clades were present by this point in time (Coiffard et al. 2013 and references therein). One unusual aspect of the flora is the array of preserved monocots (Mohr et al. 2008). This diversity is noteworthy because these angiosperm groups do not occur at midlatitudes until much later in the Turonian (94–90 Ma). Then, these plants are typical tropical rainforest (TRF) clades. Mohr et al. (2006) suggested that the low-latitude Crato region experienced seasonal drought during the Aptian–Albian. Under these conditions, angiosperms remained a minor component of the Early Cretaceous vegetation and restricted to floodplain understories and aquatic environments. The angiosperm record in younger Cretaceous sediments is sparse, with few localities identified and studied. One of these floras is the Campanian-Maastrichtian flora from the La Colonia Formation in Patagonia [U0515] (Fig. 5.24; Gandolfo et al. 2004), where aquatic angiosperms (Fig. 5.25) and water ferns are preserved and associated with palms (Cúneo et al. 2014).

Fig. 5.24 Fossil site of the Lefipan Formation, Late Cretaceous, Patagonia, Argentina

Fig. 5.25 *Lefipania padillae*, La Colonia Formation, Late Cretaceous, Chubut, Patagonia, Argentina

5.8 Cretaceous of Africa and Arabia

Jiří Kvaček and Clement Coiffard

We move to Africa, on the other side of the South Atlantic rift, in an attempt to locate the earliest angiosperm records on this continent. Here, their first appearance occurs in the form of dispersed pollen. Their pollen grains are only about 10 µm in diameter and are rare. Samples are dominated by several thousands of other sporomorphs, so patience becomes a virtue to find the angiosperm pollen. And, if we are patient enough, there is some evidence to indicate that angiosperms originated in Northern Gondwana, particularly in North Africa and the Near East. Here, angiosperms were found in late Valanginian (135 Ma) sediments from Israel (Brenner 1996) and Morocco (Gübeli et al. 1984). The same pollen type also is recorded from the Valanginian of Tuscany (Trevisan 1988), which was a part of Gondwana at the time (Brogi and Giorgetti 2010). Younger sediments record significant changes in angiosperm diversity.

The Hauterivian (133–129 Ma) is the time when sulcate pollen (these have a furrow across the middle of the grain's outer face when it was in its tetrad) is comparatively widespread in the fossil record. Again, these grains occur in samples from Israel (Brenner and Bickoff 1992) and

Fig. 5.26 *Sapindopsis* sp., Aptian–Albian, Jarash Formation, Jordan. Scale bar 10 mm

Morocco (Gübeli et al. 1984). Beginning in the Barremian, Northern Gondwana witnessed a rapid and high diversification of angiosperms across the region (Doyle 1999; Schrank and Mahmoud 2002). This diversification, though, is not restricted to Northern Gondwana, but also documented in various other parts of the world (Europe, North America; Hughes 1994; Friis et al. 2010a). Pollen diversification includes the first occurrence of tricolpate (three colpi or furrows) pollen, which marks the unequivocal presence of eudicots (Doyle et al. 1977; Penny 1991). Beginning in the Albian and continuing through the Late Cretaceous, an explosive radiation of eudicots is seen not only in Africa but also globally. Yet, there continued to be regional differences in the timing of angiosperm expansion. For example, angiosperms appear in both Antarctica and South Africa, in response to warm temperate conditions, quite late in the early Aptian (Zavada 2004). But, the record of angiosperm fossils is not equally preserved in space and time.

Angiosperm megafossils [U0518] are not as common as pollen and spores in Africa. In fact, there is a very sparse Lower Cretaceous record, limited to a few localities spread across the continent. For example, Early Cretaceous Aptian strata in Tunisia preserve aquatic plants (Barale and Ouaja 2001) represented by the genus *Klitzschophyllites*. We have encountered this plant in South America , but they also grew in the Aptian of Egypt, Tunisia, and Spain. The Aptian–Albian succession in Jordan records a coastal vegetation comprised of the salt-marsh fern, *Weichselia* and wetland forests including angiosperms of the *Sapindopsis* type (Fig. 5.26). Upland vegetation consisted of conifers, particularly araucarians, producing amber (Abu Hamad et al. 2016). These regions are interpreted as having experienced a hot climate during the Aptian. Subsequently, a change in climate is followed by a change in angiosperms.

Beginning in the Cenomanian (100–94 Ma), angiosperms dominated all of North Gondwana. Diversified floras containing flowering plants with prevailingly entire-margined leaves grew in the Cenomanian of Lebanon (Fig. 5.27; Dilcher and Basson 1990; Krassilov and Bacchia 2000) and Egypt (Lejal-Nicol and Dominik 1990). The flora is relatively rich in ferns and is evidence of a relatively high level of humidity that is needed for the group to grow and reproduce. Vegetation rich in angiosperms and virtually lacking conifers grew in Israel in the Turonian (94–90 Ma) (Dobruskina 1997; Krassilov et al. 2005) where the paleoclimate of the coastal lowlands supported mesophytic lauroids and *Debeya-Dewalquea* plants. The presence of standing bodies of water (abandoned river channels or lakes) is evidenced by a number of aquatic plant and hygrophyte monocot taxa. On the other hand, shrubby eudicots (*Eocercidiphyllites*) grew in seasonally dry uplands.

Angiosperms dominated both of these plant assemblages, which represent more than two-thirds of the plants' diversity (Krassilov et al. 2005). Late Cretaceous fossil plants from the low latitudes of North Gondwana are very scarce, but Upchurch et al. (1998) suggested that tropical rainforests existed during the Maastrichtian. This interpretation is based on the occurrence of numerous peat forests (coal deposits) in the context of an angiosperm-dominated pollen record. Other parts of Africa also are interpreted to have supported rainforests.

Fossil megafloras preserved in the Campanian (84–72 Ma) of Egypt show a high diversity of angiosperm taxa. More than 30 angiosperm species are reported in a flora with only 2 fern taxa and no conifer remains. This tropical vegetation is characterized by a rich assemblage of monocots, including members of the Araceae, which are closely related to modern taro, and philodendron (Fig. 5.28a, b; Coiffard and Mohr 2015, 2016, 2018). Both plants are common in extant tropical rainforests. But let us now explore the rest of Gondwana.

Fig. 5.27 *Sapindopsis anhauryi*, Cenomanian of Lebanon. Scale bar 20 mm

Fig. 5.28 *Afrocasia kahlertiana*, araceous leaf from the Quseir Formation, Campanian, Baris, Egypt. (**a**) Photograph; (**b**) line drawing

5.9 Cretaceous Angiosperms in Australia and Antarctica

Jiří Kvaček

The continents of Australia, New Zealand, and Antarctica and, for a shorter time, India and Madagascar were united together as one big fragment of Gondwana in the Early Cretaceous. Whether the arrival of angiosperms in Australia occurred while this part of Gondwana was conjoined with the other landmasses or via some islands from Laurasia is debated. The arrival of the group, though, was probably not via Antarctica. This is because angiosperms did not appear on the Antarctic continent until the early Albian.

Australia hosted a landscape, beginning in the latest Barremian or, possibly, as late as the Aptian, in which flowering plants grew (see also Poropat in Tschopp et al., Chap. 8). There is some debate as to whether the first basal angiosperms producing *Clavatipollenites* pollen appeared in latest Barremian, but these forms are known from at least the earliest Aptian. Soon thereafter, tricolpate pollen of higher angiosperms (eudicots) are preserved in late Aptian sediments (Dettmann in Hill 1994). Most of the Australian continent was affected by deep weathering in the mid-Cretaceous, diagenetically altering older sediments and removing evidence of fossils from this part of the record. Except in the northern part of Queensland, where wetland and swampy vegetation flourished in the late Early Cretaceous, angiosperms only occupied a position in the understory of conifer forests in most parts of Australia. This vegetational association continued into the Late Cretaceous until the evolution of two, new angiosperm families, the Nothofagaceae and Proteaceae, appeared (Dettmann in Hill 1994). With their arrival, southern beech and proteoids became important components of both the forest canopy and understory vegetation. There continue to be a number of these Mesozoic taxa (e.g., *Microcachrys*) growing in Australia and Oceania, today, although they were extirpated in the Northern Hemisphere during Cretaceous–Paleocene times. Several other plants, including members of the seed ferns (i.e., *Komlopteris*), survived in Tasmania until the early Eocene, some 13 million years after the extinction of dinosaurs (McLoughlin et al. 2008). In contrast, bennettitaleans survived in this refugium until the Oligocene (McLoughlin et al. 2011). The continent of Antarctica conserves a similar story.

Antarctica was not always covered by ice. During the Cretaceous, the continent was green, covered by coniferous forests. Flowering plants arrived here in the early Albian, with evidence of rare pollen assigned to the *Clavatipollenites* type occurring later than its appearance on other Gondwanan landmasses. There are two competing hypotheses about the way in which angiosperms arrived on Antarctica (Cantrill and Poole 2012). The first idea involves an expansion in their biogeographic range from South America across the Antarctic Peninsula. This hypothesis, though, appears to be contradicted by the fact that the earliest Antarctic angiosperms are known from the Albian, whereas the first South American angiosperms are of a much older Barremian age (Friis et al. 2010a). The second, and more plausible hypothesis, is that angiosperms came from Australia and to Antarctica via Laurasia (Burger 1981). The earliest evidence of angiosperms in Australia is reported from the latest Barremian to the early Aptian deposits (Dettmann 1986; Burger 1990) and is, thus, much closer in time to the oldest Antarctic occurrences than the South American ones. However, it is very probable that flowering plants could have extended their biogeographic range from Australia via

Antarctica to South America in the Late Cretaceous. This is because the Antarctic Peninsula acted as a gateway and bridge connecting Antarctica with South America beginning in the Cenomanian. Antarctica retained its high polar latitudinal position for most of its history, where it hosted Early Cretaceous cool temperate forests of evergreen conifers, araucarias, podocarps, and various gymnosperms. Similar to high-latitude forests in the Northern Hemisphere, these groups were adapted to life subjected to a regular half year of darkness.

Regardless of the light regime under which these forests grew, angiosperms began to play a greater role at high latitudes during the Cretaceous. Early Cretaceous vegetation was rich in podocarps and araucarians, in which a diverse bryophyte and hepatophyte flora coexisted (Cantrill and Poole 2012). These lower vascular plants, at times, account for as much as 20% of the dispersed palynomorph spectra. And, it is in podocarp, araucarian, and fern-dominated forests that the first angiosperms are reported from deposits exposed on Alexander Island (Cantrill and Poole 2012).

Forests on the Antarctica Peninsula (Fig. 5.29) grew under wet and warm paratropical conditions by the Late Cretaceous. These were comprised of conifers and angiosperms [U0519] with a rich understory of tree ferns. The youngest fossils of the Marattiaceae (*Marattiopsis*; Kvaček 2014) are preserved in these deposits. Beginning in the Coniacian, coniferous forests were replaced, in part, by arborescent angiosperms, and there is a diversification in species of the southern beech (*Nothofagus*). Today, this tree or shrub is native to the southern part of South America and Australia, as well as New Zealand and South Pacific islands including New Guinea. However, the genus has never been found in India or Africa, indicating that its evolution and biogeographic range is constrained to these other continents. In Antarctica, a number of new angiosperm groups diversified in the Late Cretaceous, with members of the Gunneraceae, Aquifoliaceae, Loranthaceae, and Arecaceae (palms) appearing for the first time (Askin 1992). There is evidence for magnoliid diversification in the Coniacian (Kvaček and Vodrážka 2016) and Santonian (Fig. 5.30),

Fig. 5.29 Fossil wood exposed in the Hidden Lake Formation, Campanian, Late Cretaceous James Ross Island, Antarctica

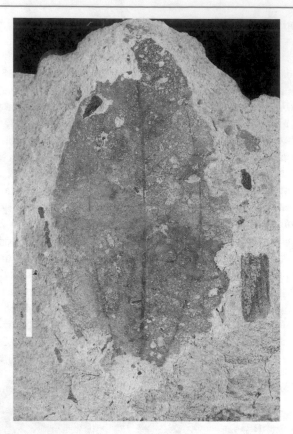

Fig. 5.30 *Cinnamomophyllum*, Santa Marta Formation, Santonian, Late Cretaceous, Antarctica. Scale bar 10 mm

with an increasing diversification of eudicot angiosperms from the Campanian to the Maastrichtian (Cantrill and Poole 2012), when subtropical vegetation was replaced by cool temperate forests (Dutra and Batten 2000).

5.10   Conclusions

Over the course of our deep-time field trip, we have seen that angiosperms diversified shortly after their first appearance in the fossil record and expanded across the planet. The origin of the group is still not completely resolved and remains partly an "abominable mystery" (Box 6.1), as it was characterized by Charles Darwin (Crepet 1998). Molecular phylogenetists frequently argue for a Jurassic appearance of angiosperms prior to physical evidence of the group in the Cretaceous (Magallón et al. 2015, but see Beaulieu et al. 2015). However, to date, there is no fossil plant of Jurassic age that can be considered to be unequivocal proof of the group's existence at that time (Herendeen et al. 2017; Saquet and Magallón 2018). The insights we have gained in this chapter are consistent with the general well-supported view that angiosperms attained ecological prominence in most terrestrial ecosystems during the latest Cretaceous (Wing and Boucher 1998). Their rise to prominence was facilitated by

their rapid diversification in the Aptian–Albian (c. 113 Ma), which also impacted the faunas at the time, as we will see in Chaps. 7 and 8.

Questions

1. What is the oldest evidence of angiosperms and where, geographically, is it found?
2. Name four characters that separate angiosperms from other seed plants, particularly the gymnosperms.
3. How many different gymnosperm groups are possible ancestors to angiosperms, and what features do each group share with flowering plants?
4. What are the three major angiosperm clades, and what characteristics separate them?
5. When do angiosperms first appear during the Cretaceous in (1) Europe, (2) High North polar latitudes, (3) Japan, (4) China, (5) Mongolia, (6) India, (7) North America, (8) Latin America, (9) Africa, and (10) Arabia?
6. What characteristics of angiosperms are used to interpret this group as being adapted to disturbed environments?
7. What ecological role(s) did angiosperms play in Early Cretaceous forests? What role(s) did they play in Late Cretaceous landscapes?
8. How does the Cretaceous angiosperm flora in the Southern (Gondwanan) Hemisphere differ from that of the paleoequatorial and Northern Hemisphere?
9. What role(s) might Cretaceous islands have played in the biogeographic distribution of this plant group?
10. What adaptions did angiosperms evolve to thrive at high paleolatitudes?

References

Abu Hamad AMB, Hamadi A, Amirehi B, Jasper A, Uhl D (2016) New palaeobotanical data from the Jarash Formation (Aptian–Albian, Kurnub Group) of NW Jordan. Palaeobotanist 65:19–29

APG (2016) An update of the Angiosperm Phylogeny Group classification for the orders and families of flowering plants: APG IV. Bot J Linn Soc 181:1–20

Archangelsky S, Barreda V, Passalia MG, Gandolfo M, Prámparo M, Romero E, Cúneo R, Zamuner A, Iglesias A, Llorens M, Puebla GG (2009) Early angiosperm diversification: evidence from southern South America. Cretac Res 30:1073–1082

Askin RA (1992) Late Cretaceous-early Tertiary Antarctic outcrops evidence for past vegetation and climates. In Kennett JP, Warnke DA, (eds.) The Antarctic paleoenvironment: a perspective on global change: part one. Antarctic Research Series, 56. https://doi.org/10.1029/AR056p0061

Axelrod DI (1972) Edaphic aridity as a factor in angiosperm evolution. Am Nat 106:311–320

Barale G, Ouaja M (2001) Découverte de nouvelles flores avec des restes à affinités angiospermiennes dans le Crétacé inférieur du Sud Tunisien. Cretac Res 22(2):131–143

Beaulieu JM, O'Meara B, Crane P, Donoghue MJ (2015) Heterogeneous rates of molecular evolution and diversification could explain the Triassic age estimate for angiosperms. Syst Biol 64(5):869–878

Brenner GJ (1996) Evidence for the earliest stage of angiosperm pollen evolution: a paleoequatorial section from Israel. In: Taylor DW, Hickey LJ (eds) Flowering plant origin, evolution and phylogeny. Chapman and Hall, New York, pp 91–115

Brenner GJ, Bickoff IS (1992) Palynology and age of the Lower Cretaceous Kurnub Group from the coastal plain to the northern Negev of Israel. Palynology 16:137–185

Brogi A, Giorgetti G (2010) The tectono-metamorphic record of the Tuscan Nappe from the Colline Metallifere region (Northern Apennines, Italy). Ital J Geosci 129(2):177–187

Burger D (1981) Observations on the earliest angiosperm development with special reference to Australia. In: Proceedings of the Fourth International Palynological Conference, Lucknow (1976–1977), vol. 3, pp 418–428

Burger D (1990) Early Cretaceous angiosperms from Queensland, Australia. Rev Palaeobot Palynol 65(1–4):153–163

Cantrill DJ, Poole I (2012) The vegetation of Antarctica through geological time. Cambridge University Press, Cambridge, 480 pp

Coiffard C, Kardjilov N, Manke I, Bernardes-de-Oliveira MEC (2019) Fossil evidence of core monocots in the Early Cretaceous. Nature Plants 5:691–696

Coiffard C, Mohr BAR (2015) Lejalia sagenopteroides gen. nov. et comb. nov.: a new tropical member of Araceae from Late Cretaceous strata of northern Gondwana (Jebel Abyad, Sudan). Taxon 64(5):987–997

Coiffard C, Mohr BAR (2016) Afrocasia kahlertiana gen. et sp. nov., a new tropical member of Araceae from Late Cretaceous strata of northern Gondwana (Baris, Egypt). Taxon 65(6):1374–1384

Coiffard C, Mohr BAR (2018) Cretaceous tropical Alismatales in Africa: diversity, climate and evolution. Bot J Linn Soc 188(2):117–131

Coiffard C, Gomez B, Kvaček J, Thevenard F (2006) Early angiosperm ecology: evidence from the Albian-Cenomanian of Europe. Ann Bot 98:495–502

Coiffard C, Mohr BAR, Bernardes-de-Oliveira MEC (2013) Jaguariba wiersemana gen. nov. et sp. nov., an Early Cretaceous member of crown group Nymphaeales (Nymphaeaceae) from northern Gondwana. Taxon 62(1):141–151

Coiro M, Doyle JA, Hilton J (2019) How deep is the conflict between molecular and fossil evidence on the age of angiosperms? New Phytol 223:83. https://doi.org/10.1111/nph.15708

Crabtree DR (1987) Angiosperms of the northern Rocky Mountains: Albian to Campanian (Cretaceous) megafossil floras. Ann Mo Bot Gard 74:707–747

Crane PR, Dilcher DL (1984) Lesqueria: an early angiosperm fruiting axis from the mid-Cretaceous. Ann Mo Bot Gard 71:384–402

Crane PR, Pedersen KJ, Friis EM, Drinnan AN (1993) Early Cretaceous (early to middle Albian) platanoid inflorescences associated with Sapindopsis leaves from the Potomac Group of eastern north america. Syst Bot 18(2):328–344

Crepet WL (1998) The abominable mystery. Science 282:1653–1654

Crepet WL, Nixon KC (1998) Fossil Clusiaceae from the Late Cretaceous (Turonian) of New Jersey and implications regarding the history of bee pollination. Am J Bot 85:1122–1133

Crepet WL, Nixon KC, Gandolfo MA (2005) An extinct calycanthoid taxon, Jerseyanthus calycanthoides, from the Late Cretaceous of New Jersey. Am J Bot 92:1475–1485

Crepet WL, Nixon KC, Weeks A (2018) Mid-Cretaceous angiosperm radiation and an asterid origin of bilaterality: diverse and extinct "Ericales" from New Jersey. Am J Bot 105(8):1412–1423

Cronquist A (1988) The evolution and classification of flowering plants, 2nd edn. New York Botanical Garden, New York, 396 pp

Cúneo NR, Gandolfo MA (2005) Angiosperm leaves from the Kachaike Formation, Lower Cretaceous of Patagonia, Argentina. Rev Palaeobot Palynol 136(1):29–47

Cúneo NR, Gandolfo MA, Zamaloa MC, Hermsen E (2014) Late Cretaceous aquatic plant world in Patagonia, Argentina. PLoS One 9(8):e104749

Dettmann ME (1986) Early Cretaceous palynoflora of subsurface strata correlative with the Koonwarra Fossil Bed, Victoria (Australia). Mem Assoc Australas Palaeontol 3:79–110

Dilcher DL, Basson PW (1990) Mid-Cretaceous angiosperm leaves from a new fossil locality in Lebanon. Bot Gaz 151:538–547

Dilcher DL, Crane PR (1984) Archaeanthus: an early angiosperm from the Cenomanian of the Western interior of North America. Ann Mo Bot Gard 71:351–383

Dilcher DL, Wang H (2009) An Early Cretaceous fruit with affinities to Ceratophyllaceae. Am J Bot 96(12):2256–2269

Dobruskina IA (1997) Turonian plants from the southern Negev, Israel. Cretac Res 18(1):87–107

Doyle JA (1992) Revised palynological correlations of the lower Potomac Group (USA) and the Cocobeach sequence of Gabon (Barremian-Aptian). Cretac Res 13(4):337–349

Doyle JA (1999) The rise of angiosperms as seen in the African Cretaceous pollen record. 3–29. In: Heine K (ed.) Proceedings of the third conference on African palynology, Johannesburg 14–19 September 1997. Palaeoecology of Africa and the surrounding islands, 26. Balkema, Rotterdam, 274 pp

Doyle JA (2012) Molecular and fossil evidence on the origin of angiosperms. Annu Rev Earth Planet Sci 40:301–326

Doyle JA, & Donoghue MJ (1985) Relationships of angiosperms and Gnetales: a numerical cladistic analysis. In: Spicer RA, Thomas BA (eds.) Systematic and taxonomic approaches in palaeobotany. Special volume no. 31. Oxford Clarendon Press. pp. 177–198

Doyle JA, Endress PK (2010) Integrating Early Cretaceous fossils into the phylogeny of living angiosperms: Magnoliidae and eudicots. J Syst Evol 48(1):1–35

Doyle JA, Hickey LJ (1976) Pollen and leaves from the mid-Cretaceous Potomac Group and their bearing on early angiosperm evolution. In: Beck CB (ed) Origin and early evolution of angiosperms. Columbia University Press, New York, pp 139–206

Doyle JA, Biens P, Doerenkamp A, Jardiné S (1977) Angiosperm pollen from the pre–Albian Cretaceous of Equatorial Africa. Bulletin des Centres de Recherches Exploration–Production Elf–Aquitaine 1:451–473

Doyle JA, Endress P, Upchurch GR (2008) Early Cretaceous monocots: a phylogenetic evaluation. Acta Musei Nationalis Pragae, Ser B - Historia Naturalis 64:59–87

Drinnan AN, Crane PR (1989) Cretaceous palaeobotany and its bearing on the biostratigraphy of Austral angiosperms. In: Taylor TN, Taylor EI (eds) Antarctic paleobiology. Springer Verlag, New York, pp 192–219

Dutra TL, Batten D (2000) Upper Cretaceous floras of King George Island, West Antarctica, and their palaeoenvironmental and phytogeographic implications. Cretac Res 21:181–209

Eklund H, Kvaček J (1998) Lauraceous inflorescences and flowers from the Cenomanian of Bohemia (Czech Republic, Central Europe). Int J Plant Sci 159:668–686

Falcon-Lang H, Kvaček J, Uličný D (2006) Mesozoic mangroves. Geoscientist 16(4):4–6

Feild TS, Arens NC, Doyle JA, Dawson TE, Donoghue MJ (2004) Dark and disturbed: a new image of early angiosperm ecology. Paleobiology 30:82–107

Fiorillo AR, McCarthy PJ, Kobayashi Y, Tomisch CS, Tykosi RS, Lee Y-N, Tanaka T, Noto CR (2018) An unusual association of hadrosaur and therizinosaur tracks within Late Cretaceous rocks of Denali National Park, Alaska. Sci Rep 8:11706

Friis EM (1983) Upper Cretaceous (Senonian) floral structures of jug-landalean affinity containing Norma-polles pollen. Rev Palaeobot Palynol 39:161–188

Friis EM (1985) *Actinocalyx* gen. nov., sympetalous angiosperm flowers from the Upper Cretaceous of southern Sweden. Rev Palaeobot Palynol 45:171–183

Friis EM, Pedersen KR (2011) *Canrightia resinifera* gen. et sp. nov., a new extinct angiosperm with *Retimonocolpites*-type pollen from the Early Cretaceous of Portugal: missing link in the eumagnoliid tree? Grana 50:3–29

Friis EM, Skarby A (1982) *Scandianthus* gen. nov., angiosperm flowers of saxifragalean affinity from the Upper Cretaceous of southern Sweden. Ann Bot New Ser 50:569–583

Friis EM, Pedersen KR, Crane PR (1999) Early angiosperm diversification: the diversity of pollen associated with angiosperm reproductive structures in Early Cretaceous floras from Portugal. Ann Mo Bot Gard 86:259–296

Friis EM, Pedersen KR, Crane PR (2000) Reproductive structure and organization of basal angiosperms from the Early Cretaceous (Barremian or Aptian) of Western Portugal. Int J Plant Sci 161:169–182

Friis EM, Pedersen KR, Crane PR (2001) Fossil evidence of water lilies (Nymphaeales) in the Early Cretaceous. Nature 410:357–360

Friis EM, Doyle JA, Endress PK, Leng Q (2003) *Archaefructus* – angiosperm precursor or specialized early angiosperm? Trends Plant Sci 8(8):369–373

Friis EM, Pedersen KR, Crane PR (2006a) Cretaceous angiosperm flowers: innovation and evolution in plant reproduction. Palaeogeogr Palaeoclimatol Palaeoecol 232:251–293

Friis EM, Pedersen KJ, Schönenberger J (2006b) Normapolles plants: a prominent component of the Cretaceous rosid diversification. Plan Syst Evol 260:107–140

Friis EM, Pedersen KR, Crane PR (2009) Early Cretaceous meso-fossils from Portugal and eastern North America related to the Bennettitales-Erdtmanithecales-Gnetales group. Am J Bot 96:252–283

Friis EM, Pedersen KJ, Crane PR (2010a) Diversity in obscurity: fossil flowers and the early history of angiosperms. Philos Trans R Soc Lond B Biol Sci 365(1539):369–382

Friis EM, Pedersen KR, Crane PR (2010b) Cretaceous diversification of angiosperms in the western part of the Iberian Peninsula. Rev Palaeobot Palynol 162:341–361

Friis EM, Pedersen KR, Crane PR (2011) Early flowers and angiosperm evolution. Cambridge University Press, Cambridge, 585 pp

Friis EM, Pedersen KR, Crane PR (2013) New diversity among chlamydospermous seeds from the Early Cretaceous of Portugal and North America. Int J Plant Sci 174:530–558

Friis EM, Grimm GW, Mendes MM, Pedersen KJ (2015a) *Canrightiopsis*, a new Early Cretaceous fossil with *Clavatipollenites*-type pollen bridge the gap between extinct *Canrightia* and extant Chloranthaceae. Grana 54(3):184–212

Friis EM, Pedersen KR, Crane PR, Stampanoni M, Marone F (2015b) Exceptional preservation of tiny embryos documents seed dormancy in early angiosperms. Nature 528:551–554

Friis EM, Crane PR, Pedersen KR (2019) The Early Cretaceous mesofossil flora of Torres Vedras (NE of Forte da Forca), Portugal: a palaeofloristic analysis of an early angiosperm community. Fossil Imprint 75:153–257

Frumin SI, Friis EM (1996) Liriodendroid seeds from the Late Cretaceous of Kazakhstan and North Carolina, USA. Rev Palaeobot Palynol 94:39–55

Gandolfo MA, Nixon KC, Crepet WL (2004) Cretaceous flowers of Nymphaeaceae and implications for complex insect entrapment pollination mechanisms in early Angiosperms. Proc Natl Acad Sci USA 101(21):8056–8060

Golovneva LB, Nosova NV (2012) Al'b-cenomanskaya flora Zapadnoi Sibiri (Albian-Cenomanian Flora of Western Siberia). Marafon, Saint-Petersburg, 436 pp. (in Russian)

Golovneva L, Alexeev P, Bugdaeva E, Volynets E (2018) An angiosperm dominated herbaceous community from the early-middle Albian of Primorye, Far East of Russia. Fossil Imprint 174(1–2):165–178

Gomez B, Daviero-Gomez V, Coiffard C, Martín-Closas C, Dilcher DL (2015) *Montsechia*, an ancient aquatic angiosperm. Proc Natl Acad Sci USA 112(35):10985–10988

Graham A (1999) Late Cretaceous and Cenozoic history of North American vegetation. Oxford University Press, New York, 350 pp

Graham A (2010) Late Cretaceous and Cenozoic history of Latin American vegetation and terrestrial environments. Missouri Botanical Garden Press, St. Louis, 618 pp

Gübeli A, Hochuli PA, Wildi W (1984) Lower Cretaceous turbiditic sediments from the Central Rif chain (northern Morocco). Palynology, stratigraphy and palaeogeographic setting. Geol Rundsch 73:1081–1114

He HY, Wang XL, Zhou ZH, Jin F, Wang F, Yang LK, Ding X, Boven A, Zhu RX (2006) 40Ar/39Ar dating of Lujiatun Bed (Jehol Group) in Liaoning, northeastern China. Geophys Res Lett 33(4):L04303

Herendeen PS, Crepet WL, Nixon KC (1993) *Chloranthus*-like stamens from the Upper Cretaceous of New Jersey. Am J Bot 80(8):865–871

Herendeen PS, Crepet WL, Nixon KC (1994) Fossil flowers and pollen of Lauraceae from the Upper Cretaceous of New Jersey. Plant Syst Evol 189:29–40

Herendeen PS, Friis EM, Pedersen KR, Crane PR (2017) Palaeobotanical redux: revisiting the age of the angiosperms. Nat Plants 3(17015):1–8

Herman AB (2002) Late early - Late Cretaceous floras of the North Pacific Region: florogenesis and early angiosperm invasion. Rev Palaeobot Palynol 122(1–2):1–11

Herman AB (2013) Albian – Paleocene flora of the North Pacific: systematic composition, palaeofloristics and phytostratigraphy. Stratigr Geol Correl 21(7):689–747

Herman AB, Spicer RA (2010) Mid-Cretaceous floras and climate of the Russian high Arctic (Novosibirsk Islands, Northern Yakutiya). Palaeogeogr Palaeoclimatol Palaeoecol 295(3–4):409–422

Herman AB, Spicer RA, Spicer TEV (2016) Environmental constraints on terrestrial vertebrate behaviour and reproduction in the high Arctic of the Late Cretaceous. Palaeogeogr Palaeoclimatol Palaeoecol 441:317–338

Heřmanová Z, Kvaček J, Friis FM (2011) *Budvaricarpus serialis* Knobloch Mai, an unusual new member of the Normapolles complex from the Late Cretaceous of the Czech Republic. Int J Plant Sci 172:285–293

Heřmanová Z, Kvaček J, Dašková J (2016) *Caryanthus* diversity in the Late Cretaceous. Rev Palaeobot Palynol 231:33–47

Heřmanová Z, Dašková J, Ekrt E, Kvaček J (2017) *Zlivifructus* gen. nov., a new member of the Normapolles complex. Rev Palaeobot Palynol 246:177–184

Hickey LJ, Doyle JA (1977) Early Cretaceous fossil evidence for angiosperm evolution. Bot Rev 43:1–104

Hill CR (1994) History of the Australian vegetation, Cretaceous to recent. Cambridge University Press, Cambridge, 433 pp

Hill CR (1996) A plant with flower-like organs from the Wealden of the Weald (Lower Cretaceous), southern England. Cretac Res 17(1):27–38

Hughes NF (1994) The enigma of angiosperm origins, Cambridge paleobiology series, vol 1. Cambridge University Press, Cambridge, 303 pp

Jud NA (2015) Fossil evidence for a herbaceous diversification of early eudicot angiosperms during the Early Cretaceous. Proc R Soc B Biol Sci 282:20151045

Jud WS, Campbell CS, Kellogg EA, Stevens PF, Donoghue MJ (2015) Plant Systematics: A Phylogenetic Approach. Fourth Edition. Sinauer Association, Sunderland 677 pp

Knobloch E, Mai DH (1986) Monographie der Früchte und Samen in der Kreide von Mitteleuropa. Rozpravy Ústředního ústavu geologického 47:1–219

Krassilov AK, Bacchia F (2000) Cenomanian florule of Nommoura, Lebanon. Cretac Res 21:785–799

Krassilov VA, Shilin PV, Vachrameev VA (1983) Cretaceous flowers from Kazakhstan. Rev Palaeobot Palynol 40:91–113

Krassilov V, Lewy Z, Nevo E, Silantieva N (2005) Late Cretaceous (Turonian) Flora of Southern Negev, Israel. Pensoft, Sophia, 252 pp

Kvaček J (2000) *Frenelopsis alata* and its microsporangiate and ovuliferous reproductive structures from the Cenomanian of Bohemia (Czech Republic, Central Europe). Rev Palaeobot Palynol 112:51–78

Kvaček J (2014) *Marattiopsis vodrazkae* sp. nov. (Marattiaceae) from the Campanian of the Hidden Lake Formation, James Ross Island, Antarctica. Sborník Národního musea v Praze. B, Přírodovědný (Acta Musei Nationalis Pragae. B, Historia Naturalis) 70(3–4): 211–218

Kvaček J, Eklund H (2003) A report on newly recovered reproductive structures from the Cenomanian of Bohemia (Central Europe). Int J Plant Sci 164(6):1021–1039

Kvaček J, Friis EM (2010) *Zlatkocarpus* gen. nov., a new angiosperm reproductive structure with monocolpate-reticulate pollen from the Late Cretaceous (Cenomanian) of the Czech Republic. Grana 49:115–127

Kvaček J, Herman AB (2004) Monocotyledons from the early Campanian (Cretaceous) of Grünbach, lower Austria. Rev Palaeobot Palynol 128(3–4):323–353

Kvaček J, Vodrážka R (2016) Late Cretaceous flora of the Hidden Lake Formation, James Ross Island (Antarctica), its biostratigraphy and palaeoecological implications. Cretac Res 58:183–201

Kvaček J, Falcon-Lang H, Dašková J (2005) A new Late Cretaceous ginkgoalean reproductive structure *Nehvizdyella* gen. nov. from the Czech Republic and its whole-plant reconstruction. Am J Bot 92(12):1958–1969

Kvaček J, Doyle JA, Endress PK, Daviero-Gomez V, Gomez B, Tekleva M (2016) *Pseudoasterophyllites cretaceus* from the Cenomanian (Cretaceous) of the Czech Republic – a possible link between Chloranthaceae and *Ceratophyllum*. Taxon 65(6):1345–1373

Lee AP, Upchurch G, Murchie EH, Lomax BH (2015) Leaf energy balance modelling as a tool to infer habitat preference in the early angiosperms. Proc R Soc B Biol Sci 282:20143052

Legrand J, Yamada T, Nishida H (2014) Palynofloras from the upper Barremian-Aptian Nishihiro Formation (Outer Zone of southwest Japan) and the appearance of angiosperms in Japan. J Plant Res 127:221–232

Lejal-Nicol A, Dominik W (1990) Sur la palaeoflore a Weichseliaceae et a angiospermes du Cenomanien de la region de Bahariya (Egypte du Sud-Oest). Berliner Geowissenschaftliche Abhandlungen Reihe A 120(2):957–992

Linder HP (1987) The evolutionary history of the Poales/Restionales: a hypothesis. Kew Bull 42:297–318

Magallón S, Gómez-Acevedo S, Sánchez-Reyes LL, Hernández-Hernández T (2015) A metacalibrated time-tree documents the early rise of flowering plant phylogenetic diversity. New Phytol 207:437–453

Manchester SR, O'Leary EL (2010) Phylogenetic distribution and identification of fin-winged fruits. Bot Rev 76:1–82

Manchester SR, Kapgate DK, Wen J (2013) Oldest fruits of the grape family (Vitaceae) from the Late Cretaceous Deccan Cherts of India. Am J Bot 100(9):1849–1859

Manchester SR, Dilcher DL, Judd WS, Corder B, Basinger JF (2018) Early Eudicot flower and fruit: *Dakotanthus* gen. nov. from the Cretaceous Dakota Formation of Kansas and Nebraska, USA. Acta Palaeobotanica 58(1):27–40

Martinez CT, Choo YS, Allevato D, Nixon KC, Crepet WL, Harbert RS, Daghlian CP (2016) *Rariglanda jerseyensis*, a new ericalean fossil flower from the Late Cretaceous of New Jersey. Botany-Botanique 94:747–758

Maslova NP, Herman AB (2006) Infructescences of *Friisicarpus* nom. nov. (Platanaceae) and associated foliage of the platanoid type from the Cenomanian of Western Siberia. Paleontol J 40(1):109–113

McLoughlin S, Carpenter RJ, Jordan GJ, Hill RS (2008) Seed ferns survived the end-Cretaceous mass extinction in Tasmania. Am J Bot 95(4):465–471

McLoughlin S, Carpenter RJ, Pott C (2011) *Ptilophyllum muelleri* (Ettingsh.) comb. nov. from the Oligocene of Australia: last of the bennettitales? Int J Plant Sci 172(4):574–585

Mendes MM, Friis EM (2018) The Nossa Senhora da Luz flora from the Early Cretaceous (early Aptian – late Albian) of Juncal in the western Portuguese Basin. Acta Palaeobotanica 58:159–174

Mendes MM, Dinis J, Pais J, Friis EM (2011) Early Cretaceous flora from Vale Painho (Lusitanian Basin, western Portugal): an integrated palynological and mesofossil study. Rev Palaeobot Palynol 166:152–162

Mendes MM, Dinis J, Pais J, Friis EM (2014a) Vegetational composition of the Early Cretaceous Chicalhão flora (Lusitanian Basin, western Portugal) based on palynological and mesofossil assemblages. Rev Palaeobot Palynol 200:65–81

Mendes MM, Grimm GW, Pais J, Friis EM (2014b) Fossil *Kajanthus lusitanicus* gen. et sp. nov. from Portugal: floral evidence for Early Cretaceous Lardizabalaceae (Ranunculales, basal eudicot). Grana 53(4):283–301

Mendes MM, Dinis P, Kvaček J (2018) Some conifers from the Early Cretaceous (late Aptian – early Albian) of Catefica, Lusitanian Basin, western Portugal. Fossil Imprint 74(3–4):317–326

Mohr BAR, Eklund H (2003) *Araripia florifera*, a magnoliid angiosperm from the Lower Cretaceous Crato Formation (Brazil). Rev Palaeobot Palynol 126:279–292

Mohr B, Bernardes-de-Oliveira MEC, Barale G, Ouaja M (2006) Palaeogeographic distribution and ecology of *Klitzschophyllites*, an Early Cretaceous angiosperm in southern Laurasia and northern Gondwana. Cretac Res 27(3):464–472

Mohr BAR, Bernardes-de-Oliveira MEC, Taylor DW (2008) *Pluricarpellatia*, a nymphaealean angiosperm from the Lower Cretaceous of northern Gondwana (Crato Formation, Brazil). Taxon 57(4):1147–1158

Nichols DJ, Matsukawa M, Ito M (2006) Palynology and age of some Cretaceous nonmarine deposits in Mongolia and China. Cretac Res 27(2):241–251

Nishida H, Legrand J (2017) Features of Cretaceous floristic changes in Japan in relation to angiosperm invasion. Fossils 101:61–67. In Japanese with English abstract

Nixon KC, Crepet WL (1993) Late Cretaceous fossil flowers of ericalean affinity. Am J Bot 80:616–623

Parish JT, Spicer RA (1988) Late Cretaceous terrestrial vegetation: a near-polar temperature curve. Geology 16:22–25

Passalia MG (2007) Nuevas evidencias de la flora creta' cica descripta por Halle (1913) en Lago San Martı'n, Santa Cruz, Argentina. Ameghiniana 44:565–595

Pedersen KR, Friis EM, Crane PR, Drinnan AN (1994) Reproductive structures of an extinct platanoid from the Early Cretaceous (latest Albian) of eastern North America. Rev Palaeobot Palynol 80(3–4):291–303

Pedersen KR, von Balthazar M, Crane PR, Friis EM (2007) Early Cretaceous floral structures and in situ tricolpate-striate pollen: new early eudicots from Portugal. Grana 46:176–196

Penny JHJ (1991) Early Cretaceous angiosperm pollen from the borehole Mersa Matruh 1, North West Desert, Egypt. Palaeontogr Abt B 222:31–88

Poinar GO, Chambers KL (2017) *Tropidogyne pentaptera*, sp. nov., a new mid-Cretaceous fossil angiosperm flower in Burmese amber. Palaeodiversity 10(1):135–140

Prasad V, Strömberg CAE, Leaché AD, Samant B, Patnaik R, Tang L, Mohabey DM, Ge S, Sahni A (2011) Late Cretaceous origin of the rice tribe provides evidence for early diversification in Poaceae. Nat Commun 2:480

Samylina VA (1968) Early Cretaceous angiosperms of the Soviet Union based on leaf and fruit remains. Bot J Linn Soc 61(384):207–218

Saquet H, Magallón S (2018) Key questions and challenges in angiosperm macroevolution. New Phytol 219(4):1170–1187

Schönenberger J, Friis EM (2001) Fossil flowers of ericalean affinity from the Late Cretaceous of southern Sweden. Am J Bot 88(3):467–480

Schrank E, Mahmoud MS (2002) Barremian angiosperm pollen and associated palynomorphs from the Dakhla Oasis area, Egypt. Palaeontology 45(1):33–56

Scotese CR (2014) Atlas of Late Cretaceous Paleogeographic Maps, PALEOMAP Atlas for ArcGIS, volume 2, The Cretaceous, Maps 16–22, Mollweide Projection, PALEOMAP Project, Evanston

Shilin PV (1986) Pozdně mělovyje flory Kazakhstana (Late Cretaceous Floras of Kazakhstan). Nauka, Alma-Ata, 136 pp

Smiley CJ (1969a) Cretaceous floras of Chandler-Colville region, Alaska: stratigraphy and preliminary floristics. Am Assoc Petrol Geol Bull 53(3):482–502

Smiley CJ (1969b) Floral zones and correlations of Cretaceous Kukpowruk and Corwin formations, Northwestern Alaska. Am Assoc Petrol Geol Bull 53(10):2079–2093

Spicer RA, Herman AB (2010) The Late Cretaceous environment of the Arctic: a quantitative reassessment based on plant fossils. Palaeogeogr Palaeoclimatol Palaeoecol 295(3–4):423–442

Spicer RA, Ahlberg A, Herman AB, Kelley SP, Raikevich M, Rees PM (2002) Palaeoenvironment and ecology of the middle Cretaceous Grebenka flora of northeastern Asia. Palaeogeogr Palaeoclimatol Palaeoecol 184(1–2):65–105

Stebbins GL (1974) Flowering plants: evolution above the species level. Harvard University Press, Cambridge, MA, 399 pp

Sun G, Dilcher DL, Zheng SL, Zhou ZK (1998) In search of the first flower: a Jurassic angiosperm, *Archaefructus*, from Northeast China. Science 282:1692–1695

Takhtajan AL (1969) Flowering plants: origin and dispersal. Smithsonian Institution Press, Washington, 310 pp

Tanaka S (2008) Early Cretaceous angiosperm pollen fossils from Hokkaido, northern Japan. Terra Nostra 2:276–277

Taylor DW, Hickey LJ (1992) Phylogenetic evidence for the herbaceous origin of angiosperms. Plant Syst Evol 180:137–156

Trevisan L (1988) Angiospermous pollen (monosulcate-trichotomosulcate phase) from the very early Lower Cretaceous of Southern Tuscany (Italy): some aspects. In: Proceedings of the 7th International Palynological Congress, Brisbane, Australia, August 29–September 3, Abstr. 165. Elsevier, Amsterdam

Trincão PR (1990) Esporos e granos de polen do Cretácico Inferior (Berriasiano- Aptiano) de Portugal. PhD thesis, Paleontología e Biostratigrafia: Lisbon, Universidade Nova de Lisboa, 312 pp

Upchurch GR, Crane PR, Drinnan AN (1994) The megaflora from the Quantico locality (upper Albian), Lower Cretaceous Potomac Group of Virginia. Virginia Museum of Natural History Memoir 4:1–57

Upchurch GR, Otto-Bliesner BL, Scotese C (1998) Vegetation–atmosphere interactions and their role in global warming during the latest Cretaceous. Philos Trans R Soc Lond B Biol Sci 353:97–112

Vakhrameev VA, Krassilov VA (1979) Reproduktivnie organi tsvetkovikh iz aliba Kazakhstana (Reproductive organs of flowering plants from the Albian of Kazakhstan). Paleontol Zh for 1979:121–128

von Balthazar M, Pedersen CJ, Friis EM (2005) *Teixeiraea lusitanica*, a new fossil flower from the Early Cretaceous of Portugal with affinities to Ranunculales. Plant Syst Evol 255(1–2):55–75

Wang H, Dilcher DL (2006) Aquatic angiosperms from the Dakota Formation (Albian, Lower Cretaceous), Hoisington III locality, Kansas, USA. Int J Plant Sci 167(2):385–401

Wang H, Dilcher DL (2018) Early Cretaceous angiosperm leaves from the Dakota Formation, Hoisington III locality, Kansas, USA. Palaeontol Electron 21.3(34A):1–49

Wang H, Dilcher DL, Schwarzwalder RN, Kvaček J (2011) Vegetative and reproductive morphology of an extinct Early Cretaceous member of Platanaceae from the Braun's ranch locality, Kansas, U.S.A. Int J Plant Sci 172(1):139–157

Wheeler EA, Baas P (1993) The potential and limitations of dicotyledonous wood. Paleobiology 19(4):487–498

Wing SL, Boucher LD (1998) Ecological aspects of the Cretaceous flowering plant radiation. Annu Rev Earth Planet Sci 26:379–421

Wu Y, You H-L, Li X-Q (2018) Dinosaur-associated Poaceae epidermis and phytoliths from the Early Cretaceous of China. Natl Sci Rev 5(5):721–727

Zavada MS (2004) The earliest occurrence of angiosperms in southern Africa. S Afr J Bot 70(4):646–653

Postcards from the Mesozoic: Forest Landscapes with Giant Flowering Trees, Enigmatic Seed Ferns, and Other Naked-Seed Plants

6

Carole T. Gee, Heidi M. Anderson, John M. Anderson, Sidney R. Ash, David J. Cantrill, Johanna H. A. van Konijnenburg-van Cittert, and Vivi Vajda

Abstract

Earth's vegetation during the 186 million years of the Mesozoic, from the Paleogene–Cretaceous boundary at 66

We wish to dedicate this chapter on Mesozoic plants to our colleague and good friend Sid Ash (1928–2019), who was a devoted and prolific Mesozoic paleobotanist. A nonagenarian with a keen intellect and immense productivity, Sid enthusiastically took the lead on writing Sect. 6.7 on the Late Triassic Petrified Forest in Arizona, but died before the final revisions could be made.

Electronic supplementary material A slide presentation and an explanation of each slide's content is freely available to everyone upon request via email to one of the editors: edoardo.martinetto@unito.it, ragastal@colby.edu, tschopp.e@gmail.com

*The asterisk designates terms explained in the Glossary.

C. T. Gee (✉)
Institute of Geosciences, Division of Paleontology, University of Bonn, Bonn, Germany
e-mail: cgee@uni-bonn.de

H. M. Anderson · J. M. Anderson
Evolutionary Studies Institute (ESI), Witwatersrand University, Johannesburg, South Africa

S. R. Ash
Department of Earth and Planetary Science, Northrop Hall, University of New Mexico, Albuquerque, NM, USA

D. J. Cantrill
Royal Botanic Gardens Victoria, Plant Sciences and Biodiversity, Melbourne, VIC, Australia
e-mail: David.Cantrill@rbg.vic.gov.au

J. H. A. van Konijnenburg-van Cittert
Laboratory of Palaeobotany and Palynology, Utrecht University, Utrecht, Netherlands

Naturalis Biodiversity Center, Leiden, Netherlands
e-mail: jtvk@kgk.nl

V. Vajda
Swedish Museum of Natural History, Department of Palaeobiology, Stockholm, Sweden
e-mail: Vivi.Vajda@nrm.se

million years ago back to the Triassic–Permian boundary at 252 million years ago, was filled with forests. Like today, the forest was the dominant terrestrial ecosystem. The trees that created the forest habitat, along with the other woody plants and ferns in the understory and groundcover, were the primary producers that powered Earth's ecosystems by converting sunlight into chemical energy through photosynthesis. Yet, the forests that flourished during the Mesozoic differed from those found on Earth today. The Mesozoic climate was generally warmer, with milder seasons, a higher sea level, and no polar ice. This resulted in evergreen forests that may have looked superficially similar to gymnosperm-dominated forests of today, but were made up of very different kinds of plants. This is because major evolutionary changes took place in the plant world during this time interval. The Cretaceous witnessed the emergence and diversification of the flowering plants, which define our global flora now. In contrast, the Jurassic and Triassic floras were dominated by gymnosperms such as conifers and cycads, as well as by other, enigmatic, naked-seed plants including seed ferns and bennettitaleans that are now extinct. Continental drift tore landmasses apart, separating Northern Hemisphere floras with ginkgoes from the Gondwana flora in the south, which also is now extinct. Geological time, biotic evolution, and plate tectonics all contributed to the making of paleobotanically unique forests in different parts of the world. In this chapter, we present a series of written postcards from the Mesozoic, each one describing a forested landscape, as we travel back in time together on a virtual plant safari.

6.1 Introduction

Greetings from the Mesozoic! It is warm and sunny today, but pleasant and shady in the Mesozoic forests. Instead of just dashing off a note "wishing you were here," we will present a

E. Martinetto et al. (eds.), *Nature through Time*, Springer Textbooks in Earth Sciences, Geography and Environment,
https://doi.org/10.1007/978-3-030-35058-1_6

series of postcards from a virtual plant safari through the Middle Life era. The postcard scenes will capture the ever-changing landscapes of forests and terrestrial vegetation as we travel back through time from 66 to 252 million years ago [U0600].

The Mesozoic is sometimes known as the "Age of Dinosaurs," but we plant people like to think of it as the "Age of Cycads." This is because the cycads are the quintessential Jurassic plant for paleobotanists. However, the Mesozoic flora embraces more than just cycads. On our plant safari, we will see monumental changes in this immense span of 186 million years, from a global flora with flowers and broad-leaved plants in the Cretaceous to the gymnosperm-dominated vegetation of the Jurassic and Triassic. Gymnosperms, or "naked-seed" plants, including seed ferns and bennettitaleans, thrived during the Mesozoic, most of which went extinct at the Mesozoic–Cenozoic boundary. Tragically, there are no survivors of these two major plant groups, as there were with the living fossils of nautilus or horseshoe crabs. However, luckily for us, some plants from our paleobotanical past did survive the end-Cretaceous extinction event. Today, *Ginkgo*, horsetails (*Equisetum*), and the cinnamon fern (formerly *Osmunda*, now known as *Osmundastrum*) are the last surviving genera of their kind. In fact, *Osmundastrum cinnamomeum* of the Early Jurassic is exactly the same species that thrives on Earth today, 180 million years later (Serbet and Rothwell 1999). Indeed, an exquisitely preserved fossil stem of the Osmundaceae family shows that the royal ferns have remained unchanged since the Early Jurassic, right down to the number and size of its chromosomes (Bomfleur et al. 2014). The fact that this fern managed to avoid extinction for 180 million years attests to the staying power of plants through geological time. However, that is a topic for another time (see Pšenička et al., Chap. 11). Instead, we will be highlighting the beauty, diversity, and uniqueness of the Mesozoic forests on our virtual field trip.

We start our plant safari just before the cusp of the Mesozoic, 66 million years ago, when a major shift in global vegetation was caused by the cataclysmic impact of a giant asteroid. Our sojourn continues back in time through the Cretaceous, Jurassic, and Triassic to look at plant life in both the Northern and Southern Hemispheres. This series of postcard scenes arising from our journey together will end in the mid-Triassic, just before the events that mark the Paleozoic–Mesozoic boundary. Hope you enjoy the trip!

6.2 Disaster and Recovery: Dramatic Vegetation Changes at the Cretaceous–Paleogene (K/Pg) Boundary

Vivi Vajda

I am sending this postcard from the early Paleocene of New Zealand [U0601]. Here, 66 million years ago, it is a fairly silent world. There are no dinosaurs but there are some ground-dwelling birds, the ones that will later adapt to a life in the trees and become our song birds (Field et al. 2018). Traces of widespread fires are visible as charcoalified logs. Ferns are the only plants you see—ferns, ferns, ferns in a sea of greenery. It is a strange world without any big trees or forests. But, let us now travel back a few years, into the Late Cretaceous, then work our way back across the Cretaceous–Paleogene boundary.

During the Cretaceous, New Zealand belonged to the *Proteacidites/Nothofagidites* Province, a flora typified by the abundant presence of pollen called *Nothofagidites* (*Nothofagus* = southern beech) and *Proteacidites* (*Banksia* family) (Fig. 6.1) [U0602] (Vajda and Raine 2003). Other typical plants in these forests were tall conifers, such as *Araucaria* (monkey puzzle tree) and podocarps (plum pines), but also Huon pine (*Lagarostrobos franklinii*) and rimu (e.g., *Dacrydium*; Vajda and Raine 2003). Flowering plants were represented mainly by dicotyledons, including species of Nothofagaceae (southern beech family), Lauraceae (laurel family), and Proteaceae, based on investigations of fossil leaves (Kennedy 2003). The genera and species preserved as leaves are in agreement with the results derived from pollen studies. It was peaceful in these New Zealand forests until the end of the Cretaceous.

Indeed, at the end of the Cretaceous, almost exactly 66 million years ago, an asteroid of 10 km in diameter penetrated Earth's atmosphere and crashed onto what is today the Yucatan Peninsula of Mexico, producing the giant Chicxulub crater and leading to global devastation. As much as 75% of all species on Earth went extinct, among them, the dinosaurs (Tschopp et al., Chaps. 7 and 8). Plants are extremely sensitive to environmental perturbations and even sudden, short-lived changes in their diversity leave traces in fossil floras. The changes in vegetation that took place the day the asteroid hit Earth can be discovered by studying the fossil pollen and spores (palynological analysis) preserved in the sediments below and above the boundary layer, that is, before and after the asteroid impact.

The end-Cretaceous asteroid hit a target rock composed mainly of limestone (calcium carbonate) and gypsum (a calcium sulfate mineral), and the asteroid itself contained elements that we currently have in the core of our Earth, such as nickel, chromium, iron, and most importantly, the rare element iridium (Vajda et al. 2001). These elements were spread over the Earth's surface by an immense dust cloud along with molten bedrock excavated from the crater. Some of it fell back onto Earth's surface as tektites (molten natural glass) [U0601] or dust particles, and was mixed with the soot from the extensive wildfires that had scorched the forests ignited by the heat of the asteroid impact. That dust-and-soot cloud resulted in an extended period of darkness around the planet (Vajda et al. 2015). This mix of material formed a layer that is preserved at various places around the globe, which is generally referred to as the Cretaceous–Paleogene (K/Pg) boundary bed [U0603, U0604] (Vajda and Bercovici 2014). Continental (non-marine) strata comprising an intact

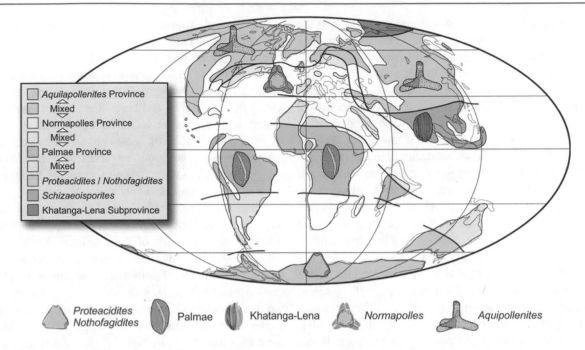

Fig. 6.1 Paleogeographic map 66 million years ago, showing main palynological provinces at the end of the Cretaceous with representative pollen morphologies. Modified from Vajda and Bercovici (2014)

boundary layer are scarce on Earth due to the erosion of sediments soon after they were deposited. Two of the best places to study how plants and land-dwelling animals responded to this disaster are the northern Great Plains of the USA and central Canada [U0605] (Vajda et al. 2013), and the western parts of New Zealand. Both places hosted swamps and wetlands with a high potential for the preservation of sediments at the time of the impact.

When the asteroid hit, dust filled the atmosphere and resulted in global darkness, which, according to physical models, lasted up to 2 years and led to the cessation of photosynthesis. Additionally, the sulfur released from the target rock gypsum combined with the water vapor in the atmosphere to form sulfuric acid (H_2SO_4) which rained down on Earth. The vegetation died abruptly as a consequence of darkness and acid rain. It seems to have been replaced completely by fungi immediately after the K/Pg event, which decomposed the last remains of the vanquished forests (Vajda and McLoughlin 2004; Vajda 2012). However, this fungal-dominated world only lasted for a brief period. As soon as light levels returned to normal values, ground-dwelling ferns and weedy angiosperms reestablished the vegetation, emerging from spores and seeds protected in the soil.

Luckily for us, a very detailed succession of spores and seeds was preserved in the geological record at Greymouth Coalfield, New Zealand [U0605] (Vajda et al. 2001). The record shows that one ground-dwelling fern species after another emerged to dominate in a succession of recovery vegetation (Vajda et al. 2001). These low-growing ferns were followed by tree ferns, such as *Cyathea* and *Dicksonia*

(Vajda et al. 2001). Many of the species with flowers became extinct, possibly owing to the prolonged darkness and the extinction of insect groups on which they had depended for pollination. However, in general, the plants in New Zealand seem to have survived with lower rates of extinction than those in regions closer to the asteroid impact site. Nevertheless, an ecosystem with a different structure appeared in New Zealand after the impact event. Most importantly for the plants, gone were the large herbivores. After a few thousands of years, the Huon pine became the dominant canopy tree, while other podocarps and *Araucaria* seem to have been outcompeted and replaced in the forest structure.

The evidence that documents the changed ecosystem mostly comes from mainland New Zealand, but when I was invited to participate in an expedition to Campbell Island by a group of New Zealand scientists, I jumped at this opportunity and just had to go! Campbell Island is an uninhabited subantarctic island located about 700 km south of New Zealand and a seemingly odd place to search for evidence of an asteroid impact that happened in North America. The core of Campbell Island is the remnant of a volcano, but Cretaceous sedimentary successions have been uplifted and preserved along the shores; these rock layers belong to the Garden Cove Formation. My colleagues and I worked hard for 2 weeks, sampling what we thought was the Cretaceous–Paleogene boundary. When we found a rusty layer in the sediments, we realized that it might indeed be the iron-rich layer laid down by the fallout from the asteroid impact, the characteristic sign of the K/Pg boundary [U0603]! Of course, we were thrilled as it meant that we had found the southernmost K/Pg boundary exposure in the world, the one furthest away from the impact site in Mexico.

Doing fieldwork in this remote place on Earth was an adventure that also turned out to be scientifically profitable. During high tide on Campbell Island, when the coastal sections were inaccessible, we enjoyed photographing the fabulous wildlife [U0601]. Our return with the *Break Sea Girl* ship was dramatic, because we were caught in a major storm and it took several days for us to reach safe harbor. Later, it turned out that the expedition had certainly been worth the trouble and seasickness. Palynological analyses confirmed that we had indeed discovered the K/Pg boundary. We found that different species of the *Banksia* family (*Proteacidites* species) had once been abundant on the islands during the Late Cretaceous, but managed to survive into the Paleocene (Wanntorp et al. 2011). Thus, life recovered after the asteroid impact and extinction disaster, but a very different world emerged afterwards. If it had not been for the Chicxulub asteroid, the dinosaurs would possibly still rule the Earth. What a postcard that would have made!

6.3   Where Have All the Flowers Gone? The Global Changeover in Vegetation During the Cretaceous, from Variations of Greens and Browns to Bright Flowery Colors

Carole T. Gee

In the Late Cretaceous, a postcard showing plants almost anywhere on Earth would have been garnished with bright flower colors and buzzing pollinators. One postcard depicting the flora some 60 million years earlier, however, would just have consisted of lush greens and browns. What happened about 125 million years ago to cause such a monumental shift in the colors of the plant world? [U0606]

One of the most exciting events happening in the Early Cretaceous that changed the course of botanical history and, thus, life on Earth, was the rise of the flowering plants. Without flowering plants, we would hardly have any of the food plants on which we subsist today, cows and horses would not be grazing on grass, giraffes would not be nibbling on acacia leaves, and the insect fauna would be vastly different, too, bereft of the many pollinators that interact with flowers. In fact, our present-day world is so strongly formed, defined, and supported by flowering plants that it would be difficult to imagine a time without them. Such a time is the Early Cretaceous. While the previous chapter (Kvaček et al., Chap. 5) discusses flowering plants, or angiosperms as they are technically known, in detail, our postcard from this part of the Mesozoic will cover the sweeping changes that occurred in the global vegetation from the Late Cretaceous,

around 70 million years ago, to the Early Cretaceous, about 133 million years ago.

By the end of the Cretaceous, before the giant asteroid hit Earth, all of the major groups of angiosperms had already evolved. These flowering plants were quite diverse in form, consisting of trees, shrubs, and herbs. The clearest fossil evidence of large trees is immense, rock-hard logs with massive amounts of wood [U0607–U0610]. More delicate forms, such as small herbaceous plants, are commonly found as surface imprints on fine-grained rock. And, by the Late Cretaceous, these plants had come to dominate some, but not all, landscapes.

There is evidence in the latest Late Cretaceous that both angiosperm- and gymnosperm-dominated floras were partitioned in different parts of the continents. For example, in what is today the American Southwest, angiosperms made up the forests in the Maastrichtian and Campanian, 83 to 66 million years ago. In New Mexico, angiosperm tree stumps are found with trunk diameters measuring up to 75 cm just above their roots (Estrada-Ruiz et al. 2012, 2018). There were also woodlands in New Mexico that were composed primarily of conifers, sometimes of even larger trees, as evidenced by their stumps with diameters up to 3 m across. Despite their success, the softwoods were being crowded out into more limited and stressed (less desirable) environments by the hardwoods at that time (Davies-Vollum et al. 2011). Thus, a postcard from the Late Cretaceous would not seem that much different from a photo taken on a plant safari to a tropical forest today, with a multitude of tall trees bearing wide leaves. By this time in Earth history, angiosperms had taken over most areas and replaced ferns and gymnosperms (Crane 1987; Friis et al. 2011). Their wily ability to adapt evolutionarily allowed them to move into a wide range of ecological niches, from the equator to the poles, from dryland habitats to freshwater setting and even into the sea. This radiation resulted in a great variety of forms and an ever-increasing number of species. However, it turns out the large broad-leaved trees of the Late Cretaceous were relative newcomers to the world of flowering plants.

The tree habit developed quite late in the evolution of angiosperms. A gigantic fossil log recently became a scientific sensation when it was reported from Turonian-aged sediments in Utah (Jud et al. 2018). The log measured nearly 2 m in diameter and is 92 million years old. This paleobotanical evidence is important because it is the oldest and largest trunk of a broad-leaved tree, which is estimated to have reached over 50 m in height. It also shows that forests of large angiosperm trees had already developed only some 33 million years after the first appearance of flowers some 125 million years ago. For paleontologists, this is a relatively short span of time.

Before the appearance of such large broad-leaved trees, flowering plants were much smaller in stature. They consisted of small trees, large shrubs, shorter shrubs, or herbaceous, weedy plants. In fact, the oldest fossil angiosperms were extremely small, measuring on the order of a few centimeters

Fig. 6.2 Relative size comparison of a modern water lily flower and the first water lily flowers from the Early Cretaceous. Based on flower dimensions given by Friis et al. (2011) and illustrated here with a photo taken by CT Gee

in height (summarized by Friis et al. 2011). To drive home how minute the first angiosperms were, let us have a look at the beautiful and charismatic water lily family and their lovely floating flowers (Fig. 6.2) [U0606, U0611].

The most common water lily in Europe, *Nymphaea alba*, produces fragrant, brilliantly white to sometimes reddish-colored flowers. The flowers are relatively small, measuring from 5 to 20 cm across. In comparison, the oldest water lily flowers from the Early Cretaceous would have been a fraction of that size, only 3 mm tall (Fig. 6.2) [U0611] (Friis et al. 2001, 2009). The significant size difference is not due to natural variation in flower dimensions between species. In this case, it is due to the fact that the earliest angiosperms were absolutely tiny.

In the last two decades, a number of small flowers, flower buds, and seeds have been described from the Early Cretaceous (summarized by Friis et al. 2011). It is actually a wonder that such diminutive, ancient flowers have been found at all. Their serendipitous discovery is due to their amazing preservation as charred plant parts. Through the process of charcoalification by wildfire, these tiny flowers and seeds have retained their three-dimensional forms with perfect cellular preservation. And, as such, it was possible to sieve these tiny bits of fossil plants out carefully from loose sediments of Early Cretaceous age, helping to explain what was considered to be a mystery for the past 140 years (Box 6.1).

Whether the elusiveness of the oldest angiosperm fossils was due to their minute sizes, non-woody habit, or preference for habitats in which they were not preserved, the intense search by plant paleontologists for them and their evolutionary precursors in the Early Cretaceous goes on today. Hence, although a Mesozoic postcard of 125 million old landscapes with the first, very inconspicuously small flowering plants may have seemed like nothing to write home about, these diminutive weeds were actually the first foot soldiers of the biotic revolution in the global flora that led to our present-day, angiosperm-dependent world.

6.4 Southern High-Latitude Forests of the Early Cretaceous in Southeastern Australia

David J. Cantrill

Conifers and ferns flourished in the global forests in both the Northern and Southern Hemispheres in the Early Cretaceous (Fig. 6.3a–e) [U0612], despite the geological upheaval that was occurring during this time. That upheaval was the continuation of the breakup of the great southern landmass of Gondwana. Large rift valleys separated land areas, parts of

Fig. 6.3 Fossil plants from the Early Cretaceous flora of southeastern Australia. (**a**) Branch with leaves of the araucarian conifer *Araucaria lanceolata*, (**b**) twig of the podocarpaceous conifer *Bellarinea*, (**c**) cone scale of *Araucarites* from an araucarian seed cone, (**d**) *Sphenopteris* fern foliage, (**e**) a frond of the osmundaceous fern *Cladophlebis australis*, (**f**) a rare thalloid liverwort, showing a darker central vascular strand. Photos by DJ Cantrill

which had been connected since the Neoproterozoic (550 Ma), and the supercontinent had split into western (Africa, South America) and eastern (Antarctica, India, Australia, New Zealand) fragments, separated by a seaway. A vast rift valley formed between India, Antarctica, and Australia, dividing eastern Gondwana into separate land masses. Rifting between Antarctica and Australia, across what is today the southern margin of Australia, started in the Jurassic (c. 144 Ma) and propagated eastward through time [U0613]. In the Early Cretaceous, much of eastern Gondwana was located further south than at present.

Set in polar latitudes (70–85° S; Torsvik et al. 2012), this region experienced several months of darkness each year. Yet, the fossil evidence points to a diverse biota that included vertebrates (Poropat et al. 2018; Poropat in Tschopp et al., Chap. 8). Our understanding of the vegetation and environ-

ments largely comes from floras preserved in the Gippsland and Otway Basins in these rift settings. The flanks of the rift valley were covered in alluvial fans with sediment derived from the hinterland (e.g., Tosolini et al. 1999), but most of the sediments that filled the basins came from contemporaneous intra-rift volcanoes (Felton 1997a, b). River systems drained these highlands and consisted predominantly of wide (kilometer-scale) channel belts with sand-rich braided channels with poorly defined banks and adjacent floodplains (Felton 1997b). The river systems in which these deposits formed may have been as deep as about 6 m or, occasionally, deeper (Felton 1997b), and experienced high variations in flow regime over time. During periods of flashy discharge and at peak flows, riverbanks were overtopped, depositing sheets of sand and silt onto the adjacent floodplains. Soil profiles developed throughout the succession and record the colonization of the floodplain and, less frequently, in the confines of the channels themselves.

The rift valley maintained terrestrial environments for 40 million years (or more) during the earliest Cretaceous before rifting was complete. Once the rift completely separated, it was flooded by the sea, isolating Australia from the rest of Gondwana. The marine incursion started earlier in the west than in the east. It is unsurprising, then, that changes in vegetation appear first throughout this interval. The youngest vegetation (Albian, c. 110 Ma) contains a rare flowering plant component (Korasidis et al. 2016). However, in contrast to the other parts of the world, angiosperms had not reached ecological dominance, as they formed only a minor component of the Aptian and Albian forests and floodplain vegetation.

The land was clothed in forests, as evidenced by the abundant woody debris entrained in river channels. The main forest trees were Araucariaceae (Fig. 6.3a, c) [U0612, U0614] (Cantrill 1991, 1992) and Podocarpaceae (Fig. 6.3b) [U0614], but other gymnosperm families were present, including the Cupressaceae and Cheirolepidiaceae (Tosolini et al. 2015). These trees formed an overstory above a fern-rich understory that also contained shrubby seed plants (Tosolini et al. 2018). Older floras also include Ginkgoales, whereas those on the basin margins were rich in Bennettitales (Douglas 1969; McLoughlin et al. 2002). The most striking aspect of this vegetation is the diversity and abundance of ferns and bryophytes (Douglas 1973; Drinnan and Chambers 1986). Ferns include tree ferns (Dicksoniaceae and Cyatheaceae), royal ferns (Osmundaceae, *Phyllopteroides*; Fig. 6.3e) [U0612, U0615] (Cantrill and Webb 1987), and a diversity of other forms such as *Sphenopteris*, *Aculea*, *Alamatus*, and *Amanda* (Fig. 6.3d) [U0616] (Douglas 1973; Drinnan and Chambers 1986). Thalloid liverworts are common in high-latitude southern floras (Cantrill 1997) and are colonizers of bare ground following flood events (Fig. 6.3f) [U0617]. Sheet-sand deposits on the floodplains often preserve mats of thalloid liverworts, and evidence of leafy liverworts from the forest is found in the Koonwarra Fish Beds

(Drinnan and Chambers 1986). Let us now travel across the Pacific and back in time, to explore Jurassic forests of the Northern Hemisphere.

6.5   The Conifer Forests of the "Jurassic Park" in Western North America

Carole T. Gee

Greetings from the Jurassic park of western North America! Here in the Late Jurassic, 150 million years ago, it is refreshingly green and resin-scented in the big-tree conifer forests. There are no broad-leaved trees in the vegetation of this epoch— the angiosperms first show up 25 million years later—just a lot of different conifers. There are also lush pockets of ferns and tree ferns, horsetails on the riverbanks, ginkgoes, cycads, and bennettitaleans in the understory, and plant-eating dinosaurs all around. Large dinosaurs are browsing on the leaves of the tall araucarian trees in the forests [U0600], whereas the younger and smaller ones seem to relish the horsetails growing along the riverbanks. Fossils of these organisms are preserved in the Upper Jurassic Morrison Formation of the western USA.

Although the Morrison Formation is deservedly world-famous for its diverse dinosaur fauna and other vertebrates (Foster et al. in Chap. 8), less fuss has been made about its fossil flora. Indeed, the giant bones of the huge sauropods tend to capture the paleontological attention of most people. Yet, a walk in the Morrison Formation today, for example, across the dry desert landscapes of eastern Utah [U0618], will reveal an abundance of fossil wood originating from the conifer trees that grew at that very same spot some 150 million years ago (Fig. 6.4a) [U0619–U0622]. Striking are also the colored bands of ancient soil horizons that characterize the upper parts of the Morrison Formation [U0618].

Located in the Western Interior of North America, the Morrison Formation is unusual because it is a laterally widespread geological unit, with outcrops occurring from what is today Montana to northern New Mexico. This broad

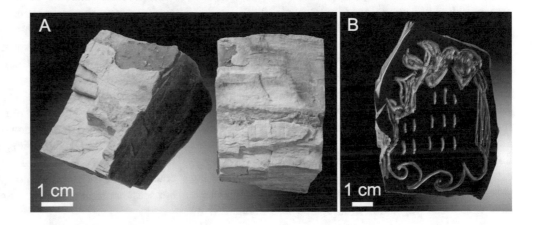

Fig. 6.4 Jurassic wood of the conifer family Araucariaceae shown in two different kinds of preservation. (**a**) Silicified wood from the Late Jurassic Morrison Formation of Utah, USA. (**b**) A piece of the black gemstone jet, carved with the Whitby Cathedral, formed from the wood of an araucarian tree growing during the Middle Jurassic in what is today Yorkshire, UK. Photos taken by Georg Oleschinski

Fig. 6.5 Woodiness in the fossil flora of the Late Jurassic Morrison Formation, USA. (**a**) Pie chart showing the relative frequencies of major plant groups. (**b**) Pie chart showing the relative frequencies of woody and non-woody (herbaceous) plant groups. Woody plant species make up two-thirds of the flora. Both charts compiled by C.T. Gee based on data reported by Ash and Tidwell et al. (1998)

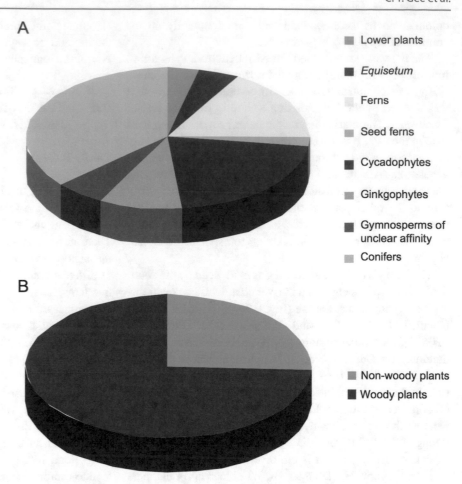

A

- Lower plants
- *Equisetum*
- Ferns
- Seed ferns
- Cycadophytes
- Ginkgophytes
- Gymnosperms of unclear affinity
- Conifers

B

- Non-woody plants
- Woody plants

latitudinal expanse means that the wide range of plant communities that thrived in the different climatic zones are found in the paleobotanical record. In the northern parts of the Morrison Formation, where the climate was cooler and moister, the flora was rich in ferns and seed ferns (Bell 1956; Brown 1972). Traveling toward the southern parts of the formation, there is a trend that shows increasing amounts of Cheirolepidiaceae conifer pollen, which likely corresponds to higher temperatures and seasonal aridity (Hotton and Baghai-Riding 2010). The now-extinct family of the Cheirolepidiaceae has long thought to be tolerant of warm, arid, or halophytic conditions (Taylor et al. 2009).

The prevalence of forests and woodlands in the Morrison Formation is reflected in the species diversity of the fossil flora. A new tabulation presented here, based on a critical review of the entire Morrison flora made 20 years ago (Ash and Tidwell 1998), shows that the vegetation was biologically diverse (Fig. 6.5a) and that three-quarters of the fossil species consisted of woody plants (Fig. 6.5b). Such abundant and consistent fossil evidence of the widespread forests flourishing in the Morrison Formation 150 million years ago contradicts a relatively recent hypothesis about the prevailing paleoenvironmental conditions of that time. This hypothesis says that fossil vegetation was sparse and dominated by mostly low-growing, non-woody plants (Parrish et al. 2004) that struggled to survive in semi-arid to arid conditions (Rees et al. 2004; Turner and Peterson 2004). As a consequence of this supposed aridity, foraging dinosaurs were unable to thrive (Engelmann et al. 2004). The extent of conifer forests now known from the Morrison Formation makes this hypothesis difficult to support.

Conifer trees formed the basic framework of the Morrison forests, creating the habitat, building the canopy, and providing food and shelter for the animals. These trees were large, with plentiful amounts of dense wood, similar to today's conifers. In northeastern Utah, for example, large fossil conifer logs are found near the city of Vernal and Dinosaur National Monument (Sprinkel et al. 2019). The largest fossil log in this area measures 11 m in length and is 127 cm wide (Fig. 6.6a) [U0621]. From the log's girth, we can calculate that this forest consisted of trees of at least 28 m in height (Gee et al. 2019). The wood of these conifers is clearly araucarian (Fig. 6.6b–d), even down to the resin plugs found in the wood's ray cells (Fig. 6.6c) [U0621]. There are no growth rings (Fig. 6.6b) [U0621], indicating that the climate was equitable. These trees did not experience any water stress, nor seasonal light or temperature fluctuations.

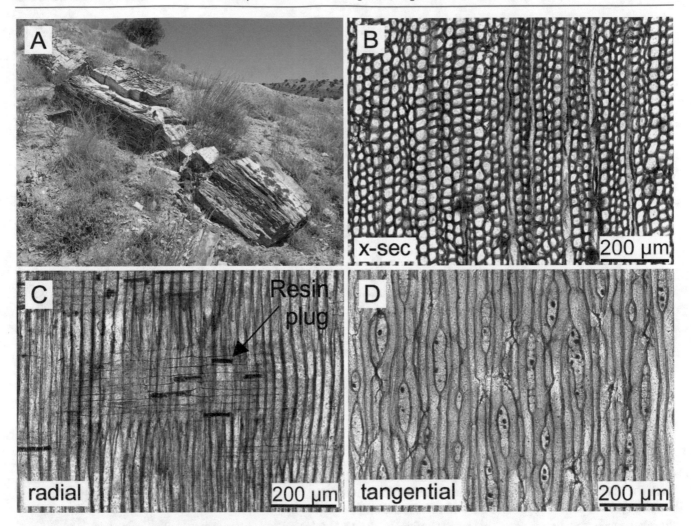

Fig. 6.6 A 150-million-year-old log from the Upper Jurassic Morrison Formation near Vernal, Utah, USA, and its anatomical structure as shown in thin-section. (**a**) The log in outcrop, with a geologic hammer for scale; its wood anatomy in (**b**) cross section (x-sec), (**c**) radial sec- tion, (**d**) tangential section. The resin plugs in the ray cells of the fossil wood are best seen in radial section (**c**). Photos (**a**) and (**b**–**d**) taken by Dale Gray and CT Gee, respectively

In the Escalante Petrified Forest State Park in southern Utah, there is an immense natural assemblage of fossil logs and pieces of wood. On display near the Visitor Center is a large fossil tree that is 15 m long. This Morrison tree is particularly interesting because it is so complete, including a hollow in the trunk and a root ball at its base [U0622]. The largest diameter fossil tree at the Escalante park known so far measures 105 cm across. This tree grew to a height at least of 22 m. The abundance of big logs and wood in this part of Utah indicates that a tall, dense conifer forest grew here 150 million years ago.

The fossil-wood flora of the entire Morrison Formation consists of eight different taxa of conifers (summarized by Gee et al. 2019). With the exception of wood identified as araucarian, most of these species cannot be assigned to any living family. However, it is clear that they all had dense coniferous wood and grew as trees (see "For deeper learning").

The diversity of the conifer flora is also reflected in its reproductive cones, which have been reported from the Morrison Formation in Utah and Wyoming (Fig. 6.7; Gee and Tidwell 2010; Gee 2013; Gee et al. 2014). Some of these cones belong to the conifer family Araucariaceae, which is well known today for the monkey puzzle tree (*Araucaria araucana*) or the Norfolk Island pine (*Araucaria heterophylla*) [U0623]. The conifer-cone flora also shows that members of the pine family (Pinaceae) and the extinct Cheirolepidiaceae family added to the forest diversity. In fact, one of the oldest seed cones of the pine family in the world comes from the Morrison Formation of Utah (Fig. 6.7b) [U0624–U0626] (Gee et al. 2014). To date, six types of seed cones and three types of pollen cones are found throughout Utah and Wyoming (Gee et al. 2014). Like the varied wood flora, the conifer-cone flora indicates that the dominant Morrison plant community consisted of mixed conifer forests.

Fig. 6.7 Rare conifer seed cones from the Upper Jurassic Morrison Formation of Utah, USA. (**a**) Araucariaceae (here in longitudinal polished section), (**b**) Pinaceae, (**c**) unknown affinity, (**d**) unknown affinity, (**e**) Cheirolepidiaceae, a now-extinct conifer family. Modified from Gee et al. (2014). Photos taken by Georg Oleschinski

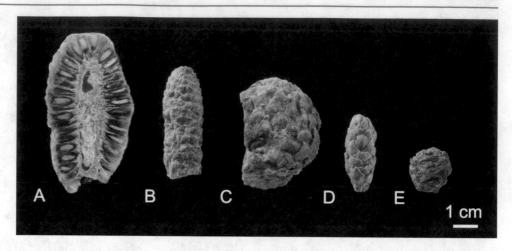

The understory in the forests, open woodlands, or clearings consisted of ferns, seed ferns, cycads and bennettitaleans, and ginkgophytes (summarized by Tidwell 1990a; Ash and Tidwell 1998; Chure et al. 2006). Moister habitats, such as around lakes, near riverbanks, or in marshes, supported the luxuriant communities of ferns and seed ferns, as well as mosses and horsetails. One common fern family was the Osmundaceae, known today as the royal ferns. These grew low on the ground, or formed a tree-like trunk and grew as a tree fern (Tidwell 1990b, 1994). Stands of cheirolepidiaceous conifers grew in drier areas. With their special leaf adaptations, such as thick cuticles and papillae around the stomata (cf. Francis 1983), the cheirolepidiaceous conifers were well suited to colonizing the more seasonally arid environments, saline soils, or even brackish-water habitats. From the abundance and broad geographic distribution of their characteristic pollen and leaf cuticle, we know that they also formed extensive forests in the Morrison Formation, mostly in the southern part of the formation (Litwin et al. 1998; Hotton and Baghai-Riding 2010).

Thus, the botanical postcards from our safari to the Upper Jurassic Morrison Formation show us a wide range of habitats with different plant communities, although most of the scenes involved conifer forests. Between the trees, smaller herbivorous dinosaurs could be spotted, nibbling on horsetails and ground-dwelling ferns, while giant sauropods were steadily browsing on conifer needles on the outskirts of the big-tree forests (Box 6.2).

Box 6.2: Preferred Food Plants of the Herbivorous Dinosaurs

The question of what the herbivorous dinosaurs fed on before the advent of the angiosperms has long been a puzzle. Plants that comprise the diet of most animal foragers and browsers today, such as grasses and dicot leaves, did not appear until the Late Cretaceous, and

extensive grasslands showed up even later, in the Cenozoic (see Saarinen et al., Chap. 3). What did the plant-eating dinosaurs eat in the age of ferns and gymnosperms during the Triassic, Jurassic, and Early Cretaceous? To answer this question, vertebrate paleontologists have looked at jaw and tooth morphology and also compared the head height and posture of various taxa to the growth heights of plants (e.g., Calvo 1994; Upchurch and Barrett 2000; Whitlock 2011).

In the last decade, a new approach was carried out that looked at the food ecology of the herbivorous sauropods from the plant perspective. To promote their fast growth and to obtain and maintain their huge body sizes, the sauropods must have been non-stop bulk feeders, but also preferred food plants offering high calorie and mineral nutrition. Thus, a laboratory study was carried out to measure and compare the calories and mineral nutrition yielded by plant groups that had survived from Mesozoic times to the present.

Using a standardized laboratory test normally designed for livestock feed, animal nutritionists and paleobotanists joined forces to analyze the nutritional content of the nearest living relatives of the Mesozoic flora. The Hohenheim Feed Test is used to analyze the digestibility of living plants such as *Equisetum*, various ferns and tree ferns, *Ginkgo*, cycads, and conifers such as *Araucaria* and *Pinus* (Hummel et al. 2008). The plant that provided the most calories in the shortest amount of time is the horsetail *Equisetum* (Fig. 6.8). The height of this low-growing plant would have been accessible to young sauropods or small-statured dinosaur species. Thus, because of its outstanding energy yield, along with its large amount of protein and mineral nutrients, *Equisetum* horsetails would have been the best food for baby and growing dinosaurs, as well (Gee 2011). Mature sauropods would have also benefited from feeding on the fast-growing colonies of *Equisetum* on the shores of ponds, lakes, and rivers.

The digestibility curve for the conifer *Araucaria* is trickier to interpret: it is slow to rise in the beginning, but in the end, it produces more calories over a longer period of time than other plant groups, with the exception of *Equisetum*. This unusual pattern of digestion of *Araucaria* foliage would have been optimal for fully grown sauropods with a long digestive retention time, for it would have allowed for the full extraction of nutrients from the leaves by the larger animals (Hummel et al. 2008; Gee 2011). In any case, the favorite food plants of the herbivorous dinosaurs—young and old—would have certainly included *Equisetum* and *Araucaria*.

In summary, the Late Jurassic Morrison flora consisted of dense, species-rich plant communities typical of the Jurassic. Widespread tall-tree forests of mixed conifers were dominated by araucarians in the mesic habitats, that is, in environments with moderate amounts of moisture [U0600]. The forest understory consisted of a rich assemblage of ferns, tree ferns, seed ferns, ginkgoes, and cycadophytes, especially where local conditions were moist or humid. In xeric, or drier, regions, the cheirolepidiaceous conifers took over the major role in building the forest and woodland communities. Colonies of horsetails and a variety of ferns thrived in

wetlands and moist areas around rivers, ponds, and lakes. The verdant landscape in western North America some 150 million years ago was, thus, strikingly different from the beautiful, but arid, countryside seen in much of the Morrison Formation in western North America today [U0654].

6.6 The Classic Mesozoic Forest of Ferns and Gymnosperms from the Middle Jurassic of Yorkshire, England

Carole T. Gee and Johanna HA. van Konijnenburg-van Cittert

In mid-Jurassic times, the rifting of the Central Atlantic Ocean basin began breaking up the northern landmass of Pangea into two parts, the Americas to the west and Eurasia to the east. At that time, about 175 million years ago, Britain was located somewhere between 30°–40° north of the equator, some 15°–25° farther south than its current position. Its location in the subtropical belt of the Northern Hemisphere meant that it experienced a warmer and moister climate than at present, and the seasons were more equable. We know about the milder Middle Jurassic climate from the fossil record because an extensive evergreen forest grew in what is today Yorkshire in northeastern England. In fact, the Yorkshire fossil flora is one of the best-known collections of Middle Jurassic plants in the entire world.

There are more than 600 horizons exposed along coastal Yorkshire in which plant fossils are preserved. This, in itself, is an amazing number of fossiliferous deposits in a single area. These are found from just north of the cathedral city of

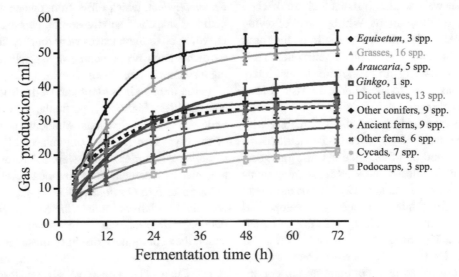

Fig. 6.8 A graph showing the digestibility of various species of plant leaves for herbivores. The plants selected for experimentation were the nearest living relatives of the Mesozoic flora, plus grasses and angiosperm leaves (in green) for comparison. The experimental trials were carried out for 3 days to simulate the length of time the plant matter would spend in a fully grown sauropod's digestive system. The plants yielding the most energy are *Equisetum* (horsetails) and *Araucaria* (trees of the monkey puzzle family), followed by gymnosperms such as *Ginkgo* (maidenhair tree) and other conifers like pines and junipers. The worst energy producers were cycads and podocarps

Whitby, southwards, and inland, as well. The fossiliferous outcrops are commonly organic-rich, gray mudstone when exposed at the seashore, whereas they are mainly claystone inland. In fact, sometimes one must pay very close attention to the tides to get to the fossil locality in the intertidal zone when collecting fossils. This is the case at Cloughton Wyke, for instance, where one has to wait for low tide before reaching the rocks and hacking through the thick cover of seaweed on them to collect the Middle Jurassic plants [U0627].

Along the North Yorkshire coast, an interbedded succession of rock preserved either the fossil evidence of marine organisms or of terrestrial plants. This interbedded character attests to the oscillating rise and fall of sea level here 170 million years ago. The non-marine intervals of the section are about 250 m thick and contain leaves and stems, as well as microscopic pollen and spores. The richest plant fossil horizon is that of the Gristhorpe Plant Bed of the Cloughton Formation, which crops out at Cloughton Wyke and at Cayton Bay near Yons Nab, the headland south of Scarborough. Other prolific plant-fossil beds are found in the underlying Saltwick Formation, in the Hayburn Wyke and Hasty Bank Plant Beds; the latter plant bed is one of the sites located inland about 50 km west of Whitby. Fossil plants from these beds have been worked on scientifically for nearly 200 years (see "For deeper learning"), yielding a breadth and depth of knowledge on this important mid-Mesozoic flora.

As elsewhere in the world, the major Middle Jurassic vegetation in Yorkshire was forest. The trees are represented in the fossil flora, for example, by conifer twigs and cones related to the living monkey puzzle tree (Araucariaceae) [U0628]. Tall araucarian trees also grew on drier soils. The araucarian forest trees were so plentiful that their trunk wood became the source of the famous Whitby jet. Following burial, millions of years of geological pressure on the fossil wood turned it into the hard, glossy, black gemstone, that is carved and sold as jewelry in Yorkshire shops today (Fig. 6.4b). There was more to the Middle Jurassic forests than just araucarians, though.

The Yorkshire forest was lush and diverse, replete with 260 species of horsetails, ferns, cycads, bennettitaleans, seed ferns, ginkgoes, and conifers. Like cycads, ginkgoes, and conifers, the bennettitaleans and seed ferns are gymnosperms (see Gastaldo et al., Chap. 12), or "naked-seed" plants. Their seeds are not tightly wrapped, or enveloped, in a protective tissue as in the angiosperms, the so-called "vessel-seed" plants. The bennettitaleans and seed ferns went extinct at the Paleogene–Cretaceous boundary, but these plant groups flourished as a common and important part of the plant world during the Mesozoic, along with cycads and ginkgoes that survived the Paleogene–Cretaceous boundary. Hence, many consider the Middle Jurassic

Yorkshire flora the "classic Mesozoic flora" for its high diversity of typically Mesozoic plants and the completeness of the assemblage.

Conifer forests dominated the lowlands, too (Fig. 6.9). These trees included members of the bald cypress family Cupressaceae and the conifer family Cheirolepidiaceae, which both formed the canopy. As found in forests in the Southern Hemisphere today, individual araucarian trees may have been the canopy emergents, projecting far into the sky above the general forest canopy. A diversity of lower-growing plants, including cycads, bennettitaleans (Fig. 6.9), and seed ferns, flourished in the understory. These cycads are not completely comparable to the living cycads, but their leaves are very similar [U0629]. The bennettitaleans are a group of plants that resemble the cycads in their leaf architecture, but have very different reproductive organs. The enigmatic gymnosperms, known as seed ferns, originally were given this name because the leaves resembled those of the ferns but bore seeds instead of sporangia. In Yorkshire, these plants are represented by the Caytoniales [U0630], Corystospermales, and Czekanowskiales (Fig. 6.10). Caytonialean leaves are palmately compound, consisting of four leaflets. Czekanowskialean leaves are often needle-like and borne in bundles called short shoots (Fig. 6.10). All of these plant groups represent woody shrubs, although some bennettitaleans may have been large enough to form small trees.

Ginkgophytes were constituent trees of the lowland forests, although these are found in low diversity. There are four genera of ginkgophyte leaves in the Yorkshire flora, including those of *Ginkgo* itself (Fig. 6.10) [U0631], which survives today as the sole member of this once larger plant group. The ginkgophyte leaves also include *Baiera* [U0631], *Sphenobaiera*, and *Eretmophyllum*, which differ from one another and *Ginkgo* in both leaf shape and subdivision of their leaf blade (Fig. 6.10). However, all of these leaves were roughly fan-shaped and bore the dichotomously∗ branching veins so characteristic of the living maidenhair tree, *Ginkgo biloba*.

Ferns were an important part of the Middle Jurassic plant communities (Fig. 6.11), particularly in the lowland forest understory and in clearings. Abundant were the large fronds of the royal fern family Osmundaceae [U0632]. Fronds of *Cladophlebis*, for example, were robust and spreading, with a regular arrangement of its pinnules∗, similar to those of the living royal fern *Osmunda* [U0632]. Other herbaceous ferns with living relatives include the Marattiaceae, Matoniaceae [U0633], Dipteridaceae, and Schizaeaceae, all of which document the evolutionary origin of these ferns deep in paleontological time, as we will see later in this book (Pšenička et al., Chap. 11). The most diverse fern family was the Dicksoniaceae [U0634], which are tree ferns native to humid forests in today's Southern Hemisphere. Represented by ten different species in the fossil flora, these ferns had distinc-

Fig. 6.9 Diversity of conifer twigs and bennettitalean pinnules in the Middle Jurassic Yorkshire flora, UK. (**a–h**) Conifer twigs: (**a**) *Pagiophyllum insigne*, (**b**) *Lindleycladus lanceolatus*, (**c**) *Geinitzia rigida*, (**d**) close-up of *Geinitzia rigida*, showing the arrangement of the tiny leaves, (**e**) *Bilsdalea dura*, (**f**) *Marskea jurassica*, (**g**) *Elatoclatus zamioides*, (**h**) *E. setosus*. (**i–m**) Bennettitales: (**i**) *Zamites*, (**j**) *Otozamites*, (**k**) *Ptilophyllum*, (**l**) *Anomozamites*, (**m**) *Pterophyllum*. All drawings by Jan van Konijnenburg, modified from *The Jurassic Flora of Yorkshire* published in 1999, and used with the permission of the Palaeontological Association, London

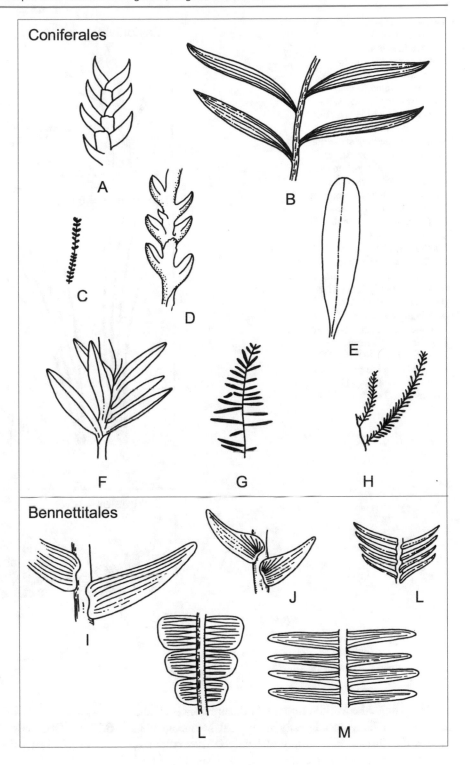

tively tiny pinnules with rounded lobes, such as *Coniopteris* (Fig. 6.11).

Bogs and freshwater marshes were dominated by *Equisetum* horsetails [U0632]. These horsetails had broad stems, up to 15 cm wide, which were massive in comparison to the slender shoots of present-day species. Similar to their living relatives, however, the Jurassic-age *Equisetum* formed dense colonies that spread by way of their underground rhizomes*. It is possible to view this growth strategy today near Hayburn Wyke along the Yorkshire coast where there is a thick, carbonaceous black bank of upright horsetail stems representing an extensive *Equisetum* marsh.

Fig. 6.10 Diversity of gymnospermous reproductive structures and leaves in the Middle Jurassic Yorkshire flora, UK. (**a–c**) Caytonialean reproductive structures: (**a**) Woody stem with a seed-bearing structure of *Caytonia nathorstii*, (**b**) A single seed-bearing structure of *Caytonia nathorstii*, (**c**) Woody stem with pollen sacs of *Caytonanthus oncodes*. (**d–h**) Czekanowskialean leaf types: (**d**) *Solenites vimineus*, (**e**) *Czekanowskia furcula*, (**f**) *C. microphylla*, (**g**) *C. blackii*, (**h**) *C. thomasii*. (**i–l**) Ginkgoalean leaf shapes and venation patterns: (**i**) *Ginkgo*, (**j**) *Baiera*, (**k**) *Sphenobaiera*, (**l**) *Eretmophyllum*. All drawings by Jan van Konijnenburg, modified from *The Jurassic Flora of Yorkshire* published in 1999, and used with the permission of the Palaeontological Association, London

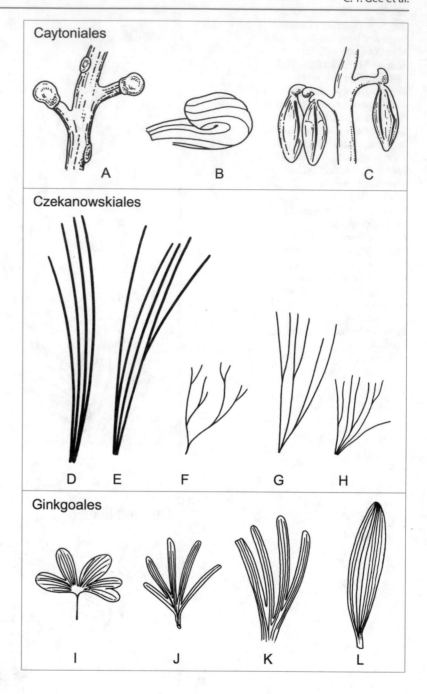

As a whole, the Middle Jurassic Yorkshire flora is characterized by a large biodiversity of typical Mesozoic plant groups that thrived in a variety of habitats. They colonized soils from the coasts to freshwater marshes and bogs, from lowland woods to slope forests. The species richness and excellent preservation of the fossil flora, as well as the two centuries of scientific work on this plant assemblage, have propelled it into being one of the best-known floras of the Mesozoic and, thus, a classic in the paleobotanical world. Let us now travel back to western North America, and into the Triassic.

6.7 The Colorful and Iconic Late Triassic Petrified Forest of Arizona, USA

Sidney R. Ash and Carole T. Gee

When driving through the southwest of the USA, it can be difficult to imagine that this arid desert was once a green and lush landscape. This ancient landscape was crisscrossed by many rivers and streams, and dotted with lakes. All of these areas were covered by groves of tall coniferous trees that grew

Fig. 6.11 Diversity of fern pinnules in the Middle Jurassic Yorkshire flora, UK. (**a–e**) Osmundaceae: (**a**) *Todites thomasii*, (**b**) *Cladophlebis denticulata*, (**c**) *Todites williamsonii*, (**d**) *Todites princeps*, (**e**) *Cladophlebis harrisii*. (**e–o**) Dicksoniaceae: (**f**) sterile *Coniopteris hymenophylloides* pinnules, (**g**) fertile *C. hymenophylloides* pinnules, (**h**) sterile *C. murrayana* pinnules, (**i**) fertile *C. murrayana* pinnules, (**j**) sterile *C. bella* pinnules, (**k**) fertile *C. bella* pinnules, (**l**) sterile *C. bella* pinnules, (**m**) fertile *C. bella* pinnules, (**n**) entire pinna of *Dicksonia mariopteris*, (**o**) Close-up of *Dicksonia mariopteris* pinnules with venation. (**p**) Thelypteridaceae, part of a frond of *Aspidistes thomasii*. (**q–s**) Matoniaceae: (**q**) *Phlebopteris polypodioides*, (**r**) *Phlebopteris woodwardii*, (**s**) *Matonia braunii*. All drawings by Jan van Konijnenburg, modified from *The Jurassic Flora of Yorkshire* published in 1999, and used with the permission of the Palaeontological Association, London

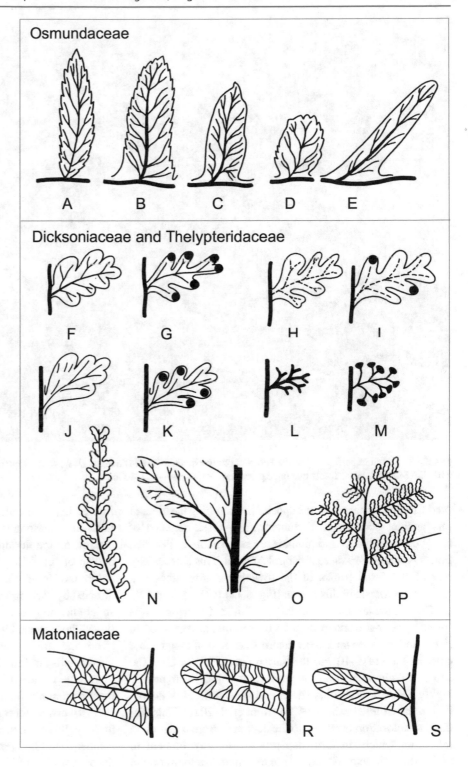

above a dense understory of seemingly familiar ferns, as well as cycads and other plants that would be unfamiliar to our eyes. A land where horsetails of all sizes up to 10 m tall colonized the margins of watercourses and lakes. A land inhabited by strange and unusual reptiles and amphibians of various shapes and sizes, as well as dinosaurs about as tall as a 10-year-old child. A land without modern birds of any kind, although there were a few small flying reptiles called pterosaurs, in addition to a variety of insects including beetles and damselflies flitting around in the warm humid air. Such was this now-desert region of Arizona about 200–230 million years ago during Late Triassic times when the strata in the intensely col-

Fig. 6.12 A picture perfect day in the Petrified Forest National Park in Arizona, USA, with a view of massive Late Triassic logs that represent the 210–230 Ma old conifer forest that once grew here. Photo taken by CT Gee

ored Chinle Formation were deposited, entombing the biota of this ancient ecosystem. Fortunately, the fossil remains of many of these plants and animals are still visible in Petrified Forest National Park in east-central Arizona. This is especially true of the fossil trunks of the towering conifer trees that formed the forests in this region (Fig. 6.12) [U0635–U0640].

The characteristically colorful Chinle Formation was formed under a tropical climate by sediments deposited in rivers and in lakes near the western shoreline of Pangea at a paleolatitude of 5–10° north (Baranyi et al. 2017). Over the course of about 30 million years, some 400 m of mostly highly colored mudstone, shale, and sandstone were deposited in the Chinle basin [U0635] (Kent et al. 2018). This distinctive coloration is related to the degree of iron oxidation in the strata which, in turn, depends on the environment in which the sediments occurred. For example, the coloration in the more somber-toned strata, like those that are bluish to gray and even black and brown in color, indicate deposition in a reducing (oxygen-starved) environment, such as in an area with a mesic climate and a high water table, or even in swampy conditions near streams and lakes (Sadler et al. 2015). Only a small thickness, perhaps the lowermost 60 m or so of the Chinle Formation, is dominantly bluish to gray in color. In contrast, red and orange rocks [U0635] generally

formed in well-drained soils with a fluctuating, deep water table in a seasonally dry climatic regime and away from streams on the floodplain. At the Petrified Forest, the upper 340 m of the Chinle Formation are dominantly reddish in color, indicating that these ancient floodplain soils were affected by varying soil moisture content under a strong seasonal climate (Therrien and Fastovsky 2000). During these times, the water table was deep, precipitation was low, and rivers and streams were small and far apart. Nevertheless, plant fossils are found throughout the entire formation.

A large variety of plants lived in the Chinle basin and are most commonly found today as long fossil logs (Fig. 6.12) [U0636] or segments of logs. They are particularly abundant in the sediments deposited during the formation of the more somber-toned beds in the lower part of the section. These deposits, formed 210–230 million years ago, reflect climatic conditions that were more favorable for preservation of plant life than in the upper, younger part of the formation (Atchley et al. 2013). Higher in the stratigraphic section, where sediments reflect climatic conditions that were drier, the variety and numbers of both plants began to decline as the red beds of the upper part of the Chinle Formation formed. Many first-time visitors to the Park are disappointed when they learn that few upright petrified trees are preserved here. Instead, they find out that the trees are nearly always represented by long lengths of their massive trunks lying prone on the desert surface at certain locations in the National Park called "forests" (Gillette et al. 1986). Sometimes the rock below a fossil log has been eroded away by wind and rain, leaving it perched on a pedestal (Fig. 6.13) [U0637].

In the Petrified Forest National Park, the permineralized wood in each forest has a distinctive coloration or other set of unique features (Fig. 6.14) [U0638]. For example, Jasper

Fig. 6.13 The whimsical "pedestal logs" at Petrified Forest National Park in Arizona, USA, which are formed when the rock below a fossil trunk has been eroded away by the weather, leaving it reclining atop a pedestal. Photos courtesy of SR Ash

Fig. 6.14 Four assemblages of fossil logs called "Forests" occur at the Petrified Forest National Park in Arizona, USA, each with its own characteristics. (**a**) Jasper Forest, (**b**) Crystal Forest, (**c**) Rainbow Forest, (**d**) Black Forest. Photos courtesy of SR Ash

Forest contains mostly reddish fossil wood; Crystal Forest contains fossil wood with small cavities called vugs that contain quartz crystals; Rainbow Forest has unusual rainbow-striped fossil wood that is highly sought after to make jewelry and various ornaments; and the Black Forest contains dense black fossil wood. The first three forests are easy to visit because they are adjacent to the main park road in the southern part of the Park, between the Puerco River and Rainbow Forest Visitors Center. In contrast, the Black Forest is located in the Painted Desert, adjacent to Lithodendron Wash in the northern part of the Park. This locality can only be visited on foot. The different colors and bright hues of the fossil logs, especially the rainbow-colored wood, are what make the plant fossils at the Petrified Forest National Park special (Box 6.3).

Most of the fossil wood in the Petrified Forest Park came from an extinct conifer commonly called *Agathoxylon arizonicum* [U0640]. Presumably, it came from trees that fell naturally into ancient channel systems at various times between 209 and 218 million years ago, and that were washed into the present National Park area where they were quickly buried and fossilized before they could rot away. There are small areas in the Park with a few short stumps of this species standing in their original position of growth, which typify the small, open groves of *A. arizonicum* trees.

In life, *A. arizonicum* trees were fairly tall, but do not appear to have been gigantic like the present-day coastal redwoods of northern California [U0640] (Ash and Creber 2000). For example, many of the fossil logs represent trees that are estimated to have reached heights of 30–40 m

Box 6.3: The Colorful World of Fossil Wood

The colors in fossil wood do not come from the original pigments in wood of the once living tree. Rather, these are produced by trace metals that enter ancient wood from the host sediment during mineralization. The colors in fossil wood can range from green, yellow, orange, red, beige, brown, to light or dark purple, or even white, gray, black, and clear. Most of these colors are produced by iron in varying amounts and different oxidation states (Mustoe and Acosta 2016). Iron-oxide minerals are, by far, the most common producer of color in silicified wood, and are also responsible for the rainbow-colored fossil woods at the Petrified Forest National Park in Arizona (Fig. 6.15) [U0639]. Some bright green fossil wood can be, however, produced by the trace element chromium. Small traces of titanium, vanadium, manganese, and copper in conjunction with iron can contribute to brown and tan colors in fossil wood, as well (Mustoe and Acosta 2016).

when living. However, there were also trees that grew even taller. The large log near the Rainbow Forest Visitors Center called "Old Faithful" [U0640] came from a tree that probably was about 59 m tall when alive. There could have been others about the same size, if not taller, in the same ancient forest.

The morphological and anatomical features of *A. arizonicum* logs offer much information on how these trees grew during Triassic times. Their root system typically consists of a broad central tap root surrounded by a ring of four to six lateral roots. Trunks do not show any evidence of having strong annual growth rings, which indicates that this region did not have well-developed seasons. Branch scars are clearly visible on most trunks [U0641] and indicate that the branches were not self-pruning. Furthermore, the scars are not arranged in whorls as in most living conifers (Ash and Creber 2000). Rather, they seem to be irregularly distributed, or perhaps arranged in a long, irregular spiral. The bark of the trees was rather thin and compares quite closely in many aspects with that of modern conifers [U0641]. These trees were prone to insect attack.

There is considerable evidence that *A. arizonicum* trees, when they were alive and following their death, were attacked by plant-eating (phytophagous) insects (Ash and Savidge 2004). There are at least two lines of evidence for this interpretation. Small tunnels in the wood resemble those produced by living powder-post beetles, as well as larger tunnels filled with cuttings made by insect larvae, similar to those of the living bark beetles (Walker 1938). Recent studies indicate that *A. arizonicum* trees were occasionally struck by lightning, as in modern forests, and the presence of small amounts of charcoal in the lower parts of the Chinle Formation shows that wildfires sometimes occurred in these ancient forests (Jones et al. 2002).

Fossil leaves of the understory plants that grew beneath these forests are not as conspicuous as the fossil wood, and have to be carefully extracted from the soft beds of mudstone in the bluish-green and brown beds in the lower part of the formation. The largest and best preserved of these plant fossils occur in overbank* deposits. Here, entire leaves, or in the case of ferns whole plants, were washed away from where they were laying or growing and entombed by a river that overflowed its banks (Daugherty 1941; Ash 2005). The small, lens-shaped deposits of dark brown shale that contain numerous small leaf fragments and other plant parts represent swamp deposits (Ash 2010b). The fossil plants usually are so well-preserved that it is possible to study their cellular features with a microscope. Investigations of these delicate fossils show that they represent about 36 genera and include the remains of living plant groups such as the lycopods (club mosses), horsetails [U0642], ferns [U0643], conifers [U0644], cycads [U0645], and ginkgophytes [U0646]. There are several extinct groups, including the bennettitaleans,

Fig. 6.15 The fossil wood at the Petrified Forest National Park in Arizona, USA, is famous for its bright, intense colors. Photo courtesy of SR Ash

seed ferns, and several other enigmatic plants, that are components of this landscape. These fossil-leaf assemblages, though, only represent part of the total plant diversity.

To date, nearly 200 genera of pollen and spores, alone, are recognized in the Chinle Formation in the Petrified Forest (Baranyi et al. 2017). In comparison, only ten genera of fossil wood and 36 genera of fossil leaves currently are known from the same area (Ash 2010a). Several of these fossil forms appear to represent evolutionary intermediates between ancient plants and their modern descendants, two of which are highlighted. One striking example of an evolutionary intermediate are some small to moderately large cones that resemble both the common Paleozoic horsetail *Calamites* and the living horsetail *Equisetum* [U0642]. As a consequence, the Chinle cones were described as *Equicalastrobus chinleana* and assigned to the horsetails (Grauvogel-Stamm and Ash 1999). The *E. chinleana* cones are distinguished by bearing long, narrow, forward-directed, leaf-like tips on the spore-bearing cone scales. A second unique plant is an unusual fern called *Cynepteris lasiophora* (Ash 1970b). This fern had net venation instead of the more typical, spreading

venation found in many living forms [U0643]. The fronds of this plant, which were large and rather sturdy, probably grew in more open areas such as the banks of streams and lakes but apparently did not survive into later Triassic times. We find no evidence of the plant in younger rocks.

There are other fossil plants found in the Park that do not appear to be closely related to any known extinct or living group of plants. One such plant fossil is called *Dinophyton spinosus* [U0646] (Ash 1970a) and is one of the most common plant fossils found in the lower part of the Chinle Formation. The entire surface of this plant was covered with narrow, needle-like leaves that bore many small, stiff, spine-like hairs, which may have deterred predation by insects. Seeds of this plant were borne in four-winged, cross-like (cruciform) structures that were distributed by the wind.

In summary, the Late Triassic conifer forests of Arizona thrived for about 14 million years near the far western edge of Pangea and a few degrees north of the equator. They consisted principally of groves of the tall *Agathoxylon arizonicum* conifer trees that reached about 60 m in height. These conifers rose above an understory of smaller plants, which included a wide variety of ferns, club mosses (lycopods), and horsetails. At the same time, several types of cycads, ginkgoes, and other small plants inhabited the more open areas and clearings in the forest. Many large horsetails grew adjacent to, and on occasion in, the many shallow lakes and streams that existed in the region.

The trees and understory plants were not the only forms of life in this forest. The common evidence of invertebrate predation on trunks and leaves illustrates the abundance of insects. Vertebrate fossils are also found in certain beds in the Park and demonstrate that a variety of amphibians and reptiles (including small dinosaurs!) were part of the forest biota. The streams also teemed with life, as fossil fishes and clam shells have been found in the stream deposits together with the remains of a crayfish.

During the time of deposition of the Chinle Formation, the climate in this region became more and more seasonally dry, and eventually the forests disappeared. With the decimation of the forests, the smaller plants of the understory vegetation and the forest animals also perished, or migrated to other regions. The area became a desert with virtually lifeless sand dunes as the Triassic Period ended about 200 million years ago.

Today, many millions of years after they existed, remnants of the Late Triassic forests of Arizona are being slowly exposed to view by erosion. In addition, the careful work of various scientists over the last 160 years is slowly revealing more and more about the composition and natural history of this ancient forest and its inhabitants. It is exciting to contemplate what the next set of paleontologists will discover about these ancient forests!

6.8 The Mid-Triassic Molteno Flora of the Karoo Basin, South Africa: Witness to the Heyday of the Gymnosperms

John M. Anderson and Heidi M. Anderson

Let us now travel southwest into Gondwana to explore the riverine forests and woodlands of the Molteno Formation at the heart of the Southern Hemisphere during the Triassic, at some 230 million years ago [U0647]. We find ourselves in an extensive lowland basin at around 50° south—in the temperate latitudes of a world without ice caps that is a good deal warmer than today. To our south occur the actively uplifting and folding Cape Fold Mountains and to our north, the stable upland hillscape of the ancient Kaapvaal Craton. The basin, a floodplain, is traversed by a series of northward-flowing braided rivers during periods of active tectonics, and by meandering rivers during tectonically stable intervals. It is the time of the Triassic diversification following the end-

Box 6.4: Field Work in the Karoo Basin

For half a century now, John and Heidi Anderson have been studying the mid-Triassic Molteno flora in the Karoo Basin of South Africa (Anderson and Anderson 1983, 1985, 1989, 1995, 2003, 2008, 2018; Anderson et al. 1999a, b, 2007). Across a span of 31 years (1967–1998), the fossil flora was intensively collected during 85 field excursions (Fig. 6.16). The Molteno flora was also sampled extensively in the sense of geographic (around the outcrop), stratigraphic (through the sequence), and paleoecologic (covering the different habitats) context. The Andersons focused on fossil populations (paleodemes), aiming to cover the range of variation of the more common species from any particular assemblage. Their collection covers 100 assemblages (taphocoenoses) from 69 localities (areas to 1 km in diameter), and includes around 27,200 curated slabs with some 300,000 identifiable fossil plant specimens. This extensive collection allowed them to identify seven major plant communities in the Molteno Beds of the Karoo Basin.

Permian extinction event, the mother of all extinctions. More on this extinction later (Delfino et al., Chap. 10). The Molteno Formation is capped by younger rocks of the Karoo Supergroup∗, most spectacularly the 1600 m thickness of flood basalts making up the Drakensberg Mountains [U0648]. The formation outcrops in a roughly trapezium-shaped area extending some 400 km north to south and 200 km west to east. It reaches a maximum thickness of 500 m in the SW and includes six fining-upwards cycles (members) reflecting pulses of uplift and erosion in, and sediment transport, from the Cape Mountains. The Molteno Formation has been the focus of many decades of dedicated field work and paleobotanical study (Box 6.4).

The mid-Triassic flora of the Molteno Formation consists of a tapestry of seven plant communities across the Karoo Basin landscape (Fig. 6.17) [U0649]. These range from diverse *Dicroidium* riverine forest to dense monospecific horsetail marsh (Cairncross et al. 1995). *Dicroidium* is an arboreal seed fern with distinctive forked leaves (Fig. 6.18a) and, along with its female (ovulate; *Umkomasia*) and male (microsporangiate; *Pteruchus*) reproductive organs, is clearly the most diverse element. The plant is known from 19 species, with growth habits ranging from trees to woody shrubs, and it is found in 75 of the 100 assemblages. *Dicroidium* is dominant here (>50% of the flora in 46 assemblages) [U0649–U0652] as it was throughout Gondwana during much of the Triassic. The genus characterizes the first three of the plant communities: the mature and immature versions of the riverine forest, and the open woodland occupying the

Fig. 6.16 Field work in the mid-Triassic Molteno Formation in the Drakensberg Mountains, South Africa. Capping the mountains is a spectacular, 1600 m thick flood basalt. Photo by JM Anderson

greater part of the landscape beyond. Several species may have been deciduous, with their short-shoot morphology similar to extant *Ginkgo* (Anderson et al. 2008). *Rissikia*, a large coniferous tree, placed in the family Podocarpaceae, forms part of the forested canopy. Its leaves resemble the bald cypress and redwoods of today, and both male and female cones are preserved in close association.

The fourth Molteno plant community, the *Sphenobaiera* woodland, encircles the lakes in the woodland. This habitat is dominated by *Sphenobaiera*, a ginkgophyte with bifurcating fronds that we encountered in the Middle Jurassic of Yorkshire (see Sect. 6.6). Leaves of nine different species of the plant dominate this lake–land habitat, which is found spread across 43 assemblages. *Ginkgoites*, represented by several species ranging from woody shrubs to trees of various sizes, appears scattered through the forested and wooded landscape. Further woody taxa include a wide diversity of other seed ferns (e.g., *Lepidopteris*; Fig. 6.18b), cycads, bennettitaleans, gnetopsids, and miscellaneous taxa that remain difficult to classify.

Heidiphyllum, a woody conifer placed in the order Voltziales, dominates fossil assemblages that constitute a major habitat of its own. The plant has elongate, broad leaves up to

10 cm in length and distinctive ovulate cones (*Telemachus*). It is very widespread and often a mono-dominant, forming a thicket to forest in areas of high water table in the floodplain or on channel sandbars (Cairncross et al. 1995).

Among the non-gymnosperms, the horsetails (sphenopsids) played a particularly significant role in the flora (Anderson and Anderson 2018). As many as eight genera and 23 vegetative species are recognized, and this group occurs in over half of the sampled assemblages. Their preferred habitat was the riverine and floodplain wetlands, where they occurred as co- or mono-dominants in a wide range of sizes. Evidence of stem fragments up to 14 cm in girth suggests bamboo-like thickets with several species reaching heights of 20 m or more. Of great interest is the discussion around the high-energy nutritional value of the horsetails in the diet of the herbivorous dinosaurs in the Late Jurassic Morrison Formation of North America (see Sect. 6.5). One wonders what symbiotic role the diverse horsetails and other lower vascular plant groups of the mid-Triassic must have played during these earliest days of dinosaur evolution in Gondwana. Lower vascular plant groups occupied positions in the undergrowth and margins of the forests. Ferns were most common and are represented by an array of genera and species

Fig. 6.17 Plants characteristic of the forest and woodland habitats in the mid-Triassic Molteno Formation, South Africa. (**a**) A riparian forest dominated by the seed fern *Dicroidium* (Umkomasiales). (**b**) The understory of the riparian forest with the shrubby seed fern *Lepidopteris* (Peltaspermales). (**c**) A mono-dominant stand of the voltzialean conifer *Heidiphyllum* (Voltziales) in an area with a high water table on the floodplain. (**d**) Large conifer trees of *Rissikia* (Coniferales) which were common in riverine areas and wetlands. (**e**) Small to large trees of *Fraxinopsis* (Fraxinopsiales), a rare to occasional gnetopsid in the riverine forest. (**f**) A meadow of *Kannaskoppia* seed ferns (Petriellales) on a sandbar in a braided river. (**g**) Mono-dominant stands of *Zonulamites* horsetails (sphenopsids) up to 20 m tall lining the marshes on the floodplain. (**h**) The herbaceous seed fern *Nataligma* (Nataligmales), which was an occasional pioneer that grew along margins of riverine forest or other water bodies. All sketches rendered by H Anderson

Fig. 6.18 Fossil seed fern foliage from the Molteno Formation, South Africa. (**a**) *Dicroidium odontopteroides* (see also Fig. 6.17a). (**b**) *Lepidopteris stormbergensis* (see also Fig. 6.17b). Photos by JM Anderson and HM Anderson

(Anderson and Anderson 2008). Lycopsids, mosses, and liverworts are more rarely found.

Considering the overall Molteno flora, the tally of vegetative taxa currently stands at 211 species placed in 61 genera. The non-gymnosperms account for 46% of the species and 56% of the genera; the gymnosperms make up 54% and 44%, respectively. Throughout the work in collecting and describing the flora, a strong focus has been placed on the affiliation of foliage and reproductive structures. Combining vegetative, female (ovulate), and male (microsporangiate) organs [U0650, U0651], it has been possible to recognize whole-plant taxa. The results of this exercise, along with a global review of gymnosperms through the fossil record until the present (Anderson et al. 2007), indicate that the Middle Triassic is witness to the heyday of the gymnosperms [U0652] (Anderson and Anderson 2003). A statistical estimate of total floral diversity rivals that of angiosperm-dominated biomes of today. And, as with extant landscapes, insects played several key roles in the Triassic ecosystems.

The insects flourishing in the varied habitats of the tree-lined rivers, lakes, and swamps reflect the Triassic explosion of life (Anderson et al. 1998). More than 40% of the plant assemblages have yielded insects, and 16 orders of insects are now identified. The dominant groups include the beetles, cockroaches, bugs, and dragonflies [U0653]. For the first time in the geological record, the beetles attain dominance and diversity. They were an integral part of the gymnosperm-dominated forests and woodlands, and presumably played a major role in pollination, as the bees and wasps do in the angiosperm-dominated floras of today. Of the approximately 2300 insect specimens, 350 species of beetles are known. It has been estimated that an astonishing 8000 or so insect species were preserved within the Molteno strata (Anderson 1999). Evidence of plant–insect interaction is common and varied, including occurrences of leaf mining in the conifer *Heidiphyllum* [U0653] (Labandeira in Anderson and Anderson 2018).

As we come to the end of our journey into the Triassic world, we have seen that the Molteno Formation at the heart of Gondwana in the Southern Hemisphere has provided us with a uniquely clear window onto the Triassic diversification of life. In our travel notes, we find the record of an extreme radiation of plants, insects, and other animals in the continental realm. The diversification that followed the end-Permian extinction was, perhaps, as extraordinary as that of the Cambrian explosion of life in the marine realm! But more on this in the next four chapters. On top of getting a first glimpse on this general diversification of life, our Triassic journey allowed us to witness the heyday of the gymnosperms, just as we are witness to the heyday of the angiosperms in today's world.

6.9 Conclusions

During our plant safari through the Mesozoic, we have experienced the beauty, diversity, and uniqueness of the Mesozoic forests. These landscapes are very different from the broad-leaved forests of the Cenozoic (Martinetto et al., Chap. 2) and the peat-forming, sporing-plant forests of the Paleozoic (Gastaldo et al., Chaps. 12 and 13). The Mesozoic forests generally consist of conifers—or in the latest Cretaceous, angiosperms—that build the basic sylvan framework and canopy with which we are familiar. Yet, gymnosperms such as cycads, bennettitaleans, and seed ferns form the mid-canopy, and ferns and fern allies reside in the understory. These are plant communities that are very different from forests of the present. Nevertheless, despite the deceptive serenity of its forests, the Mesozoic was an era marked with upheaval. The end-Cretaceous extinction, identified as the Cretaceous–Paleogene boundary, marks the catastrophic event of an asteroid impacting Earth and causing a sunless, two-year long "nuclear winter" for the globe's vegetation. In contrast, the start of the Mesozoic saw the end-Permian extinction event that was so all-encompassing that life almost died out completely (Benton 2015; Delfino et al., Chap. 10). Even during the Mesozoic, there were seismic shifts on Earth—literally. The single landmass of Pangea, which existed in the late Paleozoic, began to split into Laurasia to the north and Gondwana to the south. Then, relatively soon afterwards, during the Early Jurassic, the various tectonic plates of Gondwana began to rift from one another and drift northwards. All of these changes affected the plant world.

It is difficult to pinpoint cause and effect, but there are certain differences between the floras found throughout the Mesozoic (Gee 2010). There is, for example, some differentiation of the global flora into a characteristic Northern Hemisphere flora with the maidenhair tree *Ginkgo*, and a typical Southern Hemisphere flora with the seed fern *Dicroidium* and its distinctive Y-shaped leaves. The most dramatic change in the plant world during the Mesozoic was, however, the advent of the angiosperms, the flowering plants. They first appear as minute weeds in the Early Cretaceous, then moved rapidly into all niches of the plant world. Soon afterwards, they overtake the landscape as giant flowering trees that dominated the global flora by the end of the Cretaceous. The main thing to remember, though, is that the Mesozoic forests were actual living, breathing, and photosynthesizing ecosystems. We hope that through this series of postcards you were taken on a fascinating safari through paleobotanical time, shown a bit on how and where we collect our fossil plants and carry out our scientific work, and received a sense of our deep affection for the Mesozoic world [U0654].

For Deeper Learning

- Fossil wood flora of the Morrison Formation: Lutz 1930; Medlyn and Tidwell 1975, 1979, 2002; Tidwell and Medlyn 1993; Tidwell et al. 1998.
- Fossil flora from the Middle Jurassic of Yorkshire, England: Young and Bird 1822; Thomas 1925; Harris 1961, 1964, 1969, 1979; Harris et al. 1974; van Konijnenburg-van Cittert 1971, 1989, 1996; van Konijnenburg-van Cittert and Morgans 1999.

Questions

1. What was the dominant type of terrestrial vegetation during the Mesozoic?
2. How did the asteroid impact at the Cretaceous–Paleogene boundary affect the global vegetation?
3. Can evidence of the asteroid impact in North America 66 million years ago be found today as far away as New Zealand?
4. What happened about 125 million years ago to cause a shift in the colors of the plant world, from variations of greens and browns to bright flowery hues?
5. Were the earliest angiosperm flowers large or small?
6. What types of plants flourished in the Early Cretaceous forests of southeastern Australia?
7. What sorts of environmental and climatic conditions did the extinct conifer family Cheirolepidiaceae tolerate so well in the Morrison Formation?
8. Which two plants were probably the most preferred foods of the herbivorous dinosaurs?
9. What family of conifers is responsible for the famous Whitby jet from the Middle Jurassic of the Yorkshire Coast, England?
10. Mesozoic seed ferns have fronds and pinnules like ferns, but did they bear spores like ferns?
11. Where do the rainbow colors in the fossil wood from the Petrified Forest National Park in Arizona, USA, come from?
12. What kind of tree do the Triassic logs at the Petrified Forest National Park represent?
13. What are the seven plant communities that made up the tapestry of mid-Triassic landscapes in the Karoo Basin?
14. Which Mesozoic postcard did you like best? Why?

References

Anderson JM (1999) Towards Gondwana alive: promoting biodiversity & stemming the sixth extinction, vol 1. Gondwana Alive Society, Pretoria, p 140

Anderson JM, Anderson HM (1983) Palaeoflora of Southern Africa: Molteno Formation (Triassic), vol 1: part 1, introduction, part 2, *Dicroidium*. Balkema, Rotterdam, p 227

Anderson JM, Anderson HM (1985) Palaeoflora of southern Africa: prodromus of South African megafloras, Devonian to Lower Cretaceous. Balkema, Rotterdam, p 423

Anderson JM, Anderson HM (1989) Palaeoflora of southern Africa: Molteno Formation (Triassic), vol 2: gymnosperms (excluding *Dicroidium*). Balkema, Rotterdam, p 567

Anderson JM, Anderson HM (1995) The Molteno Formation: window onto Late Triassic floral diversity. In: Pant DD (ed) Proceedings of the international conference on global environment and diversification of plants through geological time. Society of Indian Plant Taxonomists, Allahabad, pp 27–40

Anderson JM, Anderson HM (2003) Heyday of the gymnosperms: systematics and biodiversity of the Late Triassic Molteno fructifications. Strelitzia 15. South African National Biodiversity Institute, Pretoria, p 398

Anderson HM, Anderson JM (2008) Molteno ferns: Late Triassic biodiversity in Southern Africa. Strelitzia 21. South African National Biodiversity Institute, Pretoria, p 258

Anderson HM, Anderson JM (2018) Molteno sphenophytes: Late Triassic biodiversity in southern Africa. Palaeontologia Africana 53 (Special Issue), Annals of the Evolutionary Studies Institute, University of the Witwatersrand, Johannesburg, 377 pp

Anderson JM, Kohring R, Schlüter T (1998) Was insect biodiversity in the Triassic akin to today?—A case study from the Molteno formation (South Africa). Entomol Gen Stuttgart 23(1–2):15–26

Anderson JM, Anderson HM, Archangelsky S, Bamford M, Chandra S, Dettmann M, Hill R, McLoughlin S, Rosler O (1999a) Patterns of Gondwana plant colonisation and diversification. J Afr Earth Sci 28(1):145–167

Anderson JM, Anderson HM, Macrae CS (1999b) Freezing cold to searing heat. Plant and insect life of the Karoo Basin. In: MacRae CS (ed) Life etched in stone. The Geological Society of South Africa, Johannesburg, pp 140–166

Anderson JM, Anderson HM, Cleal CJ (2007) Brief history of the gymnosperms: classification, biodiversity, phytogeography and ecology. Strelitzia 20. South African National Biodiversity Institute, Pretoria, p 280

Anderson HM, Holmes WBK, Fitness FA (2008) Stems with attached *Dicroidium* leaves from the Ipswich coal measures, Queensland, Australia. Mem Queensland Mus 52:1–12

Ash SR (1970a) *Dinophyton*, a problematical new plant from the upper Triassic of the southwestern United States. Palaeontology 13:646–664

Ash SR (1970b) Ferns from the Chinle Formation (Upper Triassic) in the Fort Wingate area, New Mexico. US Geological Survey Professional Paper 613D, 40 pp.

Ash SR (2005) Petrified forest. A story in stone. Petrified Forest Museum Association, Petrified Forest National Park, Arizona, p 54

Ash SR (2010a) Summary of the Upper Triassic flora of the newspaper rock bed and its paleoclimatic implications. In: Trendell A (ed.) Paleosols and soil surface system analogs. SEPM–NSF research conference and workshop, Petrified Forest National Park, Arizona, USA. Baylor University, Waco, pp 59–71

Ash SR (2010b) Preliminary observations on the Upper Triassic plant debris beds exposed in the road cuts along old US 180, Petrified Forest National Park, Arizona. In: Trendell A (ed.) Paleosols and soil surface system analogs. SEPM–NSF research conference and workshop, Petrified Forest National Park, Arizona, USA. Baylor University, Waco, pp 131–134

Ash SR, Creber G (2000) The Late Triassic *Araucarioxylon arizonicum* trees of the Petrified Forest National Park, Arizona, USA. Palaeontology 43:15–28

Ash SR, Savidge RA (2004) The bark of the Late Triassic *Araucarioxylon arizonicum* tree from Petrified Forest National Park, Arizona. Int Assoc Wood Anatom J 25:349–368

Ash SR, Tidwell WD (1998) Plant megafossils from the Brushy Basin member of the Morrison Formation near Montezuma Creek Trading Post, southeastern Utah. Mod Geol 22:321–339

Atchley SC, Nordt LC, Dworkin SI, Bowring J, Parker WG, Ash SR, Bowring SA (2013) A linkage among Pangean tectonism, cyclic alluviation, climate change, and biological turnover in the Late Triassic: the record from the Chinle Formation, southwestern United States. J Sediment Res 83:1147–1116

Baranyi V, Reichgelt T, Olsen PE, Parker WG, Kurschner WM (2017) Norian vegetation history and related environmental changes: new data from the Chinle Formation, Petrified Forest National Park (Arizona, SW USA). Geol Soc Am Bull 130:775–795

Bell WA (1956) Lower Cretaceous floras of western Canada. Geol Surv Can Mem 285:1–331

Benton MJ (2015) When life nearly died. Thames & Hudson, New York

Bomfleur B, McLoughlin S, Vajda V (2014) Fossilized nuclei and chromosomes reveal 180 million years of genomic stasis in royal ferns. Science 343:1376–1377

Brown JT (1972) The flora of the Morrison Formation (Upper Jurassic) of central Montana. PhD dissertation, University of Montana, Missoula, 65 pp

Cairncross B, Anderson JM, Anderson HM (1995) Palaeoecology of the Triassic Molteno Formation, Karoo Basin, South Africa—sedimentological and palaeoecological evidence. S Afr J Geol 98:452–478

Calvo JO (1994) Jaw mechanics in sauropod dinosaurs. Gaia 10:183–193

Cantrill DJ (1991) Broad leafed coniferous foliage from the Lower Cretaceous Otway Group, southeastern Australia. Alcheringa 15:177–190

Cantrill DJ (1992) Araucarian foliage from the Lower Cretaceous of southern Victoria, Australia. Int J Plant Sci 153:622–645

Cantrill DJ (1997) Hepatophytes from the Early Cretaceous of Alexander Island, Antarctica: systematics and paleoecology. Int J Plant Sci 158:476–488

Cantrill DJ, Webb JA (1987) A reappraisal of *Phyllopteroides* Medwell (Osmundaceae) and its stratigraphic significance in the Lower Cretaceous of eastern Australia. Alcheringa 11:59–85

Chure DJ, Litwin R, Hasiotis ST, Evanoff E, Carpenter K (2006) The fauna and flora of the Morrison Formation. In: Foster JR, Lucas SGRM (eds) Paleontology and geology of the Upper Jurassic Morrison Formation. New Mexico Museum of Natural History and Science Bulletin, vol 36, pp 233–249

Crane PR (1987) Vegetational consequences of the angiosperm diversification. In: Friis EM, Chaloner WG, Crane PR (eds) The origins of angiosperms and their biological consequences. Cambridge University Press, Cambridge, pp 107–144

Daugherty LH (1941) The Upper Triassic flora of Arizona, vol 526. Carnegie Institution of Washington Publication, Washington, pp 1–108

Davies-Vollum K, Boucher LD, Hudson P, Proskurowski AY (2011) A Late Cretaceous coniferous woodland from the San Juan Basin, New Mexico. PALAIOS 26:89–98

Douglas JG (1969) The Mesozoic floras of Victoria: parts 1 and 2. In: Memoir, vol 28. Department of Mines, Victoria, pp 1–310

Douglas JG (1973) The Mesozoic floras of Victoria: parts 3. In: Memoir, vol 29. Department of Mines, Victoria, pp 1–185

Doyle JA (2012) Molecular and fossil evidence on the origin of angiosperms. Ann Rev Earth Planet Sci 40:301–326

Drinnan AN, Chambers TC (1986) Flora of the Lower Cretaceous Koonwarra fossil bed (Korumburra Group), South Gippsland, Victoria. Assoc Australas Palaeontol Mem 3:1–77

Engelmann GF, Chure DJ, Fiorillo AR (2004) The implications of a dry climate for the paleoecology of the fauna of the Upper Jurassic Morrison Formation. Sediment Geol 167:297–308

Estrada-Ruiz E, Upchurch GR Jr, Wheeler EA, Mack GH (2012) Late Cretaceous angiosperm woods from the Crevasse Canyon and McRae Formations, south-central New Mexico, USA: part 1. Int J Plant Sci 173:412–428

Estrada-Ruiz E, Wheeler EA, Upchurch GR Jr, Mack GH (2018) Late Cretaceous angiosperm woods from the McRae Formation, south-central New Mexico, USA: part 2. Int J Plant Sci 179:136–150

Feild TS, Arens NC, Doyle JA, Dawson TE, Donoghue MJ (2004) Dark and disturbed: a new image of early angiosperm ecology. Paleobiology 30:82–107

Felton EA (1997a) A non-marine Lower Cretaceous rift-related epiclastic volcanic unit in southern Australia: the Eumeralla Formation in the Otway Basin. Part 1: lithostratigraphy and depositional environments. Aust Geol Surv Org J Geol Geophys 16:717–730

Felton EA (1997b) A non-marine Lower Cretaceous rift-related epiclastic volcanic unit in southern Australia: the Eumeralla Formation in the Otway Basin. Part II: fluvial systems. Aust Geol Surv Org J Geol Geophys 16:731–757

Field DJ, Bercovici A, Berv JS, Dunn R, Fastovsky DE, Lyson TR, Vajda V, Gauthier JA (2018) Early evolution of modern birds structured by global forest collapse at the end-Cretaceous mass extinction. Curr Biol 28:1825–1831

Francis JE (1983) The dominant conifer of the Jurassic Purbeck Formation, England. Palaeontology 26:277–294

Friedman WE (2009) The meaning of Darwin's "abominable mystery." Am J Bot 96:5–21

Friis EM, Pedersen KR, Crane PR (2001) Fossil evidence of water lilies (Nymphaeales) in the Early Cretaceous. Nature 410:357–360

Friis EM, Pedersen KR, von Balthazar M, Grimm GW, Crane PR (2009) *Monetianthus mirus* gen. et sp. nov., a nymphaealean flower from the Early Cretaceous of Portugal. Int J Plant Sci 170:1086–1101

Friis EM, Crane PR, Pedersen KR (2011) Early flowers and angiosperm evolution. Cambridge University Press, Cambridge, p 585

Gee CT (2010) (ed) Plants in Mesozoic time: morphological innovations, phylogeny, ecosystems. Indiana University Press, Bloomington, p 373

Gee CT (2011) Dietary options for the sauropod dinosaurs from an integrated botanical and paleobotanical perspective. In: Klein N, Remes K, Gee CT, Sander PM (eds) Biology of the sauropod dinosaurs: understanding the life of giants. Indiana University Press, Bloomington, pp 34–56

Gee CT (2013) Applying microCT and 3D visualization to Jurassic silicified conifer seed cones: a virtual advantage over thin-sectioning. Appl Plant Sci 1:1300039

Gee CT, Tidwell WD (2010) A mosaic of characters in a new whole-plant *Araucaria*, *A. delevoryasii* Gee sp. nov., from the Late Jurassic Morrison Formation of Wyoming, U.S.A. In: Gee CT (ed) Plants in Mesozoic time: morphological innovations, phylogeny, ecosystems. Indiana University Press, Bloomington, pp 67–94

Gee CT, Dayvault RD, Stockey RA, Tidwell WD (2014) Greater palaeobiodiversity of conifer seed cones in the Upper Jurassic Morrison Formation of Utah, USA. Palaeobiodiv Palaeoenviron 94:363–365

Gee CT, Sprinkel DA, Bennis MB, Gray DE (2019) Silicified logs of *Agathoxylon hoodii* (Tidwell et Medlyn) comb. nov. from Rainbow Draw, near Vernal, Utah, and their implications for araucariaceous conifer forests in the Upper Jurassic Morrison Formation. Geol Intermountain West 6:77–92

Gillette DD, Ash SR, Long RA (1986) Paleontology of the Petrified Forest National Park, Arizona. In: Nations JD, et al. (eds.) Geology of central and northern Arizona: Geological Society America, Rocky Mountain Section, Field trip guidebook, pp 59–69

Grauvogel-Stamm L, Ash SR (1999) "*Lycostrobus*" *chinleana*, an equisetalean cone from the Upper Triassic of the Southwestern United States and its phylogenetic implications. Am J Bot 86:1391–1405

Harris TM (1961) The Yorkshire Jurassic Flora I. Thallophyta–Pteridophyta. British Museum (Natural History), London, p 212

Harris TM (1964) The Yorkshire Jurassic Flora II. Caytoniales, Cycadales and Pteridosperms. British Museum (Natural History), London, p 191

Harris TM (1969) The Yorkshire Jurassic Flora III. Bennettitales. British Museum (Natural History), London, p 186

Harris TM (1979) The Yorkshire Jurassic Flora V. Coniferales. British Museum (Natural History), London, p 166

Harris TM, Millington W, Miller J (1974) The Yorkshire Jurassic Flora IV. Ginkgoales and Czekanowskia. British Museum (Natural History), London, p 150

Hotton CL, Baghai-Riding NL (2010) Palynological evidence for conifer dominance within a heterogeneous landscape in the Late Jurassic Morrison Formation. In: Gee CT (ed) Plants in Mesozoic time: morphological innovations, phylogeny, ecosystems. Indiana University Press, Bloomington, pp 295–328

Hummel J, Gee CT, Südekum KH, Sander PM, Nogge G, Clauss M (2008) *In vitro* digestibility of fern and gymnosperm foliage: implications for sauropod feeding ecology and diet selection. Proc R Soc Ser B 275:1015–1021

Jones TP, Ash SR, Figueiral I (2002) Late Triassic charcoal from Petrified Forest National Park, Arizona, USA. Palaeogeogr Palaeoclimatol Palaeoecol 188:127–139

Jud NA, D'Emic MD, Williams SA, Mathews JC, Termaine KM, Bhattacharya J (2018) A new fossil assemblage shows that large angiosperm trees grew in North America by the Turonian (Late Cretaceous). Sci Adv 4(9):eaar8568

Kennedy EM (2003) Late Cretaceous and Paleocene terrestrial climates of New Zealand: leaf fossil evidence from South Island assemblages. N Z J Geol Geophys 46:295–306

Kent DV, Olsen PE, Rasmussen C, Lepre C, Mundil R, Irmis RB, Gehrels GE, Giesler D, Geissman JW, Parker WG (2018) Empirical evidence for stability of the 405-kiloyear Jupiter–Venus eccentricity cycle over hundreds of millions of years. Proc Nat Acad Sci USA 115:6153–6158

Korasidis VA, Wagstaff BE, Gallagher SJ, Duddy IR, Tosolini AM, Cantrill DJ, Norvick M (2016) Early angiosperm diversification in the Albian of southeast Australia: implications of flowering plant radiation across eastern Gondwana. Rev Palaeobot Palynol 232:61–80

Litwin RJ, Turner CE, Peterson F (1998) Palynological evidence on the age of the Morrison Formation, Western Interior U.S. Mod Geol 22:297–319

Lutz HJ (1930) A new species of *Cupressinoxylon* (Goeppert) Gothan from the Jurassic of South Dakota. Bot Gaz 90:92–107

McLoughlin S, Tosolini A-MP, Nagalingum NS, Drinnan AN (2002) Early Cretaceous (Neocomian) flora and fauna of the lower Strzelecki Group, Gippsland Basin, Victoria. Assoc Aust Palaeontol Mem 26:1–144

Medlyn DA, Tidwell WD (1975) Conifer wood from the Upper Jurassic of Utah. Part I. *Xenoxylon morrisonense* sp. nov. Am J Bot 62:203–208

Medlyn DA, Tidwell WD (1979) A review of the genus *Protopiceoxylon* with emphasis on North American species. Can J Bot 57:1451–1463

Medlyn DA, Tidwell WD (2002) *Mesembrioxylon obscurum*, a new combination for *Araucarioxylon*? *obscurum*, from the Upper Jurassic Morrison Formation. Western North Am Nat 62:210–217

Mustoe G, Acosta M (2016) Origin of petrified wood color. Geosciences 6(2):25

Parrish JT, Peterson F, Turner CE (2004) Jurassic "savannah"—plant taphonomy and climate of the Morrison Formation (Upper Jurassic, Western USA). Sediment Geol 167:137–162

Poropat SF, Martin SK, Tosolini A-MP, Wagstaff BE, Bean LB, Kear BP, Vickers-Rich PV, Rich TH (2018) Early Cretaceous polar biotas of Victoria, southeastern Australia—an overview of research to date. Alcheringa 42:157–229

Rees PM, Noto CR, Parrish JM, Parrish JT (2004) Late Jurassic climates, vegetation, and dinosaur distributions. J Geol 112:643–653

Sadler C, Parker W, Ash SR (2015) Dawn of the dinosaurs. The Late Triassic in the American Southwest. Petrified Forest Museum Association, Petrified Forest National Park, Arizona. 124 pp

Serbet R, Rothwell GW (1999) *Osmunda cinnamomea* (Osmundaceae) in the Upper Cretaceous of Western North America: additional evidence for exceptional species longevity among filicalean ferns. Int J Plant Sci 160:425–433

Sprinkel DA, Bennis MB, Gray DE, Gee CT (2019) Stratigraphic setting of fossil log sites in the Morrison Formation (Upper Jurassic) near Dinosaur National Monument, Uintah County, Utah. Geology of Intermountain West 6:61–76

Taylor DW, Hickey LJ (1990) An Aptian plant with attached leaves and flowers: implications for angiosperm origin. Science 247:702–704

Taylor DW, Hickey LJ (1996) Evidence for and implications of an herbaceous origin for angiosperms. In: Taylor DW, Hickey LJ (eds) Flowering plant origin, evolution & phylogeny. Chapman & Hall, New York, pp 232–266

Taylor TN, Taylor EL, Krings M (2009) Paleobotany: the biology and evolution of fossil plants. Academic Press, San Diego, p 1230

Therrien F, Fastovsky DE (2000) Paleoenvironments of early theropods, Chinle Formation (Late Triassic), Petrified Forest National Park, Arizona. PALAIOS 15:194–211

Thomas HH (1925) The Caytoniales, a new group of angiospermous plants from the Jurassic rocks of Yorkshire. Philos Trans R Soc Lond Ser B 213:299–363

Tidwell WD (1990a) Preliminary report on the megafossil flora of the Upper Jurassic Morrison Formation. Hunteria 2(8):1–11

Tidwell WD (1990b) A new arborescent osmundaceous species (*Osmundacaulis lemonii* n. sp.) from the Upper Jurassic Morrison Formation. Hunteria 2(7):3–11

Tidwell WD (1994) *Ashicaulis*, a new genus for some species of *Millerocaulis* (Osmundaceae). Sida Contrib Bot 16:253–261

Tidwell WD, Medlyn DA (1993) Conifer wood from the Upper Jurassic of Utah, part II: *Araucarioxylon hoodii* sp. nov. Palaeobotanist 42:70–77

Tidwell WD, Britt BB, Ash SR (1998) Preliminary floral analysis of the Mygatt-Moore Quarry in the Jurassic Morrison Formation, west-central Colorado. Mod Geol 22:341–378

Torsvik TH, van der Voo R, Preeden U, MacNiocall C, Steinsberger B, Doubrovine PV, van Hinsbergen DJJ, Domeier M, Gaina C, Tohver E, Meert JG, McCausland PJA, Cocks LRM (2012) Phanerozoic polar wander, palaeogeography and dynamics. Earth Sci Rev 114:325–368

Tosolini A-MP, McLoughlin S, Drinnan AN (1999) Stratigraphy and fluvial sedimentary facies of the Neocomian lower Strzelecki Group, Gippsland Basin, Victoria. Aust J Earth Sci 46:951–970

Tosolini A-MP, McLoughlin S, Wagstaff BE, Cantrill DJ, Gallagher SJ (2015) Cheirolepidiacean foliage and pollen from Cretaceous high-latitudes of southeastern Australia. Gondwana Res 27:960–977

Tosolini A-MP, Korasidis VA, Wagstaff BE, Cantrill DJ, Gallagher SJ, Norvick MS (2018) Palaeoenvironments and palaeocommunities from Lower Cretaceous high-latitude sites, Otway Basin, southeastern Australia. Palaeogeogr Palaeoclimatol Palaeoecol 496:62–84

Turner CE, Peterson F (2004) Reconstruction of the Upper Jurassic Morrison Formation extinct ecosystem—a synthesis. Sediment Geol 167:309–355

Upchurch P, Barrett PM (2000) The evolution of sauropod feeding mechanisms. In: Sues H-D (ed) Evolution of herbivory in terrestrial vertebrates: perspectives from the fossil record. Cambridge University Press, Cambridge, pp 79–122

van Konijnenburg-van Cittert JHA (1971) In situ gymnosperm pollen from the Middle Jurassic of Yorkshire. Acta Bot Neerlandica 20:1–96

van Konijnenburg-van Cittert JHA (1989) Dicksoniaceous spores in situ from the Middle Jurassic of Yorkshire, England. Rev Palaeobot Palynol 61:273–301

van Konijnenburg-van Cittert JHA (1996) Two *Osmundopsis* species from the Middle Jurassic of Yorkshire and their sterile foliage. Palaeontology 39:719–731

van Konijnenburg-van Cittert JHA, Morgans HS (1999) The Jurassic Flora of Yorkshire. Palaeontol Assoc Field Guides Fossils 8:1–134

Vajda V (2012) Fungi, a driving force in normalization of the global carbon cycle following the end-Cretaceous extinction. In: Talent JA (ed) Earth and life global biodiversity, extinction intervals and biogeographic perturbations through time. Springer, Dordrecht, pp 132–144

Vajda V, Bercovici A (2014) The global vegetation pattern across the Cretaceous–Paleogene mass-extinction interval—an integrated global perspective. Glob Planet Chang 12:29–49

Vajda V, McLoughlin S (2004) Fungal proliferation at the Cretaceous–Tertiary boundary. Science 303:1489

Vajda V, Raine I (2003) Pollen and spores in marine Cretaceous/Tertiary boundary sediments at mid-Waipara River, North Canterbury, New Zealand. N Z J Geol Geophys 46:255–273

Vajda V, Raine JI, Hollis CJ (2001) Indication of global deforestation at the Cretaceous–Tertiary boundary by New Zealand fern spike. Science 294:1700–1702

Vajda V, Lyson TR, Bercovici A, Doman JH, Pearson DA (2013) A snapshot into the terrestrial ecosystem of an exceptionally well-preserved dinosaur (Hadrosauridae) from the Upper Cretaceous of North Dakota, USA. Cretac Res 46:114–122

Vajda V, Ocampo A, Ferrow E, Bender Koch C (2015) Nano particles as the primary cause for long-term sunlight suppression at high southern latitudes following the Chicxulub impact—evidence from ejecta deposits in Belize and Mexico. Gondwana Res 27:1079–1088

Walker MV (1938) Evidence of Triassic insects in the Petrified Forest National Monument, Arizona. US National Mus Proc 85:137–141

Wanntorp L, Vajda V, Raine JI (2011) Past diversity of Proteaceae on subantarctic Campbell Island, a remote outpost of Gondwana. Cretac Res 32:357–367

Whitlock JA (2011) Inferences of diplodocoid (Sauropoda: Dinosauria) feeding behavior from snout shape and microwear analyses. PLoS One 6(4):e18304

Young G, Bird J (1822) A geological survey of the Yorkshire Coast. Clark, Whitby, p 332

Dinosaurs, But Not Only: Vertebrate Evolution in the Mesozoic

Emanuel Tschopp, Ricardo Araújo, Stephen L. Brusatte,
Christophe Hendrickx, Loredana Macaluso,
Susannah C. R. Maidment, Márton Rabi, Dana Rashid,
Carlo Romano, and Thomas Williamson

Abstract

If we imagine walking through Mesozoic lands, we would be able to observe vertebrates with peculiar combinations of morphological traits, some of which would seem to be intermediary to animals seen today. We would witness a terrestrial vertebrate fauna dominated by dinosaurs of various sizes and diversity, accompanied by many other animal groups that often are overlooked. Current research suggests that many of the main vertebrate clades existing today originated or diversified sometime in the Triassic or Early to Middle Jurassic. Herein, we profile some of the major transformations in both terrestrial and aquatic vertebrate evolution during the Mesozoic. We highlight: the appearance of features that allowed sauropod dinosaurs to become the largest animals to ever walk on Earth's continents, the appearance of herbivory among the usually carnivorous theropod dinosaurs, and we follow the specific changes that led to the evolution of avian flight. Our Mesozoic tour across the globe will allow us to see how different evolutionary forces led to convergent shifts to quadrupedality in ornithischian dinosaurs and to an aquatic lifestyle in turtles, crocodiles, and plesiosaurs. Last, but not least, we examine changes in the Mesozoic fauna linked to the rise of mammals, and the diversification patterns in several clades of fishes after the End-Permian Mass Extinction.

Electronic supplementary material A slide presentation and an explanation of each slide's content is freely available to everyone upon request via email to one of the editors: edoardo.martinetto@unito.it, ragastal@colby.edu, tschopp.e@gmail.com

*The asterisk designates terms explained in the Glossary.

E. Tschopp (✉)
Division of Paleontology, American Museum of Natural History, New York, NY, USA
e-mail: etschopp@amnh.org

R. Araújo
Instituto de Plasmas e Fusão Nuclear, Instituto Superior Técnico, Universidade de Lisboa, Lisbon, Portugal

S. L. Brusatte
School of Geosciences, University of Edinburgh, Edinburgh, UK

C. Hendrickx
Unidad Ejecutora Lillo, CONICET-Fundación Miguel Lillo, San Miguel de Tucumán, Argentina

L. Macaluso
Department of Earth Sciences, University of Turin, Turin, Italy

S. C. R. Maidment
Department of Earth Sciences, Natural History Museum, London, UK

M. Rabi
Central Natural Science Collections, Martin-Luther University Halle-Wittenberg, Halle (Saale), Germany

D. Rashid
Department of Cell Biology and Neuroscience, Montana State University, Bozeman, MT, USA

C. Romano
Paläontologisches Institut und Museum, Universität Zürich, Zürich, Switzerland

T. Williamson
New Mexico Museum of Natural History, Albuquerque, USA

7.1 Introduction

Traveling back to the Mesozoic, which spans nearly 185 million years of time, brings us to ecosystems that looked quite different from what we know today, not just in terms of plant diversity (see Kvaček et al. and Gee et al., Chaps. 5 and 6). Arriving at the end of the Mesozoic, we pass from one geological era, the Cenozoic, back into another. As we have seen while analyzing the plants of New Zealand (Vajda in Gee et al., Chap. 6), these two eras are separated by the most recent of the big five mass extinctions in Earth's history. The Cretaceous–Paleogene (K/Pg) mass extinction was the result of a combination of factors including an extended period of strong volcanism in the Indian subcontinent and an asteroid

© Springer Nature Switzerland AG 2020
E. Martinetto et al. (eds.), *Nature through Time*, Springer Textbooks in Earth Sciences, Geography and Environment,
https://doi.org/10.1007/978-3-030-35058-1_7

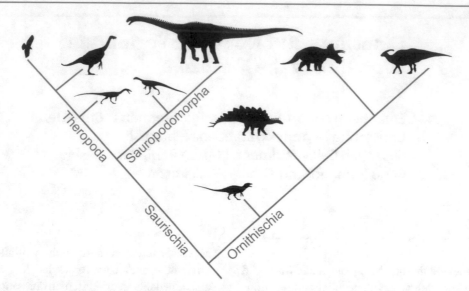

Fig. 7.1 A cladogram showing the major subclades of Dinosauria, the dominant terrestrial vertebrates during most of the Mesozoic Era. Note that early members of each clade are bipedal and have relatively similar body shapes, whereas derived members vary greatly in morphology. Silhouettes in the tree represent *Coelophysis*, *Therizinosaurus*, and *Buteo* within Theropoda (from bottom to top), *Thecodontosaurus* and *Cetiosaurus* within Sauropodomorpha (from bottom to top), and *Lesothosaurus*, *Stegosaurus*, *Triceratops*, and *Parasaurolophus* in Ornithischia (from bottom to top). Silhouettes are all taken from PhyloPic.org, and were produced by Raven Amos (*Triceratops*, shared under CC-BY 3.0 unported license), Lauren Anderson (*Buteo*, public domain), Scott Hartman (*Coelophysis*, *Stegosaurus*, shared under CC-BY 3.0 unported license), Jamie Headden (*Lesothosaurus*, *Thecodontosaurus*, shared under CC-BY 3.0 unported license), Martin Kevil (*Therizinosaurus*, shared under CC-BY 3.0 unported license), Jack Mayer Wood (*Parasaurolophus*, public domain), Michael P. Taylor (*Cetiosaurus*, public domain)

impact in Central America (Brusatte et al. 2015a; Sprain et al. 2019). Among terrestrial animals, the K/Pg event had its worst effect on large-sized species, many of them vertebrates. This is probably because large animals could not effectively hide from radiation and forest fires (Field et al. 2018), and because they relied on the primary productivity of plants in terrestrial ecosystems (Sheehan and Hansen 1986; Brusatte et al. 2015a). It is thought that a general darkening of the atmosphere, caused by ejecta* from the volcanism and the bolide impact, affected photosynthesis and, thus, production of plant biomass would have been slowed down or even halted for some period of time (Sheehan and Hansen 1986). During the event, large terrestrial herbivores went extinct, and consequently, carnivores suffered the same fate. This biosphere collapse liberated many ecological niches that were filled subsequently by the survivors of the mass extinction. Many of these survivors belonged to groups of vertebrates that either originated or significantly radiated throughout the Mesozoic. In contrast, one of the major terrestrial vertebrate clades that experienced near total extinction is Dinosauria.

Dinosaurs were (and still are) one of the most diverse groups of Mesozoic terrestrial vertebrates. They commonly are divided into three main clades, the Sauropodomorpha, Theropoda, and Ornithischia. All of these groups evolved in the Triassic, with their early members being mostly bipedal, probably omnivorous animals (Baron et al. 2017). During the Late Triassic and Early Jurassic, all three clades under-

went important radiation*, which produced a variety of subclades* with diverse body sizes and shapes, feeding strategies, styles of locomotion, and ecological functions (see also Roghi et al., Chap. 9). This series of multiple radiations resulted in the evolution of the largest land animals that ever existed on Earth, the sauropods—the most derived members of Sauropodomorpha. Within Theropoda, several clades lost the teeth and developed beaks and feathers. Last, but not least, some ornithischians became quadrupedal, and evolved spectacular skull ornamentation and body armor for display and defense (Fig. 7.1). Some of these radiations, as well as major evolutionary transformations in other Mesozoic vertebrate clades, are the focus of short field trips throughout this time interval.

7.2 Thunder-Lizards: Gigantism in Sauropods

Emanuel Tschopp

One of the most spectacular sights when traveling in the Mesozoic would be the enormous sauropod dinosaurs. These animals reached lengths of 40–50 m from head to tail, with body masses estimated to be up to 100 tons (Carballido et al. 2017). Sauropods were the largest terrestrial animals that ever existed on this planet. Their size, and other features in

Fig. 7.2 Mounted skeletons of three distinct subgroups of Sauropoda. (**a**) *Cetiosaurus*, a basal member of Eusauropoda from the Middle Jurassic of the UK, mounted in the Leicester New Walk Museum and Art Gallery, Leicester, UK (photo courtesy of Jeff J Liston). (**b**) *Brontosaurus*, a member of Diplodocidae from the Late Jurassic of the USA, mounted at the Yale Peabody Museum, New Haven, USA (photo by Matt Wedel, shared under a CC-BY 4.0 International license). (**c**) *Brachiosaurus*, a member of Titanosauriformes from the Late Jurassic of the USA, mounted at the Field Museum in Chicago, USA (photo by Matt Wedel, shared under a CC-BY 4.0 International license)

the bones, led early workers to name some of them "whale-lizard" (*Cetiosaurus* from the Middle Jurassic of England; Owen 1842) or "thunder-lizard" (*Brontosaurus* from the Late Jurassic of the USA; Marsh 1879). Sauropods are known from the Late Triassic to the latest Cretaceous, and from all continents including Antarctica. Several groups are recognized; the one that includes the largest species is called Titanosauriformes (Fig. 7.2).

Traveling back to the Late Cretaceous, in what is now known as Argentina, it would be impossible to miss individuals of a group of titanosauriforms. These sauropods are known to have dug colonial nests with their hind feet and laid up to 35 eggs per nest in a geothermally* active area (Grellet-Tinner and Fiorelli 2010). What we observe here is a crucial, energy-saving strategy. This behavioral pattern is part of a suite of features that allowed sauropods to evolve such astonishing body sizes. In fact, a body size as that attained by gigantic sauropods comes with a series of biological problems [U0702], the most critical of which is energy. Evolving energy-efficient physical (morphological) traits and behaviors allowed sauropods to grow to such immense sizes.

Sauropods needed to funnel most of the energy into growth (as juveniles) and, once they reached sexual maturity, body maintenance. To overcome these obstacles, they optimized their reproduction strategy, lung structure, body cooling mechanism, food intake, and digestion (Fig. 7.3) [U0702]. Reproduction is among the most energy-consuming activities during an organism's life. In fact, growth generally slows down when sexual maturity is reached. This phenomenon is seen in the developmental records of sauropod bone as shown by histological* thin sections of their internal structure (Sander et al. 2011). Before reaching sexual maturity, sauropods grew significantly faster than any reptile

known today, nearly reaching mammalian growth rates. As reptiles, but unlike most mammals, sauropods continued to grow as adults, although at a slower rate (Griebeler et al. 2013). Such continuous growth was likely possible because sauropods did not invest a significant amount of energy into raising their young. Egg size was comparatively small in relation to their body size (Sander et al. 2008) and, given their body mass, it is unlikely that individuals engaged in brooding activities; they may have used other strategies instead. As noted above, some sauropods took advantage of the heat generated through geothermic activity to incubate their eggs. Others buried them, possibly together with plant material that released heat while decomposing (Vila et al. 2010). Once hatched and growth began, another energy-efficient feature employed by sauropods is the inferred structure of their lungs.

The bird-like lung structure of sauropods (Fig. 7.3), with air sacs connected to a predicted multichambered lung, increased efficiency while, at the same time, reducing the risk of overheating. Air sacs in birds allow for a cross-current gas exchange, which is more efficient than the unidirectional gas exchange that occurs in mammalian lungs (Perry and Sander 2004). The highly pneumatic axial skeleton of sauropods, as indicated by the complex arrangement of thin bone struts* and cavities in their vertebrae (Fig. 7.3) and, sometimes, even girdle and long bones, are features used to infer that these animals had an equally efficient lung (Wedel 2003). In addition, such an extensive air-sac system would have significantly increased the surface area in contact with air. This is seen as a feature that greatly contributed to heat exchange (Perry et al. 2009). Thus, excessive heat produced in these enormous bodies through muscle activity, fermentation, and digestion could be released not only through the skin, but also internally into the air-sac system [U0703].

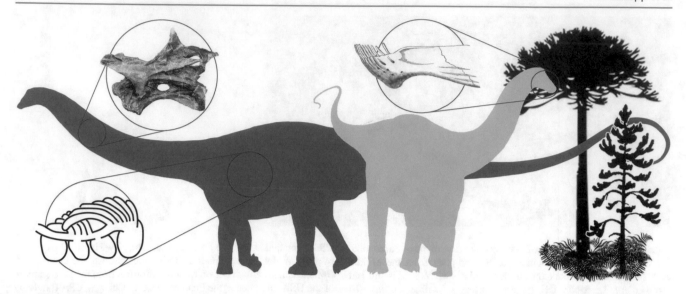

Fig. 7.3 Features that allowed sauropods to grow to gigantic body sizes. In a clock-wise direction, starting in the lower left corner: multi-chambered lung with high efficiency (modified from Perry et al. 2011), and with air-sacs that invade nearly the entire vertebral column in derived taxa (see the highly pneumatized cervical vertebra of *Galeamopus pabsti* with complex arrangement of thin bone struts and cavities; shown in left lateral view, modified from Tschopp and Mateus 2017). The absence of teeth for mastication (as demonstrated by the lower jaw of *Diplodocus* sp., modified from Marsh 1896), and a large feeding envelope thanks to the long neck, indicated by the different vegetation. The silhouettes are based on reconstructions of *Galeamopus pabsti* by Davide Bonadonna (Milan, Italy)

Energy is acquired through the digestion of food [U0704]. Sauropods were herbivores, living off a diet of various plant groups (Gee 2011; Gee et al. Chap. 6). Their long necks allowed them to browse a large area before having to spend energy to move to a new feeding location (Wedel et al. 2000). Unlike elephants, which masticate upwards of 20 h a day (Hummel and Clauss 2011), sauropods did not chew their food (Fig. 7.3), thereby saving a lot of time for browsing (Sander 2013). This time-saving device was possible due to a combination of factors. There is evidence that at least some sauropod species ingested small stones, the so-called gastroliths, that acted as kind of a mill in the moving stomach, grinding down plant material and increasing the efficiency of digestive enzymes (Sanders et al. 2001). In addition, sauropods likely had a hind-gut fermentation system similar to many mammals and some birds. With their enormous gut volume, these animals could retain food in their digestive tract for several days (Hummel and Clauss 2011). This additional retention time allowed for the combination of chemical and mechanical treatments to extract as many nutrients as possible from the food resource. These adaptations to increase energy-efficiency and reach enormous body size are reasons for their evolutionary success.

Sauropods dominated the landscape throughout the entire Jurassic and Cretaceous, a time interval spanning 135 million years! They were one of the most speciose groups among the major clades of non-avian dinosaurs (Mannion et al. 2011). As such, they were an impressive evolutionary success. A combination of morphological features and behavioral adaptations allowed them to evolve gigantic sizes. In addition, fast growth was probably also fueled by a fierce evolutionary arms race [U0705], during which juvenile or generally smaller individuals had to escape predators unless they reached a body size that was difficult to attack (Sander 2013). Only once in evolutionary history, early in the Mesozoic, did all these factors converge perfectly, allowing terrestrial animals to attain and, then, maintain such extreme body masses.

7.3 The Biggest Claws Do Not Hurt: Herbivorous Theropods

Christophe Hendrickx

We trek to the other side of the planet, in what is now Mongolia, during the Late Cretaceous. Here, sauropods were also abundant but were less diverse when compared to other continents including South America (Currie et al. 2018). Instead, another animal would likely catch most of our attention. This animal would be a large bipedal feathered dinosaur with extremely long claws on its hand and a beak. This beast is called *Therizinosaurus* and belongs to the theropod dinosaurs (Fig. 7.1). It holds the record for the longest hand claws of all time, attaining a length of up to 70 cm (Maleev 1954). Evidence including a downturned lower jaw, a large gut region, and an elongated neck indicates that the entire group of therizinosaurs was strictly herbivorous, probably feeding on leaves [U0706] (Zanno et al. 2009). The very long manual claws of *Therizinosaurus* were likely used for pulling down and browsing vegetation. Maybe surprisingly, although large predators may have preyed on this animal, there is no strong evidence that the hypertrophied (excessively developed) claws were used as a defense mechanism (Lautenschlager 2014). Theropod herbivory, once recognized, was a bit of a surprise among dinosaur researchers, given that theropods were known generally as the carnivorous dinosaurs [U0707].

Allosaurus, *Tyrannosaurus*, and *Velociraptor* are the best-known examples of all the ferocious meat-eating theropods that lived in the Mesozoic (and which you would prefer to avoid as a time-traveler). However, the discovery of toothless dinosaurs at the end of the nineteenth century (Marsh 1890) and the beginning of the twentieth century (Osborn 1924), as well as some strange looking theropods with beaks and leaf-shaped teeth in the 1970s (Perle 1979), led researchers to

suggest that some theropods were actually herbivorous (e.g., Cracraft 1971). *Therizinosaurus* is not the only herbivorous theropod from the Cretaceous of Mongolia (Barta in Tschopp et al., Chap. 8). This animal co-occurred with herbivore members of two additional theropod clades, known as oviraptorosaurs and ornithomimosaurs. These dinosaurs were also mostly toothless, bipedal, and feathered. In fact, recent studies have shown that many theropod groups were actually not carnivorous but omnivorous or even strictly herbivorous [U0708] (Zanno and Makovicky 2011). These diets are reflected best in the morphology of the teeth (Fig. 7.4) or in their complete absence in the jaw [U0709].

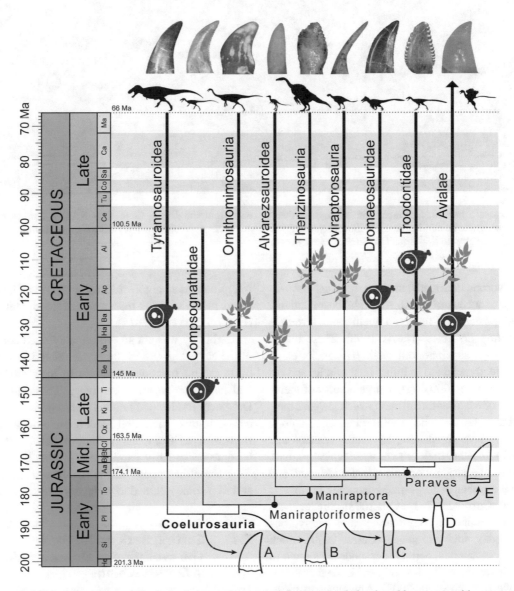

Fig. 7.4 Tooth evolution and diversity, and diet in coelurosaur theropods. Teeth evolved from a ziphodont (i.e., distally recurved and blade-shaped) tooth with serrated anterior and posterior edges (**a**) in basal coelurosaurs (e.g., Tyrannosauroidea), to a smaller and more simplified ziphodont tooth with a smooth anterior edge (**b**) in basal neocoelurosaurs (e.g., Compsognathidae). Maniraptoriform teeth, such as those of ornithomimosaurs, are no longer ziphodont but folidont (leaf-shaped) or conidont (cone-shaped), with minute or no denticles at all (**c**), and

maniraptoran teeth developed long roots with convex margins (**d**) as in therizinosaurs. Finally, unserrated teeth with a ziphodont morphology (**e**) are typical for paravians such as dromaeosaurids, troodontids, and the first true birds (Avialae). Tyrannosauroids, compsognathids, and dromaeosaurids were strictly carnivorous, whereas ornithomimosaurs, alvarezsauroids, therizinosaurs, and oviraptorosaurs were omnivorous and/or herbivorous. Carnivorous and herbivorous theropods are found among troodontids and birds

Fig. 7.5 *Deinocheirus*, a beaked, feathered, ornithomimosaur theropod. This genus is thought to be omnivorous. It is shown here in a floodplain environment together with a specimen of *Tarchia*, an ankylosaur, also from the Upper Cretaceous Nemegt Formation in Mongolia. Artwork copyright by Andrey Atuchin, used with permission

Typical carnivorous theropods, such as *Acrocanthosaurus* from the Early Cretaceous of North America, bear relatively large-sized and posteriorly recurved teeth with serrated∗ edges made of tiny denticles (Harris 1998). These blade-shaped teeth, known as ziphodont teeth (Hendrickx et al. 2015b), were perfectly adapted to inflict fatal injuries and cut through flesh (D'Amore 2009). Yet, a large number of theropods in the group of Maniraptoriformes (which includes therizinosaurs, oviraptorosaurs, and ornithomimosaurs, among others) did not have blade-shaped teeth (Zanno and Makovicky 2011). Instead, their crowns were typically small, constricted at the base, straight or only slightly recurved, and had only a small number of large-sized denticles, or none at all (Fig. 7.4; e.g., Currie 1987; Ji et al. 2003). Similar tooth morphologies are found in omnivorous and herbivorous extant animals, today, such as iguanas (Barrett 2000). Much the same as some sauropods, certain theropods are known to have possessed a gastric mill—a strong evidence for an herbivorous lifestyle (e.g., Ji et al. 2003). Analogous to birds, several theropod groups also completely or partly replaced their teeth by a horny beak (Louchart and Viriot 2011). This is the case of the tooth-less ornithomimosaur *Deinocheirus* from the Late Cretaceous of Mongolia. This large-sized theropod, with giant forelimbs and a long spatulated beak, is interpreted as an omnivore, eating fishes and browsing herbaceous plants in rivers that traversed floodplain environments (Fig. 7.5; Lee et al. 2014).

Herbivory was not an evolutionary dead end. Indeed, two of the closest groups of theropods related to the ancestors of birds exhibit functional changes in their tooth morphology that indicate a return to carnivory. These groups are called Dromaeosauridae (the famous *Deinonychus* and *Velociraptor* belong to this group) and Troodontidae (including *Troodon*). Both evolved from omnivorous or herbivorous animals to become secondarily carnivorous (Hendrickx et al. 2015a). Why some theropods became herbivorous in the first place, and why some returned to carnivory later, remains a mystery.

7.4 Getting Back on All Four: The Evolution of Quadrupedality in Ornithischians

Susannah C. R. Maidment

It is time to move from Asia to North America, and witness the changes in the fauna while traveling back in time from the Late Cretaceous to the Jurassic. One visual aspect of the megafauna that would be immediately recognizable during this time travel is the variability and diversity of four-legged (quadrupedal) dinosaurs. Not all of them were sauropods (Fig. 7.6). Late Cretaceous North American quadrupedal ornithischians include horned and frilled ceratopsians (as, for example, *Triceratops*, 71–66 Ma), gracile, duck-billed ornithopods (for instance, *Parasaurolophus*, 85–71 Ma), and heavily armored ankylosaurs (such as *Scolosaurus*, 85–71 Ma). During the Late Jurassic (151–145 Ma), the famous *Stegosaurus* with plates on its back and spikes at the end of its tail was one of the most common quadrupedal ornithischians living alongside the sauropods. This high diversity of quadrupeds is surprising because the first dinosaurs were bipedal, with forelimbs modified for grasping (Fig. 7.1).

The transition from bipedal to quadrupedal stance is exceptionally rare in the history of tetrapod evolution. It only occurred once outside of Dinosauria, in a Triassic reptile group called Silesauridae, which is closely related to dinosaurs (Fig. 7.7; Barrett and Maidment 2017). Among dinosaurs, quadrupedality evolved on at least four occasions independently. It is documented at least once in Sauropodomorpha (during the evolution of sauropods), and three times in Ornithischia. These include once in the armored Thyreophora, at least once in Ornithopoda, and at least once in Ceratopsia [U0710]. The transition in locomotion from bipedal ancestors to quadrupedal descendants appears to have underpinned major radiations for each of the groups in which it occurred. It was also associated with a series of physical changes, especially, of course, in the limbs (Box 7.1).

Fig. 7.6 Mounted skeletons of quadrupedal ornithischians. (**a**) *Triceratops* (a ceratopsid) at the Los Angeles Museum of Natural History, USA. By Allie Caulfield, shared under a CC BY-SA 3.0 copyright license (https://creativecommons.org/licenses/by-sa/3.0/deed.en). (**b**) *Parasaurolophus* (a hadrosaur) at the Field Museum of Natural History, Chicago, USA. By Zissoudisctrucker, shared under a CC BY-SA 4.0 copyright license (https://creativecommons.org/licenses/by-sa/4.0/). (**c**) *Scolosaurus* (an ankylosaur) at the Royal Tyrrell Museum of Palaeontology, Canada. By I.J. Reid, shared under a CC BY-SA 4.0 copyright license. (**d**) *Stegosaurus* (a thyreophoran) at the Natural History Museum, London, UK. Copyright The Natural History Museum

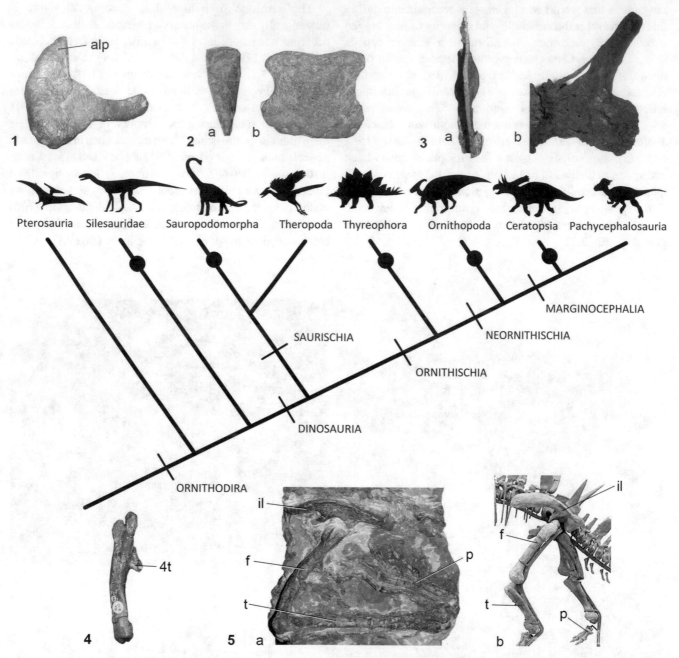

Fig. 7.7 Simplified phylogeny of Ornithischia and outgroups. Black dots on branches indicate groups in which quadrupedality has evolved. Photographs around the edge indicate features characteristic of quadrupedal ornithischians, convergently evolved in each major clade. (**1**) Proximal view of the left ulna of the quadrupedal hadrosaur *Brachylophosaurus canadensis* (CMN 8893) showing the development of an anterolateral process (alp). This process is absent in bipedal ornithischians. The transverse width of the ulna is 12 cm. (**2**) Terminal toe bones in bipedal and quadrupedal ornithischians contrasted. (**2a**) The bipedal neornithischian *Othnielosaurus consors* (SMA 0010), with typical "claw-like" bones. The phalanx is 3 cm long. (**2b**) The quadrupedal stegosaur *Kentrosaurus aethiopicus* (MB R.2951) with "hoof-like" bones typical of quadrupeds. The phalanx is 5 cm long. (**3a**) A dorsal view of the left ilium (il) of the bipedal basal ornithischian *Lesothosaurus diagnosticus* (NHMUK PV RUB 17), which is transversely compressed, typical of bipeds. Anterior is to the top of the image; the ilium is 7.9 cm long. (**3b**) Dorsal view of the right ilium of the quadrupedal stegosaur *Stegosaurus stenops* (NHMUK PV R36730), showing transverse expansion of the ilium. Anterior is toward the top of the image; the ilium is 82 cm long. (**4**) Lateral view of the left femur (f) of the bipedal basal ornithischian *Lesothosaurus diagnosticus* (NHMUK PV RUB 17), showing a prominent, pendant fourth trochanter (4t). The fourth trochanter is greatly reduced or lost in quadrupedal ornithischians. The femur is 10.4 cm long. (**5a**) The hind limb of the bipedal heterodontosaurid *Abrictosaurus consors* (NHMUK PV RUB 54; femur is 7.7 cm long); (**5b**) the hind limb of the quadrupedal stegosaur *Huayangosaurus taibaii* (ZDM T7001; femur is 51 cm long). In (**5a**), the femur is shorter than the tibia (t), where the reverse is true in (**5b**); in (**5a**), the foot (pes; p) is large and elongate, but is much smaller in (**5b**); in (**5a**), the femur is anteriorly curved along its length in lateral view, but is straight in (**5b**). All photos by the author. Institutional abbreviations: *CMN* Canadian Museum of Nature, Ottawa, Canada, *MB* Museum für Naturkunde, Berlin, Germany, *NHMUK* Natural History Museum, London, UK, *SMA* Sauriermuseum, Aathal, Switzerland, *ZDM* Zigong Dinosaur Museum, Sichuan Province, People's Republic of China

Box 7.1: Morphological Features Indicating Quadrupedality

Quadrupedal ornithischian dinosaurs are characterized by a suite of anatomical features that indicate their locomotion strategy (Maidment and Barrett 2014). These include the lengthening of the forelimb (Galton 1970), and the re-orientation of the lower forelimb bones (ulna and radius) and the forefoot (manus; e.g., Bonnan 2003). There were accompanying changes in the hip bone (ilium), which became wider (Fig. 7.7; Maidment and Barrett 2014). Changes in the hind limb mostly affected the thigh bone, which became longer than the shin bone (Norman 1980) and straight when seen from the side, and also developed reduced attachment sites for certain muscles (Yates and Kitching 2003). Additionally, the foot became relatively short and developed hoof-like toe bones (Galton 1970). To date, though, it is not always clear what might have been the functional significance of these skeletal adaptations. Some changes are undoubtedly related to modifications of the limbs to accommodate greater weight-bearing. For example, the rotation of the lower forelimb bones is related to pronation* of the wrist, which means that the palm faced downwards rather than inwards, providing more stability (Bonnan 2003). And, changes in the hind limb made it more columnar, which also increased its weight-bearing capacity (Maidment et al. 2014b). Some of these changes in bone structure related to the evolution of quadrupedalism in Ornithischia are also observed in Sauropodomorpha (Yates and Kitching 2003).

There are several shared morphological changes among different groups that evolved quadrupedality. This suggests that there are a limited number of biomechanical pathways along which this locomotion strategy can be achieved. Indeed, limb-bone morphology is highly convergent in many four-legged ornithischians. For example, the head of the humerus* is restricted to the back side of this bone in all quadrupedal groups. As such, the humerus was habitually retracted and could not have been moved forward past a vertical position [U0711]. This limited movement of the humerus indicates that the animal could only move slowly (Barrett and Maidment 2017). Despite a high degree of similarity in the bones, recent studies have demonstrated that there are also significant differences between quadrupedal ornithischian groups.

The advent of computer modeling has allowed scientists to use muscle reconstructions in combination with quantitative computational analyses to show that ornithischians exhibited a diversity of stances and gaits [U0712, U0713]

(Maidment et al. 2014a). While stegosaurs and ceratopsids appear to have held their elbows out from their bodies, ankylosaurs and hadrosaurs held them parallel to the plane that divides the body into right and left halves (Maidment and Barrett 2012). Ceratopsids, with relatively wide bodies, had a "wide-gauged" stance. In contrast, hadrosaurs, with narrower torsos, placed their feet on the midline during walking and running (Maidment et al. 2014b). Differing proportions of the humeri in ceratopsids and stegosaurs may indicate differences in where each animal's center-of-mass was located, perhaps related to the very large skulls of ceratopsids (Maidment et al. 2014b). Such a difference would, no doubt, have impacted locomotor style.

The reasons why all major groups of ornithischian dinosaurs evolved quadrupedality have proven difficult to resolve. Some authors suggested that quadrupedality is a consequence of large body size (e.g., Sereno 1997). But, this explanation is unsupportable in light of evidence that several small-bodied ornithischians were quadrupedal (e.g., the ceratopsian *Leptoceratops*) and that some very large theropods retained bipedalism (e.g., *Tyrannosaurus*). A confounding factor in our understanding of the phenomenon may be that different groups evolved quadrupedality for different reasons. It is possible that quadrupedality in ceratopsids was a consequence of a change in the animal's center-of-mass related to the evolution of very large skulls, elaborate horns, and frills, all of which may have been used for intra- or interspecific display [U0714] (Maidment et al. 2014b). This simple argument cannot be invoked for ornithopods or thyreophorans. It is known that all groups in which quadrupedality evolved, including Sauropodomorpha, are herbivorous. Thus, quadrupedality may be related to the need for a longer, larger, or more complex gut for efficient digestion of high-fiber plant material (Barrett and Maidment 2017). This hypothesis, however, has remained difficult to test in the absence of soft tissue data on dinosaurian gut structure. Hence, although we now know many aspects of how quadrupedality evolved, many more questions remain to be solved.

The evidence indicates that different clades of ornithischians attained the characteristics necessary for quadrupedality at different times in Earth history and along different pathways. The timing and diversity of acquisition of this trait indicates that ornithischians were not significantly constrained by their bipedal ancestry to develop a four-legged stance and locomotion in any particular way. Transitions in movement underpin some of the most important radiations in the history of tetrapod evolution. The rise to dominance of the quadrupedal ornithischian dinosaurs is no exception. Future work elucidating how quadrupedal dinosaurs moved, and testing hypotheses about why quadrupedality evolved, will no doubt shed significant light on the paleobiology and paleoecology of Mesozoic terrestrial ecosystems. It is now time to move to the islands.

7.5 Toothed, Feathered Dragons: Birds Are Dinosaurs

Dana Rashid, Loredana Macaluso, and
Emanuel Tschopp

Exploring a Late Jurassic archipelago in what is now Germany, with forests of conifers and gymnosperms filled with numerous invertebrates, lizards, turtles, pterosaurs, and dinosaurs (Röper 2005), we spot a very peculiar flying, feathered animal. At first sight, it seems to be a bird, but looking more closely, we observe that it has teeth instead of a beak; a long, bony tail; and claws on its wings. This is *Archaeopteryx*, the first and oldest fossil bird ever found [U0715]. If it lacked feathers and wings, we would most likely have mistaken it for a small theropod dinosaur. Although it remains unclear if *Archaeopteryx* was able to fly actively, or if it just glided from one tree to another, its discovery was the first sign that birds were derived from dinosaurs [U0716, U0717]. That fact is now generally accepted and supported by numerous other discoveries of feathered dinosaurs or toothed birds, especially from the Cretaceous of China (Fig. 7.8; Brusatte et al. 2015b; Cau 2018). But, how exactly did birds evolve from theropod dinosaurs?

Genetics and developmental studies provide some clues about the evolution of birds from dinosaurs [U0718]. Recent studies on tail development in mice have shown that mouse mutants* with fusion in their caudal vertebrae usually also have short, truncated tails. This feature mirrors the tail morphology of modern birds [U0719] (Rashid et al. 2014). Thus, significant changes in anatomy and morphology sometimes can result from very few mutations and can happen in very short time frames. Those rapid modifications are too short to

be detectable in the fossil record. Another recent study looked at the backward rotation (retroversion) of the pubis, a hip bone, in birds [U0720] (Macaluso and Tschopp 2018). Here, this bone is connected to the tail through muscles that control tail movement and help during lift off from the ground (Baumel et al. 1990). Such pubis retroversion is unique among living animals but happened five times in dinosaurs. And, four such retroversions are found among the Maniraptora, the clade of theropod dinosaurs from which birds evolved [U0721] and which also includes many herbivorous theropods (see Sect. 7.3). The evolution of a retroverted pubis seems to be correlated with the evolution of the peculiar ventilation mode of modern birds (Macaluso and Tschopp 2018). Modern birds use ossified hooked (uncinate) processes fused with the ribs as a lever to breathe more efficiently (Claessens 2015). Similar processes are present in cartilaginous form in crocodilians and were probably cartilaginous in early dinosaurs, too. But, they are found to have been ossified only in maniraptoran theropods and finally fused to the ribs in more derived birds (Codd et al. 2008). This ossification and fusion increased the lever arm for the muscles used for breathing and, thus, also the efficiency of the main ventilation mode that uses the ribs to increase and contract the lung volume during respiration. Hence, the importance of having an accessory respiratory system was reduced [U0722, U0723].

The accessory respiratory system that is inferred to be basal for dinosaurs and crocodilians is called cuirassal ventilation. Cuirassal ventilation worked with a muscle connecting two hip bones, the ischium and pubis, to the "belly ribs" [U0722]. This muscle, the Musculus ischiotruncus, increased the size of the abdominal cavity when contracted which, in turn, supported the usual costal respiration in inflating the animal's lungs. With the ossification of the hooked uncinate

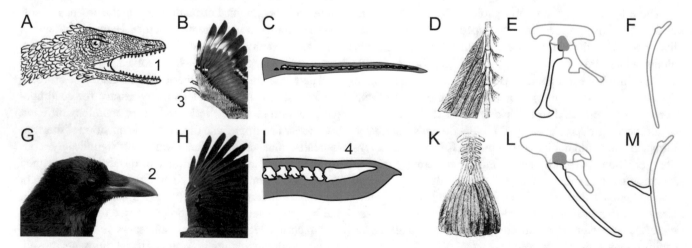

Fig. 7.8 Selected major changes in traits during the evolution of theropod dinosaurs (**a–f**) to birds (**g–m**): Loss of teeth and evolution of a beak (1,2; **a, g**); loss of clawed wings (3; **b, h**); reduction of tail length and appearance of pygostyle (4; **c, i**); changing feather arrangement on tail from frond to fan (**d, k**); retroversion of the pubis (**e, l**); ossification of uncinate processes on dorsal ribs (**f, m**). Images (**a**) Laura Garcés Gómez, shared under a CC-BY-SA 3.0 Unported License; (**c, i**) by D Rashid; (**d, k**) unknown author, Public Domain; (**e, f, l, m**) modified from Macaluso and Tschopp (2018)

processes [U0723], the presence of an accessory system such as the cuirassal ventilation became less important. As a consequence, the orientation of the hip bones was not tied to the Musculus ischiotruncus anymore, and the hip could be restructured. The pubis was free to rotate backwards, and to serve as an insertion point for a set of (pubocaudalis) muscles that control tail movement (Macaluso and Tschopp 2018). The hip and tail were not the only features that changed during the evolution of birds from theropods.

There are many additional traits that we generally associate with modern birds, other than the orientation of the pubis and the short tail. Nearly all of these additional traits actually evolved in non-avian dinosaurs, before *Archaeopteryx* ever took to the air [U0724, U0725]. Feathers, for example, predate birds, and have been discovered on a multitude of more basal, non-avian dinosaurs (Prum 1999). Hollow bones, which are so advantageous for bird flight, find their origins in Mesozoic dinosaurs (O'Connor and Claessens 2005). Another anatomical feature seen in modern birds is the reduction or complete loss of a muscle that connects the hind limbs with the tail (the Musculus caudofemoralis) and coordinates their movements for locomotion in reptiles. In birds, however, the reduction of this muscle has allowed the tail to act independently of the hind limbs. As such, the tail is its own locomotory unit. And, its independence greatly contributes to flight maneuverability through its novel connection to the hip through the pubocaudalis muscles (Gatesy and Dial 1996). The slow, progressive reduction and, subsequent, disappearance of the M. caudofemoralis has been documented in a long line of dinosaurs, since before the emergence of Maniraptora (Allen et al. 2013). All in all, birds share somewhere between 100 and 200 specific traits with their dinosaur ancestors (Cau 2018). Hence, these shared morphologies have substantiated the hypothesis, first proposed by Huxley (1868), that birds are the evolutionary legacy of Mesozoic dinosaurs.

7.6   Not Just Mouse-Like: Mammalian Radiations among Dinosaurs

Thomas Williamson and Stephen L. Brusatte

Mammals would clearly not be the first vertebrates you would see when traveling from Europe east across the Eurasian continent during the Late Jurassic. However, contrary to popular belief, Mesozoic mammals were not just small, insignificant animals, living in the shadows of the dinosaurs [U0726]. In a forested area with numerous lakes in what is now China, an attentive time-traveler could spot a wide variety of mammalian body types in the understory, on trees, and even gliding across the skies and swimming in water bodies (Fig. 7.9). Such a high diversity of shapes and ecological disparity implies a dynamic early evolutionary

history, which has been confirmed with recent discoveries all around the planet [U0727] (Luo 2007).

A number of anatomical/morphological, physiological, and developmental features separate early stem mammal-like animals (mammaliaforms) from other Late Jurassic animals and their ancestors. These include a high metabolism and insulation allowing activity in the cool of night, the evolution of chewing mechanisms, larger brains that helped to process sharpened senses and coordination, and fast growth helped by a novel way of nourishing offspring. The evolution of fur stabilized body temperature by providing insulation and helping to radiate excess heat (Crompton et al. 1978). The earliest mammaliaforms developed a simple jaw joint, enlarged muscles, and complicated teeth for chewing to maximize feeding efficiency (Luo 2007). As adaptations for chewing evolved in early mammaliaforms, the ear region became isolated from the jaw and hearing sensitivity was enhanced (Luo 2007). Brains in these early forms were large and embellished in areas that processed hearing, smell, and touch, all of which were transmitted through hairs and whiskers (Rowe et al. 2011). These super senses may have made early mammals and their relatives keenly aware of their surroundings in the dark of night. The earliest mammals possessed mammary glands that provided nourishment for fast growing young. Tooth replacement changed from a pattern of constant turnover throughout an individual's life to only a single episode of replacement. This innovation allowed the young to suckle and ensured that increasingly complicated cheek teeth of the upper and lower jaws precisely locked together (occluded) for better chewing (Luo et al. 2004). When did this complex interplay of features evolve?

Animals that are considered to be stem, or basal, mammaliaforms survived the End-Permian Mass Extinction (EPME), the most severe mass-extinction event in history (Benton 2003; Delfino et al., Chap. 10). By the Late Triassic, these stem mammaliaforms showed diverse feeding adaptations for eating a variety of different insect prey (e.g., *Morganucodon*, *Kuehneotherium*; Gill et al. 2014). Soon thereafter, in a crisis at the Triassic–Jurassic boundary, life on Earth was impacted by another Mesozoic mass extinction (see Roghi et al., Chap. 9). But, unlike other tetrapods, these mammaliaforms survived and underwent an evolutionary burst. The Multituberculata was one of the first major clades that appeared. These mammals are characterized by their distinctive enlarged incisors that are set apart by a large gap from the blade-shaped lower premolars and multi-cusped∗ grinding molars. By the Middle Jurassic, another group of mammals, the Tribosphenida, had evolved a tooth morphology that enabled a mortar-and-pestle chewing style. This evolutionary innovation, in turn, permitted a broader diet for these animals. Tribosphenids gave rise to therians, which contain metatherians (including modern marsupials) and eutherians (including the dominant group of mammals today, the Placentalia) to which we humans belong. Therian mam-

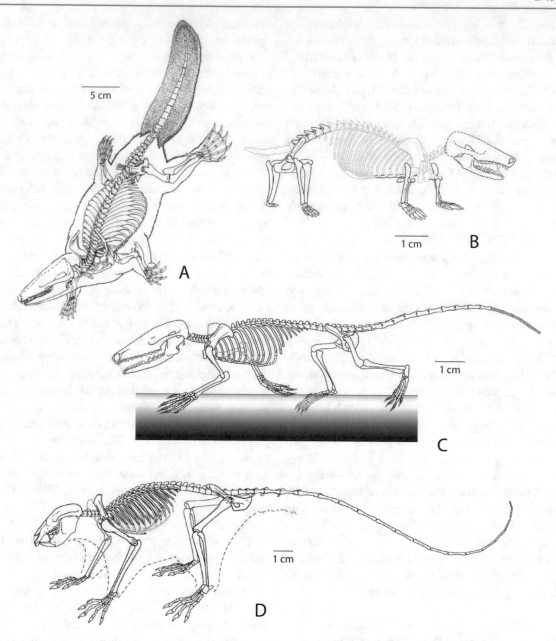

Fig. 7.9 Middle Jurassic mammals from China. (**a**) *Castorocauda lutrasimilis* (modified from Ji et al. 2006). (**b**) *Docofossor brachydacty-lus* (modified from Luo et al. 2015). (**c**) *Agilodocodon scansorius* (mod-ified from Meng et al. 2015). (**d**) *Maiopatagium furculiferum* (modified from Meng et al. 2017)

mals give birth to live young, with different strategies among the two main groups. Metatherians give birth to poorly devel-oped young which, in extant marsupials, develop in a pouch. Eutherians have offspring that are more developed at birth. Thanks to these, and other innovations, subsequently evolv-ing mammalian groups were able to increase in size and eco-logical breadth. Indeed, both metatherian and eutherian mammals underwent rapid diversification through the Late Cretaceous and expanded their biogeographic range, with possible members of Placentalia (e.g., *Protungulatum*) pres-ent in latest Cretaceous rocks of North America. However, let us return to the Middle and Late Jurassic of China, again, and have a closer look at the mammaliaforms that lived there.

The consequences of the post-extinction evolutionary burst are witnessed in the spectacular fossils found in the Jurassic and Cretaceous of China (Fig. 7.9). Here, Jurassic mammals, from about 160-million-year-old lake deposits, show specializations for swimming (e.g., the stem mammal *Castorocauda*; Fig. 7.9a; Ji et al. 2006), digging (the mole-like *Docofossor*; Fig. 7.9b; Luo et al. 2015), tree-living (the shrew-sized *Agilodocodon*; Fig. 7.9c; Meng et al. 2015), and gliding (such as *Maiopatagium*, which looked a bit like today's flying squirrel; Fig. 7.9d; Luo et al. 2017). Staying put in the same area throughout the Cretaceous would allow us to spot some of the earliest therians such as the small, tree-living *Eomaia* [U0728] (Ji et al. 2002). Metatherians developed a range of dietary specializations that included insectivory, carnivory, opportunistic feeding, and possibly frugivory (eating fruits and seeds) (Wilson 2013; Williamson et al. 2014). Eutherians were less diverse but also included insectivorous and carnivorous members and, possibly, some representatives that also ate seeds and fruits. Multituberculates, which probably had a rather herbivorous diet, underwent a major expansion that may be related to the rise of angiosperms (Wilson et al. 2012; Kvaček et al., Chap. 5). An unusual group of Multituberculata from the Southern Hemisphere, gondwanatheres, evolved ever-growing, high-crowned cheek teeth, possibly for chewing tough and fibrous plant material such as early grasses (Prasad ct al. 2005; see also Saarinen et al., Chap. 3). A European multituberculate subgroup with a square-like skull morphology developed small brains and other insular features due to their isolation on Mesozoic islands (Csiki-Sava et al. 2018). Yet, not all mammals were of small stature. Some mammals increased in body size during the Cretaceous.

Around the planet, some Cretaceous mammals had evolved into relatively large, carnivorous animals, further increasing the diversity of mammals that lived with dinosaurs. Examples are the Early Cretaceous *Repenomamus* [U0729] found in China and the Late Cretaceous *Didelphodon* from North America. *Repenomamus* reached about a meter in length and had slicing teeth that were well-adapted for carnivory. Indeed, a specimen was found with the remains of an individual of the small ceratopsian dinosaur *Psittacosaurus* in its gut (Hu et al. 2005). The metatherian *Didelphodon* grew to about 5 kg, and had teeth adapted for bone crushing (Wilson et al. 2016). And, there probably were other large mammals for which we have, yet, to discover their remains. As the events that ended the Cretaceous period occurred, eutherian, metatherian, multituberculate, and more primitive mammals were diverse and globally distributed. Most of their fossil record consists of small animals, no larger than a badger, but these were successful while living under foot of the dinosaurs.

7.7 Growing Fins: Adaptation to Aquatic Lifestyles in Mesozoic Reptiles

Márton Rabi and Ricardo Araújo

Exploring the rivers, lakes, and oceans by raft or dug-out canoe during the Late Jurassic reveals a completely different vertebrate fauna than we saw on land. Aquatic crocodiles were present, as well as a variety of strictly marine reptiles including ichthyosaurs, mosasaurs, and plesiosaurs. Turtles would have been the easiest to spot, given that they were the second most abundant and diverse animals, behind fishes, inhabiting both marine and freshwater habitats.

7.7.1 Turtles

Turtles have a unique morphology among vertebrates, with a shell partly formed by their ribs and shoulder girdles [U0730] (Lyson and Joyce 2012). This unique body plan obviously hinders an understanding of their origin [U0731]. Indeed, turtles are the single last major vertebrate group without a consensual position on the tree of life. How did their ancestors look?

Identifications of turtle ancestors in the fossil record, as well as the ecology of these enigmatic species, remain ambiguous (Joyce 2015). Today, turtles inhabit a wide range of ecosystems ranging from deserts to wetlands and marine habitats. Their extensive fossil record, combined with molecular data from living species, reveals that there were a number of repetitive, highly comparable ecological transitions between terrestrial and aquatic lifestyles during their 250-million-year-old evolutionary history. The rare, early turtles from the Late Triassic exhibit a fully formed shell, and appear to be terrestrial in habitat, judging from the anatomy of their limbs, neck, and tail osteoderms (bony deposits that form scales, plates, or other dermal structures) [U0732, U0733]. By the Middle Jurassic, turtles adapted to a freshwater lifestyle and became a characteristic and dominant element of wetlands (Rabi et al. 2010). From the Late Jurassic onwards, fossils of turtles now begin to appear in marine deposits [U0734]. The slow break-up of Jurassic Pangea contributed to this ecological radiation, and coincided with a split into the two major turtle clades to which all extant species belong, today (Joyce et al. 2016).

Presently, turtles are grouped into side-neck and hidden-neck turtles. Living side-neck turtles are all freshwater and restricted to the continents that formed the southern continent of Gondwana in the Mesozoic. Back then, and until the Paleogene, these turtles were adapted to a coastal marine lifestyle and, thus, had a near-global distribution. However, side-neck turtles never developed advanced marine adaptations. Instead, such adaptations may have occurred twice in hidden-neck turtles during the Cretaceous, resulting in the lineage of modern-day sea turtles (Cadena and Parham 2015). These adaptations include enlarged paddles, girdles, salt glands, and a reduced armor, all of which were key morphological transformations that allowed their acquisition of a marine lifestyle (Hirayama 1994). Interestingly, similar adaptations occurred in different reptile groups in similar ways, as we will see below in crocodiles (Sect. 7.7.2) and plesiosaurs (Sect. 7.7.3). However, hidden-neck turtles did not only colonize the oceans.

In contrast to side-neck turtles, hidden-neck turtles evolved into semi-terrestrial to terrestrial forms multiple

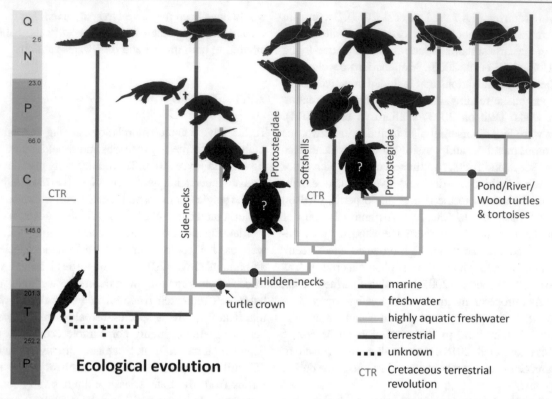

Ecological evolution

Fig. 7.10 Phylogeny and the evolution of habitat ecology of extant and extinct turtles (phylogeny after Joyce et al. 2016). Note the repeated evolution of marine and terrestrial lifestyles. Adaptations to a life on land occurred in tandem with the Cretaceous Terrestrial Revolution, when flowering plants appeared. *P* Permian, *T* Triassic, *J* Jurassic, *C* Cretaceous, *P* Paleogene, *N* Neogene, *Q* Quaternary

times. The first such event occurred in tandem with the spread of flowering plants during the Cretaceous. Thus, turtles could be another example for the wide-reaching impact of the evolution of flowering plants, just like multituberculate mammals (see Sect. 7.6 and Kvaček et al., Chap. 5). A second time hidden-neck turtles evolved terrestrial forms happened during the Paleogene, among the group that includes modern-day pond/river/wood turtles and tortoises. Some freshwater taxa, on the other hand, became highly specialized for an aquatic lifestyle and evolved into flatfish-like forms during the Cretaceous (softshell turtles; Fig. 7.10; Brinkman et al. 2017).

7.7.2 Crocodyliformes

Diving among marine turtles in the Late Jurassic oceans would have been dangerous. Several groups of carnivorous reptiles adapted to a marine and sometimes even a pelagic lifestyle, too. Crocodyliformes are only one example. In this group, metriorhynchoid crocodyliforms (the only archosaur group to evolve a marine lifestyle) were among the largest and most advanced clades [U0735]. These animals had a streamlined skull with dorsally shifted nostrils and laterally placed eye sockets (orbits), paddles, a tail fin, salt glands, and reduced bony armor. In these aspects, marine crocodyliforms

shared many traits with marine turtles, and even more with extant toothed whales. Just as in modern toothed whales, metriorhynchoid crocodyliforms also developed a wide variety of body sizes (Young et al. 2010). Indeed, the best modern-day analogs for these Late Jurassic crocodyliforms are dolphins and killer whales.

As a group, Crocodyliformes originated during the Late Triassic. Their extant descendants include crocodiles, alligators, caimans, and gavials. Initially, crocodyliforms were all terrestrial but they evolved into coastal marine forms by the Early Jurassic [U0736, U0737]. It remains unclear, however, whether these animals lived in freshwater prior to their transition to a saltwater habit. Early marine crocodiles superficially resembled extant gavials, the long-snouted, fish-eating forms of the present-day Indomalayan realm (Young et al. 2016). Pelagic forms had also evolved by the Early Jurassic, implying rapid diversification of the open-marine lineage of metriorhynchoids [U0738]. A rudimentary tail fin evolved early in this group and subsequent specializations during the Middle and Late Jurassic made them particularly adapted for a swimming lifestyle [U0739] (Ősi et al. 2018). Such adaptations for an open-marine lifestyle only occurred once during the evolutionary history of crocodyliforms, and did not last too long. Near the Jurassic/Cretaceous boundary, the Metriorhynchoidea declined in diversity, and went extinct by the second half of the Early Cretaceous (Chiarenza et al. 2015).

Other marine crocodyliforms also experienced extinction in the Early Cretaceous. These include a lineage of the gavial-like forms that evolved a massive skull, robust dentition, and giant body size (c. 7 m; Fanti et al. 2016). On the other hand, the gavial-like body plan was a recurring strategy for adaptation to a coastal marine lifestyle throughout the Mesozoic and similar body plans evolved independently in multiple lineages (Lee and Yates 2018).

7.7.3   Plesiosaurs

Amid the turtles and crocodyliforms occupying Mesozoic oceans were other strangely looking forms of marine reptiles—animals that have no extant analog whatsoever. These are the plesiosaurs [U0740]. As Cuvier once posed, plesiosaurs are "the most heteroclite (abnormal or irregular) inhabitants of the ancient world." They were part of a highly morphologically variable clade of mostly marine reptiles (Box 7.2), with rare specimens found in freshwater fluvial–lacustrine deposits (Sato et al. 2003). Plesiosaurs had four subequal, paddle-like limbs, streamlined bodies, and a relatively short tail (Fig. 7.11) [U0741]. They were obligate swimmers. Histological analyses of their inner bone structure reveal that these animals' skeletons experienced widespread hardening and increased bone density (osteosclerosis), a feature known to occur in living whales and other marine vertebrates. However, their bizarre anatomy with four subequally sized flippers is unparalleled among vertebrate species and poses a hydrodynamic paradox that has not been completely resolved yet [U0742-U0744]. Various motion combinations of the limb-swimming cycle have been proposed [U0745], but there is no clear consensus on how plesiosaurs propelled themselves through water (Araújo and Correia 2015). Their diet was predominantly of fishes (piscivores), as their jaw adduction muscles indicate [U0746] (Araújo and Polcyn 2013), but stomach contents have also revealed the inclusion of invertebrates including ammonites and belemnites. One plesiosaur, *Aristonectes*, seems to have used its numerous, needle-like teeth to perform filter-feeding (O'Keefe et al. 2017). As an even more complete adaptation to an aquatic lifestyle, plesiosaurs were viviparous∗ and K-selectors∗. This mode of reproduction indicates that they had few offspring but possibly increased the time of parental care for the young (O'Keefe and Chiappe 2011).

Box 7.2: Plesiosaur Diversity and Systematics

Plesiosaurs have traditionally been grouped into long-headed, short-necked forms (pliosauromorphs), and short-headed, long-necked forms (plesiosauromorphs). However, recent phylogenetic studies dispute this historic separation that was based on the proportions of the head and neck (Benson and Druckenmiller 2014). Instead, Plesiosauria now includes three groups. The Rhomaleosauridae is the most basal clade of plesiosaurs comprising Early to Mid-Jurassic forms mostly from Europe. The Pliosauridae includes longirostrine∗ forms, such as *Hauffiosaurus*, from the late Early Jurassic from England and Germany, as well as very large genera like *Liopleurodon* from the Late Jurassic from Europe. Mid-Cretaceous forms, called Brachaucheninae, are reported from the Western Interior Seaway and Australia. Finally, the Plesiosauroidea comprises the Microcleididae from the Early to Middle Jurassic of Europe, the Cryptoclididae from the Late Jurassic of Europe and North America, and the highly diverse Xenopsaria, which includes the famous long-necked Elasmosauridae that existed from the Early to the latest Cretaceous as well as the short-necked Leptocleidia from the Early to mid-Cretaceous.

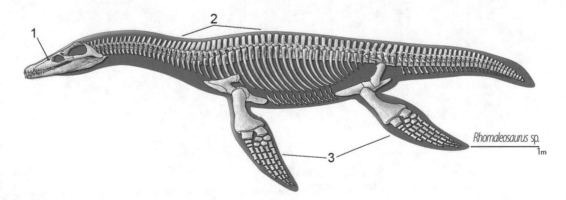

Fig. 7.11 Features acquired during the evolution to a completely marine lifestyle in the plesiosaur *Rhomaleosaurus*. Features that were convergently acquired in other marine reptile groups (e.g., turtles and crocodyliforms) are highlighted with an asterisk. Nasal openings are retracted (1), highly streamlined body shape∗ (2), flippers∗ with an increased number of phalanges and a reduced and simplified morphology of the arm bones (3). Drawing © Fernando JS Correia, Laboratório de Ilustração Científica -dbio/UA, used with permission

Plesiosauria are part of the sauropterygian lineage, a clade of marine reptiles whose relationships within Reptilia and within Diapsida are still debated (Scheyer et al. 2017). All members of Sauropterygia were adapted to an aquatic lifestyle, exhibiting shortening of the upper arm and leg bones and histological adaptations, such as bone densities similar to modern penguins (pachyostosis). However, aquatic adaptations among Sauropterygia are most conspicuous in plesiosaurs [U0741]. Divers in the Mesozoic oceans could have encountered these bizarre creatures almost throughout the entire era: they first appear near the end of the Triassic and are known up until the K/Pg mass extinction (Benson and Druckenmiller 2014). Their fossil remains occur in rocks from all over the planet. Something they all had in common were well-developed paddles, just as turtles and marine crocodyliforms. Like the latter, plesiosaurs also had streamlined bodies (Fig. 7.11). No member of the group escaped the end-Cretaceous mass-extinction event.

7.8   Starting a New Wave: Diversification Patterns among Mesozoic Fishes

Carlo Romano

While diving through Mesozoic lakes, rivers, and oceans, you would see many more fishes than aquatic reptiles (see Sect. 7.7). But what is, in fact, a fish? Fishes, as we commonly understand the group today, comprise a variety of distantly related, aquatic vertebrate species (Box 7.3; Fig. 7.12). Here, we focus especially on the ray-fins, coelacanths, and lungfishes, which collectively belong to the bony fishes, and which contribute a large percentage of today's "fish" diversity [U0747]. Interestingly, Mesozoic mass-extinction events did not impact fishes to the degree they affected other vertebrates. Hence, to observe important evolutionary changes in the group, we have to travel all the way back to the Triassic, which both began and ended with a mass-extinction event.

Looking back, there is an apparent long-term increase in bony fish diversity beginning in the Triassic, leading to the remarkable species richness of this group today (Friedman and Sallan 2012). In the wake of the K/Pg extinction, the so-called spiny-finned teleosts (e.g., perches, cods, oarfish) diversified. Spiny-finned teleosts comprise nearly 30% of all living vertebrate species and have a stunning array of body shapes (Friedman 2010). As with the K/Pg event, no significant crisis is evident during the Triassic–Jurassic boundary extinction event (Smithwick and Stubbs 2018), which preceded the diversification of mammals on land, for example (see Sect. 7.6). Nevertheless, some diversification events happened after this biological crisis. The evolution of the first rays in Chondrichthyes occurred in the Early to Middle Jurassic (Guinot and Cavin 2015), and the appearance of novel adaptations for durophagy* is found in teleostean actinopterygians (Smithwick and Stubbs 2018).

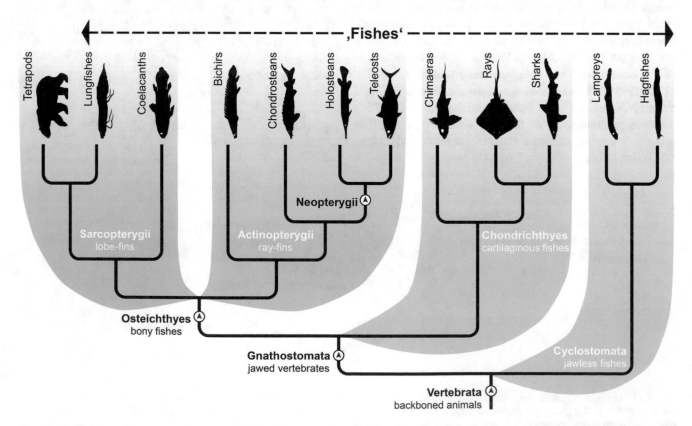

Fig. 7.12 Phylogenetic tree of extant backboned animals (Vertebrata). Osteichthyes (bony fishes) shown on the left side (Sarcopterygii in pink, Actinopterygii in purple), Chondrichthyes (cartilaginous fishes) in blue, and extant jawless fishes (Cyclostomata) in beige (simplified from Friedman and Sallan 2012). Note that with over 32,000 species, teleosts comprise about 50% of today's vertebrate diversity, whereas all other groups of "fishes" represent less speciose or depauperate lineages

Box 7.3: Systematics of Extant 'Fishes'

Most present-day fishes are classified as teleosts, whereas all other living groups belong to either less speciose or depauperate lineages [U0747] (Sallan 2014). Therefore, most of the commonly known fishes are teleosts, including tuna, salmon, pikes, catfishes, flatfishes, eels, seahorses, and anglerfishes. Teleosts are part of the bony fishes (Osteichthyes; Fig. 7.12), which comprise >32,000 living fish species (which are about 50% of all extant backboned animal species!). As indicated by their name, the skeleton of bony fishes is mostly ossified; although there are exceptions (e.g., sturgeons, whose skeletons are mostly cartilaginous). Another successful clade also takes its name from the general condition of the skeleton. Cartilaginous fishes (Chondrichthyes) include sharks, rays, skates and chimaeras (ghost sharks), and comprise over 1100 species today (Nelson et al. 2016). A third fish lineage are the jawless fishes (Cyclostomata), but these are much less diverse – both today and in the fossil record. Osteichthyes, Chondrichthyes, and Cyclostomata represent separate evolutionary lineages since the early Paleozoic. Nevertheless, when we commonly talk about "fishes", we usually mean all of these lineages – but we exclude one group of their descendants: tetrapods.

The term "fish" does not encompass a so-called monophyletic clade: a natural group that has a single common ancestor and includes all its descendants. It all comes down to one osteichthyan subgroup. Bony "fishes" are divided into two main subgroups, sarcopterygians (lobe-fins) and actinopterygians (ray-fins). Ray-fins can be further divided into Polypteriformes (bichirs, reedfishes), Chondrostei (sturgeons, paddlefishes), and Neopterygii, the latter of which includes Holostei (bowfin, gars) and Teleostei (the group encompassing over 96% of present-day fishes). Extant lobefins, on the other hand, include Actinistia (coelacanths) and Dipnoi (lungfishes), as well as the forms that gave rise to Tetrapoda (the terrestrial "four-limbs"). Hence, lobe-finned "fishes", and with them Osteichthyes and all its more-inclusive clades, include terrestrial vertebrates, as well. "Fishes", excluding us terrestrial forms, is therefore a so-called paraphyletic group. They have a common ancestor but some of this ancestor's descendants are excluded. Because scientists prefer names that describe monophyletic clades, the term "fish" is not used in formal systematic literature.

Traveling further back in time, we get close to Earth's most severe biotic crisis, the End-Permian Mass Extinction (EPME). Similar to the other mass-extinction events, this one does not seem to have impacted fishes significantly [U0748]. Analysis of fish diversity and body size before and after the crisis revealed comparably low extinction rates and no size selectivity∗ during the EPME (Puttick et al. 2017). Subsequent smaller extinction events during the Early Triassic, like the Smithian–Spathian boundary extinction (SSBE), severely decimated some groups (e.g., conodonts), but left others seemingly unaffected. How did the diversity of bony fishes change spatially and temporally after these extinctions?

Early Triassic bony fish assemblages prior to the SSBE were globally homogenous, with the same genera and families occurring on opposite sides of the supercontinent Pangea, and as distant as Greenland and Madagascar (Romano et al. 2016). Early Triassic fishes were also relatively large on average, mostly because small-sized species with body lengths of less than 10 cm were missing or not preserved (Fig. 7.13). A change in global bony fish communities is evident between the Early Triassic and Middle Triassic, when these cosmopolitan faunas were replaced by regionally contrasted associations (Romano et al. 2016). On a global scale, this trend resulted in a radiation of bony fishes [U0748], which can first be observed in large, predatory rayfins belonging to Palaeopterygii and smaller-sized Subholostei [U0749]. The average body size significantly decreased in the Middle Triassic, driven by a diversification of small-bodied taxa, most of which were neopterygians, including early teleosts (Teleosteomorpha) [U0750]. During this first neopterygian radiation event (Tintori et al. 2014), increasing functional and body-shape disparity, as well as novel feeding specializations, emerged [U0751] (Smithwick and Stubbs 2018). The question begs as to whether the Triassic radiation of bony fishes, and of neopterygians, in particular, was triggered by the extinction of Paleozoic chondrichthyan clades (Fig. 7.14).

Using molecular clocks, the origin of teleosts is thought to have occurred sometime in the Paleozoic (Near et al. 2012), but fossils of the earliest unambiguous members of this clade are known from the Triassic. Extant teleost clades do not appear before the Jurassic (Sallan 2014). Triassic-aged rocks also yield archaic-looking, stem ray-fin fishes. Their phylogenetic relationships often remain enigmatic, although some of them probably belong to extant non-teleost clades (Giles et al. 2017). Thus, Paleozoic and early Mesozoic fish assemblages were characterized mostly by non-teleosts (see also Tinn et al., Chap. 14). Even though these forms are only distantly related to modern species, you would recognize body shapes that are common in today's fishes, too. These body shapes include large-finned over-water gliders (*Thoracopterus*), elongate ambush predators (*Saurichthys*), fast swimmers (e.g., *Birgeria*, *Rebellatrix*), and unhurried deep-bodied forms (e.g., *Kyphosichthys*, *Cleithrolepis*, *Bobasatrania*, *Sargodon*, *Dapedium*) [U0751]. The conver-

Fig. 7.13 Examples of Early Triassic ray-fins (Actinopterygii). (**a**) Partial skull of *Birgeria americana* from Nevada (USA), with an estimated body length of almost 2 meters. The large eyes and snout to the right of the fossil are not preserved. (**b**) Skull of a medium-sized individual of *Saurichthys* sp. from Idaho (USA). (**c**) Complete fossil of *Watsonulus eugnathoides* from Madagascar, an early neopterygian. *Birgeria* and *Saurichthys* were the top-predators among Triassic ray-fins. Photos by C Romano

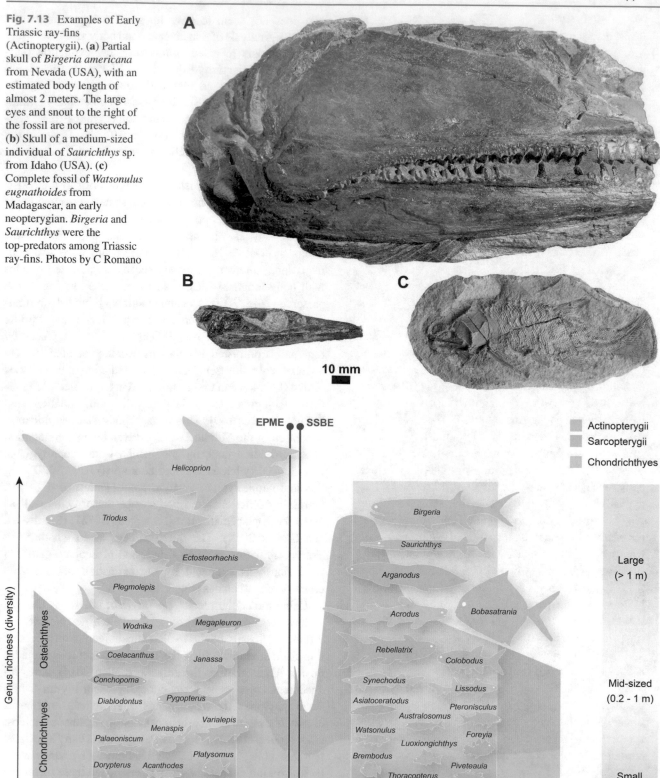

Fig. 7.14 The background diagram shows the stacked diversity of bony fishes (pink/purple shaded area; tetrapods excluded) and cartilaginous fishes (blue) during the Permian and Triassic, implying a turnover in diversity at the beginning of the Mesozoic (modified from Romano et al. 2016). Foreground: Overview of body shapes (silhouettes) and relative sizes (not to scale) of selected Permian and Triassic taxa. Body size correlates with trophic level in fishes (see Romano et al. 2016). The aquatic top predators during the Triassic and later were tetrapods (see Sect. 7.7, and Roghi et al., Chap. 9). *EPME* End-Permian Mass Extinction, *SSBE* Smithian–Spathian boundary extinction

gence in body plan suggests that these groups had evolved various lifestyles much before modern teleosts acquired them (Fletcher et al. 2014).

7.9 Conclusions

Our short journey during the Mesozoic has shown that, not surprisingly, Earth's biosphere witnessed a lot of variation and evolutionary experimentation over its 185-million-year history [U0752]. Not only did most of the vertebrate groups we recognize, today, originate or at least radiate throughout this era, many groups that are now extinct evolved peculiar features for which we do not have any modern analogs. Feathers appeared among reptiles, and were later adapted for flight, together with a series of additional changes in morphology that led to the evolution of modern birds. The largest terrestrial animals that ever existed, the sauropods, evolved during the Triassic, thanks to their unique body plan with elongated necks and highly efficient lungs, as well as strongly favorable environments and ecosystems. Carnivorous, bipedal dinosaurs became herbivorous, and several acquired a quadrupedal habit. In contrast, other terrestrial reptiles, including turtles, crocodiles, and lizards, adapted many times in convergent ways to an aquatic lifestyle. Even though relatively small, mammals experienced an impressive diversification, evolving forms adapted for swimming, digging, climbing, and gliding. Interestingly, fishes were not affected significantly by any Mesozoic mass-extinction event. Nevertheless, the first unambiguous records of the dominant group of fishes today, the teleosts, are known from the Triassic. Their origin was followed by a fast and vast radiation especially of small-sized species. All of these windows into Mesozoic life, and many more, demonstrate that it was a period of evolutionary radiation as well as experimentation possibly unparalleled in Earth history.

Questions

1. What is convergent evolution? Provide three examples from vertebrates from the Mesozoic.
2. Dinosaurs were ubiquitous and very diverse throughout the Mesozoic. Which environment did they not colonize? Any ideas why that could be?
3. Three mass-extinction events happened just before, during, and at the end of the Mesozoic. Name them, and discuss why they were important in shaping today's vertebrate diversity.
4. What factors allowed gigantism in sauropod dinosaurs, and why are there no terrestrial vertebrates today that reach similar sizes?
5. How did paleontologists recognize that some groups of the long-thought carnivorous theropod dinosaurs were, in fact, omnivorous/herbivorous? List three adaptations towards herbivory seen in some theropods.
6. Name five traits that modern birds share with Mesozoic dinosaurs but not with mammals or other reptiles.
7. Explain in your own words why the use of the term "fishes" is inadequate in an evolutionary sense.

Acknowledgments We thank RA Gastaldo and E Martinetto for highly useful reviews on an early draft of this chapter. ET thanks the Richard Gilder Graduate School and AMNH for funding and Mark Norell for the freedom to use his time for this book project.

References

Allen V, Bates KT, Li Z, Hutchinson JR (2013) Linking the evolution of body shape and locomotor biomechanics in bird-line archosaurs. Nature 497(7447):104–107

Araújo R, Correia F (2015) Plesiosaur pectoral myology. Palaeontol Electron 18(1):1–32

Araújo R, Polcyn MJ (2013) A biomechanical analysis of the skull and adductor chamber muscles in the late Cretaceous plesiosaur *Libonectes*. Palaeontol Electron 16(2):1–25

Baron MG, Norman DB, Barrett PM (2017) A new hypothesis of dinosaur relationships and early dinosaur evolution. Nature 543(7646):501–506

Barrett PM (2000) Prosauropod dinosaurs and iguanas: speculations on the diets of extinct reptiles. In: Sues H-D (ed) Evolution of Herbivory in terrestrial vertebrates. Cambridge University Press, Cambridge, pp 42–78

Barrett PM, Maidment SCR (2017) The evolution of ornithischian quadrupedality. J Iber Geol 43(3):363–377

Baumel JJ, Wilson JA, Bergren DR (1990) The ventilatory movements of the avian pelvis and tail: function of the muscles of the tail region of the pigeon (*Columba livia*). J Exp Biol 151(1):263–277

Benson RBJ, Druckenmiller PS (2014) Faunal turnover of marine tetrapods during the Jurassic–Cretaceous transition. Biol Rev 89(1):1–23

Benton MJ (2003) When life nearly died: the greatest mass extinction of all time. Thames & Hudson, New York

Bonnan MF (2003) The evolution of manus shape in sauropod dinosaurs: implications for functional morphology, forelimb orientation, and phylogeny. J Vertebr Paleontol 23(3):595–613

Brinkman D, Rabi M, Zhao L (2017) Lower Cretaceous fossils from China shed light on the ancestral body plan of crown softshell turtles (Trionychidae, Cryptodira). Sci Rep 7(1):6719

Brusatte SL, Butler RJ, Barrett PM, Carrano MT, Evans DC, Lloyd GT et al (2015a) The extinction of the dinosaurs. Biol Rev 90(2):628–642

Brusatte SL, O'Connor JK, Jarvis ED (2015b) The origin and diversification of birds. Curr Biol 25(19):R888–R898

Cadena EA, Parham JF (2015) Oldest known marine turtle? A new protostegid from the Lower Cretaceous of Colombia. PaleoBios 32(1):1–42

Carballido JL, Pol D, Otero A, Cerda IA, Salgado L, Garrido AC et al (2017) A new giant titanosaur sheds light on body mass evolution among sauropod dinosaurs. Proc R Soc B 284(1860):20171219

Cau A (2018) The assembly of the avian body plan: a 160-million-year long process. Bollettino della Società Paleontologica Italiana 57(1):2

Chiarenza AA, Foffa D, Young MT, Insacco G, Cau A, Carnevale G, Catanzariti R (2015) The youngest record of metriorhynchid crocodylomorphs, with implications for the extinction of Thalattosuchia. Cretac Res 56:608–616

Claessens LPAM (2015) Anatomical transformations and respiratory innovations of the archosaur trunk. In: Dial KP, Shubin N, Brainerd EL (eds) Great transformations in vertebrate evolution. University of Chicago Press, Chicago, pp 47–62

Codd JR, Manning PL, Norell MA, Perry SF (2008) Avian-like breathing mechanics in maniraptoran dinosaurs. Proc R Soc B Biol Sci 275(1631):157

Cracraft J (1971) Caenagnathiformes; Cretaceous birds convergent in jaw mechanism to dicynodont reptiles. J Paleontol 45(5): 805–809

Crompton AW, Taylor CR, Jagger JA (1978) Evolution of homeothermy in mammals. Nature 272(5651):333

Csiki-Sava Z, Vremir M, Meng J, Brusatte SL, Norell MA (2018) Dome-headed, small-brained island mammal from the Late Cretaceous of Romania. Proc Natl Acad Sci 115(19):4857–4862

Currie PJ (1987) Bird-like characteristics of the jaws and teeth of troodontid theropods (Dinosauria, Saurischia). J Vertebr Paleontol 7(1):72–81

Currie PJ, Wilson JA, Fanti F, Mainbayar B, Tsogtbaatar K (2018) Rediscovery of the type localities of the Late Cretaceous Mongolian sauropods *Nemegtosaurus mongoliensis* and *Opisthocoelicaudia skarzynskii*: stratigraphic and taxonomic implications. Palaeogeogr Palaeoclimatol Palaeoecol 494:5–13

D'Amore DC (2009) A functional explanation for denticulation in theropod dinosaur teeth. Anat Rec Adv Integr Anat Evol Biol 292(9):1297–1314

Fanti F, Miyashita T, Cantelli L, Mnasri F, Dridi J, Contessi M, Cau A (2016) The largest thalattosuchian (Crocodylomorpha) supports teleosaurid survival across the Jurassic-Cretaceous boundary. Cretac Res 61:263–274

Field DJ, Bercovici A, Berv JS, Dunn R, Fastovsky DE, Lyson TR et al (2018) Early evolution of modern birds structured by global forest collapse at the end-Cretaceous mass extinction. Curr Biol 28(11):1825–1831.e2

Fletcher T, Altringham J, Peakall J, Wignall P, Dorrell R (2014) Hydrodynamics of fossil fishes. Proc R Soc B 281(1788): 20140703

Friedman M (2010) Explosive morphological diversification of spiny-finned teleost fishes in the aftermath of the end-Cretaceous extinction. Proc R Soc Lond B Biol Sci 277(1688):rspb20092177

Friedman M, Sallan LC (2012) Five hundred million years of extinction and recovery: a phanerozoic survey of large-scale diversity patterns in fishes. Palaeontology 55(4):707–742

Galton PM (1970) The posture of hadrosaurian dinosaurs. J Paleontol 44(3):464–473

Gatesy SM, Dial KP (1996) Locomotor modules and the evolution of avian flight. Evolution 50(1):331–340

Gee CT (2011) Dietary options for the sauropod dinosaurs from an integrated botanical and paleobotanical perspective. In: Klein N, Remes K, Sander PM, Gee CT (eds) Biology of the sauropod dinosaurs: understanding the life of giants. Indiana University Press, Bloomington, pp 34–56

Giles S, Xu G-H, Near TJ, Friedman M (2017) Early members of 'living fossil' lineage imply later origin of modern ray-finned fishes. Nature 549(7671):265–268

Gill PG, Purnell MA, Crumpton N, Brown KR, Gostling NJ, Stampanoni M, Rayfield EJ (2014) Dietary specializations and diversity in feeding ecology of the earliest stem mammals. Nature 512(7514):303–305

Grellet-Tinner G, Fiorelli LE (2010) A new Argentinean nesting site showing neosauropod dinosaur reproduction in a Cretaceous hydrothermal environment. Nat Commun 1:32

Griebeler EM, Klein N, Sander PM (2013) Aging, maturation and growth of sauropodomorph dinosaurs as deduced from growth

curves using long bone histological data: an assessment of methodological constraints and solutions. PLoS One 8(6):e67012

Guinot G, Cavin L (2015) 'Fish' (Actinopterygii and Elasmobranchii) diversification patterns through deep time. Biol Rev 91(4): 950–981

Harris JD (1998) A reanalysis of *Acrocanthosaurus atokensis*, its phylogenetic status, and paleobiogeographic implications, based on a new specimen from Texas. N M Mus Nat Hist Sci Bull 13:1–75

Hendrickx C, Hartman SA, Mateus O (2015a) An overview on non-avian theropod discoveries and classification. PalArch's J Vertebr Palaeontol 12(1):1–73

Hendrickx C, Mateus O, Araújo R (2015b) A proposed terminology of theropod teeth (Dinosauria, Saurischia). J Vertebr Paleontol 35(5):e982797

Hirayama R (1994) Phylogenetic systematics of chelonioid sea turtles. Island Arc 3(4):270–284

Hu Y, Meng J, Wang Y, Li C (2005) Large Mesozoic mammals fed on young dinosaurs. Nature 433(7022):149–152

Hummel J, Clauss M (2011) Sauropod feeding and digestive physiology. In: Klein N, Remes K, Gee CT, Sander PM (eds) Biology of the sauropod dinosaurs: understanding the life of giants. Indiana University Press, Bloomington, pp 11–33

Huxley TH (1868) On the animals which are most nearly intermediate between birds and reptiles. Ann Mag Nat Hist 2:66–75

Ji Q, Luo Z-X, Yuan C-X, Wible JR, Zhang J-P, Georgi JA (2002) The earliest known eutherian mammal. Nature 416(6883):816–822

Ji Q, Norell MA, Makovicky PJ, Gao K-Q, Ji S-A, Yuan C (2003) An early ostrich dinosaur and implications for ornithomimosaur phylogeny. Am Mus Novit 3420:1–19

Ji Q, Luo Z-X, Yuan C-X, Tabrum AR (2006) A swimming mammaliaform from the middle Jurassic and ecomorphological diversification of early mammals. Science 311(5764):1123–1127

Joyce WG (2015) The origin of turtles: a paleontological perspective. J Exp Zool B Mol Dev Evol 324(3):181–193

Joyce WG, Rabi M, Clark JM, Xu X (2016) A toothed turtle from the Late Jurassic of China and the global biogeographic history of turtles. BMC Evol Biol 16(1):236

Lautenschlager S (2014) Morphological and functional diversity in therizinosaur claws and the implications for theropod claw evolution. Proc R Soc Lond B Biol Sci 281(1785):20140497

Lee MSY, Yates AM (2018) Tip-dating and homoplasy: reconciling the shallow molecular divergences of modern gharials with their long fossil record. Proc R Soc B 285(1881):20181071

Lee Y-N, Barsbold R, Currie PJ, Kobayashi Y, Lee H-J, Godefroit P et al (2014) Resolving the long-standing enigmas of a giant ornithomimosaur *Deinocheirus mirificus*. Nature 515:257–260

Louchart A, Viriot L (2011) From snout to beak: the loss of teeth in birds. Trends Ecol Evol 26(12):663–673

Luo Z-X (2007) Transformation and diversification in early mammal evolution. Nature 450(7172):1011–1019

Luo Z-X, Kielan-Jaworowska Z, Cifelli RL (2004) Evolution of dental replacement in mammals. Bull Carnegie Mus Nat Hist 2004(36):159–175

Luo Z-X, Meng Q-J, Ji Q, Liu D, Zhang Y-G, Neander AI (2015) Evolutionary development in basal mammaliaforms as revealed by a docodontan. Science 347(6223):760–764

Luo Z-X, Meng Q-J, Grossnickle DM, Liu D, Neander AI, Zhang Y-G, Ji Q (2017) New evidence for mammaliaform ear evolution and feeding adaptation in a Jurassic ecosystem. Nature 548(7667):326–329

Lyson TR, Joyce WG (2012) Evolution of the turtle bauplan: the topological relationship of the scapula relative to the ribcage. Biol Lett 8(6):1028–1031

Macaluso L, Tschopp E (2018) Evolutionary changes in pubic orientation in dinosaurs are more strongly correlated with the ventilation system than with herbivory. Palaeontology 61(5):703–719

Maidment SCR, Barrett PM (2012) Does morphological convergence imply functional similarity? A test using the evolution of quadrupedalism in ornithischian dinosaurs. Proc R Soc Lond B Biol Sci 279(1743):3765–3771

Maidment SCR, Barrett PM (2014) Osteological correlates for quadrupedality in ornithischian dinosaurs. Acta Palaeontol Pol 59(1):53–70

Maidment SCR, Bates KT, Falkingham PL, VanBuren C, Arbour V, Barrett PM (2014a) Locomotion in ornithischian dinosaurs: an assessment using three-dimensional computational modelling. Biol Rev 89(3):588–617

Maidment SCR, Henderson DM, Barrett PM (2014b) What drove reversions to quadrupedality in ornithischian dinosaurs? Testing hypotheses using centre of mass modelling. Naturwissenschaften 101(11):989–1001

Maleev EA (1954) New turtle-like reptile in Mongolia. Priroda 3:106–108

Mannion PD, Upchurch P, Carrano MT, Barrett PM (2011) Testing the effect of the rock record on diversity: a multidisciplinary approach to elucidating the generic richness of sauropodomorph dinosaurs through time. Biol Rev 86(1):157–181

Marsh OC (1879) Principal characters of American Jurassic dinosaurs. Part II. Am J Sci (Series 3) 17:86–92

Marsh OC (1890) Description of new dinosaurian reptiles. Am J Sci (Series 3) 39:81–86

Marsh OC (1896) The dinosaurs of North America. US Geol Surv Ann Rep 16:143–244

Meng Q-J, Ji Q, Zhang Y-G, Liu D, Grossnickle DM, Luo Z-X (2015) An arboreal docodont from the Jurassic and mammaliaform ecological diversification. Science 347(6223):764–768

Meng Q-J, Grossnickle DM, Liu D, Zhang Y-G, Neander AI, Ji Q, Luo Z-X (2017) New gliding mammaliaforms from the Jurassic. Nature 548(7667):291–296

Near TJ, Eytan RI, Dornburg A, Kuhn KL, Moore JA, Davis MP et al (2012) Resolution of ray-finned fish phylogeny and timing of diversification. Proc Natl Acad Sci 109(34):13698–13703

Nelson JS, Grande TC, Wilson MVH (2016) Fishes of the world. Wiley, Hoboken

Norman DB (1980) On the ornithischian dinosaur *Iguanodon bernissartensis* from the Lower Cretaceous of Bernissart (Belgium). Memoir de l'Institut R des Sci Nat de Belgique 178:1–105

O'Connor PM, Claessens LPAM (2005) Basic avian pulmonary design and flow-through ventilation in non-avian theropod dinosaurs. Nature 436(7048):253–256

O'Keefe FR, Chiappe LM (2011) Viviparity and K-selected life history in a Mesozoic marine plesiosaur (Reptilia, Sauropterygia). Science 333(6044):870–873

O'Keefe FR, Otero RA, Soto-Acuña S, O'gorman JP, Godfrey SJ, Chatterjee S (2017) Cranial anatomy of *Morturneria seymourensis* from Antarctica, and the evolution of filter feeding in plesiosaurs of the Austral Late Cretaceous. J Vertebr Paleontol 37(4):e1347570

Osborn HF (1924) Three new theropoda, *Protoceratops* zone, Central Mongolia. Am Mus Novit 144(7):1–12

Ősi A, Young MT, Galácz A, Rabi M (2018) A new large-bodied thalattosuchian crocodyliform from the Lower Jurassic (Toarcian) of Hungary, with further evidence of the mosaic acquisition of marine adaptations in Metriorhynchoidea. PeerJ 6:e4668

Owen R (1842) Report on British fossil reptiles Pt. II. Rep Br Assoc Adv Sci 1841:60–204

Perle A (1979) Segnosauridae – a new family of theropods from the Late Cretaceous of Mongolia. Trans Joint Soviet–Mongolian Palaeontol Exped 8:45–55

Perry SF, Sander PM (2004) Reconstructing the evolution of the respiratory apparatus in tetrapods. Respir Physiol Neurobiol 144(2):125–139

Perry SF, Christian A, Breuer T, Pajor N, Codd JR (2009) Implications of an avian-style respiratory system for gigantism in sauropod dinosaurs. J Exp Zool A Ecol Genet Physiol 311A(8):600–610

Perry SF, Breuer T, Pajor N (2011) Structure and function of the sauropod respiratory system. In: Klein N, Remes K, Gee CT, Sander PM (eds) Biology of the sauropod dinosaurs: understanding the life of giants. Indiana University Press, Bloomington, pp 83–93

Prasad V, Strömberg CAE, Alimohammadian H, Sahni A (2005) Dinosaur coprolites and the early evolution of grasses and grazers. Science 310(5751):1177–1180

Prum RO (1999) Development and evolutionary origin of feathers. J Exp Zool 285(4):291–306

Puttick MN, Kriwet J, Wen W, Hu S, Thomas GH, Benton MJ (2017) Body length of bony fishes was not a selective factor during the biggest mass extinction of all time. Palaeontology 60(5):727–741

Rabi M, Joyce WG, Wings O (2010) A review of the Mesozoic turtles of the Junggar Basin (Xinjiang, Northwest China) and the paleobiogeography of Jurassic to Early Cretaceous Asian testudinates. Palaeobiodivers Palaeoenviron 90(3):259–273

Rashid DJ, Chapman SC, Larsson HC, Organ CL, Bebin A-G, Merzdorf CS et al (2014) From dinosaurs to birds: a tail of evolution. EvoDevo 5:25

Romano C, Koot MB, Kogan I, Brayard A, Minikh AV, Brinkmann W et al (2016) Permian–Triassic Osteichthyes (bony fishes): diversity dynamics and body size evolution. Biol Rev 91(1):106–147

Röper M (2005) Field trip C: lithographic limestones and plattenkalk deposits of the Solnhofen and Mörnsheim formations near Eichstätt and Solnhofen. Zitteliana 26:71–85

Rowe TB, Macrini TE, Luo Z-X (2011) Fossil evidence on origin of the mammalian brain. Science 332(6032):955–957

Sallan LC (2014) Major issues in the origins of ray-finned fish (Actinopterygii) biodiversity. Biol Rev 89(4):950–971

Sander PM (2013) An evolutionary cascade model for sauropod dinosaur gigantism - overview, update and tests. PLoS One 8(10):e78573

Sander PM, Peitz C, Jackson FD, Chiappe LM (2008) Upper Cretaceous titanosaur nesting sites and their implications for sauropod dinosaur reproductive biology. Palaeontogr Abt A 284:69–107

Sander PM, Christian A, Clauss M, Fechner R, Gee CT, Griebeler E-M et al (2011) Biology of the sauropod dinosaurs: the evolution of gigantism. Biol Rev Camb Philos Soc 86(1):117–155

Sanders F, Manley K, Carpenter K (2001) Gastroliths from the Lower Cretaceous sauropod *Cedarosaurus weiskopfae*. In: Mesozoic vertebrate life. Indiana University Press, Bloomington, pp 166–180

Sato T, Li C, Wu X-C (2003) Restudy of *Bishanopliosaurus youngi* Dong 1980, a freshwater plesiosaurian from the Jurassic of Chongqing. Vertebr PalAsiatica 41(1):18–33

Scheyer TM, Neenan JM, Bodogan T, Furrer H, Obrist C, Plamondon M (2017) A new, exceptionally preserved juvenile specimen of *Eusaurosphargis dalsassoi* (Diapsida) and implications for Mesozoic marine diapsid phylogeny. Sci Rep 7(1):4406

Sereno PC (1997) The origin and evolution of dinosaurs. Annu Rev Earth Planet Sci 25(1):435–489

Sheehan PM, Hansen TA (1986) Detritus feeding as a buffer to extinction at the end of the Cretaceous. Geology 14(10):868–870

Smithwick FM, Stubbs TL (2018) Phanerozoic survivors: Actinopterygian evolution through the Permo-Triassic and Triassic-Jurassic mass extinction events. Evolution 72(2):348–362

Sprain CJ, Renne PR, Vanderkluysen L, Pande K, Self S, Mittal T (2019) The eruptive tempo of Deccan volcanism in relation to the Cretaceous-Paleogene boundary. Science 363(6429):866–870

Tintori A, Hitij T, Jiang D, Lombardo C, Sun Z (2014) Triassic actinopterygian fishes: the recovery after the end-Permian crisis. Integrat Zool 9(4):394–411

Tschopp E, Mateus O (2017) Osteology of *Galeamopus pabsti* sp. nov. (Sauropoda: Diplodocidae), with implications for neurocentral closure timing, and the cervico-dorsal transition in diplodocids. PeerJ 5:e3179

Vila B, Galobart À, Oms O, Poza B, Bravo AM (2010) Assessing the nesting strategies of Late Cretaceous titanosaurs: 3-D clutch geometry from a new megaloolithid eggsite. Lethaia 43(2):197–208

Wedel MJ (2003) The evolution of vertebral pneumaticity in sauropod dinosaurs. J Vertebr Paleontol 23(2):344–357

Wedel MJ, Cifelli RL, Sanders RK (2000) Osteology, paleobiology, and relationships of the sauropod dinosaur *Sauroposeidon*. Acta Palaeontol Pol 45(4):343–388

Williamson TE, Brusatte SL, Wilson GP (2014) The origin and early evolution of metatherian mammals: the Cretaceous record. ZooKeys 2014(465):1–76

Wilson GP (2013) Mammals across the K/Pg boundary in northeastern Montana, U.S.A.: dental morphology and body-size patterns reveal extinction selectivity and immigrant-fueled ecospace filling. Paleobiology 39(3):429–469

Wilson GP, Evans AR, Corfe IJ, Smits PD, Fortelius M, Jernvall J (2012) Adaptive radiation of multituberculate mammals before the extinction of dinosaurs. Nature 483(7390):457–460

Wilson GP, Ekdale EG, Hoganson JW, Calede JJ, Linden AV (2016) A large carnivorous mammal from the Late Cretaceous and the North American origin of marsupials. Nat Commun 7:13734

Yates AM, Kitching JW (2003) The earliest known sauropod dinosaur and the first steps towards sauropod locomotion. Proc R Soc Lond B Biol Sci 270(1525):1753–1758

Young MT, Brusatte SL, Ruta M, de Andrade MB (2010) The evolution of Metriorhynchoidea (Mesoeucrocodylia, Thalattosuchia): an integrated approach using geometric morphometrics, analysis of disparity, and biomechanics. Zool J Linnean Soc 158(4):801–859

Young MT, Rabi M, Bell MA, Foffa D, Steel L, Sachs S, Peyer K (2016) Big-headed marine crocodyliforms and why we must be cautious when using extant species as body length proxies for long-extinct relatives. Palaeontol Electron 19(3):1–14

Zanno LE, Makovicky PJ (2011) Herbivorous ecomorphology and specialization patterns in theropod dinosaur evolution. Proc Natl Acad Sci 108(1):232–237

Zanno LE, Gillette DD, Albright LB, Titus AL (2009) A new North American therizinosaurid and the role of herbivory in 'predatory' dinosaur evolution. Proc R Soc B Biol Sci 276(1672):3505–3511

How to Live with Dinosaurs: Ecosystems Across the Mesozoic

Emanuel Tschopp, Daniel E. Barta, Winand Brinkmann,
John R. Foster, Femke M. Holwerda,
Susannah C. R. Maidment, Stephen F. Poropat,
Torsten M. Scheyer, Albert G. Sellés, Bernat Vila,
and Marion Zahner

Abstract

We continue our trip back in time through the Mesozoic, visiting several different ecosystems across the planet. Each of these was strongly influenced by the continental breakup from a single landmass into several tectonic plates and associated landmasses during this period. We will visit localities on several continents, observe how their vertebrate faunas changed over time, and what external factors might have contributed to these differences.

During the Cretaceous, we visit the Iberian Peninsula, where hadrosauroids replaced titanosaurs as the most abundant dinosaur taxon. On the other side of the planet, a succession of geologic formations in Australia shows a gradual change from aquatic to terrestrial faunas resulting from sea-level changes of a now non-existent inland ocean. A visit to two polar ecosystems indicates possible mutual exclusion between amphibians (temnospondyls) and reptiles (crocodylomorphs), because they occupied similar ecological niches. Observing the record of Cretaceous landscapes in what is now Mongolia shows how changes in environment and climate correlate with changes in faunal composition.

Heading back, we check if there are distinct differences in vertebrate diversity in space and time in the Late Jurassic of North America. Then we move south, to Argentina, and back to the Middle and Early Jurassic. Here, we will try to understand where these Late Jurassic faunas originated and what influence the fragmentation of the supercontinent Pangea had on their evolution and diversity. Finally, we will stop our trip in the Late Triassic of Central Europe, examining a typical vertebrate fauna from the time when dinosaurs began their domination of the planet.

Electronic supplementary material A slide presentation and an explanation of each slide's content is freely available to everyone upon request via email to one of the editors: edoardo.martinetto@unito.it, ragastal@colby.edu, tschopp.e@gmail.com

*The asterisk designates terms explained in the Glossary.

E. Tschopp (✉)
Division of Paleontology, American Museum of Natural History, New York, NY, USA
e-mail: etschopp@amnh.org

D. E. Barta
Department of Anatomy and Cell Biology Oklahoma State University College of Osteopathic Medicine at the Cherokee Nation Tahlequah, Oklahoma, NY, USA

W. Brinkmann · T. M. Scheyer · M. Zahner
Universität Zürich, Paläontologisches Institut und Museum, Zürich, Switzerland

J. R. Foster
Utah Field House of Natural History State Park Museum, Vernal, UT, USA

F. M. Holwerda
Staatliche Naturwissenschaftliche Sammlungen Bayerns, Bayerische Staatssammlung für Paläontologie und Geologie, Munich, Germany

S. C. R. Maidment
Department of Earth Sciences, Natural History Museum, London, UK

S. F. Poropat
Swinburne University of Technology, Hawthorn, VIC, Australia

A. G. Sellés · B. Vila
Departament de Recerca de Faunes del Mesozoic, Institut Català de Paleontologia, Barcelona, Spain

Museu de la Conca Dellà, Lleida, Spain

8.1 Introduction

Ecosystems of the Mesozoic were very different than today's, a consequence of the mass extinction at the end of the Cretaceous, which eliminated most of the commanding vertebrates across the planet (Brusatte et al. 2015). On land, these dominant vertebrates were dinosaurs. Dinosaurs were, and still are, among the most diverse vertebrate groups, with combined

E. Martinetto et al. (eds.), *Nature through Time*, Springer Textbooks in Earth Sciences, Geography and Environment, https://doi.org/10.1007/978-3-030-35058-1_8

estimates of extinct and extant forms (birds) reaching approximately 14,000 species (Wang and Dodson 2006; Davis and Page 2014). However, other vertebrates populated Earth, as well. In fact, all of today's major vertebrate groups have been around since the Mesozoic or, in the case of several groups, since the Paleozoic (see Chaps. 10 and 14). After having focused on specific vertebrate clades and how they evolved through time in the previous chapter (Chap. 7), we now look at the taxa that lived together throughout the approximately 180 million years of the Mesozoic Era. We start our imaginary field trip about 66 million years ago, just before the asteroid hit Earth in the latest Cretaceous. From there, we will travel back to the Triassic stopping in seven localities scattered across both hemispheres (Fig. 8.1).

8.2 From Super-Rich to Nothing: The End-Cretaceous Extinction in Spain

Bernat Vila and Albert G. Sellés

Sixty-six million years have passed since one of the most catastrophic episodes in Earth's history. It was a devastating impact caused by a ten-kilometers-wide asteroid that crashed into the ocean at high speed and caused the extinction of 60% of the biosphere (Schulte et al. 2010; see also Vajda in Chap. 6). The day after the impact, global climate changed, the environments were no longer the same, and many organisms had vanished forever. Among the victims were most of the species of a clade of vertebrates that had ruled the planet for

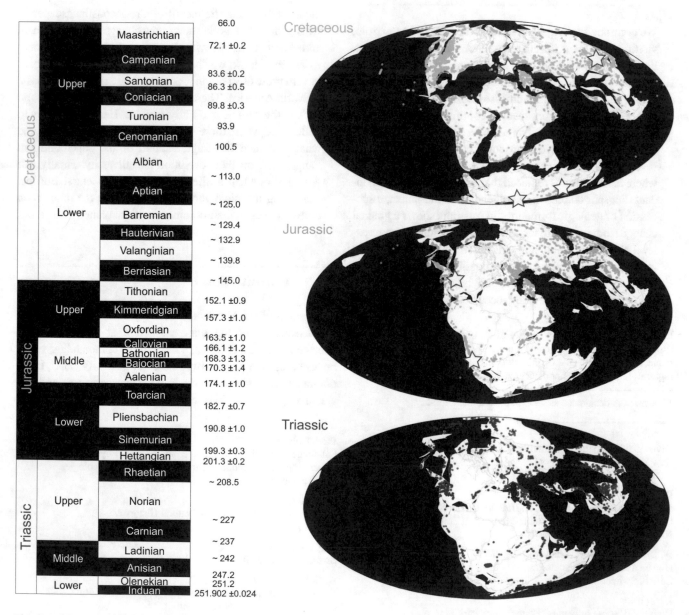

Fig. 8.1 Mesozoic timeline and paleogeographic maps with fossil occurrences. The localities we visit throughout this chapter are indicated by stars. Geological periods (e.g., Cretaceous), epochs (e.g., Lower Jurassic), and stages (e.g., Norian) are indicated on the timeline, together with their most current date estimated in Ma (based on the International Chronostratigraphic Chart, v2018/08; Cohen et al. 2013). Occurrence maps are from the PaleobiologyDatabase (www.paleobiodb.org)

165 million years: Dinosauria. The most famous area in which this dramatic scenario is recorded is western North America, where thousands of fossil remains of dinosaurs, plants, and a plethora of other organisms are collected every year, providing new data on fossil ecosystems before the global catastrophe (Brusatte et al. 2015). However, much less is known about species that lived on other landmasses at the time of the impact and, more importantly, about changes in ecosystems in other regions during the latest Maastrichtian, the latest geological stage in the Cretaceous (72–66 Ma; Fig. 8.1).

Among the few regions of the planet geographically distant from the impact zone, which reveal secrets of the crisis, are New Zealand, which we visited in Chap. 6 (Vajda in Chap. 6), and the Ibero-Armorican domain, to where we now are heading. During the Late Cretaceous, the Ibero-Armorican domain was the largest island of the European archipelago (Fig. 8.2) [U0802]. Today, the rocks in the southern Pyrenees record how dinosaur faunas evolved in this particular island laboratory. There, in the present-day regions of Catalonia and Aragón, hundreds of fossils recovered from the badlands of the Tremp Formation not only provide a window into the terminal Mesozoic ecosystems but also provide a valuable test of whether ecological and extinction patterns recorded elsewhere are a global phenomenon [U0803]. Investigations on this region conducted over the last few decades have yielded an extraordinary record of its latest Cretaceous dinosaur communities (e.g., López-Martínez 2001; Vila et al. 2006).

Recent analyses that explain the temporal distribution of these communities are based on a comprehensive database of all dinosaur fossil occurrences from the late Campanian to the latest Maastrichtian (77–66 Ma) in the Ibero-Armorican domain [U0804] (Vila et al. 2016; Fondevilla et al. 2019). And, the results are clear. Vertebrate communities were extremely dynamic and changed through time. Dinosaur assemblages experienced a major turnover right before the end of the Cretaceous, which was accompanied by environmental change (Fig. 8.2). The pre-turnover communities were composed of various dinosaur clades. There were several different species of derived titanosaurs (a group of sauropods that also includes the nesting animals we observed in Argentina in Chap. 7), which were characteristically gracile and relatively small in size (e.g., *Lirainosaurus*; Díez Díaz et al. 2013). Other herbivorous dinosaurs living alongside these small giants were ornithischians, including basal iguanodontians (*Rhabdodon* spp.; Pereda Suberbiola et al. 1999), basal hadrosauroids, and the rare, armored nodosaurids (a group of ankylosaurs), such as the small-sized *Struthiosaurus*. They both represent a small part of the ecological diversity in these pre-turnover communities. Theropod dinosaurs, including large abelisauroids (Tortosa et al. 2014) and small dromaeosaurs (Torices et al. 2015), were also present [U0805].

The composition of the vertebrate communities changed during the beginning of the early Maastrichtian, about 71 million years ago (Fig. 8.2). New immigrants arrived from the Americas, Eurasia, and Gondwana, as land bridges probably were created in response to a drop in sea level (Csiki-Sava et al. 2015). For a period of approximately two million years, the pre-turnover dinosaur taxa coexisted with the new arrivals, and their replacement was somewhat progressive with most of the species likely competing for environmental space and resources (Sellés et al. 2017). The most successful newcomers were lambeosaurine hadrosaurids, a group that rapidly occupied all available environments and displaced the cohort of herbivorous clades throughout the region. Thus, the post-turnover dinosaur faunas at the beginning of the late Maastrichtian were composed of new and moderately larger titanosaurs, lambeosaurine hadrosaurids, together with basal hadrosauroids, and flocks of large abelisauroid and small dromaeosaurid theropods [U0806]. Numbers of nodosaurids and rhabdodontids were in decline, and these groups were pushed almost to extinction in these ecosystems [U0805] (Prieto-Márquez et al. 2013; Vila et al. 2016). The new communities inhabited the island for the last part of the Cretaceous, and no major loss of biodiversity is observed before the catastrophic extinction event (Canudo et al. 2016).

The terminal dinosaur record in the Pyrenean region is based on well-dated chronostratigraphic successions and consists of nearly 100 sites. This fossil record contains skeletal remains of hadrosauroid ornithopods, titanosaurs, and theropods, as well as traces of their activities including egg sites and trackways (Vila et al. 2016). Fossil remains of dinosaurs are abundant in rocks of the latest Cretaceous but disappear abruptly near the top of the Tremp Formation, although the type of sediment and, thus, fossilization potential remained similar [U0807]. Even more unexpectedly, lower Paleocene strata are not only virtually devoid of dinosaurs but of any vertebrate fossils. The rapid disappearance of vertebrates at the end of the Cretaceous indicates that their extinction was nearly immediate in the southern Pyrenees and, probably, in the entire Ibero-Armorican region. Additionally, the general rarity of vertebrate fossils throughout the earliest Paleogene implies that these ecosystems took relatively long to recover (Canudo et al. 2016). On the other side of the planet, Cretaceous ecosystems looked quite different.

8.3 Sea, Land, and Darkness: Australia Throughout the Cretaceous

Stephen F. Poropat

Australia has not always been a sunburnt country. During the Cretaceous, central Queensland was a floodplain – as it is now. However, this Cretaceous floodplain was covered in a lush conifer forest and was not semi-arid grassland as it is

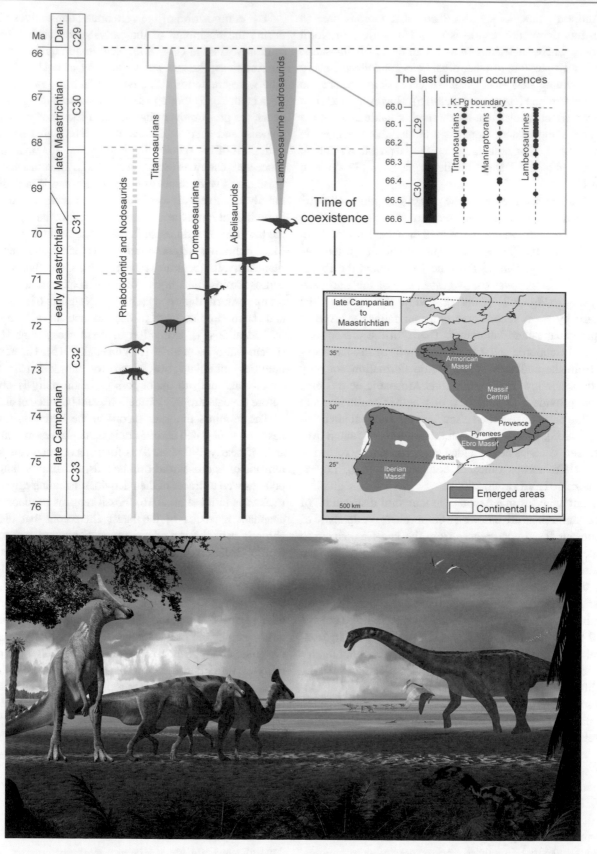

Fig. 8.2 Temporal distribution of dinosaur faunas during the latest Cretaceous in the Ibero-Armorican island, with indications of the last occurrences right before the Cretaceous-Paleogene boundary (based on Vila et al. 2016; Canudo et al. 2016; and Fondevilla et al. 2019). Below, a late Maastrichtian floodplain environment in the Ibero-Armorican island, depicting the herds of lambeosaurine hadrosaurids, titanosaurs, and dromaeosaurids, among other vertebrates (image courtesy of Oscar Sanisidro)

today (Cantrill in Gee et al., Chap. 6). Where kangaroos now hop, dinosaurs once walked and ran. Similarly, Australia has not always been an island continent. As we have seen during our visit to the Australian conifer forests, the continent was part of East Gondwana during the Cretaceous, with a southern connection to Antarctica and an eastern connection with a continental landmass called Zealandia (which is now largely submerged). Antarctica's intermittent connection with South America facilitated the passage of terrestrial animals between East and West Gondwana until about 50 million years ago, when rifting finally detached it completely from adjoining landmasses (Box 8.1). Although Australia has been steadily drifting northward since then, it has deserved the nickname "Down Under" since the Cretaceous, when its southeast corner was situated within the Antarctic Circle.

The polar fossil floras from the Early Cretaceous (c. 125–105 Ma) of southeastern Australia are well documented [U0812, U0813] (Tosolini et al. 2018). By contrast, Cretaceous floras in northeastern Australia are only known from the mid-Cretaceous Winton Formation (Eromanga Basin) and from the Early Cretaceous of the Styx and Maryborough basins. The plants that thrived in these landscapes vary in composition through time, but several similarities are apparent. Conifers were generally dominant, with ferns, ginkgoes, cycads, bennettitales, and horsetails making up various proportions of the understory and groundcover vegetation. Importantly, these floras document the rapid rise of angiosperms during the mid-Cretaceous. Flowering plants

went from relative obscurity in southeastern Australia at about 120 million years ago to co-dominating northeastern Australian floras with conifers by about 95 million years ago (McLoughlin et al. 2010). It is in these vegetated landscapes that the vertebrates thrived.

Terrestrial tetrapods occupied the forests, plains, rivers, and lakes of a varied geomorphic nature. Ornithopod dinosaurs reigned supreme in southeastern Australia during the late Early Cretaceous (Poropat et al. 2018). Their remains are more abundant than those of any other tetrapod, and five taxa (*Leaellynasaura*, *Atlascopcosaurus*, *Qantassaurus*, *Diluvicursor*, *Galleonosaurus*) have been named, to date. These small herbivores shared their environment with ankylosaurs, megaraptorid theropods, turtles (e.g., *Otwayemys*, *Chelycarapookus*), flying pterosaurs, lizards, and numerous mammals. Currently, the mammalian component includes monotremes (*Teinolophos*, *Kryoryctes*), a multituberculate (*Corriebaatar*), and ausktribosphenids (*Ausktribosphenos*, *Bishops*). The rocks exposed in the upper Strzelecki Group (c. 125–115 Ma) of eastern Victoria have produced abundant evidence of the latest-surviving temnospondyls (considered a primitive amphibian group) known anywhere in the world (*Koolasuchus*). Yet, these rocks have yielded no evidence of crocodylomorph fossils [U0812, U0813]. By contrast, in western Victoria, the slightly younger Eumeralla Formation (c. 115–105 Ma) has yielded crocodylomorph remains but no temnospondyls [U0814]. This faunal change is attributed to a warming trend during the late Early Cretaceous. Cool climates might have favored temnospondyls, whereas rising temperatures might have left them vulnerable to displacement [U0815]. Another major vertebrate clade that is conspicuously absent from the Victorian Cretaceous is the sauropods.

Relatively diverse sauropod faunas are known from mid-Cretaceous localities in Queensland [U0809, U0810]. The geologically oldest specimens are preserved in the marine Toolebuc Formation (c. 106–104 Ma) and the Allaru Mudstone (c. 104–102 Ma), which have produced partial skeletons of *Austrosaurus* (Poropat et al. 2017). Remains of other dinosaurs have also been found in these marine units, particularly from the Mackunda Formation (c. 102–101 Ma). These include the bird *Nanantius*, the ankylosaur *Kunbarrasaurus*, and the large ornithopod *Muttaburrasaurus*. Whereas pterosaurs are relatively rare in the marine units of the Eromanga Basin (*Mythunga*, *Aussiedraco*; Pentland and Poropat 2019), remains of marine reptiles are common. Ichthyosaurs (*Platypterygius*), several types of plesiosaurs (including pliosaurs, elasmosaurs, and polycotylids, see also Rabi and Araújo in Chap. 7), and turtles (*Cratochelone*, *Notochelone*, *Bouliachelys*) are often recovered [U0808] (Kear 2016; Kear et al. 2018). With the gradual retreat of the inland sea, terrestrial fossils, including those of sauropods, become more abundant (Fig. 8.3) [U0809].

The most diverse sauropod fauna known is that from the Winton Formation (c. 101–93 Ma) exposed in central-

Box 8.1: Australian Paleogeography in the Cretaceous

Our understanding of Early Cretaceous tetrapod faunas in southeastern Australia (Victoria) is hosted in the Gippsland and Otway basins, which formed during the early stages of Australian – Antarctic rifting (Duddy 2003). The river systems that flowed through this rift valley carried huge volumes of contemporaneous volcanogenic sediment from the east, and it is in these sediments that tetrapod fossils have been found. By contrast, the Eromanga Basin of northeastern Australia (Queensland) was tectonically stable throughout the Mesozoic. Most of the sediments deposited in the Eromanga Sea, which initially covered these inland basins, were sourced from the Whitsundays Volcanic Province to the northeast. During the Late Cretaceous, this inland sea gradually receded [U0808] (Cook et al. 2013) and was succeeded by a vast floodplain (which formed the Winton Formation). Erosion of these sediments over the past 95 million years has resulted in the succession of marine to terrestrial strata we can now observe and explore for fossils (Fig. 8.3) [U0809].

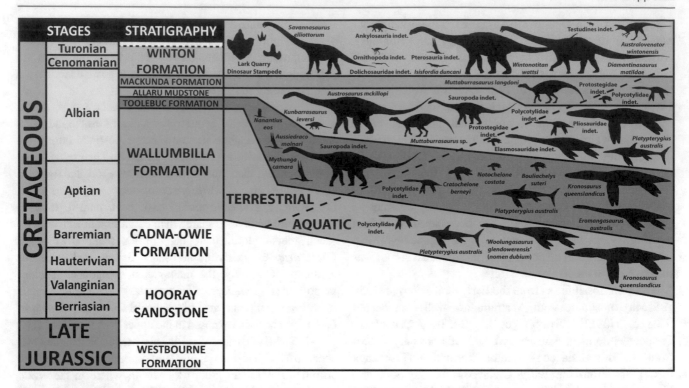

Fig. 8.3 Mid-Cretaceous tetrapods of the Eromanga Basin. Silhouettes of tetrapods from the Eromanga Sea at its maximum aerial extent (Wallumbilla Formation), its contraction (Toolebuc, Allaru, and Mackunda), and its final recession. A vast floodplain took the place of the former Eromanga Sea, and the sediments deposited therein are known as the Winton Formation

western Queensland. Here, three sauropod taxa are currently described (*Wintonotitan, Diamantinasaurus, Savannasaurus*) [U0810]. Alongside these behemoths lived the megaraptorid theropod *Australovenator*, small ornithopods, and ankylosaurs. But, the diversity in this site is not based solely on megafossil remains. Trace fossils of footprints in the rocks exposed at Lark Quarry indicate the presence of larger ornithopods and tiny theropods. In addition, the tetrapod fauna comprises pterosaurs, crocodylomorphs (*Isisfordia*), turtles, and lizards (Hocknull et al. 2009; Poropat et al. 2015).

The record of sauropods might be over-represented in the fossil collections from Queensland. Most specimens found in this region are made by local graziers, and the sheer size of sauropod bones means that they are more likely to remain recognizable when lying in a field rather than other, smaller tetrapod remains [U0811]. Thus, as shown by the diversity of the Lark Quarry footprints, other tetrapods are probably under-represented by body fossils. By contrast, Cretaceous strata in coastal Victoria are subjected to continual erosion and have been systematically sampled for decades by paleontologists specifically searching for small tetrapod fossils (Rich and Vickers-Rich 2000). However, not even a sauropod tooth has been found here. Perhaps these giants were genuinely absent from Victoria between about 125 and 105 million years ago. Sauropods might only have entered this region later in the Cretaceous when temperatures increased

(Poropat et al. 2016). In sum, while much of Australia languished underwater during the Early Cretaceous, ornithopods and megaraptorids thrived in the conifer-dominated rift valleys of the southeast. Once the Eromanga Sea receded, sauropods strode across the conifer-forested floodplains of the northeast, with burgeoning angiosperms underfoot.

8.4   Changing Climates and Faunas: The Case of Mongolia

Daniel E. Barta

The Gobi Desert of Mongolia contains some of the richest Cretaceous vertebrate localities in the world. These sites document changes in climate and faunas from the Early Cretaceous (>125 Ma) to nearly the end of the Late Cretaceous (c. 70 Ma; Fig. 8.1) [U0816, U0817]. Unique dinosaur and other vertebrate faunas characterize each formation deposited during this time span. However, an absence of absolute radiometric ages for many key strata currently prohibits the construction of a detailed chronostratigraphic framework such as that obtained for the Upper Cretaceous rocks of western North America. Therefore, relative dating of formations usually depends on lithostratigraphic evidence and correlations of invertebrate and vertebrate fossils (e.g., Khand

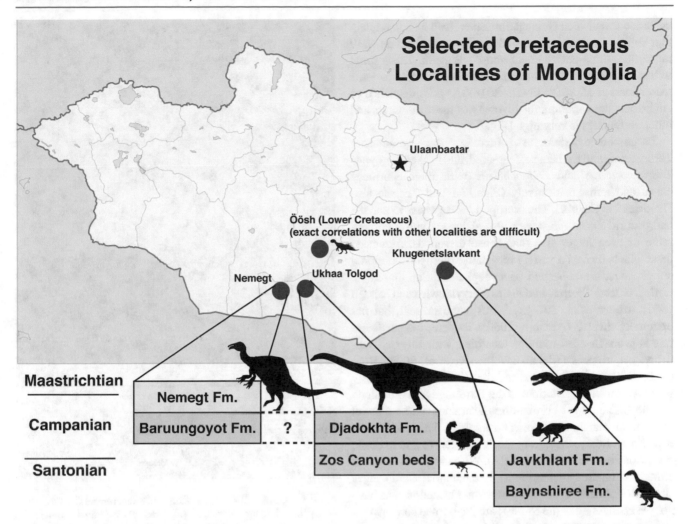

Fig. 8.4 Selected Cretaceous localities of Mongolia and their representative dinosaur taxa, with schematic stratigraphic columns and tentative correlations indicated. Dinosaur silhouettes by Daniel E. Barta (*Citipati*), Conty (vectorized by T. Michael Keesey) (*Tyrannosaurus*, CC BY 3.0 https://creativecommons.org/licenses/by/3.0/), Andrew A. Farke (*Protoceratops*, CC BY 3.0), FunkMonk (*Oryctodromeus*, *Segnosaurus*, Public Domain), Jaime Headden (*Psittacosaurus*, CC BY 3.0), T. Michael Keesey (*Deinocheirus*, Public Domain), and Slate Weasel (*Opisthocoelocaudia*, Public Domain). *Oryctodromeus* and *Tyrannosaurus* silhouettes represent the related Mongolian taxa *Haya* and *Tarbosaurus*, respectively

et al. 2000). Recent fossil-based correlation between rocks in the eastern and western Gobi reveals a composite Upper Cretaceous stratigraphic succession that may be equivalent with the Cenomanian to Campanian interval (Fig. 8.4; Norell and Barta 2016). The sedimentological and fossil evidence contained in these extensively exposed rocks, in combination with stable-isotope data, tell a tale of fluctuating environments and changes in faunal composition, the links between which are not currently well understood. Nevertheless, some patterns emerge as we journey through time. Our trip starts in the uppermost (and, therefore, youngest) geological layers deposited during the Maastrichtian and Campanian stages in the Cretaceous (see Fig. 8.1) and continues back almost to the Jurassic.

8.4.1 The Late Campanian to Maastrichtian Nemegt and Baruungoyot Formations

The Nemegt and Baruungoyot formations were deposited during the late Campanian to Maastrichtian under relatively wet to rather arid conditions. The Nemegt Formation is generally younger than the Baruungoyot Formation, but the oldest layers of the former seem to be of about the same age as the uppermost portions of the latter [U0818]. We will, therefore, start our journey in the Nemegt Formation.

Nemegt rocks represent predominantly river and lake deposits, signifying a landscape in which generally wetter conditions prevailed—very different from today's desert. Sedimentological and stable carbon-and-oxygen isotopic evidence reveals that Nemegt dinosaurs lived on a floodplain that shifted from meandering channels to a mixed alluvial/marsh/lacustrine environment and back again. These shifts reflect changes in tectonic uplift and/or rainfall patterns (Montanari et al. 2013; Eberth 2018). As a consequence of environmental variation, the diversity of these dinosaurs and other vertebrates is very high (Fig. 8.4).

Large-bodied dinosaurs include the tyrannosaur *Tarbosaurus* and the bizarre, feathered, and beaked theropods *Therizinosaurus* and *Deinocheirus* with their enormous claws and arms, respectively (Figs. 7.1 and 7.5; see also Hendrickx in Chap. 7). The sauropod *Nemegtosaurus* and the hadrosaurid *Saurolophus* were also abundant, and of comparable or even larger size than those theropods. Alongside these giants thrived a variety of smaller theropods, armored ankylosaurs, dome-headed pachycephalosaurian dinosaurs, turtles, crocodyliforms, and fishes (Jerzykiewicz and Russell 1991; Arbour et al. 2014). Lizards were present, but not extremely diverse (Alifanov 2000). To date, mammals are rare at probable Maastrichtian localities, with only two published occurrences (Kielan-Jaworowska et al. 2000). Both the composition of the fauna and the climate change slightly when we travel back in time to the Baruungoyot Formation.

The Baruungoyot Formation is characterized by alluvial fan, eolian, marsh, and lake environments. These sediments reveal a cyclical pattern from mesic (wetter) to arid to mesic conditions over time (Eberth 2018). With the somewhat drier climate compared to the Nemegt, the fauna also changes slightly. Animals of the Baruungoyot Formation that have not yet been found in the Nemegt are the horned ceratopsians (Funston et al. 2017). Pachycephalosaurs, small theropods, crocodyliforms, and lizards thrived there too (Fanti et al. 2018). Interestingly, some of the genera known from the Baruungoyot Formation also occur in the correlative or underlying Djadokhta Formation (Dashzeveg et al. 1995), where we will travel next.

8.4.2 The Campanian Djadokhta Formation

The Campanian (c. 75–71 Ma; Dashzeveg et al. 2005) Djadokhta Formation is one of the most fossiliferous in Mongolia, but exact stratigraphic correlations with other formations remain debated. The Djadokhta Formation may either correlate to, or underlie, parts of the Baruungoyot Formation (Dashzeveg et al. 1995). Sedimentological and stable carbon-and-oxygen isotopic evidence (Dingus et al. 2008; Montanari et al. 2013) indicates that the Djadokhta Formation was deposited under hot and arid desert-like conditions [U0819]. Deposits include eolian and dune-adjacent sediments, similar to some parts of the Baruungoyot Formation. The known vertebrate fauna is very diverse and

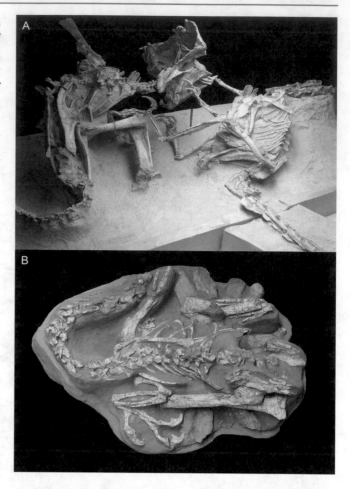

Fig. 8.5 Djadokhta Formation sediments reveal remarkable direct evidence of dinosaur behavior. (**a**) This *Protoceratops* (left) and *Velociraptor* (right) were entombed in the act of fighting. (**b**) This *Citipati* was buried while protecting its nest of eggs and shows a typical avian-style brooding posture

abundantly preserved but is concentrated in only a few depositional environments.

Evidence of large dinosaurs in the eolian and neighboring deposits is very rare and consists mainly of tracks and teeth of theropods, sauropods, and hadrosaurs (Osmólska 1980; Dingus et al. 2008). Indeed, large dinosaurs, such as the hadrosauroid *Plesiohadros*, are much better known from localities where more fluvial sediments are exposed (Hasegawa et al. 2009; Tsogtbaatar et al. 2014). In contrast, sand dune collapses entombed numerous small to medium-sized animals, capturing many in the act of nesting (Fig. 8.5b; Norell et al. 1995, 2018) and even fighting (Fig. 8.5a; Kielan-Jaworowska and Barsbold 1972). Theropod dinosaur diversity was high and included representatives of alvarezsaurs (possible insect-eaters with shortened arms with a single claw), beaked, tooth-less oviraptorids, dromaeosaurs (as for example the famous *Velociraptor*), troodontids, and birds (Dashzeveg et al. 2005; Dingus et al. 2008). Theropod eggs are commonly found. The most abundant single dinosaur taxon is the sheep-sized ceratopsian *Protoceratops*. Numerous localities also preserve remains of the ankylosaur *Pinacosaurus* (Hill et al. 2003). Crocodyliforms are represented by the neosuchian *Shamosuchus* (Pol et al. 2009). Cretaceous lizards are highly abundant and diverse, with more than 30 species of Iguania (Iguanas), Gekkota (geckos), Scincomorpha (a group of lizard), and Anguimorpha (alligator, glass, and legless lizards, and galliwasps) from Djadokhta-age localities. These species include many carnivorous, insectivorous, and herbivorous forms with a variety of dental specializations reflecting their diets (Gao and Norell 2000). Many lizards from the Ukhaa Tolgod locality were likely preserved in burrows, or struggling to escape the collapsing dunes (Gao and Norell 2000). Campanian mammals are also extraordinarily abundant and diverse and include dozens of multituberculate, metatherian, and eutherian species (Kielan-Jaworowska et al. 2000). Traveling back in time, the known diversity decreases once again.

8.4.3 The Santonian to Campanian Javkhlant Formation

The next formation we visit, the Javkhlant Formation, is Santonian to Campanian in age (c. 86–75 Ma). It is thought to be temporally equivalent to the Zos Canyon beds further to the west (Norell and Barta 2016). The Zos Canyon beds, in turn, conformably underlie the Djadokhta Formation we just visited (Fig. 8.4). Evidence from the sedimentary environments in the Javkhlant Formation indicates the presence of a seasonally wet-dry alluvial plain setting that, over time, transitioned to an alluvial fan setting (Eberth et al. 2009). The fauna inhabiting this environment [U0820] was poorly known until recently.

A spate of discoveries revealed the presence of a varied dinosaur fauna. These finds include skeletons of the early ceratopsian *Yamaceratops* (Makovicky and Norell 2006), an enantiornithine bird embryo (Balanoff et al. 2008), the diminutive bird-like alvarezsaur *Albinykus* (Nesbitt et al. 2011), and the most common animal in the formation, the early-diverging ornithischian dinosaur *Haya* (Makovicky et al. 2011). The latter is also present in the Zos Canyon beds further to the west (Norell and Barta 2016).

The shared presence of *Haya* is what led Norell and Barta (2016) to suggest that the Javkhlant Formation and the Zos Canyon Beds are time-equivalent. If this correlation is correct, it seems that some sections of exposed rocks in Mongolia comprise a conformable succession spanning at least 25 million years of the Cretaceous. This succession begins at the base of the Baynshiree Formation, which is the next stop on our visit, and extends to the top of the Djadokhta Formation (Fig. 8.4).

8.4.4 The Cenomanian to Santonian Baynshiree Formation

Traveling deeper in time, the climate in this part of Mongolia becomes less seasonal. The Baynshiree Formation (Cenomanian-Santonian, c. 100–86 Ma) [U0821] is characterized by gray to multicolored sandstone and claystone (Jerzykiewicz and Russell 1991). These rocks were deposited in generally humid riverside environments (Hicks et al. 1999; Shuvalov 2000) and preserved fossils of a more varied fauna when compared to our last stop.

The Baynshiree vertebrate fauna is diverse, although the environment is very different than that of the similarly fossiliferous Djadokhta Formation. The riverside environments of the Cenomanian to Santonian were inhabited by a number of medium- to large-sized dinosaurs, as well as other vertebrates. Dinosaurs include the ornithomimosaur *Garudimimus* (Barsbold 1981), the bizarre therizinosaur *Erlikosaurus* (Barsbold and Perle 1980), hadrosaurs (Jerzykiewicz and Russell 1991), the sauropod *Erketu* (Ksepka and Norell 2006), and the armored ankylosaur *Talarurus* (Maleev 1952). Other Baynshiree vertebrates include pterosaurs (Watabe et al. 2009), the neosuchian crocodyliform *Paralligator* (Turner 2015), turtles (Jerzykiewicz and Russell 1991), and the deltatheroidian mammal *Tsagandelta* (Rougier et al. 2015).

8.4.5 The Berriasian to Barremian Formations

Our last stop on this tour through Mongolia's Mesozoic past is in the Lower Cretaceous rocks (Berriasian to Barremian). Many of the localities from this time (c. 141–126 Ma; Shuvalov 2000), such as Öösh and Ondai Sair, consist of sandstone, freshwater limestone, and organic-rich, black paper shale. These sedimentary rocks generally indicate the presence of alluvial fan, deep lake, and lakeside environments [U0822]. The presence of gypsum in some lake deposits at Öösh may indicate a period during which the locality experienced a semiarid climate (Berkey and Morris 1927). Fossils of both terrestrial and aquatic animals, as well as plants, are found in these shale, limestone, and sandstone beds.

The fossil record of the Berriasian to Barremian of Mongolia is very diverse. Early reports from the 1920s already identified the presence of dinosaurs, fishes, crustaceans, insects, and occasionally plant fragments (Osborn 1923; Berkey and Granger 1923; Berkey and Morris 1927). A number of other vertebrates have been found since then. Interestingly, notable dinosaurs tend to be small-bodied. They include dromaeosaurids and the ubiquitous primitive ceratopsian *Psittacosaurus* (Turner et al. 2007; Prieto-Márquez et al. 2012). Pterosaurs, gecko-like lizards, and mammals also thrived in these environments (Rougier et al. 2001; Andres and Norell 2005; Conrad and Norell 2006).

8.4.6 The Gobi Region as a Model for Regional Changes During the Cretaceous?

As we have seen during this part of our trip, both climate and fauna changed considerably throughout the Cretaceous in Mongolia. However, in-depth studies of the links between faunal and climate change in the Gobi Cretaceous rocks are just beginning. These will benefit from future efforts to better assign relative and absolute ages to formations and identify links between changes in climate and changes in preservational modes at the localities. Ongoing field work and detailed analyses continuously add to what is, already, one of the richest sources of data for reconstructing Cretaceous life on Earth. A very similar case, although more limited in time, is the Upper Jurassic Morrison Formation of the USA, where we will head now.

8.5 A Jurassic Diversity Hotspot? The Morrison Formation of the USA

John R. Foster, Emanuel Tschopp, and
Susannah C. R. Maidment

About 150 million years ago, North America in the Late Jurassic was about 650 km farther south than its present latitude. The western half of the continent was an enormous floodplain in which one of the richest vertebrate paleofaunas we know was preserved. Here, over an interval of approximately eight to ten million years, what we now know as the Morrison Formation was deposited [U0823]. Being further south than today, lower in altitude, and with a generally warmer global climate, the local conditions at the time of sediment deposition was very different from what paleontologists experience when they are exploring these areas today (see Chap. 6).

Reconstruction of the paleoclimate of western North America comes from several different proxies including fossil plants, ancient soils (paleosols) and sediments, and computer models. Interpretations of the prevailing climate during Morrison times differ. Often, the southern parts of the formation are considered to have been deposited in a semi-arid setting (Valdes 1994), but as we have seen in Chap. 6, paleobotanical studies and general circulation models have indicated that wet (mesic) conditions prevailed, at least locally, at sites ranging from southern Utah to central Montana [U0824] (Tidwell et al. 1998). Sedimentological studies found that there were

times when the floodplains were waterlogged and times when the groundwater level was low and calcic (limestone-rich) paleosols formed (Dodson et al. 1980; Owen et al. 2015). All this combined evidence likely indicates that the climate of the Morrison basin varied from seasonally wet to seasonally dry or semi-arid and back, on a cyclical basis. We have already explored the forests in this ecosystem in Chap. 6. Now, we will have a closer look at the vertebrates calling them home.

The Morrison Formation is best known for its dinosaurs (Fig. 8.6c) [U0825] (Dodson et al. 1980). The unit has a relatively low diversity of ornithischians, compared to other major clades. Ornithischians include the small-bodied, likely omnivorous *Fruitadens*, the armored ankylosaurs *Mymoorapelta* and *Gargoyleosaurus*, several plated stegosaurs including the famous *Stegosaurus*, possibly several small, bipedal forms like *Drinker* and *Othnielosaurus*, and the somewhat larger *Dryosaurus* and *Camptosaurus*. Theropod dinosaurs are relatively diverse, including large-bodied taxa like *Ceratosaurus*, *Torvosaurus*, *Allosaurus*, and *Saurophaganax*. Smaller theropods were the poorly known *Marshosaurus*, *Ornitholestes*, *Coelurus*, the early tyrannosauroid *Stokesosaurus*, and the bird-like troodontid *Koparion*. In addition, several unnamed species await detailed study, among which are members of theropod groups from which there have not been any record in the Morrison Formation before. The diversity of the sauropods, as a group, is the most impressive, given their size. They include *Dystrophaeus* from the stratigraphically oldest Tidwell Member along with *Haplocanthosaurus* from low stratigraphic levels. Other known sauropods have only been found in the upper layers. These sauropods include *Camarasaurus* and *Brachiosaurus* among macronarians, and a number of diplodocoids (*Suuwassea*, *Diplodocus*, *Barosaurus*, *Supersaurus*, *Kaatedocus*, *Galeamopus*, *Apatosaurus*, *Brontosaurus*; Foster 2007; Whitlock et al. 2018).

Although dinosaurs would have been the most visible components of the Morrison fauna, there are an impressive number of non-dinosaurian taxa. Numerous sites have yielded various aquatic or semi-aquatic animals including fishes, frogs, salamanders, freshwater turtles, and crocodyliforms. In addition, several terrestrial crocodyliforms are known, along with a number of lizards, a snake, omnivorous and herbivorous rhynchocephalians (similar to the living tuatara), a choristodere (*Cteniogenys*), several species of the flying pterosaurs, and a number of mammal groups (including many of those we have seen earlier in China; Williamson and Brusatte in Chap. 7). What is remarkable about the vertebrates in the Morrison Formation, which goes unrecognized by many, is that there are approximately as many species of mammals known from here as there are dinosaurs (Kielan-Jaworowska et al. 2004; Foster 2009). Fossilized bones of all these vertebrates have been recovered from several hundred localities, and their diversity is also evidenced in at least 65 sites where tracks and traces are preserved (Fig. 8.7; Foster

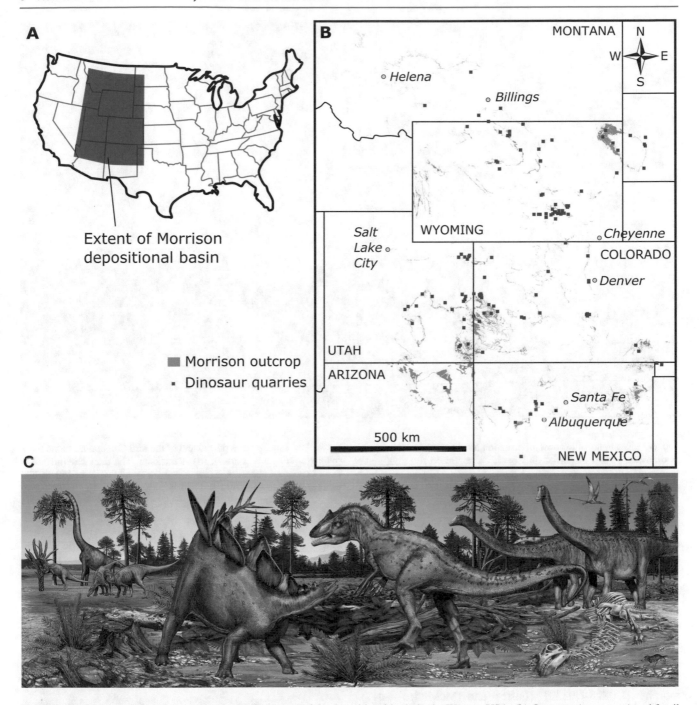

Fig. 8.6 The Morrison Formation ecosystem. (**a**) The extent of the depositional basin in the Western USA. (**b**) Outcrops (grey areas) and fossil localities (red squares). (**c**) A reconstruction of the ecosystem by Davide Bonadonna (Milan, Italy). Used with permission

and Lockley 2006). Such a large diversity of vertebrates requires an even larger food supply.

The diets of all Morrison vertebrates obviously depended on an abundant and diverse flora and invertebrate fauna. Plant fossils are common elements preserved in the formation (see also Chap. 6). Representatives of green algae, mosses, horsetails, many ferns, cycads, bennettitales, ginkgoes (*Ginkgo* and *Czekanowskia*), and a diversity of conifers are well documented (Ash and Tidwell 1998; Gee et al. 2014). In addition, the unit preserves abundant specimens of

mollusks (unionid bivalves and various species of gastropods), arthropods (crayfish, ostracods, conchostracans, insects), and freshwater sponges (Chure et al. 2006). Trace fossils indicate a greater diversity of insects and other arthropods than is known from body fossils, alone (Hasiotis 2004).

The overall high biodiversity might be surprising, in particular, for large vertebrates like dinosaurs. However, there are some indications of geographical and, potentially, temporal segregation among various animals [U0826-U0830]. Recent studies indicate that stegosaurs from northern

Fig. 8.7 Examples of Morrison Formation localities and its fossils. (**a**) A dinosaur quarry. A partial diplodocid sauropod on the right and a theropod in the back on the left (copyright by Mike Eklund, used with permission). (**b**) A microfossil site (photograph by J.R. Foster). (**c**) A fossil plant locality and a partial leaf of the fern *Coniopteris* found there (photograph by J.R. Foster). (**d**) A tracksite. The trackway (arrow) is from a sauropod that walked toward the photographer about 150 million years ago (photograph by J.R. Foster)

exposures of the formation are different from those found further south (Maidment et al. 2018). Several sauropods seem to follow a similar pattern, with *Suuwassea* and *Kaatedocus* restricted to north-central Wyoming and Montana (Whitlock et al. 2018), whereas *Brachiosaurus* and *Camarasaurus* are more widespread (Woodruff and Foster 2017; Maltese et al. 2018). Turtles and semi-aquatic crocodyliforms generally occur in more northern and eastern areas, which correspond to wetter environments, whereas the more arid southwest of the Morrison basin was inhabited by cursorial* crocodylomorphs [U0828-U0830] (Foster and McMullen 2017). Preliminary studies also imply that there is some temporal segregation of sauropods (Tschopp et al. 2016), but more research is needed to confirm stratigraphic correlation among fossil sites [U0827]. Furthermore, the impact of sampling biases on both temporal and spatial segregation of Morrison vertebrates remains to be better understood.

Cropping out over 1.2 million km² (Box 8.2), and having been extensively sampled since the discovery of its diverse and well-preserved dinosaurian fauna in the later part of the nineteenth century, the Morrison Formation is among the best studied Upper Jurassic terrestrial ecosystem on Earth. However, many questions about the spatial and temporal diversity of its fauna, the prevailing climate or oscillations

Box 8.2: Morrison Formation Paleogeography
The Morrison Formation of the western USA extends from central New Mexico and Arizona north to Montana, and from central Utah to the western Great Plains of South Dakota (Fig. 8.5). These rocks represent deposition in rivers and across floodplains and wetlands that lay east of the ancient Sevier mountain range (Turner and Peterson 2004). Surface runoff from the western mountains flowed in perennial and ephemeral streams and rivers toward the east, depositing channel sand and overbank mud in fan-like bodies known as distributive fluvial systems (Owen et al. 2015). A shallow inland seaway existed to the north of the Morrison basin, in what is now Canada (Peterson 1988).

in climate, and the paleogeography of the area remain unanswered. Research efforts continue unabated, and new discoveries are made every year. Hence, the Morrison continues to be the best-available case study for the Late Jurassic world, a glimpse of which can be seen while visiting Dinosaur National Monument in Jensen, Utah.

8.6 Underexplored Treasures: The Middle Jurassic of Argentina

Femke M. Holwerda

Sauropods from the Late Jurassic, including *Brontosaurus*, *Diplodocus*, and *Brachiosaurus* from the Morrison Formation, are generally well known, as are the world's largest dinosaurs, such as *Dreadnoughtus* and *Patagotitan* (see Sect. 8.5, and Chap. 7). However, the origins of these large, long-necked herbivores, after their initial emergence in the Late Triassic (Buffetaut et al. 2000; Lallensack et al. 2017), are still not well understood. This is mainly due to limited exposures of Early and early Middle Jurassic strata in which vertebrate fossils are preserved (Mannion et al. 2011). Relatively abundant sauropod material is known from Laurasia (e.g., *Shunosaurus*; Zhang 1988). However, most taxa from the Early and Middle Jurassic hail from Gondwana [U0831]. Many Gondwanan sauropods are from North Africa (e.g., *Tazoudasaurus*; Allain et al. 2004), India (e.g., *Barapasaurus*; Bandyopadhyay et al. 2010), and Argentina (e.g., *Patagosaurus*; Bonaparte 1979). Here, we

travel to the Early to Middle Jurassic of what is now Argentina to see how far sauropod evolution has proceeded, and which other organisms thrived alongside these giants.

The fossil record of Argentina is particularly good, with abundant sauropod material preserved in the early Middle Jurassic Cañadón Asfálto Formation in west-central Chubut, Patagonia (Fig. 8.8) [U0832]. Since its discovery, over 20 species of different taxonomic groups (including sauropod, theropod, and ornithischian dinosaurs, pterosaurs, sphenodonts, mammals, fishes, frogs, turtles, and crocodylomorphs) have been discovered. These fossils demonstrate the presence of a rich ecosystem that existed in a lush conifer-dominated forest, in contrast to the prevailing aridity currently characterizing the region (Escapa et al. 2008; Pol et al. 2013; Olivera et al. 2015). Two sauropods, *Patagosaurus* and *Volkheimeria*, have been named from these beds, to date, but more taxa are likely to be present. *Patagosaurus* is the more common sauropod, with over eight specimens found in two localities, Cerro Condor Norte and Cerro Condor Sur, adjacent to the Chubut River (Fig. 8.8). However, several additional studies have since shown that some material ascribed to this genus may, in fact, belong to a different taxon (Rauhut 2003; Holwerda et al. 2015). Moreover, additional finds, and

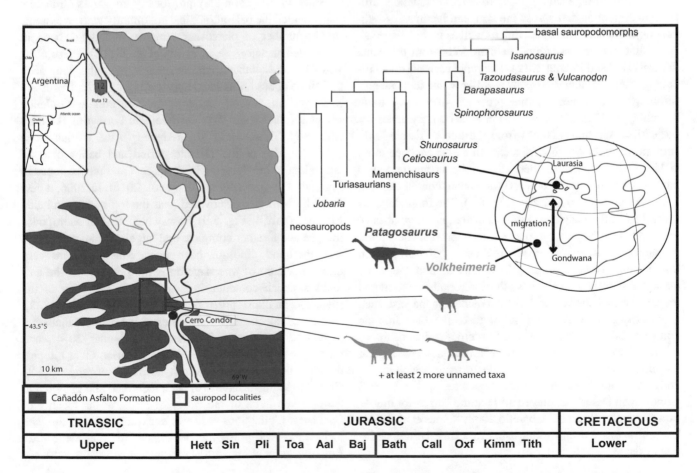

Fig. 8.8 The Toarcian-Bajocian Cañadón Asfalto Fm., Chubut, Patagonia, Argentina. Here, four types of sauropods, including *Patagosaurus* and *Volkheimeria*, are reported. The phylogenetic posi-

tions of these taxa are indicated along with the possible faunal connections across Gondwana and Laurasia

more detailed research on isolated sauropod teeth, indicate the presence of at least four, if not more, sauropod taxa here [U0833] (Becerra et al. 2017; Carballido et al. 2017). This is a higher diversity than expected considering that these species are from a time early in the evolution of the clade.

The Cañadón Asfálto Formation has been redated from what had been an assumed Callovian age to a much older Toarcian-Bajocian age (c. 183–168 Ma; Cúneo et al. 2013), which coincides with the early evolution of sauropods. Hence, the presence of *Patagosaurus* and its allies in these rocks places them right in the middle of this crucial time of early evolution and geographic dispersal. According to latest research, the various named and unnamed sauropod taxa mentioned above occupy highly divergent positions in the sauropod family tree (Fig. 8.8; Holwerda and Pol 2018). *Volkheimeria* is considered to hold a relatively basal position, *Patagosaurus* having an intermediate position, and some yet-unnamed material being identified as a rather derived neosauropod. Hence, various distinct clades co-occurred in what is now Argentina early in the evolutionary history of sauropods. At the same time, the Argentinian taxa are very closely related to species from the UK and Niger.

The relationships among Early to Middle Jurassic sauropods in Argentina and across the Pangean landmasses indicate rapid regional diversification as well as global dispersal. A similar pattern was found in theropods from the same deposits (Pol and Rauhut 2012). The rapid regional diversification of Argentinian sauropods was likely due to niche partitioning as indicated by the presence of several tooth morphotypes [U0834] (Holwerda et al. 2015). Evidence for the earliest sauropods comes from the latest Triassic-Early Jurassic rocks of South Africa (Raath 1972; McPhee et al. 2015). Once established there, the group is interpreted to have dispersed toward other Gondwanan continents and, finally, migrated into Laurasia (Fig. 8.8). The fossil deposits preserved in the Cañadon Asfálto Formation, in what once was southwest Gondwana, make it an important checkpoint for sauropod dispersal. Indeed, the diversity of sauropods in this locality (*Patagosaurus, Volkheimeria,* and the other unnamed sauropod taxa) shows that sauropod evolution and dispersal established several distinct clades. In the past, such a diversity usually has been ascribed to the Late Jurassic (Barrett and Upchurch 2005). The close relationship of *Patagosaurus* with sauropods from Laurasia shows a great mobility of sauropods in the Middle Jurassic, which apparently was not hindered by the presence of the Central Gondwanan Desert, an alleged arid natural barrier for migration (Rauhut and Lopez-Arbarello 2008; Remes et al. 2009). Let's now travel even further back, to a time before sauropods existed.

8.7 An Early Dinosaur Community in a Triassic Ecosystem from Switzerland

Torsten M. Scheyer, Marion Zahner, and Winand Brinkmann

When Triassic dinosaurs and other, now extinct, vertebrate animals roamed the area of Frick, Switzerland, this region was not a valley nestling between gently rolling hills, covered with luscious meadows, green forests, and shady gardens with fruit trees. In the Late Triassic (c. 210 Ma; Fig. 8.1), a rather flat, semiarid coastal landscape sprawled as far as the eye could see [U0835]. Where the Alps now stand, a light-drenched, shallow sea with warm water existed, and monsoon-like rains led to periodic floods across the landscape. Rivers flowed in meandering channels toward the sea. In some parts of the floodplain, water bodies may have remained for a time and small, temporary ponds possibly developed. During humid periods and in humid areas, local plant life flourished. In this scenery, early dinosaurs thrived; their remains are now entombed in large numbers in the Late Triassic sediments of Frick.

Since the 1960s, a clay pit, locally known as Gruhalde, has provided the paleontological community with an extraordinary number of plateosaur remains, including abundant articulated skeletons (e.g., Foelix et al. 2011). *Plateosaurus* was a bipedal, plant-eating dinosaur with a rather long neck and sharp claws on its hand and feet [U0837]. Its fossils are preserved in several horizons in the middle Gruhalde Member (parts 3 and 4) of the Klettgau Formation [U0836] (Fig. 8.9). These rocks are assigned to the Norian stage (227–209 Ma) of the Triassic period and believed to be equivalent to the Trossingen Formation and the Löwenstein Formation in Germany (Jordan et al. 2016). In 2006, a new fossil horizon was discovered near the top of the Gruhalde Member (part 4; Fig. 8.9) from which several sauropodomorphs and a rather complete and largely articulated predatory theropod dinosaur have been collected. One very unusual feature of this skeleton is the presence of the animal's stomach contents, in which bones and teeth of the rhynchocephalian reptile *Clevosaurus* are identified [U0838]. Additional spectacular discoveries at the clay pit include a juvenile *Plateosaurus* (Hofmann and Sander 2014), and more, yet-to-be-described theropod material. Over the past decade, the dinosaur-bearing horizons of the middle Gruhalde Member have also been identified and explored at the opposite flank of the valley of Frick, at the locality of Frickberg. Collections made in excavation pits at this site include a well-preserved specimen of the stem-turtle *Proganochelys* along with plateosaur remains (Fig. 8.10; see

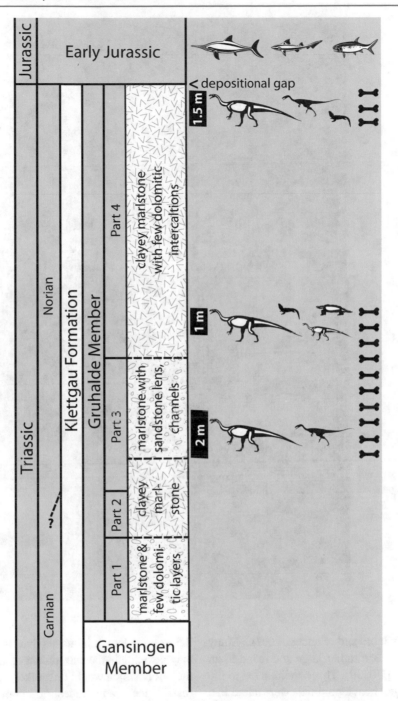

Fig. 8.9 Stratigraphic profile at the clay pit Gruhalde at Frick, Switzerland, with important vertebrate fossil findings (based on Jordan et al. 2016, Reisdorf et al. 2011; personal observations; animal sketches by Beat Scheffold, PIMUZ, used with permission). Black bones = Vertical range of horizons of the Late Triassic plateosaur bone bed with vertebrate remains; black boxes = positions and thicknesses of the Lower (plateosaurs, and new, larger neotheropod remains), Middle (plateosaurs, juvenile plateosaur, stem-turtle, remains of rhynchocephalians, and other reptiles, as well as fishes), and Upper (sauropodomorphs, smaller neotheropod *Notatesseraeraptor frickensis*, and rhynchocephalian remains) Dinosaur Layer containing articulated vertebrate material. From the Early Jurassic of Frick ichthyosaurs and fishes are known

also Foelix et al. 2011). The uppermost rocks exposed in Frick are of Early Jurassic age and contain marine reptiles (e.g., Reisdorf et al. 2011), recording the transgression of the Jurassic sea in the area (Fig. 8.9).

The genus *Plateosaurus* is a representative of a basal grade of sauropodomorph dinosaurs, which first appear in the Late Triassic. These animals reached 5–10 m in body length and had a lifespan of over 20 years, as shown by bone

Fig. 8.10 Fossils from the Klettgau Fm. at the Sauriermuseum Frick. (**a**) Reconstructed stem-turtle *Proganochelys*. (**b**) One of the most complete *Plateosaurus quenstedti* in typical sitting position. (**c**) Skull of *Plateosaurus*. (**d**) Well-preserved and articulated hand (on the left) and foot (on the right) of *Plateosaurus*. (**e**) Skeleton of the small theropod *Notatesseraeraptor frickensis*. (**f**) Reconstruction of a small rhynchocephalian, which was found as stomach content in the theropod skeleton. (**g**) Reconstruction of the new theropod dinosaur *Notatesseraeraptor frickensis*. The models were built by Beat Scheffold (University of Zurich, Palaeontological Institute and Museum) for the Sauriermuseum Frick (all images T Scheyer, except images (**e**) and (**f**) taken by Beat Scheffold; used with permission)

histological analyses (Klein and Sander 2007). Many specimens of *Plateosaurus* are rather large and found in an upright "sitting position" [U0839]. This condition led to the hypothesis that these larger, heavier animals were mired in a mud-hole trap and were, thus, preserved in situ (Sander 1992). Younger, less heavy individuals and other faunal elements, including small reptiles, on the other hand, represent only rare components in the fossil assemblage. This taphonomic explanation for the position and orientation of large skeletons observed in the field fits well with modern environmental reconstructions. The Triassic landscape around Frick was an extensive and open, vegetated floodplain occupied by large river systems, and not, as traditionally hypothesized, desert-like environments (e.g., Huene 1928; Sander 1992). The recovery of smaller skeletons of neotheropod dinosaurs, the stem-turtle, and a juvenile *Plateosaurus* of just over

2.5 m body length indicates that the mud-hole trap scenario may not explain the entrapment of all specimens in the accumulation, which would be unlikely. In addition to the macrofossils, teeth and bones of hybodont sharks, lungfishes, ray-finned fishes, rhynchocephalians (beak-head lizards), armored aetosaurs, carnivorous saurians, and mammal-like reptiles have also been recovered.

The conclusions drawn from the fossil assemblages from the Klettgau Formation are important for not only a deeper understanding of *Plateosaurus*, itself, but they also allow for a glimpse into the early evolution of two important dinosaur lineages that diversified during this time. These Late Triassic lineages are the sauropodomorphs (which later gave rise to the sauropods) and the neotheropods (which are part of the larger clade Theropoda, and later gave rise to birds). Other fossil evidence allows for the reconstruction of

Fig. 8.11 The clay pit Gruhalde at Frick, Switzerland. (**a**) View along the active quarry up to the excavation site in 2010. The start of the fossiliferous horizons in the stratigraphic sequence is marked by white asterisks. (**b**) Excavation site in the lower fossiliferous layers in 2011, where the bones of a large *Plateosaurus* are extracted from the matrix (image A by T. Scheyer; image B by Grabungsteam Frick under supervision by Ben Pabst, used with permission)

the ecosystems in which these animals lived. To date, no ornithischian dinosaur remains are known from Frick, but members of this clade were extremely rare worldwide during the Triassic (Barrett et al. 2014). Nevertheless, the tremendous geographic scope of the Norian dinosaur-bearing horizons and the relatively dense accumulations of vertebrate fossils make the valley of Frick a genuine dinosaur El Dorado. Within Europe, the Gruhalde clay pit has been one of the most productive dinosaur localities for decades (Fig. 8.11), due to frequent commercial mining of the Norian clay deposits used as aggregate for brick production. The continuous production of this raw material by the company that owns the clay pit is accompanied by scientific explorations and excavations, supervised and coordinated by a permanent commission of the community of Frick. This fruitful collaboration is expected to yield many more spectacular finds in future.

8.8 Conclusions

Going back in time, we realize that ecosystems and the organisms inhabiting them become less and less familiar [U0840]. This is especially true after passing one of the big five mass extinction events such as the K/Pg and the one at the end of the Triassic. Moving across these event boundaries, we see the effects of the disappearance of dominant species and the fast recolonization of the new open ecological niches. During the Mesozoic, other abiotic and biotic changes also had a big influence on biodiversity and ecosystem organization. Biotic changes include the first angiosperm appearance in the Early Cretaceous and the group's radiation throughout that period, coupled with the split of the supercontinent Pangea into several smaller continents and their movement toward their current position on the globe. Even though the direct impact of these changes on vertebrate communities and ecologies is not clear yet, a changing food resource definitely had an influence on the environments that were inhabited by vertebrates. Herein, we visited several places around the planet to see how Earth and its biota changed over time, and what kind of vertebrates managed to live there. A Triassic environment in Switzerland was dominated by the plant-eating sauropodomorph dinosaur, *Plateosaurus*. These early dinosaurs were subsequently replaced by more derived members of their own clade, the gigantic sauropods that can be observed all over the planet beginning in the earliest Jurassic and surviving until the latest Cretaceous. However, their diversity declined and these giants were replaced by hadrosaurs during the Cretaceous as the dominant herbivores in some areas. Simultaneously with sauropods, carnivorous dinosaurs also evolved larger body sizes. Most other vertebrate clades diversified significantly throughout the Triassic and Jurassic. During the Cretaceous, ecosystems began to look a bit more like those of the present, with the extinction of the giant salamander-like temnospondyls, the arrival of birds and derived turtles, as well as the diversification of flowering plants seen in our stop in Australia. However, the organisms that continued to dominate the terrestrial fauna were the dinosaurs until the very end of the era.

Questions

1. Where can we study the extinction of non-avian dinosaurs? And what kind of evidence do we see in the fossil record that there was a severe and sudden change?
2. Faunal changes do not only happen at mass extinction events. Provide examples of faunal turnovers that are not linked to mass extinction, and discuss what may have been their cause.
3. What are the difficulties in reconstructing ecosystems based on the fossil record? Why is it important that paleontologists from various different disciplines collaborate?
4. Most of the places where we excavate now were not at that same spot on Earth during the Mesozoic. Why is that so, and what does this mean for us when we attempt to reconstruct the ecosystem these animals lived in?
5. Ecosystems today seem fairly stable, without a lot of change in species composition. Why are these changes more apparent in the fossil record?

Acknowledgments We thank RA Gastaldo and E Martinetto for their extremely thoughtful reviews of an early draft of this chapter. DE Barta thanks the staff of the Mongolian Academy of Sciences, Mark Norell and Mike Novacek (AMNH), Dave Varricchio (MSU), Konstantin Mikhailov (PIN), and Adam Halamski (ZPAL) for discussion and the opportunities they have provided to study Mongolian fossils in the field and collections. The following persons and institutions were involved with the excavations in Frick, Switzerland: Jasmina Hugi, René Kindlimann, Ben Pabst, Esther Premru, Monica Rümbeli, Beat Scheffold, Iwan Stössel, Lui Unterrassner, Martin Sander, Nicole Klein, the Arbeitskreis Frick and the community of Frick, Hochschule Aalen, Canton Aargau, the MNF of the University of Zurich, the Paul Scherrer Institute Villigen, the Saurierkommission Frick, the University of Bonn, and the Universitätsspital Zürich. The Swiss National Science Foundation (grant nos. 205321_162775 to TMS and 31003A_163346 to WB), the Ministerio de Ciencia e Innovación, the Ministerio de Economía y Competitividad [grant numbers CGL2011-30069-C02-01,02, CGL2016-73230-P], the Departament de Cultura de la Generalitat de Catalunya [grant number 20014/100927], the Richard Gilder Graduate School and the Division of Paleontology (AMNH, to ET and DEB) are acknowledged for funding.

References

Alifanov VR (2000) The fossil record of Cretaceous lizards from Mongolia. In: Benton MJ, Shishkin M, Unwin DM, Kurochkin EN (eds) The age of dinosaurs in Russia and Mongolia. Cambridge University Press, Cambridge, pp 368–389

Allain R, Aquesbi N, Dejax J, Meyer C, Monbaron M, Montenat C et al (2004) A basal sauropod dinosaur from the Early Jurassic of Morocco. C R Palevol 3(3):199–208

Andres B, Norell MA (2005) The first record of a pterosaur from the Early Cretaceous strata of Öösh (Övörkhangai; Mongolia). Am Mus Novit 3472:1–6

Arbour VM, Currie PJ, Badamgarav D (2014) The ankylosaurid dinosaurs of the Upper Cretaceous Baruungoyot and Nemegt Formations of Mongolia. Zool J Linnean Soc 172(3):631–652

Ash SR, Tidwell WD (1998) Plant megafossils from the Brushy Basin member of the Morrison Formation near Montezuma Creek trading post, southeastern Utah. Mod Geol 22:321–340

Balanoff AM, Norell MA, Grellet-Tinner G, Lewin MR (2008) Digital preparation of a probable neoceratopsian preserved within an egg, with comments on microstructural anatomy of ornithischian eggshells. Naturwissenschaften 95(6):493–500

Bandyopadhyay S, Gillette DD, Ray S, Sengupta DP (2010) Osteology of *Barapasaurus tagorei* (Dinosauria: Sauropoda) from the Early Jurassic of India. Palaeontology 53(3):533–569

Barrett PM, Butler RJ, Mundil R, Scheyer TM, Irmis RB, Sánchez-Villagra MR (2014) A palaeoequatorial ornithischian and new constraints on early dinosaur diversification. Proc R Soc Lond B Biol Sci 281(1791):20141147

Barrett PM, Upchurch P (2005) Sauropodomorph diversity through time. In: Curry Rogers K, Wilson J (eds) The Sauropods: evolution and paleobiology. University of California Press, Berkeley, pp 125–156

Barsbold R (1981) Toothless carnivorous dinosaurs of Mongolia. Trudy Sovmestnoi Sovetsko-Mongol'skoi Paleontologicheskoi Ekspeditsii 15:28–39

Barsbold R, Perle A (1980) Segnosauria, a new infraorder of carnivorous dinosaurs. Acta Palaeontol Pol 25(2):187–195

Becerra MG, Gomez KL, Pol D (2017) A sauropodomorph tooth increases the diversity of dental morphotypes in the Cañadón Asfalto Formation (Early – Middle Jurassic) of Patagonia. C R Palevol 16(8):832–840

Berkey CP, Granger W (1923) Later sediments of the desert basins of central Mongolia. Am Mus Novit 77:1–16

Berkey CP, Morris FK (1927) Geology of Mongolia: a reconnaissance report based on the investigations of the years 1922-1923, vol 2. American Museum of Natural History, New York

Bonaparte JF (1979) Dinosaurs: a jurassic assemblage from Patagonia. Science 205(4413):1377–1379

Brusatte SL, Butler RJ, Barrett PM, Carrano MT, Evans DC, Lloyd GT et al (2015) The extinction of the dinosaurs. Biol Rev 90(2):628–642

Buffetaut E, Suteethorn V, Cuny G, Tong H, Loeuff JL, Khansubha S, Jongautchariyakul S (2000) The earliest known sauropod dinosaur. Nature 407(6800):72–74

Canudo JI, Oms O, Vila B, Galobart À, Fondevilla V, Puértolas-Pascual E et al (2016) The upper Maastrichtian dinosaur fossil record from the southern Pyrenees and its contribution to the topic of the Cretaceous–Palaeogene mass extinction event. Cretac Res 57:540–551

Carballido JL, Holwerda FM, Pol D, Rauhut OWM (2017) An early Jurassic sauropod tooth from Patagonia (Cañadón Asfalto Formation): implications for sauropod diversity. Publ Electr Asoc Paleontol Argent 17(2):50–57

Chure DJ, Litwin R, Hasiotis ST, Evanoff E, Carpenter K (2006) The fauna and flora of the Morrison Formation: 2006. New Mexico Museum of Natural History and. Sci Bull 36:233–249

Cohen KM, Finney SC, Gibbard PL, Fan J-X (2013) The ICS international chronostratigraphic chart. Episodes 36:199–204. (updated)

Conrad JL, Norell MA (2006) High-resolution X-ray computed tomography of an Early Cretaceous gekkonomorph (Squamata) from Öösh (Övörkhangai; Mongolia). Hist Biol 18(4):405–431

Cook AG, Bryan SE, Draper J (2013) Post-orogenic Mesozoic basins and magmatism. In: Jell P (ed) Geology of Queensland. Geological Survey of Queensland, Brisbane

Csiki-Sava Z, Buffetaut E, Ősi A, Pereda-Suberbiola X, Brusatte SL (2015) Island life in the Cretaceous - faunal composition, biogeography, evolution, and extinction of land-living vertebrates on the Late Cretaceous European archipelago. ZooKeys 469:1–161

Cúnco R, Ramezani J, Scasso R, Pol D, Escapa I, Zavattieri AM, Bowring SA (2013) High-precision U–Pb geochronology and a new chronostratigraphy for the Cañadón Asfalto Basin, Chubut, central Patagonia: implications for terrestrial faunal and floral evolution in Jurassic. Gondwana Res 24(3):1267–1275

Dashzeveg D, Dingus L, Loope DB, Swisher CC, Dulam T, Sweeney MR (2005) New stratigraphic subdivision, depositional environment, and age estimate for the Upper Cretaceous Djadokhta Formation, southern Ulan Nur Basin, Mongolia. Am Mus Novit 3498:1–31

Dashzeveg D, Novacek MJ, Norell MA, Clark JM, Chiappe LM, Davidson A et al (1995) Extraordinary preservation in a new vertebrate assemblage from the Late Cretaceous of Mongolia. Nature 374(6521):446–449

Davis KE, Page RDM (2014) Reweaving the tapestry: a supertree of birds. PLoS Curr 6. https://doi.org/10.1371/currents.tol. c1af68dda7c999ed9f1e4b2d2df7a08e

Díez Díaz V, Tortosa T, Le Loeuff J (2013) Sauropod diversity in the Late Cretaceous of southwestern Europe: the lessons of odontology. Ann Paléontol 99(2):119–129

Dingus L, Loope DB, Dashzeveg D, Swisher CC, Minjin C, Novacek MJ, Norell MA (2008) The geology of Ukhaa Tolgod (Djadokhta Formation, Upper Cretaceous, Nemegt Basin, Mongolia). Am Mus Novit 3616:1–40

Dodson P, Behrensmeyer AK, Bakker RT, McIntosh JS (1980) Taphonomy and paleoecology of the dinosaur beds of the Jurassic Morrison Formation. Paleobiology 6(2):208–232

Duddy LR (2003) Mesozoic. In: Birch WD (ed) Geology of Victoria, 3rd edn. Geological Society of Australia (Victoria Division), Melbourne, pp 239–286

Eberth DA (2018) Stratigraphy and paleoenvironmental evolution of the dinosaur-rich Baruungoyot-Nemegt succession (Upper Cretaceous), Nemegt Basin, southern Mongolia. Palaeogeogr Palaeoclimatol Palaeoecol 494:29–50

Eberth DA, Kobayashi Y, Lee Y-N, Mateus O, Therrien F, Zelenitsky DK, Norell MA (2009) Assignment of *Yamaceratops dorngobiensis*

and associated redbeds at Shine Us Khudag (eastern Gobi, Dorngobi Province, Mongolia) to the redescribed Javkhlant Formation (Upper Cretaceous). J Vertebr Paleontol 29(1):295–302

Escapa IH, Sterli J, Pol D, Nicoli L (2008) Jurassic tetrapods and flora of Cañadón Asfalto Formation in Cerro Cóndor area, Chubut province. Rev Asoc Geol Argent 63(4):613–624

Fanti F, Cantelli L, Angelicola L (2018) High-resolution maps of Khulsan and Nemegt localities (Nemegt Basin, southern Mongolia): stratigraphic implications. Palaeogeogr Palaeoclimatol Palaeoecol 494:14–28

Foelix RF, Pabst B, Kindlimann R (2011) Die Saurier in Frick. Nat Aargau 37:59–76

Fondevilla V, Riera V, Vila B, Sellés AG, Dinarès-Turell J, Vicens E et al (2019) Chronostratigraphic synthesis of the latest Cretaceous dinosaur turnover in south-western Europe. Earth Sci Rev 191:168–189

Foster JR (2007) Jurassic west: the dinosaurs of the Morrison Formation and their world. Indiana University Press, Bloomington and Indianapolis

Foster JR (2009) Preliminary body mass estimates for mammalian genera of the Morrison Formation (Upper Jurassic, North America). PaleoBios 28(3):114–122

Foster JR, Lockley MG (2006) The vertebrate ichnological record of the Morrison Formation (Upper Jurassic, North America). New Mexico Museum of Natural History and. Sci Bull 36:203–216

Foster JR, McMullen SK (2017) Paleobiogeographic distribution of Testudinata and neosuchian Crocodyliformes in the Morrison Formation (Upper Jurassic) of North America: evidence of habitat zonation? Palaeogeogr Palaeoclimatol Palaeoecol 468:208–215

Funston GF, Mendonca SE, Currie PJ, Barsbold R (2017) A dinosaur community composition dataset for the Late Cretaceous Nemegt Basin of Mongolia. Data Brief 16:660–666

Gao K-Q, Norell MA (2000) Taxonomic composition and systematics of Late Cretaceous lizard assemblages from Ukhaa Tolgod and adjacent localities, Mongolian Gobi Desert. Bull Am Mus Nat Hist 249:1–118

Gee CT, Dayvault RD, Stockey RA, Tidwell WD (2014) Greater palaeobiodiversity in conifer seed cones in the Upper Jurassic Morrison Formation of Utah, USA. Palaeobiodivers Palaeoenviron 94(2):363–375

Hasegawa H, Tada R, Ichinnorov N, Minjin C (2009) Lithostratigraphy and depositional environments of the Upper Cretaceous Djadokhta Formation, Ulan Nuur basin, southern Mongolia, and its paleoclimatic implication. J Asian Earth Sci 35(1):13–26

Hasiotis ST (2004) Reconnaissance of Upper Jurassic Morrison Formation ichnofossils, Rocky Mountain region, USA: paleoenvironmental, stratigraphic, and paleoclimatic significance of terrestrial and freshwater ichnocoenoses. Sediment Geol 167(3):177–268

Hicks JF, Brinkman DL, Nichols DJ, Watabe M (1999) Paleomagnetic and palynologic analyses of Albian to Santonian strata at Bayn Shireh, Burkhant, and Khuren Dukh, eastern Gobi Desert, Mongolia. Cretac Res 20(6):829–850

Hill RV, Witmer LM, Norell MA (2003) A new specimen of *Pinacosaurus grangeri* (Dinosauria: Ornithischia) from the Late Cretaceous of Mongolia: ontogeny and phylogeny of ankylosaurs. Am Mus Novit 3395:1–29

Hocknull SA, White MA, Tischler TR, Cook AG, Calleja ND, Sloan T, Elliott DA (2009) New mid-Cretaceous (Latest Albian) dinosaurs from Winton, Queensland, Australia. PLoS One 4(7):e6190

Hofmann R, Sander PM (2014) The first juvenile specimens of *Plateosaurus engelhardti* from Frick, Switzerland: isolated neural arches and their implications for developmental plasticity in a basal sauropodomorph. Peer J 2:e458

Holwerda FM, Pol D (2018) Phylogenetic analysis of Gondwanan basal eusauropods from the Early-Middle Jurassic of Patagonia, Argentina. Span J Palaeontol 33(2):289–298

Holwerda FM, Pol D, Rauhut OWM (2015) Using dental enamel wrinkling to define sauropod tooth morphotypes from the Cañadón Asfalto Formation, Patagonia, Argentina. PLoS One 10(2):e0118100

von Huene F (1928) Lebensbild des Saurischier-Vorkommens im obersten Keuper von Trossingen in Württemberg. Palaeobiologica 1:103–116

Jerzykiewicz T, Russell DA (1991) Late Mesozoic stratigraphy and vertebrates of the Gobi Basin. Cretac Res 12(4):345–377

Jordan P, Pietsch JS, Bläsi H, Furrer H, Kündig N, Looser N et al (2016) The middle to late Triassic Bänkerjoch and Klettgau Formations of northern Switzerland. Swiss J Geosci 109(2):257–284

Kear BP (2016) Cretaceous marine amniotes of Australia: perspectives on a decade of new research. Mem Mus Vic 74:17–28

Kear BP, Fordyce RE, Hiller N, Siversson M (2018) A palaeobiogeographical synthesis of Australasian Mesozoic marine tetrapods. Alcheringa 42(4):461–486

Khand Y, Badamgarav D, Ariunchimeg Y, Barsbold R (2000) Cretaceous system in Mongolia and its depositional environments. In: Developments in palaeontology and stratigraphy, vol 17. Elsevier, Amsterdam, pp 49–79

Kielan-Jaworowska Z, Barsbold R (1972) Narrative of the Polish-Mongolian palaeontological expeditions 1967-1971. Palaeontol Pol 27:5–13

Kielan-Jaworowska Z, Cifelli RL, Luo Z-X (2004) Mammals from the age of dinosaurs: origins, evolution, and structure. Columbia University Press, New York

Kielan-Jaworowska Z, Novacek MJ, Trofimov BA, Dashzeveg D (2000) Mammals from the Mesozoic of Mongolia. In: Benton MJ, Shishkin M, Unwin DM, Kurochkin EN (eds) The age of dinosaurs in Russia and Mongolia. Cambridge University Press, Cambridge, pp 573–626

Klein N, Sander PM (2007) Bone histology and growth of the prosauropod dinosaur *Plateosaurus engelhardti* von Meyer, 1837 from the Norian bonebeds of Trossingen (Germany) and Frick (Switzerland). Spec Pap Palaeontol 77:169–206

Ksepka DT, Norell MA (2006) *Erketu ellisoni*, a long-necked sauropod from Bor Guvé (Dornogov Aimag, Mongolia). Am Mus Novit 3508(1):16

Lallensack JN, Klein H, Milàn J, Wings O, Mateus O, Clemmensen LB (2017) Sauropodomorph dinosaur trackways from the Fleming Fjord Formation of East Greenland: evidence for Late Triassic sauropods. Acta Palaeontol Pol 62(4):833–843

López-Martínez N (2001) La extinción de los dinosaurios y su registro en los Pirineos Meridionales. In: Actas de las II Jornadas sobre Dinosaurios y su Entorno. Colectivo Arqueológico-Paleontológico Salense, Burgos, pp 71–89

Maidment SCR, Woodruff DC, Horner JR (2018) A new specimen of the ornithischian dinosaur *Hesperosaurus mjosi* from the Upper Jurassic Morrison Formation of Montana, U.S.A., and implications for growth and size in Morrison stegosaurs. J Vertebr Paleontol 38(1):e1406366

Makovicky PJ, Kilbourne BM, Sadleir RW, Norell MA (2011) A new basal ornithopod (Dinosauria, Ornithischia) from the Late Cretaceous of Mongolia. J Vertebr Paleontol 31(3):626–640

Makovicky PJ, Norell MA (2006) *Yamaceratops dorngobiensis*, a new primitive ceratopsian (Dinosauria: Ornithischia) from the Cretaceous of Mongolia. Am Mus Novit 3530:1–42

Maleev EA (1952) A new ankylosaur from the Upper Cretaceous of Mongolia. Dokl Akad Nauk SSSR 87:273–276

Maltese A, Tschopp E, Holwerda FM, Burnham D (2018) The real bigfoot: a pes from Wyoming, USA is the largest sauropod pes ever reported and the northern-most occurrence of brachiosaurids in the Upper Jurassic Morrison Formation. Peer J 6:e5250

Mannion PD, Upchurch P, Carrano MT, Barrett PM (2011) Testing the effect of the rock record on diversity: a multidisciplinary approach to elucidating the generic richness of sauropodomorph dinosaurs through time. Biol Rev 86(1):157–181

McLoughlin S, Pott C, Elliott D (2010) The Winton Formation flora (Albian–Cenomanian, Eromanga Basin): implications for vascular plant diversification and decline in the Australian Cretaceous. Alcheringa 34(3):303–323

McPhee BW, Bonnan MF, Yates AM, Neveling J, Choiniere JN (2015) A new basal sauropod from the pre-Toarcian Jurassic of South Africa: evidence of niche-partitioning at the sauropodomorph–sauropod boundary? Sci Rep 5:13224

Montanari S, Higgins P, Norell MA (2013) Dinosaur eggshell and tooth enamel geochemistry as an indicator of Mongolian Late Cretaceous paleoenvironments. Palaeogeogr Palaeoclimatol Palaeoecol 370:158–166

Nesbitt SJ, Clarke JA, Turner AH, Norell MA (2011) A small alvarezsaurid from the eastern Gobi Desert offers insight into evolutionary patterns in the Alvarezsauroidea. J Vertebr Paleontol 31(1):144–153

Norell MA, Balanoff AM, Barta DE, Erickson GM (2018) A second specimen of *Citipati osmolskae* associated with a nest of eggs from Ukhaa Tolgod, Omnogov Aimag, Mongolia. Am Mus Novit 3899:1–44

Norell MA, Barta DE (2016) A new specimen of the ornithischian dinosaur *Haya griva*, cross-Gobi geologic correlation, and the age of the Zos Canyon Beds. Am Mus Novit 3851:1–20

Norell MA, Clark JM, Chiappe LM, Dashzeveg D (1995) A nesting dinosaur. Nature 378(6559):774–776

Olivera DE, Zavattieri AM, Quattrocchio ME (2015) The palynology of the Cañadón Asfalto Formation (Jurassic), Cerro Cóndor depocentre, Cañadón Asfalto Basin, Patagonia, Argentina: palaeoecology and palaeoclimate based on ecogroup analysis. Palynology 39(3):362–386

Osborn HF (1923) Two Lower Cretaceous dinosaurs from Mongolia. Am Mus Novit 95:1–10

Osmólska H (1980) The Late Cretaceous vertebrate assemblages of the Gobi Desert, Mongolia. Mem Soc Geol Fr 139:145–150

Owen A, Nichols GJ, Hartley AJ, Weissmann GS, Scuderi LA (2015) Quantification of a distributive fluvial system: the salt wash DFS of the Morrison Formation, SW U.S.A. J Sediment Res 85(5):544–561

Pentland AH, Poropat SF (2019) Reappraisal of *Mythunga camara* Molnar & Thulborn, 2007 (Pterosauria, Pterodactyloidea, Anhangueria) from the upper Albian Toolebuc Formation of Queensland, Australia. Cretac Res 93:151–169

Pereda Suberbiola X, Murelaga X, Baceta JI, Corral JC, Badiola A, Astibia H (1999) Nuevos restos fósiles de vertebrados continentales en el Cretácico Superior de Álava (Región Vasco-Cantábrica): sistemática y posición estratigráfica. Geogaceta 26:79–82

Peterson F (1988) Stratigraphy and nomenclature of middle and upper Jurassic rocks, Western Colorado Plateau, Utah and Arizona. US Geol Surv Bull 1633-B:B13–B56

Pol D, Carballido JL, Rauhut OWM, Rougier GW, Sterli J (2013) Biogeographic distribution patterns of tetrapods during the Jurassic: new information from the Cañadón Asfalto Basin, Patagonia, Argentina. J Vertebr Paleontol 33:192A

Pol D, Rauhut OWM (2012) A Middle Jurassic abelisaurid from Patagonia and the early diversification of theropod dinosaurs. Proc R Soc B Biol Sci 279(1741):3170–3175

Pol D, Turner AH, Norell MA (2009) Morphology of the Late Cretaceous crocodylomorph *Shamosuchus djadochtaensis* and a discussion of neosuchian phylogeny as related to the origin of Eusuchia. Bull Am Mus Nat Hist 324:1–103

Poropat SF, Mannion PD, Upchurch P, Hocknull SA, Kear BP, Elliott DA (2015) Reassessment of the non-titanosaurian somphospondylan *Wintonotitan wattsi* (Dinosauria: Sauropoda: Titanosauriformes) from the mid-Cretaceous Winton Formation, Queensland, Australia. Pap Palaeontol 1(1):59–106

Poropat SF, Mannion PD, Upchurch P, Hocknull SA, Kear BP, Kundrát M et al (2016) New Australian sauropods shed light on Cretaceous dinosaur palaeobiogeography. Sci Rep 6:34467

Poropat SF, Martin SK, Tosolini A-MP, Wagstaff BE, Bean LB, Kear BP et al (2018) Early Cretaceous polar biotas of Victoria, southeastern Australia–an overview of research to date. Alcheringa 42(2):157–229

Poropat SF, Nair JP, Syme CE, Mannion PD, Upchurch P, Hocknull SA et al (2017) Reappraisal of *Austrosaurus mckillopi* Longman, 1933 from the Allaru Mudstone of Queensland, Australia's first named Cretaceous sauropod dinosaur. Alcheringa 41(4):543–580

Prieto-Márquez A, Bolortsetseg M, Horner JR (2012) A diminutive deinonychosaur (Dinosauria: Theropoda) from the Early Cretaceous of Öösh (Övörkhangai, Mongolia). Alcheringa 36(1):117–136

Prieto-Márquez A, Dalla Vecchia FM, Gaete R, Galobart À (2013) Diversity, relationships, and biogeography of the lambeosaurine dinosaurs from the European Archipelago, with description of the new aralosaurin *Canardia garonnensis*. PLoS One 8(7):e69835

Raath MA (1972) Fossil vertebrate studies in Rhodesia: a new dinosaur (Reptilia: Saurischia) from near the Trias-Jurassic boundary. Arnoldia (Rhod) 7:1–7

Rauhut OWM (2003) A dentary of *Patagosaurus* (Sauropoda) from the Middle Jurassic of Patagonia. Ameghiniana 40(3):425–432

Rauhut OWM, Lopez-Arbarello A (2008) Archosaur evolution during the Jurassic: a southern perspective. Rev Asoc Geol Argent 63(4):557–585

Reisdorf AG, Wetzel A, Schlatter R, Jordan P (2011) The Staffelegg Formation: a new stratigraphic scheme for the Early Jurassic of northern Switzerland. Swiss J Geosci 104(1):97–146

Remes K, Ortega F, Fierro I, Joger U, Kosma R, Ferrer JMM et al (2009) A new basal sauropod dinosaur from the Middle Jurassic of Niger and the early evolution of Sauropoda. PLoS One 4(9):e6924

Rich THV, Vickers-Rich P (2000) Dinosaurs of darkness. Indiana University Press, Bloomington

Rougier GW, Davis BM, Novacek MJ (2015) A deltatheroidan mammal from the Upper Cretaceous Baynshiree Formation, eastern Mongolia. Cretac Res 52:167–177

Rougier GW, Novacek MJ, McKenna MC, Wible JR (2001) Gobiconodonts from the Early Cretaceous of Oshih (Ashile), Mongolia. Am Mus Novit 3348:1–30

Sander PM (1992) The Norian *Plateosaurus* bonebeds of central Europe and their taphonomy. Palaeogeogr Palaeoclimatol Palaeoecol 93(3):255–299

Schulte P, Alegret L, Arenillas I, Arz JA, Barton PJ, Bown PR et al (2010) The Chicxulub asteroid impact and mass extinction at the Cretaceous-Paleogene boundary. Science 327(5970):1214–1218

Sellés AG, Vila B, Galobart À (2017) Evidence of reproductive stress in titanosaurian sauropods triggered by an increase in ecological competition. Sci Rep 7(1):13827

Shuvalov VF (2000) The Cretaceous stratigraphy and palaeobiogeography of Mongolia. In: Benton MJ, Shishkin M, Unwin DM, Kurochkin EN (eds) The age of dinosaurs in Russia and Mongolia. Cambridge University Press, Cambridge, pp 256–278

Tidwell WD, Britt BB, Ash SR (1998) Preliminary floral analysis of the Mygatt-Moore Quarry in the Jurassic Morrison Formation, West-Central Colorado. Mod Geol 22:341

Torices A, Currie PJ, Canudo JI, Pereda-Suberbiola X (2015) Theropod dinosaurs from the Upper Cretaceous of the South Pyrenees Basin of Spain. Acta Palaeontol Pol 60(3):611–626

Tortosa T, Buffetaut E, Vialle N, Dutour Y, Turini E, Cheylan G (2014) A new abelisaurid dinosaur from the Late Cretaceous of southern France: palaeobiogeographical implications. Ann Paléontol 100(1):63–86

Tosolini A-MP, Korasidis VA, Wagstaff BE, Cantrill DJ, Gallagher SJ, Norvick MS (2018) Palaeoenvironments and palaeocommunities from Lower Cretaceous high-latitude sites, Otway Basin, southeastern Australia. Palaeogeogr Palaeoclimatol Palaeoecol 496:62–84

Tschopp E, Giovanardi S, Maidment SCR (2016) Temporal distribution of diplodocid sauropods across the Upper Jurassic Morrison Formation (USA). J Vertebr Paleontol Prog Abstr 2016:239

Tsogtbaatar K, Weishampel DB, Evans DC, Watabe M (2014) A new hadrosauroid (*Plesiohadros djadokhtaensis*) from the Late Cretaceous Djadokhtan fauna of southern Mongolia. In: Eberth DA, Evans DC, Ralrick PE (eds) Hadrosaurs. Indiana University Press, Bloomington, pp 108–135

Turner AH (2015) A review of *Shamosuchus* and *Paralligator* (Crocodyliformes, Neosuchia) from the cretaceous of Asia. PLoS One 10(2):e0118116

Turner AH, Hwang SH, Norell MA (2007) A small derived theropod from Öösh, Early Cretaceous, Baykhangor Mongolia. Am Mus Novit 3557:1–27

Turner CE, Peterson F (2004) Reconstruction of the Upper Jurassic Morrison Formation extinct ecosystem–a synthesis. Sediment Geol 167(3–4):309–355

Valdes P (1994) Atmospheric general circulation models of the Jurassic. In: Allen JRL, Hoskins BJ, Sellwood BW, Spicer RA, Valdes PJ (eds) Palaeoclimates and their modelling: with special reference to the Mesozoic era. Springer, Dordrecht, pp 109–118

Vila B, Gaete R, Galobart A, Oms O, Peralba J, Escuer J (2006) Nuevos hallazgos de dinosaurios y otros tetrápodos continentales en los Pirineos sur-centrales y orientales: resultados preliminares. In: Actas de las III Jornadas sobre Dinosaurios y su Entorno. Salas de los infantes, Burgos, pp 365–378

Vila B, Sellés AG, Brusatte SL (2016) Diversity and faunal changes in the latest Cretaceous dinosaur communities of southwestern Europe. Cretac Res 57:552–564

Wang SC, Dodson P (2006) Estimating the diversity of dinosaurs. Proc Natl Acad Sci 103(37):13601–13605

Watabe M, Tsuihiji T, Suzuki S, Tsogtbaatar K (2009) The first discovery of pterosaurs from the Upper Cretaceous of Mongolia. Acta Palaeontol Pol 54(2):231–242

Whitlock JA, Trujillo KC, Hanik GM (2018) Assemblage-level structure in Morrison Formation dinosaurs, Western Interior, USA. Geol Intermt West 5:9–22

Woodruff DC, Foster JR (2017) The first specimen of *Camarasaurus* (Dinosauria: Sauropoda) from Montana: the northernmost occurrence of the genus. PLoS One 12(5):e0177423

Zhang Y (1988) The Middle Jurassic dinosaur fauna from Dashanpu, Zigong, Sichuan, vol. 1: sauropod dinosaur (I): *Shunosaurus*. Sichuan Publishing House of Science and Technology, Chengdu

Early Mesozoic Nature In and Around Tethys

Guido Roghi, Ricardo Araújo, Massimo Bernardi,
Fabrizio Bizzarini, Mirco Neri, Fabio Massimo Petti,
Rossana Sanfilippo, and Edoardo Martinetto

Abstract

At the onset of the Mesozoic Era, all of the landmasses were assembled into a single supercontinent (Pangea) that enclosed a smaller ocean, called Tethys, and was surrounded by a single vast ocean (Panthalassa). Carnivorous archosaurs and crocodile-like reptiles, together with amphibians, populated the land, but these groups witnessed a dramatic change that began during the Late Triassic. This change was the rise of the dinosaurs. During the Jurassic, dinosaurs evolved to enormous body sizes and expanded their range over all parts of a gradually more fragmented supercontinent. The break-up of Pangea resulted in extensive coastlines and embayments in which ammonites, other mollusks, and swimming reptiles thrived. The marine reptiles include groups known as ichthyosaurs, placodonts, and nothosaurs, all of which are predatory. Once the invertebrate communities recovered from the end-Permian extinction event, shallow marine settings witnessed the development of Triassic reefs. These were initially built by several different invertebrate and plant groups (calcareous algae), only to be replaced by a new group of corals, the Scleractinia, in the Late Triassic. Marine biodiversity increased when a seasonal climate prevailed along the coastal zones. These conditions affected upwelling and nutrient supplies in the oceans and altered the atmospheric circulation pattern across the land's surface. The interiors of the continents were propelled into a semi-arid or arid climate, during which time the long and warm dry period was interrupted by several humid pulses in the Late Triassic. The most significant humid pulse is called the Carnian Pluvial Episode. At about the same time, extensive basalts, known as the Wrangellia Traps, formed through effusive eruptions. It is possible that this high volcanic activity influenced climate and triggered the Carnian Pluvial Episode. During this humid interval, high extinction rates affected marine groups (ammonoids, crinoids, bryozoans, conodonts), and changes in ocean chemistry resulted in the evolution of calcareous marine nannoplankton, forming chalk. Simultaneously, on land, the number of wetland plants increased along with the expansion of conifers, among which dinosaurs and pterosaurs diversified.

Electronic supplementary material A slide presentation and an explanation of each slide's content is freely available to everyone upon request via email to one of the editors: edoardo.martinetto@unito.it, ragastal@colby.edu, tschopp.e@gmail.com

*The asterisk designates terms explained in the Glossary.

G. Roghi (✉)
Istituto di Geoscienze e Georisorse - CNR, c/o Dipartimento di Geoscienze, Universita' di Padova, Padua, Italy
e-mail: guido.roghi@igg.cnr.it

R. Araújo
Instituto de Plasmas e Fusão Nuclear, Instituto Superior Técnico, Universidade de Lisboa, Lisbon, Portugal

M. Bernardi · F. M. Petti
MUSE–Museo delle Scienze, Trento, Italy
e-mail: Massimo.Bernardi@muse.it; fabio.petti@socgeol.it

F. Bizzarini
Cannaregio 1269/a, Venezia, Italy
e-mail: fabrizio.bizzarini@alice.it

M. Neri
Museo Civico di Storia Naturale Città di Vignola,
Vignola, MO, Italy
e-mail: mirco.lias@gmail.com

R. Sanfilippo
Department of Biological, Geological and Environmental Sciences (BiGeA), University of Catania, Catania, Italy
e-mail: sanfiros@unict.it

E. Martinetto
Department of Earth Sciences, University of Turin,
Turin, Italy
e-mail: edoardo.martinetto@unito.it

9.1 Introduction

Massimo Bernardi and Guido Roghi

The reader who arrived to this chapter has traveled back in time for more than 200 million years, visiting numerous terrestrial sites along the way, which exemplify past natural systems outside the marine environment. This chapter explores the seashores and we will dive into the early Mesozoic seas. In fact, we need to take a glimpse both on land and into the sea to understand what happened to the biosphere at that time. At the onset of the Mesozoic Era, during the Triassic, the paleogeographic setting [U0901] was one in which a single vast ocean (Panthalassa) surrounded a single supercontinent (Pangea) to the south, west, and north (Fig. 9.1). The eastern shore of Pangea faced a smaller ocean called Tethys. The Tethys shores are particularly intriguing because a thick succession of sedimentary rocks preserves a rich fossil heritage. These successions are precious archives of past environments and their life forms, as we will observe in our trip. However, to understand what we will see through our postcard window at each stop, we will need some additional information about two main events that we would not be able to perceive by visiting single localities, alone.

Decades of geo-paleontological research, which has integrated an enormous data set from many localities, reveal that the Triassic was marked by three major extinction events. The more recent extinction occurred at the boundary between the Triassic and the Jurassic, about 201 million years ago (Cohen et al. 2015). Both sea (reef ecosystems, bivalves, ammonites) and land (terrestrial tetrapods) diversity declined dramatically (Bond and Gresby 2017) as result of the global warming, ocean acidification and anoxia caused by the eruption of the Central Atlantic Magmatic Province (Preto et al. 2010). Going back in time, the second event occurred about 232 million years ago, during the Late Triassic, and it is the so-called Carnian Pluvial Episode (CPE; Simms and Ruffell 1989). As the name implies, the CPE was characterized by a strongly humid climate and stands out as a major global environmental perturbation that triggered biotic turnover (Dal Corso et al. 2015). Even though it probably lasted only about 2 million years or less (Miller et al. 2017), extinction rates in several groups of marine (e.g., ammonoids, crinoids, conodonts) and terrestrial (e.g., rhychosaurs) animals, as well as plants, were high. A rapid floral turnover is documented by an increase in water-loving (hygrophytic) plants and conifers (Roghi et al. 2010). Alongside the diversification of conifers was the radiation of various vertebrate groups during or soon after the main humid pulses. These include the expansion of the dinosaurs, crocodiles, turtles, and mammals (Furin et al. 2006). But what triggered all these events?

The main cause for the climatic perturbation during the CPE is thought to be volcanism. The event was likely triggered by the eruption of the Wrangellia magmatic province (Dal Corso et al. 2015), which crops out along the western coast of the USA and Canada. The expulsion of basaltic lava and the gasses trapped in that magma pushed global ecosystems towards a mass-extinction scenario. As the relative amount of atmospheric CO_2 increased, temperature rose and the ocean acidified. It is thought that denudation of the continents followed, which resulted in a global biodiversity crisis and subsequent recovery (Dal Corso et al. 2012). However, the CPE was a minor event compared to what happened 20 million years earlier.

The most dramatic biodiversity crisis in Earth's history took place at the boundary between the Permian and the Triassic periods, which also mark the boundary between the Paleozoic and the Mesozoic Eras (252 Ma). This crisis is the End-Permian Mass Extinction (EPME), the second outstanding event considered in this chapter. We already learned about its (surprisingly small) impact on fish diversity (Romano in Tschopp et al., Chap. 7) and will go into more detail in the following chapter of this book (Delfino et al., Chap. 10). However, before getting there, we will gain some more insights into the Mesozoic natural systems by visiting

Fig. 9.1 Triassic paleogeographic setting of Pangea showing the location of the Tethys Ocean

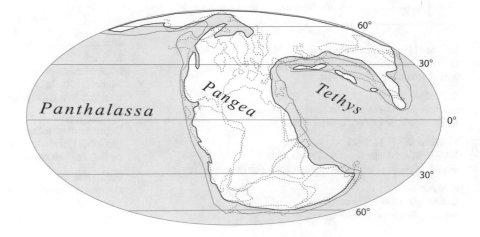

various localities along the shores of the Tethys Ocean. We will start our journey in the Early Jurassic, and work our way back in time, through the CPE, to the beginning of the Triassic, ending in a late Permian shallow Tethys ecosystem.

9.2 Early Jurassic Environment in Central-Eastern Pangea

Mirco Neri

A visit to central-eastern Pangea along the westernmost part of the Tethys margin, at about 30° N latitude in the northern part of Gondwana, would appear like a visit to an island paradise. Here, during the Early Jurassic, deep blue ocean waters surrounded numerous islands and the setting is believed to have been similar to the present-day Bahamas (Bosellini et al. 1981). But this island paradise did not last

long. In the eastern part of the southern Alps (northeastern Italy), we can find the stratigraphic evidence for the break-up of Pangea, linked to the rifting of the westernmost opening of the North Atlantic Ocean. The so-called Trento Platform (Fig. 9.2), bordered to the east by the Belluno basin and to the west by the Lombardy basin (Masetti et al. 1998), retains the evidence of the regional paleogeographic setting in this part of the Tethyan margin [U0902] (Zempolich 1993). The Lower Jurassic rocks that make up the Trento Platform, namely the Calcari Grigi Group, are limestones that formed in shallow water [U0903-U0904]. Fossils here are from both marine and continental organisms. Marine animals include bivalves, foraminifera, ostracods, and corals [U0905]. Here we can observe the famous *Lithiotis* coquina*, a rock that is nearly entirely composed of fossil shells. In this particular coquina, large shells, about 50–70 cm in length, belong to the genera *Lithiotis*, *Cochlearites,* and *Lithioperna* (Posenato and Masetti 2012). These were large bivalves with distinctive

Fig. 9.2 Paleogeographic setting of the Italian southern Alps in the Lower Jurassic. The Trento Platform is bordered by the Belluno and the Lombardy basins

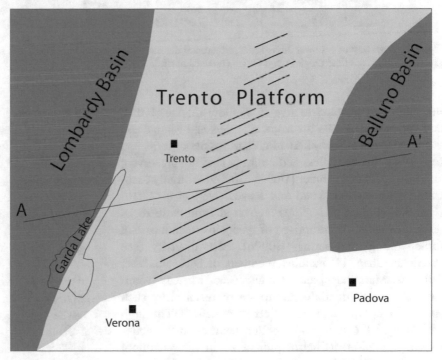

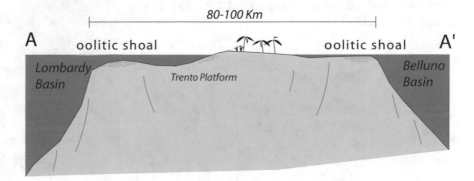

Fig. 9.3 Trackway of a Lower Jurassic large theropod dinosaur (*Kayentapus* isp.; right) and seed ferns *Dichopteris visianica* (left) from the famous macroflora of the Calcari Grigi Group (Lower Jurassic)

morphologies adapted to live in shallow marine waters, and widespread across the tropical Tethys and Panthalassa oceans (Franceschi et al. 2014). Plant fossils of ferns and fern allies, bennettitales, and conifers also are preserved; and Lower Jurassic amber [U0906-U0909] is strictly associated with conifer wood and leaves (Neri et al. 2016) (Fig. 9.3). Pollen and spores recovered from these rocks reflect a vegetation comprised of lycopsids (club mosses), ferns, and gymnosperms [U0906]. Fern diversity (see Psenicka, Chap. 11) includes members of Dicksoniaceae, Filicales, Marattiales, and Osmundaceae, whereas lycopsids of the Selaginellales group are represented by such spores as *Lycopodiacidites* (Neri et al. 2018). The high percentage of *Circumpolles* pollen testifies to the abundance of conifers (Cheirolepidiaceae) in the paleoflora (Neri et al. 2016). In several cases, these plant fossils are associated with dinosaur footprints (Box 9.1) (e.g., Avanzini et al. 2006).

Box 9.1: Trento Platform Track Megasites

Locked away on the surfaces of Early Jurassic coastal sediments are dinosaur trackways. The track megasites* (extensively exposed and broad surfaces on which tracks are preserved [U0909]) on the Trento Platform are special because they record the presence of a typical dinosaur-dominated tetrapod assemblage (Lucas et al. 2005a). Few of these sites exist in Europe. The trace fossils are preserved in limestone units once thought to be entirely marine in origin. The most abundant trackways are of theropods; other taxa represent early sauropods and several unidentified bipedal dinosaurs. Even small ornithischians walked across these surfaces, and, when flooded, bipedal dinosaurs and non-dinosaurian reptiles left swimming tracks. However, the composition of single track sites can vary.

The differences in trackway densities and species diversities of the Trento megasites seem to be related to prevailing climatic and environmental conditions during the Early Jurassic. The younger (Pliensbachian, 191–183 Ma) track sites record more humid conditions. In contrast, older megasites (Hettangian-Sinemurian, 201–191 Ma) evidently formed in a semi-arid to arid, upper intertidal to supratidal environment. This change in climate is recorded in the palynological assemblages that reflect a shift from warm and rather arid climate to humid environments through time (Fig. 9.4). The proportion of sauropods versus theropods also reflects this climate shift. Under the more humid conditions, the number of sauropods increased and the number of theropod species decreased (Avanzini et al. 2006).

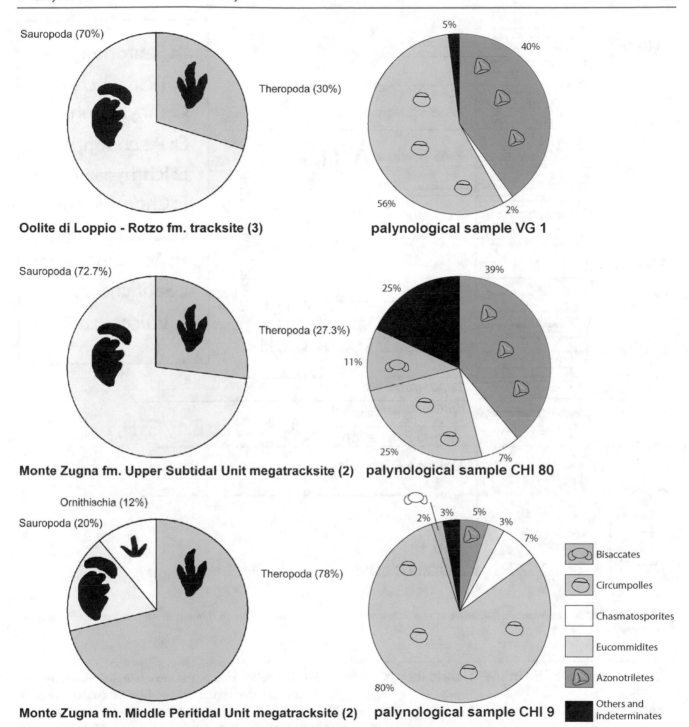

Fig. 9.4 Diagrams of quantitative palynological data plotted and compared to the quantitative dinosaur data. Three main dinosaur track levels are compared to the paleobotanical component of the environments. Sample CHI 9 (first track level), gymnospermous *Circumpolles* pollen represents the main group (80%) and indicates a subtropical, warm, and rather arid climate. In the samples CHI80 and VG1 (next two track lev-els), an interval of typical *Azonotriletes* spores indicates a change toward a more humid climate with freshwater influence. The basal track level is dominated by theropod traces with fewer imprints from sauro-pods and ornithopods, whereas in levels 2 and 3, sauropod-like tracks represent the dominant ichnofauna. Ornithischian-like footprints are absent at higher levels

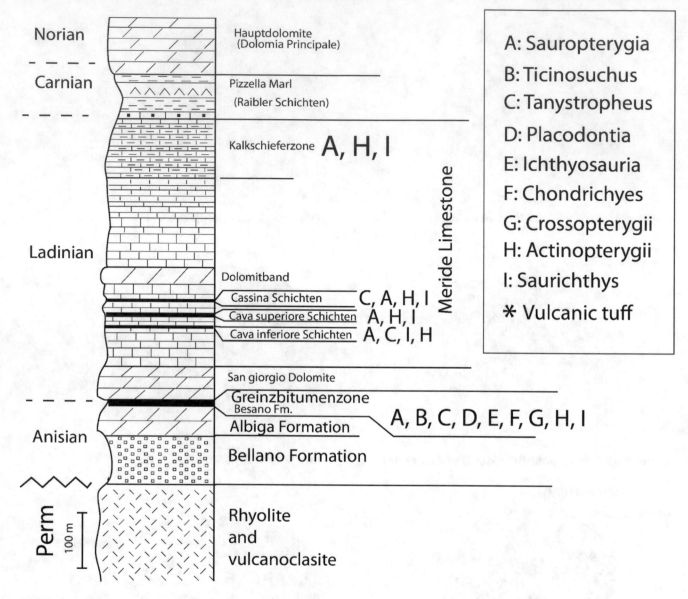

Fig. 9.5 Stratigraphic succession and distribution of vertebrates in the Grenzbitumenzone and Meride Limestone at Monte San Giorgio (da Furrer 2004, modif.)

9.3 Vertebrate Life in the Triassic Tethys

Ricardo Araújo and Guido Roghi

A little bit further west from the Trento Platform, at the border between Italy and Switzerland, we visit a small green mountain above a beautiful subalpine lake. This mountain is Monte San Giorgio (MSG) and has been one of the best places in the world to find Triassic marine vertebrates [U0910]. Collections have been made, here, for the last two centuries. Excavated quarries have yielded thousands of articulated skeletal remains (Felber et al. 2004). Paleontological research on MSG began in the middle of the nineteenth Century, when the first specimens were discov-

ered on the Italian side of the mountain. Subsequently, fossils were also found on the Swiss side of the border (Fig. 9.5). These Middle Triassic animals lived in and around the Tethys beginning in the late Anisian (247–241 Ma) to the late Ladinian (247–237 Ma).

During the Middle Triassic, the depositional environment now exposed at the MSG site is interpreted as an undisturbed tropical lagoon. Major radiation events are seen in both reptiles and actinopterygian fishes at this time. This diversification is recorded by the more than 10,000 fossil reptile specimens collected from quarries around the mountain. Many of these consist of completely articulated skeletons, of a number of taxa including marine ichthyosaurs, nothosaurs, placodonts, and aerial protorosaurs. Enigmatic finds at MSG

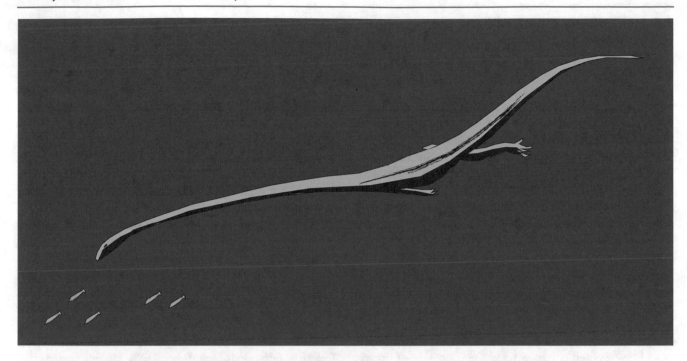

Fig. 9.6 Reconstruction of the long-necked protorosaur *Tanystropheus longobardicus* (Monte San Giorgio)

include several specimens of the long-necked protorosaur *Tanystropheus longobardicus* (Fig. 9.6) [U0911], and a 6-m-long skeleton of the ichthyosaur *Besanosaurus*, which is the largest complete specimen found here, to date. The only major marine reptile group so far missing at MSG is the plesiosaurs, which we observed in unit 7 as swimming forms in the Jurassic and Cretaceous oceans (Rabi and Araújo in Tschopp et al., Chap. 7).

The absence of plesiosaurs at MSG may be explained by their general rarity during the Middle Triassic. According to the most recent phylogenetic analysis and age dates of these rocks, there is a significant sampling gap between the youngest pistosauroids (the group thought to include the ancestors of plesiosaurs) and the basal-most plesiosaurs (Benson et al. 2010). Although *Bobosaurus* from the lower Carnian (c. 237 Ma) sediments of Italy was previously interpreted as a basal plesiosaur (e.g., Fabbri et al. 2014), more focused and comprehensive phylogenetic analyses consider this taxon as a pistosauroid (Benson and Druckenmiller 2014). The multiple basal plesiosaur taxa from the Blue Lias Formation (Street, UK), which were thought to have lived near the Triassic-Jurassic boundary (e.g., Storrs and Taylor 1996), are, instead, from the lowermost Hettangian, and are of definite Jurassic age (Benson et al. 2012). In sum, we currently do not know of any plesiosaur from before the Jurassic. Thus, the MSG is no exception from global trends. Despite the absence of plesiosaurs, the MSG localities have yielded 30 reptile spe-

cies, 80 fish species, some 100 macro-invertebrates (such as bivalves, ammonites, echinoderms, crustaceans), as well as abundant marine microorganisms. Although the vast majority of these species are marine, a few terrestrial plants and animals are also encountered.

Members of the continental biome include reptiles, insects, and the base of the food chain, plants. One of the most exceptional terrestrial vertebrate fossils is a complete skeleton of the crocodile-like archosaur *Ticinosuchus ferox*. This is the oldest complete specimen of an archosaur discovered in the Northern Hemisphere. Noteworthy insects include a wide range of terrestrial groups, with phytophagous (plant eating) and predatory habits, as well as aquatic groups, preserved in both their larval (e.g., Plecoptera) and adult stages (e.g., Ephemeroptera). Some insects are new to science, such as the beetle *Praedodromeus sangiorgensis* (Strada et al. 2014). The diversity of the insect fauna from one horizon, together with the high biodiversity of the fish fauna (Lombardo et al. 2012), supports the interpretation that the region was composed of a complex of paleoenvironments. Monte San Giorgio was likely a shallow lagoon adjacent to a carbonate platform, with permanent freshwater pools or small streams entering the ocean. There was limited interchange between the open ocean and deeper offshore areas (Strada et al. 2014). Unfortunately, MSG only provides a limited fossil record of the terrestrial biota of Europe, and to look for more diversified assemblages, we need to travel to elsewhere on the continent.

9.4 The Early Mesozoic Biota of the Grès à Voltzia

Guido Roghi

In Middle Triassic times, the area along today's border between France and Germany was occupied by an epicontinental∗ sea that covered the so-called Germanic Basin. This sea was, at times, connected further south to the Tethys and, more rarely, to the Panthalassic Ocean through a passage to the north. The sediments deposited here built up a geological formation known as Grès à Voltzia, which mainly consists of sandstone, but also contains small silty-clayey lenses in which an abundance of terrestrial and marine organisms are preserved. These fine-grained lenses are interpreted as pools in a coastal deltaic system, which appeared during the wet season and disappeared in the dry season (Selden and Nudds 2004). It is believed that the climate was probably seasonal. The drying of ephemeral pools [U0912] resulted in the mass death of the aquatic plants and animals which, in turn, promoted low oxygen conditions and the formation of organic films coating their remains and facilitating fossilization

(Selden and Nudds 2004). There is a marked stratification to the fossils in these deposits.

The various layers of the fine-grained lenses preserve distinct fossil assemblages. In combination with the inferred salinity requirements of each assemblage, it is possible to reconstruct the history of the pond. The lower part of the lenses contains only an aquatic biota and include arthropods (such as horseshoe crabs, crustaceans, insects, myriapods, scorpions, spiders), cnidarids (jellyfish, corals), brachiopods, annelids, mollusks, and fishes. Fully marine salinities are required by many of the invertebrates. The upper layers of these clay lenses preserve remains of the conifer *Voltzia heterophylla*, the plant on which the name of the formation is based. Voltzian conifers are reconstructed as shrubs or small trees growing around the edges of pools. They are often associated with another conifer, *Aethophyllum*, which was herbaceous and grew to a maximum height of less than 1 m (Grauvogel-Stamm 1978; Gall and Grauvogel-Stamm 2005) (Fig. 9.7). Remains of terrestrial arthropods are often preserved with the plant fossils but are less diverse than in the lower level assemblages. Terrestrial arthropods include scorpions, spiders, and insects. Dessication cracks, reptile footprints, and remains of land

Fig. 9.7 Reconstruction of the paleoenvironments of the Grès à Voltzia

plants in upright, growth position are preserved at the top of each clay lens. Freshwater or, possibly, brackish salinities are needed to sustain these organisms. As such, these distinct layers nicely show how the seasons may have changed from wet to dry during the time these fossil assemblages formed (Selden and Nudds 2004). After this short visit to a coastal delta, it is time to return back into the oceans.

9.5 Shallow Marine Life in the Triassic of the Dolomites

Fabrizio Bizzarini and Guido Roghi

One of the most attractive and characteristic landscapes of the Alps can be found in the Dolomites. Here, the link between present and past is so strong that modern geomorphology is highly controlled by the ancient environment of about 240–230 million years ago. The massive carbonate platforms of the Dolomites, formed in shallow marine settings [U0913], are exposed in some of the best outcrops of the world (Fig. 9.8). These isolated carbonate bodies resist subaerial erosion much better than the clay-rich and soft marine sediments that formed around them in deep sea environments. Due to their prominence, the region has attracted geologists since the early nineteenth century and has been the object of continuous stratigraphic and sedimentologic research over the past 200 years. The stratigraphic framework of the Dolomites includes Permian to Cenozoic units,

but rocks formed in the Triassic dominate the area. The stratigraphic basement on which these occur is a succession of deformed and metamorphosed, lower Paleozoic rocks. In the early Permian, a thick volcanic succession was deposited (Avanzini et al. 2013), followed by late Permian fluvial conglomerate and sandstone (Massari and Neri 1997). Overlying these deposits are late Permian shallow marine evaporites and carbonates (Bellerophon Formation) which, in turn, are overlain by Early Triassic shallow marine carbonate and continental (terrigenous) deposits. By the time the Middle Triassic began, carbonate deposits (Gianolla et al. 1998) were tabular-shaped platforms (Anisian, 247–241 Ma). These were replaced by wide, mainly microbial isolated platforms in the Ladinian (237–241 Ma) and, at the onset of the Upper Triassic (Carnian, 228–237 Ma), coral reefs appeared. The Dolomites are thick, a fact that was used as early as 1860 as proof that long-term subsidence influenced the buildup of carbonate reefs (Richthofen 1860; Schlager and Keim 2009).

Studies of these carbonate platforms have identified abrupt facies changes, reflecting the spatial and temporal variation in this part of the Tethyan ocean. For example, back-reef, shallow marginal basins, slope, and central basin are particularly well represented in the Cassian platform and San Cassiano Formation (Upper Triassic) in the central Dolomites (Fig. 9.9). Structures conserved in these rocks are comparable to processes operating in modern reefs and atolls, leading to the conclusion that massive gravity flows from the platform to the basin floor occurred then, as they do now. Each environment preserved a typical fossil assem-

Fig. 9.8 The Lower and Middle Triassic succession below the carbonate platform of Catinaccio/Rosengarten, Dolomites

Fig. 9.9 Cyclicity of Carnian sediments in the Richthofen Reef and Set Sass. Third-order sea-level cycles are found in the Lower Cassian Dolomite (LCD) and Lower St. Cassian/San Cassiano Formation (LSC), the Richthofen Reef and Upper St. Cassian/San Cassiano Formation (USC), and Upper Cassian Dolomite (UCD). The Lower and Upper Cassian Dolomites represent the highstand phases when sediments prograded into the basin of the St. Cassian/San Cassiano. The Upper St. Cassian/San Cassiano (USC) succession onlaps the slope of the Lower Cassian Dolomites, indicating a rise in sea level

blage including algae and foraminifera, along with calcareous sponge or coral-patch reefs with a highly diverse fauna of frame-builders and reef-dwellers. Algal meadows are often encountered where they actually grew more than 200 million years ago (they are autochthonous∗). These meadows are associated with gastropods and bivalves in the shallow marginal basins. Boulders that originated from the edge of the carbonate platform can now be found in slope deposits, transported there by gravity flows (Russo et al. 1997). These carbonate blocks commonly preserve worn shells of the bivalve *Pachycardia rugosa* and echinoderm debris. Further seaward and away from the carbonate platform, in the deeper basins, pelagic assemblages are dominated by the bivalves *Daonella* and *Posidonia*, but ammonites are commonly found. The sedimentary environment, though, changed dramatically in the Late Triassic.

Purely carbonate, peritidal (within the tidal zone) deposits can be found on the platform during the Norian (227–209 Ma), and these characteristic rocks can be traced over an extended area of about 200 km. This formation, known in the Dolomites as the Hauptdolomit/Dolomia Principale [U0914]

(Bosellini and Hardie 1988), is represented by the dolomitic cliffs that now top the underlying carbonate platform units (Fig. 9.10). These calcareous sediments are characterized by a peculiar paleoenvironmental cyclicity of subtidal dolomite and inter-supratidal stromatolitic layers. The thicker subtidal layers feature gastropods and giant megalodontid bivalves, which reach diameters of up to half a meter (Fig. 9.11). But the Dolomites do not only provide a window into Triassic marine life, these mountains also yielded globally relevant records of the Triassic terrestrial life: amber and dinosaur tracks.

9.6 Amber, Dinosaurs, and the Carnian Pluvial Episode in Europe

Massimo Bernardi and Guido Roghi

Amber was an essential gemstone used in a famous series of science-fiction films that started in the 1990s (Jurassic Park) to bring extinct dinosaurs back to life. In our imaginary field

Fig. 9.10 The Dolomia Principale Formation at the Tre Cime di Lavaredo mountain group in the Dolomites

Fig. 9.11 Megalodontid bivalve from the Upper Triassic of the Dolomites

trip to the past, strictly based on sound scientific evidence, amber and dinosaurs are somehow connected. Bernardi et al. (2018) studied dinosaur trackways, especially from the Italian Dolomites, and combined their new data with global information from skeletal remains of the same time period. They found evidence for a distinctive and rapid diversification of dinosaurs (in particular saurischians, the group that includes sauropodomorphs and theropods, see Tschopp

et al., Chaps. 7 and 8) in the early late Carnian, around 232 million years ago. This is when several amber-bearing layers are found across the Dolomites (Fig. 9.12). The volume of amber seems to be linked to one of the major paleoclimatic events of the CPE (see Sect. 9.1). Carnian fossil resin is known from Europe to Africa and from North to South America (Fig. 9.13). One of the most important amber finds was the recovery of several thousands of small (2–4 mm

Fig. 9.12 Upper Triassic amber drops from the Dibona Section in the Tofana Group (Dolomites)

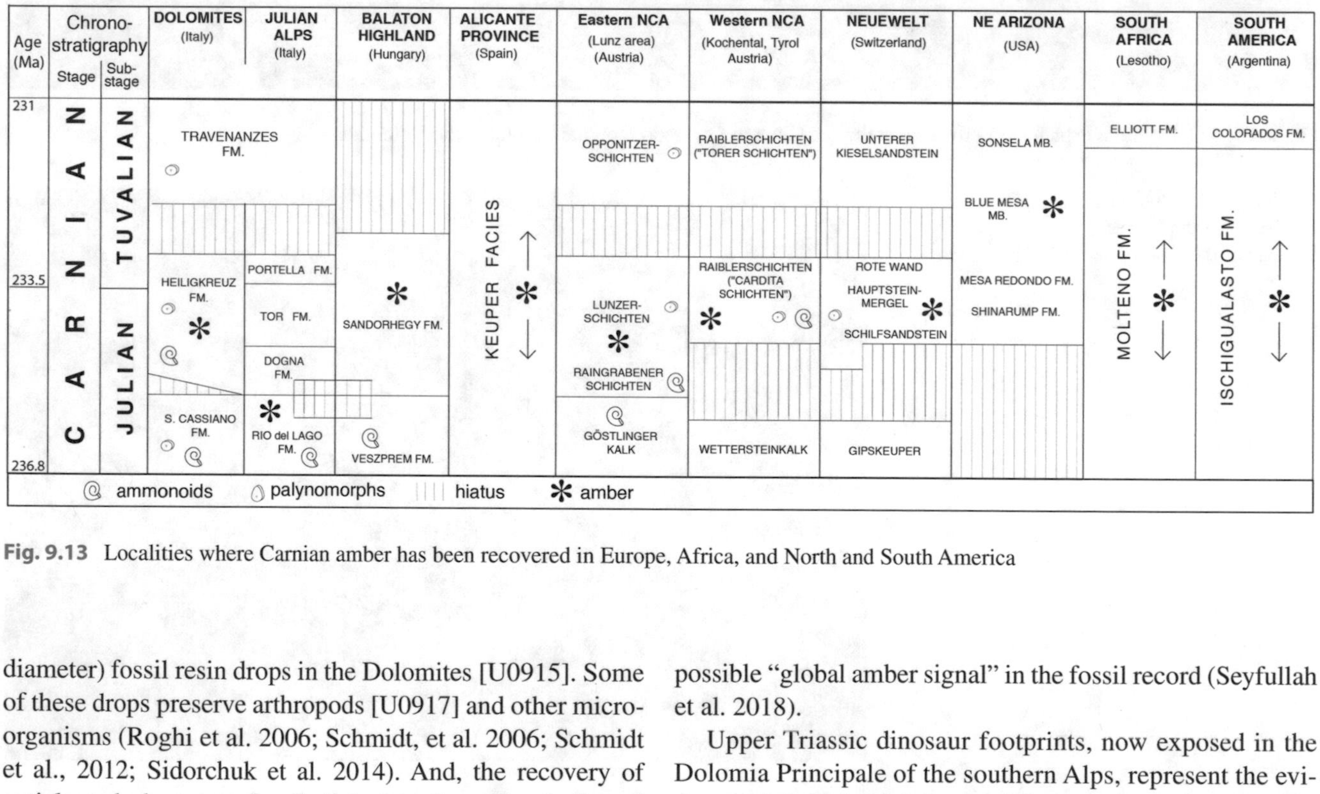

Age (Ma)	Chrono-stratigraphy		DOLOMITES (Italy)	JULIAN ALPS (Italy)	BALATON HIGHLAND (Hungary)	ALICANTE PROVINCE (Spain)	Eastern NCA (Lunz area) (Austria)	Western NCA (Kochental, Tyrol Austria)	NEUEWELT (Switzerland)	NE ARIZONA (USA)	SOUTH AFRICA (Lesotho)	SOUTH AMERICA (Argentina)
	Stage	Sub-stage										
231	C A R N I A N	T U V A L I A N	TRAVENANZES FM.			KEUPER FACIES	OPPONITZER-SCHICHTEN	RAIBLERSCHICHTEN ("TORER SCHICHTEN")	UNTERER KIESELSANDSTEIN	SONSELA MB.	ELLIOTT FM.	LOS COLORADOS FM.
233.5			HEILIGKREUZ FM. ✳	PORTELLA FM.	SANDORHEGY FM. ✳	✳	LUNZER-SCHICHTEN ✳	RAIBLERSCHICHTEN ("CARDITA SCHICHTEN")	ROTE WAND	BLUE MESA MB. ✳	MOLTENO FM.	ISCHIGUALASTO FM.
				TOR FM.					HAUPTSTEIN-MERGEL ✳	MESA REDONDO FM.		
		J U L I A N		DOGNA FM.			RAINGRABENER SCHICHTEN		SCHILFSANDSTEIN	SHINARUMP FM.		
236.8			S. CASSIANO FM. ✳	RIO del LAGO FM. ✳	VESZPREM FM.		GÖSTLINGER KALK	WETTERSTEINKALK	GIPSKEUPER			

ammonoids   palynomorphs ||| hiatus   ✳ amber

Fig. 9.13 Localities where Carnian amber has been recovered in Europe, Africa, and North and South America

diameter) fossil resin drops in the Dolomites [U0915]. Some of these drops preserve arthropods [U0917] and other micro-organisms (Roghi et al. 2006; Schmidt, et al. 2006; Schmidt et al., 2012; Sidorchuk et al. 2014). And, the recovery of cuticle and plant megafossils linked to the amber indicated that the source plants were conifers of the extinct family Cheirolepidiaceae (see Gee et al., Chap. 6). These conifers were part of xerophytic plant communities growing under climatically stressed conditions. When damaged, these plants exuded a resin as a protection against insects, pathogens, and wildfire; upon the loss of its volatile chemicals, the resin turned into amber. The abundance of amber during the CPE, linked to climate-stressed ecosystems [U0916], is used as a possible "global amber signal" in the fossil record (Seyfullah et al. 2018).

Upper Triassic dinosaur footprints, now exposed in the Dolomia Principale of the southern Alps, represent the evidence of their domination in the proximal marine tidal flat environments. More than ten trackway (ichno) sites, of late Tuvalian (225–216.5 Ma) to Norian (227–208 Ma) age, are evidence for the abundant presence of dinosaurs in the region. Numerous dinosaur footprints are described from the Mt. Pelmetto ichnosite (Fig. 9.14), which also yielded a single *Brachychirotherium*-like footprint. One trampled horizon occurs in the lowermost part of the Dolomia Principale directly overlying the Travenanzes Formation

Fig. 9.14 Dinosaur footprints from the Dolomia Principale of the Mt. Pelmetto ichnosite (Dolomites)

and is dated to the late Tuvalian. A *Eubrontes-Grallator* association is reported from the Tre Cime site, and dinosaur-only associations are also known in the Friuli and Carnic Prealps. All these localities are dated to the late Carnian-Norian by means of lithostratigraphy. An association of *Evazoum, Eubrontes,* and *Grallator* is also found in the Pasubio Massif, whose Norian (Alaunian, 215.6–212 Ma) age is established by conodont biostratigraphy. A mixed chirotheroid–dinosaur footprint association occurs in the Val Pegolera outcrop, and a dinosaur-dominated *Atreipus-Grallator* association was described from the Moiazza ichnosite, both in the Province of Belluno. All Late Triassic ichnosites of late Tuvalian to Norian age provide evidence for dinosaur dominance in this coastal area, with more than 90% of tracks being assignable to dinosaurian track makers [U018].

9.7   Early–Middle Triassic Trackways on Tethys Shores

Fabio Massimo Petti

We have reached a stage in our long excursion when it is time to lie on our backs of a Tethyan beach near the beginning of the Triassic. Which large terrestrial animals would

we be our cohabitants? If we want an answer to this question, we should look, as usual in this book, to the sedimentary record and the biological evidence preserved, therein. Just as we would leave footprints in the sand, fossil trackways are one of the most apparent and abundant records of the organisms that walked these Tethys shorelines. The most common Lower to Middle Triassic footprints and trackways (the so called "chirotherian footprints") are attributed to archosaur-like (archosauriform) vertebrates. These are well known from Argentina (Melchor and De Valais 2006), the North American southwest (Klein and Lucas 2010), Africa (Klein et al. 2011), and southern China (Xing et al. 2013). But the most abundant and diversified assemblages are reported from siliciclastic* (sand and mud) and carbonate (limestone) sediments of Europe. Nearly identical trace-fossil taxa and trace-fossil assemblages are known from the Lower to Middle Triassic of Great Britain (King et al. 2005) and Germany (e.g., Klein and Haubold 2007), Lower Triassic rocks of Poland (Klein and Neidzwiedzki 2012), and Middle Triassic of France (e.g., Gand 1979) and of the Iberian Range in Spain (Dìaz-Martinez et al. 2015). Curiously, archosauriform tracks are almost absent in the Early–Middle Triassic Alpine rocks, along with those of amphibians and synapsids. However, the southern Alps are the place of the historically oldest reports of Lower Triassic vertebrate tracks.

9.7.1 Early Triassic Trackways in the Alps

In 1926, the Austrian paleontologist Othenio Abel (1875–1946) reported tracks found by Julius von Pia (1887–1943) in the Braies Dolomites. He erected a new ichnospecies *Rhynchosauroides tirolicus* (Abel 1926) to describe them. Today, those trackways are attributed to a group of reptiles comprising all diapsids closer to the lizards than to archosaurs, a clade called Lepidosauromorpha. Archosauriform tracks were found later in the twentieth Century. For many years, the only Early Triassic archosauriform track from the southern Alps was the one found in the 1950s by Piero Leonardi (1908–1998). That track occurs in the uppermost part of the Werfen Formation (Leonardi 1968). Subsequent discoveries indicated that the Lower Triassic tetrapod-track record of the southern Alps does not mirror the global body fossil record. One explanation may be linked to a paleoenvironmental bias implicit in the depositional environment in which these trackways are found. The upper part of the Werfen Formation is interpreted as a gently dipping shallow marine surface (ramp*) on which both carbonate and terrigenous sedimentation occurred. This setting may also explain why there is an absence of fossil plants, the remains of which may have decayed before fossilization could have happened in this oxygen-rich substrate (Petti et al. 2013). However, it may also be possible that the local paleoclimatic conditions impacted the diversity of plants and trackmakers. The track-bearing horizons of the late Olenekian (251–247 Ma) rocks are marked by clear sedimentological features of a dry interval. In contrast, a somehow different paleoenvironment seems to have been present in the Lower Triassic of the western Alps.

No vertebrate-track sites were known in the Lower Triassic of the western Alps at the beginning of the twenty-first Century. New trackways with archosauriform footprints were only recently discovered at two sites in the past few years. The first site, with footprints tentatively assigned to the ichnogenus *Chirotherium,* is preserved in lagoonal mudrock of the uppermost Olenekian. These deposits are now exposed in the Ligurian Alps of northwestern Italy (Santi et al. 2015). However, the most relevant and promising track site is located about 50 km north in the Gardetta Plateau [U0919]. Today, this location is a pleasant, treeless landscape [U0920] where hundreds of cows mow flowering grasslands during the short Alpine summers. These meadows grow on very interesting geological substrates (Fig. 9.15a). Below a pure quartz sandstone (arenite) is a bed with ripple marks (Fig. 9.15b). This bed was deposited during the Early Triassic (?Olenekian, c. 250 Ma) and covers a trampled surface. Tracks and trackways are preserved as concave epirelief* on a steeply tilted bedding plane, a consequence of Neogene Alpine tectonism. The track-bearing succession contains cross-bedded quartz-arenite and quartz-siltstone,

interpreted as having been deposited in a fluvial to mudflat setting. The Gardetta assemblage includes archosauriform footprints, assigned to *Chirotherium* and *Isochirotherium*, which are typical of the upper Olenekian–lower Anisian interval. Archosaur-like reptiles (possibly from a clade named Erythrosuchidae), up to 5 m long, are the most likely trackmakers, but footprint morphology (Fig. 9.15c) and trackway patterns [U0921] do not exclude the possible presence of crown*-archosaurs, either.

The Gardetta site, together with evidence from a few other localities, unequivocally demonstrates the presence of archosauriforms [U0922] at low latitudes during the Lower Triassic. Several authors (e.g., Sun et al. 2012) have suggested that this vertebrate clade had not yet appeared in the time immediately following the end-Permian event. However, this idea is based solely on the absence of skeletal remains, which most often are not preserved unless specific geochemical conditions are present when bones are buried. The ichnological record helps to fill in our understanding of deep-time landscapes in the absence of body fossils. In fact, there is abundant trace-fossil evidence in support of the presence of archosauriforms in the aftermath of the EPME. Their trackways are common in the intertidal carbonate mudflats of the tropical coastal areas bordering Tethys and Pangea. Diedrich (2015) reported on numerous archosauriform trackways (*Isochirotherium*) in the Germanic Muschelkalk basin and suggested that these tetrapods were at the top of the terrestrial food chain, feeding on smaller vertebrates. This coastal habitat probably allowed archosauriforms to migrate long distances across the single supercontinent, spreading from Europe to what is now Northern America, Africa, and China [U0923].

9.8 The Lost Permian Eden of the Tethys

Rossana Sanfilippo

We now travel to the ocean basin that preceded the Tethys at the end of the Paleozoic Era, the Paleo-Tethys [U0924]. This ocean was located on the eastern margin of the supercontinent Pangea near to the equator (Scotese 2014), and it teemed with life, looking like a real "Eden". In this warm, ancient sea, sponge-coral-crinoid reefs thrived in shallow waters, providing shelter for a wide diversity of fishes and cephalopods. These reef frameworks offered hard substrates colonized by an abundant and diversified invertebrate fauna. The westernmost end of the Paleo-Tethys, the area where Sicily now is positioned [U0925], was submerged and covered with late Paleozoic reef communities [U0927]. At deeper depths, along the slope adjacent to the reefs (Catalano et al. 1991), lived trilobites and ostracods. Highly fossiliferous rocks known as

Fig. 9.15 Gardetta site, Lower Triassic of the western Alps. (**a**) Study of archosaurifom trackways in the field. (**b**) One of the archosauriform trackway assigned to the ichnogenus *Isochirotherium*. (**c**) The modern landscape of the Gardetta plateau, where the highest visible mountain is made up by vertical Triassic beds (photos M Defino and FM Petti)

the Sosio Limestone (Box 9.2), cropping out in western Sicily, preserve these ancient marine ecosystems (Flügel et al. 1991; Di Stefano and Gullo 1997). The Sosio Limestone, discovered in 1887 (Gemmellaro 1887–1889) and described by Mascle (1979), is unique. Unlike the geographically extensive carbonate platforms of the Triassic we visited earlier in our journey, the fossil-bearing sediments are limestone megablocks (the most famous one called "Pietra di Salomone") enclosed in debris flow and turbidite deposits. These sediments of the middle-to-late Permian (c. 268–251 Ma) were moved from a carbonate platform into the deepest parts of the ancient ocean by gravity.

Box 9.2: Fossils of Sosio Megablocks

The Sosio megablocks are highly fossiliferous because they preserve the remains of animals that lived on the platform before the blocks were moved downslope to their final resting place. Interestingly, these blocks contain numerous forms with bizarre shapes that do not resemble anything we know living in today's ocean. Although we can place the animals into a variety of major clades based on several of their features, their unusual shapes indicate that they belong to extinct groups. These unknown animals are classified as *incertae sedis* ("uncertain placement"), and generally constitute organisms that have not been known previously to science. Consequently, hundreds of new invertebrate species have been described from the Sosio Limestones [U0926]; new species continue to be discovered. These include planktonic forms, such as radiolarians and foraminifera (like *Colaniella* and the wheat grain–shaped fusulinids, e.g., *Neoschwagerina)*, but also microscopic crustaceans, bryozoans, and parts (sclerites and thorns) of sea slugs (holothurians). Larger previously unknown invertebrate species are assigned to sponges (*Steinmannia*, *Corynella,* and *Sollasia)*, crinoids (*Palermocrinus*, *Heterobrachiocrinus,* and *Plocostoma*), and brachiopods, as well as corals, gastropods, bivalves, nautiloid and ammonoid cephalopods, trilobites, crustaceans, echinoids, and blastoids (Fig. 9.16). In fact, all the known invertebrate groups have representatives in the bizarre animal assemblages preserved in the Sosio Limestones. Vertebrates are represented by conodonts, and plants include calcareous algae. Recently (Sanfilippo et al. 2016, 2018), five genera of sabellid and serpulid tube worms [U0926] were discovered (*Paleotubus*, *Filograna*, *Serpula*, *Propomatoceros*, and *Pyrgopolon*; Sanfilippo et al. 2017). These worms are uncommon in many Paleozoic and Mesozoic marine assemblages before the Late Triassic. Their presence in this unique, end-Permian fossil assemblage from Sicily demonstrates that tube worms existed before the "Great Dying" and survived the extinction event. Hence, the taxonomic richness of the Sosio Limestones continues to be unique in the world. These fossils inform us about a little-known part of the Tethys Ocean, a kind of metazoan Eden of marine life in a very remote era, the Paleozoic [U0927].

9.9   Discussion

Massimo Bernardi and Guido Roghi

Our imaginary field trip backward in time is not based on our imagination. What we have seen is recorded in the rich terrestrial and marine records along the Tethyan coasts and in the Tethys Ocean. However, visiting just a small number of localities, each of which is of a different age, makes it difficult to understand how dramatic the events were that shaped the Mesozoic biosphere. Indeed, the dramatic events that altered Earth systems can only be recognized and detailed due to the collaboration of hundreds of scientists and the integration of vast amounts of data. Looking "forward" from the Permian to the Jurassic now allows us to witness the calamities in Earth history that altered life on our planet.

Shortly after the deposition of the Sosio Limestones, which we visited in Sicily, the Tethys biota experienced the End-Permian Mass Extinction (EPME). This extinction event is the greatest biodiversity crisis in the history of our planet. Very few shallow-marine organisms of late Permian ecosystems survived this disastrous event about 252 million years ago. The brachiopods, coral groups, and cephalopods that dominated Paleozoic oceans were nearly extinguished; and as we have seen in the Dolomites (Sect. 9.5), recovery of marine life during the Early and Middle Triassic resulted in the evolution and diversification of mollusks, crinoids, and new reef-building invertebrate animals (Erwin 2006). Most importantly, calcareous algae and bryozoans that were builders of late Paleozoic reef communities were replaced by a new group of corals, the scleractinian corals. This group continues to dominate modern reefs (Stanley Jr 2006). On land and in the oceans, as seen at Monte San Giorgio (Sect. 9.3), there was a Middle Triassic radiation of diapsid reptiles (Benton 2015). Marine placodonts, nothosaurs, and ichthyosaurs occupied different niches, feeding on nearly every invertebrate and vertebrate group regardless of their own growth strategy. And the presence at this site of the terrestrial archosaur and close relative of the dinosaur clade, *Ticinosuchus*, signals the prelude to the arrival of true dinosaurs sometime in the Carnian.

One of the most recent advances in the understanding of the Mesozoic tetrapod evolution is the discovery of the interdependence between the Carnian Pluvial Event and the diversification of dinosaurs, or what some workers call the Dinosaur Diversification Event (DDE). Dinosaurs (see Tschopp et al., Chaps. 7 and 8) probably originated in the late Early or early Middle Triassic (248–245 Ma; Brusatte

Fig. 9.16 Paleoenvironmental reconstruction of the shallow-reef bottoms from the middle-to-upper Permian European Paleo-Tethys. Marine life includes dominant sponges, crinoids, brachiopods, corals, mollusks, and polychaetes. All taxa known from the area, to date, are figured: (**a**) *Glomerula* spp., (**b**) *Filograna* sp., (**c**) *Propomatoceros permianus*, (**d**) '*Serpula*' *distefanoi*, (**e**) *Pyrgopolon gaiae*, (**f**) *Paleotubus sosiensis*, and (**g**, **h**, **i**) three recently described species of Serpulidae. (Ideation: R Sanfilippo, A Rosso, and A Reitano; graphic design: Giuseppe Navarria)

et al. 2011; Nesbitt et al. 2012), during the aftermath of the devastating EPME (see Delfino et al., Chap. 10). Recently, Bernardi et al. (2018) showed that dinosaurs radiated in the early late Carnian, at or slightly after the CPE. These authors found evidence for a synchronous burst in dinosaur abundance across Pangea, based on vertebrate assemblages not only from the Dolomites (see Sect. 9.6) but also from Argentina (Ischigualasto Basin), Brazil (Santa Maria Formation), and central Europe (Central European Basin). Benton et al. (2018) applied a statistical analysis to the same data set and demonstrated a major macroecological shift of tetrapod faunas at this time. These authors suggested that the event was one of the most important in the history of life. It not only witnessed the dawn of the "age of dinosaurs", but also boosted the diversification of most key terrestrial clades that now form the modern fauna, including lizards, turtles, crocodiles, and mammals. Hence, the CPE profoundly reshaped the history of life.

The Late Triassic climate along the Tethys coast was seasonal with abundant precipitation, whereas interior parts of the supercontinent experienced considerably lower precipitation rates and more semi-arid conditions (Stefani et al. 2010). This long and warm latest Triassic interval was interrupted by several pulses where a humid climate prevailed. The CPE, in particular, has been characterized as a series of five episodes of worldwide pluvial conditions over a 2-million-year interval (234–232 Ma). These fluctuations in the hydrological cycle affected both the terrestrial and marine biosphere. Changes in vegetation triggered tetrapod diversification across the continents. A modern conifer-dominated landscape resulted in the occurrence of a new preservational medium, resin (amber), produced by conifers and other gymnosperms. Conifers characteristic of Triassic landscapes are also found dominant in Early Jurassic plant communities. Their presence has been used to infer another long-lasting "pluvial" interval from 191 to 183 million years ago. It is around this time that dinosaurs became abundant, as evidenced by skeletal remains and abundant trackways (see Sect. 9.2). In fact, around 201 million years ago, the end-Triassic mass extinction event occurred, and it was during this crisis when the main ecological competitors of dinosaurs, a group of archosaur reptiles, the crurotarsans, went extinct. Brusatte et al. (2008) demonstrated that morphological diversity (disparity) in the dinosaur clade rapidly increased across the Triassic–Jurassic boundary interval and that dinosaurs only became dominant vertebrates after the end-Triassic event. The end-Triassic crisis also affected the oceans, where marine sediments record profound changes and perturbations in the carbon cycle. These perturbations severely impacted marine ecosystems.

9.10 Conclusions

Our virtual visit, beginning along the Tethys margin in the Jurassic (190 Ma) and ending back in the Permian (260 Ma), has allowed us to glimpse postcards of ecosystems and the organisms that inhabited them when Earth, in many of its facets, was very different than at present. Along the easternmost part of Pangea, environments ranged from coastal plains and shallow waters of epicontinental seas to carbonate platforms and reefs, down to deep marine basins. This physical aspect of our journey is not much different than the organization of our current planet. Most of our time involved travel in tropical seas, around numerous islands similar to the present-day Bahamas. What is different is the extent to which these oceans surrounded a single supercontinent. Tethys, where much of the record presented in the chapter originates, was a central gulf and a part of the Panthalassic Ocean that surrounded the eastern part of the supercontinent of Pangea. Complex shorelines separated by carbonate platforms or emerged lands marked the ocean's westernmost extension. In this paleogeographic context, biological systems experienced the largest and most severe extinction of both marine and terrestrial faunas in history, the End-Permian Mass Extinction, about 252 million years ago. After a recovery period and diversification of several terrestrial and marine groups, the Carnian Pluvial Episode (CPE, c. 232 Ma) marked another major biological turnover event. And, again, that turnover event is recorded both in the oceans and on land after this Middle to Upper Triassic crisis.

We have experienced the change in the planet's biosphere as a consequence of the CPE. Middle Triassic marine vertebrate assemblages are exceptionally well documented in a few Fossil-Lagerstätten. Several of these now can be found on the southern slope of the European Alps, such as at Monte San Giorgio, and provide outstanding evidence for the evolution of several marine lineages. Sites in which terrestrial organisms are found are less common in Europe. But the Grès à Voltzia, another locality in which exceptional fossils exist, shows us that conifer forests supported a diversified arthropod fauna. The Heiligkreutz Formation, with its Triassic amber drops, documents the presence of microorganisms most often escaping from the fossil record. And the Calcari Grigi deposits provide insight into the Lower Jurassic floras that repopulated land after the CPE. Remnants of the vertebrates that roamed through and used these plants as a resource across time are often found as traces.

The most continuous record of terrestrial vertebrates is provided by footprints preserved in several sites of late Permian to Early Jurassic age. The extensive carbonate platforms of the Late Triassic, cyclically submerged and then exposed, have recorded migrations of terrestrial reptiles across mudflats and other coastal environments. Their preservation in the form of footprints demonstrates a rich record of their evolution and lifestyle. These footprints testify to the first burst of dinosaur evolution in the Upper Triassic of Argentina, where the oldest skeletal remains of dinosaurs are also found. However, soon after the appearance of dinosaur footprints in South America, trackways became common in Europe, reflecting an additional step towards the global conquest of the land by dinosaurs. During the Early Jurassic, the opening and widening of the Central Atlantic Ocean triggered the development of shallow-water platforms separated by deep basins to the west of Tethys. These areas correspond to the southern Alps, today. Some Jurassic Fossil-Lagerstätte, such as Holzmaden, show that life was teeming in Tethys, and a prevailing humid climate promoted the growth of terrestrial forests in which vertebrate clades diversified.

For Deeper Learning

- Calcari Grigi paleontological content: Bosellini and Broglio Loriga (1971), Clari (1975), Avanzini et al. (2006), De Zigno (1856, 1885), Wesley (1956, 1958).
- Carbonate platforms, middle Triassic start: Gianolla et al. (1998), Stefani et al. (2010).
- Dinosaur-dominated Early Jurassic tetrapod track assemblage: Lucas (1998), Lucas and Huber (2003), Lucas et al. (2005a, b).
- Permian volcanic package: Avanzini et al. (2013), Morelli et al. (2007).
- Sosio Limestone, Sicily, late Permian fossils: Kozur and Mostler (1989), Flügel et al. (1991), Ernst (2000), Jenny-Deshusses et al. (2000), Jones et al. (2015).
- Subsidence and sedimentation on Jurassic passive continental margins: Winterer and Bosellini (1981)

Questions

1. How can you account for the presence of terrestrial plant fossils preserved along with marine invertebrates in the Trento Platform?
2. Perturbation to the hydrological cycle occurred during the Carnian Pluvial Extinction (CPE). What mechanisms and causes are proposed to explain this event?
3. What are several possible reasons for the apparent absence of Triassic plesiosaurs in marine rocks?
4. Why is the succession of limestone in the Dolomites of Italy important to our understanding of marine life?

5. Which animal groups dominated the marine ecosystems during the Triassic?
6. What is amber, how does it form, and what relationship might its appearance have with the CPE?
7. Provide an explanation as to how dinosaur footprints and trackways might form and be preserved in coastal tidal deposits?
8. Explain how the ichnological record of trackways fills in gaps in the fossil record of skeletal-bearing stratigraphies.
9. What makes the fossil assemblages preserved in the Sosio Limestone megablocks unique?
10. What is the Dinosaur Diversification Event (DDE) and what evidence do we have in support of this hypothesis?

References

Abel O (1926) Der erste Fund einer Tetrapodenfährte in der unteren alpinen Trias. Paläontol Z 7:22–23

Avanzini M, Pibelli D, Mietto P, Roghi G, Romano R, Masetti D (2006) Lower Jurassic (Hettangian-Sinemurian) dinosaur track megasites, southern Alps, Northern Italy. In: Harris JD, Lucas SG, Spielmann JA, Lockley MG, Milner ARC, Kirkland JI (eds) The Triassic – Jurassic terrestrial transition. New Mexico Museum of Natural History and Science Bulletin, vol 37. The University of New Mexico, Albuquerque, pp 207–216

Avanzini M, Bargossi GM, Borsato A, Selli L (2013) Note Illustrative – Foglio 060 Trento, Carta Geologica d'Italia alla Scala 1:50000. ISPRA, Roma, p 244

Benson RB, Druckenmiller PS (2014) Faunal turnover of marine tetrapods during the Jurassic–Cretaceous transition. Biol Rev 89(1):1–23

Benson RB, Butler RJ, Lindgren J, Smith AS (2010) Mesozoic marine tetrapod diversity: mass extinctions and temporal heterogeneity in geological megabiases affecting vertebrates. Proc R Soc Lond B Biol Sci 277(1683):829–834

Benson RB, Evans M, Druckenmiller PS (2012) High diversity, low disparity and small body size in *Plesiosaurs* (Reptilia, Sauropterygia) from the Triassic–Jurassic boundary. PLoS One 7(3):e31838. https://doi.org/10.1371/journal.pone.0031838

Benton MJ (2015) When life nearly died, 2nd edn. Thames & Hudson, London

Benton MJ, Bernardi M, Kinsella C (2018) The Carnian Pluvial Episode and the origin of dinosaurs. J Geol Soc 175:1019–1026

Bernardi M, Gianolla P, Petti FM, Mietto P, Benton MJ (2018) Dinosaur diversification linked with the Carnian Pluvial Episode. Nat Commun 9:1499

Bond DPG, Grasby ES (2017) On the causes of mass extinctions. Palaeogeo, Palaeoclima, Palaeoecol 478:3–29

Bosellini A, Broglio Loriga C (1971) I "Calcari Grigi" di Rotzo (Giurassico Inferiore, Altopiano di Asiago) e loro inquadramento nella paleogeografia e nella evoluzione tettonico-sedimentaria delle Prealpi venete. Annali dell'Università di Ferrara (Sezione Scienze Geologiche e Paleontologiche) 5:1–61

Bosellini A, Hardie LA (1988) Facies e cicli della Dolomia Principale delle Alpi Venete. Mem Soc Geol Ital 30:245–266

Bosellini A, Masetti D, Sarti M (1981) A Jurassic "Tongue of the Ocean" infilled with oolitic sands: the Belluno Trough, Venetian Alps, Italy. Mar Geol 44:59–95

Brusatte SL, Benton MJ, Ruta M, Lloyd GT (2008) Superiority, competition, & opportunism in the evolutionary radiation of dinosaurs. Science 321:1485–1488

Brusatte SL, Niedzwiedzki G, Butler RJ (2011) Footprints pull origin and diversification of dinosaur stem lineage deep into early Triassic. Proc Biol Sci. 278:1107–1113

Catalano R, Di Stefano P, Kozur H (1991) Permian circumpacific deep-water faunas from the western Tethys (Sicily, Italy), new evidences for the position of the Permian Tethys. Palaeogeogr Palaeoclimatol Palaeoecol 87:75–108

Clari P (1975) Caratteristiche sedimentologiche e paleontologiche di alcune sezioni dei Calcari Grigi del Veneto. Memorie degli Istituti di Geologia e Mineralogia dell'Università di Padova 31:1–63

Cohen KM, Finney SC, Gibbard PL, Fan J-X (2015) The ICS International Chronostratigraphic Chart. Episodes 36:199–204

Dal Corso J, Mietto P, Newton RJ, Pancost RD, Preto N, Roghi G, Wignall PB (2012) Discovery of a major negative δ13C spike in the Carnian (late Triassic) linked to the eruption of Wrangellia flood basalts. Geology 40:79–82

Dal Corso J, Gianolla P, Newton RJ, Franceschi M, Roghi G, Caggiati M, Preto N (2015) Carbon isotope records reveal synchronicity between carbon cycle perturbation and the 'Carnian pluvial event' in the Tethys realm (late Triassic). Glob Planet Chang 127:79–90

De Zigno A (1856) Flora fossilis formationis oolithicae, vol 1. Tip. Del seminario, Padova

De Zigno A (1885) Flora fossilis formationis oolithicae. Vol. Tip. Del seminario, Padova, p 2

Di Stefano P, Gullo M (1997) Permian deposits of sicily. A review. Geodiversitas 19:193–202

Dìaz-Martinez I, Castanera D, Gasca JM, Canudo JI (2015) A reappraisal of the Middle Triassic chirotheriid Chirotherium ibericus Navas, 1906 (Iberian Range, NE Spain), with comments on the Triassic tetrapod track biochronology of the Iberian Peninsula. PeerJ 3:1044. https://doi.org/10.7717/peerj.1044

Diedrich C (2015) *Isochirotherium* trackways, their possible trackmakers (?*Arizonasaurus*) intercontinental giant archosaur migrations in the Middle Triassic tsunami-influenced carbonate intertidal mud flats of the European Germanic Basin. Carbonates Evaporites 30:229–252

Ernst AK (2000) Permian Bryozoans of the NW-Tethys. Facies 43: 79–102

Erwin DH (2006) Extinction: how life on earth nearly ended 250 million years ago. Princeton University Press, Princeton

Fabbri M, Dalla Vecchia FM, Cau A (2014) New information on Bobosaurus forojuliensis (Reptilia: Sauropterygia) implications for plesiosaurian evolution. Hist Biol 26(5):661–669

Felber M, Furrer H, Tintori A (2004) The Triassic of Monte San Giorgio in the World Heritage list of UNESCO: an opportunity for science, the local people and tourism. Eclogae Geol Helv 97:1–2

Flügel E, Di Stefano P, Senowbari-Daryan B (1991) Microfacies and depositional structure of allochthonous carbonate base-of-slope deposits: the late Permian Pietra di Salomone megablock, Sosio Valley (western Sicily). Facies 25:147–186

Franceschi M, Dal Corso J, Posenato R, Roghi G, Masetti D, Jenkyns HC (2014) Early Pliensbachian (Early Jurassic) C-isotope perturbation and the diffusion of the Lithiotis fauna: insights from the western Tethys. Palaeogeogr Palaeoclimatol Palaeoecol 410:255–263

Furin S, Preto N, Rigo M, Roghi G, Gianolla P, Crowley JL, Bowring SA (2006) High- precision U–Pb zircon age from the Triassic of Italy: implications for the Triassic time scale and the Carnian origin of calcareous nannoplankton and dinosaurs. Geology 34:1009–1012

Gall JC (1971) Faunes et paysages du Grès à Voltzia du Nord des Vosges. Essai paléoécologique sur le Buntsandstein supérieur Mémoire du Service de la Carte Géologique d'Alsace et de Lorraine 34, Bibliothèque de l'Institut de géologie de l'Université de Strasbourg, Paris

Gall JC, Grauvogel-Stamm L (2005) The early Middle Triassic Gres a Voltzia formation of eastern France: a model of environmental refugium. Comptes Rendus Palevol 4(6–7):637–652

Gand G (1979) Description de deux nouvelles traces d'*Isochirotherium* observées dans le grès du Trias moyen de Bourgogne. Bulletin de la Société d'Histoire Naturelle du Creusot 37:13–25

Gemmellaro GG (1887-88-90) La Fauna dei calcari con Fusulina della valle del fiume Sosio nella Provincia di Palermo, Fasc. 1–2. Cephalopoda, Ammonoidea, Nautioidea, Gastropoda. Giornale di Scienze Naturali ed Economiche di Palermo 19:1–106; 20: 1–26, 97–182

Gianolla P, De Zanche V, Mietto P (1998) Triassic sequence stratigraphy in the Southern Alps (Northern Italy) definition of sequences and basin evolution. In: de Graciansky PC, Hardenbol J, Jacquin T, Vail PR (eds) Mesozoic-Cenozoic sequence stratigraphy of European Basins, vol 60. SEPM Special Publication, Tulsa, pp 723–751

Grauvogel-Stamm L (1978) La flore du Grès à Voltzia (Buntsandstein supérieur) des Vosges du Nord (France). Morphologie, anatomie, interprétations phylogénétique et paléogéographique. Sciences Géologiques, bulletins et mémoires Strasbourg 50

Jenny-Deshusses C, Marrini R, Zaninerri L (2000) Découverte du foraminifère *Colaniella* Likharev dans le Permien supérieur de la vallée du Sosio (Sicile). Comtes Rendus de l'Académie des Sciences 330:799–804

Jones WT, Feldmann RM, Schweitzer CE, Reitano A, Insacco G (2015) A new pygocephalomorph (Peracarida) from the Permian of the Sosio Valley (Sicily Italy). J Crustac Biol 35:627–632

King MJ, Sarjeant WAS, Thompson DB, Tresise G (2005) A revised systematic ichnotaxonomy and review of the vertebrate footprint ichnofamily Chirotheriidae from the British Triassic. Ichnos 12:241–299

Klein H, Haubold H (2007) Archosaur footprints – potential for biochronology of Triassic continental sequences. NM Mus Nat Hist Sci Bull 41:120–130

Klein H, Lucas SG (2010) Review of the tetrapod ichnofauna of the Moenkopi Formation/Group (Early-Middle Triassic) of the American Southwest. NM Mus Nat Hist Sci Bull 50:1–67

Klein H, Neidzwiedzki G (2012) Revision of the Lower Triassic Tetrapod Ichnofauna from Wiory, Holy Cross Mountains, Poland. NM Mus Nat Hist Sci Bull 56:1–62

Klein H, Voigt S, Saber H, Schneider JW, Hminna A, Fischer J et al (2011) First occurrence of a Middle Triassic tetrapod ichnofauna from the Argana Basin (Western High Atlas, Morocco). Palaeogeogr Palaeoclimatol Palaeoecol 307:218–231

Kozur H, Mostler H (1989) Echinoderm remains from the Middle Permian (Wordian) from Sosio Valley (Western Sicily). Neues Jahrbuch fur Geologie und Paleontologie (Abhandlungen) 132:677–685

Leonardi P (1968) Le Dolomiti, Geologia dei Monti tra Isarco e Piave. Rovereto, Edizioni Manfrini. 1019 pp

Lombardo C. Tintori A. Tona D. (2012) A new species of Sangiorgioichthys (Actinopterygii, Semionotiformes) from the Kalkschieferzone of Monte San Giorgio (Middle Triassic; Meride, Canton Ticino, Switzerland). Bollettino della Societa Paleontologica Italiana 51(3):203–212

Lucas SG (1998) Global Triassic tetrapod biostratigraphy and biochronology. Palaeogeogr Palaeoclimatol Palaeoecol 143:347–384

Lucas SG, Huber P (2003) Vertebrate biostratigraphy and biochronology of the nonmarine Late Triassic. In: LeTourneau PM, Olsen PE (eds) The great rift valleys of Pangea in eastern North America, v. 2: sedimentology, stratigraphy and paleontology. Columbia University Press, New York, pp 143–191

Lucas SG, Tanner LH, Heckert AB (2005a) Tetrapod biostratigraphy and biochronology across the Triassic-Jurassic boundary in northeastern Arizona. NM Mus Nat Hist Sci Bull 29:84–94

Lucas SG, Heckert AB, Tanner LH (2005b) Arizona's Jurassic fossil vertebrates and the age of the Glen Canyon Group. NM Mus Nat Hist Sci Bull 29:95–104

Mascle GH (1979) Etude geologique des Monts Sicani. Memorie Rivista Italiana di Paleontologia e Stratigrafia 16:1–431

Masetti D, Claps M, Giacometti A, Lodi P, Pignatti P (1998) I Calcari Grigi della Piattaforma di Trento (Lias inferiore e medio, Prealpi Venete). Atti Ticinensi di Scienze della Terra 40:139–183

Massari F, Neri C (1997) The infill of a supradetachment (?) basin: the continental to shallow-marine Upper Permian in the Dolomites and Carnia (Italy). Sediment Geol 110:181–221

Melchor RN, De Valais S (2006) A review of Triassic tetrapod track assemblages from Argentina. Palaeontology 49(2):355–379

Miller CS, Peterse F, da Silva AC, Baranyi V, Reichart GJ, Kürschner W (2017) Astronomical age constraints and extinction mechanisms of the Late Triassic Carnian crisis. Sci Rep 7:2557

Morelli C, Bargossi GM, Mair V (2007) The Lower Permian volcanics along the Etsch Valley from Meran to Auer (Bozen). Mitteilungen der Österreichischen Mineralogischen Gesellschaft 153:195–218

Neri M, Roghi G, Ragazzi E, Papazzoni CA (2016) First record of Pliensbachian (Lower Jurassic) amber and associated palynoflora from the Monti Lessini (northern Italy). Geobios 50:49–63

Neri M, Kustatscher E, Roghi G (2018) Megaspores from the Lower Jurassic (Pliensbachian) Rotzo Formation (Monti Lessini, northern Italy) and their paleoenvironmental implications. Palaeobiodivers Palaeoenviron 98(1):97–110

Nesbitt SJ, Barrett PM, Werning S, Sidor CA, Charig AJ (2012) The oldest dinosaur? A Middle Triassic dinosauriform from Tanzania. Biol Lett 9:20120949

Petti FM, Bernardi M, Kustatscher E, Renesto S, Avanzini M (2013) Diversity of continental tetrapods and plants in the Triassic of the Southern Alps: Ichnological, paleozoological and paleobotanical evidence. NM Mus Nat Hist Sci Bull 61:458–484

Posenato R, Masetti D (2012) Environmental control and dynamics of Lower Jurassic bivalve build-ups in the Trento Platform (Southern Alps, Italy). Palaeogeogr Palaeoclimatol Palaeoecol 361–362:1–13

Preto N, Kustatscher E, Wignall PB (2010) Triassic climates - state of the art and perspectives. Palaeogeo, Palaeoclima, Palaeoeco, 290: 1–10

Richthofen F (1860) Geognostische Beschreibung der Umgebung von Predazzo Sanct Cassian und Seisseralpen in Südtirol. Perthes, Gotha, p 327

Roghi G, Ragazzi E, Gianolla P (2006) Triassic amber of the Southern Alps (Italy). Palaios, 21:143–154

Roghi G, Gianolla P, Minarelli L, Pilati C, Preto N (2010) Palynological correlation of Carnian humid pulses throughout western Tethys. Palaeogeogr Palaeoclimatol Palaeoecol 290:89–106

Russo F, Neri C, Mastandrea A, Baracca A (1997) The mud-mound nature of the Cassian platform margins of the Dolomites. A case history: the Cipit boulders from Punta Grohmann (Sasso Piatto Massif, northern Italy). Facies 36:25–36

Sanfilippo R, Reitano A, Insacco G, Rosso A (2016) A new tubeworm of possible serpulid affinity from the Permian of Sicily. Acta Palaeontol Pol 61(3):621–626

Sanfilippo R, Rosso A, Reitano A, Insacco G (2017) First record of sabellid and serpulid polychaetes from the Permian of Sicily. Acta Palaeontol Pol 62(1):25–38

Sanfilippo R, Rosso A, Reitano A, Insacco G (2018) A further new species of serpulidae polychaete from the Permian of Sicily. Acta Palaeontol Pol 63(3):579–584

Santi G, Lualdi A, Decarlis A, Nicosia U, Ronchi A (2015) Chirotheriid footprints from the Lower-Middle Triassic of the Briançonnais Domain (Pelite di Case Valmarenca, Western Liguria, NW Italy). Bollettino della Società Paleontologica Italiana 54(2):82

Schlager W, Keim L (2009) Carbonate platforms in the Dolomites area of the Southern Alps – historic perspectives on progress in sedimentology. Sedimentology 56:191–204

Schmidt AR, Ragazzi E, Coppellotti O Roghi, G (2006) A microworld in Triassic amber. Nature 444:835

Schmidt AR, Jancke S, Lindquist EE, Ragazzi E, Roghi G, Nascimbene PC, Schmidt K, Wappler T, Grimaldi DA (2012) Arthropods in amber from the Triassic Period. PNAS 109:14796–14801

Scotese CR (2014) Atlas of Middle & Late Permian and Triassic Paleogeographic maps, maps 43-48 from volume 3 of the PALEOMAP Atlas for ArcGIS (Jurassic and Triassic) and maps 49-52 from volume 4 of the PALEOMAP PaleoAtlas for ArcGIS (Late Paleozoic). Mollweide Projection, PALEOMAP Project, Evanston

Selden PA, Nudds JR (2004) Evolution of fossil ecosystems. The University of Chicago Press, Chicago. 160 p

Seyfullah LJ, Roghi G, Dal Corso J, Schmidt A (2018) The Carnian pluvial episode and the first global appearance of amber. J Geol Soc 175:1012–1018

Sidorchuk EA, Schmidt AR, Ragazzi E, Roghi G, Lindquist EE (2014) Plant-feeding mite diversity in Triassic amber (Acari: Tetrapodili). J Syst Palaeontol 13(2):129–151

Simms MJ, Ruffell AH (1989) Synchroneity of climatic change and extinctions in the Late Triassic. Geology 17:265–268

Stanley GD Jr (2006) Photosymbiosis and the evolution of modern coral reefs. Science 312:857–858

Stefani M, Furin S, Gianolla P (2010) The changing climate framework and depositional dynamics of Triassic carbonate platforms from the Dolomites. Palaeogeogr Palaeoclimatol Palaeoecol 290:43–57

Storrs GW, Taylor MA (1996) Cranial anatomy of a new plesiosaur genus from the lowermost Lias (Rhaetian/Hettangian) of Street, Somerset, England. J Vertebr Paleontol 16(3):403–420

Strada L, Montagna M, Tintori A (2014) A new genus and species of the family Trachypachidae (Coleoptera, Adephaga) from the upper Ladinian (Middle Triassic) of Monte San Giorgio. Riv Ital Paleontol Stratigr 120:183–190

Sun YD, Joachimski MM, Wignall PB, Yan CB, Chen YL, Jiang HS, Wang LN, Lai XL (2012) Lethally hot temperatures during the early Triassic greenhouse. Science 338:366–370

Wesley A (1956) Contribution to the knowledge of the flora of the Grey limestone of Veneto. Part I. Memorie degli Istituti di Geologia e Mineralogia dell'Università di Padova 19:1–69

Wesley A (1958) Contribution to the knowledge of the flora of the Grey Limestone of Veneto, Part. II. Memorie degli Istituti di Geologia e Mineralogia dell'Università di Padova 21:1–57

Winterer EL, Bosellini A (1981) Subsidence and sedimentation on Jurassic Passive Continental Margin, Southern Alps Italy. AAPG Bull 65:394–421

Xing LD, Klein H, Lockley MG, Li J, Zhang J, Matsukawa M, Xiao J (2013) *Chirotherium* trackways from the Middle Triassic of Guizhou, China. Ichnos 20:99–107

Zempolich WG (1993) The drowning succession in Jurassic carbonates of the Venetian Alps, Italy: a record of supercontinent breakup, gradual eustatic rise, and eutrophication of shallow-water environments. In: Loucks RG, Sarg JF (eds) Carbonate sequence stratigraphy – recent developments and applications, vol 57. AAPG Memories, Tulsa, pp 63–105

The End-Permian Mass Extinction: Nature's Revolution

10

Massimo Delfino, Evelyn Kustatscher, Fabrizio Lavezzi, and Massimo Bernardi

Abstract

Several localities around the world expose successions of rocks that straddle the Permian–Triassic boundary documenting a common pattern of environmental change. This change testifies to a large-scale event that led to the extinction of a significant portion of biodiversity, the most severe mass extinction of all times. This event is called the End-Permian Mass Extinction (EPME) or the Permian–Triassic Mass Extinction and was likely triggered by extensive volcanism. It not only affected the biodiversity of the marine realm, but also the terrestrial environments where faunas showed a marked reduction in diversity, whereas evidence for a mass extinction among plants is less robust. Even if the synchronicity of the extinctions in terrestrial and marine environments is still controversial, it seems clear that the event itself was rather fast, and that it took several million years for life to recover completely from the crisis. In fact, it seems likely that massive volcanic eruptions not only caused a chain of reactions that led to the extinction but also hindered the recovery of most of the surviving taxa. The EPME changed life forever, and the following recovery saw the evolution and radiation of many modern taxa that still characterize our planet today.

10.1 Introduction

The End-Permian Mass Extinction (EPME) (also known as Permian–Triassic Mass Extinction, PTME) is one of the most studied geobiological events of the past. It is the most severe mass extinction of all life—"the mother of all extinctions"—and promoted the evolution of modern ecosystems (e.g., Raup and Sepkoski 1982; Erwin 1993, 2006). Hundreds of scientific papers have been written on the topic (see Fig. 1 in Twitchett 2006), and the conclusions of these studies have been summarized in popular books with apocalyptic titles like "When life nearly ended: The greatest mass extinction of all time" (Benton 2005) and "Extinction: How life on Earth nearly ended 250 million years ago" (Erwin 2006). Popular books and TV documentaries are a clear sign that this topic is not only the "intellectual exercise" for a large group of scientists, but that it conveys so many deep implications about our world today that it has the potential of attracting the general public. This is especially true in a time of global warming and, hopefully, increasing environmental awareness.

Electronic supplementary material A slide presentation and an explanation of each slide's content is freely available to everyone upon request via email to one of the editors: edoardo.martinetto@unito.it, ragastal@colby.edu, tschopp.e@gmail.com

*The asterisk designates terms explained in the Glossary.

M. Delfino (✉)
Department of Earth Sciences, University of Turin, Turin, Italy

Institut Català de Paleontologia Miquel Crusafont, Universitat Autònoma de Barcelona, Barcelona, Spain
e-mail: massimo.delfino@unito.it

E. Kustatscher
Museum of Nature South Tyrol, Bozen/Bolzano, Italy

Department für Geo- und Umweltwissenschaften, Paläontologie und Geobiologie, Ludwig-Maximilians-Universität, Munich, Germany

Bayerische Staatssammlung für Paläontologie und Geobiologie, Munich, Germany
e-mail: evelyn.kustatscher@naturmuseum.it

F. Lavezzi
Department of Earth Sciences, University of Turin, Turin, Italy
e-mail: fabriziolavezzi@yahoo.it

M. Bernardi
MUSE–Museo delle Scienze, Trento, Italy

School of Earth Sciences, University of Bristol, Bristol, UK
e-mail: massimo.bernardi@muse.it

© Springer Nature Switzerland AG 2020
E. Martinetto et al. (eds.), *Nature through Time*, Springer Textbooks in Earth Sciences, Geography and Environment,
https://doi.org/10.1007/978-3-030-35058-1_10

The vast amount of data available for the EPME and the numerous congruent hypotheses developed in the last decades to explain its dynamics provide us today with an excellent model to describe what happened when a global ecosystem crisis pushed life close to complete annihilation, about 252 million years ago (Chen and Benton 2012). In an excellent overview, still broadly valid even if written about 15 years ago (Benton and Twitchett 2003), three aspects were highlighted as key for understanding the EPME: the scale and pattern of species loss, what combination of environmental changes could possibly have had such a devastating effect, and the pattern of the recovery. Here, we will provide an updated account of such key event for the evolution of life as currently known.

10.2 The Magnitude of Species Loss

The rate and the pattern of species loss is one of the main questions about this event. This is not an easy question to answer actually, because it is time-dependent and, therefore,

needs precise dating of the rocks and fossil associations preserved therein. According to the most recent version of the International Chronostratigraphic Chart (v2019/5; Cohen et al. 2018), the boundary between the late Permian and the Early Triassic is set at a CA-ID-TIMS Uranium-Lead derived age of 251.902 ± 0.024 million years ago. Such CA-ID-TIMS analyses of a population of single-crystal zircons from a horizon or bed provide the highest resolution age estimate for its emplacement. The age of the end-Permian event has been determined in China, at the reference locality for the Permian–Triassic Boundary (PTB). The locality is called Meishan and is placed in Changxing County, Zhejiang Province (Fig. 10.1). It is one of several Global Stratotype Sections and Points (GSSP), where a "golden spike" was literally fixed into the sedimentary succession to mark the exact position of the boundary. In Meishan, each of the limestone and mudstone beds in the succession has been given a number in ascending stratigraphic order. In this succession, the PTB is placed at the base of Bed 27c, which records a deep marine environment from where the first occurrence of the conodont species *Hindeodus parvus* is documented (Yin et al. 2001). However,

Fig. 10.1 The Global Stratotype Section and Point (GSSP) of the Permian–Triassic boundary is located at Meishan (Changxing County, Zhejiang Province) where a monumental garden has been created to celebrate the relevance of the outcrop in Earth history. Photo by Robert A Gastaldo

the extinction is not a one-time or one-level event. It has a sort of a prelude at the base of Bed 24e, a main episode corresponding to Bed 25, and an epilogue at Bed 28 (Yin et al. 2007a). Thanks to the extremely precise dating available in these rocks, we now know that the extinction peak occurred about 40,000 years before the horizon marking the PTB (Burgess et al. 2014; Burgess and Bowring 2015). The whole event was extremely fast in geological terms, with a duration of about 60,000 years (e.g., Brand et al. 2012; Burgess et al. 2014; Shen et al. 2018). Data from other sections, as in the Dolomites of northeast Italy, indicate that the event could have been even more rapid, with a duration of a few millennia (see Brand et al. 2012; Posenato 2019). According to the interpretations of various results of different studies, the extinction events in continental ecosystems occurred over a time span as short as 10,000 years or across a much longer interval of up to 200,000 years (Smith and Botha-Brink 2014, and literature therein). In any case, the rock record clearly indicates that sedimentary strata, in which marine or terrestrial organisms are preserved, are rich in fossils. These fossil assemblages testify to the presence of well-structured communities for most of the late Permian (Lopingian) (Erwin 1993; Bernardi et al. 2017) when the action of marine communities that lived on and in the sediment is registered by intense bioturbation of the ocean floor, indicating oxic conditions (Benton and Twitchett 2003). However, close to the end of the Permian, the oceans around the world record sediments that reflect dysoxic∗ or anoxic∗ conditions. These resulted in the production of non-bioturbated, laminated∗, dark, and pyrite-rich sediments that are scarcely fossiliferous in the marine realm (Wignall and Twitchett 2002; Benton and Twitchett 2003). Degraded terrestrial environments from the same time interval testify to erosion following the reported loss of vegetation (Benton and Newell 2014). Such rocks indicate that, before the PTB, the environmental conditions began to deteriorate, resulting in increasing stress in the biotic communities. Was the end-Permian event limited to the marine realm or did it affect also continental organisms?

According to some research (e.g., Smith and Botha-Brink 2014; Cascales-Miñana et al. 2016), the EPME is the only one of the five major extinction events in the marine realm that "coincided" with a clear and abrupt decrease in continental biodiversity. It is also the only mass extinction thought to have had devastating effects on vegetation at a global scale (but see Nowak et al. 2019). In fact, peat-accumulating wetlands (see Gastaldo et al., Chap. 12) were absent on the landscape for more than ten million years; and, therefore, Early Triassic coals are unknown, except in western China (Thomas et al. 2011), and Middle Triassic coals are rare and thin. Their absence has been termed the "coal gap" by Retallack et al. (1996). But is the concept of coincident diversity loss in the oceans and on land accurate?

The terrestrial vertebrate-fossil record in the Karoo Basin, South Africa (Fig. 10.2), has acted as the cornerstone on which the end-Permian extinction event is recognized on land. Having been considered the standard for understanding the end-Permian turnover in terrestrial ecosystems, any new data that can be used to constrain "time" on land in the Karoo Basin helps to tie together global events. Smith and Botha-Brink (Smith and Botha-Brink 2014; but see also Gastaldo and Neveling 2016) proposed that there were three phases of drought-induced extinctions lasting, respectively (from the oldest), 21,000, 33,000, and 8000 years, separated by periods of 7000 and 50,000 years, over a timeframe equivalent, and coincident with, the marine event (see Fig. 12 in Smith and Botha-Brink 2014). The model has been extrapolated to other parts of the globe (e.g., Benton and Newell 2014) where this polyphased extinction resulted in the decimation of most vertebrate taxa. In the Karoo Basin, the most recognizable vertebrates that went extinct are probably the medium-sized dicynodont therapsid *Dicynodon* and the large gorgonopsid therapsid *Rubidgea* (see also, among others, Retallak et al. 2003; Ward et al. 2005). But, the term "coincident" loss may not be appropriate because, as remarked by Padian (2018) and Fielding et al. (2019), there is growing evidence for a series of events on land that are not synchronous with those that occurred in the oceans. For example, Gastaldo et al. (2015, p. 939) presented the first high precision age on strata close to the inferred PTB in the Karoo Basin, South Africa. These authors concluded that the turnover in vertebrate taxa at this biozone boundary is older than the marine extinction event and "probably does not represent the biological expression of the terrestrial end-Permian mass extinction". Additionally, a more recent analysis in the Karoo Basin likewise concluded that the turnover from the *Daptocephalus* (formerly *Dicynodon*) to *Lystrosaurus* Assemblage Zones (Fig. 10.3), once used to define the PTB on land, is not coincident with the end-Permian marine event (Gastaldo et al. 2018, 2019a). Recently, Gastaldo et al. (2020) published a CA-ID-TIMS Uranium-Lead age date from a pristine ash-fall bed in the base of the *Lystrosaurus* Assemblage Zone that is more than 300,000 years prior to the marine crisis. Because of these and other reasons, the expression "the terrestrial Permian–Triassic boundary event [bed] is a nonevent" recently appeared in the literature (Gastaldo et al. 2009; Gastaldo and Neveling 2012, 2016; Ward et al. 2012).

Due to scarce and impoverished fossil megafloras from the earliest Triassic on the different continents, the common perception has been that land plants suffered a mass extinction like the terrestrial and marine animals, but doubts always remained (Nowak et al. 2019). Early Triassic megafloras are, indeed, often markedly impoverished and dominated by opportunistic taxa such as the iconic lycophyte *Pleuromeia* (Box 10.1; Looy et al. 1999; Grauvogel-Stamm and Ash 2005; McElwain and Punyasena 2007), although some megafloras were more diverse and gymnosperm-dominated (Feng et al. 2018). This apparent loss in species is probably based on a strong taphonomic bias (Gastaldo et al.

Fig. 10.2 The Karoo Basin, South Africa, has provided a key fossil record for evaluating the impact of the EPME on terrestrial vertebrate communities. This photo by Robert A. Gastaldo shows the Bethel Farm in Bethulie Valley

Fig. 10.3 Several medium-sized dicynodontian therapsids went extinct in the Karoo Basin at the end of the Permian. In this artistic interpretation by Fabrizio Lavezzi, the body of *Dicynodon* is covered with hair as possibly indicated by bone histology and tentatively suggested in a recent study of Permian coprolites (Bajdek et al. 2016)

2005) and/or on the fact that the macrofossil record (especially the gymnosperm one) is considerably undersampled for the Early Triassic, giving the impression of an increased gymnosperm extinction during the latest Permian (Nowak

et al. 2019). The recent discovery of nearly all major post-extinction plant groups, including bona fides Cycadales, Corystospermales, Bennettitales, Czekanowskiales, Podocarpales, and Araucariaceae in the Permian (e.g. DiMichele et al. 2001; Abu Hamad et al. 2008; Blomenkemper et al. 2018; Kustatscher et al. 2019) and the presence of "mixed" floras (e.g. *Glossopteris-*

Box 10.1: *Pleuromeia sternbergii*

Pleuromeia sternbergii (Fig. 10.4) is the most characteristic plant of the biotic recovery during the Early Triassic. The first megafossil was discovered in 1836 at the Magdeburger cathedral, when weathered sandstone fell down from the tower during repair works, burst, and released an impression of a yet-unknown fossil plant. During the same year, the specimen was brought to Georg Graf zu Münster (1776–1844), also a famous paleobotanist living in the city. He described it in 1839 as *Sigillaria sternbergii* in honor of paleobotanist Kaspar Maria Graf von Sternberg (1761–1844), one of the founders of paleobotany. Later, the species

Fig. 10.4 *Pleuromeia sternbergii* (Münster) Corda is the most characteristic plant megafossil of the biotic recovery during the Early Triassic. (**a**) Fossil from Sollingen, Germany (BSC445, 10 × 14 cm; picture courtesy Léa Grauvogel-Stamm); (**b**) artistic rendering of the live plant by Fabrizio Lavezzi

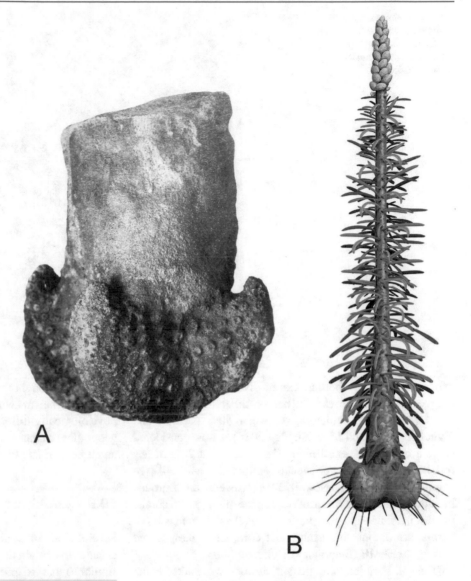

A

B

was transferred by Corda to the new genus *Pleuromeia* (originally spelled *Pleuromeya*).

The individual plants have an undivided stem with a maximum height of two meters and a terminal, heterosporous∗ cone-like structure composed of oval sporophylls∗. The four-lobed plant base (rhizophore) shows a typical sequence of root scars. The structure of the rhizophores and the ultrastructure of the micro- and megaspores∗ indicate a strong relationship of this genus with the extant lycophyte *Isoetes*. Several species of *Pleuromeia* colonized all continents during the Early Triassic. *Pleuromeia* is considered a highly adaptive pioneering plant. It experienced a quick radiation during the earliest Triassic where it formed monotypic stands or was part of low-diversity communities that were subsequently replaced by more complex communities in the Anisian.

Dicroidium floras; Schneebeli-Hermann et al. 2015) indicates that these groups did not "evolve" in response to the EPME, but rather had their major radiation after it.

Although the palynological record (Fig. 10.5) reflects a series of ecological disturbances and climatic changes around the PTB (Hochuli et al. 2010, 2016), the microscopic plant remains show less apparent changes during the extinction event than the macroscopic ones. This is possibly because of the higher preservation potential of pollen and spores (Fig. 10.6; Nowak et al. 2019). Nonetheless, the magnitude of the extinction among plants is still a matter of harsh debate (see Nowak et al. 2019).

The estimate of loss in the marine ecosystem, on the other hand, has been calculated several times. As we have seen earlier (Romano in Chap. 7), "fishes" were not too severely affected. However, estimates vary according to which database is used and the methods applied to analyze the data. Most studies agree that up to 49% of marine animal families went extinct, with foraminifers, bryozoans, and brachiopods

Fig. 10.5 *Kraeuselisporites apiculatus*, a spore from the Permian–Triassic boundary interval of the Finnmark platform (Norway)

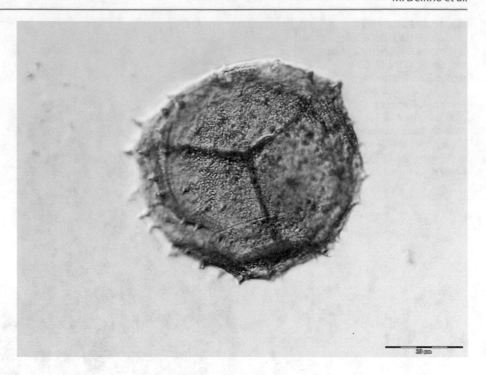

experiencing severe bottlenecks in diversity. Extinctions at the marine-species level have been estimated up to 96% by Raup (1979), but something between 80% (Benton and Twitchett 2003) and 85% (Stanley 2016) is now considered to be a more reliable estimate. On land, about 70% of terrestrial vertebrates experienced extinction, as well (see details in Benton and Newell 2014). Several large groups including fusulinid foraminifera, rugose and tabulate corals, the last trilobites and eurypterids, as well as blastoid echinoderms, are among the groups that completely disappeared during the EPME (Benton and Twitchett 2003).

The end-Permian biological crisis not only resulted in the largest mass extinction of all times but also influenced the biogeographic range and the body size of the lineages that survived on both land and in the oceans (as we have seen in neopterygians; Romano in Chap. 7). Bernardi et al. (2018), for example, found evidence for forced range expansion among tetrapods that moved to higher-latitudinal, cooler regions, in an attempt to escape from the superhot climate that developed in the equatorial belt in the earliest Triassic (see below; Sun et al. 2012). Schaal et al. (2016) recorded the maximum size of 11,224 mostly invertebrate specimens belonging to 2743 species of gastropods, bivalves, calcitic and phosphatic brachiopods, ammonoids, ostracods, conodonts, and foraminiferans, spanning the late Permian through the Middle to Late Triassic. They concluded that the EPME induced more size reduction (the so-called "Lilliput effect"; Urbanek 1993) among species than any other interval in Earth history. The decrease of size in specific lineages not only resulted from the extinction of larger taxa, but also because of the evolution of smaller sizes in the surviving lin-

eages. It is thought that the post-extinction world must have favored organisms with small body sizes in ecologically and physiologically different clades. But what caused such biogeographic changes, size reduction, and the disappearance of about 80% of late Permian biodiversity?

10.3 Looking for the Smoking Gun

Several lines of evidence have been presented in the last decades to explain the EPME. They include a bolide impact, similar to that experienced at the end of the Cretaceous, and extensive volcanism. The role of a bolide impact for the EPME has been thoroughly discussed in several papers (Benton and Twitchett 2003 and references therein). But, remarkably, data from the GSSP at Meishan, as well as other sections in south China and in central and western Tethyan, do not support an instantaneous calamity. The classical impact-related features (such as bolide ejecta or the crater itself) have never been found in any area. The evidence in these localities indicate that the marine biotic crisis started approximately 50,000 years before the possible impact of any bolide (Yin et al. 2007a, b; see also Bottjer et al. 2008). Therefore, as concluded by Twitchett (2006, p. 199), most of the evidence for the bolide impact "is seriously flawed and impact is not favored as a cause for this event" (see also Benton and Twitchett 2003). However, volcanism might have played a major role in the crisis.

The onset of the extensive volcanism that resulted in the emplacement of the Siberian traps (Fig. 10.7), a Large Igneous Province (LIP) in Russia, is now considered as the

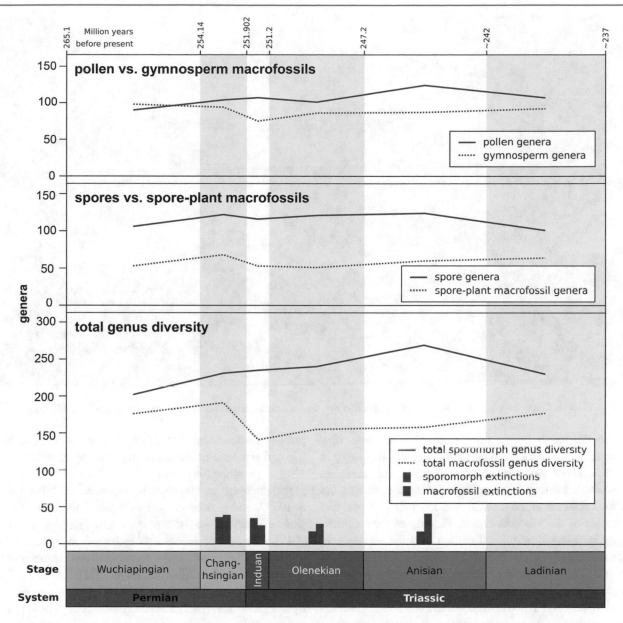

Fig. 10.6 Genus-level diversity curves for land-plant macrofossils and sporomorphs (spores and pollen) from the late Permian to the Middle Triassic (after Nowak et al. 2019). Macrofossil diversity decreases between the late Permian and the Early Triassic, but only less than 20% of the genera go extinct – significantly less than among animals. Sporomorph diversity even increases across the Permian–Triassic boundary and during the Early Triassic, showing no signs of a mass extinction. Interestingly, the diversity trends for spores and macrofossils of spore-producing plants (mosses, club mosses, ferns, horsetails) are quite similar, whereas those for pollen and macrofossils of gymnosperms (the seed plants that produced the pollen) are very different. This is probably because pollen and gymnosperm remains had different limitations affecting transport and preservation. Courtesy Hendrik Nowak

most likely trigger of the EPME (among others, Svensen et al. 2009; Chen and Benton 2012). This prolonged, multiphase volcanic activity, which started in the latest Permian and continued for about five million years through most of the Early Triassic, produced immense volumes of effusive volcanic rocks over an extensive area of what is now Siberia. Estimates of the volume of basalts, tuffs, and intrusive igneous rock injected into the crust are up to four million km³ and activated the cascade of events that produced the so-called "lethal triade": extreme global warming, ocean anoxia, and acidification (e.g., Wignall and Hallam 1992; Saunders and Reichow 2009). The magmatic event was characterized by multiple episodes of CO_2 and CH_4 release. The events leading up to, and following, the extinction crisis are associated with a negative excursion in stable-carbon isotope values, indicating the addition of "light" carbon to the planet. The isotopic signature locked in these rocks of −4 to −6‰ is thought to be the evidence that explains the negative excursion of $\delta^{13}C$ across the PTB (Wignall 2001; Payne and Kump 2007; Bond and Wignall 2014 and literature therein).

Fig. 10.7 Siberian Traps cropping out in the area of Talnakh (the Red Stones locality, Krasnoyarsk Krai, Russia). Photo by Cindy Looy

Because the basalt floods were recurrent over time, they impacted the host rock exposed on the surface and deeper in the crust. Those rocks included both coals and limestone (carbonates). Contact between the igneous magma and the host rock resulted in the thermal metamorphism of coal and carbonates of the Siberian craton. Heating of the sedimentary rocks released additional carbon into the atmosphere, which had repercussions in both terrestrial and marine environments. These multiple episodes of lava flood and carbon-rich gas release lasted into the Triassic and are, at least, partly related to the Early Triassic negative and positive $\delta^{13}C$ excursions recorded in these rocks (Benton and Twitchett 2003; Benton 2005; Erwin 2006 and literature therein). However, the magnitude of the negative excursions of $\delta^{13}C$ is so conspicuous that the volcanic CO_2 and CH_4 release, even if coupled with a decrease in biological productivity, cannot explain the phenomenon in its entirety. There must be additional factors or reasons.

It is likely that as a consequence of the rapid global warming (Sun et al. 2012; but see also Goudemand et al. 2013), methane clathrates trapped on shallow oceanic floors and/or high latitude permafrost melted and released light methane (with a $\delta^{13}C$ signature of $-65‰$) into the atmosphere (Benton and Twitchett 2003; Twitchett 2006). A similar scenario might have happened at the end of Paleocene, during another event of global warming we visited earlier on our journey (Dickens et al. 1995; see also Chap. 4). The amount of CO_2 and CH_4 released in a sort of "runaway greenhouse", with a positive-feedback loop established between clathrate melting and temperature increase triggered by Siberian volcanism, determined the extent of global warming. The average global increase in temperature has been calculated to be 8–10 °C, and geochemical data indicate that the warming was very rapid (see the summary and Fig. 2 in Lai et al. 2018). The rate of increase is estimated to have occurred over about 2000 years and resulted in ocean temperatures in the intertropical belt of about 39–42 °C (Brand et al. 2012; Sun et al. 2012). The direct release of large quantities of hydrochloric acid (HCl) and chloromethane (CH_3Cl) into the atmosphere, as a result of venting and basin-scale metamorphism, resulted in significant stratospheric ozone depletion. And this modification of the stratosphere increased the incidence of harmful ultraviolet-B radiation to an extent that significantly affected life (Beerling et al. 2007; Benca et al. 2018). It is thought that increasing concentrations of greenhouse gasses resulted in increasing aridity, as well as a short-term production of acid rains that negatively affected terrestrial vegetation (but see Li et al. 2017, Fielding et al. 2019, and Gastaldo et al. 2019b for evidence against abrupt aridification in Australia and South Africa). A loss in plants, due to the disturbance of the terrestrial ecosystems across the land, may have increased soil erosion and, in this model, resulted in a higher sedimentation rate in both freshwater and saltwater basins (Chen and Benton 2012). Shortly after the extreme warming phase, ocean basins experienced widespread anoxic/dysoxic and euxinic (anoxia in waters rich of hydrogen sulfide) conditions: ocean waters soon became stratified and increasingly acid (e.g. Isozaki 1997; Clarkson et al. 2015). It is the interaction of these factors across Earth Systems that likely explains the marine ecological and environmental turmoil that eclipsed the end-Permian ecosystems (Erwin 1993, 2006; Benton and Newell 2014). And if the establishment of hostile conditions had been a geologically rapid phenomenon, a cascade of geochemical events determined their persistence for a long time thereafter. These conditions shaped the prolonged recovery that followed the most catastrophic mass extinction in Earth history.

10.4   The Recovery

The pattern of recovery is attracting more and more interest of researchers, because it may serve as a model for today's crisis, similar to the Paleocene-Eocene Thermal Maximum (see DeVore and Pigg, Chap. 4). The paleoenvironments and fossil assemblages recorded in the post-extinction strata represent an excellent laboratory for studying evolutionary trajectories and the origin of modern ecosystems. This is because many of the lineages that characterize the life on our planet, today, date back to the Triassic recovery (Chen and Benton 2012). Many new families, genera, and species, in fact, appeared or rapidly diversified in the Early and Middle Triassic. As such, they are technically not part of a real "recovery", if the term "recovery" is thought of as a return to a previous state (Chen and Benton 2012). But these groups are interesting proxies of how ecosystems and trophic levels are restructured as the biosphere returned to a higher level of complexity.

The post-extinction recovery was a slow process for many of the animal groups that survived. This is because hostile conditions prevailed for a few million years. Among other things, hostile conditions across the globe are evidenced by: (1) the presence of multiple stable-carbon isotopic excursions that followed the one found at the PTB (Payne et al. 2004); (2) peaks in temperature (Sun et al. 2012); and (3) changes in the biogeochemical sulfur cycle, which resulted in widespread euxinic conditions and posed a sustained threat to marine life (Schobben et al. 2017; see also Benton et al. 2013). Early Triassic ecosystems were highly unstable (Roopnarine et al. 2007; Roopnarine and Angielczyk 2015), and it has been repeatedly suggested that the aftermath of the crisis extended into the Middle Triassic, about ten million years after the EPME (e.g., Lau et al. 2016; Martindale et al. 2018).

The change of the reef ecosystems before and after the PTB is particularly interesting. Reefs of the late Permian, constructed by sponges, algae, bryozoans, large foramin-

ifera, microbes, and rugose corals, hosted benthic invertebrates, including brachiopods, mollusks, foraminifera, sponges, and bryozoans (Martindale et al. 2018). Most of these reef-forming groups never made it past the crisis. After the mass extinction, stromatolites (layered cyanobacterial structures), calcimicrobes (calcareous colonial microfossils), and thrombolites (clotted, cyanobacteria mats) were common at relatively low latitudes. Microbial mats filled the ecological niche previously occupied by metazoan reefs (Heindel et al. 2018). Recent evidence indicates that these microbialites were associated with small-sized, oxygen-dependent metazoans, including ostracods, microconchids, brachiopods, gastropods, bivalves, crinoids, echinoids, and conodonts but no corals (e.g., Martindale et al. 2018). The concept of a "reef gap" has been modified into "reef eclipse" because microbial reefs were present, although severely altered, in the aftermath of the extinction. The first metazoan biostromes (horizontally bedded fossil assemblages) from the Early Triassic (Olenekian) are relatively small structures devoid of corals. These assemblages were made by bivalves and sponges in association with stromatolitic and thrombolitic microbialites. Therefore, a long coral reef eclipse characterized the aftermath of the EPME (Martindale et al. 2018).

In the depauperate Early Triassic marine ecosystems, the presence of some disaster taxa, especially among brachiopods and mollusks, is remarkable (Fig. 10.8). The inarticulate brachiopod *Lingula*, rare in pre-extinction sediments, is an example of a very common and globally widespread taxon in subtidal environments immediately after the crisis (Twitchett 2006). Bivalve disaster taxa are also widespread and numerically abundant in Early Triassic sediments and include well-known taxa like *Claraia*, *Eumorphotis*, *Promyalina,* and *Unionites* (e.g., Bottjer et al. 2008).

The rate of recovery varied in different groups. Both slow and delayed recovery rates are recorded for benthic groups such as bivalves and gastropods (with an initial opportunistic proliferation of microgastropods). In contrast, explosive and rapid recovery is found in ammonoid cephalopods (in less than 2 My after the PTB, diversity levels were higher than in the Permian), conodonts, and some groups of foraminifera (Bottjer et al. 2008; Brayard et al. 2009; Stanley 2009; Song et al. 2011). Rapid diversification also occurred in marine vertebrates. According to Benton et al. (2013), bony (actinopterygian) fishes explosively diversified in the Early Triassic (but see Friedman and Sallan 2012 and Romano in Chap. 7), while the diversification of marine reptiles only occurred in the latest Early–Middle Triassic (Motani 2009). This latter diversification involved ichthyosaurs, thalattosaurs, pachypleurosaurs, nothosaurs, and placodonts (Chen and Benton 2012; Benton et al. 2013).

Fig. 10.8 Examples of disaster taxa that flourished after the EPME. Mollusks: (**a**) *Claraia*; (**b**) *Eumorphotis*; (**c**) *Unionites*; (**d**) *Promyalina*. Brachiopod: (**e**) *Lingula*. Artwork by Fabrizio Lavezzi. (after Benton and Harper 2009)

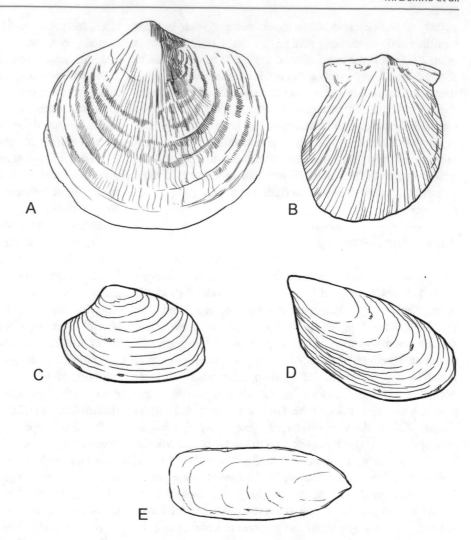

Following the lingering effects of the EPME, when environmental conditions became favorable again for the development of well-structured communities, new groups radiated. For example, Dai et al. (2018) reported high levels of benthic and nektonic faunas in the late Induan, soon after the last carbon-cycle perturbations, with the appearance of 26 new species of mollusks, brachiopods, foraminifers, conodonts, ostracods, and echinoderms. The most dramatic Early Triassic marine radiations are found in the scleractinian corals, bivalves, and crinoids (e.g., Martindale et al. 2018, and literature therein). On land, several groups of ferns and conifers appear for the first time (Cascales-Miñana et al. 2016), and insects are represented by an evolutionary modern fauna with derived clades of orthopteroids and hemipteroids and, especially, holometabolous taxa that replaced the previous Paleozoic faunas (Labandeira 2005). Among terrestrial vertebrates, several tetrapods including lissamphibians (frogs, salamanders) (Ascarrunz et al. 2016), squamate reptiles (lizards, snakes) (Simões et al. 2018), and several archosauriforms, comprising the dinosaurs (Brusatte et al. 2011; Ezcurra and Butler 2018), appear (Box 10.2).

Box 10.2: Archosauromorph Diversity and the EPME

A recent analysis studied archosauromorph diversity from the middle Permian to the early Late Triassic (Ezcurra and Butler 2018). This clade of reptiles likely originated in the middle Permian and may not have been affected by the EPME. The morphological disparity (variation) in the group, depending on the analytical method used, does not change significantly into the earliest Triassic (Induan). In fact, morphological disparity in this group increased significantly in the Olenekian or in the Anisian. Peaks of phylogenetic diversity in both the Induan and Olenekian correspond to very high evolutionary rates. The Induan forms are characterized by a low disparity of a sort considered to characterize a disaster fauna, dominated by proterosuchids and other lineages that were rather homogeneous in terms of their morphology. These forms underwent a major phylogenetic radiation in the Olenekian, which led to the origin or diversification of major clades. These clades include

Fig. 10.9 The archosauriforms *Chasmatosuchus* (**a**), *Erythrosuchus* (**b**), and *Tanystropheus* (**c**) are among the taxa that underwent a major phylogenetic radiation after the EPME in the Triassic. Artwork by Fabrizio Lavezzi

the rhynchosaurs, archosaurs, erythrosuchids, and tanystropheids (Fig. 10.9). During the Anisian, the archosauromorphs underwent an increase in terms of the number of species, abundance, ecomorphological disparity, and also size. In fact, the first part of the Middle Triassic witnessed the appearance of large hypercarnivore taxa, bizarre and highly specialized herbivores, long-necked marine predators, and gracile and agile dinosauromorphs (Ezcurra and Butler 2018). Their appearance corresponds with the return to stable ecosystems, as indicated by the end of the carbon-isotope excursions and the structuring of conifer-dominated forests (Sun et al. 2012). According to Bernardi et al. (2018), the emergence and the radiation of entirely new vertebrate groups, such as the archosaurs (including the dinosaurs) and others, might have been prompted by the forced biogeographic shift induced by the rapidly changing conditions (e.g., global warming) that characterized the aftermath of the EPME.

ful among the marine biota, although the various groups were affected differently depending on their life style (our knowledge is biased by their skeletal type and, therefore, the probability of retrieving information). Among the marine organisms, nektonic groups including the ammonoids and fishes were less affected than benthic forms such as bivalves, brachiopods, or reef-building organisms. On land, tetrapods and insects were more affected by either a loss in plant diversity or ecological turnover, although the composition and richness of the communities changed in the latter group as well. The prolonged unstable climatic conditions in the aftermath of the mass extinction affected the biotic recovery of the groups that were affected by the EPME. But, at the same time, it was a cradle for the evolution and especially radiation of a series of animals (archosaurs, squamates) and plants (corystosperms, Cycadales) that became important elements of Triassic ecosystems. Although the latest Permian and the earliest Triassic are among the most studied periods of time in Earth history, the dynamics and timeline of the extinction, biotic recovery, and radiation are still incompletely understood and further studies in the next tens of years will likely shed more light on "the mother of all extinctions".

10.5   Conclusions

As shown above and summarized in Fig. 10.10, the end of the Permian is a drastic moment in Earth history with the most severe mass extinction of all life. This was most impact-

For Deeper Learning

- Early, general works dealing with the EPME: Newell (1973); Raup (1979); Gould and Calloway (1980); Sepkoski (1981).

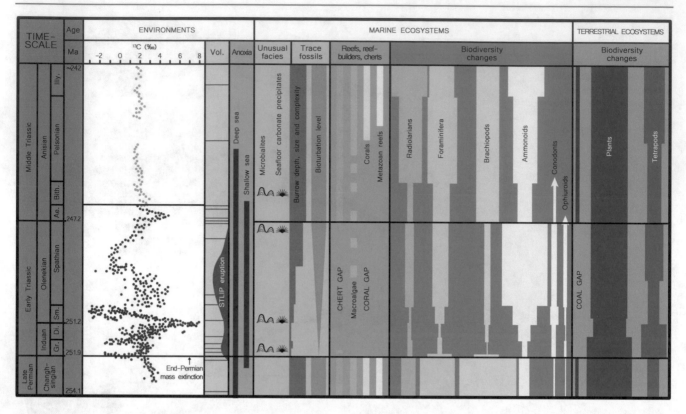

Fig. 10.10 A graphic summary of the environmental changes (oscillation of δ13C, volcanism = Vol., and anoxia), coupled with peculiar events in marine ("unusual" facies, trace fossils, reefs, reef-builders, chert, biodiversity changes in radiolarians, foraminiferans, brachiopods, ammonoids, conodonts, and ophiuroids) and terrestrial (coal gap and biodiversity changes in plants and vertebrates) ecosystems. *Ae.* Aegean, *Bith.* Bithynian, *Di.* Dienerian, *Gr.* Griesbachian, *Illy.* Illyrian, *Sm.* Smithian. Modified from Chen and Benton (2012; Fig. 1) after Nowak et al. (2019) by Fabrizio Lavezzi

- Siberian traps and the "lethal triade" (global warming, ocean anoxia, acidification): Wignall (2001); Algeo and Twitchett (2010); Algeo et al. (2011); Clapham and Payne (2011); Brand et al. (2012); Sun et al. (2012); Clarkson et al. (2015).
- Ocean waters stratification and acidification: Wignall and Hallam (1992); Wignall (2001); Wignall and Twitchett (2002); Algeo and Twitchett (2010); Clapham and Payne (2011); Brand et al. (2012); Winguth and Winguth (2012); Silva Tamayo et al. (2018); Shen et al. (2018).
- Stromatolites, microbialites, and trace fossils across the Permian–Triassic boundary: Schubert and Bottjer (1992); Twitchett and Wignall (1996); Kershaw et al. (2012).
- Plant evolution across the EPME: DiMichele et al. (2001); Hochuli et al. (2010, 2016); Blomenkemper et al. (2018); Nowak et al. (2019).
- Insect evolution across the EPME: Shcherbakov (2008); Labandeira (2005); Ponomarenko (2016).
- Vertebrate evolution across the EPME: Friedman and Sallan (2012); Viglietti et al. (2016).

Questions

1. Why is the EPME considered the worst of the Big Five mass extinctions over geologic time?
2. Which groups of organisms were particularly affected, and which ones suffered relatively low extinction rates?
3. Over what duration of time has the EPME been estimated to have lasted in the oceans and on land? How might you explain the disparity in these estimates?
4. What evidence exists in marine sediments for reduced oxygen (dysoxia/anoxia) at the sea floor during the EMPE?
5. How might a non-synchronous series of events, between the events on land and those in the oceans, be explained?
6. What is the "Lilliput effect" and in which groups of Early Triassic organisms is it found?
7. The emplacement of the Siberian Traps is considered a possible "smoking gun" that pushed ecosystem instability. What are the Siberian Traps, how might their emplacement have resulted in a lethal triade, and what lines of evidence are used to support this concept?

8. What evidence exists for a hostile planet in the aftermath of the EPME?

9. Why might different marine invertebrate groups have undergone different rates of recovery in the Early Triassic?

Acknowledgments The Editors, Edoardo Martinetto, Robert A Gastaldo, and Emanuel Tschopp, are thanked not only for having invited us to contribute to this volume, but also for their continuous support and guidance. Robert A Gastaldo and Cindy Looy kindly provided the photographs of relevant localities.

References

Abu Hamad A, Kerp H, Vörding B, Bandel K (2008) A late Permian flora with *Dicroidium* from the Dead Sea region, Jordan. Rev Palaeobot Palynol 149:85–130

Algeo TJ, Twitchett RJ (2010) Anomalous Early Triassic sediment fluxes due to elevated weathering rates and their biological consequences. Geology 38:1023–1026

Algeo TJ, Chen ZQ, Fraiser L, Twitchett RJ (2011) Spatial variation in sediment fluxes, redox conditions, and productivity in the Permian–Triassic Panthalassic Ocean. Palaeogeogr Palaeoclimatol Palaeoecol 308:65–83

Ascarrunz E, Rage J-C, Legreneur P, Laurin M (2016) *Triadobatrachus massinoti*, the earliest known lissamphibian (Vertebrata: Tetrapoda) re-examined by μCT scan, and the evolution of trunk length in batrachians. Contrib Zool 85(2):201–234

Bajdek P, Qvarnström M, Owocki K, Sulej T, Sennikov AG, Golubev VK, Niedźwiedzki G (2016) Microbiota and food residues including possible evidence of pre-mammalian hair in Upper Permian coprolites from Russia. Lethaia 49(4):455–477

Beerling DJ, Harfoot M, Lomax B, Pyle JA (2007) The stability of the stratospheric ozone layer during the end-Permain eruption of the Siberian Traps. Philos Trans R Soc A Math Phys Eng Sci 365:1843–1866

Benca JP, Djiunstee IAP, Looy CV (2018) UV-B–induced forest sterility: implications of ozone shield failure in Earth's largest extinction. Sci Adv 4:e1700618

Benton MJ (2005) When life nearly ended. The greatest mass extinction of all time. Thames & Hudson, London. 336pp

Benton MJ, Harper DAT (2009) Introduction to paleobiology and the fossil record. Wiley-Blackwell, Oxford. 592pp

Benton M, Newell AJ (2014) Impacts of global warming on Permo-Triassic terrestrial ecosystems. Gondwana Res 25:1308–1337

Benton MJ, Twitchett RJ (2003) How to kill (almost) all life: the end-Permian extinction event. Trends Ecol Evol 18(7):358–365

Benton MJ, Zhang Q, Hu S, Chen Z-Q, Wen W, Liu J, Huang J, Zhou C, Xie T, Tong J, Choo B (2013) Exceptional vertebrate biotas from the Triassic of China, and the expansion of marine ecosystems after the Permo-Triassic mass extinction. Earth Sci Rev 125:199–243

Bernardi M, Petti FM, Kustatscher E, Franz M, Hartkopf-Fröder C, Labandeira CC, Wappler T, van Konijnenburg-van Cittert JHA, Peecook BR, Angielczyk KD (2017) Late Permian (Lopingian) terrestrial ecosystems: a global comparison with new data from the low latitude Bletterbach Biota. Earth Sci Rev 175:18–43

Bernardi M, Petti FM, Benton MJ (2018) Tetrapod distribution and temperature rise during the Permian-Triassic mass extinction. Proc R Soc B 285(1870):20172331

Blomenkemper P, Kerp H, Abu Hamad A, DiMichele WA, Bomfleur B (2018) A hidden cradle of plant evolution in Permian tropical lowlands. Science 362:1414–1416

Bond DPG, Wignall PB (2014) Large igneous provinces and mass extinctions: an update. Geol Soc Am Spec Pap 505:29–55

Bottjer DJ, Clapham ME, Fraiser ML, Powers CM (2008) Understanding mechanisms for the end-Permian mass extinction and the protracted Early Triassic aftermath/recovery. Geol Soc Am Today 18(9):4–10

Brand U, Posenato R, Came R, Affek H, Angiolini L, Azmy K, Farabegoli E (2012) The end-Permian mass extinction: a rapid volcanic CO2 and CH4-climatic catastrophe. Chem Geol 322–323:121–144

Brayard A, Escarguel G, Bucher H, Monnet C, Brühwiler T, Goudeman N, Galfetti T, Guex J (2009) Good genes and good luck: ammonoid diversity and the end-Permian mass extinction. Science 325(5944):1118–1121

Brusatte SL, Niedźwiedzki G, Butler RJ (2011) Footprints pull origin and diversification of dinosaur stem lineage deep into Early Triassic. Proc R Soc B Biol Sci 278(1708):1107–1113

Burgess SD, Bowring SA (2015) High-precision geochronology confirms voluminous magmatism before, during, and after Earth's most severe extinction. Sci Adv 1:e1500470

Burgess SD, Bowring S, Shen S-Z (2014) High-precision timeline for Earth's most severe extinction. Proc Natl Acad Sci 111:3316–3321

Cascales-Miñana B, Diez JB, Gerrienne P, Cleal CJ (2016) A palaeobotanical perspective on the great end-Permian biotic crisis. Hist Biol 28(8):1066–1074

Chen Z-Q, Benton MJ (2012) The timing and pattern of biotic recovery following the end-Permian mass extinction. Nat Geosci 5(6):375–383

Clapham MF, Payne JL (2011) Acidification, anoxia, and extinction: a multiple logistic regression analysis of extinction selectivity during the middle and late Permian. Geology 39(11):1059–1062

Clarkson MO, Kasemann SA, Wood RA, Lenton TM, Daines SJ, Richoz S, Ohnemueller F, Meixner A, Poulton SW, Tipper ET (2015) Ocean acidification and the Permo-Triassic mass extinction. Science 348(6231):229–232

Cohen KM, Finney SC, Gibbard PL, Fan J-X (2013; updated 2018) The ICS international chronostratigraphic chart. Episodes 36: 199–204. http://www.stratigraphy.org/ICSchart/ChronostratChart2018-08.pdf

Dai X, Song H, Wignall PB, Jia E, Bai R, Wang F, Chen J, Tian L (2018) Rapid biotic rebound during the late Griesbachian indicates heterogeneous recovery patterns after the Permian-Triassic mass extinction. Bull Geol Soc Am 130(11–12):2015–2030

Dickens GR, O'Neil JR, Rea DK, Owen RM (1995) Dissociation of oceanic methane hydrate as a cause of the carbon isotope excursion at the end of the Paleocene. Paleoceanography 10(6):965–971

DiMichele WA, Mamay SH, Chaney DS, Hook RW, Nelson J (2001) An early Permian Flora with late Permian and Mesozoic affinities from north central Texas. J Palaeontol 75:449–460

Erwin DH (1993) The great Paleozoic crisis, life and death in the Permian. 327pp. Colombia University Press, New York

Erwin DH (2006) Extinction: how life nearly died 250 million years ago. Princeton University Press, Princeton. 296pp

Ezcurra MD, Butler RJ (2018) The rise of the ruling reptiles and ecosystem recovery from the Permo-Triassic mass extinction. Proc R Soc B 285:20180361

Feng Z, Wei H, Guo Y, Bomfleur B (2018) A conifer-dominated Early Triassic flora from Southwest China. Sci Bull 63:1462–1463

Fielding CR, Frank TD, McLoughlin S, Vajda V, Mays C, Tevyaw AP, Winguth A, Winguth C, Nicoll RS, Bocking M, Crowley JL (2019) Age and pattern of the southern high-latitude continental end-Permian extinction constrained by multiproxy analysis. Nat Commun 10:385

Friedman M, Sallan LC (2012) Five hundred million years of extinction and recovery: a Phanerozoic survey of large-scale diversity patterns in fishes. Palaeontology 55(4):707–742

Gastaldo RA, Neveling J (2012) The terrestrial Permian–Triassic boundary event is a nonevent. Geology 40(3):e257

Gastaldo RA, Neveling J (2016) Comment on: "Anatomy of a mass extinction: sedimentological and taphonomic evidence for drought-induced die-offs at the Permo-Triassic boundary in the main Karoo Basin, South Africa" by R.M.H. Smith & J. Botha-Brink, Palaeogeography, Palaeoclimatology, Palaeoecology 396:99–118. Palaeogeogr Palaeoclimatol Palaeoecol 447:88–91

Gastaldo RA, Adendorff R, Bamford M, Labandeira CC, Neveling J, Sims H (2005) Taphonomic trends of macrofloral assemblages across the Permian-Triassic boundary, Karoo Basin, South Africa. PALAIOS 20:479–497

Gastaldo RA, Neveling J, Clark CK, Newbury SS (2009) The terrestrial Permian–Triassic boundary event bed is a non-event. Geology 37(3):199–202

Gastaldo RA, Kamo SL, Neveling J, Geissman JW, Bamford M, Looy CV (2015) Is the vertebrate-defined Permian–Triassic boundary in the Karoo Basin, South Africa, the terrestrial expression of the end-Permian marine event? Geology 43:939–942

Gastaldo RA, Neveling J, Geissman JW, Kamo SL (2018) A lithostratigraphic and magnetostratigraphic framework in a geochronologic context for a purported Permian-Triassic boundary section at Old (West) Lootsberg Pass, Karoo Basin, South Africa. Bull Geol Soc Am 130(9–10):1411–1438

Gastaldo RA, Neveling J, Geissman JW, Looy CV (2019a) Testing the *Daptocephalus* and *Lystrosaurus* Assemblage Zones in a Lithostratigraphic, Magnetostratigraphic, and Palynological Framework in the Free State, South Africa. PALAIOS 34:542–561

Gastaldo RA, Neveling J, Geissman JW, Li JW (2019b) A multidisciplinary approach to review the vertical and lateral facies relationships of the purported vertebrate-defined terrestrial boundary interval at Bethulie, Karoo Basin, South Africa. Earth-Sci Rev 189: 220–243

Gastaldo RA, Kamo SL, Neveling J, Geissman JW, Looy CV, Martini AM (2020) The base of the *Lystrosaurus* Assemblage Zone, Karoo Basin, predates the end-Permian marine extinction. Nat. Comm

Goudemand N, Romano C, Brayard A, Hochuli PA, Bucher H (2013) Comment on "Lethally hot temperatures during the Early Triassic greenhouse". Science 339(6123):1033

Gould SJ, Calloway CB (1980) Clams and brachiopods–ships that pass in the night. Paleobiology 6(4):383–396

Grauvogel-Stamm L, Ash SR (2005) Recovery of the Triassic land flora from the end-Permian life crisis. C R Palevol 4:593–608

Heindel K, Foster WJ, Richoz S, Birgel D, Roden VJ, Baud A, Brandner R, Krystyn L, Mohtat T, Koşun E, Twitchett RJ, Reitner J, Peckmann J (2018) The formation of microbial-metazoan bioherms and biostromes following the latest Permian mass extinction. Gondwana Res 61:187–202

Hochuli PA, Vigran JO, Hermann E, Bucher H (2010) Multiple climatic changes around the Permian-Triassic boundary event revealed by an expanded palynological record from mid-Norway. Geol Soc Am Bull 122:884–896

Hochuli PA, Sanson-Barrera A, Schneebeli-Hermann E, Bucher H (2016) Severest crisis overlooked – worst disruption of terrestrial environments postdates the Permian–Triassic mass extinction. Sci Rep 6:28372

Isozaki Y (1997) Permo-Triassic boundary superanoxia and stratified superocean: records from lost deep sea. Science 276(5310):235–238

Kershaw S, Crasquin S, Li Y, Collin PY, Forel MB, Mu X, Baud A, Wang Y, Xie S, Maurer F, Guo L (2012) Microbialites and global environmental change across the Permian–Triassic boundary: a synthesis. Geobiology 10:25–47

Kustatscher E, Visscher H, Van Konijnenburg-van Cittert JHA (2019) Did the Czekanowskiales already exist in the late Permian? PalZ 93:465

Labandeira CC (2005) The fossil record of insect extinction: new approaches and future directions. Am Entomol 51:14–29

Lai X, Jiang H, Wignall PB (2018) A review of the late Permian – Early Triassic conodont record and its significance for the end-Permian mass extinction. Rev Micropaleontol 61:155–164

Lau KV, Maher K, Altiner D, Kelley BM, Kump LR, Lehrmann DJ, Silva-Tamayo JC, Weaver KL, Yu M, Payne JL (2016) Marine anoxia and delayed earth system recovery after the end-Permian extinction. Proc Natl Acad Sci 113(9):2360–2365

Li JW, Gastaldo RA, Neveling J, Geissman JW (2017) Siltstones across the *Daptocephalus* (*Dicynodon*) and *Lystrosaurus* assemblage zones, Karoo Basin, South Africa, show no evidence for aridification. J Sediment Res 87:653–671

Looy CV, Brugman A, Dilcher DL, Visscher H (1999) The delayed resurgence of equatorial forests after the Permian–Triassic ecologic crisis. Proc Natl Acad Sci 96:13857–13862

Martindale RC, Foster WJ, Velledits F (2018) The survival, recovery, and diversification of metazoan reef ecosystems following the end-Permian mass extinction event. Palaeogeogr Palaeoclimatol Palaeoecol 513:100–115

McElwain JC, Punyasena SW (2007) Mass extinction events and the plant fossil record. Trends Ecol Evol 22:548–557

Motani R (2009) The evolution of marine reptiles. Evol Educ Outreach 2:224–235

Newell ND (1973) The very last moment of the Paleozoic era. In: Logan A, Hill LV (eds) Permian and Triassic systems and their mutual boundary, vol 2. Canadian Society of Petroleum Geologists Memoirs, Calgary, pp 1–10

Nowak H, Schneebeli-Hermann E, Kustatscher E (2019) No mass extinction for land plants at the Permian–Triassic transition. Nat Commun 10:384

Padian K (2018) Measuring and comparing extinction events: reconsidering diversity crises and concepts. Integr Comp Biol 58(6): 1191–1203

Payne JL, Kump LR (2007) Evidence for recurrent Early Triassic massive volcanism from quantitative interpretation of carbon isotope fluctuations. Earth Planet Sci Lett 256(1–2):264–277

Payne JL, Lehrmann DJ, Wei J, Orchard MJ, Schrag DP, Knoll AH (2004) Large perturbations of the carbon cycle during recovery from the end-Permian extinction. Science 305:506–509

Ponomarenko AG (2016) Insects during the time around the Permian–Triassic crisis. Paleontol J 50(2):174–186

Posenato R (2019) The end-Permian mass extinction (EPME) and the Early Triassic biotic recovery in the western Dolomites (Italy): state of the art. Boll Soc Paleontol Ital 58:11–34

Raup DM (1979) Size of the Permo-Triassic bottleneck and its evolutionary implications. Science 206:217–218

Raup DM, Sepkoski JJ Jr (1982) Mass extinctions in the marine fossil record. Science 215:1501–1503

Retallack GJ, Veevers JJ, Morante R (1996) Global coal gap between Permian-Triassic extinction and Middle Triassic recovery of peat-forming plants. Geol Soc Am Bull 108(2):195–207

Retallak GJ, Smith RMH, Ward PD (2003) Vertebrate extinction across Permian-Triassic boundary in Karoo Basin, South Africa. Geol Soc Am Bull 115(9):1133–1152

Roopnarine PD, Angielczyk KD (2015) Community stability and selective extinction during the Permian-Triassic mass extinction. Science 350:90–93

Roopnarine PD, Angielczyk KD, Wang SC, Hertog R (2007) Trophic network models explain instability of Early Triassic terrestrial communities. Proc R Soc B 274:2077–2086

Saunders A, Reichow M (2009) The Siberian Traps and the end-Permian mass extinction: a critical review. Chin Sci Bull 54(1):20–37

Schaal EK, Clapham ME, Rego BL, Wang SC, Payne JL (2016) Comparative size evolution of marine clades from the late Permian through Middle Triassic. Paleobiology 42(1):127–142

Schneebeli-Hermann E, Kürschner WM, Kerp H, Bomfleur B, Hochuli PA, Bucher H, Ware D, Roohi G (2015) Vegetation history across the Permian–Triassic boundary in Pakistan (Amb section, Salt Range). Gondwana Res 27:911–924

Schobben M, Stebbins A, Algeo TJ, Strauss H, Leda L, Haas J, Struck U, Korn D, Korte C (2017) Volatile earliest Triassic sulfur cycle: a consequence of persistent low seawater sulfate concentrations and a high sulfur cycle turnover rate? Palaeogeogr Palaeoclimatol Palaeoecol 486:74–85

Schubert JK, Bottjer DJ (1992) Early Triassic stromatolites as post-mass extinction disaster forms. Geology 20:883–886

Sepkoski JJ Jr (1981) A factor analytic description of the Phanerozoic marine fossil record. Paleobiology 7:36–53

Shcherbakov DE (2008) Insect recovery after the Permian/Triassic crisis. Alavesia 2:125–131

Shen S-Z, Ramezani J, Cao C-Q, Erwin DH, Zhang H, Xiang L, Schoepfer SD, Henderson CM, Zheng Q-F, Bowring SA, Wang Y, Li X-H, Wang X-D, Yuan D-X, Zhang Y-C, Mu L, Wang J, Wu Y-S (2018) A sudden end-Permian mass extinction in South China. Bull Geol Soc Am 131(1/2):205–223. https://doi.org/10.1130/B31909.1

Silva Tamayo J-C, Lau KV, Jost AB, Payne JL, Wignall PB, Newton RJ, Eisenhauer A, Depaolo DJ, Brown S, Maher K, Lehrmann DJ, Altiner D, Yu M, Richoz S, Paytan A (2018) Global perturbation of the marine calcium cycle during the Permian-Triassic transition. Geol Soc Am Bull 130(7–8):1323–1338

Simões TR, Caldwell MW, Tałanda M, Bernardi M, Palci A, Vernygora O, Bernardini F, Mancini L, Nydam RL (2018) The origin of squamates revealed by a Middle Triassic lizard from the Italian Alps. Nature 557(7707):706–709

Smith R, Botha-Brink J (2014) Anatomy of a mass extinction: sedimentological and taphonomic evidence for drought induced die-offs at the Permo-Triassic boundary in the main Karoo Basin, South Africa. Palaeogeogr Palaeoclimatol Palaeoecol 396:99–118

Song H, Wignall PB, Chen Z-Q, Tong J, Bond DPG, Lai X, Zhao X, Jiang H, Yan C, Niu Z, Chen J, Yang H, Wang Y (2011) Recovery tempo and pattern of marine ecosystems after the end-Permian mass extinction. Geology 39(8):739–742

Stanley SM (2009) Evidence from ammonoids and conodonts for multiple Early Triassic mass extinctions. Proc Natl Acad Sci 106(36):15264–15267

Stanley SM (2016) Estimates of the magnitudes of major marine mass extinctions in earth history. Proc Natl Acad Sci 113(42): E6325–E6334

Sun YD, Joachimski MM, Wignall PB, Yan CB, Chen YL, Jiang HS, Wang LN, Lai XL (2012) Lethally hot temperatures during the Early Triassic greenhouse. Science 338:366–370

Svensen H, Planke S, Polozov AG, Schmidbauer N, Corfu F, Podladchikov YY, Jamtveit B (2009) Siberian gas venting and the end-Permian environmental crisis. Earth Planet Sci Lett 277: 490–500

Thomas SG, Tabor NJ, Yang W, Myers TS, Yang Y, Wang D (2011) Paleosol stratigraphy across the Permo-Triassic boundary, Bogda Mountains, NW China: implications for palaeoenvironmental transition through Earth's largest mass extinction. Palaeogeogr Palaeoclimatol Palaeoecol 308:41–64

Twitchett RJ (2006) The palaeoclimatology, palaeoecology and palaeoenvironmental analysis of mass extinction events. Palaeogeogr Palaeoclimatol Palaeoecol 232(2–4):190–213

Twitchett RJ, Wignall PB (1996) Trace fossils and the aftermath of the Permo–Triassic mass extinction: evidence from northern Italy. Palaeogeogr Palaeoclimatol Palaeoecol 124:137–152

Urbanek A (1993) Biotic crises in the history of Upper Silurian graptoloids: a palaeobiological model. Hist Biol 7:29–50

Viglietti PA, Smith RMH, Angielczyk KD, Kammerer CF, Fröbisch J, Rubidge BS (2016) The *Daptocephalus* Assemblage Zone (Lopingian), South Africa: a proposed biostratigraphy based on a new compilation of stratigraphic ranges. J Afr Earth Sci 113: 153–164

Ward PD, Botha J, Buick R, Dekock MO, Erwin DH, Garrison G, Kirschvink J, Smith RHM (2005) Abrupt and gradual extinction among late Permian land vertebrates in the Karoo Basin. Science 307:709–714

Ward PD, Retallack GJ, Smith RHM (2012) The terrestrial Permian–Triassic boundary event bed is a nonevent. Geology 40(3):e256

Wignall PB (2001) Large igneous provinces and mass extinctions. Earth Sci Rev. 53(1):1–33

Wignall PB, Hallam A (1992) Anoxia as a cause of the Permian/Triassic extinction: facies evidence from northern Italy and the western United States. Palaeogeogr Palaeoclimatol Palaeoecol 93:21–46

Wignall PB, Twitchett RJ (2002) Extent, duration, and nature of the Permian–Triassic superanoxic event. Geol Soc Am Spec Pap 356: 395–413

Winguth C, Winguth AME (2012) Simulating Permian–Triassic oceanic anoxia distribution: implications for species extinction and recovery. Geology 40:127–130

Yin H, Zhang K, Tong J, Yang Z, Wu S (2001) The global stratotype section and point (GSSP) of the Permian-Triassic boundary. Episodes 24(2):102–114

Yin H, Feng Q, Lai X, Baud A, Tong J (2007a) The protracted Permo-Triassic crisis and multi-episode extinction around the Permian-Triassic boundary. Glob Planet Chang 55(1–3):1–20

Yin H, Feng Q, Baud A, Xie S, Benton MJ, Lai X, Bottjer DJ (2007b) The prelude of the end-Permian mass extinction predates a postulated bolide impact. Int J Earth Sci 96(5):903–909

Long-Lasting Morphologies Despite Evolution: Ferns (Monilophytes) Throughout the Phanerozoic

Josef Pšenička, Jun Wang, Ronny Rößler,
Mihai Emilian Popa, and Jiří Kvaček

Abstract

Successful evolutionary forms are characterized by their longevity in the fossil record. There are many plant groups that exhibit these traits; here we have selected the ferns as one acknowledged evolutionary model. Ferns are the most successful cryptogamic plants in geologic history and are known from nearly all fossil floras since their first appearance in the Devonian. Ferns, autotrophic organisms, have colonized nearly all types of continental environments over time [U1102]. Fern groups experienced rapid radiation at the familial level in the Carboniferous and are well documented in the fossil record thereafter. Their megaphyllous leaves are easy to recognize and have reproductive organs (sporangia filled by spores) borne on the lower side of pinnules. Although growth forms show great variability in morphology and size, all have retained this simple reproductive organization for over 400 million years from when first recognized in Devonian floras.

Electronic supplementary material A slide presentation and an explanation of each slide's content is freely available to everyone upon request via email to one of the editors: edoardo.martinetto@unito.it, ragastal@colby.edu, tschopp.e@gmail.com

*The asterisk designates terms explained in the Glossary.

J. Pšenička (✉)
Centre of Palaeobiodiversity, West Bohemian Museum in Pilsen, Pilsen, Czech Republic
e-mail: jpsenicka@zcm.cz

J. Wang
State Key Laboratory of Paleobiology and Stratigraphy, Nanjing Institute of Geology and Palaeontology, Chinese Academy of Sciences, Nanjing, PR China
e-mail: jun.wang@nigpas.ac.cn

11.1 Introduction

The previous chapters of this book presented a nature walk through the Cenozoic and Mesozoic landscapes, populated by the survivors of the End-Permian Mass Extinction (EPME) (Chap. 10). Before going deeper into the pre-extinction scenarios, we may be curious to know if any landscape-forming organisms were important and had a similar aspect before and after the EPME. Yes, we have an example, and this is ferns! Hence, we will follow their history in this chapter. The ferns are very successful vascular plants that have been a significant part of the biosphere throughout their geologic history. They currently grow in mountains, deserts, aquatic environments, open fields, and all types of forest habitats [U1106]. The first fern or fern-like plants (Chap. 15) have been identified in Devonian floras (Taylor et al. 2009), and, presently, more than 10,000 living species are known. Thanks to the fact that most of their mineral uptake occurs through the surface of the plant, ferns are able to colonize habitats where soil is not developed (e.g., on the rocks). This adaption, under the proper climatic conditions, has allowed ferns to be typical pioneer plants and to colonize the majority of environments subjected to erosion, including highland settings, where fossilization is mostly impossible. However, what is, in fact, a fern?

R. Rößler
Museum für Naturkunde, Chemnitz, Germany

TU Bergakademie Freiberg, Geological Institute, Department of Palaeontology/Stratigraphy, Freiberg, Germany
e-mail: roessler@naturkunde-chemnitz.de

M. E. Popa
University of Bucharest, Faculty of Geology and Geophysics, Department of Geology, Bucharest, Romania
e-mail: mihai@mepopa.com

J. Kvaček
Department of Palaeontology, National Museum, Prague, Praha 1, Czech Republic
e-mail: jiri_kvacek@nm.cz

Ferns are plants formed by rhizomes (subterranean), roots (similar to those of seed plants), stems (aboveground level), and leaves (fronds) with spore-producing reproductive organs, known as sporangia, that develop on the lower side of pinnules. Ferns belong to cryptogamic plants (plants that do not produce seeds) with alternating asexual (Box 11.1) and sexual (Box 11.2) generations. Both generations are characterized by having dissimilar body habits and function. Generally, ferns exhibit a wide variety of growth forms including trees, lianas, scramblers, epiphytes, and herbaceous groundcover plants [U1103, U1107]. Regardless of geologic age or growth form, the one significant diagnostic

---

Box 11.1: Fern Sporophyte Generation [U1105]

The sporophyte generation is characterized by the development of a megaphyllous plant body. The plant size varies from centimeters (herbaceous form) up to several meters in height (liana/scrambling-like or tree-like forms). Herbaceous ferns are the most widespread forms in nature, growing on mineral soils, bedrock exposures, or on top of other plants as epiphytes. They are characterized by shortly-erect or creeping/climbing rhizomes bearing leaves of variable size. Tree-like forms have a distinctly developed, tall, trunk-like rhizome bearing a palm-like set of terminal fronds in a crown. These growth strategies can be less than 1 m tall, but they may reach a height of up to 21 m. Some tree-like ferns produce large fronds, up to 5 m in length.

The sporophyte generation produces reproductive organs called sporangia. Based on the character of sporangia, two major groups of ferns are distinguished: eusporangiate and leptosporangiate. Eusporangiate ferns are characterized by a sporangial wall that consists of several cell layers and develops from the division of a group of periclinal (parallel to the surface of the meristem) superficial initial cells. The sporangia are "large" (more than 1 mm), contain hundreds of spores, and usually are grouped into synangia (fused group of sporangia). A typical member of the eusporangiate group includes ferns assigned to the Marattiales. Leptosporangiate ferns have a sporangial wall that consists of a single cell layer and develops from a slightly oblique division of a single superficial initial cell. The rather small sporangium* (less than 1 mm) contains only tens of spores. The vast majority of living ferns belong to leptosporangiate groups. Nevertheless, some ferns (e.g., Osmundaceae) are intermediate in their characteristics between eusporangiate and leptosporangiate ferns where sporangia arise from a single or multiple initials. These intermediate states show that the classification into eusporangiate or leptosporangiate forms is more or less arbitrary. The vast majority of fern groups are isosporous, producing only one type of spore, whereas there is one small group, the Salviniales, that is heterosporous, producing two distinctly different sized spores that are released to the environment.

Box 11.2: Fern Gametophyte Generation

Spores that are released from sporangia and land on a suitable substrate germinate and develop freely into a gametophyte plant. The gametophyte (N) generation develops from a small group of tissues composed of the millimeter-sized, heart-shaped prothallus. Isosporous ferns (vast majority of fern groups) develop a prothallus on which both eggs from archegonia and sperm from antheridia come from a common gametophyte. Heterosporous ferns (Salviniales), on the other hand, develop a micro-gametophyte body that grows from a microspore* and on which sperm cells develop. Similarly, mega-gametophytes develop separately from the germination of a megaspore, and eggs are produced in reproductive structures of these plants. As in other plant groups, fertilization of an egg by sperm results in a diploid zygote which, in turn, develops into another sporophyte generation.

---

feature of the group is leaf morphology [U1107]. All leaves are planar compound fronds with reproductive organs (sporangia) located on the abaxial (lower) side of the subdivided laminae known as pinnules. This sporangial placement is interpreted as a persistent advantage as the sporangia are "protected" by the leaf lamina*, which retains a photosynthetic function. The type of sporangia has traditionally been used to divide ferns into two major groups: the eusporangiate and leptosporangiate forms [U1104] (Box 11.1). However, these categories do not seem to reflect phylogeny.

Many current studies have abandoned the terms eusporangiate and leptosporangiate ferns because sporangial development in modern plants shows a high degree of variability in all sporangium-related features (Taylor et al. 2009). Given that fossils provide limited information about sporangial development, these two terms are now used in a rather descriptive sense. Many of the oldest plants remaining from the Paleozoic do not show the presence of planar fronds, but they have unique characters associated with their reproductive organs (e.g., sporangia have several cell layers and a distinct annulus), or they produced secondary tissue in the rhizome (a feature known from seed plants). It is often difficult to place these early plants into either the eusporangiate and leptosporangiate ferns. Yet, many of them show some of the same characteristics with ferns and, therefore, are placed

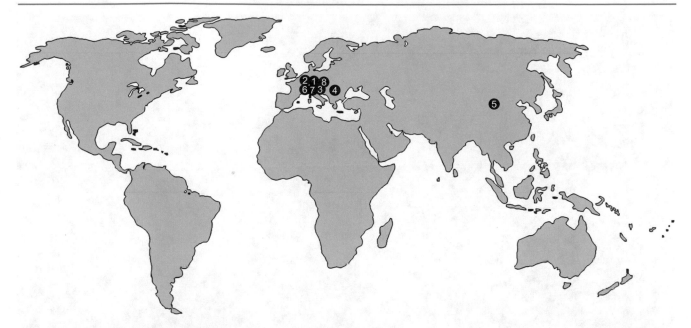

Fig. 11.1 Map of localities described in this chapter. (1) Svatý Jan pod Skalou—Quaternary (Czech Republic), (2) Bílina—Cenozoic (Northern Czech Republic), (3) Grünbach, Austria, (4) Anina—Jurassic (Romania), (5) Wuda—Permian (Inner Mongolia, China), (6) Flöha—Carboniferous and Chemnitz—Permian (Germany), (7) Radnice (Ovčín)—Carboniferous (Czech Republic), (8) Hostim—Devonian (Czech Republic)

among so-called fern-like plants. One such group is the zygopterids. These early fern-like plants are restricted to the Paleozoic, whereas both the eusporangiate and leptosporangiate ferns are found ranging from the Paleozoic to Recent.

Genetic methods used for the classification of modern ferns can obviously not be applied to fossil forms, which are classified based on phenotypes. One primary morphological character used to classify fossil ferns is their leaf structure. The compound leaves of ferns can be complex and variable in both pinna* and pinnule morphology. This complexity often is intensified in the fossil record because of the fragmentary nature of specimens, which has resulted in complications of classification and taxonomy in the absence of reproductive or anatomical structures. Thus, fern systematics and the interpretation of fern evolution is frequently reinterpreted as new specimens are discovered. However, their important role in ecosystems around the planet remains undisputed.

This chapter focuses on fern-dominated floras from nine localities (Fig. 11.1) of various geologic ages to examine the role that ferns have played in the landscape. These localities [U1108, U1109] were selected based on the interpretation of their floras having been preserved in original growth position (autochthonous) or in the environment where they grew (parautochthonous*; DiMichele and Gastaldo 2008), allowing the study of a phytocoenosis (the plants in any environmental setting). The postcards from these nine localities describe the composition of the flora, the growth habit and leaf morphology of ferns, as well as their role in the biotope.

11.2 The Holocene Svatý Jan pod Skalou Locality Barrandien, Czech Republic

Svatý Jan pod Skalou is in the Barrandian area (between Pilsen and Prague) of the Czech Republic [U1110]. Traveling to the fossil site we encounter a large karst spring, 70 m long, 80 m wide, and 17 m deep (Fig. 11.2). The fossil-plant assemblage includes ferns and is preserved in a freshwater limestone (travertine) (Němejc 1927). The deposit has produced many imprints of plants assignable to angiosperms including *Acer*, *Ulmus*, *Tilia*, and *Sambucus* (elderberry), and the herbaceous fern *Asplenium scolopendrium* (syn. *Phyllitis scolopendrium*) (Němejc 1927). This freshwater deposit also preserves more than 50 species of mollusks, mostly snails. The fern *Asplenium scolopendrium* is easily identified by its linear sorus (a cluster of sporangia) and an overlapping venation pattern (Fig. 11.3). Today, *A. scolopendrium* is known as a frost-resistant "whole-margin" fern with large and long glossy leaves. The plant thrives in semi-shaded to fully shaded, high humidity habitats growing on permeable moist alkaline (slightly calcareous) substrates. Svatý Jan pod Skalou represents a postglacial environment that experienced a warmer and more humid climate following deglaciation in the region (Chlupáč et al. 2002). As such, more thermophilic plant species, including *A. scolopendrium*, colonized this locality. To date, no other ferns have been reported from this deposit.

Fig. 11.2 Svatý Jan pod Skalou locality with large karst spring. Photo by Josef Pšenička

11.3 The Miocene Bílina Locality of the Most Basin, Northern Czech Republic

A little over an hour drive north of Svatý Jan pod Skalou is the locality known as Bílina (see 2.2.3), where there is a large lignite open-pit mine [U1111] in the Most Basin of northern Bohemia (Fig. 11.4). The Most Basin is one of the several structural low-lying areas in the Krušné hory Piedmont, which extends for over 1000 km. The stratigraphy records six distinct phases influenced by the local movement of the Earth's crust and environmental change. These are: (1) a central river, (2) floodplain and first moors, (3) a whole basin swamp, (4) localized lakes, (5) a whole basin lake, and (6) a swamp rehabilitation phase (Mach et al. 2014). The first phase of environmental change is linked to the tectonic development of an internally drained basin at the end of the Oligocene. During this time, with the onset of a second syn-rift interval, the Eger graben developed. A river transported sand and mud (siliciclastics) into the Most Basin, filling it with fluvial and floodplain sediments. In phases 2–4, the basin was filled partially with sediment during a climate interval that was ever wet. These conditions resulted in the development of peat swamps which, after burial, were converted to brown coals over time. In the final stages, lacustrine environments prevailed and clastic sediments were no longer transported from the Bohemian Massif to Saxony. The final stage of Cenozoic

deposition occurred at the end of the early Miocene (Mach et al. 2014) during the onset of the Mid-Miocene Climatic Optimum (sensu Zachos et al. 2001), which pre-dated the widespread expansion of grasslands (Chap. 3).

The early Miocene in the region is typified by a gradual increase in temperature and precipitation, which peaked 17–15 million years ago (Teodoridis and Kvaček 2015). Five generalized environments can be identified in these deposits. These environments include: (1) better drained soils on slopes of the basin margin, (2) riparian settings that developed along the river margins, (3) wetland soils found in back swamps, (4) mires, and (5) lakes (Kvaček et al. 2004b). Ferns grew in all five settings and, to date, ten fern types have been described from this locality. These taxa include *Osmunda parschlugiana* (Royal fern), *Asplenium* sp., *Woodwardia muensteriana, Blechnum dentatum, Pronephrium stiriacum, Lygodium kaulfusii, Salvinia reussii* (Box 11.3), *Azolla* aff. *nana, Azolla* aff. *rossica*, and *Azolla* aff. *ventricosa* (Kvaček et al. 2004a, b).

Box 11.3: The Aquatic Ferns of Bílina

Open-water, aquatic vegetation [U1112] of the Bílina paleolake is represented by the presence of the heterosporous aquatic ferns *Salvinia reussii* (Fig. 11.5) and *Azolla* sp. Both *Salvinia* and *Azolla* are ferns that grow

Several of these fossil forms show great similarity to modern, living taxa. The fossil herbaceous fern *Osmunda parschlugiana* shows a morphological relationship with the extant forms *O. palustris* or *O. regalis* [U1111] (Bůžek 1971; Kvaček and Bůžek 1983). Presently, *Osmunda* lives across a wide range of habitats growing in tropical to temperate zones. In the deposits of the Most Basin, *Osmunda* is found in the riparian floras indicating that it inhabited river banks. Sakala (2000) identified herbaceous *Asplenium* sp. represented by a small pinnate* leaf (=frond). Poor preservation of the specimen does not allow for any comparison with extant taxa, making its relationship to living forms problematic. The fossil taxon *Woodwardia muensteriana* (a herbaceous fern) is a typical element of mire floras. This species shares many characters with the somewhat smaller and extant species, *W. virginica*. The fossil herbaceous fern *Blechnum dentatum* grew in the peat marshes as undergrowth of a wetland vegetation. This genus is known in the tropics and subtropics but it is not found in the Czech Republic, today. *Blechnum dentatum* is similar to *B. serrulatum* presently known from southeastern United States (Kvaček et al. 2004a). The fossil *Pronephrium stiriacum*, a swamp or wetland fern (Kvaček et al. 2004a), occupied moist soils as part of the wetland vegetation; this taxon has no known modern relative. In contrast, *Lygodium kaulfusii* is an interesting paleofloristic element from the Most Basin. Recent *Lygodium*, a member of the family Schizaeaceae, is a thermophilic fern and restricted to tropical to subtropical zones. Species of this genus grow in both dry and wet habitats. Thus, the presence of this genus seems to confirm that temperatures were relatively high at the time. Our next stop will be in an equally warm if not warmer environment and time.

Fig. 11.3 *Asplenium scolopendrium* (syn. *Phyllitis scolopendrium*) with linear sorus overlapping veins. Scale bar 20 mm. Photo by Jiří Kvaček; collection of National Museum Prague

and float directly on the water's surface. They represent an interesting group that develop a unique reproductive strategy. Microsporangia (male) and megasporangia (female) develop separately as sporocarps on a common leaf. Microspores are released in a bubbly gelatinous substance (massula) having numerous hooked structures (glochidia) which attach to megaspore. This ensures that both the mega-gametophyte and microgametophyte develop in close proximity to one another, and the sexual reproduction phase can take place more easily. Macrofossils of *Salvinia reussii* from Bílina preserve a very rare case, in that both megasporangia and microspore massulae are found in organic connection to a parent plant. The megasporangia of *S. reussii* yield megaspores named *Salvinia cerebrata*, which are common in the fossil assemblage and found isolated in the lake sediment. *Azolla* species also are represented in the fossil record by spores and macrofossils (Teodoridis 2007). These aquatic fern groups have reached the modern taxonomic structure during the Late Cretaceous and Tertiary (Collinson 2002).

11.4 Cretaceous Ferns from Grünbach, Austria

A five-hour train ride south, with several platform changes, brings us to the Grünbach locality (Fig. 11.6) in the Austrian Alps [U1113]. The rocks exposed at Grünbach belong to the Gosau Formation (Campanian, Late Cretaceous) and represent one of the coal-mining areas in Europe. During the Late Cretaceous, Gosau was a tectonically active island located in a climate suitable for peat accumulation (see Chap. 12).

The vegetation that grew in these Cretaceous "coal forests" was luxuriant, with both pandans and palms (Herman and Kvaček 2010). Ferns were present in the understory and grew as epiphytes on various trees. Both archaic and modern ferns co-occurred in these swamp forests, growing alongside horsetails with stems reaching 3 cm in diameter. Various types of ferns are preserved in the Grünbach flora. These include tree ferns (Dicksoniaceae—*Coniopteris*) with large fronds of *Cladophlebis*-type morphologies, typical wetland

Fig. 11.4 Bílina locality—large lignite open pit mine in the Most Basin (North Bohemia). Photo by Zdeněk Dvořák

dwellers (Marattiaceae and Matoniaceae) of small stature, and the pioneer ferns *Korallipteris* (Gleicheniaceae). In addition, there are several common leaf forms that are traditionally assigned to ferns, among them *Sphenopteris gruenbachiana* (Fig. 11.7) [U1113]. These ferns grew in light breaks of the virgin swamp forest created by fallen trees (Kvaček and Herman 2004).

Ferns in the Cretaceous were influenced by the expansion of flowering plants. As angiosperms diversified and expanded into nearly all environments (see Chap. 5), diversity in the fern families characteristic of the Jurassic and Early Cretaceous declined and new families overtook their role in the environment. There is a marked reduction in ferns assigned to Matoniaceae, Dicksoniaceae, and Marattiaceae, and an increase in the evolution and diversification of Gleicheniaceae, Schizaeaceae, and Polypodiaceae. This turnover in fern groups is typical for the period. Traveling back to the Jurassic will make this turnover even more evident.

11.5 The Early Jurassic Ferns of Anina, Romania

Anina is a Romanian fossil-Lagerstätte (Fig. 11.8) known for both a high biodiversity and excellent preservation of the fossil material [U1114]. It is a key Lower Jurassic plant locality in the Eurosinian paleofloristic context (Popa 1997, 2005, 2009, 2014) and is only a ten-hour drive southeast from Grünbach, through Hungary. Anina is located in the center of the Reșița Basin (also known as the Reșița-Moldova Nouă sedimentary zone), in the Getic Nappe of the south Carpathians. Anina, formerly known as Steierdorf, has a long tradition of coal mining. Coal-mining activities began in 1792 and ended in 2006, hosting both underground and open-cast mines. The underground mines were among the most complex European structures, reaching 1300 m in depth and extending over more than 100 km of active galleries toward the end of the twentieth Century. The intricate underground mine works offer a unique window to the geo-

Fig. 11.5 Fossil water fern *Salvinia reussii*. Scale bar 10 mm. Photo by Zlatko Kvaček

logical past, permitting a highly detailed, three-dimensional study of the floral distribution in these Jurassic "coal forests" (Popa 1992, 1997, 2000, 2005, 2009, 2014). Today, opencast mines and overburden dumps are conserved areas because of their rich fossiliferous content, which includes both plants and animals. The flora is Early Jurassic (Hettangian-Sinemurian) in age, confined to the coal bearing, continental Steierdorf Formation (Bucur 1997; Popa and Kedzior 2008). This succession yields eight bituminous, coking coal seams. The typical compression-impression coal floras preserve an array of bryophytes, pteridophytes, gymnosperms, and of course ferns.

The Anina ferns include vegetative structures, such as rhizomes and leaves, as well as fertile structures with in situ spores. The fern groups include the Marattiales and Filicales. Marattialean ferns are represented by *Marattia intermedia*, a ubiquitous species that is mainly Sinemurian in age (c. 199.3–190.8 Ma). This species with elongated pinnules is very similar to the living tropical species of *Marattia*. The Filicales are represented by families including Osmundaceae, Dipteridaceae, Matoniaceae, and Dicksoniaceae. Last, but not least, there are fragments of sterile fern leaves, for which we do not know their systematic affinity yet. Let us examine representatives of the Filicales in a bit more detail.

There are several Osmundaceae ferns [U1115] preserved in the Anina locality including fertile fragments of *Osmundopsis sturii* associated with leaves and rhizomes of *Cladophlebis denticulata* (Fig. 11.9). When found isolated, *Cladophlebis denticulata* often could not be assigned to any

known group based on morphology. However, in some cases, leaves of this species are associated with *Osmundopsis sturii*, confirming its osmundaceous affinities. A unique site with in situ Sinemurian Osmundaceae ferns occurs in the Ponor Quarry, also in Anina, where a structural surface is covered with their rhizomes in growth position (Popa 2001). Here, some of their remains are preserved as charcoal.

The Dipteridaceae ferns [U1115] are particularly diverse at Anina. They are mainly Hettangian in age (201–199 Ma) and are preserved in the Steierdorf Formation. This family includes *Thaumatopteris brauniana*, which is considered the biostratigraphic marker for the Hettangian floral assemblage in Europe and Asia. In addition, *Dictyophyllum nilssonii*, *D. nervulosum*, *D. irregularis* [U1115], *Hausmannia buchii*, *H. ussuriensis*, and *Clathropteris meniscoides* are common elements of the flora. These taxa display leaf characters similar to modern Dipteridaceae including anastomosed* venation, overall leaf shape, and spore morphology. Other Jurassic fern families also appear similar to their modern counterparts.

Members of the Matoniaceae from the Reșița Basin are both Hettangian and Sinemurian in age. Recognized species include *Matonia braunii*, *Phlebopteris formosa* (with in situ spores; Popa and Van Konijnenburg-Van Cittert 1999), *P. woodwardii* (also with in situ spores), and *P. angustiloba*. These plants are very similar to living Matoniaceae with respect to leaf morphology, reproductive structures, and spore structure.

Representatives of Dicksoniaceae are mainly Hettangian in age. These ferns exhibit small-sized, lobed pinnules and

Fig. 11.6 Historic photograph of the Grünbach locality. Photo by Jiří Kvaček

large fronds, and are similar to living Dicksoniaceae. Representative taxa include *Kylikipteris arguta* (with in situ spores; Popa and Van Konijnenburg-Van Cittert 1999) and *Coniopteris murrayana*. All members of this group, and those described above, played a similar ecological role in the "coal forests."

All ferns of the Steierdorf Formation in the Reşiţa Basin, as well as from all other Lower Jurassic basins of the south Carpathians and Europe, were secondary plants in peat-accumulating forests. Ferns are preserved in association with bennettitaleans and ginkgoaleans (Popa 1998, 2014; Barbacka et al. 2016), and grew as part of the riparian and floodplain vegetation. Ferns occupied open spaces and grew, as well, under various canopies; the plants appear to have been opportunists. Their remains, commonly, are found near the top of swamp deposits interpreted as fern marshes and represent the final phases of peat accumulation before burial by siliciclastic sediments. A similar growth strategy in another deep-time peat forest can be found in Inner Mongolia. It's time to travel back in time to China.

11.6 The Permian Wuda Locality, Inner Mongolia, China

There is an extensive man-made exposure extending kilometers in all directions in the Wuda Coalfield of Inner Mongolia [U1116] in which ferns dominated parts of the landscape. The site originally was located on the northern margin of the North China Plate, when this terrane was a large island located somewhere in the tropical Tethys Ocean. The North China Plate collided with the Mongolian Plate during the latest Permian (Pfefferkorn and Wang 2007), and active volcanism was associated with the convergence of these two tectonic regimes. Explosive volcanism and a thick layer of ash resulted in the preservation of autochthonous coal forests (see Chap. 12) in the area. Here, plant remains are preserved in a volcanic tuff enveloped in the coal bed. This locality represents a T⁰ assemblage (Gastaldo et al. 1995), which is known as either the Chinese "Vegetational Pompeii" (Wang et al. 2012) or the Wuda Tuff Flora (Wang et al. 2013). The flora represents a peat-forming plant assemblage situated on

Fig. 11.7 Common leaf form of *Sphenopteris gruenbachiana* traditionally placed in the ferns. Photo by Jiří Kvaček

a Permian coastal delta plain. Its preservation has allowed for a full characterization of the area at the time of burial.

The flora has been investigated using quantitative methods in several sites, along with the examination of random outcrops extending over a 20 km² area. Two types of fern-dominated forests are recognized. One is a *Cordaites-Psaronius* (C-P) forest in the southern part of the basin, and the other is a *Sigillaria-Psaronius-Paratingia* (S-P-P) forest (Fig. 11.10) [U1119] in the northern part of the coal basin. All fern-like plants, including leptosporangiate and eusporangiate taxa, are found in various parts of the flora. Ferns occupy positions in the groundcover, understory, and canopy.

The common groundcover is the zygopterid fern-like plant *Nemejcopteris haiwangii* Pšenička et al. (2020), which created a dense undergrowth [U1118]. *N. haiwangii* developed erect phyllophores that attained heights of 1–2 m and grew from rhizomes. The leaves of the plant are about 0.5 m long and organized in four series of primary pinnae. Phillips and Galtier (2005) note that zygopterids adapted to dry and high light-intensity environments. However, the Wuda plants

Fig. 11.8 A paleosol (fossil soil) exposed in the Anina Mine as a structural surface with in situ rhizomes and short pseudostems of Osmundaceous ferns. Photo by Mihai Popa

Fig. 11.9 Herbaceous fern *Cladophlebis denticulata* from Ponor Quarry, Anina. Scale bar 10 mm. Photo by Mihai Popa

Fig. 11.10 Reconstruction of the forest from Wuda locality—*Sigillaria-Psaronius-Paratingia* forest. 1—*Sigillaria*, 2—*Psaronius*, 3—*Paratingia*, 4—*Nemejcopteris*, 5—*Eocycas*, Drawing and photo by Jiří Svoboda

grew both on peat substrates and on clastic floodplain deposits in wet, occasionally flooded settings. It is well known for more than a century that xerophytic characters occur in plants growing under ever-wet regimes (Cleal and Shute 2012). Other fern groups also are found as groundcover in these forests.

The leptosporangiate ferns are known to have been lianas or herbaceous forms and represent the predominant groundcover across the forest floor. One of the most frequently encountered leptosporangiate fern species is *Chansitheca wudaensis* (Fig. 11.11) [U1118]. The general growth stature of *C. wudaensis* in still unknown, because this fern is pre-

Fig. 11.11 The most frequent leptosporangiate fern species collected in the Wuda coal field is *Chansitheca wudaensis*. Scale bar 10 mm. Photo by Josef Pšenička

served only as isolated pinnae, which are more than 20 cm long and attached to about 3.5 mm wide rachises*. Their character is interpreted to indicate a herbaceous (He et al. 2016) or scrambling form. *Chansitheca wudaensis* is often accompanied by the foliage of the leptosporangiate fern, *Cladophlebis*, a form we have seen in the Mesozoic. These plants are preserved in the lowermost part of the tuff layer, indicating that they were likely groundcover at the time when ash covered and buried the forest. The leptosporangiate ferns in the Wuda locality are still under investigation. But, to date, at least eight species are recognized and it is assumed that additional new species will be identified in future.

Overall, it is clear that the ferns played an important role in the coal forests of the Permian. Their in situ preservation in volcanic ash allows us to reconstruct their whole-plant biology, as well as those of other taxa. To date, whole plant reconstructions include the spore-producing plants *Sphenophyllum angustifolium* (Yan et al. 2013), *Sigillaria* cf. *ichthyolepis* (Wang et al. 2009b), *Paratingia wudensis* (Wang et al. 2009a), and *Tingia unita* (Wang 2006). In addition, the intraspecific variation of the frond of *Pecopteris lativenosa* has been documented by Zhang et al. (2008), and an abnormal leaf, possibly referable to *Pecopteris norinii*, has been given the name *Aphlebia hvistendahliae* (Wang et al. 2014). A nearly complete reconstruction [U1118] of all life stages of *Nemejcopteris haiwangii* has been undertaken, demonstrating what is possible for other Wuda coalfield taxa (Pšenička et al. 2020). What is remarkable is the continuity of the fern form from older coal forests of the Carboniferous.

All fern taxa from Wuda show more or less evolutionary continuity from the coal forests of Carboniferous age (see Chap. 13). Many fern groups interpreted to be archaic in

nature, such as the zygopterid fern–like plants or eusporangiate marattialean ferns, were abundant in the Permian landscape. Nevertheless, at the same time, ferns from the coastal-deltaic sediments of the Wuda locality also show a newly derived reproductive strategy found in modern leptosporangiate ferns [U1119]. Did similar fern-dominated Permian landscapes occur on the other side of the hemisphere?

11.7   The Permian Chemnitz and Carboniferous Flöha Localities (Germany)

Two localities, Chemnitz and Flöha, are situated in Saxony (southestern Germany) and exquisitely preserve ferns in volcanic ash deposits, similar to the Wuda coalfield. Chemnitz and Flöha are parts of the post-Variscan intramontane basin structures developed in Central Europe during the latest Paleozoic. Here, some pioneering paleobotanical work was undertaken by Frenzel (1759), Sprengel (1828), and Cotta (1832). A common character of both Flöha and Chemnitz sites is that each forest ecosystem was entombed instantaneously, preserving significant T^0 assemblages. These forests were buried by pyroclastic flows and accompanying airfall ash deposits from very local volcanoes [U1120]. Analyses of isolated zircon crystals using U-Pb geochronological dating methods indicate that the Flöha site was buried about 310 million years ago and is Bolsovian (Carboniferous) in age (Löcse et al. 2019). In contrast, the Chemnitz Fossil Lagerstätte is approximately 290.6 million years old, and assigned to a Sakmarian-Artinskian (Permian) age (Rößler et al. 2009). These fossil assemblages are preserved most often as compressions/impressions, which rarely preserve cellular or anatomical detail. What is unusual about both localities is the presence of permineralization, which often are referred to as petrifactions. The mode of preservation known as petrifaction is where the internal structures are replaced, or filled, by silica, carbonate, various iron oxides, or fluorite minerals. These minerals preserve cellular anatomical structures in three dimensions. Each locality is unique, and the setting for each flora differs.

The Pennsylvanian Flöha assemblage is found in grey-colored, coal-bearing sediments overlain by volcanic pyroclastic deposits, whereas the Chemnitz locality consists of alluvial red-beds. The Flöha flora is interpreted to have flourished under equable wet tropical conditions. In contrast, the sedimentology at Chemnitz indicates the area experienced more of a seasonal climate with alternating wet and dry phases (Luthardt et al. 2016, 2017). The forests preserved at both Flöha and Chemnitz are dense and multi-aged, with water-loving (hygrophilous) elements dominated by conservative Carboniferous lineages. Whereas the Chemnitz ecosystem developed in a seasonally dry alluvial plain setting with sparse vegetation cover, and likely represented a "wet spot" sensu DiMichele et al. (2006) characterized by a sub-humid local paleoclimate, the Flöha site represented a typical peat-forming alluvial swamp.

11.7.1 Permian Chemnitz Locality

The flora preserved at Chemnitz consists of arborescent plants at all growth (ontogenetic) stages but also preserves a rich terrestrial fauna consisting of invertebrates (arthropods) and vertebrates (Rößler et al. 2012; Dunlop et al. 2016; Spindler et al. 2018) (see Chap. 12). Fossil trees are commonly in situ, upright, and rooted in the underlying mineral-rich paleosol [U1120] (Rößler et al. 2014). The root systems show adaptation to specific substrates, and seasonal growth interruptions are recorded in the stems, branches, and roots of long-lived woody plants (Luthardt et al. 2017). Tree ferns play a significant role in the forest.

There are different tree ferns in the Chemnitz assemblage. The tree fern *Psaronius* (Fig. 11.13) is dominant and accounts for approximately 22% of the arborescent vegetation [U1121]. The growth forms of the taxon (Box 11.4) include various distichous∗ (leaves borne on opposite sides of the trunk) and polystichous (spirally arranged phyllotaxy) forms. In contrast, smaller tree forms, such as *Asterochlaena ramosa* or *A. laxa*, are rare [U1122] (Rößler and Galtier 2002). These trunks are characterized by a central actinostele (the central vascular core is lobed or fluted) surrounded by a mantle of spirally arranged, rather loose semi-erect persistent petioles that bore the pinnae and pinnules, rarely preserved in growth position. These trees also developed small adventitious

Box 11.4: The *Psaronius* Plant

The tree fern *Psaronius* [U1117], which could have grown to more than 4 m in height, is a member of the eusporangiate ferns and is part of the forest understory. Together with zygopterid fern-like plants, they formed the major components of both the C-P and S-P-P forest communities. It is possible to gain a better understanding of the group's ecology from the exquisitely preserved 5–30 cm wide, 3-dimensional *Psaronius* trunks and leaves (fronds). The erect trees grew 3–5 m apart from each other, indicating that they probably formed a rather open understory layer. Leaves of *Psaronius*, referred to as the genus *Pecopteris*, are often found in organic connection with the parental trunk, and nine different species have been identified, to date. In addition, several trees were preserved during a reproductive phase. The fertile fronds of many *Pecopteris* species have been collected and likely belong to one of the several fossil-genera of reproductive structures (e.g., *Scolecopteris* or *Eoangiopteris*).

The association or, in rare cases, the organic attachment of *Psaronius* and *Pecopteris* can be found in several localities other than the Permian Wuda (China) coal. These other sites include Chemnitz (Germany) (Fig. 11.12) or the Carboniferous Ovčín locality near Radnice (Czech Republic). The relationship between *Psaronius* and *Pecopteris* has been inferred for a long time, based on leaves associated or attached to compressed trunks on which elliptical leaf scars occur on the surface of the stem [U1117]. Trunks show different size, shape, and organization of leaf scars and, therefore, different morpho-genera were established to accommodate the variation (e.g., *Caulopteris*, *Megaphyton*). Such adpression trunks traditionally were referred to as *Psaronius*. Clear evidence that these leaf-scar types were, in fact, *Psaronius* was discovered on petrified [U1121] or partly petrified (3D preserved) trunks preserved in both the Wuda and Chemnitz localities. Thus, we have conclusive evidence that elliptical leaf-scar patterns on the surface of adpression trunks correspond with leaf scars discovered on petrified material. Ultimately, a natural reconstruction of the plant can be made using *Psaronius* trunks and *Pecopteris* leaves.

roots∗. Stelar and foliar anatomy provide reliable diagnostic characteristics that place *Asterochlaena* in the zygopterid ferns. However, to date, the phylogenetic position of *Asterochlaena* is still not established because the fertile organs of the plant are unknown. Zygopterid and other fern groups also are found as parts of the groundcover.

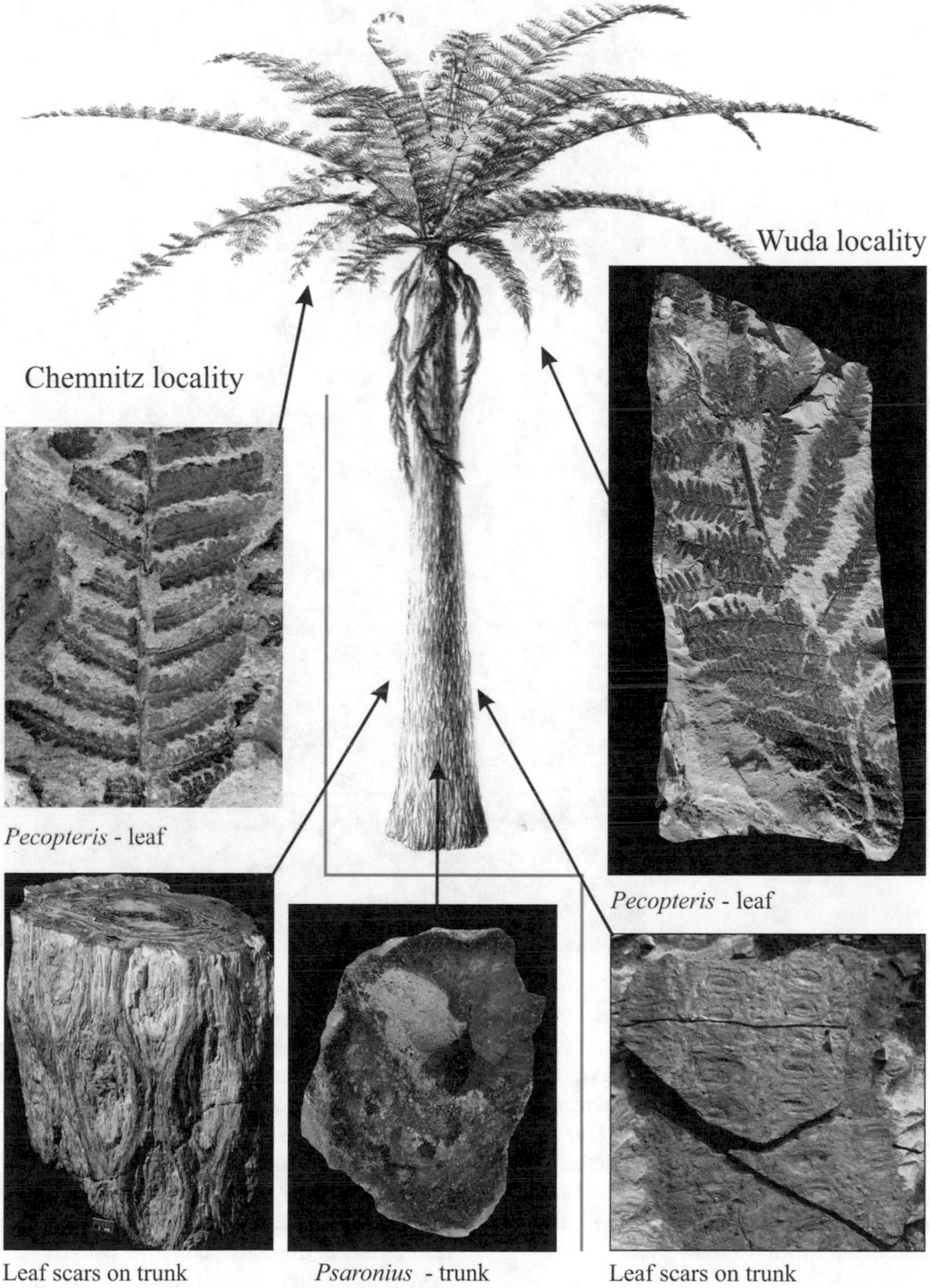

Fig. 11.12 Leaves and trunks of three *Psaronius* ferns from the Chemnitz and Wuda localities. Drawing by Jiří Svoboda, photos by Josef Pšenička and Ronny Rößler

Herbaceous zygopterid or anachoropterid (filicalean) ferns are quite common at Chemnitz, and these non-arborescent forms represent either climbing or epiphytic ferns associated with *Psaronius* trunks. They often are preserved in the dense mantle of adventitious aerial roots of other tree ferns (Rößler 2000). The liana-like zygopterid, *Ankyropteris brongniartii*, and the epiphytic anachoropterids, *Tubicaulis grandeuryi* and *T. bertheri*, are frequently encountered. In

Fig. 11.13 Three dominant ferns assigned to *Psaronius* at Chemnitz. *S*—stele, *IRM*—inner root mantle, *ORM*—outer root mantle. Scale bar 100 mm. Photo by Ronny Rößler

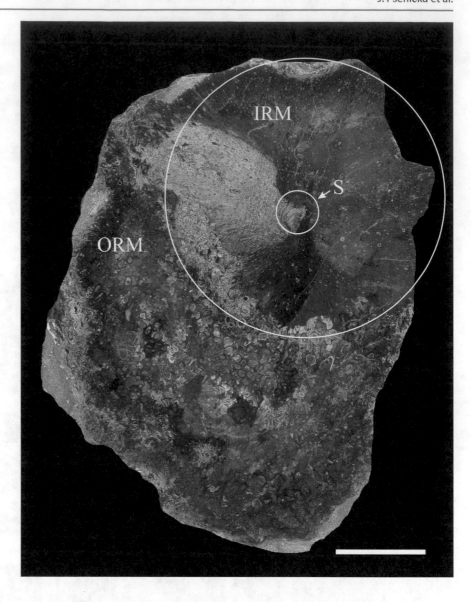

most instances, several axes of both climbers and epiphytes have been found on the same *Psaronius* trunk. And, in one instance, the root mass of *A. brongniartii* is preserved encircling a stem of a standing specimen of *Calamites striata*. Easily recognizable by their diarch roots (lateral roots develop between the conducting strands), small ferns are not only identifiable in the voluminous root mantles of *Psaronius* near the trunk base. In some instances, though, they also have been found in the crowns of these trees.

11.7.2 Carboniferous Flöha Locality

The Flöha flora has yielded several unique tree ferns and fern-like plants that extend the biological ranges of several Permian taxa back into the Pennsylvanian. These plants include *Tubicaulis solenites*, *Zygopteris primaria*, and distichous *Psaronius* sp. [U1123] (Löcse et al. 2013, 2017). And,

the new observations may challenge traditional ideas of evolutionary development of zygopterid and anachoropterid ferns in the future. Compression fossils, referable to *Alloiopteris*, a semi-self-supporting zygopterid, are found in the ash deposits and show that this plant grew as part of the understory [U1124].

Entombment by volcanic ash allows for excellent preservation of in situ plant communities, providing a snapshot of the ecological diversity across the landscape. Recent, large road construction projects have resulted in new outcrops where both field work and radiometric dating of the volcanic host rocks are now underway (Löcse et al. 2015, 2019). Petrified fossil ferns from Chemnitz and Flöha have significantly enriched our understanding of Permian and late Carboniferous paleoenvironments in which they grew, anatomical and morphological characters of the taxa that inhabited these forests and, thus, provided additional information about the evolutionary puzzle of ferns. A quick return to the

Czech Republic provides a window into the role that these fern groups played in the Middle Pennsylvanian.

11.8 The Carboniferous Ovčín Locality (Czech Republic), Whetstone Horizon

The Ovčín locality, situated in the Radnice Basin of the Czech Republic, is a part of a larger geological complex of continental basins that extends from western to northeastern Bohemia. The only Pennsylvanian strata of early Moscovian (Late Duckmantian and middle Bolsovian, 315.2–307 Ma) age in this region are found in the Radnice Basin [U1125] (Opluštil et al. 2016). Four sedimentary facies are recognized in the succession of rocks, with fluvial strata accounting for the majority of the stratigraphic section. The other three facies—colluvial, lacustrine, and coal—are a minor part of the succession (Opluštil 2005). The productive coals are known as the Lower and Upper Radnice Coals that are separated by the widespread and thick silica-cemented volcanic ash bed, known as the Whetstone Horizon. The "Bělka" bed, in the lowermost part of the Whetstone Horizon, has been the focus of paleobotanical studies for many years. The "Bělka" is interpreted as an ash-fall deposit that covered an extensive area in the western part of the Czech Republic (Opluštil et al. 2014). According to Opluštil et al. (2016), the entombed fossils grew on a landscape about 313.4 million years ago; the age estimate is based on U-Pb isotopic analyses of magmatic zircon crystals. An in situ Carboniferous forest was buried by volcanic ash in this bed (see Chap. 12). In addition to the foliage of the canopy, coalified and upright stems are preserved rooted in mudstone overlying the lower Radnice Coal. Due to the rapid burial of the forest, the spatial distribution and vertical stratification of the plants are preserved allowing for an understanding of the forest's synecological relationships in extraordinary detail.

Volcanic ash buried a lepidodendrid-cordaitalean forest with a diverse fern component (Fig. 11.14) [U1127] (Opluštil et al. 2014). The ferns are represented by taxa with a variety of growth forms including low trees and herbaceous groundcover found in the understory and as lianas. In the exposures studied, to date, *Psaronius* trunks are associated with a single megaphyllous frond, *Lobatopteris aspidioides* [U1126]. Most of specimens of *L. aspidioides* are sterile, but occasionally fertile pinnae are recovered. Fertile organs are represented by synangia, a character used to assign these plants to the eusporangiate ferns. This 2–4 m tall tree was the dominant tree fern in the lower stories of the forest. The occurrence of only a single Marattialean species at Ovčín is in sharp contrast with other Carboniferous-Permian localities in the Northern Hemisphere where several Marattialeans co-occur. But Marratialeans were not the only understory ferns.

Fern-like zygopterids are the most dominant plant at the Ovčín locality. They are represented by two species: the fossil-genus* "*Corynepteris*" *angustissima* [U1126] is common (Fig. 11.15) whereas *Desmopteris longifolia* is rare. Abundant fossil remains of "*Corynepteris*" *angustissima* demonstrate that this plant developed a 1–2 m erect phyllophore, bearing about 0.4 m long biseriate primary pinnae, growing from rhizomes. In contrast, because of its rarity, we have little knowledge of *Desmopteris longifolia* due to its fragmentary nature. Both ferns were part of the understory, each may have grown in small patches, and each is interpreted to have shared more open (less shaded) areas of the peat swamp (Opluštil et al. 2009a, b).

Ferns growing as either groundcover or with a vine-habit also are present. "Small" leptosporangiate ferns were subdominant in the forest. These are represented by the herbaceous form *Hymenotheca* sp. or the lianas *Oligocarpia lindsaeoides* (Pšenička and Bek 2001), *Senftenbergia plumosa* [U1126] (Bek and Pšenička 2001; Pšenička and Bek 2003), and *Sonapteris bekii*.

The Carboniferous forests and Ovčín, in particular, represent a part of biological history when fern-like plants evolved together with the eusporangiate ferns, and accompanied by a variety of leptosporangiate ferns. In many instances, though, fossil material of leptosporangiate ferns often is highly degraded and difficult to assess. Their poor record may be due to decay or breakdown during transport to a site of burial away from where they grew, or rapid decay based on their anatomical organization. It is in contrast, then, that the Ovčín locality preserved a relatively high diversity of leptosporangiate ferns. But, when we understand that this locality represents an in situ forest that had been instantaneously buried by volcanic ash (represented T⁰ plant assemblage), their presence is not much of a surprise. The Ovčín phytocoenosis, more or less, corresponds with the original plant composition at the time of burial. Due to this fact, the Ovčín locality plays an important role in the evaluation of fern evolution.

11.9 The Devonian Hostim Locality Near Srbsko (Barrandian, Czech Republic)

Only an hour's drive to the southwest from Ovčín, through Prague and the surrounding countryside, are outcrops of marine sediments of the Srbsko Formation (Givetian, Middle Devonian, 387–383 Ma) [U1128]. These are exposed near Beroun in the Czech Republic. Few in situ fossil-plant assemblages of Devonian age are reported from around the world, and many localities only preserve transported (allochthonous*) plant remains. The well-preserved fossil plants at Srbsko include *Protopteridium hostimense*, *Pseudosporochnus*, *Protolepidodendron scharianum* (probably a club moss [lycophyte]), and *Barrandeina dusliana*.

Fig. 11.14 View across a Pennsylvanian peat-forming forest. 1—*Psaronius* (tree fern), 2—*Lepidodendron* (tree lycopsid), 3—*Lepidophloios* (tree lycopsid), 4—*Calamites* (horsetail), 5—"*Corynepterys*" (fern-like plant), 6—*Sphenophyllum* (horsetail), 7—*Sphenopteris mixta* (seed fern) 8—*Spencerites* (herbaceous lycopsid). Images by Jiří Svoboda

Fig. 11.15 An example of a zygopterid fern, "*Corynepteris*" *angustissima*, from the understory of the Ovčín fossil-forest locality. Scale bar 20 mm. Photo by Josef Pšenička

Fig. 11.16 *Pseudosporochnus* sp. from Hostim, Czech Republic. Scale bar 20 mm. Photo by Jiří Kvaček; collection of National Museum in Prague

The first two fossil-plant taxa are considered fern-like plants (Kenrick and Crane 1997) and placed into a special "fern" group, the Cladoxylales. Cladoxylalean ferns attained an unusual tree-like growth habit. Compression specimens of the cladoxylalean *Pseudosporochnus* (Box 11.5) collected from the Hostim locality exhibit dichotomizing ultimate segments that also appear to be planar. This feature may be due to the orientation of the axes in the rock or may be a result of compression during burial. The assemblage preserves a large number of one species, *Pseudosporochnus krejcii* (Fig. 11.16), and it is believed that this plant was a dominant element of the Devonian coastal "forests" (Fig. 11.17).

Several genera/species were established as fern-like plants from Devonian-aged rocks over the course of paleobotanical history. These forms usually are placed into a group called Cladoxylopsida, which also includes *Pseudosporochnus*. The onset of forms that appear fern-like and may signal the group's evolution is typified by a

Box 11.5: The *Pseudosporochnus* Plant
The cladoxylalean fern *Pseudosporochnus* is interpreted to have been a small tree or shrub attaining a height of 1–4 m. What is unusual about the plant is

Fig. 11.17 General view of a Devonian "forest". 1—*Protopteridium*, 2—*Pseudosporochnus*, 3—*Barrandeina*, 4—*Protolepidodendron*. By Jiří Svoboda

that it is "rooted" with a bulbous base [U1129]. The plant did not produce a leaf in the strict sense of a fern leaf. The "leaves" of these plants are subdivided branches that developed at the upper (distal) end of a robust and tapering main "trunk." These terminal branching units, forming a crown, were the photosynthetic organs. Cladoxylaleans reproduced via sporangia attached to the end of these ultimate branching units. Sterile and fertile ultimate units are slightly different. Berry and Fairon-Demaret (2002) have demonstrated that second-order branches equally forked (dichotomized) numerous times, resulting in a three-dimensional terminal branching architecture. Yet, there is evidence in some anatomically preserved specimens of *Pseudosporochnus* from the Devonian of eastern New York (Matten 1974) that the ultimate appendages may have been similar to planar, two-dimensional fronds of modern ferns. When sporangia developed, these were large (c. 3 mm long) and ellipsoidal. To date, the spores that developed in the sporangia are unknown. It is likely that the evolution of the ferns may have begun in the Givetian, or sometime prior to 383 million years ago. However, the systematic placement of cladoxylaleans remains enigmatic (see Chap. 15), because both the anatomical and morphological features of the group are not associated with any typical fern.

large number of homoplasies (traits lost or gained independently in separate evolutionary lineages). And, because of these, the earliest representatives have been treated with a varied taxonomic history. Currently, it is not often clear which Devonian fossil species represent true fern taxa. We have chosen the Hostim window [U1130] as a representative of the beginning of fern evolution but realize that new data may change our understanding of the group.

11.10 A Brief Evolution of Ferns

As our virtual trip highlighted using various localities over geologic time, the group we know as ferns established themselves sometime in the Devonian and have continued to be important in ecosystems until today [U1131]. These plants have maintained, more or less, the same major characteristics that they attained during the early evolution of the group (Fig. 11.18). These are the presence of planar leaves and compound fronds with reproductive organs located on the underside of the pinnules, an evolutionary strategy that ensures their protection against desiccation. All ferns reproduce by spores that require a wet substrate for germination. Nevertheless, early fern-like plants grew as small-to-moderate size tree-like architectures or as rhizomatous* forms, with reproductive organs often situated at the end of ultimate appendages, restricted to wetlands. One early fern-like plant from the Devonian Hostim locality near Srbsko in

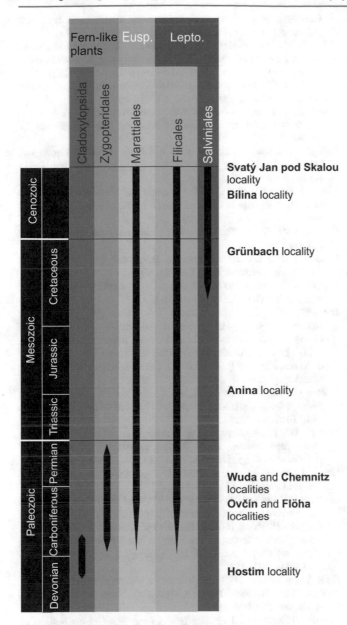

Fig. 11.18 Simplified evolutionary history of ferns from the Devonian to the Cenozoic. *Eusp.* eusporangiate ferns, *Lepto* leptosporangiate ferns. By Josef Pšenička

dichotomous* branching system. More recently, Galtier (2010) proposed the origin from a modification of large lateral branch systems as found in *Pseudosporochnus*. The diversification of the group during the Carboniferous exhibits features not found today.

By the time of the "coal forests", several early Carboniferous fern-like plants (e.g., zygopterids) had developed three-dimensional fronds with planar penultimate pinnae. But, one unusual feature of these plants was the appearance of secondary (woody) tissue originating in horizontal rhizomes. These features are unknown in extant ferns. At the same time, early Carboniferous eusporangiate and leptosporangiate ferns probably also had evolved (Pryer et al. 2004). This important and interesting phase of fern evolution is captured at both the Ovčín locality (Czech Republic) and Flöha locality (Germany). Diversification continued during the Late Paleozoic Ice Age.

Large arborescent Marattialean ferns evolved in the Carboniferous and Permian with eusporangiate representatives, including *Psaronius*, having grown to heights of about 8 m. These forms appear more familiar because they established a crown of large, pinnate leaves that were either spirally arranged at the top of, or alternately spaced along, the trunk. Generally, eusporangiate ferns were abundant in the Pennsylvanian and Permian paleotropical wetlands of Euramerica and Cathaysia. Although greatly diminished in diversity, they have persisted to the present. Localities at Chemnitz (Germany) and Wuda (China) offer a window into this phase of fern evolution.

The largest group of living monilophytes are leptosporangiate ferns, which are known from the early Carboniferous to Recent. The more-derived polyploid ferns, comprising more than 80% of living fern species, diversified in the Cretaceous. Their diversification has been interpreted as an ecological opportunistic response to the diversification of angiosperms (Schneider et al. 2004). Their expansion also impacted the group as a whole. Fern families typical of the Jurassic and Early Cretaceous, including Matoniaceae, Dicksoniaceae (both Filicales), and Marattiaceae (Marattiales), experienced a decrease in diversity. In contrast, fern families including Gleicheniaceae, Schizaeaceae, and Polypodiaceae (all Filicales) underwent a radiation. The "modernization" of aquatic ferns also happened during the Cretaceous, indicating that this time interval is another important phase in fern evolution. This evolutionary step is captured at Anina (Romania) and Grünbach (Austria). By the time we arrived at Bílina (Czech Republic) and Svatý Jan pod Skalou (Czech Republic), both of which are of stratigraphically youngest age, the fern groups we encounter appear most similar to taxa living in today's world.

In general, ferns have maintained a great level of diversity beginning in the early Carboniferous and continuing to the Recent. The diversification of the group has allowed these

the Czech Republic has a branching system without a clear stem/leaf organography. According to Pryer et al. (2004), the first clade that diverged from the crown-group of ferns consisted of fern-like plants together with ophioglossoids, whereas marattioids, eusporangiate, and leptosporangiate ferns diverged during the Late Devonian. The appearance of large, planar fronds (megaphylly) is an important evolutionary step that provided a characteristic aspect to ferns, as we know them today. Holmes (1989) studied the branching pattern of some middle Carboniferous planar filicalean fern fronds (e.g., Psalixochlaenaceae or Botryopteridaceae) and proposed their evolution from a preceding three-dimensional

plants to adapt to varied physical conditions. Therefore, it is not surprising to find that ferns occupy almost every conceivable environment around the world.

Questions

1. Ferns have been around for hundreds of millions of years. What are some of the key features of ferns that make them successful?

2. What cellular features of sporangia are used to differentiate fossil eusporangiate and leptosporangiate ferns, and why might these criteria not apply to the classification of living forms?

3. What is the concept of "alternation of generations," and what role does this life strategy play in ferns?

4. Why might the Osmundaceae be considered as "living fossils"?

5. Name several benefits to the investigation of fossil floras having been preserved in volcanic ash.

6. What evidence exists in the Wuda Coal forest for the tiering (multi-level development) of ferns?

7. Which physical characteristics of tree-fern trunks (aerial rhizomes) allow you to differentiate different taxa in the fossil record?

8. Describe the differences in growth forms of arborescent (tree habit) Late Devonian Cladoxylaleans and Marattialean tree ferns of the late Carboniferous and early Permian.

9. Why can the Cretaceous be considered a critical phase in the evolution of ferns?

Acknowledgments The authors would like to acknowledge Mathias Merbitz of the Museum für Naturkunde of Chemnitz for research on the Chemnitz locality; Zlatko Kvaček, Jakub Sakala of the Faculty of Natural Science, Charles University of Prague, for providing specimens from some localities (Bílina) and improving the text that deals with Cenozoic ferns from the Bílina locality; Jiří Dvořák of Severočeské doly a.s. of Bílina, for providing a photograph of the Bílina Mine; Jiří Svoboda for providing paintings of all reconstructions. Special thanks to Dan Chaney for constructive remarks and grammar correction of the text. This chapter was partly supported by Gr. No. 2016VEA004 (Chinese Academy of Sciences Visiting Professorship for Senior International Scientists) for J. Pšenička, National Natural Science Foundation of China (Grant 41530101), the Strategic Priority Research Program (B) of the Chinese Academy of Sciences (XDB18030404) for J. Wang, and Grant Agency of the Czech Republic (Grants 19-06728S and 17-102333S).

References

Barbacka M, Popa ME, Mitka J, Bodor E, Puspoki Z, McIntosh RW (2016) A quantitative approach for identifying plant ecogroups in the Romanian early Jurassic terrestrial vegetation. Palaeogeogr Palaeoclimatol Palaeoecol 446:44–54

Bek J, Pšenička J (2001) *Senftenbergia plumosa* (Artis) emend. and its spores from the Carboniferous of the Kladno and Pilsen Basins, Bohemian Massif, and some related and synonymous taxa. Rev Palaeobot Palynol 116:213–232

Berry CM, Fairon-Demaret M (2002) The architecture of *Pseudosporochnus nodosus* Leclercq et Banks: a Middle Devonian Caldoxylopsid from Belgium. Int J Plant Sci 163:699–713

Bucur II (1997) Formaţiunile mesozoice din zona Reşiţa-Moldova Nouă. Presa Universitară Clujeană, Cluj-Napoca, 214pp

Bůžek Č (1971) Tertiary flora from the northern part of the Pětipsy area (North-Bohemian basin). Rozpr Ústředního ústavu geologického 36:1–118

Chlupáč I, Brzobohatý R, Kovanda J, Stráník Z (2002) Geologická minulost České republiky (in Czech). Akademie Věd České Republiky, Academia. 432 pp

Cleal CJ, Shute CH (2012) The systematic and palaeoecological value of foliage anatomy in Late Palaeozoic medullosalean seed-plants. J Syst Palaeontol 10(4):765–800

Collinson ME (2002) The ecology of Cainozoic ferns. Rev Palaeobot Palynol 119:51–68

Cotta B (1832) Die Dendrolithen in Bezug auf ihren inneren Bau. Arnoldische Buchhandlung, Leipzig, 89pp

DiMichele WA, Gastaldo RA (2008) Plant paleoecology in deep time. Ann Mo Bot Gard 95(1):144–198

DiMichele WA, Tabor NJ, Chaney DS, Nelson WJ (2006) From wetlands to wet spots: environmental tracking and the fate of Carboniferous elements in Early Permian tropical floras. In: Greb SF, DiMichele WA (eds) Wetlands through time: Geological Society of America Special Paper, vol 399. Geological Society of America, Boulder, pp 223–248

Dunlop JA, Legg DA, Selden PA, Vet V, Schneider JW, Rößler R (2016) Permian scorpions from the Petrified Forest of Chemnitz, Germany. BMC Evol Biol 16:72

Frenzel D (1759) Zuverlässige Nachricht von einem zu Steine gewordenen Baume, nebst dessen eigentlicher Abbildung. In: Dresdnisches Magazin, vol 1. Michael Gröll, Dresden, pp 39–47

Galtier J (2010) The origins and early evolution of the megaphyllous leaf. Int J Plant Sci 171(6):641–661

Gastaldo RA, Pfefferkorn HW, DiMichele WA (1995) Taphonomic and sedimentologic characterization of "roof-shale" floras. In: Lyons P, Wagner RH, Morey E (eds) Historical persepctive of early twentieth century Carboniferous paleobotany in North America: in memory of William Culp Darrah, vol 185. Geological Society of America, Boulder, pp 341–352

He X, Wang S, Wang J (2016) *Chansitheca wudaensis* (Gleicheniaceae, fern) from the early Permian WudaTuff Flora, Inner Mongolia. Palaeoworld 25:199–211

Herman AB, Kvaček J (2010) Late Cretaceous Grünbach flora of Austria. Naturhistorisches Museum Wien, Vienna. 215pp

Holmes J (1989) Anomalous branching patterns in some fossil *Filicales*: implications in the evolution of the megaphyll and the lateral branch, habit and growth pattern. Plant Syst Evol 165:137–158

Kenrick P, Crane PR (1997) The origin and early diversification of land plants: a cladistic study. Smithsonian Institution, Washington, DC

Kvaček Z, Bůžek Č (1983) Třetihorní rostlinná společenstva severočeské hnědouhelné pánve ve vztahu k litofaciálnímu vývoji (In Czech). Výzkumná zpráva, Ústav geologie a geotechniky ČSAV, Ústřední ústav geologický, Praha, 46pp

Kvaček J, Herman AB (2004) The Campanian flora of Lower Austria: palaeoecological interpretations. Ann Naturhist Mus Wien 106A:91–101

Kvaček Z, Dvořák Z, Mach K, Sakala J (2004a) Třetihorní rostliny severočeské hnědouhelné pánve. Severočeské doly a.s., Chomutov, Granit, s.r.o., Praha, 158pp

Kvaček Z, Böhme M, Dvořák Z, Konzalová M, Mach K, Prokop J, Rajchl M (2004b) Early Miocene freshwater and swamp ecosystems of the Most Basin (northen Bohemia) with particular reference to the Bílina Mine section. J Czech Geol Soc 49(1-2):1–40

Löcse F, Meyer J, Klein R, Linnemann U, Weber J, Rößler R (2013) Neue Florenfunde in einem Vulkanit des Oberkarbons von Flöha - Querschnitt durch eine ignimbritische Abkühlungseinheit. Veroff Mus Naturkunde Chem 36:85–142

Löcse F, Linnemann U, Schneider G, Annacker V, Zierold T, Rößler R (2015) 200 Jahre *Tubicaulis solenites* (Sprengel) Cotta - Sammlungsgeschichte, Paläobotanik und Geologie eines oberkarbonischen Baumfarn-Unikats aus dem Schweddey-Ignimbrit vom Gückelsberg bei Flöha. Veroff Mus Naturkunde Chem 38:5–46

Löcse F, Zierold T, Rößler R (2017) Provenance and collection history of *Tubicaulis solenites* (Sprengel) Cotta – a unique fossil tree fern and its 200-year journey through the international museum landscape. J Hist Collections 30:241

Löcse F, Linnemann U, Schneider G, Merbitz M, Rößler R (2019) First U-Pb LA-ICP-MS zircon ages assessed from a volcano-sedimentary complex of the mid-European Variscids (Pennsylvanian, Flöha Basin, SE Germany). Int J Earth Sci 108:713

Luthardt L, Rößler R, Schneider JW (2016) Palaeoclimatic and site-specific conditions in the early Permian fossil forest of Chemnitz: sedimentological, geochemical and palaeobotanical evidence. Palaeogeogr Palaeoclimatol Palaeoecol 441:627–652

Luthardt L, Rößler R, Schneider JW (2017) Tree-ring analysis elucidating palaeo-environmental effects captured in an in situ fossil forest – the last 80 years within an Early Permian ecosystem. Palaeogeogr Palaeoclimatol Palaeoecol 487:278–295

Mach K, Teodoridis V, Grygar MT, Kvaček Z, Suhr P, Standke G (2014) Evaluation of palaeogeography and palaeoecology in the Most Basin (Czech Republic) and Saxony (Germany) from Late Oligocene to Early Miocene. N Jb Geol Paläont 272(1):13–45

Matten LC (1974) The Givetian flora from Cairo, New York: *Rhacophyton, Triloboxylon* and *Cladoxylon*. Bot J Linn Soc 68:303–318

Němejc F (1927) Paleobotanická studie o fossilních travertinových sedimentech v kraji mezi Prahou a Berounem (In Czech). Rozpravy II. Třídy České akademie 36(22):1–9

Opluštil S (2005) Evolution of the Middle Westphalian river valley drainage systém in central Bohemia (Czech Republic) and its palaeogeographic implication. Palaeogeogr Palaeoclimatol Palaeoecol 222(3–4):223–258

Opluštil S, Pšenička J, Libertín M, Bek J, Dašková J, Šimůnek Z, Drábková J (2009a) Composition and structure of an in situ Middle Pennsylvanian peat-forming plant assemblage in volcanic ash, Radnice Basin (Czech Republic). Palaios 24:726–746

Opluštil S, Pšenička J, Libertín M, Bashforth AR, Šimůnek Z, Drábková J, Dašková J (2009b) A Middle Pennsylvanian (Bolsovian) peat-forming forest preserved in situ in volcanic ash of the Whetstone Horizon in the Radnice Basin, Czech Republic. Rev Palaeobot Palynol 155:234–374

Opluštil S, Pšenička J, Bek J, Wang J, Feng Z, Libertín M, Šimůnek Z, Bureš J, Drábková J (2014) T⁰ peat-forming plant assemblage preserved in growth position by volcanic ash-fall: a case study from the Middle Pennsylvanian of the Czech Republic. Bull Geosci 89(4):773–818

Opluštil S, Schmitz M, Cleal CJ, Martínek K (2016) A review of the Middle–Late Pennsylvanian west European regional substages and floral biozones, and their correlation to the Geological Time Scale based on new U–Pb ages. Earth Sci Rev 154:301–335

Pfefferkorn HW, Wang J (2007) Early Permian coal-forming floras preserved as compressions from the Wuda District (Inner Mongolia, China). Int J Coal Geol 69(1-2):90–102

Phillips TL, Galtier J (2005) Evolutionary and ecological perspectives of Late Paleozoic ferns: part I. Zygopteridales. Rev Palaeobot Palynol 135:165–203

Popa ME (1992) The Early Liassic of Anina: new paleobotanical aspects. Doc Nat 1-3:1–9

Popa ME (1997) Liassic ferns from the Steierdorf Formation, Anina, Romania, 4th European Palaeobotanical and Palynological Conference. Heerlen: Mededelingen Nederlands Instituut voor Toegepaste Geowetenschappen TNO, pp 139–148

Popa ME (1998) The Liassic continental flora of Romania: Systematics, Stratigraphy and Paleoecology. Acta Bot Horti Bucurestensis 1997-1998:177–184

Popa ME (2000) Early Jurassic land flora of the Getic Nappe, Faculty of Geology and Geophysics. University of Bucharest, Bucharest, p 258

Popa ME (2001) Ponor SSSI (Site of Special Scientific Interest). Lower Jurassic Paleoflora. In: Bucur II, Filipescu S, Săsăran E (eds) Algae and carbonate platforms in western part of Romania. Field trip guidebook. Babes-Bolyai University, Cluj-Napoca, pp 167–171

Popa ME (2005) Aspects of Romanian Early Jurassic palaeobotany and palynology. Part VI. Anina, an exceptional locality. Acta Palaeontol Rom 5:375–378

Popa ME (2009) Late Palaeozoic and Early Mesozoic continental formations of the Reșița Basin. Editura Universității din București, Bucharest, 197pp

Popa ME (2014) Early Jurassic bennettitalean reproductive structures of Romania. Palaeobio Palaeoenv 94:327–362

Popa ME, Kedzior A (2008) High resolution paleobotany and sedimentology of the Steierdorf Formation, Reșița Basin. In: Bucur II, Filipescu S (eds) Annual scientific session "Ion Popescu Voitești". Cluj University Press, Cluj-Napoca, pp 57–59

Popa ME, Van Konijnenburg-Van Cittert JHA (1999) Aspects of Romanian Early Jurassic palaeobotany and palynology. Part I. In situ spores from the Getic Nappe, Banat, Romania. Acta Palaeobot Suppl 2:181–195

Pryer KM, Schuettpelz E, Wolf PG, Schneider H, Smith AR, Cranfill R (2004) Phylogeny and evolution of ferns (monilophytes) with a focus on the early leptosporangiate divergences. Am J Bot 91:1582–1598

Pšenička J, Bek J (2001) *Oligocarpia llndsaeoides* (Ettingshausen) Stur and its spores from the Westphalian of Central Bohemia (Czech Republic). Acta Mus Nat Prague Ser B Hist Nat 57(3-4):57–68

Pšenička J, Bek J (2003) Cuticles and spores of *Senftenbergia plumosa* (Artis) Bek and Pšenička from the Carboniferous of Pilsen Basin, Bohemian Massif. Rev Palaeobot Palynol 125:299–312

Pšenička J, Wang J, Bek J, Pfefferkorn H, Opluštil S, Zhou V, Frojdová J, Scanu GG, Libertín M (2020) A zygopterid fern with fertile and vegetative parts in anatomical and compression preservation from the earliest Permian of Inner Mongolia, China. Rev Palaeobot Palynol

Rößler R (2000) The late Palaeozoic tree fern *Psaronius* – an ecosystem unto itself. Rev Palaeobot Palynol 108:55–74

Rößler R, Galtier J (2002) *Dernbachia brasiliensis* gen. nov. et sp. nov. – a new small tree fern from the Permian of NE Brazil. Rev Palaeobot Palynol 122:239–263

Rößler R, Kretzschmar R, Annacker V, Mehlhorn S (2009) Auf Schatzsuche in Chemnitz – Wissenschaftliche Grabungen '09. Veroff Mus Naturkunde Chem 32:25–46

Rößler R, Zierold T, Feng Z, Kretzschmar R, Merbitz M, Annacker V, Schneider JW (2012) A snapshot of an early Permian ecosystem preserved by explosive volcanism: new results from the Chemnitz Petrified Forest, Germany. Palaios 27:814–834

Rößler R, Merbitz M, Annacker V, Luthard L, Noll R, Neregato R, Rohn R (2014) The root systems of Permian arborescent sphenopsids: evidence from the Northern and Southern hemispheres. Palaeontographica B 290(4-6):65–107

Sakala J (2000) Flora and vegetation of the roof of the main lignite seam in the Bilina Mine (Most Basin, Lower Miocene). Acta Mus Nat Pragae Ser B Hist Nat 56(1-2):49–84

Schneider H, Schuettpelz E, Pryer KM, Cranfill R, Magallón S, Lupia R (2004) Ferns diversified in the shadow of angiosperms. Nature 428:553–557

Spindler F, Werneburg R, Schneider JW, Luthardt L, Annacker V, Rößler R (2018) First arboreal pelycosaurs (Synapsida: Varanopidae) from the early Permian Chemnitz Fossil Lagerstätte, SE-Germany, with a review of varanopid phylogeny. Palaeontol Z 92:315

Sprengel A (1828) Commentatio de Psarolithis, ligni fossilis genere. Libraria Antoniana, Halle, 38pp

Taylor TN, Taylor EL, Krings M (2009) Paleobotany: the biology and evolution of fossil plants. Academic Press, Amsterdam, 1230pp

Teodoridis V (2007) Revision of *Potamogeton* fossils from the Most Basin and their palaeoecological significance (Early Miocene, Czech Republic). Bull Geosci 82(4):409–418

Teodoridis V, Kvaček Z (2015) Palaeoenvironmental evaluation of Cainozoic plant assemblages from the Bohemian Massif (Czech Republic) and adjacent Germany. Bull Geosci 90(3):695–720

Wang J (2006) *Tingia unita* sp. nov. (Noeggerathiales) with strobilus from the Lower Permian of Wuda, Inner Mongolia, China. Chin Sci Bull 51(21):2624–2633

Wang J, Pfefferkorn HW, Bek J (2009a) *Paratingia wudensis* sp. nov., a whole noeggerathialean plant preserved in an air fall tuff of earliest Permian age (Inner Mongolia, China). Am J Bot 96(9):1676–1689

Wang J, Feng Z, Yi Z, Wang S (2009b) Confirmation of *Sigillaria* Brongniart as a coal forming plant in Cathaysia: occurrence from an Early Permian autochthonous peat-forming flora in Inner Mongolia. Geol J 44:480–493

Wang J, Pfefferkorn HW, Zhang Y, Feng Z (2012) Permian vegetational Pompeii from Inner Mongolia and its implications for landscape paleoecology and paleobiogeography of Cathaysia. Proc Natl Acad Sci U S A 109(13):4927–4932

Wang J, He X, Pfefferkorn HW, Wang J (2013) A constrain on the compaction rate of Early Permian Tuff Bed from Wuda Coalfield of Inner Mongolia. Acta Geol Sin 87(5):1242–1249

Wang J, Wan M, Pfefferkorn HW (2014) *Aphlebia hvistendahliae* sp. nov. from the early Permian Wuda Tuff Flora, Inner Mongolia. Rev Palaeobot Palynol 210:69–76

Yan MX, Libertin M, Bek J, Wang J (2013) Morphological reconstruction and ecological habit of *Sphenophyllum angustifolium* (Germar) Goeppert from Early Permian of Wuda, Inner Mongolia. Acta Palaeontol Sin 52(4):467–483. (in Chinese with English abstract)

Zachos J, Pagani M, Sloan L, Thomas E, Billups K (2001) Trends, rhythms, and aberrations in global climate 65 Ma to present. Science 292:686–693

Zhang Y, Wang J, Liu LJ (2008) Reconstruction of pinna of *Pecopteris lativenosa* Halle based on statistic analysis. Acta Palaeontol Sin 47(1):21–38. (in Chinese with English abstract)

The Non-analog Vegetation of the Late Paleozoic Icehouse–Hothouse and Their Coal-Forming Forested Environments

Robert A. Gastaldo, Marion Bamford, John Calder, William A. DiMichele, Roberto Iannuzzi, André Jasper, Hans Kerp, Stephen McLoughlin, Stanislav Opluštil, Hermann W. Pfefferkorn, Ronny Rößler, and Jun Wang

Abstract

A walk in the Carboniferous-and-Permian woods of the Late Paleozoic, a time known as the Late Paleozoic Ice Age (LPIA), would not be a walk in the woods comparable to today's Holocene forests. The vegetation that colonized and inhabited the landscapes during glacial and interglacial episodes are non-analogs with the world we witness around us. Unlike continents covered in seed-bearing forests, the systematic affinities of the largest trees, and many shrubs, groundcover, vines (lianas), and epiphytes lie with the spore-producing ferns and fern allies. These ferns and fern allies, including the club mosses (lycopsids) and horsetails (sphenopsids), dominated both organic-rich (peat) and mineral-substrate soils from the Mississippian until the latest Pennsylvanian. Even the gymnosperm groups, which commonly grew in mineral-rich soils, are unfamiliar and subdominant components of these landscapes. The extinct pteridosperms and cordaitaleans, and the extant ginkgoalean, cycad, and conifer clades, ultimately diversify and occupy better drained soil conditions that developed in response to global climate change from icehouse to hothouse conditions. Beginning in the latest Pennsylvanian and increasing their dominance in the Permian, seed-producing clades expanded their biogeographic ranges, displacing the former fern and fern-ally giants. This change in diversity occurs during a unique interval in the history of Earth's biosphere.

The LPIA is the only time, other than the late Cenozoic, since the evolution and colonization of terrestrial plants,

Electronic supplementary material A slide presentation and an explanation of each slide's content is freely available to everyone upon request via email to one of the editors: edoardo.martinetto@unito.it, ragastal@colby.edu, tschopp.e@gmail.com

*The asterisk designates terms explained in the Glossary.

R. A. Gastaldo (✉)
Department of Geology, Colby College, Waterville, ME, USA
e-mail: robert.gastaldo@colby.edu

M. Bamford
Evolutionary Studies Institute, University of the Witwatersrand, Johannesburg, South Africa
e-mail: Marion.Bamford@wits.ac.za

J. Calder
Geological Survey Division, Nova Scotia Department of Energy and Mines, Halifax, NS, Canada
e-mail: John.H.Calder@novascotia.Ca

W. A. DiMichele
Department of Paleobiology, Smithsonian Institution, United States National Museum, Washington, DC, USA
e-mail: DIMICHEL@si.edu

R. Iannuzzi
Instituto de Geociências, Universidade Federal do Rio Grande do Sul, Porto Alegre, Brazil
e-mail: roberto.iannuzzi@ufrgs.br

A. Jasper
Universidade do Vale do Taquari - Univates, Lajeado, RS, Brazil
e-mail: ajasper@univates.br

H. Kerp
Institute of Geology and Palaeontology – Palaeobotany, University of Münster, Münster, Germany
e-mail: kerp@uni-muenster.de

S. McLoughlin
Department of Palaeobiology, Swedish Museum of Natural History, Stockholm, Sweden
e-mail: steve.mcloughlin@nrm.se

S. Opluštil
Institute of Geology and Paleontology, Charles University in Prague, Prague, Czech Republic
e-mail: stanislav.oplustil@natur.cuni.cz

H. W. Pfefferkorn
Department of Earth and Environmental Science, University of Pennsylvania, Philadelphia, PA, USA
e-mail: hpfeffer@sas.upenn.edu

R. Rößler
Museum für Naturkunde, Chemnitz, Germany
e-mail: roessler@naturkunde-chemnitz.de

J. Wang
Nanjing Institute of Geology and Palaeontology, Chinese Academy of Sciences, Nanjing, People's Republic of China
e-mail: jun.wang@nigpas.ac.cn

© Springer Nature Switzerland AG 2020
E. Martinetto et al. (eds.), *Nature through Time*, Springer Textbooks in Earth Sciences, Geography and Environment,
https://doi.org/10.1007/978-3-030-35058-1_12

when the planet experienced prolonged icehouse and greenhouse conditions. Extensive tropical peat swamps, similar in physical properties to current analogs in Southeast Asia, accumulated in coastal plain lowlands. These forests extended over thousands of square kilometers during periods when global sea level was low in response to the development of extensive Gondwanan glaciation at the southern pole. When these ice sheets melted and sea-level rose, the tropical coastal lowlands were inundated with marine waters and covered by near-shore to offshore ocean sediments. The waxing and waning of glacial ice was influenced by short- and long-term changes in global climate that were, in turn, controlled by extraterrestrial orbital factors. As the LPIA came to a close, a new forested landscape appeared, more familiar but, still, distant.

12.1 Introduction

The Industrial Revolution began in the middle of the eighteenth century and progressed rapidly after the invention of the steam engine in 1776. The technological advances that marked this time were powered by coal, which is the byproduct of peat accumulation in extensive tropical wetland forests that covered equatorial regions hundreds of millions of years ago in what is often referred to as "Deep Time." From the end of the eighteenth century to the middle of the twentieth century, as exploitation of this natural resource expanded, coal-mining operations uncovered countless troves of fossilized plants representing ancient forests. But, the plant groups comprising these forests were not the same as the angiosperms and gymosperms that have dominated the landscapes of the Paleogene or Neogene (Martinetto et al., Chaps. 1 and 2), or even back in the Mesozoic (Gee et al., Chap. 6). Rather, a hard look at them tells us that the systematic affinities of the largest trees, and many shrubs, groundcover, vines, and epiphytes lie, mainly, with the spore-producing ferns and fern allies. Imagine standing under a grove of trees, 30 m in height, where a broken canopy casts little shade, and the release of spores turns the vista into a yellowish color. When fossilized seeds were first discovered, these, too, had features very different from modern groups (Oliver and Scott 1903). And, when seeds were found attached to their parental plants, it was recognized that these gymnosperms were also very different from modern forms, although several growth architectures look familiar (DiMichele et al. 2005a).

12.2 LPIA Tropical Forests: The Players

The coal forests of the Carboniferous were dominated by entirely different plant groups from those contributing most of the biomass in modern ecosystems [U1201]. DiMichele et al. (2005a) note that four Linnaean classes of vascular plants–lycopsids, sphenopsids, ferns, and seed plants–were co-equal components of Carboniferous–Permian peat forests. Most of these plants reproduced exclusively by spores–the lycopsids, sphenopsids, ferns, and enigmatic progymnosperms–whereas the gymnosperms reproduced by seeds. Many of the fossil taxa are unique to this time interval, but several subgroups in each broad clade persist to the present in similar ecological settings. The lycopsids were confined primarily to wetlands, the soils of which ranged from purely mineral to purely organic matter (peat), and constitute the majority of biomass contribution to the paleotropical peat swamps (=Carboniferous and Permian coals). One taxon, *Sigillaria*, is known from sites in which the soil moisture conditions were better drained and, probably, seasonally dry (DiMichele et al. 2005a). After the demise of tree lycopsids in the Late Pennsylvanian of the paleotropics, representatives of the clade occupied wetlands in the North and South China Blocks, which persisted until the end of the Permian (see Gastaldo et al., Chap. 13). Calamitean sphenopsids are similar in gross structural organization to living Equisetales, except for the presence of secondary xylem (wood), which allowed them to grow to the size of trees. This group occupied a narrow range of habitats, those of disturbed settings and floodplains where sedimentation built up land surfaces. The clonal growth habit of some taxa permitted regeneration following burial in many instances (Gastaldo 1992). Several groups of ferns are known in the fossil record, some of which are extant (see Pšenička et al., Chap. 11). The most conspicuous group is the marattialean ferns. Whereas modern members of this clade remain understory forms, some Paleozoic species grew to be trees during the LPIA. These plants, inexpensively constructed in terms of carbon-biomass allocation, dominated tropical wetlands in the latest Carboniferous and were opportunistic taxa. In contrast, the progymnosperm group, which is a holdover from the latest Devonian, is more prominent in Mississippian floras with few recognized individuals in younger forests. These plants produced woody stems with conifer-like wood, some of which grew to tall trees but reproduced all by spores. The progymnosperm group is a transitional mosaic to the true seed-producing gymnosperms (see Gensel et al., Chap. 15). With the advent of seeds, true gymnospermous plants came to dominate terra ferma habitats but were also widespread in wetlands. Several extinct groups are found in the Permo–Carboniferous swamps including medullosan and lyginopterid pteridosperms (seed ferns) in the wet tropics and the cordaites, a sister group of the conifers, in both the wet and seasonally dry tropics and the north temperate zone. The fossil record of the conifers and other gymnospermous groups is less common in Carboniferous paleoequatorial forests and, in general, found in the seasonally dry tropics. These groups are encountered more commonly in the south temperate zone and become dominant forest elements in the Permian (DiMichele et al. 2005a). One gymnospermous group, the glossopterids, first appears in abundance in the south temperate regions following the deglaciation of Gondwana. These plants dominated landscapes until the end of the era. As one might anticipate, a walk through these "woods" at different times across different continents would encompass the same, or greater, landscape diversity than we envision for post-Paleozoic worlds.

12.2.1 Club Mosses (Lycopsids)

Lycopsida is a group of vascular plants that originated in the late Silurian. They are one of two major lineages of vascular plants, the other encompassing virtually all the plants that dominate modern landscapes and most landscapes of the past (the ferns, sphenopsids, and seed plants; Bateman et al. 1998). The common ancestor of these two major lineages lacked roots, leaves, wood, and bark (secondary tissues), and reproduced by spores. Later, both the lycopsids and the other plant groups evolved these features independently (leaves, roots,

wood) and also evolved more complex reproductive systems, including seeds and seed-like organs (Phillips 1979). The earliest appearing lycopsids, and their immediate ancestors, the zosterophylls, appear to have been ecologically centered in wetlands, more so than the other lineages of vascular plants. Their colonization and occupation of wetlands is a pattern that continues today in some of the living groups.

Slogging through wetlands of the Late Devonian, we can encounter several distinct evolutionary lineages of lycopsids, three of which are still represented in the modern landscape [U1202]. These orders are called Lycopodiales, Selaginellales, and Isoetales. Members of the first two of these were small-bodied and had, for the most part, a sprawling, groundcover habit throughout their evolutionary history. In contrast, Isoetales evolved centrally rooted, upright forms, and tree habits (Fig. 12.1). Selaginellales and Isoetales are united by several features. Two of these are the presence of ligules*, tiny spine-like appendages borne on leaves, near their point of attachment to stems, and heterosporous reproduction. In heterospory, the parent plants produce two sizes of spores. Large megaspores contain the female reproductive organs bearing egg cells, whereas small microspores pro-

Fig. 12.1 Carboniferous lycopsid trees. (**a**) Standing lycopsid at the UNESCO Joggins World Heritage site; hammer for scale. (**b**) *Diaphrodendron* stem cross section showing major tissues. Note that the center of stem is root penetrate Scale = 1 cm. (**c**) Trunk of polycarpic tree growth form with two, opposite rows of scars marking former position of branches; hand for scale. (**d**) Crown branch of monocarpic growth form with associated cones (Images by (**a**) J Calder, (**b–d**) WA DiMichele)

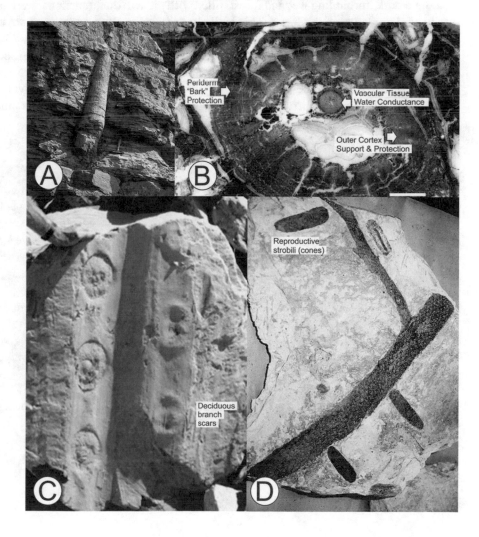

duce sperm. To complete the life cycle, sperm must be released in an aquatic medium where, through chemical signals, they detect and swim to the megasporangium, and fertilize the waiting eggs. Members of Isoetales are further united, as a group, by their unique central rooting systems and production of secondary tissues, such as wood and protective bark (periderm) tissues (Boyce and DiMichele 2016).

Sometime in the Late Devonian or Early Mississippian, we find that an evolutionary lineage of isoetaleans diverged significantly in their morphology and ecological importance from the main group. This lineage is often referred to as the lepidodendrids; they are members of the order Isoetales but are clearly a distinct group in that order (Bateman et al. 1992). Colloquially, these plants are the "arborescent lycopsids" of the Coal-Age [U1203]. If you have visited one or more of the world's natural history museums, dioramas reconstruct this iconic cameo group that serves to illustrate just how strange the Coal Age was, in comparison to today. These were the dinosaurs of the plant world (Fig. 12.1a, c). Lepidodendrids are usually shown as giant trees in swampy environments, dominating the canopy, which is most often reconstructed as dense, creating a dark, forbidding wetland forest, filled with giant insects and predatory amphibians. This is likely an incorrect representation (see Gastaldo et al., Chap. 13). The growth habits and architectures of arborescent lycopsids evolved in several different lineages and were found throughout the world of the Carboniferous. We concentrate here on the tropical forms that dominated Pennsylvanian wetlands because they are the best known and are illustrative of the larger group in their ecology and growth forms.

The Pennsylvanian tropical arborescent lycopsids were large trees, some forms exceeding 30 m in height and 2 m in basal girth. They had unique root systems, known as *Stigmaria*, that extended out many meters from the base of the tree at shallow depths. The main root axes were surrounded by a cloud of thin, branching rootlets firmly anchoring the plant to the ground (Hetherington et al. 2016). Despite their great height, the trees were determinate, having a programmed branching pattern and death that occurred after a certain interval of growth had been attained (Bateman 1994). Their large trunks were not woody, in contrast to most modern flowering plant and gymnospermous trees. Rather, the trunk was supported by a rind of bark, reaching thicknesses of perhaps 10 cm, which was chemically enriched in compounds such as suberin (Fig. 12.1b). Hence, the bark was both water-and-decay resistant; this bark accounts for much of the biomass that contributed to Lower and Middle Pennsylvanian coal beds. Unlike in any living tree, support and water-conduction functions were separated in arborescent lycopsids. With aerial support taken care of by the bark rind, the movement of water occurred in a highly specialized and extremely efficient, but small, woody cylinder in the central parts of the stem (Fig. 12.1b) and main root axes. Several main growth forms are recognized in this group's determinate development (Box 12.1).

Box 12.1: Lycopsid Growth Architectures

The most commonly reconstructed tree's habit consists of a tall trunk, capped by a crown of equally forking branches; a typical example, *Lepidodendron mannabachense*, has been reconstructed in detail from specimens preserved in volcanic ash (Fig. 12.2) [U1204] (Opluštil 2010). The crown developed late in the plant's growth strategy and reproductive organs were borne at the tips of branches, ending the life cycle (Fig. 12.1d). These plants spent most of their lives growing as unbranched poles, partially covered in leaves and with reproductive organs confined to the crowns (DiMichele et al. 2013). Thus, their reproductive period was relatively short compared to the overall life of the tree. They grew, they reproduced, and they died. The Early and Middle Pennsylvanian Coal-Age forests, in which these trees were dominant, were likely not the dense, dark landscapes of most reconstructions. Rather, these forests were relatively open to sunlight penetration, at least until crowns began to develop. They would have been dark only if the final growth phases were somehow synchronized among all the trees on the landscape. The lepidodendrids with these growth habits are classified in the family Lepidodendraceae. There are other growth strategies in the group.

Three other major growth forms are recognized in the lepidodendrids (Fig. 12.2a). The first is typical of the Diaphorodenraceae and the Ulodendraceae and may be the primitive growth habit of the lineage. These trees developed a main trunk, along which were borne, in two opposite vertical rows, relatively small deciduous lateral branches. The cones were borne in these lateral branch systems, which were shed, leaving a row of circular scars on either side of the trunk (DiMichele et al. 2013). This, of course, resulted in an enormous amount of litter and contributed to peat accumulation in some wetland settings (see Sect. 12.5). The other growth form was that of the Sigillariaceae, a widespread and important group. These arborescent plants are survivors and remain ecologically important in the Late Pennsylvanian and Permian, after other groups experience extinction. Sigillarians had sparsely branched trunks, with the branches again developing only late in the life of a tree. Reproductive organs were borne in whorls on the branches (DiMichele and Phillips 1985). A third growth form, represented by the genus *Hizemodendron*, is a sprawling growth habit that is interpreted as a developmentally stunted arborescent form (Fig. 12.2b; Bateman and DiMichele 1991).

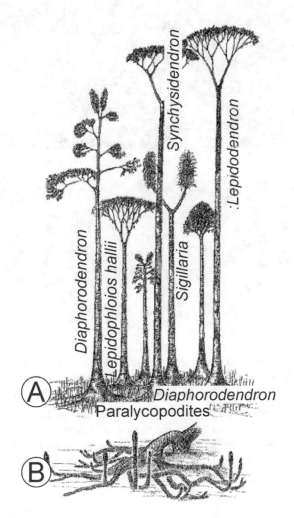

Fig. 12.2 Growth forms of Permo–Carboniferous lycopsids. (**a**) Reconstructed arborescent growth architectures. (**b**) Sprawling groundcover growth form of *Hizemodendron*. (reproduced with permission: Bateman et al. (1992))

12.2.2 Horsetails (Sphenopsids)

The progenitors of our living horsetails first occur in the Late Devonian times where two distinct lineages of sphenopsids are preserved [U1205]. One is the tree-sized calamitaleans and the other is the scrambling and climbing sphenophylls. Both orders become fully established with the onset of the Northern Hemisphere equatorial forest ecosystems in the Carboniferous. Their success parallels that of the lycopsids, having acquired a striking evolutionary burst and diversification in wetland plant communities. These groups persisted for more than 60 million years. Accordingly, they are common in fossil assemblages of both the classical Pennsylvanian Coal Measures and in restricted Permian wetlands (Grand'Eury 1877). Calamiteans would be the conspicuous elements of the forest canopy and subcanopy and are the focus of this section.

A panoramic scan of the forest makes it immediately apparent that the growth architecture of calamitaleans differs dramatically from the lycopsids. This growth form has been interpreted to represent an archaic strategy [U1206]. These spore-producing plants are traditionally reconstructed as enlarged examples of the extant horsetail *Equisetum*. Although this model of an ecologically well-adapted plant has been adopted in many publications, the model combines unique developmental features of both a rhizomatous and arborescent nature in wetlands. Calamitean growth in tropical peat-accumulating swamps and in mineral-soil floodplain habitats resulted in tall trunks, attaining heights of up to 20 m, with a diverse branching architecture (Fig. 12.3; Rößler et al. 2012). The most common fossils of these plants include sediment-casts of their hollowed stems, in some cases even preserved in growth position (Gastaldo 1992), whorled leafy twigs, and sporangia-bearing organs (strobili), all of which reveal the characteristic appearance of distinctive horizontal nodes and internodes (Fig. 12.4). Both pith casts* and compressions of stems exhibit a variety of branching patterns (DiMichele and Falcon-Lang 2011; Thomas 2014). Details of these plants are known from anatomically preserved specimens.

Specimens that preserve plant anatomy are known from coal balls* or "petrified" individuals. These fossils indicate that calamitaleans (Box 12.2) displayed a range of secondary developmental patterns (e.g., Wang et al. 2005). This is particularly the case in Permian plants where woody stems attained diameters of up to 60 cm! Changes in wood production as a function of age are interpreted to indicate that individual plants responded to seasonal climate and environmental change (Rößler and Noll 2006). Their wood differs from gymnosperms in that it consisted of up to 50% soft tissue (parenchyma), representing an enormous water-storage capacity (Fig. 12.3d) [U1207]. This feature indicates that

Fig. 12.3 Early Permian calamitalean trees. (**a**) Cross section anatomy of arborescent trunk (*Arthropitys*) showing pith surrounded by secondary xylem (wood). (**b**) Cast of aerial trunk demonstrating the articulated nature of the pith composed of internodes and nodes, from which lateral branches and/or leaves originated. Small branch scars are arranged alternately. (**c**) Rooting system of calamitalean tree in the paleosol beneath volcanic ash deposits at Chemnitz, Germany. (**d**) Anatomical root-cross section (*Astromylon*) showing the abundant secondary xylem along with a reduced central pith structure. (Images by R. Rößler)

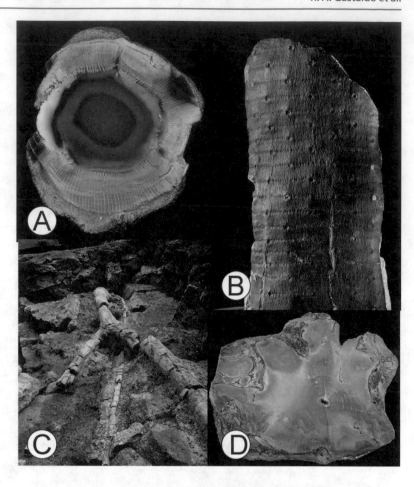

Box 12.2: Sphenopsid Growth Architectures

There is a long-held idea that calamitalean aerial axes originated either from a clonal underground system of rhizomes (Hirmer 1927) or laterally from neighboring stems (Pfefferkorn et al. 2001). This view has recently been modified based on new fossil material (Fig. 12.4c). Although first reported from the late nineteenth century, hypotheses inferring alternative growth architectures, which included free-standing calamitaleans from the Pennsylvanian of France (e.g., Grand'Eury 1877) or England (Maslen 1905), were usually met with doubt. The skepticism of free-standing growth strategies persisted into the mid-twentieth century, although such growth models were presented by Leistikow (1962) and Barthel (1980) from the Euramerican Permo–Carboniferous paleotropics. Since then, there has been the discovery of free-stemmed calamitaleans. This evidence comes from in-situ rooted trunk bases in Brazil and Germany, in which multiple organic connections exist between stems and roots (Rößler et al. 2014). Sizable trees are preserved anchored in soils by numerous stem-borne secondary roots (Fig. 12.3c), which arose from differ-

ent nodes of the trunk base. These roots branched several times while tapering on their oblique geopedal, downward course. As such, these recently discovered fossil species followed growth strategies of more modern trees, differing from what was previously thought for calamitaleans, and underline the considerable adaptive control of this plant group.

some Permian calamitaleans may have been able to survive short seasonal episodes of dryness, or drought, during which the plants reduced water uptake and shed their leafy twigs. Nevertheless, these highly specialized functional features were not sufficient to prevent the group's eventual demise in the latest Permian, as increasing seasonal aridity reduced populations and drove them to extinction. This may have been due to a non-clonal growth strategy.

12.2.3 Ferns (Pteridophytes)

Ferns are familiar to botanists and non-botanists, alike, as they grow in habitats ranging from mangroves at sea level to

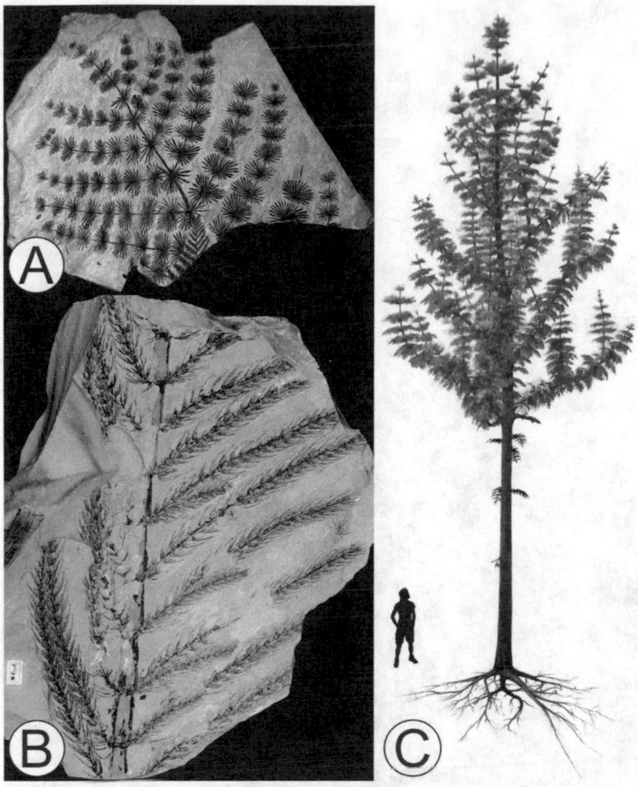

Fig. 12.4 Foliage and reconstruction of calamitean growth architecture at Chemnitz, Germany. (**a**) Whorled, tongue-shaped leaves assigned to *Annularia*. (**b**) Whorled, linear-lanceolate leaves assigned to *Asterophyllites*, (**c**) Reconstruction of free-standing arborescent calamitean. (Images by R Rößler)

alpine biomes above tree line, temperate-and-tropical forests to arctic tundra, and from wetlands to deserts. The group is the second most diverse plant group on the planet, with only the seed plants being more diverse, and have been featured in other units of this book (see Pšenička et al., Chap. 11). As such, ferns display a wide diversity in functions and habitats, which are reflected in a similar diversity in the size and shape of the group's megaphyllous leaf (Fig. 12.5a). As we have seen previously, fern leaves are described as consisting of a central axis, termed a rachis, from which lateral pinnae or pinnules are alternately or oppositely arranged [U1208]. These compound, divided, or dissected leaves are called fronds, which are arranged in various phyllotaxis* around either a subterranean (underground) axis (rhizome) in groundcover taxa or an aerial (trunk) axis (Fig. 12.5b) as in tree ferns. The plant group reproduces by spores, and two groups exhibit different developmental reproductive patterns. Most taxa produce one type of reproductive spore (homosporous), but a small number of taxa evolved a reproductive strategy in which both a male-and-female spore is produced separately (heterospory). LPIA ferns are known from compression-impressions and permineralizations, and display the same range in growth architectures and habitats

as do modern fern groups. Paleozoic ferns are discussed in unit 11, and the reader is directed to that chapter for details of their diversity over time.

12.2.4 Gymnosperms

Gymnosperms consist of seed plants, many groups of which are familiar to anyone walking through present-day woods. Wood in these plants is produced by a two-sided (bifacial) vascular cambium that accretes annual growth rings. The presence of a megaphyllous leaf, a photosynthetic lamina with several or many veins arranged either in a branching or parallel pattern, has been considered as a criterion to separate clades from those without a leafy structure [U1209]. Both characteristics have been used to interpret a phylogenetic relationship between these clades (Crane ct al. 2004).

12.2.4.1 Seed Ferns (Pteridosperms)
The terms "seed fern" and "pteridosperm" are widely used in the paleobotanical literature for plants with foliage that, superficially, appears to be fern-like (Figs 12.6b and 12.7b, c; Taylor et al. 2009). It would be difficult to distinguish these plants from tree ferns wandering through the forests of the LPIA. However, if you were to fell one, you would see that the stems that bore this foliage were woody. And, if you were there at the right time, you would see that the plants reproduced via pollen and seeds rather than spores (Figs 12.6c and 12.7d). Many disparate plant groups of late Paleozoic and Mesozoic ages have been assigned to the "seed ferns," making the term essentially meaningless. At best, the group represents a 'grade' of gymnosperm evolution [U1210]. What we can say is that the group encompasses numerous lineages, with varying degrees of secondary wood in stems on which fern-like leaf architectures developed. And, reproductive structures evolved, through time, into more complex organs enclosing and protecting both seeds and pollen. Hence, the group is a collection of early seed plants in which a wide variety of experimentation with different stem, leaf, and reproductive architectures evolved to adapt to specific niches in the late Paleozoic forests.

Currently, nine orders of late Paleozoic 'seed ferns' are identified based on anatomical, morphological, and reproductive features. These include the Calamopityales, Buteoxylonales, Lyginopteridales, Medullosales, Callistophytales, Gigantopteridales, Glossopteridales, Peltaspermales, and Corystospermales. Both Calamopityales and Buteoxylonales are known exclusively from stem anatomy, and neither their growth habits nor ecologies are documented. Hence, these two groups that play a minor role in LPIA forests are not treated in this chapter. Several other groups, especially the Medullosales, Gigantopteridales, and Glossopteridales, were major biomass contributors to the

Fig. 12.5 Modern and ancient tree ferns. (**a**) Arborescent fern canopy in a New Zealand rainforest. (**b**) Permineralized stem of Pennsylvanian tree fern, *Psaronius*, showing the trunk's internal anatomy, divergence of c-shaped leaf traces, and surrounding, and supporting, root mantle. See Unit 11 for details of the group. (Images by (**a**) H Kerp, (**b**) R Rößler)

coal-forming swamps of the Carboniferous and Permian. These, and seed ferns occupying better drained soils, are treated below.

12.2.4.1.1 Lyginopteridales

Lyginoperidales is a heterogenous group that occurred predominantly in the Carboniferous of Euramerica, but reports of several taxa extend its biogeographic range worldwide. Representative plants are known from both anatomically preserved and compression-impression specimens (Fig. 12.6) [U1211]. Slender stems consist of a relatively broad pith, modest secondary vascular tissue development, and prominent radial bands of fibers in the outer cortex (Fig. 12.6a). Adventitious roots emerged from the stem above ground. Typical genera include *Lyginopteris* and *Heterangium*, both of which are interpreted to have been scrambling groundcover or climbing lianas (Masselter et al. 2007). A characteristic feature of both leaves and young stems of many taxa is the presence of multicellular hairs and

glands, and some genera developed hook-like appendages for climbing (e.g., *Karinopteris*; Krings et al. 2003). Bi- to tri-pinnate leaves of this group, including *Lyginopteris* (Fig. 12.6b), *Sphenopteris*, *Cardiopteris*, *Sphenopteridium*, *Rhacopteris*, *Eusphenopteris*, *Eremopteris* and *Polycalyx*, were forked (Y-shaped), with each division bearing highly dissected leaflets. The pollen organ, *Crossotheca*, was a disc- or cup-shaped structure with fused, or tightly bound, pollen sacs on its lower surface (Fig. 12.6d, e). The female reproductive structures are unique with ovules surrounded by cup-shaped structures, the partial fusion of leaf-like appendages, to form a protective coat. These cupulate seeds, generally called *Lagenostoma*, typically had elaborate funnel-like apical structures that functioned to capture pollen and seal the pollen chamber* after pollination (Fig. 12.6c). Lyginopterids had a similar distribution to Medullosales in the mid-Carboniferous but declined in the late Carboniferous to be largely replaced by the latter group (DiMichele et al. 2005a).

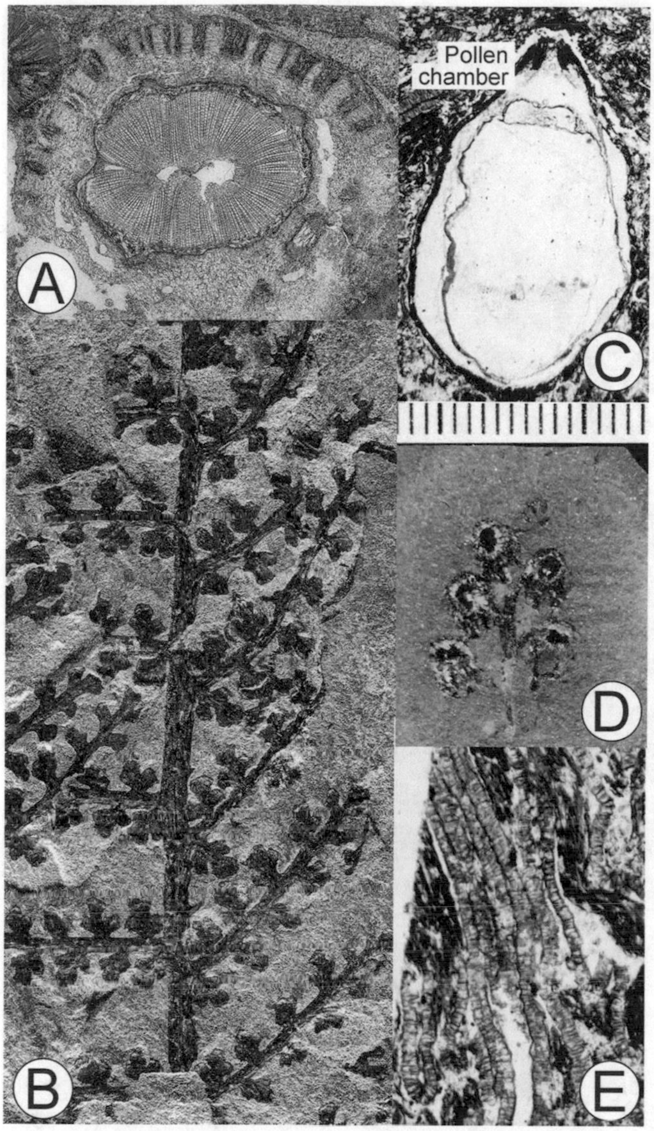

Fig. 12.6 Representative seed fern organs of the Lyginopteridales. (**a**) Permineralized transverse section of a stem of *Lyginopteris oldhamia* showing prominent fiber bands in cortex, which appear as longitudinal striations along the stem of compression specimens. (**b**) Foliage of the form genus *Lyginopteris*. (**c**) Permineralized longitudinal section of the lyginopterid seed, *Sphaerostoma,* in which the specialized pollen-receiving structure, the lagenostome, can be seen. (**d**) Pollen-bearing organs assigned to *Crossotheca*. (**e**) Permineralized longitudinal section of pollen sacs. (Images H Kerp)

12.2.4.1.2 Medullosales

Representatives of Medullosales were, physically, the largest of the Northern Hemisphere seed ferns and ranged in age from the Late Mississippian into the Permian. Their permineralized remains are well known from coal balls of Europe and North America, and sizeable stems preserved in volcanic ash, as well as numerous compressions-impressions of their aerial parts in both sandstone and mudstone. You would encounter this group of important trees wandering through any of the late Carboniferous peat-accumulating swamps or coastal floodplains. The wood in the trunk was peculiar,

composed of multiple vascular segments each surrounded by rings of secondary xylem, and functioned similar to that of modern angiosperms (Fig. 12.7A) [U1212] (Wilson 2016). Woody stems attained diameters in excess of 50 cm, contained a thick bark, with trunk wood assigned to several genera (e.g., *Medullosa, Colpoxylon, Quaestoria,* and *Sutcliffia*) differing in the arrangement of the vasculature and architecture of the leaf bases (Rößler 2001). These trees (Box 12.3) bore very large leaves, reaching up to 7 m in length, which were asymmetrically forked and had finely divided leaflets [U1213]. Pinnate leaves are assigned to numerous genera including *Odontopteris* (Fig. 12.7b),

---

Box 12.3: Medullosan Pteridosperm Growth Architectures

Growth architectures of Carboniferous pteridosperms are not as well known as either the lycopsids or sphenopsids. This is, in part, because they occupied either an understory position in these forests, of which few have been preserved in an upright orientation (Falcon-Lang 2009; Rößler et al. 2012), or were lianas (Krings et al. 2003). Most commonly, trunks of understory trees with attached leaves and/or leaf bases are found lying flat, parallel to bedding. Their preservation was a consequence of storm blow-downs (e.g., Pfefferkorn et al. 1984). Medullosan trunks are encountered more often due to their unique anatomy, consisting of several separate woody vascular bundles* surrounded by fibrous tissues. Where examples are rooted in a paleosol, vertically oriented tap roots extend downward from which lateral roots and rootlets developed. Basal trunk diameters vary, ranging from 10 cm (Wnuk and Pfefferkorn 1984) to nearly 50 cm, and some display the development of a root mantle that surrounds the stem. Medullosans developed a single erect axis from which robust leaves were born in a helical, or spiral, arrangement. When leaves became non-functional as new leaves developed at the growing apex, the leafy pinnules decayed but the petioles remained attached to the monoaxial trunk. This resulted in a "skirt" of abundant, closely spaced and downward-recurved axes around the lower part of the trunk, a feature found, today, in some angiosperms (e.g., palms; Wnuk and Pfefferkorn 1984). It has been suggested that these forms, in particular, may not have been "free standing." Rather, such architectures reflect a flexuous habit that required support from adjacent plants, which may have resulted in mono- or polyspecific stands. Such an interpretation is supported by paleoecological studies (see DiMichele, in Gastaldo et al., Chap. 13). Other species, though, appear to have been self-supporting (DiMichele et al. 2006; Rößler et al. 2012). Medullosan

Fig. 12.7 Representative seed fern organs of the Medullosales. (**a**) Transverse section of a permineralized stem of *Medullosa leuckartii* exhibiting numerous secondary vascular tissue bundles, characteristic of the genus. (**b**) Typical medullosan pinnate foliage assigned to *Odontopteris*. (**c**) Typical medullosan foliage (pinna) assigned to *Neuropteris*. (**d**) A cluster of large seeds assigned to *Trigonocarpus noeggerathii*. (Images (**a**) R Rößler, (**b–d**) S McLoughlin)

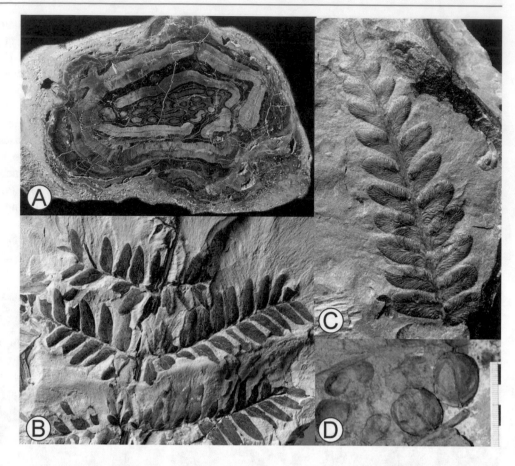

leaf architecture, in general, consists of a proximal (near the stem) stout petiole that, at some distance from the stem, divides into two or four main axes. Circular or subdivided leaflets (pinnules), known as *Aphlebia*, may have emerged from along the undivided petiole, and individual pinnae or pinnate leaflets were organized along laterals that developed from the bifurcated* main axes. Leaves attained lengths of several meters, depending on the taxon. Estimates of at least 5 m in height have been made for these understory shrubs. In contrast, several taxa are known to have leaves that grew to lengths of up to 7 m, and there is no evidence for an erect aerial stem where these are preserved as monocultures. It is possible that these forms had a subterranean or short emergent stem from which the leaves grew, similar in growth habit to *Nipa* palm found in today's tropics.

Neuropteris (Fig. 12.7c), and others, based on leaflet shape and attachment to the rachis. Medullosans produced large (0.7–10 cm long; Gastaldo and Matten 1978) wingless seeds that, structurally, replaced some leaflets on frond margins. Seeds mostly exhibit radial symmetry–a character that has

been used to suggest distant affinities with cycads–and are, again, assigned to various genera (e.g., *Trigonocarpus*, *Pachytesta*, and *Stephanospermum*) based on preservational state and modifications of the integument (Fig. 12.6d). The pollen organs (e.g., *Halletheca*, *Bernaultia*, *Potoniea*) were large (up to several centimeters in diameter) and mostly compound inverted cup-shaped (bell-like) structures composed of numerous fused pollen sacs that yielded large, monolete pollen that, in some cases, had weakly developed wings.

12.2.4.1.3 Callistophytales

Callistophytales was a group of woody scrambling or climbing, possibly understory, plants known from the Middle–Upper Pennsylvanian of North America and Europe, early Permian of Europe, China, and possibly Brazil. The best known examples of its permineralized wood (*Callistophyton*) are from, typically, less than 3 cm diameter stems exhibiting a modest-sized pith and extensive secondary xylem (Rothwell 1981). The flattened (platyspermic) seeds, called *Callospermarion*, could be as large as 5 mm long. Pollen was borne in fused pollen sacs attached to modified leaflets, and the winged pollen was of the *Vesicaspora* type–some of which preserve unequivocal pollen-tube development.

Fig. 12.8 Representative foliar organs of the Gigantoperidales. (**a**) Entire leaf of *Evolsonia* morphotype. (**b**) Leaf morphotype assigned to *Cathaysiopteris* / *Gigantopteris whitei*. (**c**) Leaf morphotype assigned to *Zelleropteris* exhibiting a networked venation pattern reminiscent of modern angiosperms. (Images (**a**, **c**) WA DiMichele, (**b**) S McLoughlin)

12.2.4.1.4   Gigantopteridales

All previous non-analog groups would have appeared odd or weird to anyone traversing forested areas in which they grew. One group that might have looked a bit more familiar is a heterogenous group of Permian plants, Gigantopteridales, where broad leaves are characterized by a complex network (reticulate) vein pattern [U1214]. Superficially, some resemble the leaves of angiosperms but the pattern, in others, is either forked or a basically pinnate architecture (Fig. 12.8b, c). The leaves, attaining lengths of 0.5 m, could be lobed or entire (Fig. 12.8a). The group is known primarily from East and southern Asia, and western North America. Numerous genera have been established for the leaves (e.g., *Gigantopteris*, *Gigantonoclea*, *Cathaysiopteris*, *Zeilleropteris*, *Delnortea*) based on variations in shape and vein architecture (Glasspool et al. 2004). In some instances, leaves were borne on thin, fibrous stems, which, when combined with the evidence of spines and hooks on the leaves and stem surface, has been used to argue that many in this group were climbing or scrambling plants (Seyfullah et al. 2014). As much as we know about

the leaves of these plants, the reproductive organs are poorly understood. They possibly bore seeds in rows at the ends of major veins on either side of the leaf midvein. Elongate sporophylls bearing fused pollen sacs (synangia) have been found in association with the foliage and may represent the male organs of this plant group. Some authors have suggested possible phylogenetic relationships to Lyginopteridales and Callistophytales based on the stem-and-leaf structure (Taylor et al. 2009).

12.2.4.1.5   Glossopteridales

Forests in the Southern Hemisphere of the supercontinent–South America, Africa, Antarctica, India, and Australia–might have appeared familiar; but, somehow, a feeling of uncertainty also may have prevailed. It appears that one plant group dominated these landscapes, similar to how modern angiosperms dominate our own [U1215]. Glossopteridales was a remarkably successful seed-plant group that dominated the vegetation of Gondwana for almost 40 million years through the Permian. It is named after the compression-leaf genus *Glossopteris* (*Glosso* = tongue-shaped and

Fig. 12.9 Glossopterid leaves and reproductive structures. (**a**) *Glossopteris* leaf showing typical venation of a broad central midrib and secondary veins diverging to the margin, with a few bifurcations (splits) and anastomoses (joins). (**b**) Male fructification (*Glossotheca*) developed in the axil of a leaf, with scale leaves bearing clusters of microsporangia. (**c**) Female fructification (*Ottokaria*) consisting of a large capitulum (compact head) surrounded by many seeds. (**d**) Female fructification (*Rigbya*) with seed-bearing scales developed at the apex of a slender stalk, which is expanded into a flattened head. (Images M Bamford)

pteris = fern; although they are not related to true ferns; Fig. 12.9a). Typical of many fossil plant groups, the various organs have been given separate names because of their disarticulation prior to burial, preservation, and collection. But several of their features are well known.

These trees had the basic wood and trunk structure of modern conifers. The fossil wood consists of well-defined 'softwood' rings made up of secondary xylem tracheids with narrow radial rays and rare, to absent, parenchyma (Fig. 12.10). The roots, called *Vertebraria*, had an unusual segmented structure superficially resembling a vertebral column of animals. The wood is composed of radial wedges and partitions of normal xylem tissue surrounding alternating chambers, or sections, of soft tissue adapted to gas exchange (aerenchyma; Decombeix et al. 2009). On the basis of this unusual structure, glossopterids are interpreted to have thrived in waterlogged peat-accumulating environments in the extensive lowlands of the middle- to high-latitude cool temperate regions of Gondwana (McLoughlin 1993).

The leaves of *Glossopteris* are the most varied organ of this group with a familiar shape, similar to those found in several angiosperm families (McLoughlin 2011). *Glossopteris* leaves were more or less elliptical or spatula-shaped, with a broad midrib composed of a cluster of veins; secondary veins arise from the midrib and arch outward to the margin, regularly branching and anastomosing along their course (Fig. 12.9; Pigg and McLoughlin 1997). To date, at least 200 'species' have been described (Anderson and

Anderson 1989). More importantly, the plants were deciduous as evidenced by dense layers of mature leaves occurring at regular intervals between layers of fine sediments, and leaf abscission scars on the short shoots. Leaf size varied from a few centimeters to over 50 cm in different "species," but most are around 10–30 cm long. At times, and most likely during short winter months, forest floors were covered in a leaf mat over which towered a barren canopy. It is unknown, though, if these plants reproduced annually or episodically.

The reproductive structures were complex and diverse, with over 40 genera described. Male and female structures were borne separately arising from the surface of modified leaves and perhaps were even produced on different plants [U1216]. The male (microsporangiate) organs consisted of scale-like bracts arranged in clusters or loose cones (Fig. 12.9b), each bearing pairs of finely branched filaments with terminal pollen sacs on one surface (Surange and Chandra 1974). Typical glossopterid pollen has a central body (corpus) bearing thickened transverse strips (taeniae). Similar to some conifer pollen, there are two opposite air sacs attached to the sides of the corpus that represent adaptations for wind dispersal. The female (seed-bearing) reproductive structures can be divided into four main families based on broad differences in their shape and position of the seeds (Anderson and Anderson 1989). Arberiaceae developed seeds on one side of the tips (apices) of loosely branched structures, whereas Dictyopteridiaceae is characterized by flattened shield-shaped organs with seeds on one surface sur-

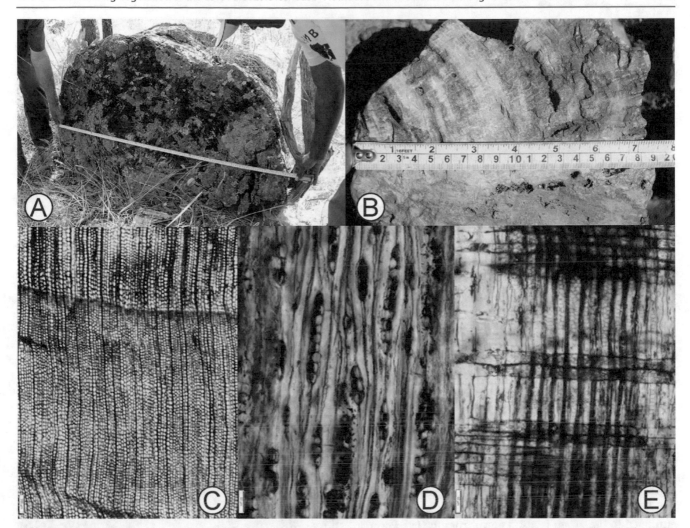

Fig. 12.10 Silicified glossopterid tree trunks from southern Africa. (**a**) Examples of huge silicified trunks (>2 m length × 1.5 m diameter) lying between trees of a modern savanna woodland (hardwoods) Luangwa Valley, Zambia. (**b**) Transverse break of silicified trunk showing growth rings of complacent character. (**c**) Transverse thin section showing wide growth rings indicating that yearly growth was rapid. (**d**) Tangential longitudinal section in which uniseriate ray parenchyma appear as vertical, lenticular features. (**e**) Radial longitudinal section in which the vertical lines are the outlines of the tracheids (water-conducting cells) and horizontal lines are the outlines of the ray cells. Circles on the tracheid walls are bordered pits. (Images M Bamford)

rounded by a lateral wing. Fan-shaped or lobed organs with a seed at the base of each lobe are placed in Rigbyaceae. In contrast, seeds in Lidgettoniaceae are represented by pairs of minute, umbrella-shaped organs with seeds on the lower surface of the hood. All reproductive structures are borne on stalks attached to the midrib, petiole, or in the axil of the leaf (Fig. 12.9c, d). The reproductive structures, themselves, have a flattened receptacle with seeds attached to one surface; and, typically, there is a marginal wing folded over that surface during early the development to protect the ovules. From whence these features evolved remains a mystery.

The evolutionary origins of glossopterids are obscure, but they are believed to have arisen in areas peripheral to the ice sheets during the peak of the LPIA (Cuneo et al. 1993). They diversified through the remainder of the Permian and were major biomass contributors to the vast peat-accumulating swamps in the Southern Hemisphere. Although plants as a whole do not seem to have been affected too much by the End-Permian Mass Extinction event (see Delfino et al., Chap. 10), Glossopteridales, in its entirety, appears to have been one of the major casualties of the event (Rees 2002). It is unclear whether they left any descendants, but it has been claimed that these plants may have been distant links to the flowering plants, which emerged in the Cretaceous (Retallack and Dilcher 1981).

12.2.4.1.6 Peltaspermales

The Peltaspermales is an order of gymnospermous plants with a shrub-like growth habit, creeping to short upright stems, and a crown of medium-sized, bipinnate leaves (DiMichele et al. 2005b). There is a considerable variety in foliage types, with compound forked leaves with highly dissected or lobed pinnae

[U1217]. Leaf cuticles are thick and resistant to decay, resulting in the naming of new taxa often distinguished based on epidermal features preserved as cuticular impressions. Female strobili (cones) consist of helically arranged, fan- or umbrella-shaped megasporophylls∗ with relatively small, flattened, and bilaterally symmetrical (platyspermic) seeds attached to their lower surfaces. The pollen organs are simple and consist of basally fused pollen sacs attached to the lower surfaces of modified pinnules (Kerp 1988). Peltasperms emerged in the latest Pennsylvanian and grew in drier, extrabasinal∗ environments (Pfefferkorn 1980). They were common and very widespread in the entire Northern Hemisphere in the Permian, where they were the dominant group of seed ferns. Although very successful in the Northern Hemisphere during the Permian, they did not colonize Gondwana until the Triassic (Blomenkemper et al. 2018).

12.2.4.1.7 Corystospermales

The Corystospermales are another order of arborescent gymnosperms that reached its maximum diversity in the Triassic of the Southern Hemisphere, where they were often dominant (Rees 2002). Early forms have been recognized in the upper Permian of the paleotropics (Kerp et al. 2006; Blomenkemper et al. 2018). Most typical and most common is the foliage known as *Dicroidium*. The leaves were small to medium-sized and bifurcated (Y-shaped) [U1218]. Female organs consist of cupules in which a single seed developed. The pollen organs are rather simple and composed of clusters of basally fused, spindle-shaped pollen sacs attached to the lower surfaces of modified leaves. Dispersed pollen attributed to the group is also known from the late Permian of Gondwana (e.g., Gastaldo et al. 2017).

12.2.4.1.8 Cycadales

Cycadales are an obscure and systematically limited plant group in today's world, appearing as minor constituents of tropical and sub-tropical regions. A very limited number of genera grow in isolation in South and Central America, southeastern United States, Australia and islands of the South Pacific, Japan, China, Southeast Asia, India, Sri Lanka, Madagascar, and southern and tropical Africa. Finding representatives of these plants in LPIA landscapes also might have been difficult. The group appeared in the Permian, or may have been present in the latest Pennsylvanian (Zeiller 1906), and became more common during the Mesozoic (see Gee et al., Chap. 6). It is recognized based on the remains of leaves and reproductive ovulate structures [U1219]. The oldest cycad leaf with well-preserved cuticle is from the upper Permian of South China (Feng et al. 2017), whereas the oldest unequivocal fertile material comes from the lower Permian of China (*Crossozamia*; Gao and Thomas 1989) and Texas (*Dioonitocarpidium*; DiMichele et al. 2004). These plants were very successful in the Mesozoic, and most of the

Fig. 12.11 Modern and LPIA ginkgophyes. (**a**) The distinctive fan-shaped leaf of *Ginkgo biloba*, the only living species. (**b**) An early Permian (Asselian) dichotomous and planated ginkgophyte leaf assigned to *Ginkgophyllum grassetii*. (**c**) A Permian dichotomous ginkgophyte leaf morphotype assigned to *Sphenobaiera*. (**d**) A Permian dichotomous ginkgophyte leaf morphotype assigned to *Sphenobaiera*. (Images (**a**) J Wang, (**b, d**) R Rößler, (**c**) H Kerp)

few living representatives have a short unbranched stem with a whorl of stiff pinnate leaves. The plants are dioecious. Ovules are attached to the base of free, leafy megasporophylls or may be organized in cones in which the sporophylls are strongly reduced and woody. Like conifers, early cycads grew in drier extrabasinal habitats.

12.2.4.2 Ginkgophytes, Cordaites, and Conifers

The most striking innovations in crown-group seed plants occur in the coniferophytes, components of all forests visited up until this section of the book. And, as seen previously, coniferophytes differ markedly from other gymnosperms in their simple leaves. These leaves may be fan- or strap-shaped with dichotomous venation in ginkgophytes and cordaitaleans, or scale- or needle-like with a single vein in most conifers.

12.2.4.2.1 Ginkgoales

Five lineages of Mesozoic ginkgophytes are known as whole plants (Zhou 1997), with the Ginkgoales best described and characterized. Today, only a single species–*Ginkgo biloba*–of the group is living and it is a dioecious tree with fan-shaped leaves (Fig. 12.11a). Although the tree has been a cultivated plant for some time, initially in China and Japan and in Europe since the eighteenth century, its natural occurrence in Southeast China was not discovered until 1956 where it was confined to the flanks of a single mountain (Crane 2015). These relatively slow-growing plants have leaves that can show considerable variation, even within a single tree. The seeds develop in pairs of two on a long stalk and are noteworthy for a thick fleshy and odoriferous (stinking) outer coat. *Ginkgo* pollen is produced on microsporophylls* that are organized in catkin-like strobili.

The earliest ginkgoaleans, described from the Permian, are dissected leaves [U1220]. Several taxa, including *Ginkgophyllum, Sphenobaiera,* and *Baiera* are recognized (Fig. 12.11b, c, d). Leaves of *Ginkgo* first appear in the Middle Jurassic, making it the oldest extant gymnosperm taxon. Ginkgoaleans were very widespread during the Mesozoic to the Neogene, ranging biogeographically from the Antarctic to the Arctic. Late Cenozoic cooling that culmi-

nated in the Pleistocene ice ages is believed to be why *Ginkgo*'s range contracted and was ultimately restricted to a very small refugium in China. Evolutionary trends show changes in the plant's reproductive structures, which include a reduction of individual stalks, a decrease in the number of ovules, and an increase in the size of the ovules. These changes were accompanied by an increase in the width of leaf segments. These evolutionary trends are roughly consistent with the ontogenetic sequence of the living species (Zheng and Zhou 2004).

12.2.4.2.2 Cordaitales

Cordaites are reconstructed as tall, 30–40 m high, riparian woody trees, medium-sized mangroves with a strongly branched root system, and smaller shrub-like plants with creeping woody axes. All members of the group have strap to lancet-shaped leaves with parallel venation (Fig. 12.12c), which might make their overall features a familiar sight in a forest setting. Although the foliage looks rather uniform, it represents a wide variety of natural species as is evidenced by cuticular analysis (Šimůnek 2007) and by the variety of the reproductive structures found associated with the plants. The stems of arborescent *Cordaites* have a well-developed secondary growth that is not well differentiated (Fig. 12.12a) [U1221], similar to the woody character of the plant's roots (Fig. 12.12b). The wood is, often, hardly distinguishable from the wood of early conifers. The stems have a central cavity with horizontal septae. Both male and female repro-

Fig. 12.12 LPIA cordaitalean gymnosperms. (**a**) Permineralized woody root assigned to *Amyelon*. (**b**) Permineralized woody trunk of *Cordaites* showing well-defined growth rings. (**c**) Long, strap-like leaf of *Cordaites* in which parallel venation is preserved. (Images (**a, c**) R Rößler, (**b**) J Wang)

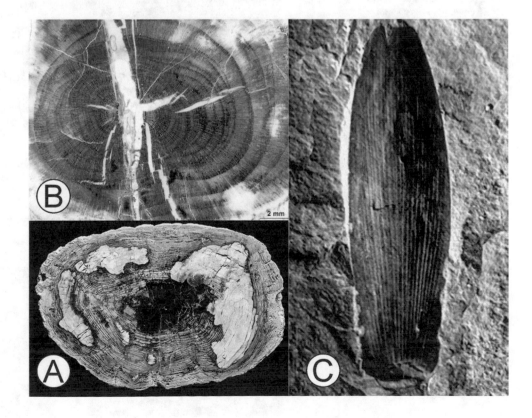

ductive structures are organized into weak cone-like structures (strobili) that consist of an axis with, usually, two rows of bracts and dwarf-shoots in their axils. The dwarf-shoots consist of a short axis with spirally arranged scales and strobili that bore either pollen sacs or ovules. The apical scales of the dwarf-shoots in the male strobili carry pollen sacs containing prepollen*, usually monosaccate with a single air bladder (saccus) surrounding the entire body (corpus). This feature is interpreted as an adaption for wind pollination. The dwarf shoots of the female strobili each hold several seeds (usually 2–3). These seeds were attached to long stalks and projected outside the edge of the strobili, and are anatomically differentiated into a fleshy seed coat (sarcotesta) and a hard inner layer (sclerotesta).

The earliest *Cordaites* are known from the Mississippian paleoequatorial regions of Euramerica. They were more common in humid Pennsylvanian peat-forming swamps but are also reported from hinterland (outside of the coastal lowlands) regions. In the Permian, representatives of the group replaced lycopsids as the major coal-forming plants. These conifer relatives became rare and are absent in the upper Permian of Euramerica but continued with a biogeographic range restricted to Gondwana, Cathaysia (a microcontinent block extending from the paleoequator into the temperate zone), and Angara (a Siberian craton located in the mid-latitudes). Cordaitales is considered to be a direct-line ancestor of the conifers, although they have a more-or-less coeval appearance in the geological record (Taylor et al. 2009). Regardless, a close relationship between the groups is clear, and both probably evolved from a common ancestor.

12.2.4.2.3 Coniferales

Once we arrive in our travels to forests where conifers dominate, landscapes begin to look more familiar and resemble those of the Mesozoic and Cenozoic we have visited in earlier chapters. We might be at home where conifers, characterized, in general, by small needle-like, grow to tree-sized stature (Fig. 12.13A, B) [U1222]. The female cones of most modern plants consist of an axis with spirally arranged woody scales with winged seeds in their axils. These are homologous to a branching system consisting of last order branchlets and an axis of penultimate (next-to-last) order. The male cones are much smaller, consist of an axis with spirally arranged microsporophylls that carry pollen sacs, and are homologous with the last order branchlets. Modern conifers have pollen grains commonly with sacci (air blad-

Fig. 12.13 Permian conifers. (**a**) *Walchia pinniformis* preserved in redbeds by hematite. Scale = 10 cm. (**b**) A Permian conifer leafy shoot. (**c**) A Permian conifer female cone. (**d**) Leafy shoot of *Dicranophyllum hallei* with several male and one female cone (upper right) from the lower Permian. (Images (**a**) WA DiMichele, (**b**, **c**) J Wang, (**d**) R Rößler)

ders). The pollen grains are caught by a pollination droplet on top of the micropyle and sucked into the pollen chamber. The pollen grains then develop a pollen tube at the distal side of the pollen grain that carries the male nuclei (sperm) to the egg cells.

The early fossil record of conifers is very scanty owing to their very limited preservation potential. We may have been able to see them in the forests, but their remains decayed in the litter scattered across the floor without a trace. Where we will see evidence of their presence is in the pollen record. The occurrence of typical conifer prepollen indicates that this group must have been in the paleotropics beginning in the Mississippian. However, the earliest body fossils of the group are found in the Pennsylvanian of North America (Lyons and Darrah 1989), and macroscopic remains, including cones, are well documented during a short interval in the Middle Pennsylvanian (Hernández-Castillo et al. 2001). And, although not common, conifer fossils start to appear regularly in the smaller intramontane basins* (e.g., France; Rothwell et al. 1997), during the Late Pennsylvanian, when these floras are still dominated by humid elements found elsewhere (e.g., calamitaleans and marattialeans). It was not until the Permian that conifer remains became common in the fossil record. The earliest conifers had typical coniferous wood, nearly indistinguishable from cordaitalean secondary xylem, and pinnate branching systems (Fig. 12.13a) like modern *Araucaria* (monkey-puzzle tree). These earliest representatives were probably trees, but the height to which they grew is uncertain. Changes in the reproductive structures, particularly in the female cones, have played a prominent role in the evolutionary history of the group (Fig. 12.13c).

Prepollen cones and ovuliferous cones may have developed on a single tree or separate trees. All Paleozoic conifers were apparently zoidiogamous, meaning these plants produced prepollen that released free-swimming sperm through a ruptured tetrad mark. The presence of a tetrad mark (monolete or trilete) on the proximal side and the lack of a thinning on the distal side of the grain suggest zoidiogamy*. The earliest siphonogamous pollen grains (with a pollen tube) are known from the Triassic. The female cones of the earliest conifers consist of an axis with spirally arranged bracts, with dwarf-shoots in their axils (Fig. 12.13c, d). The bracts are often forked and similar to the leaves of the vegetative axes of penultimate (next-to-last) order. The axis carrying the female cones makes an angle of approximately 120° with the cone axis. The dwarf-shoots consist of a very short axis with small, in early forms spirally arranged scales, one or more of which bearing ovules. The morphology of the dwarf-shoots of the female cones is of primary importance for the systematics and phylogeny. They are often found isolated, because cones easily disintegrated after having shed the seeds. The dwarf-shoots are flattened, and the number of scales and the degree of fusion of the sterile-and-fertile scales are diagnos-

tic features. The general evolutionary trend in conifers can be characterized as an increasing reduction of the number of scales per dwarf-shoot, and an increasing fusion of the individual scales (Taylor et al. 2009).

There are several well-known and common LPIA conifers. *Walchia piniformis* (Fig. 12.13a) is one of the best known and most common Paleozoic conifers of which we know its trunk, foliated branches, and ovuliferous and pollen cones with cuticle (Florin 1939). Conifers with strongly decurrent leaves are usually assigned to either this genus or *Utrechtia*. Other representative genera include: (1) *Thucydia*, the earliest well-known conifer, from the Pennsylvanian of the USA; (2) *Emporia*, from the famous Hamilton Lagerstätte (Kansas), slightly younger; (3) *Otovicia*, with very small leaves, from the Permian of Europe; (4) *Majonica*, late Permian, Europe, with winged seeds similar to modern *Acer* seeds; (5) *Pseudovoltzia*, late Permian, Europe; (6) *Ortiseia*, thick fleshy leaves, from the upper Permian of the Southern Alps; and (7) *Ullmannia* with a single, large, rounded scale that carries a large winged seed.

12.3 Extraterrestrial Control on the Late Paleozoic Ice Age

The LPIA is unique in Earth history because of several factors. The assembly of the supercontinent, Pangea, during the Carboniferous and Permian witnessed the assembly of all major tectonic plates that coalesced into a single landmass [U1223]. It was oriented in a north-south direction, with parts of the present-day continents of South America, Africa, India, Antarctica, and Australia located at a high southern paleolatitude (Scotese and McKerrow 1990). The formation of extensive mountain chains at the boundaries of continental collisions and the contiguous landscape affected atmospheric patterns over land as well as oceanic circulation. These, in conjunction with extraterrestrial factors, promoted the build-up and loss of glacial ice in the Southern Hemisphere which, in turn, affected the relative position of sea level and coastal plains (Fielding et al. 2008a).

The extraterrestrial factors [U1224] responsible for changes in climate over the Phanerozoic and, most likely, into the Precambrian, occurred in response to how our planet orbits the sun. Long-term effects occurred then, and now, on the scales of tens of thousands of years, but also on shorter time scales. Long- and short-term oscillations in climate were most pronounced during the LPIA (Fielding et al. 2008a, b). In combination, three orbital parameters are responsible for the flux from icehouse-to-hothouse climates that were first identified and mathematically described by Milanković (1998). Milanković's model explains how variations in our planet's position and orientation, relative to the sun, alter global climate (Box P3). In combination, in- and out-of-phase Milanković orbital factors

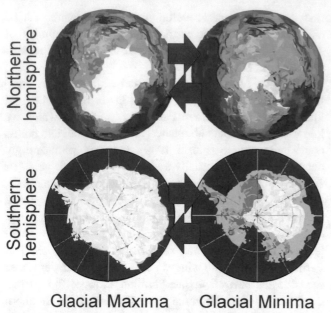

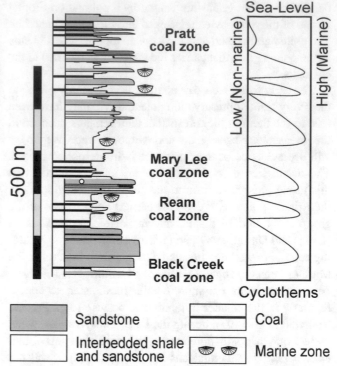

Fig. 12.14 Reconstruction of late Cenozoic continental glaciation in the Northern and Southern Hemispheres. The plots show the extent of glacial ice during both maxima and minima, along with the change in sea level as a consequence of build-up and melting. Nearshore marine sediments deposited during glacial minima are exposed and colonized by terrestrial plants during glacial maxima, which resulted in coastal plain soils and peat swamps. Following deglaciation, coastal lowlands are covered in marine waters and sediments, resulting cyclical successions known as cyclothems

Fig. 12.15 An example of the repetitive nature of LPIA cyclothems from the Lower Pennsylvanian record of the Black Warrior Basin, Alabama, demonstrating the relationship between coal-bearing coastal deposits and overlying marine rocks. (modified from Gastaldo et al. 2009)

influence short (10,000 year) to long (100,000–400,000 year) climate cycles on Earth [U1205]. These, in turn, promote the formation and advance of ice sheets when the (paleo)geographic position of continents are at high latitudes near the poles (Fig. 12.14). Recently, in a geologic sense, our planet has experienced oscillations in icehouse-to-hothouse conditions over the past 23 million years, beginning in the Oligocene (Zachos et al. 2001). The advances and retreats in polar glaciers we have seen during our trip through the late Cenozoic (Fig. 12.14; Martinetto et al., Chap. 1) are not unlike what the planet experienced during the LPIA [U1226].

12.4 Features of the Late Paleozoic Ice Age

Our ideas about the scope and duration of the LPIA [U1227] have been transformed since first proposed more than 100 years ago. Early in the twentieth century, the model used to explain the expansion of Carboniferous "coal-forming" forests and the reasons behind their success centered on the idea that Earth experienced one, very long icehouse period (e.g., Frakes et al. 1992). That icehouse was thought to have lasted more than 30 million years. The first significant ice sheets advanced in the latest Mississippian (Serpukovian, 331–323 Ma) and retreated in the Early

Permian (Kungurian, 284–273 Ma), with an acknowledged warming interlude during the Late Pennsylvanian (c. 307 Ma). It was thought that polar regions experienced continued cold climates whereas land masses located near the equator, as the supercontinent of Pangea formed, remained warm and wet (Fielding et al. 2008a). These warm-and-wet conditions promoted the expansion of extensive tropical forests. Both subterranean (rooting) and aerial (stems, branches, leaves, reproductive structures) plant parts of successive generations of plants accumulated in geographically extensive tropical peat swamps across low-lying coastal plain topographies. At the time, geoscientists acknowledged the fact that sea levels fell (due to glacial ice advances) and rose (during ice sheet melting) multiple times, affecting the coastal forests across the entire planet (Wanless and Shepard 1936). When sea levels were low, soils of various nature, including the organic, peat-rich soils, formed. During subsequent periods of high sea levels, these soils were buried by nearshore and fully marine sediments, in which fossilized shells of marine invertebrates are preserved. The alternation between a succession of marine rock, bearing invertebrate fossils, overlain by one or more coal (peat)- bearing intervals of rock constituted a cyclothem* (Fig. 12.15) [U1228] (Wanless and Weller 1932). There are several hundred cyclothems, and potential

Box 12.4: One Long Icehouse Interval or Multiple Icehouse-to-Hothouse Transitions

Global studies of the sediments associated with the coal-bearing equatorial localities correlated with glacial deposits, preserved at high paleolatitudes in South America, southern Africa, Antarctica, India, and Australia [U1229], demonstrate that the LPIA is a more complex interval than once thought. It is now recognized that several discrete icehouse climates occurred beginning in the Late Mississippian (Serpukhovian) and ending in the Late Middle Pennsylvanian (lower Moscovian). Each cold-climate interval was punctuated by a warm interval during which time the extent of ice sheet dynamics was minimal (Fig. 12.16) [U1230] (Fielding et al. 2008b). Widespread continental glaciation began, once again, in the Late Pennsylvanian and continued into the middle Permian (Fig. 12.17). But, unlike earlier models that interpreted a prolonged icehouse, another four discrete icehouse-to-hothouse transitions are now known, with the deglaciation of the planet occurring in the middle Permian.

fossil-plant assemblages, from which we reconstruct vegetation across the paleotropics and high latitude land masses in space and time. For example, there are at least 54 cyclothems recorded in the Ostrava Formation of Late Mississippian (Serpukhovian) age in the Silesian Basin of the Czech Republic and Poland (Gastaldo et al. 2009; Jirásek et al. 2018); at least 35 cycles are reported for the Early Pennsylvanian (Baskirian, 323–315 Ma) and around 30 cycles for the Middle Pennsylvanian (Moscovian, 315–307 Ma) of the Donets Basin in the Ukraine (Eros et al. 2012); and a total of 60 minor, intermediate, and major Middle Pennsylvanian cyclothems (Moscovian = mid-Desmoinesian to Missourian; Heckel 2008) are recognized in the Mid-Continent of the United States. The Late Pennsylvanian is reported to contain at least 30 cycles (Eros et al. 2012), many of which are constrained by U-Pb radiometric age dates. With the advances in our resolution of major global changes in climate states during the LPIA (Box 12.4), it has been possible to evaluate the vegetational dynamics in the paleotropics of these peat- and non-peat-accumulating swamps (Wilson et al. 2017).

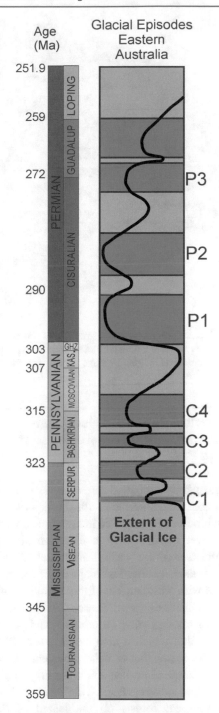

Fig. 12.16 The time scale of the LPIA against which the Carboniferous (C1–C4) and Permian (P1–P4) glacial and interglacial stages, based on the sedimentologic and stratigraphic record of eastern Australia, are shown (modified from Fielding et al. 2008a, b). Spore-bearing plants dominate Pennsylvanian coal forests which, subsequently, are replaced by seed-bearing groups in much of the Permian

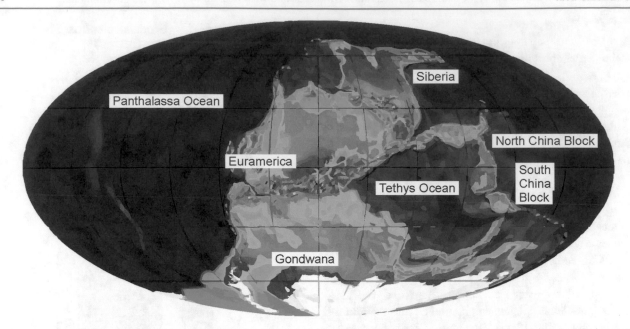

Fig. 12.17 A paleogeographic reconstruction of the supercontinent, Pangea, on which major continental blocks and oceans are identified. The illustration depicts a glacial stage where the polar Southern Hemisphere continents, including South America, South Africa, Antarctica, and Australia, were ice covered. (modified from Scotese Paleomap Project: http://www.scotese.com/)

12.5 Tropical Peat- and Non-peat-Accumulating Forests: Present and Past

Today, angiosperm-dominated forests cover more than 30% of the available land's surface with the densest vegetation and highest biodiversity centered in the tropics. These forests are spread over coastal to high elevation sites [U1231]. Lowland forests colonize various substrates including mineral and organic-rich soils. The majority of forests grow on mineral soils with less than 5% of this biome established on organic-rich substrates (peat; Pearce 2017). Yet, the plant systematics of taxa growing on mineral- and peat-substrate forests are similar, which is also the case for LPIA vegetation. One major reason that explains why generations of trees grow on either mineral or peat soil is the clay mineralogical content of the soil substrate itself. This is best exemplified by peat- and non-peat forests on the island of Borneo in Southeast Asia, both of which have similar forests dominated by the angiosperm family Dipterocarpaceae (Fig. 12.18a).

There are several major drainage systems that transport water and sediment from the central mountains of the Borneo Massif to the ocean under the same ever-wet climate. Two of these, the Mahakam and Rajang rivers, drain opposite sides of the island; the Mahakam discharges into the Makassar Strait to the east, and the Rajang discharges into the South China Sea to the west [U1232]. The type of sediment transported and deposited in these coastal zones has different origins, resulting in physical conditions that prevent or promote the accumulation of thick peat that blankets the land's surface (Gastaldo 2010). The sediment originating from the eastern side of the Borneo Massif comes from Tertiary-aged continental and shallow marine sediments. These are sands and silts, with a low proportion of clay minerals. Soils formed on these coastal deposits allow for water to drain through them establishing a subsurface water table, the depth of which is controlled by rainfall. Roots and other biological activities promote the decay of aerial plant parts that fall to the soil's surface, constantly recycling the organic matter. In contrast, the sediment originating from the western side of the Borneo Massif comes from Tertiary-aged deep marine sediments of silt and clay minerals. The clay minerals in these sediments are expandable (swelling) and mixed-layered clay species, including illite–smectite, illite, and kaolinite (Staub and Gastaldo 2003). Deposits in which swelling clays occur absorb water and, essentially, block its downward transport through the soil horizon. As a consequence, water ponds at the surface of the clay-rich soil. Rooting into the clay-rich soil first establishes the vegetation (Fig. 12.18c). Subsequently, aerial plant debris begins to accumulate at the soil's surface, where it rots but is not completely recycled. As the rate of organic matter accumulation exceeds the rate of recycling, peat begins to accumulate and establish the organic-rich soil. Subsequent generations of plants add more organic matter to the peat, ultimately increasing the surface of the deposit above the original clay-rich soil [U1233]. The water

Fig. 12.18 The Rajang River delta, Sarawak, is a modern tropical analog setting in which thick peat deposits accumulate in raised forests. (**a**) An oblique aerial image showing a dense canopy of Dipterocarpaceae forest rooted in peat. (**b**) Peat swamps drain into coastal plain rivers that are tea-colored, as a consequence of humic acids, and sediment starved. (**c**) Vibracore through the peat swamp and underlying mineral-substrate soil showing 4 m of peat cross cut by rooting structures. The base of the peat has a ^{14}C age constraint of c. 7.5 ka, whereas the underlying mineral soil formed during Oxygen-isotope stage 5, c. 120 ka (Staub and Gastaldo 2003). (Images RA Gastaldo)

table is maintained above the land's surface (rain-fed, ombrotrophic[*] mire). This is because the peat acts similar to a sponge, retaining water, and the peat body thickens as more subterranean and aerial plant debris is added. In the Rajang River delta, centers of peat domes have attained a thickness of 16 m *above* the clay-rich soil in less than 7000 years of forest-litter accumulation (Staub and Gastaldo 2003). These deposits blanket more than 6500 km² of area in the coastal lowland. Since the last rise in sea level in response to Northern Hemisphere deglaciation, thick peat deposits are found across 250,000 km² of Southeast Asia. The areas include Peninsular Malaysia, Indonesia, and eastward into Papua New Guinea, although the proportion of these peatlands has decreased significantly over the past few decades by more than 50% (77% to 36%, Miettinen et al. 2012) due to anthropogenic activities.

12.5.1 Tropical Forests in Deep Time

Peat- and mineral-substrate forests in today's tropics are analogs for those preserved in the Carboniferous–Permian rock record. We understand the most about LPIA forests that colonized coastal lowland environments. This record is a direct consequence of their preservation in the stratigraphic record of an interplay between the long- and short-term Milanković cycles, which controlled the advances and retreats of continental glaciers of Gondwana and the lowering and raising of

global sea level, respectively, and geological events responsible for subsidence (lowering) of the land's surface in response to plate tectonic activities. Geoscientists recognize a myriad of tectonic basins based on the tectonic regime under which each formed; our discussion will omit these complexities, and the reader is directed elsewhere for details on the topic. For the sake of simplicity, the following will focus on the generalized effects of the fall and rise of sea level on the distribution of LPIA forests (Fig. 12.19).

Continental glaciers "grow" and expand their area over the time scales of several tens of thousands of years in a stepwise pattern. As continental glaciers build up and advance, the position of eustatic sea level is lowered, also in a stepwise manner. In contrast, deglaciation and the rate of sea-level rise is significantly more rapid. It is estimated that LPIA sea levels were as much as 125 m lower at glacial maxima than at highstand (Haq and Shutter 2008). When a drop in sea level occurred, what once were nearshore coastal marine sediments were exposed to the atmosphere. These deposits of sand, silt, and mixtures of sand-and-silt were subjected to soil-forming processes and colonized, first, by wetland taxa. If these sediments were poor in expandable-clay minerals, forests grew on mineral-substrate soils and blanketed these areas (Fig. 12.19a). If these sediments contained a proportion of expandable-clay minerals and prevented soil formation and promoted the ponding of water, organic matter accumulated, resulting in peat accumulation. Both mineral- and peat-substrate forests could coexist across

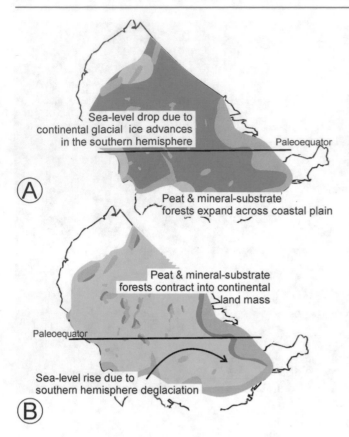

Fig. 12.19 Paleogeographic reconstruction of equatorial North America during the Pennsylvanian. (**a**) Extent of coastal lowlands and peat-accumulating forests during a glacial maximum in the Southern Hemisphere, showing the build-out of the continent into what, once, was ocean. (**b**) Coastline, lowland, and peat-accumulation forests retreat during times of glacial minima, when sea level rises in response to deglaciation in the Southern Hemisphere. (reproduced with permission: Pfefferkorn et al. (2017))

the coastal plain landscape at any point in time, depending on the origin of the sediment transported to the coast. In effect, LPIA plants colonized any, and all, available soil substrates, similar to Holocene forests in Southeast Asia (Fig. 12.19a, b). The difference in such forests between now and several hundred million years ago is the systematic composition of the deep-time equivalents. At each glacial maximum during one Milanković-controlled cyclothem, peat- and mineral-substrate forests expanded geographically across the sediments of what had been previously marine shelf. During this time interval, many of the thickest and geographically extensive peat forests formed. As continental glaciers melted, sea level rose, and a change in climate state resulted in increased sediment supply to the coastal zone. At these times, peat forests were buried in deposits of sand and mud. These new lands were stabilized for shorter durations but, nevertheless, also were colonized (Fig. 12.19B). Their geographical extent was less than the area of the landmass at glacial maxima, though. As Southern Hemisphere glaciers continued to melt, ultimately these coastal deposits were

covered in shallow-to-deep ocean waters, the sea floor colonized by marine taxa dominated by the invertebrate group of brachiopods (see Tinn et al., Chap. 14), and the world's coastlines were pushed inland to the interior of continents [U1234]. This cyclicity was accompanied, at times, by dramatic climate perturbation, allowing for the dismantling and reorganization of forest composition and their structure when critical thresholds were surpassed during the LPIA (Pfefferkorn et al. 2017).

The physical conditions under which the coal forests of the Carboniferous and Permian grew, and peat deposits accumulated, were not significantly different than those our planet has experienced over the last few million years (see Martinetto et al., Chap. 1). Changes in both latitudinal patterns of vegetation and its long-term temporal patterns of turnover, replacement, and extinction under a transition from icehouse-to-hothouse conditions, have been proposed as a deep-time analog to what may portend for our biosphere under the unidirectional global warming now underway (Gastaldo et al. 1996). The Permo–Carboniferous ice sheets waxed and waned across Gondwana at the high southern latitudes. Evidence of their movement is recorded in near-field paleoclimate proxies in both Antarctica (Isbell et al. 2008) and Australia (Fielding et al. 2008b) along with contemporary glacial deposits in Asia (Isbell et al. 2016). This cyclicity promoted intervals of high rainfall in the paleotropics of Euramerica when ice sheets were nearing and retreating from their maximum geographical extent. In response to high precipitation, clay-rich soils developed primarily in coastal plain settings, retarding drainage and enhanced the sequestration of forest biomass in extensive peat swamps. Geological activities, including sea-level rise, as well as earthquake and volcanic activities associated with tectonic (mountain building) processes, often buried standing, in situ coastal forests in an "instant" (see Gastaldo et al., Chap. 13). These events preserved not only those plants and animals that inhabited the landscape, but also the spatial heterogeneity of these forests. What is significantly different about LPIA forests from those of either the Mesozoic or the Cenozoic is their systematic composition: the first appearance of angiosperms occurred more than 125 million years in the future (see Kvaček et al., Chap. 5).

12.6 Conclusions

The proliferation of wetland forests across continents during the LPIA occurred during the wet phases of glacial–interglacial cycles. Ice sheets and mountain glaciers periodically covered the high latitude, southern polar continents, alternating with times during which ice sheets retreated or, temporarily, disappeared from the land surface. The driving force for the expansion and contraction of glacial ice, and the response of the biosphere to periodic changes in Earth Systems (i.e., lithosphere, hydrosphere, cryosphere, and atmosphere), was not controlled by any Earth-bound mechanism. Rather, significant changes in global climate states, from icehouse to hothouse and back, again, across time scales of more or less 100,000 to 400,000 years, are a consequence of extraterrestrial factors. It is climatic conditions and changes in them across these grand time scales that, ultimately, are responsible for the formation of the Carboniferous–Permian coals. These coals continue to constitute one of the major fuel resources responsible for the world as we know it today (coal-fired power plants, as of this writing, still account for nearly 40% of all electricity generated globally and, in addition, provide the energy source for most smelting operations of iron and steel). These deep-time coal forests have Recent analogs in the peat-accumulating and mineral substrate tropical swamps of Southeast Asia. Yet, the plant communities and biomass that accumulated in peat swamps, today, are vastly different from the plants that occupied the landscapes of the LPIA.

The systematic affinities of the plants that dominated the Carboniferous to early Permian were not seed-bearing groups. Rather, the spore-bearing plant groups that are subdominant or cryptic in today's flora were the giants of the LPIA. These groups–club mosses (lycopsids), horsetails (sphenopsids), and ferns (pteridophytes)–occupied nearly all available habitats where the physical conditions were met for reproduction. The growth architectures of these clades ranged from groundcover and scrawling plants to towering trees, attaining basal-trunk diameters of 2 m and heights approaching 50 m. Several taxa of these spore-bearing groups developed woody trunks, whereas others attained their structural integrity and height through the development of a unique bark (periderm) configuration. Plants that reproduced by seeds are assigned to the umbrella group of gymnosperms and were more common in mineral-substrate settings until the latest Pennsylvanian and early Permian. And, although most of these clades are now extinct [U1235], ginkgophytes and conifers remain successful elements of today's landscape.

Questions

1. Unlike today, which Linnaean plant classes dominated the landscapes of the Late Paleozoic? How does each Class differ from angiosperms?
2. What was the reproductive strategy of the dominant Carboniferous vegetation that tied these plants to wetlands of various physical characters? Why does this reproductive strategy require moisture?
3. Club mosses found in temperate, subtropical, and tropical forests across the planet are small, diminutive plants that grow up to 30 cm in height. What anatomical features of Permo–Carboniferous club mosses allowed some of these plants to grow to tree heights exceeding 30 meters?
4. What are the typical growth architectures found in Pennsylvanian "lepidodendrids?"
5. The horsetail group colonizes soils from the high, polar latitudes to the equator, with modern representatives growing to heights of 2 m. What physical and anatomical features of Carboniferous relatives allowed these plants to grow to tree-sized heights?
6. What characteristics of calamitean growth architecture set these plants apart from the other dominant Carboniferous groups?
7. How do the major groups of Pennsylvanian and Permian gymnosperms differ from one another?
8. What variation in growth architecture do Medullosan and Lyginopterid pteridosperms display?
9. In 1915, Alfred Wegener proposed that the Southern Hemisphere continents once were united into a single land mass. Evidence he used in support of his continental drift (today's plate tectonic) theory included the distribution of the gymnosperm, *Glossopteris*. How does this plant group differ from other Carboniferous clades, and why did Wegener think its paleogeographic distribution supported his idea?
10. Cordaites and conifers are considered to be related as sister groups. What features of their reproductive strategies do they share and would support this relationship?
11. Of the plants found in the Permo–Carboniferous, which groups maintain a foothold in the landscapes of today, and what role(s) do they continue to play?
12. What is a cyclothem, and how might these records reflect the influence of extraterrestrial (Milanković orbital parameters) factors on climate?
13. Today, peat-accumulating equatorial forests accumulate thick and extensive peat deposits act as organic soils for a variety of vegetation. Why are these settings considered analogs for deep-time coal deposits?

Acknowledgments The authors would like to acknowledge the array of grant-funding agencies that have supported their research over the decades, which has resulted in the compilation of case studies presented herein. Those agencies are, in alphabetical order: Alexander von Humboldt Stiftung, Germany; American Chemical Society, Petroleum Research Fund; Brazilian National Council for Scientific and Technological Development (CNPq); Conselho Nacional de Desenvolvimento Científico e Tecnológico, Brazil; Coordenação de Aperfeiçoamento de Pessoal de Nível Superior, Brazil; Deutsche Forschungsgemeinschaft, Bonn, Germany; Fulbright Scholars Program, USA; Grant Agency of the Czech Republic (project 16-24062S); Jogins Fossil Institute; National Research Foundation of South Africa– African Origins Platform; National Natural Science Foundation of China (Grant No.41530101); National Science Foundation of the United States of America; Nova Scotia Department of Natural Resources; The Research Support Foundation of the State of Rio Grande do Sul (FAPERGS); Strategic Priority Research Program of

Chinese Academy of Sciences (Grant No. XDB26000000); The Swedish Research Council (Vetenskapsrådet); The United States National Museum, Smithsonian Institution, Washington DC USA; Volkswagen Foundation, Hannover, Germany.

References

Anderson JM, Anderson HM (1989) Palaeoflora of southern Africa. Molteno Formation (Triassic). Volume 2: Gymnosperms (excluding *Dicroidium*). A.A. Balkema, Rotterdam. 423 pp

Barthel M (1980) Calamiten aus dem Oberkarbon und Rotliegenden des Thüringer Waldes. In: Vent W (ed) 100 Jahre Arboretum. Akad. Verlag, Berlin, pp 237–258

Bateman RM (1994) Evolutionary-developmental change in the growth architecture of fossil rhizomorphic lycopsids: Scenarious constructed on cladistics foundations. Biol Rev 69:527–597

Bateman RM, DiMichele WA (1991) *Hizemodendron*, gen. nov., a Pseudoherbaceous segregate of *Lepidodendron* (Pennsylvanian): phylogenetic context for evolutionary changes in lycopsid growth architecture. Syst Bot 16:195–205

Bateman RM, DiMichele WA, Willard DA (1992) Experimental cladistic analysis of anatomically preserved arborescent lycopsids from the Carboniferous of Euramerica: an essay in paleobotanical phylogenetics. Ann Mo Bot Gard 79:500–559

Bateman RM, Crane PR, DiMichele WA, Kenrick P, Rowe NP, Speck T, Stein WE (1998) Early evolution of land plants: phylogeny, physiology, and ecology of the primary terrestrial radiation. Ann Rev Ecol Syst 29:263–292

Blomenkemper P, Kerp H, Abu Hamad A, DiMichele WA, Bomfleur B (2018) A hidden cradle of plant evolution in Permian tropical lowlands. Science 362:1414–1416

Boyce CK, DiMichele WA (2016) Arborescent lycopsid productivity and lifespan: constraining the possibilities. Rev Palaeobot Palynol 227:97–110

Crane PR (2015) *Ginkgo:* the tree that time forgot. Yale Univ Press, New Haven. 408 p

Crane PR, Herendeen P, Friis EM (2004) Fossils and plant phylogeny. Am J Bot 91:1683–1699

Cuneo NR, Isbell J, Taylor EL, Taylor TN (1993) The *Glossopteris* flora from Antarctica: Taphonomy and paleoecology: Compt. Rendu XII ICC-P 2:13–40

Decombeix AL, Taylor EL, Taylor TN (2009) Secondary growth in *Vertebraria* roots from the Late Permian of Antarctica: a change in developmental timing. Int J Plant Sci 170:644–656

DiMichele WA, Falcon-Lang HJ (2011) Pennsylvanian 'fossil forests' in growth position (T⁰ assemblages): origin, taphonomic bias and palaeoecological insights. J Geol Soc Lond 168:1–21

DiMichele WA, Phillips TL (1985) Arborescent lycopod reproduction ad paleoecoogy in a coal-swamp environment of Late Middle Pennsylvanian age (Herrin Coal, Illinois, USA). Rev Palaeobot Palynol 44:1–26

DiMichele WA, Hook RW, Nelson WJ, Chaney DS (2004) An unusual Middle Permian flora from the Blaine Formation (Pease River Group: Leonardian-Guadalupian Series) of King County, West Texas. J Paleontol 78:765–782

DiMichele WD, Gastaldo RA, Pfefferkorn HW (2005a) Biodiversity partitioning in the late Carboniferous and its implications for ecosystem assembly. Proc Calif Acad Sci 56(suppl. 1):24–41

DiMichele WA, Kerp H, Krings M, Chaney DS (2005b) The peltasperm radiation: evidence from the southwestern United States. New Mexico Museum of Natural History and. Sci Bull 30:67–79

DiMichele WA, Phillips TL, Pfefferkorn HW (2006) Paleoecology of Late Paleozoic pteridosperms from tropical Euramerica. J Torr Bot Soc 133:83–118

DiMichele WD, Elrick SD, Bateman RM (2013) Growth habit of the late Paleozoic rhizomorphic tree-lycopsid family Diaphorodendraceae: phylogenetic, evolutionary, and paleoecological significance. Am J Bot 100:1604–1625

Eros JM, Montañez IP, Osleger DA, Davydov V, Nemyrovska TI, Poletaev VI, Zhykalyak MV (2012) Sequence stratigraphy and onlap history, Donets Basin, Ukraine: Insight into Late Paleozoic Ice Age dynamics. Palaeogeogr Palaeoclimatol Palaeoecol 313-314:1–25

Falcon-Lang HJ (2009) A *Macroneuropteris scheuchzeri* tree in growth position in the Middle Pennsylvanian Sydney Mines Formation, Nova Scotia, Canada. Atl Geol 45:74–80

Feng Z, Lv Y, Guo Y, Wei HB, Kerp H (2017) Leaf anatomy of a late Paleozoic cycad. Biol Lett 13:20170456

Fielding CR, Frank TD, Isbell J (2008a) Resolving the Late Paleozoic Gondwanan ice age in time and space. Geol Soc Am Spec Pap 441:1–354

Fielding CR, Frank TD, Birgenheier LP, Rygel MC, Jones AT, Roberts J (2008b) Stratigraphic imprint of the Late Paleozoic ice age in eastern Australia: a record of alternating glacial and nonglacial climate regime. J Geol Soc 165:129–140

Florin R (1939) The morphology of the female fructifications in *Cordaites* and conifers of Palaeozoic age. Bot Notiser 36:547–565

Frakes LA, Francis JE, Syktus JI (1992) Climate modes of the Phanerozoic. Cambridge University Press, Cambridge. 274 p

Gao Z, Thomas BA (1989) A review of fossil cycad megasporophylls, with new evidence of *Crossozamia* Pomel and its associated leaves from the lower Permian of Taiyuan, China. Rev Palaeobot Palynol 60:205–223

Gastaldo RA (1992) Regenerative growth in fossil horsetails (*Calamites*) following burial by Alluvium. Hist Biol 6:203–220

Gastaldo RA (2010) Peat or no peat: why do the Rajang and Mahakam Deltas Differ? Int J Coal Geol 83:162–172

Gastaldo RA, Matten LC (1978) *Trigonocarpus leeanus*, a new species from the middle Pennsylavanian of southern Illinois. Am J Bot 65:882–890

Gastaldo RA, DiMichele WA, Pfefferkorn HW (1996) Out of the icehouse into the greenhouse: a Late Paleozoic analogue for modern global vegetational change. GSA Today 10:1–7

Gastaldo RA, Purkyňová E, Šimůnek Z, Schmitz MD (2009) Ecological Persistence in the Late Mississippian (Serpukhovian – Namurian A) Megafloral Record of the Upper Silesian Basin, Czech Republic. PALAIOS 24:336–350

Gastaldo RA, Neveling J, Looy CV, Bamford MK, Kamo SL, Geissman JW (2017) Paleontology of the Blaauwater 67 Farm, South Africa: testing the *Daptocephalus/Lystrosaurus* Biozone boundary in a stratigraphic framework. PALAIOS 34:349–366

Glasspool I, Hilton J, Collinson ME, Wang SJ (2004) Defining the gigantopterid concept: a reinvestigation of *Gigantopteris*. A reinvestigation of *Gigantopteris* (*Megalopteris*) *nicotianaefolia* Schenck and its taxonomic implications. Palaeontology 47:1339–1361

Grand'Eury FC (1877) Mémoire sur la flore carbonifère du Départment de la Loire et du centre de la France, étudiée aux trois points de vue botanique, stratigraphique et géognostique. Mém. Academie de Science Institut France, Science Mathematiques et Physiques 24:1–624

Haq BU, Shutter SR (2008) A chronology of Paleozoic sea-level changes. Science 322:64–68

Heckel PH (2008) Pennsylvanian cyclothems in midcontinent North America as far-field effects of waxing and waning of Gondwana ice sheets. In: Fielding CR, Frank TD, Isbell JL (eds) Resolving the Late Paleozoic Gondwanan ice age in time and space, vol 441. Geological Society of America, Boulder, pp 275–289

Hernández-Castillo GR, Rothwell GW, Mapes G (2001) Thucydiaceae fam. nov., with a review and reevaluation of Paleozoic walchian conifers. Int J Plant Sci 162:1155–1185

Hetherington AJ, Berry CM, Dolan L (2016) Networks of highly branched stigmarian rootlets developed on the first giant trees. Proc Natl Acad Sci USA 113:6695–6700

Hirmer M (1927) Handbuch der Paläobotanik, Band I: Thallophyta–Bryophyta–Pteridophyta. R. Oldenbourg, Berlin. 708 pp.

Isbell JL, Koch ZJ, Szablewski GM, Lenaker PA (2008) Permian glacigenic deposits in the Transantarctic Mountains, Antarctica. In: Fielding CR, Frank TD, Isbell JL (eds) Resolving the Late Paleozoic Ice Age in Time and Space, vol 441. Geological Society of America, Boulder, pp 59–70

Isbell JL, Biakov AV, Verdernikov I, Davydov VI, Gulbranson EL, Fedorchuk ND (2016) Permian diamictites in northeastern Asia: their significance concerning the bipolarity of the late Paleozoic ice age. Earth-Sci Rev 154:279–300

Jirásek J, Opluští S, Sivek M, Schmitz MD, Abels HA (2018) Astronomical forcing of Carboniferous paralic sedimentary cycles in the Upper Silesian Basin, Czech Republic (Serpukhovian, latest Mississippian): new radiometric ages afford an astronomical age model for European biozonations and substages. Earth-Sci Rev 177:715–741

Kerp JHF (1988) Aspects of Permian palaeobotany and palynology. X. The West- and Central European species of the genus *Autunia* Krasser emend. Kerp (Peltaspermaceae) and the form-genus *Rhachiphyllum* Kerp (callipterid foliage). Rev Palaeobot Palynol 54:249–360

Kerp H, Abu Hamad A, Vörding B, Bandel K (2006) Typical Triassic Gondwanan floral elements in the Upper Permian of the paleotropics. Geology 34:265–268

Krings M, Kerp H, Taylor TN, Taylor EL (2003) How Paleozoic vines and lianas got off the ground: on scrambling and climbing Late Carboniferous-Early Permian pteridosperms. Bot Rev 69:204–224

Leistikow KU (1962) Die Wurzeln der Calamitaceae. Ph.D. Thesis, Bot. Inst., Univ. Tübingen, 67pp.

Lyons PC, Darrah WC (1989) Earliest conifers of North America: upland and/or paleoclimate indicators? PALAIOS 4:480–486

Maslen AJ (1905) The relation of root to stem in *Calamites*. Ann Bot 19:61–73

Masselter T, Rowe NP, Speck T (2007) Biomechanical reconstruction of the Carboniferous seed fern *Lyginopteris oldhamia*. Implications for growth form reconstruction and habit. Int J Plant Sci 168:1177–1189

McLoughlin S (1993) Plant fossil distributions in some Australian Permian non-marine sediments. Sediment Geol 85:601–619

McLoughlin S (2011) *Glossopteris* – insights into the architecture and relationships of an iconic Permian Gondwanan plant. J Bot Soc Bengal 65:93–106

Miettinen J, Shi C, Liew SC (2012) Two decades of destruction in Southeast Asia's peat swamp forests. Front Ecol Environ 10:124–128

Milanković M (1998) Canon of Insolation and the Ice-Age Problem. Zavod Za Udžbenike I Nastavna Sredstva, Beograd. 634 p

Oliver F, Scott DH (1903) On *Lagenostoma lomaxi*, the seed of *Lyginodendron*. Proc Roy Soc Lond 71:477–481

Opluštil S (2010) Contribution to knowledge on ontogenetic developmental stages of *Lepidodendron mannabachense* Presl, 1838. Bull Geosci 85:303–316

Pearce F (2017) Can we find the world's remaining peatlands in time to save them? Yale Environment 360. https://e360.yale.edu/features/can-we-discover-worlds-remaining-peatlands-in-time-to-save-them

Pfefferkorn HW (1980) A note on the term "upland flora". Rev Palaeobot Palynol 30:157–158

Pfefferkorn HW, Gillespie WH, Resnik DA, Scheining MH (1984) Reconstruction and architecture of medullosan pteridosperms (Pennsylvanian). Mosasaur 2:1–8

Pfefferkorn HW, Archer AW, Zodrow EL (2001) Modern tropical analogs for Carboniferous standing forests: comparison of extinct *Mesocalamites* with extant *Montrichardia*. Hist Biol 15:235–250

Pfefferkorn HW, Gastaldo RA, DiMichele WA (2017) Impact of an icehouse climate interval on tropical vegetation and plant evolution. Stratigraphy 14:365–376

Phillips TL (1979) Reproduction of heterosporous arborescent lycopods in the Mississippian–Pennsylvanian of Euramerica. Rev Palaeobot Palynol 27:239–289

Pigg KB, McLoughlin S (1997) Anatomically preserved *Glossopteris* leaves from the Bowen and Sydney basins, Australia. Rev Palaeobot Palynol 97:339–359

Rees PM (2002) Land-plant diversity and the end-Permian mass extinction. Geology 30:827–830

Retallack G, Dilcher DL (1981) Arguments for a glossopterid ancestry of angiosperms. Paleobiology 7:54–67

Rößler R (2001) The petrified Forest of Chemnitz. Museum für Naturkunde, Chemnitz. 250 pp

Rößler R, Noll R (2006) Sphenopsids of the Permian (I): the largest known anatomically preserved calamite, an exceptional find from the petrified forest of Chemnitz, Germany. Rev Palaeobot Palynol 140:145–162

Rößler R, Zierold T, Feng Z, Kretzschmar R, Merbitz M, Annacker V, Schneider JW (2012) A snapshot of an Early Permian ecosystem preserved by explosive volcanism: new results from the petrified forest of Chemnitz, Germany. PALAIOS 27:814–834

Rößler R, Merbitz M, Annacker V, Luthardt L, Noll R, Neregato R, Rohn R (2014) The root systems of Permian arborescent sphenopsids: evidence from the Northern and Southern hemispheres. Palaeontogr Abt B 291:65–107

Rothwell GW (1981) The Callistophytales (Pteridospermopsida): reproductively sophisticated Paleozoic gymnosperms. Rev Palaeobot Palynol 32:103–121

Rothwell GW, Mapes G, Mapes RH (1997) Late Paleozoic conifers of North America: structure diversity and occurrences. Rev Palaeobot Palynol 95:95–113

Scotese CR, McKerrow WS (1990) Paleozoic paleogeography and biogeography, vol 12. Geological Society of London Memoirs, London, pp 1–435

Seyfullah L, Glasspool I, Hilton J (2014) Hooked: habits of the Chinese Permian gigantopterid *Gigantonocle*. J Asian Earth Sci 83:800–890

Šimůnek Z (2007) New classification of the genus *Cordaites* from the Carboniferous and Permian of the Bohemian Massif, based on cuticle micromorphology. Acta Musei Nationalis Pragae, Series B – Historia Naturalis 62:97–210

Staub JR, Gastaldo RA (2003) Late Quaternary Incised-valley Fill and Deltaic Sediments in the Rajang River Delta. In: Sidi HF, Nummedal D, Imbert P, Darman H, Posamentier HW (eds) Tropical Deltas of Southeast Asia – Sedimentology, Stratigraphy, and Petroleum Geology, vol 76. SEPM Special Publication, McLean, pp 71–87

Surange KR, Chandra S (1974) Some male fructifications of Glossopteridales. Palaeobotanist 21:255–266

Taylor TN, Taylor EL, Krings M (2009) Palaeobotany: the biology and evolution of fossil plants, 2nd edn. Elsevier, Oxford, pp 1–1230

Thomas BA (2014) *In situ* stems: preservation states and growth habits of the Pennsylvanian (Carboniferous) calamitaleans based upon new studies of *Calamites* Sternberg, 1820 in the Duckmantian at Brymbo, North Wales, UK. Palaeontology 57:21–36

Wang Q, Geng BY, Dilcher DL (2005) New perspective on the architecture of the Late Devonian arborescent lycopsid *Leptophloeum Rhombicum* (Leptophloeaceae). Am J Bot 92:83–91

Wanless HR, Shepard FP (1936) Sea level and climate changes related to late Paleozoic cycles. Geol Soc Am Bull 47:1177–1206

Wanless HR, Weller JM (1932) Correlation and extent of Pennsylvanian cyclothems. Geol Soc Am Bull 43:1003–1016

Wilson JP (2016) Evolutionary trends in hydraulic conductivity after land plant terrestrialization: from *Psilophyton* to the present. Rev Palaeobot Palynol 227:65–76

Wilson JP, Montañez IP, White JD, DiMichele WA, McElwain JC, Poulsen CJ, Hren MT (2017) Dynamic Carboniferous tropical forests: new views of plant function and potential for physiological forcing of climate. New Phytol 215:1333–1353

Wnuk C, Pfefferkorn HW (1984) The life habits and paleoecology of Middle Pennsylvanian medullosan pteridosperms based on an in situ assemblage from the Bernice Basin (Sullivan County, Pennsylvania, U.S.A.). Rev Palaeobot Palynol 41(3–4):329–351

Zachos J, Pagani M, Sloan L, Thomas E, Billus K (2001) Trends, rhythms, and aberrations in global climate 65 Ma to present. Science 292:686–693

Zeiller R (1906) Études sur la flore fossile du bassin houiller et permien de Blanzy et du Creusot. Études Gîtes Minérlogie France, Paris. 514 p

Zheng SL, Zhou ZY (2004) A new Mesozoic Ginkgo from western Liaoning, China and its evolutionary significance. Rev Palaeobot Palynol 131:91–103

Zhou ZY (1997) Mesozoic ginkgoalean megafossils: a systematic review: *Ginkgo biloba* a global treasure. Springer, Tokyo, pp 183–206

The Coal Farms of the Late Paleozoic

13

Robert A. Gastaldo, Marion Bamford, John Calder,
William A. DiMichele, Roberto Iannuzzi, André Jasper,
Hans Kerp, Stephen McLoughlin, Stanislav Opluštil,
Hermann W. Pfefferkorn, Ronny Rößler, and Jun Wang

Abstract

The assembly of the supercontinent Pangea resulted in a paleoequatorial region known as Euramerica, a northern mid-to-high latitude region called Angara, and a southern high paleolatitudinal region named Gondwana. Forested peat swamps, extending over hundreds of thousands of square kilometers, grew across this supercontinent during the Mississippian, Pennsylvanian, and Permian in response to changes in global climate. The plants that accumulated as peat do not belong to the plant groups prominent across today's landscapes. Rather, the plant groups of the Late Paleozoic that are responsible for most of the biomass in these swamps belong to the fern and fern allies: club mosses, horsetails, and true ferns. Gymnosperms of various systematic affinity play a subdominant role in these swamps, and these plants were more common outside of wetland settings. It is not until the Permian when these seed-bearing plants become more dominant. Due to tectonic activity associated with assembling the supercontinent, including earthquakes and volcanic ashfall, a number of these forests were buried in their growth positions. These instants in time, often referred to as T^0 assemblages, provide insight into the paleoecological relationships that operated therein. Details of T^0 localities through the Late Paleozoic demonstrate that the plants, and plant communities, of the coal forests are non-analogs to our modern world. Analysis of changing vegetational patterns from the Mississippian into the Permian documents the response of landscapes to overall changes in Earth Systems under icehouse to hothouse conditions.

Electronic supplementary material A slide presentation and an explanation of each slide's content is freely available to everyone upon request via email to one of the editors: edoardo.martinetto@unito.it, ragastal@colby.edu, tschopp.e@gmail.com

*The asterisk designates terms explained in the Glossary.

R. A. Gastaldo (✉)
Department of Geology, Colby College, Waterville, ME, USA
e-mail: robert.gastaldo@colby.edu

M. Bamford
Evolutionary Studies Institute, University of the Witwatersrand, Johannesburg, South Africa
e-mail: Marion.Bamford@wits.ac.za

J. Calder
Geological Survey Division, Nova Scotia Department of Energy and Mines, Halifax, NS, Canada
e-mail: John.H.Calder@novascotia.ca

W. A. DiMichele
Department of Paleobiology, Smithsonian Institution, United States National Museum, Washington, DC, USA
e-mail: DIMICHEL@si.edu

R. Iannuzzi
Instituto de Geociências, Universidade Federal do Rio Grande do Sul, Porto Alegre, Brazil
e-mail: roberto.iannuzzi@ufrgs.br

A. Jasper
Universidade do Vale do Taquari - Univates, Lajeado, RS, Brazil
e-mail: ajasper@univates.br

H. Kerp
Institute of Geology and Palaeontology – Palaeobotany, University of Münster, Münster, Germany
e-mail: kerp@uni-muenster.de

S. McLoughlin
Department of Palaeobiology, Swedish Museum of Natural History, Stockholm, Sweden
e-mail: steve.mcloughlin@nrm.se

S. Opluštil
Institute of Geology and Paleontology, Charles University in Prague, Prague, Czech Republic
e-mail: stanislav.oplustil@natur.cuni.cz

H. W. Pfefferkorn
Department of Earth and Environmental Science, University of Pennsylvania, Philadelphia, PA, USA
e-mail: hpfeffer@sas.upenn.edu

R. Rößler
Museum für Naturkunde, Chemnitz, Germany
e-mail: roessler@naturkunde-chemnitz.de

J. Wang
Nanjing Institute of Geology and Palaeontology, Chinese Academy of Sciences, Nanjing, People's Republic of China
e-mail: jun.wang@nigpas.ac.cn

© Springer Nature Switzerland AG 2020
E. Martinetto et al. (eds.), *Nature through Time*, Springer Textbooks in Earth Sciences, Geography and Environment,
https://doi.org/10.1007/978-3-030-35058-1_13

13.1   Introduction

Over the course of the Late Paleozoic Ice Age (LPIA, 359–273 Ma), and in conjunction with oscillations of both physical and chemical conditions operating on Earth at that time (see Gastaldo et al., Chap. 12), thick and geographically expansive "coal forests" formed. Carboniferous coals are found primarily across North America, Europe, and China, whereas the younger, Permian coals accumulated in the high-paleolatitude continents of South America, South Africa, India, Australia, and in paleoequatorial China (Fig. 13.1). The complement of Carboniferous and Permian [U1301] plants that thrived for almost 50 million years are not familiar to us because they play a minor role in today's landscape. Their rise, expansion, and ultimate demise as the dominant vegetation were a consequence of global climate change when Earth moved from a glaciated icehouse to a non-glaciated hothouse state over several phases of increasingly warming conditions. Carboniferous and Permian fossil plants occur in both continental and coastal sandstone and mudstone deposits, often preserved in wetland settings where they grew (Greb et al. 2006). Most of the biomass in these coastal deltaic lowlands was recycled, leaving us with a sporadic fossil record at any one point in time. In contrast, the majority of biomass generated by forests in wetland settings resulted in one of the greatest carbon sequestration events in Earth's history. We have exploited these resources since the Industrial Revolution, and it is these coals that continue to be responsible for many of our current energy needs. Fossil plants are preserved in the coals, themselves, as well as in the

mudrocks below and above these seams. Under unusual and, generally, short-term events, such as seismic activity associated with earthquakes or ash fall as a consequence of local or regional volcanic eruptions, these coal forests were buried "alive." These in situ fossil assemblages provide us with snapshots in time, referred to as T^0 windows (Gastaldo et al. 1995). It is from these entombed assemblages that we can examine plant architectures, community structure, and ecosystem partitioning over spatial scales similar to modern ecology [U1302].

This chapter diverges from the book's theme of tracing the history of our biosphere back in time. Wandering through the coal forests forward in time, the approach taken in this unit, is important, and critical, to understand their development and change in a temporal context. Plants, unlike animals, are not able to "migrate" from one locality to another in response to any chemical or physical change operating on Earth. Plants are fixed to a soil substrate, although current computer graphic animation may have us believe otherwise. And, as such, plants either expand or contract their biogeographic range(s) as the conditions of their substrate or environment change or are altered. Biogeographic range expansion is accomplished through the spread of their reproductive propagules (spores or seeds) to a similar site where the conditions are favorable for germination, establishment and growth, and continued reproduction. Range contraction, regional or hemispherical extirpation (site-specific loss of a plant or plant group), or outright extinction of any group occurs when conditions for its growth and reproduction no longer can be met. These patterns play a prominent role in

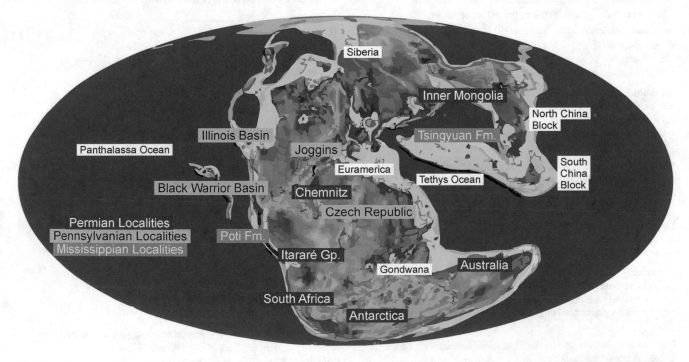

Fig. 13.1 Permian paleogeographic map of the supercontinent Pangea illustrating the main physiographic provinces and the localities of T^0 forests presented in this unit. (adapted from the Paleomap Project, http://www.scotese.com/)

the history of LPIA vegetation. Here, we present postcards of in situ forests from the Late Mississippian (early Carboniferous) to the middle Permian, focused on the Euramerican paleoequatorial region, the Northern Hemisphere mid-latitudes, and the high latitudes of the Southern Hemisphere. As we've seen in the previous unit, changes in terrestrial environments, ocean and atmospheric circulation, and glacial ice all are controlled by short- and long-term cycles in climate. Climate drives change.

13.2   Late Mississippian Bolivian, Peruvian, and Brazilian Forests

Roberto Iannuzzi and Hermann W. Pfefferkorn

Late Mississippian forests are not well known because very few areas expose rocks of this age on a global scale. Therefore, it is not possible to, literally, walk through these forested coastlines because fossil plants are preserved in sedimentary environments away from where the plants grew. Localities where insights into these landscapes are known include the Poti Formation deposits in northeastern Brazil (Fig. 13.2) [U1303]. Here, sandstones, mudrock, and a small proportion of carbonaceous shales were deposited in coastal plain to nearshore marine settings in which fossil plants, palynomorphs, and invertebrate remains (bivalves) are preserved (Santos and Carvalho 2009). Assigning the rocks in this part of the Parnaíba basin to the Late Mississippian is based on palynology from equivalent areas in Bolivia (Suárez Soruco and Lobo Boneta 1983), which are comparable to those in Brazil (Melo and Loboziak 2003). Currently, these rocks are considered to be late Visean (c. 334 Ma) in Brazil and Serpukhovian (331–323 Ma) in Peru and Bolivia (di Pasquo and Iannuzzi 2014). In the Precordillera of Argentina, Pérez Loinaze (2007) defined the equivalent palynological zone as middle Visean in age based on a geochronometric age date of approximately 336 million years ago. Hence, we've got a glimpse into the Late Mississippian forests that lined rivers and waterways.

The flora of the Poti Formation first was reported by Oliveira (1934) with new material added by Dolianiti (1954) in a locality known as "Curral de Pedra." Iannuzzi and Pfefferkorn (2002) revised the flora and documented the presence of older Mississippian-age plants in the assemblage. These older forms include: stems of lycopsid (*Lepidodendropsis* and *Cyclostigma*-type stems) [U1304], sphenopsid (*Archaeocalamites*), and basal leafless fern (*Paulophyton*) taxa; primitive seed fern foliage, including *Aneimites, Diplothmema, Fryopsis, Nothorhacopteris, Sphenopteridium,* and *Fedekurtzia* [U1305] (Iannuzzi et al. 2006); and reproductive structures (*Kegelidium, Stamnostoma*; Fig. 13.2d, e) [U1306]. The plant assemblage is similar to

other Late Mississippian palynofloras and macrofloras from Gondwanan regions that include western South America, north-central Africa, the northern Indian subcontinent, and eastern Australia [U1307]. The plant assemblage was confined to a paleolatitudinal belt (Box 13.1) that extended approximately from between 30° and 60° S (di Pasquo and Iannuzzi 2014). A different story is recorded in the rocks of the North China block.

Box 13.1: The Paraca Floral Realm

The Paraca floral realm [U1308] first was recognized by Alleman and Pfefferkorn (1988) because it differed from the plant assemblages preserved in Euramerica to the north and those in Gondwanan to the south. The transported branches, leaves, and reproductive structures come from a landscape characterized by a low-diversity flora with only a few typical lycopsids (*Tomiodendron*), sphenopsids (*Archaeocalamites*), and seed ferns (e.g., *Nothorhacopteris, Triphyllopteris, Fryopsis, Sphenopteridium,* and *Diplothmema*). This plant assemblage, characteristic of a warm temperate climate, existed for only a few million years (early late Visean into the Serpukhovian) when glacial ice began to form in Argentina (Césari et al. 2011). Its persistence over only a relatively short time span was a function of the paleogeographic setting in Gondwana, during the Mississippian when the planet experienced a hothouse climate [U1308] (e.g., Pfefferkorn et al. 2014).

The Paraca floral realm extended over a vast geographic extent of Gondwana, ranging from Peru, Bolivia, and Brazil, to Niger (Africa), and India, to Australia (Iannuzzi and Pfefferkorn 2002). And, due to its distribution, differences in the plant community composition between the distinct Gondwana regions should be expected in response to local climate conditions. This seems to be true in the Poti Formation. Here, endemic elements include *Kegelidium lamegoi* (Iannuzzi and Pfefferkorn 2014) and *Diplothmema gothanica* (Iannuzzi et al. 2006). In combination with the short stature of the plants and low diversity, these observations have been used as evidence to interpret a vegetation restricted and constrained by a more seasonally dry climate. In fact, northern Brazil in the mid-Mississippian has been reconstructed as having been in a zone of semiarid conditions (Iannuzzi and Rösler 2000). As such, the paleoflora is somewhat distinct from those preserved in Peru, Bolivia, and Argentina, interpreted as more humid regions. This is evidenced in differences found in both the palynological and macrofossil plant composition (Iannuzzi and Pfefferkorn 2002).

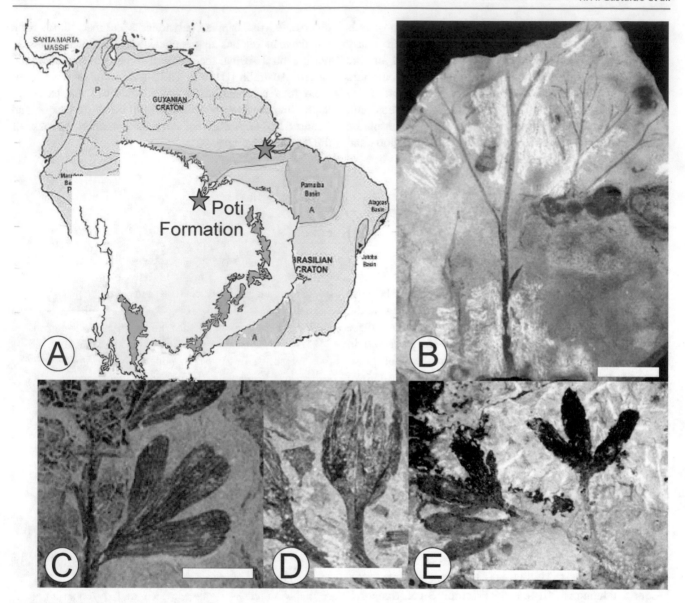

Fig. 13.2 Plants characteristic of the Poti Formation, Brazil, during a short hothouse interval in the Late Mississippian. (**a**) Current (base map) and Carboniferous (insert map) geographies showing the location of the fossil-plant assemblage (star). (**b**) Leafless fern, *Paulophyton sommeri*, showing features of basal members of the clade. Scale = 1 cm. (**c**) Leaflet of *Fedekurtzia* cf. *argentina*, a seed-bearing gymnosperm. Scale = 5 mm. (**d**) The pre-ovule, or seed, *Calymmatotheca* sp. Scale = 5 mm. (**e**) Pollen organ assigned to *Kegelidium lamegoi* consisting of a leafless branching system bearing terminally unfused, paired terminal sporangia. Scale = 5 mm. (Images R. Iannuzzi)

13.3 Late Mississippian Coastal Vegetation in China

Jun Wang

Hiking through Mississippian coastal vegetation of China, preserved and now exposed in rocks of the North Qilian Mountain, Gansu, and Qinghai provinces, impart a very different landscape than the Gondwana floras of the Paraca biome (Fig. 13.3) [U1309]. Here, we encounter our first well-developed "coal forests" with plant remains found as peat. Late Mississippian plant fossils are restricted to coal-bearing intervals, as both permineralized peat and compression-impressions, and are common in the rocks of the Serpukhovian aged Tsingyuan Formation. The floristic components are similar to those identified in other peat swamps of central Europe (Gastaldo et al. 2009). Lycopsids (including *Lepidodendron* and *Stigmaria*) comprise the greatest proportion of peat biomass (up to 75%), a clue to their dominance in the peat swamp vegetation. Two other plant groups, calamitean horsetails and cordaitalean gymnosperms, comprised canopy elements but are found in lower proportions (Li et al. 1995). In contrast, ferns and seed ferns

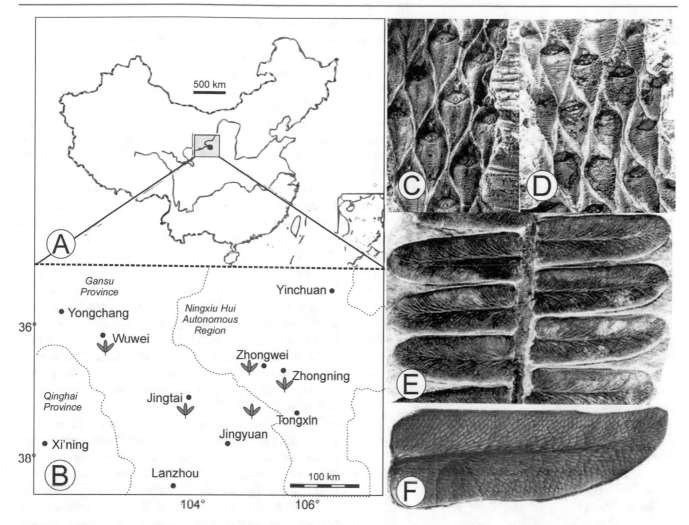

Fig. 13.3 Mississippian coastal floras of North China. (**a**) Map of China showing the geographic position of the North Qilian Mountain localities. (**b**) Mississippian-age fossil-plant assemblages are found over a distance of >200 km in Gansu Province. (**c, d**) Specimens of *Lepidodendron* cf. *wortheni*, the most abundant tree that dominated peat swamps. (**e**) The common pteridosperm, *Neuralethopteris* cf. *schlehanii*, found in mineral-substrate assemblages. (**f**) The common pteridosperm, *Linopteris lepida*, also an element of mineral-substrate assemblages. (Images by J. Wan)

occupied a position in the understory or sporadically grew as epiphytes in the peat swamp forest. Plant fossils are more abundant in the clastic deposits found either beneath or overlying the coals [U1310].

The fossil plants preserved in sandstone and mudrock represent vegetation that lived on the coastal/littoral plain and delta plain, mimicking the plant group composition found in the peat. The megaflora consists of every major plant group in varying proportions, all represented by fossil genera or species based on individual organs (Li et al. 1993). Lycopsid stem diversity is restricted to two genera, *Lepidodendron* (Fig. 13.3c, d) and *Bothrodendron,* along with aerial leaves (*Eleutherophyllum*), cones and cone scales (e.g., *Lepidostrobus, Lepidostrobophyllum*), and common rooting structures (*Stigmaria*). Sphenopsid genera are the same as those reported in other parts of the paleotropics and include both stems of *Calamites* and *Archaeocalamites,* along

with their leaves and cones (*Asterophyllites, Calamostachys*), and the sphenophyllalean scrambler *Sphenophyllum*. A group commonly found in other parts of the Euramerican province, the Noeggerathiales of uncertain systematic affinity, are represented by both leaf (*Noeggerathia*) and reproductive structures (*Archaeonoeggerathia*). The highest systematic diversity of coastal plants is found in the ferns and seed ferns (pteridosperms) where 25 genera and 64 species, mostly leaves with a minor number of pollen organs/sporangia, and ovuliferous organs, are reported to account for more than 70% of the total flora. Most of these fern-like foliage groups are interpreted to represent seed ferns, including both lyginopterids (e.g., *Lyginopteris, Lyginodendron*) and medullosans (e.g., *Neuropteris, Paripteris, Linopteris,* and *Neuralethopteris*; Fig. 13.3e, f). Other taxa with forked (Y-shaped) fronds, a characteristic of pteridosperm leaf architectures, include *Eusphenopteris, Karinopteris,* and

Diplothmema, whereas some fronds may be referable to ferns (*Alloiopteris*, *Sphenopteris*) which are much less abundant. The cosmopolitan genus, *Cordaites*, is common.

The paleogeographic setting of China during the Mississippian is interpreted to have consisted of numerous tropical islands or microcontinents, with coastal plain deposits draping their margins. Carboniferous rocks now exposed in the North Qilian Mountain are considered as having accumulated in this setting, as they represent coastal plain deposits associated with a part of the southwestern North China Block [U1311]. It is in these isolated regions that the typical Carboniferous forest elements are interpreted to have originated and spread globally. This hypothesis is based on several systematic studies of the paleobiogeographic distribution of these plants during the Mississippian and Pennsylvanian. Several taxa have been shown to have extended their biogeographic range, slowly, between China in the east and Euramerica in the west of the Paleo-Tethys Ocean, along the paleotropical coastal line (Li et al. 1993; Laveine 1997). Although several of these plants appear in Early Pennsylvanian floras of Euramerica, coal forests of the Black Warrior and Appalachian basins in North America were overgrown by endemic taxa. It's now time to head across the ocean to North America.

13.4 Early Pennsylvanian Forests of the Black Warrior Basin, United States

Robert A. Gastaldo

Several million years forward in time, coastal plains that extended along the western side of the Appalachian Mountains in North America were forested with members of each plant clade growing on both mineral and peat substrates. Walking from coastal plain communities established on a mineral soil type to forests growing on peat wouldn't present you with a different view because all "the players" would be there in a varying mosaic of vegetation. We can visualize these relationships because in situ, erect forests, with groundcover, understory, and canopy forms in growth position, are commonly preserved in the Black Warrior Basin, southeastern United States [U1312] (Pashin and Gastaldo 2009). The combination of processes responsible for the preservation of many of these standing forests is unusual. The reason for this is because the preservation of these forests involves mechanisms that operated on time scales much shorter than Milanković parameters. The series of these short-term events was the result of high-magnitude earthquakes that affected the Black Warrior Basin, as tectonic activity built the mountain chains to the east (Appalachian mountains) and southwest (Ouachita mountains). In a near instant, high-magnitude earthquakes lowered parts of the coastal plain, resulting in

subsidence of the land's surface to positions several meters below sea level. This sudden downward shift of the land's surface left the vegetation upright and the forest structure intact. These submerged coastal forests, then, were buried rapidly by tidal sedimentation that filled in the area over a period of only a few years [U1323] (Gastaldo et al. 2004b). As a consequence, we have gained insight into their systematic composition and organization and, where exposed in coal mines, have been able to admire these Early Pennsylvanian (Baskirian, 323–315 Ma) forests.

Unlike closed-canopy tropical forests of the present (Fig. 12.18a), Early Pennsylvanian forests displayed an open, although tiered, structure. Canopy elements included various lycopsid (*Lepidodendron*, *Lepidophloios*, and *Sigillaria*) [U1313], sphenopsid (*Calamites*), and rare cordaitalean (*Cordaites*) taxa (Fig. 13.4; Gastaldo et al. 2004a). Juvenile "pole trees" of lycopsid affinity dominated parts of the forest, whereas mature individuals were spaced sporadically across the landscape. Both juvenile and mature lycopsid trees grew coevally, indicating that the canopy was not comprised of an even-aged stand of individuals. The distance between lycopsid trees was wide (Gastaldo 1986a, 1986b), leaving sufficient space for the growth of an understory and significant groundcover plants, which appear to have produced the majority of biomass at times. One unique aspect of calamitean plants buried alive in coastal sediments of tidal origin was their ability to regenerate new rooting structures and aerial shoots from the upright axes (Gastaldo 1992). Understory groups included scattered tree ferns, with fronds organized either in a spiral (Gastaldo 1990) or opposite (distichous) arrangement, along with a low-diversity pteridosperm assemblage. The number of seed-fern taxa was low, with most leaves assigned either to *Neuralethopteris* or *Alethopteris*. In many cases, large fragments of these fronds, which attained lengths of several meters, are preserved at the peat/sediment interface. Large parts of the forests were covered in creeping and liana taxa of sphenopsid, gymnosperm (seed fern), and fern affinity [U1314]. *Sphenophyllum*, a sister group of the calamitalean trees, is preserved with its whorls of wedge-shaped leaves with a forked venation pattern. Lyginopterid seed ferns, assigned to the cosmopolitan Euramerican genus *Lyginopteris* are common, as are various *Alloiopteris* fern species. The groundcover at this time in the Appalachian basin was dominated by an endemic plant, *Sphenopteris pottsvillea* (Gastaldo 1988), with fronds attaining an estimated 2 m and more in length, originating from either a rhizome or short vertical stem. The affinity of this plant is unknown, because neither sporangia nor seeds have been recovered from preserved specimens. Gastaldo et al. (2004a) noted that the number of groundcover taxa was equal to the diversity of understory taxa in the Blue Creek peat swamp. A reconstruction of this Early Pennsylvanian age forest, based on biomass contribution to the swamp,

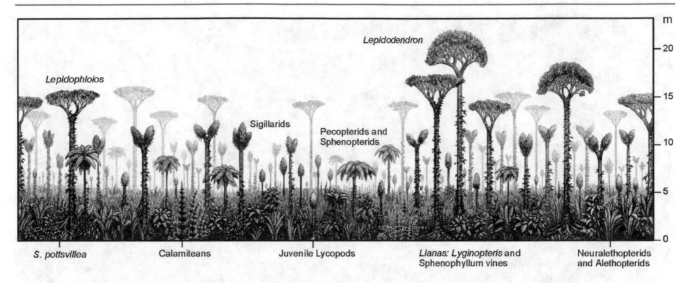

Fig. 13.4 Reconstruction of in situ peat swamp forest of the Early Pennsylvanian, Blue Creek Coal in the Black Warrior Basin, USA. (reproduced with permission: Gastaldo et al. (2004a))

shows these landscapes to have been covered by densely packed, upright fern and pteridosperm groundcover attaining heights of several meters, through which isolated (tree fern and cordaitean) or clumped (calamitean) shrubs/trees comprised an understory. Emergent juvenile lycopsids, with unbranched trunks enveloped by meter-long, linear microphyllous leaves, were interspersed with mature individuals. The growth habit of mature lycopsids varied from the development of a wide to reduced dichotomous crown. In the first growth strategy, branches were terminated by reproductive cones encircled by small, short microphylls (*Lepidodendron* and *Lepidophloios*). In contrast, *Sigillaria* produced a reduced dichotomous crown with opposite rows of reproductive cones spaced along the terminal branches [U1315]. You could easily have seen the sky if you were taller than the groundcover. Let us travel to the other side of the Pennsylvanian mountain belt, today further north, to see if and how these forests changed geographically.

13.5 Early Pennsylvanian Forests: A UNESCO World Heritage Site at Joggins, Nova Scotia

John Calder

The Lower Pennsylvanian (Baskirian) strata exposed in the coastal cliffs at Joggins, Nova Scotia, Canada, hold a prominent place in the history of science, paleontology, and paleoecology, and are inscribed on the list of UNESCO World Heritage sites [U1316]. In 1852, Charles Lyell and William Dawson discovered standing fossil forests and tetrapod bones in the cast of a fallen lycopsid tree fossil (Lyell and Dawson 1853), which later were recognized as the earliest

reptile (amniote) in the fossil record, *Hylonomus lyelli* (Carroll 1964). Darwin (1859) drew upon the descriptions of Dawson and Lyell, incorporating them in his seminal work *On the Origin of Species*, which has led creationists to Joggins to try to debunk the standing trees as in situ forests drowned by successive events rather than by the one Biblical "Deluge" (Calder 2012). The forests preserved in the Joggins succession are very similar to those further south in the Black Warrior Basin. Unlike the fossil forests in Alabama exposed as a consequence of coal-mining activities, the Joggins cliffs are a natural laboratory in which new parts of these forests are constantly being unearthed.

Fossilized, erect lycopsid trees occur throughout the approximate 1630 m-thick succession of "coal measures" comprising the Joggins and Springhill Mines formations of the Cumberland Group (Fig. 13.5; Davies et al. 2005; Rygel et al. 2014). Erosion of the coastal cliffs by the world's highest tides along the Bay of Fundy continually brings new exposures of fossil trees to view, each tree surviving in the cliff face for 3–5 years. This slow "unveiling" of the floral components confirms that they, indeed, are remains of forest stands. The setting of these Joggins forests has been interpreted as coastal wetlands (Davies and Gibling 2003), although no unequivocal open marine fauna exists within the section.

Most lycopsid trees are rooted in organic-rich beds and coals, ranging from centimeter- to meter-scale in thickness [U1317]. The Joggins coals have been interpreted as the product of plant parts accumulating in groundwater-influenced (rheotrophic*) swamps that struggled to maintain equilibrium with the rapidly subsiding basin and accumulating sediment carried by seasonal rainfall events (Waldron et al. 2013). It is unlikely, given this environmental scenario, that the standing trees represent a mature forest of these

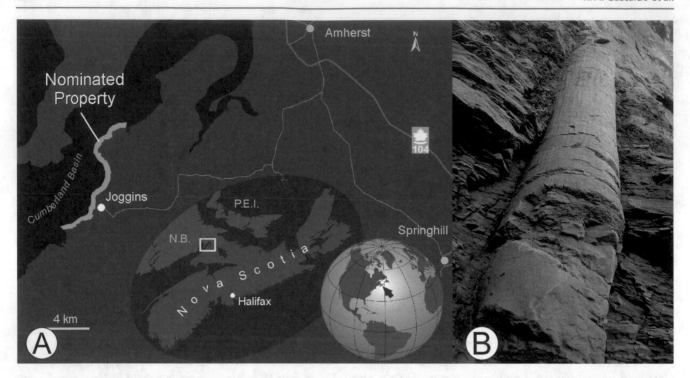

Fig. 13.5 The Joggins Cliffs UNESCO World Heritage Site, Nova Scotia, Canada. (**a**) Map showing the geographic position of the Joggins fossil cliffs along the eastern side of the Bay of Fundy. (**b**) An example of one in situ, standing lycopsid (club moss) tree preserved at nine discrete stratigraphic horizons in the succession. (Images by J Calder)

struggling peatlands, instead reflecting the most disturbance-adapted members of the lycopsids. Hence, these are unlike those in the Black Warrior Basin where forest plants maintained continued cycles of regeneration. Regardless of the substrate on which trees are rooted, the trees at Joggins commonly have a diameter at breast height (1.4 m) of 45 cm, which indicates that they attained such a diameter quickly, and experienced little secondary growth thereafter. Lycopsid stands whose stigmarian rootstocks became overwhelmed by the depth of accumulating sediment were succeeded by groves of *Calamites*, which demonstrate a regenerative and adventitious growth strategy (Gastaldo 1992).

Identification of the standing trees can be problematic because of the loss of diagnostic leaf scar patterns due to bark loss, either by physiological or decay functions, and disruption by secondary growth. Logs representing more aerial pieces of these plants, and lying in close proximity to the standing trees, are dominated by *Sigillaria*. Their ongoing, polycarpic* reproductive strategy (DiMichele and Phillips 1994), where cone development occurred throughout the life of the tree, enabled them to succeed in the disturbed habitats at Joggins. Other lycopsid components, identified in the compression flora at Joggins, include *Lepidodendron*, *Lepidophloios*, and *Paralycopodites* (Calder et al. 2006). It is within hollowed trunks of these plants that the early tetrapods discovered by Lyell and Dawson (1853) are found (Box 13.2).

Box 13.2: Wildfire and a Tetrapod Tree Fauna

It is not known if lycopsids or *Sigillaria*, specifically, evolved adaptation to wildfire, but evidence of recurring wildfire is commonplace in dispersed charcoal clasts, charred logs, and accumulation of charcoal clasts in the interior fill of once hollow standing trees (Calder et al. 2006). Standing trees that exhibit basal charring, or that contain an accumulation of charcoal, commonly contain disarticulated skeletal remains of amphibians and early reptiles (Carroll 1964; Milner 1987). Their co-occurrence is suggestive that wildfire played a role in their preservation [U1318]. Charred, notched bases of some trees indicate the development of fire scars that modern vertebrates use as "doorways" to standing trees (Calder 2012). Although the "pitfall" hypothesis, that the tetrapods fell victim to gaping, partially buried tree hollows, has been the most widely accepted explanation for the occurrence of the tree hollow fauna, modern analogs point to the use of such tree hollows as dens for animals. The tree hollow tetrapods have long been held as unique to Joggins, but it is likely that future search strategies based on the taphonomy* of the Joggins trees will show that other fossil lycopsid forests hosted their own tree fauna. The unique preservation of primary producers, aquatic and terrestrial invertebrates, and sharks, cartilagenous ray-like, and true bony fishes (Box 7.3), along with some of the earliest reptiles, provides insight into a fully functioning ecosystem in this coastal setting [U1319].

13.6 Middle Pennsylvanian Forests of the Herrin No. 6 Coal, Illinois Basin

William A. DiMichele

The late Middle-Pennsylvanian-age Herrin (No. 6) Coal is the most intensively studied coal, paleobotanically, in the United States. Because it is widespread and of mineable thickness over a broad area, including large deposits of low sulfur content, this Moscovian-aged (315–307 Ma) coal has been, and continues to be, of great economic importance. The result has been intensive mining, in the course of which, coal balls, in large numbers have been exposed, reported, and collected by paleobotanists [U1320]. Coal balls are original peat stages of the coal that were entombed by mineral matter, usually CaCO$_3$, generally very early in the post-burial history of the peat body. They capture, in anatomical detail, the plant components of the original peat swamp, both aerial organs and roots (Phillips et al. 1976). And, although we may not be able to "walk" through the peat swamps over lateral distances, as in the earlier case studies, we can "travel" through these peat swamps over time. This "time travel" is made possible because coal balls often are preserved in a stratigraphic succession, vertically through the swamp.

Hence, coal-ball profiles through the ancient peat body document the changes in community structure over hundreds to thousands of years.

Much of our understanding of the anatomy and morphology of peat/coal-forming plants comes from a few well-collected and studied sites in Euramerica and, more recently, in China. One of these, the Sahara Mine No. 6, in southern Illinois, has been the source of thousands of coal balls extracted from the Herrin Coal (Fig. 13.6). The plants entombed in these coal balls have been described over the past 75 years, such that the taxonomic and morphological literature based on them is very large, and very taxonomically particular, and thus will not be referred to here. This information can be found easily by searching the scientific literature online. An overview of plant groups present at the time was given in Chap. 12 (Gastaldo et al., Chap. 12).

Studies on the paleoecology of the Herrin Coal are of specific interest when discussing coal forests of the LPIA. These studies were done mainly by Tom Phillips and his students and collaborators. The earliest of these (Phillips et al. 1977) developed a quantitative sampling method for characterizing the composition of coal-ball floras, a method that has since been used by other workers, either directly or modified to suit their particular research problem (e.g., Pryor 1988). This

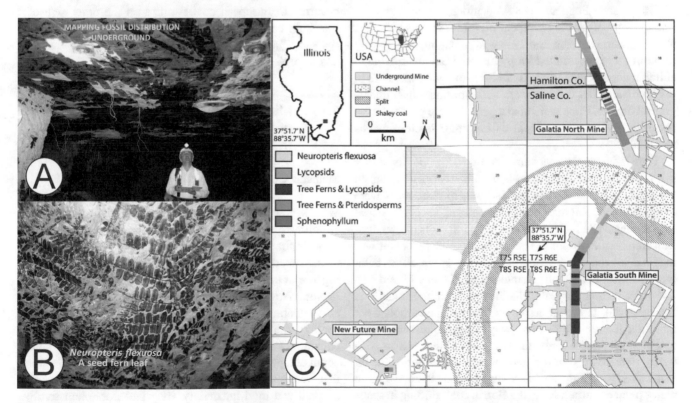

Fig. 13.6 The Herrin No. 6 Coal of the Illinois Basin is the most intensely studied peat forest in the United States. (**a**) Middle Pennsylvanian fossil plants are preserved in the "roof shale" siltstone above the coal, allowing for them to be mapped over large distances. (**b**) A complete leaf (frond) of *Neuropteris flexuosa* showing the quality of preservation. (**c**) A distribution map of vegetation in a transect away from one main river system, showing the heterogeneity of the peat swamp forest. ((**a, b**) images by WA DiMichele; (**c**), from DiMichele et al. (2017). (reproduced with permission)

early study was based on coal balls from the Sahara No. 6 Mine and was followed by a more detailed analysis aimed at characterizing the ecology in some detail (Phillips and DiMichele 1981). Later, a study was carried out in the Old Ben (No. 24) Mine (DiMichele and Phillips 1988), situated close to the Walshville paleochannel*, the remains of a large river system that ran through the Herrin peat swamp. Winston (1986) expanded our understanding of the macroflora by comparing coal balls to polished blocks of coal, collected in the Peabody Coal Company Camp 11 Mine, in western Kentucky, and used Herrin coal balls, among others, to investigate peat compaction.

In combination, these studies indicate several broad patterns in the Herrin peat swamp community and the resulting peat body [U1321]. The most abundant plants are the arborescent lycopsids, accounting in aggregate for about 58% of the peat biomass, including both roots and stems. Second in abundance are pteridosperms, at 16%, followed by ferns, consisting mainly of marattialean tree ferns, at 15%, and sphenopsids, at 5%. Cordaitaleans account for only trace amounts of peat biomass, which is true for most coals of late Middle Pennsylvanian age. Prior to the development of the Herrin swamp, cordaitaleans were more abundant during the Middle Pennsylvanian as evidenced by their proportional remains in slightly older coals (Phillips and Peppers 1984). Unidentifiable elements account for 6% of the total peat biomass. Of this biomass, 69% was aerial tissues and 27% was roots; the discrepancy (around 4%) is accounted for by material not identifiable to either category. This yields a shoot-to-root ratio of 2.5, which means that there is more aerial biomass than underground organs. Fires were also a part of these peat-forming swamps; 5.9% of Herrin coal-ball biomass is preserved as charcoal.

Translation of the peat composition and proportional biomass into the standing forest is a matter for discussion. In the studies that have been carried out, to date, roots were eliminated and biomass was normalized to 100% aerial components. The major issue here are the roots of marattialean tree fern *Psaronius*. The aerial stem of this plant was mantled by roots, which permitted it to attain a tree habit (see Pšenička et al., and Gastaldo et al., Chaps. 11 and 12). The thin, morphologically distinctive inner root mantle is clearly aerial. However, aerial and subterranean roots from the outer mantle are nearly always indistinguishable, with notable exceptions. Thus, the solution applied was to arbitrarily remove 50% of the tree fern roots and normalize on that basis. Resulting analyses suggest that there are three major plant communities in the Herrin coal, shifting in space through time.

Most abundant were parts of the swamp dominated by the arborescent lycopsid *Lepidophloios hallii*, a monocarpic* form (reproduces once and then dies) that occurs most com-

monly in low diversity plant assemblages, often in high dominance [U1322]. The inference is that this plant dominated those areas of the swamp with the longest periods of standing or high water table. Other assemblages were dominated by, or rich in, the arborescent lycopsid *Diaphorodendron scleroticum* or, less commonly, its relative *Synchysidendron resinosum*. And, finally, numerous assemblages were dominated by, or rich in, medullosan seed ferns. In many cases, these latter assemblages were preserved as thick accumulations of medullosan leaves and stems, possibly representing storm blow downs. Tree ferns were very widespread in their occurrence, generally intermixed with other plants, most rarely with *Lepidophloios*. This local "catholic" occurrence of *Psaronius* is in keeping with its general distribution throughout the tropical regions of Euramerican Pangea throughout the Pennsylvanian and Permian (Fig. 13.6c). *Psaronius* came to dominate Late Pennsylvanian peat-forming environments but also reached the far ends of western Pangea, occurring in wet habitats even in places where the surrounding landscapes were strongly moisture stressed seasonally. This was, no doubt, attributable to both its capacity for wide dispersal via small isospores (both sexes, permitting self-fertilization) and an ability to tolerate a wide spectrum of physical conditions, even if tilted toward microhabitats with higher levels of soil moisture. These plants also occurred in European coal swamps on the other side of a mountain belt known as the Variscan orogeny, and is best seen in T⁰ assemblages in the Czech Republic, our next LPIA stop.

13.7   Middle Pennsylvanian Forests of Central Europe Buried in Volcanic Ash

Stanislav Opluštil

What is now central Europe, a region experiencing a temperate climate (see Preface), was a flourishing tropical landscape in the Middle Pennsylvanian along the eastern margin of the supercontinent. Endless peat-forming swamps covering tens of thousands of square kilometers, coeval with those in North America, spread from extensive coastal lowlands in north Germany and Poland several hundred kilometers southward along the river valleys into the hilly interior of equatorial Pangea [U1323]. In the interior parts of the supercontinent, peat swamps and clastic wetlands covered valleys that were surrounded by hilly topographies. The deposition of sand and mud in, mainly, river and floodplain settings preserved excellent records of both fossil and climate proxies. This evidence, when combined, documents oscillations between ever-wet and seasonal climates, on several time scales and intensities, during the Middle and Late

Fig. 13.7 A Middle
Pennsylvanian (Moscovian
age) peat swamp buried in
volcanic ash. (**a**) Map of the
Czech Republic showing two
contemporaneous localities,
Štilec and Ovčín, where
excavations have
demonstrated differences in
the peat swamp communities.
(**b**) Reconstruction of the
Ovčín locality in which a
tiered forest, consisting of c.
40 biological species, is
preserved. (Opluštil et al.
(2009). reconstructions
reproduced with permission)

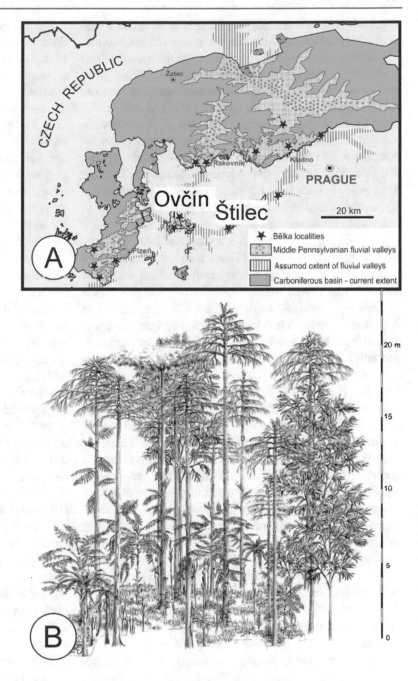

Pennsylvanian that resulted in biotic changes (DiMichele 2014). In the central and western parts of the Czech Republic (Fig. 13.7a), Middle Pennsylvanian peat-forming tropical swamps covered valleys transected by rivers flowing to shallow seaways in north Germany (Opluštil 2005). Deposition was accompanied by volcanic activity from several volcanic centers, the largest one situated around the Czech–German border. The Teplice-Altenberg Caldera exploded and rained volcanic ash that covered tens of thousands of square kilometers of landscape and buried all inhabitants alive (Opluštil et al. 2016). In the central and western Czech Republic, over 110 km from the eruption center, the resultant volcanic ash

(tuff) bed is about 1.5 m thick, preserving the peat swamp of the Lower Radnice Coal. It bears an in situ buried single-aged (T^0; Gastaldo et al. 1995) peat-accumulating forest rooted in the underlying coal. It is here that we can meander, literally, through a Middle Pennsylvanian peat forest that has been unearthed since its burial 314 million years ago.

Although the thickness of the tuff (Bělka) bed is insufficient to have buried the forest in its entirety [U1324], the bases of stems in upright/growth position, ranging from less than 1 cm to over 1 m in diameter, are preserved. In addition, the herbaceous groundcover is conserved in situ, along with elements of the forest canopy that fell as a consequence of damage by volcanic ash loading. Their broken branches and/or fronds are concentrated around the parent stumps, together with foliage of vine-like plants climbing on the trees (Opluštil et al. 2009). The unique preservational mode allows for a better understanding and detailed reconstruction of the Coal Farm plants (e.g., Šimůnek et al. 2009; Opluštil 2010). Distribution of plant compressions in the Bělka is laterally and vertically irregular. Lateral variations in plant composition reflect the original structure and density of the vegetation cover, whereas the vertical distribution mostly corresponds to plant habit (Opluštil et al. 2014). Compression of the peat swamp under the weight of the volcanic ash led to a change in landscape topography.

Overlying the Bělka bed is a complex of laminated lake muds called the Whetstone. These lacustrine deposits are composed of redeposited volcaniclastics, mixed with sand and mud, which were washed down into the valley from the surrounding paleo-highlands along valley margins. The Whetstone is up to 10 m thick, preserving upright stems protruding from the underlying tuff. Several stems are more than 6 m tall, indicating rapid deposition of sediments in a shallow lake generated by compaction of underlying peat (Opluštil et al. 2014). Besides these in situ trunks, the Whetstone contains drifted plant fragments, either irregularly scattered or concentrated on discrete bedding planes, or in thin beds, and common in the lower part of the Whetstone interval. Although both the Whetstone and Bělka floras are derived from the same generation of the pre-eruption vegetation, their species composition differs somewhat. The Whetstone flora is enriched by fragments of plant taxa transported from clastic wetlands and well-drained Piedmont to upland settings located on adjacent paleohighs covered by regional (or zonal) vegetation (Opluštil et al. 2014).

Two contrasting plant assemblages have been identified in the Bělka tuff bed at two localities 24 km apart. The assemblage of the Štilec locality [U1325] consists of herbaceous and shrubby plants dominated by small zygopterid ferns and calamiteans with subdominant lycopsids (Libertín et al. 2009). The groundcover was composed of small-stature ferns (*Kidstonia heracleensis*, *Dendraena pinnatilobata*, *Desmopteris alethopteroides*, and *Sphenopteris cirrhifolia*) accompanied by 1–2 m dwarf *Calamites* sp. growing in clumps. In lesser quantities and scattered across the study area was the small, shrubby lycopsid *Spencerites leismanii* reconstructed as having been less than 1.5 m tall (Bek et al. 2009). There was only minor spatial variability in vegetational density, cover, and abundance of individual taxa over an area of about 900 m². At the time of burial, all the species were mature with fertile organs in which ripened spores are preserved. This low-diversity assemblage is interpreted as a pioneering recovery flora (Libertín et al. 2009). Palynological records from the roof of the Lower Radnice Coal, however, show a more diverse flora in the peat swamp. Its systematic composition shows a close relationship between this low-diversity herbaceous recovery assemblage with the underlying, higher diversity lepidodendrid lycopsid forest.

In addition to a diverse fern assemblage (see Pšenička et al., Chap. 11) the Ovčín locality [U1326] exposes a lepidodendrid-cordaitalean forest preserved in the Bělka tuff (Fig. 13.7b). The "Ovčín coal forest," studied in detail over an area of about 200 m², consisted of 32 biological species (only an additional four species were identified in a collection made from approximately 10 ha of the same opencast mine). This minor difference in diversity between areas suggests that the plants found in a few hundred square meters are well representative of structure and diversity of the overall "Ovčín coal forest." This forest was structured into well-developed tiers. The canopy was dominated by lycopsid (*Lepidodendron (Paralycopodites) simile*, *L. lycopodioides*, and *Lepidophloios acerosus*) and cordaitalean (*Cordaites borassifolius*) trees. They formed a relatively dense canopy, locally interrupted by significant gaps allowing for the development of a rich groundcover composed mainly of ferns, sphenophylls, and juvenile calamites. Groundcover, together with liana-like plants of fern and lyginopterid pteridosperm affinities, represent the most diverse part of the forest. An epiphyte-life strategy was suggested for one *Selaginella*-like plant that was found attached to tree top branches of the arborescent lycopsid *Lepidodendron* (Pšenička and Opluštil 2013). The understory was comprised of a less diverse flora including calamiteans, medullosan pteridosperms (*Laveineopteris*-type foliage), and marattialean tree ferns (*Psaronius* with *Pecopteris* foliage), displaying a patchy distributional pattern presumably related to the density of the canopy. There are (paleo)ecological methods that can be used to understand the diversity in the swamp. One technique, known as a species-area curve, indicates that only 200 m² need to be surveyed to accurately represent the vegetational complexity of this forest. Results from this analysis indicate low partitioning of the dominant taxa (*Cordaites* vs. lepidodendrid lycopsids), resulting in some slight heterogeneity in diversity. A smaller quadrat of only about 60 m² is needed to characterize both the groundcover and shrubby stories (Opluštil et al. 2014). Other areas buried in volcanic ash show a different pattern.

Examination of collections from old mines exploiting the Lower Radnice Coal in the central and western Czech Republic indicate that regional, basin-scale diversity of the Bělka coal forest is higher than at the Ovčín locality. Here, the biological species is estimated to be about 90–95 taxa (Opluštil et al. 2007). This diversity is comparable with that of Early Pennsylvanian roof shale floras of the Ruhr Valley in Germany and the Pennines in the United Kingdom. Diversity estimates in these sites reach approximately 100 species (Cleal et al. 2012) and contrasts with the diversity of coal-ball floras of slightly older coals (late Early Pennsylvanian) in the United Kingdom and Belgium (c. 40–45 species; Galtier 1997). This number is comparable with the 40–50 species of Middle Pennsylvanian-aged (Asturian) floras identified in coals balls from North American coalfields (e.g., DiMichele and Phillips 1996).

13.8 Early Permian Forests of Inner Mongolia

Jun Wang and Hermann W. Pfefferkorn

Travel through LPIA forests buried in volcanic ash is not restricted to the Carboniferous of central Europe. There are several examples of these Coal Farms throughout the supercontinent Pangea. We visited the early Permian (c. 300 Ma) Wuda Tuff flora [U1327] previously in Sect. 11.6 (Pšenička ct al., Chap. 11) where the diversity of ferns is highlighted. The locality is on the northwest margin of the Helanshan mountain chain, an isolated desert mountain range in Inner Mongolia (Fig. 13.8a), and these rocks are placed in the Taiyuan Formation (Wang et al. 2012). The volcanic tuff bed separated two coals, which occur in a syncline* of about 20 km², and the preserved plants represent the peat swamp forest of Coal No. 7. During the Permian, Wuda was located on the northwest sector of the North China Block (Fig. 13.1), which is interpreted either as a large island or microcontinent in the tropical zone of the Paleo-Tethys Ocean. The peat-forming forest was preserved in a manner similar to Štilec and Ovčín in the Czech Republic. It was smothered by volcanic event, which now is represented by a 66-cm thick bed after compaction and lithification. The ash bed can be traced over a north-south distance of 10 km (Fig. 13.8b) and its original thickness only can be estimated. But, based on compaction features of the fossil plants, the original thickness of the ash is interpreted to have been about 1.50 m (Wang et al. 2013). The event that emplaced the ash buried plants, broke off twigs and leaves, toppled trees, and preserved the forest remains in place. Systematic excavation of the volcanic tuff in quadrats at three different sites has allowed for the reconstruction of the spatial distribution of trees, groundcover, and other plant parts (Fig. 13.8c, d).

Six plant groups comprise the part of the forest studied in detail [U1328]. In contrast to previous case studies, the most abundant plants were the marattialean tree ferns, with subordinate numbers of pole-tree lepidodendraleans assigned to *Sigillaria*. Herbaceous ferns preserved as groundcover include *Nemejcopteris*, *Cladophlebis*, and *Sphenopteris* (Pšenička et al., Chap. 11), along with the scrambler/liana sphenopsid, *Sphenophyllum*. Growth forms of calamitalean include a dwarf shrub and small, possibly juvenile plants. Similar to floras of the Czech Republic, the enigmatic spore-bearing Noeggerathiales, an extinct group of uncertain sys-

tematic affinity, is represented by species of *Tingia* and *Paratingia* (Wang 2006; Wang et al. 2009). Although *Cordaites* is a component of the forest structure, the complement of other gymnosperms is different. Two forest genera, *Taeniopteris* and *Pterophyllum*, are possible early representatives of the cycads.

The structure of the forest also appears to be different from those in the Pennsylvanian (Box 13.3). *Sigillaria* and *Cordaites* were tall trees that grew to mature heights exceeding the general canopy. These emergents attained heights of 25 m or more and towered over a canopy of Marattialean tree ferns, which reached heights of up to 10–15 m. Smaller stature trees included the Noeggerathiales and possible early cycads. Climbing vines were rare but may include only one species of *Sphenopteris*. In contrast, the groundcover was

---

Box 13.3: Reconstructing Forest Spatial Heterogeneity

Reconstructions of the Wuda Tuff vegetation are based on excavation of three sites [U1329] in which diversity counts of entombed macroflora followed modern ecological methods. Gazing at the vegetational patterns in these three sites could disorient the visitor. These forests don't look modern. Small tree- or shrub-sized ferns are dominant in most areas, whereas semi-emergent *Sigillaria* and Noeggerathiales alternate, spatially, between being subdominant or locally dominant elements. All other components, including the potential early cycads, *Cordaites*, sphenopsids, and herbaceous ferns, are patchy in their distribution. It is likely that the same type of vegetation would have covered the extensive mire with changes in community heterogeneity and ecological gradients over time. The basic patterns are clearly identifiable in the T⁰ assemblage. Of particular note, though, is the existence of Noeggerathiales throughout the forest where the group can be dominant locally. This is an enigmatic group of spore-bearing plants that is known incompletely and is poorly defined. Their occurrence is unique because the group is not known as a major biomass contributor in any other LPIA peat-forming communities. Hence, its common occurrence is distinct from those species reported from the scanty fossil record of extrabasinal settings in Euramerica (Leary and Pfefferkorn 1977). The canopy trees *Cordaites* and *Sigillaria* occur together in only one part of the forest, but do not co-occur in other parts of the excavated areas, to date. These two taxa, one a seed plant and the other as a spore-producing plant, show spatial differentiation and a pattern of co-occurrence characteristic of an ecotone [U1330].

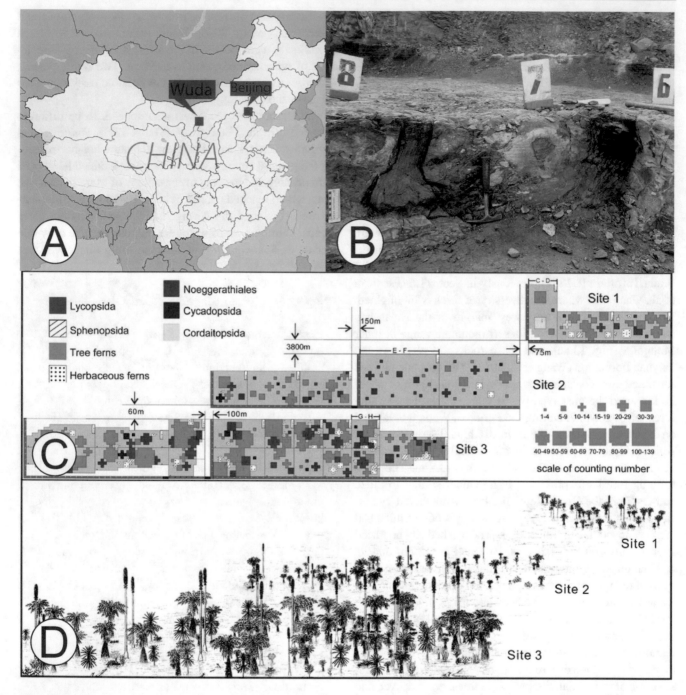

Fig. 13.8 The early Permian peat swamp of Inner Mongolia preserved in a volcanic ashfall. (**a**) Originally located on a tropical island associated with the North China tectonic block, the Wuda coalfield is now exposed over 20 km² of Inner Mongolia. (**b**) Excavation of the volcanic ash bed found separating two coal seams. Casts of standing lycopsids are identified by the numbers. Scale = hammer. (**c**) A spatial map of the major peat-forming plants in three localities. (**d**) A reconstruction of vegetation at these three sites showing a heterogenous forest community. (reproduced with permission: Wang et al. (2012))

composed of the fern *Nemejcopteris* and the small-stature sphenopsids, *Sphenophyllum* and *Asterophyllites*. These clonal horsetails developed only in small patches as is typical in tropical swamp forests today. A different early Permian vegetation grew on mineral soils where better drainage prevented the accumulation of peat. It's time to return to Central Europe and see it.

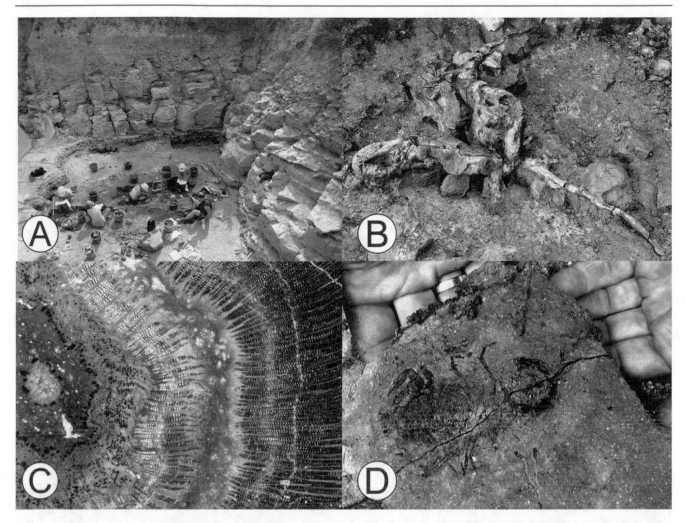

Fig. 13.9 The early Permian (291 ± 2 Ma) forest of Chemnitz, Germany, buried in ashfall. (**a**) Excavation site showing the thickness, up to 35 m in some locations, of the ashfall event. (**b**) In situ upright basal stump with rooting structure. (**c**) Permineralized trunks preserve annual growth rings that evidence variations in climate and solar insolation, affecting growth. (**d**) A scorpion preserved in the paleosol beneath the ash deposit, along with terrestrial gastropods, amphibians, and reptiles, attests to a well-established ecosystem with an advanced trophic structure. (Images by R Rößler)

13.9 The Early Permian Forest at Chemnitz

Ronny Rößler

The most striking witness of how ancient ecosystems responded to environmental and climatic changes lies in rock archives. Wandering through an early Permian forest growing in another volcanic landscape, where both invertebrates and vertebrates are entombed, gives us a deeper understanding of the ecosystem that surrounds us. A well-preserved fossil forest with both a long history of investigation (e.g., Sterzel 1875) and a multidisciplinary approach (e.g., Luthardt et al. 2016, 2017) lies beneath the city of Chemnitz in southeast Germany (Fig. 13.9a). We caught a glimpse of a part of the ferns that occupied this early Permian landscape in unit 11. Here, we will focus on the most completely fossilized forest ecosystems in the late Paleozoic of the Northern Hemisphere in tropical Pangea.

The fossil site is located in the eastern part of the Chemnitz Basin, which represents a post-Variscan intra-montane basin [U1331]. Sediments that filled this basin consist of floodplain red-bed deposits and various volcanics that span parts of the Permian. Few organic-rich deposits are known. The first record of the fossil forest dates back to the early eighteenth century when outstanding anatomically preserved trees provided the basis for the introduction of fossil plant names, including *Psaronius* (Box 11.3, Pšenička et al., Chap. 11), *Medullosa*, and *Calamitea* (Cotta 1832). As the city developed through the late nineteenth and the early twentieth centuries, excavations uncovered more of the buried forest. It wasn't until 2008 when a permanent display of the Chemnitz fossil forest was established based on two excavations. One is known as Chemnitz-Hilbersdorf (2008–2011) in which more than 2000 petrifactions, molds, casts, and adpressions were collected from different volcanic units (Rößler et al. 2012). Zircon grains recovered from the host rock that buried the forest, the Zeisigwald Tuff pyroclastics, has been U-Pb isotopically dated at about 291 million years ago, indicating an early Permian age (Sakmarian/Artinskian; Luthardt et al. 2018). The volcanic event is dominated by a series of eruptions, which initially deposited wet, cool, and highly fragmented ash tuffs promoting the detailed preservation of the ancient ecosystem. The last eruptive pulse deposited massive hot and dry ignimbrites*. The buried forest has yielded significant insights into the fossil plants, animals, paleoecological interactions [U1332], and paleoclimatic conditions (e.g., Dunlop et al. 2016; Spindler et al. 2018).

Trunk bases still standing in their places of growth and rooted in the underlying immature (entisol) paleosol (Fig. 13.9b) characterize this fossil Lagerstätte as an outstanding T⁰ assemblage (Gastaldo et al. 1995). This "Permian Pompeii" forested lowland sheltered a dense wetland vegetation dominated by medullosan seed ferns, cordaitaleans, calamitaleans, and tree ferns (Rößler et al. 2012, 2014). Trees commonly are found broken or erect, permineralized by silica or fluorite, and anatomically well preserved (Fig. 13.9c). The forest grew under seasonally dry conditions as evidenced by the wood anatomy (Box 13.4), similar to modern trees, and the co-occurrence and intergrowth of carbonate and hematite glaebules in the paleosol. The regional paleoclimate is interpreted as having been monsoonal (Roscher and Schneider 2006) under seasonally dry conditions with probably prolonged and severe dry phases (Luthardt et al. 2016). Its position in the basin appears as a spatially restricted and taphonomically favored "wet spot" (e.g., DiMichele et al. 2006) characterized by a subhumid local paleoclimate with an estimated annual precipitation of 800–1100 mm (Luthardt et al. 2016).

A diverse community of arborescent plants grew on an immature mineral soil that lacks features of intense chemical weathering, probably due to a relatively short time of formation. Fifty-three upright standing petrified trees, still anchored in the paleosol, are preserved at one site together with a variety of minimally transported logs and twigs. *Psaronius* tree ferns and calamiteans are common and are plants adapted to seasonally variable water availability as evidence by their production of shallow and voluminous root systems. Their rooting systems only penetrate the upper soil horizons. In contrast, taproots of cordaitalean gymnosperms and medullosan seed ferns penetrate to deeper depths of the soil profile. The excellent preservation of both woody trunks and their rooting systems provides a means to understand how the forest grew over time.

Besides a diverse plant community, a diverse macrofauna inhabited the forest and colonized the paleosol [U1344]. The array of animals recovered from Chemnitz excavations includes terrestrial gastropods (snails) and arthropods and various vertebrates (Fig. 13.9d). Together, these data show

Box 13.4: Tree Rings, Paleoclimate, and Evidence for the Influence of Sun Spots

Dendrological studies allow for insights into the fourth dimension of this three-dimensionally preserved forest ecosystem (Luthardt et al. 2017), unraveling its development over a period of several decades in the scale of years [U1333]. Tree-ring sequences can be correlated among the coeval trees as a consequence of their instantaneous burial, and living versus deadwood has been recognized among the broken fossil logs. An analysis of these tree rings shows that the ecosystem was environmentally stressed, controlled by the major limiting factor of water supply. The presence of different tree-ring types, used to evaluate a tree's sensitivity and protection strategy, demonstrates that different plant groups likely show variable adaptations to seasonal droughts. Medullosans and calamitaleans exhibit a high mean-wood sensitivity and can be used as good indicators for paleoenvironmental changes and events. In contrast, the less sensitive cordaitaleans and conifers have the best tree-ring record offering the highest dendrochronological resolution. So-called event rings mark distinct environmental perturbations on plant growth, induced by a variety of factors including severe droughts and accidental events, such as lightning strike. Tree growth appears also to have been affected by extraterrestrial factors.

The dendrochronological record of Chemnitz trees has provided evidence for the influence of the solar variation in the forest's growth. Tree-ring analysis has demonstrated the effects of sunspot activity, known as the 11-year solar cycle, along with the first statistical evidence of sunspot periodicity in deep time (Luthardt and Rößler 2017). A periodicity of 10.62 years, spanning a time frame of up to 80 years, exhibits a slightly higher frequency of sunspot activity in the early Permian compared to the modern average periodicity of 11.12 years.

that the forest was a relatively young but well-established ecosystem with a strikingly advanced trophic structure. Fossilized leaves and plant molds co-occur with various arachnid, amphibian, and synapsid (pelycosaur) remains (e.g., Dunlop and Rößler 2013; Spindler et al. 2018), reflecting the role of primary producers as well as primary and secondary consumers. The presence of invertebrate detritivores and fungi attests to a fully functional ecosystem. A very different ecosystem is found in the high southern paleolatitudes, which is our next LPIA stop.

13.10 The Early-Mid Permian Brazilian Forest

André Jasper

While ice sheets melted and retreated to the high southern latitudes of the Gondwanan continents, these Late Paleozoic glacial landscapes were gradually replaced by a lush vegetation that appears in stages [U1335]. The interval over which these continents witnessed deglaciation is recorded in strata of the Paraná Basin, Brazil. The Paraná Basin is an extensive intracratonic basin* covering about 1,500,000 km² of southeastern and southern Brazil (Milani et al. 2007). The changes in depositional setting and paleoenvironmental conditions are reflected as two informal but successive stages based on plant-fossil assemblages (Christiano-de-Souza and Ricardi-Branco 2015). The first stage in vegetational turnover is preserved by Late Pennsylvanian floras in which the plants represent pioneers and are replaced, subsequently, by glossopterid assemblages. Our understanding of this vegetational phase comes, mostly, from the Itararé Group. Sediments in this succession were deposited in response to glacial melting of areas that were previously covered by ice sheets beginning in the mid-Carboniferous. Plant colonization of these newly emergent land surfaces occurred under a post-glacial climate that oscillated between cold and temperate conditions (Iannuzzi 2010). The second stage is represented by floras preserved in association with the coal-bearing strata of the Rio Bonito Formation (Guatá Group) of early Permian age (c. 290 Ma; Cagliari et al. 2014). These represent the *Glossopteris* flora that occupied the humid lowland paleoenvironments under a milder climate (Guerra-Sommer et al. 2008). The outcrop at Quitéria, Rio Grande do Sul (Fig. 13.10a, b), displays a unique example of this second vegetational phase.

The outcrop at Quitéria exposes a 6.4 m vertical section on the southern border of the Paraná Basin (Jasper et al. 2006, 2008). A thick, massive siltstone layer (between 2.6 and 3.0 m in the profile) preserves an erect forest at the upper contact. That forest is dominated by the small tree or shrub lycopsid *Brasilodendron pedroanum* (Fig. 13.10b) [U1336] (Chaloner et al. 1979). The growth strategy of this lycopsid differs from those we have seen in other parts of our journey. Here, the shrubby trees have a round, cormose base, similar to an onion, from which thin roots grew downwards. Although these trees are common elements in coal-bearing parts of the Paraná Basin (e.g., Morro do Papaléo outcrop – see Spiekermann et al. 2018), the assemblage preserved at Quitéria is unique because it occurs in situ. Other well-preserved taxa representative of Gondwanan assemblages occur along with the erect, upright trunks. An unusual aspect of the flora is the presence of glossopterid leaves (*Glossopteris browniana* and *Gangamopteris buriadica*) along with the forest lycopsid *Brasilodendron pedroanum*. Understory plants include herbaceous lycopsids (*Lycopodites riograndensis*), leaves of two species of the seed fern *Botrychiopsis* and, possible fern, *Rhodeopteridium*, together with fertile and sterile leafy conifer shoots (*Coricladus quiteriensis*; Jasper et al. 2006; Iannuzzi and Boardman 2008). Scanning the forest gives one a very different impression of its structure than anything visited, previously, in this unit. The presence of macroscopic remains of charcoal [U1337] (Jasper et al. 2008) includes bark and wood related to *Agathoxylon* and confirms the occurrence of paleowildfires in the surrounding areas during the preservational event (Da Costa et al. 2016). Wildfire may have played a role in the forest ecology, similar to the role it plays in modern coniferous forests [U1338] on coastal floodplains.

The outcrop succession at Quitéria traditionally has been interpreted to represent deposition in a coastal microtidal environment, associated with a restricted lagoon protected by a barrier island (Jasper et al. 2006). The level at which the *Brasilodendron* forest is preserved was considered as a roof shale flora, which originated as a consequence of overbank deposits (crevasse splays) that covered the swamp (between 0.0 and 2.6 m in the profile). However, more extensive, ongoing studies that are integrating paleofloristic, taphonomic, and sedimentological data will probably change the paleoenvironmental interpretations made, to date [U1339].

13.11 Permian Forests of the Youngest Late Paleozoic Ice Age : Australia and South Africa

Stephen McLoughlin and Marion Bamford

Throughout the Late Paleozoic, the Australian paleocontinent rotated progressively southwards into higher paleolatitudes. As such, the area was subjected to glacial and interglacial intervals for which there is scant paleobotanical evidence of the plants that grew during these times [U1340]. As a consequence, the existing evidence shows that the low-diversity, lycopsid, and progymnosperm-dominated floras of Mississippian age became successively more impoverished, and those lycopsids that persisted show increasing evidence of seasonality in their rhythmic production of leaves along the stem. Maximum glaciation of the continent began near the Pennsylvanian-Permian boundary, when Australia was located in near-polar latitudes on the southeastern margin of Gondwana. Here, LPIA deposits are manifest in the geological record by tillites* (consolidated moraine deposits), diamictites* (rocks consisting of two distinct grain sizes—normally isolated pebbles to boulders set in fine muds—caused by glacial rafting of rocks into quiet marine settings), glendonites (radiating crystals of calcite replacing ikaite,

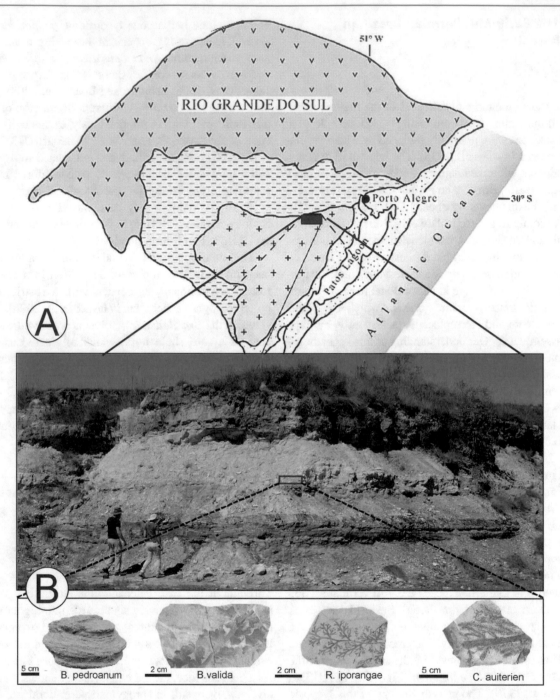

Fig. 13.10 The early–middle Permian forest at Quitéria, Brazil. (**a**) Map of Rio Grande do Sul province, Brazil, on which the locality is indicated. (**b**) Standing cormose, arborescent lycopsids, *Brasilodendron*, are preserved erect in a massive siltstone along with herbaceous groundcover. *B. perdroanum* = *Brasilodendron; B. valida* = *Botrychiopsis; R.* = *Rhacopteris; C.* = *Callipteris*. (Images by A Jasper)

which typically forms in cold water marine settings), varved sediments, and an almost complete absence of plant fossils (Fielding et al. 2008). Better paleobotanical records accompany the phases of deglaciation.

As Australia began to emerge from the LPIA in the late early Permian, a low diversity vegetation colonized emergent landscapes. The plants included a few sphenopsids, ferns, a relict pteridosperm (*Botrychiopsis*), and the appearance of a new evolutionary clade of gymnosperms, Glossopteridales

(Fig. 13.11) [U1341]. This latter group came to dominate the lowland vegetation of not only the Australian continent but extended its biogeographic range into the middle to high latitudes of the rest of Gondwana. It dominated the Southern Hemisphere floras for the remainder of the Permian (Anderson et al. 1999). The earliest Permian vegetation of Australia has been interpreted as an analog of modern tundra because of its monotonous herbaceous groundcover interspersed with sparse, dwarfed woody shrubs.

Fig. 13.11 An early Permian "tundra" vegetation in Australia. (**a**) *Botrychiopsis plantiana*, a pteridosperm holdover from the Carboniferous. (**b**) *Bergiopteris reidsdomae*, foliage of a presumed pteridosperm associated with the *Glossopteris* flora. (**c**) *Gangamopteris angustifolia*, a typical glossopterid leaf taxon in post-glacial floras. (**d**) *Gangamopteris spatulata*, a typical glossopterid leaf taxon in post-glacial floras. (Images by S McLouglin)

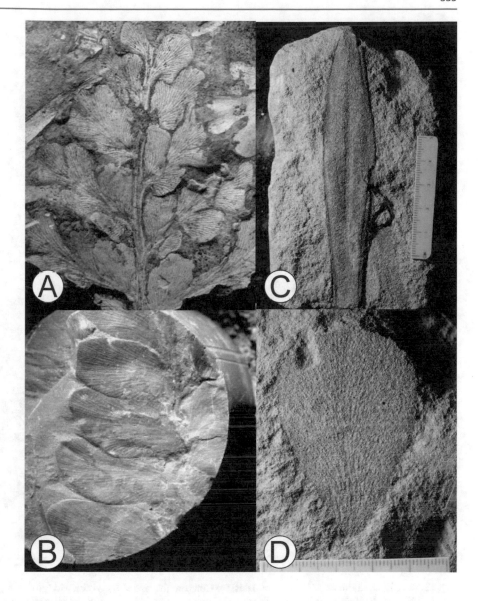

As the pulses of LPIA glaciation gradually waned through the early to middle Permian, woody vegetation became more prominent. Foremost among the woody riparian plants were the glossopterids [U1342]. These gymnosperms rapidly diversified and came to dominate the extensive lowlands of all sedimentary basins following the retreat of the ice sheets. Members of the group became specialists adapted to high water tables and swampy conditions—their segmented roots contained air chambers that helped maintain the underground parts growing in waterlogged, dysoxic peaty substrates. Their geographic range extended from about 30° S to essentially the high polar latitudes in Antarctica (McLoughlin 2011). Even in the absence of continental glaciers, the polar high latitudes, in particular, must have experienced very low temperatures and intervals of several months of darkness or twilight each year. Glossopterids appear to have adapted to these conditions by adopting a deciduous habit—shedding

their leaves in the autumn and surviving over winter by going into dormancy. A modern analog might be *Betula* (birch), which includes species that thrive in high-latitude swampy environments in the modern Northern Hemisphere vegetation. Like *Betula*, glossopterids might have evolved a conical growth form to optimize the interception of low angle sunlight through much of the growing season.

Southern Africa was positioned around 35° S during the early Permian and, as such, experienced a warmer temperate climate than Australia or Antarctica. A diverse flora soon became established after the glaciers had melted in the middle Permian (c. 273 Ma), with meltwaters flowing into the large inland Karoo sea. Glossopterids were the dominant plants and formed peats in the uneven topography that was left behind by the receding ice sheets. These peat bodies were later buried and compressed into coal seams. Large silicified logs are exposed and scattered in some parts of South Africa

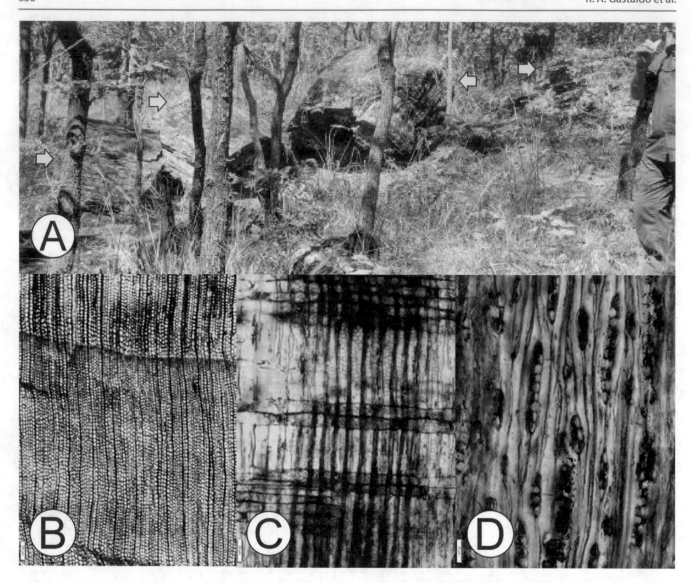

Fig. 13.12 Silicified trunks of Glossopteridales. (**a**) Compare the size and diameter of Permian-aged trunks with modern savanna-woodland trees in the north Luangwa Valley, Zambia. (**b**) Transverse thin sections showing a wide growth ring, reflecting rapid growth in the middle pale-olatitudes of Permian Gondwana. (**c**) Radial thin section. (**d**) Tangential thin section. (Images by M Bamford)

today, but there are some deposits of numerous tree trunks of extraordinary size farther north (Fig. 13.12). These are found in southern Zambia and northern Mozambique on the north and south banks of today's Zambezi River. Trunks of over 2 m diameter and broken lengths of more than 20 m are testimony of impressive forests of Permian plants that are now extinct.

Evidence pertaining to the structure of glossopterid forests is scarce. Hence, it is difficult to provide the reader with a "visual" of the tree density, understory, groundcover vegetation, and ecosystem composition as has been possible in other parts of this unit. There are a few permineralized peats from the Transantarctic Mountains and Prince Charles Mountains in Antarctica, and from the Sydney-Bowen Basin complex in Australia. These sparse sites record the three-dimensionally entombed remains of plant parts accumulating in the Permian swamps. Unfortunately, the record is neither equivalent to, nor as extensive as, data from the Pennsylvanian-aged coal-ball floras of the paleotropics nor are many forests buried in volcanic ash as in other parts of the globe. High latitude, Southern Hemisphere peats are commonly rich in *Vertebraria*, the roots of the glossopterid plants, indicating that the coals comprised the in situ (autochthonous to parautochthonous) remains of glossopterid trees (Slater et al. 2015). The relatively low ash and low sulfur contents of these coals, together with their low floristic diversity and high levels of decayed biomass, have been used to infer that many peats accumulated in raised forest mires akin to those found in present-day Siberia. Unlike the Holocene raised peat swamps of Borneo (see Gastaldo et al., Chap. 12), Siberian raised swamps form under high rainfall and low evaporation conditions, which enable the peat surface to accumulate well above the regional water table over very extensive areas of a subdued landscape. Additional data on the structure of these forests come from a few examples of glossopterid-stump horizons preserved in situ by volcanic ash deposits in eastern Australia and the Transantarctic Mountains, Antarctica. Although work is incomplete on these in situ forests, they appear to represent immature communities with trunk diameters of generally less than 20 cm and spacing between individuals of only a few meters (Taylor and Ryberg 2007). Very high latitude forests would be expected to have relatively open-canopied vegetation and widely spaced trees to optimize the interception of low angle light in the growing season. Hence, the Permian forests may be an anomaly.

Glossopterids reached their acme in the middle to late Permian when, by this time, they had diversified into four distinct families and formed vast peat-producing forests across the Gondwanan lowlands. The bituminous coals that are extensively mined in India, Australia, and South Africa, currently constitute about 20% of world coal production. It is impressive to know that these resources are derived largely from fossilized glossopterid biomass [U1343]. In a few instances, there is evidence of the invertebrates and vertebrates that lived in the forests. A few fish fossils along with aquatic reptiles and amphibians are known from freshwater deposits, and a few mammal-like reptiles are reported from southern Africa. But, the vertebrate faunas were relatively impoverished compared to the succeeding Triassic faunas in this region. However, then, as now, arthropods and fungi were the most important herbivores and saprotrophs*, respectively, in these deciduous forests. There is extensive evidence of leaf-feeding, galling, and wood-boring by insects and mites documented from Permian high southern latitude glossopterids [U1344] (Prevec et al. 2010; Slater et al. 2015). A diverse array of saprotrophic (extracellular digestion of organic matter) and parasitic (nutrition derived directly from the host) fungi also has been identified in permineralized remains of glossopterid wood and leaves and in the fossils of associated plants [U1345] (Slater et al. 2015). Abundant charcoalified plant remains are reported throughout the Australian and Antarctic Permian coals, attesting to the regular occurrence of wildfires perturbating these landscapes. Although not always the case, a common difference between the Carboniferous coals of Euramerica and the Permian coals of Gondwana is the higher proportion of charcoalified woody components in coals from the latter region. Combining all available data, it is possible to reconstruct an energy flow diagram for these communities [U1346].

Box 13.5: Paleoclimate at High Southern Latitudes

The presence of growth rings in glossopterid wood attests to the persistence of strongly seasonal conditions in Australia until the latest Permian (Fig. 13.12b–d). Australia remained in a high southern latitudinal position throughout this period, and growth conditions were amenable at these latitudes [U1349]. Several months of winter darkness were compensated by several months of continuous sunlight throughout the growing season. In response to changes in light regime, glossopterid trees commonly show very thick (up to 1 cm) annual growth rings in trunks that attained diameters of up to 1 m (McLoughlin 1993). Woods from more temperate areas, such as southern Africa, sometimes have even wider growth rings, up to 16 mm, but the types of growth rings are complex and not easy to decipher (Bamford 2016).

Plants that constituted the understory in Australian glossopterid swamp forests, or that grew in mineral-substrate soils along lake and river margins, included a modest range of herbaceous lycopsids, sphenopsids, and osmundaceous ferns [U1347]. As in other parts of the globe, cordaitalean gymnosperms were secondary trees in these forests, and only a few other plants are known to have been forest components (Hill et al. 1999). Additional forest components include relict lyginopterid seed ferns, cycad-like plants, and scale-leafed (voltzialean) conifers [U1348]. These latter plant groups tend to become more common elements of fossil floras in the late Permian, as the climate ameliorated (Box 13.5) and a wider array of depositional sites developed along coastal areas of the large sedimentary basins (Shi et al. 2010). The pattern of an increasing proportion of seed-bearing plants is indicative that many of these secondary groups occupied more seasonally dry (extrabasinal or upland) sites during the dominance of the glossopterids, and extended their range into the lowlands in response to a more seasonal climate, as documented in other parts of the planet (Looy et al. 2014).

The glossopterid clade experienced an abrupt extinction across their biogeographical range very close to the Permian-Triassic transition. Similar to the response of gigantopterids in the Cathaysian tropical wet forests, and cordaitaleans in the high northern latitudes of Siberia, glossopterids were one of the major casualties of the southern moist temperate broad-leafed forests at the close of the Permian. The precise timing and causes of their demise are still matters of great debate, but a unidirectional, progressive shift toward seasonally drier climates may have contributed to their demise

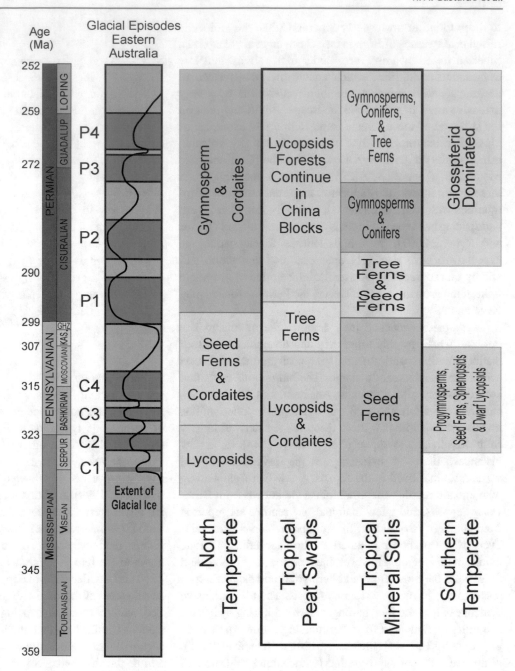

Fig. 13.13 Summary diagram on which the icehouse intervals (Carboniferous 1–4; Permian 1–3) and hothouse interglacials are illustrated against the International Chronostratigraphic Chart. Changes in major vegetational biomes in the North Temperate, Tropics, and Southern Temperate zones are mapped. (adapted from Gastaldo et al. 1996)

(McLoughlin et al. 1997). In most of the world, the first 5 million years of the succeeding Triassic period are notable for an absence of economic coals. This "coal gap" is significant in representing the only interval in the past 350 million years during which little or no peat accumulated anywhere on the planet (Retallack et al. 1996). The broader geographic distribution of red-bed facies in the Early Triassic, extending almost to polar latitudes by the late Early Triassic, may be a reflection of intensification of the "Gondwana monsoon" climate system and prominent seasonality of much of the interior of the vast supercontinent of Pangea. The *Glossopteris* flora was replaced in eastern Australia by a short-lived vegetational association dominated by other gymnosperms that included Peltaspermalean seed ferns and Voltzialean coni-

fers. Both groups produced small leaves with a thick cuticle and sunken, or otherwise, protected stomata. Combined, these physiognomic features indicate adaptations to water stress. A shift in climate and the loss of extensive peat-forming habitats may have signaled the Death Knell of glossopterids at the close of the Paleozoic.

13.12 Synopsis

The coal forests of the Carboniferous were dominated by entirely different plant groups from those that comprise most of the biomass in modern ecosystems. We have seen through our excursions that, at the highest systematic level, five groups of vascular plants (see Gastaldo et al., Chap. 12) were important components of these ecosystems over space and time. Four of these reproduce exclusively by spores—the lycopsids, sphenopsids, ferns, and enigmatic progymnosperms (see Gensel et al., Chap. 15)—whereas the fifth group reproduces by seeds, the gymnosperms. Many of the taxa recognized in the fossil assemblages were unique to this time interval, but several subgroups in each broad clade persist to the present. Patterns of vegetational stasis, turnover, replacement, and extinction throughout the Carboniferous and Permian are complex, controlled by regionally (e.g., Gastaldo et al. 2009) and temporally (e.g., Pfefferkorn et al. 2008) constrained factors including paleogeography and climate cycles on various time scales. It is beyond the scope of the current chapter to explore these patterns, and the reader is directed to other literature sources for greater depth on the topic (e.g., DiMichele et al. 2001; Montañez et al. 2007, 2016; Cleal et al. 2012). Based on the postcards presented herein, several broad and generalized statements can be made about changes witnessed in the "coal forests" of the Mississippian, Pennsylvanian, and Permian.

Over a period of almost 50 million years, from the Late Mississippian (330 Ma) to middle Permian (283 Ma), there was a significant shift of plant group dominance in the peat accumulating forests, the "coal forests" (Fig. 13.13). Plants growing in Mississippian forests were dominantly of spore-producing clades, with one group, progymnosperms, being a holdover from the latest Devonian wetlands. The proportion of seed plants as a component of these landscapes is low and, seemingly, restricted to understory tiers. The rise to dominance by the spore-producing lycopsids established them as the principal group responsible for biomass production and accumulation in peat swamps, from the latest Mississippian to the Middle Pennsylvanian in the equatorial paleotropics. Other spore-producing clades, including the ferns and calamitalean horsetails, coinhabited these forests as near equals in diversity and density. Seed ferns of various systematic affinities continued to play a subdominant role in the structure of forests growing on peat substrates but were more common in mineral-substrate soils where drainage was better. Unlike wetland forests of today where either gymnosperms or angiosperms dominate, the Carboniferous coal forests are the only time in Earth's history where four different plant groups were equal in dominance and diversity. Earth Systems associated with the Euramerican paleotropics experienced a short and, seemingly, rapid perturbation in the Middle Pennsylvanian, resulting in the demise of the lycopsid swamp forests and their replacement by tree fern-dominated forests. The causes of this demise have been debated, and are generally attributed to some combination of events, related to consecutive pulses of extreme warming and cooling, resulting loss of habitat area, and accompanying periods of widespread tropical moisture deficits (Montañez et al. 2016, Wilson et al. 2017). Yet, although lycopsid-dominated forests were extirpated in the paleotropical belt, they continued to thrive in both the North and South China blocks into the Permian when the tree lycopsids finally experienced extinction.

Mid- to high paleolatitudes in the early to middle Permian witnessed the onset of seed-producing groups occupying the wetlands, earlier dominated by elements of the Carboniferous flora. The spore-producing lycopsids, sphenopsids, and ferns now found themselves growing in the shadows of large, woody gymnosperms of various systematic affinities. But, the extensive peat forests that became established across the high paleolatitudes of Gondwana all were dominated by a single seed-bearing group, the Glossopteridales. Similar to angiosperms, today, that dominate most latitudinal belts, glossopterids covered the high paleolatitudes. Other gymnospermous groups, including various seed ferns, cycads, and conifers that are found rarely, or in low proportions of several late Carboniferous localities, take on a more prominent role in these, and mineral substrate, forests. These plants mostly evolved outside of the preservational window in Carboniferous times. Their rise to prominence is seen as a consequence of increasing seasonality across the planet (Looy et al. 2014), with subsequent range expansion, and sometimes radiation into the preservational window following environmental change. Hence, following complete deglaciation of the Gondwanan subcontinent, seed-bearing groups, more tolerant of seasonally dry conditions, ultimately expanded their biogeographic range into landscapes in which preservation potential was higher, providing a fossil record of their existence. These gymnospermous groups radiated into all inhabitable environments and came to dominate the Mesozoic floras, at least until the arrival of angiosperms (see Kvaček et al., Chap. 5, and Gee et al., Chap. 6).

Questions

1. What characteristics define an LPIA interval as either icehouse or hothouse?

2. What distinguishes T^0 fossil-plant assemblages from other fossiliferous deposits?

3. The Paraca floral realm is considered unique in space and time. What are the characteristics of these plant assemblages, during which part of the Carboniferous did the Paraca floral realm exist, and across which latitudinal (or hemispherical) gradient is it restricted?

4. How do the fossil plants of the Mississippian of China compare with those of the Early Pennsylvanian of Alabama, United States, and Joggins, Nova Scotia? How might you explain their similarities or differences knowing their reconstructed paleogeographic positions?

5. Evidence of Early Pennsylvanian tetrapods is uncommon in the fossil record. What physical conditions may have played a role in their preservation at the UNESCO World Heritage site at Joggins, Nova Scotia?

6. Coal is a combustible sedimentary rock formed from ancient vegetation that is consolidated and transformed by both microbial decay and prolonged burial (increasing pressure and heat over millions of years). Most often, original plant parts decay into either fibrous tissues or amorphous organic muck. Explain how data from coal-ball studies provide insight into the contribution of plants to the peat.

7. Using the Bělka bed exposed at Ovčin and Štilec in the Czech Republic, discuss the spatial heterogeneity in forest structure of this Middle Pennsylvanian forest. What factors may have controlled these relationships?

8. Why are both the Wuda Tuff and Chemnitz forests described as "deep time" sites equivalent to the Roman city of Pompeii? What features do each preserve that allows for the analogy?

9. Although roughly contemporaneous in time, the early Permian floras preserved in Chemnitz and Inner Mongolia are very different in plant composition. Compare and contrast these two Fossil Lagerstätten, and provide an explanation as to why these forests are so different.

10. How does the growth architecture of the early Permian lycopsid, *Brasilodendron*, differ from other Carboniferous lycopsid trees?

11. Dendrochronology* is the study of tree rings in which (paleo)environmental information is recorded. What environmental signals are caught in any tree-ring record, and what do dendrochronological studies tell us about the Permian?

12. The Permian *Glossopteris* flora grew at high southern latitudes following the deglaciation of Gondwana. What features did these plants evolve to live under conditions that included cold winter temperatures, low (or limited) light conditions during winter months, and limited rainfall?

13. Explain the major changes in global vegetational patterns recorded in the Carboniferous to Permian rock record. What impact(s) did these have on the trajectory of plants into the Mesozoic?   •

Acknowledgments The authors would like to acknowledge the array of grant funding agencies that have supported their research over the decades, which has resulted in the compilation of case studies presented herein. Those agencies are, in alphabetical order: Alexander von Humboldt Stiftung, Germany; American Chemical Society, Petroleum Research Fund; Brazilian National Council for Scientific and Technological Development (CNPq); Conselho Nacional de Desenvolvimento Científico e Tecnológico, Brazil; Coordenação de Aperfeiçoamento de Pessoal de Nível Superior, Brazil; Deutsche Forschungsgemeinschaft, Bonn, Germany; Fulbright Scholars Program, USA; Grant Agency of the Czech Republic (project 16-24062S); Joggins Fossil Institute; National Research Foundation of South Africa—African Origins Platform; National Natural Science Foundation of China (Grant No.41530101); National Science Foundation of the United States of America; Nova Scotia Department of Natural Resources; The Research Support Foundation of the State of Rio Grande do Sul (FAPERGS); Strategic Priority Research Program of Chinese Academy of Sciences (Grant No. XDB26000000); The Swedish Research Council (Vetenskapsrådet); The United States National Museum, Smithsonian Institution, Washington DC USA; Volkswagen Foundation, Hannover, Germany.

References

Alleman V, Pefferkorn HW (1988) Licopodos de Paracas: Significación geológica y paleo-climatológica. Bolletin Sociedad Geologoie Perú 78:131–136

Anderson JM, Anderson HM, Achangelsky S, Bamford M, Chandra S, Dettmann M, Hill R, McLoughlin S, Rösler O (1999) Patterns of Gondwana plant colonisation and diversification. J Afr Earth Sci 28:145–167

Bamford MK (2016) Fossil woods from the Upper Carboniferous to Lower Jurassic Karoo Basin and the environmental interpretation. In: Linol B, de Wit M (eds) Origin and evolution of the Cape Mountains and Karoo Basin. Regional geology reviews. Springer, Berlin, pp 158–167. https://doi.org/10.1007/978-3-319-40859-0_16

Bek J, Libertín M, Drábková J (2009) *Spencerites leismanii* sp. nov., a new sub-arborescent compression lycopsid and its spores from the Pennsylvanian of the Czech Republic. Rev Palaeobot Palynol 155:116–132

Cagliari J, Lavina ELC, Philipp RP, Tognoli FMW, Basei MAS, Faccini UF (2014) New Sakmarian ages for the Rio Bonito formation (Paraná Basin, southern Brazil) based on LA-ICP-MS U-Pb radiometric dating of zircons crystals. J S Am Earth Sci 56:265–277

Calder JH (2012) The Joggins fossil cliffs: coal age Galápagos. Nova Scotia Department of Natural Resources, Halifax. 96 p

Calder JH, Gibling MR, Scott AC, Hebert BL (2006) A fossil lycopsid forest succession in the classic Joggins section of Nova Scotia: paleoecology of a disturbance-prone Pennsylvanian wetland. In: Greb SF, DiMichele WA (eds) Wetlands through time, vol 399. Geological Society of America, Boulder, pp 169–195

Carroll RL (1964) The earliest reptiles. Journal of the Linneann Society Zoology 45:61–83

Césari SN, Limarino CO, Gulbranson EL (2011) An upper Paleozoic bio-chronostratigraphic scheme for the western margin of Gondwana. Earth Science Reviews 106:149–160

Chaloner WG, Leistikow KU, Hill A (1979) *Brasilodendron* gen. nov. and *B. pedroanum* (Carruthers) comb. nov., a permian lycopod from Brazil. Rev Palaeobot Palynol 28:117–136

Christiano-de-Souza IC, Ricardi-Branco FS (2015) Study of the West Gondwana Floras during the Late Paleozoic: a paleogeographic approach in the Paraná Basin – Brazil. Palaeogeogr Palaeoclimatol Palaeoecol 426:159–169

Cleal CL, Uhl D, Cascales-Miñana B, Thomas BA, Bashforth AR, King SC, Zodrow EL (2012) Plant biodiversity changes in Carboniferous tropical wetlands. Earth-Sci Rev 114:124–155

Cotta B (1832) Die Dendrolithen in Bezug auf ihren inneren Bau. Arnoldische Buchhandlung, Leipzig. 89 pp

Da Costa L, Kionka DCO, Périco E, Japser A (2016) Identificação de carvão vegetal macroscópico no nível de roof-shale do Afloramento Quitéria, Formação Rio Bonito, permiano inferior da Bacia do Paraná. Geosul, Florianópolis 31(61):133–154

Darwin C (1859) The origin of species by means of natural selection. John Murray, London. 513 p

Davies SJ, Gibling MR (2003) Architecture of coastal and alluvial deposits in an extensional basin: the Carboniferous Joggins Formation of eastern Canada. Sedimentology 50:415–439

Davies SJ, Gibling MR, Rygel MC, Calder JH, Skilliter DM (2005) The Pennsylvanian Joggins Formation of Nova Scotia: sedimentological log and stratigraphic framework of the historic fossil cliffs. Atl Geol 41:115–142

di Pasquo M, Iannuzzi R (2014) New palynological information from the Poti Formation (upper Visean) at the Roncador creek, Parnaíba Basin, northeastern Brazil. Bolletin Geologie y Mineralogie 125:405–435

DiMichele WA (2014) Wetland-dryland vegetational dynamics in the Pennsylvanian ice age tropics. Int J Plant Sci 175:123–164

DiMichele WA, Phillips TL (1988) Paleoecology of the Middle Pennsylvanian age Herrin coal swamp (Illinois) near a contemporaneous river system, the Walshville paleochannel. Rev Palaeobot Palynol 56:151–176

DiMichele WA, Phillips TL (1994) Paleobotanical and paleoecological constraints on models of peat formation in the Late Carboniferous of Euramerica. Palaeogeogr Palaeoclimatol Palaeoecol 106:39–90

DiMichele WA, Phillips TL (1996) Climate change, plant extinctions, and vegetational recovery during the Middle-Late Pennsylvanian transition: the case of tropical peatforming environments in North America. In: Hart ML (ed) Biotic recovery from mass extinction events, vol 102. Geological Society Special Publication, London, pp 201–221

DiMichele WA, Pfefferkorn HW, Gastaldo RA (2001) Response of Late Carboniferous and Early Permian plant communities to climate change. Annu Rev Earth Planet Sci 29:461–487

DiMichele WA, Tabor NJ, Chaney DS, Nelson WJ (2006) From wetlands to wet spots: environmental tracking and the fate of Carboniferous elements in Early Permian tropical floras. Geol Soc Am Spec Pap 399:223–248

DiMichele WA, Elrick SD, Nelson WJ (2017) Vegetational zonation in a swamp forest, Middle Pennsylvanian, Illinois Basin, U.S.A., indicates niche differentiation in a wetland plant community. Palaeogeogr Palaeoclimatol Palaeoecol 487:71–92

Dolianiti E (1954) A flora do Carbonífero Inferior de Teresina, Piauí. Depart Nacional Prod Mineral Div Geol Mineral, Boletim 148:1–56

Dunlop JA, Rößler R (2013) The youngest trigonotarbid Permotarbus schuberti n. gen., n. sp. from the Permian Petrified Forest of Chemnitz in Germany. Foss Rec 16:229–243

Dunlop JA, Legg DA, Selden PA, Fet V, Schneider JW, Rößler R (2016) Permian scorpions from the Petrified Forest of Chemnitz, Germany. BMC Evol Biol 16:72

Fielding CR, Frank TD, Birgenheier LP, Rygel MC, Jones AT, Roberts J (2008) Stratigraphic imprint of the Late Paleozoic Ice Age in eastern Australia: a record of alternating glacial and nonglacial climate regime. J Geol Soc 165:129–140

Galtier J (1997) Coal-ball floras of the Namurian–Westphalian of Europe. Rev Palaeobot Palynol 95:51–72

Gastaldo RA (1986a) Implications on the paleoecology of autochthonous Carboniferous lycopods in clastic sedimentary environments. Paleogeogr Paleoclimatol Paleoecol 53:191–212

Gastaldo RA (1986b) An explanation for lycopod configuration, 'Fossil Grove' Victoria Park, Glasgow. Scott J Geol 22:77–83

Gastaldo RA (1988) The frond architecture of Sphenopteris pottsvillea (White) Gastaldo and Boersma. J Paleontol 62:982–991

Gastaldo RA (1990) Earliest evidence for helical crown configuration in a Carboniferous tree of uncertain affinity. J Paleontol 64:146–151

Gastaldo RA (1992) Regenerative growth in fossil horsetails (Calamites) following burial by Alluvium. Hist Biol 6:203–220

Gastaldo RA, Pfefferkorn HW, DiMichele WA (1995) Taphonomic and sedimentologic characterization of "roof-shale" floras. In: Lyons P, Wagner RH, Morey E (eds) Historical persepctive of Early Twentieth Century Carboniferous Paleobotany in North America, vol 185. Geological Society of America Memoirs, Boulder, pp 341–352

Gastaldo RA, DiMichele WA, Pfefferkorn HW (1996) Out of the icehouse into the greenhouse: a late paleozoic analogue for modern global vegetational change. GSA Today 10:1–7

Gastaldo RA, Stevanovic-Walls I, Ware WN, Greb SF (2004a) Community heterogeneity of Early Pennsylvanian peat mires. Geology 32:693–696

Gastaldo RA, Stevanovic-Walls I, Ware WN (2004b) In Situ, Erect Forests Are Evidence for Large-Magnitude, coseismic base-Level Changes within Pennsylvanian cyclothems of the Black Warrior Basin, USA. In: Pashin JC, Gastaldo RA (eds) coal-bearing strata: sequence stratigraphy, paleoclimate, and tectonics, vol 51. AAPG Studies in Geology, Tulsa, pp 219–238

Gastaldo RA, Purkyňová E, Šimůnek Z, Schmitz MD (2009) Ecological Persistence in the Late Mississippian (Serpukhovian – Namurian A) Megafloral Record of the Upper Silesian Basin, Czech Republic. PALAIOS 24:336–350

Greb SF, DiMichele WD, Gastaldo RA (2006) Evolution of wetland types and the importance of wetlands in earth history. In: DiMichele WA, Greb S (eds) Wetlands Through Time, vol 399. Geological Society of America Special Publication, Boulder, pp 1–40

Guerra-Sommer M, Cazzulo-Klepzig M, Jasper A, Kalkreuth W, Menegat R, Barboza EG (2008) Palaeoecological patterns at the coal-roof shale transition in an outcrop of the Permian Brazilian Gondwana. Revista Brasil de Paleontologie 11:11–26

Hill RS, Truswell EM, McLoughlin S, Dettmann ME (1999) The evolution of the Australian flora: fossil evidence. In: Orchard AE (ed) Flora of Australia, 1 (Introduction), 2nd edn. CSIRO, Melbourne, pp 251–320

Iannuzzi R (2010) The flora of early Permian coal measures from the Paraná Basin in Brazil: a review. Int J Coal Geol 83:229–247

Iannuzzi R, & Boardman DR (2008) Novas ocorrências de Botrychiopsis plantiana (Carr.) Arch. and Arrondo no Afloramento Quitéria, Permiano Inferior, Rio Grande do Sul: implicações bioestratigráficas. In: 12° Simpósio de Paleobotânicos e Palinólogos (Florianópolis), Boletim de Resumos, p 95

Iannuzzi R, Pfefferkorn HW (2002) A pre-glacial, warm-temperate floral belt in Gondwana (late Visean, Early Carboniferous). PALAIOS 17:571–590

Iannuzzi R, Pfefferkorn HW (2014) Re-interpreting Kegelidium lamegoi Dolianiti, a microsporangiate organ from the Poti Formation (Late Visean), Parnaíba Basin, northeastern Brazil. Comunication Geológie 101:451–453

Iannuzzi R, Rösler O (2000) Floristic migration in South America during the Carboniferous: phytogeographic and biostratigraphic implications. Palaeogeogr Palaeoclimatol Palaeoecol 161:71–94

Iannuzzi R, Pfefferkorn HW, Rösler O (2006) Reavaliação da flora da Formação Poti: Diplothmema gothanica (Dolianiti) Iannuzzi. Revista Brasil Paleontologie 9:9–20

Jasper A, Menegat R, Guerra-Sommer M, Cazzulo-Klepzig M, Souza PA (2006) Depositional cyclicity and paleoecological variability in an outcrop of Rio Bonito formation, Early Permian, Paraná Basin, Rio Grande do Sul, Brazil. J S Am Earth Sci 21:276–293

Jasper A, Uhl D, Guerra-Sommer M, Mosbrugger V (2008) Palaeobotanical evidence of wildfires in the late Palaeozoic of South America – Early Permian, Rio Bonito formation, Paraná Basin, Rio Grande do Sul, Brazil. J S Am Earth Sci 26:435–444

Laveine JP (1997) Synthetic analysis of the Neuropterids. Their interest for the decipherment of Carboniferous palaeogeography. Rev Palaeobot Palynol 95:155–189

Leary RL, Pfefferkorn HW (1977) An early pennsylvanian flora with megalopteris and noeggerathiales from west Central Illinois USA. Ill State Geol Surv Circ 500:1–78

Li XX, Wu XY, Shen GL, Liang XL, Zhu HC, Tong ZS, Li L (1993) The Namurian and its biota in the east sector of north Qilian Mountain. Shandong Science and Technology Press, Jinan, pp 1–482. (in Chinese with English abstract)

Li XX, Shen GL, Tian BL, Wang SJ, Ouyang S (1995) Some notes on Carboniferous and Permian floras in China. In: Li XX, Zhou ZY, Cai CY, Sun G, Ouyang S, Deng LH (eds) Fossil floras of China through the geological ages. Guangdong Science and Technology Press, Guangzhou, pp 244–304

Libertín M, Opluštil S, Pšenička J, Bek J, Sýkorová I, Dašková J (2009) Middle Pennsylvanian pioneer plant assemblage buried *in situ* by volcanic ash-fall, central Bohemia, Czech Republic. Rev Palaeobot Palynol 155:204–233

Looy CV, Kerp H, Duijnstee IAP, DiMichele WA (2014) The late Paleozoic ecological-evolutionary laboratory, a land-plant fossil record perspective. Sediment Rec 12(4):4–10

Luthardt L, Rößler R (2017) Fossil forest reveals sunspot activity in the early Permian. Geology 45:279–282

Luthardt L, Rößler R, Schneider JW (2016) Palaeoclimatic and site-specific conditions in the early Permian fossil forest of Chemnitz – sedimentological, geochemical and palaeobotanical evidence. Palaeogeogr Palaeoclimatol Palaeoecol 441:627–652

Luthardt L, Rößler R, Schneider JW (2017) Tree-ring analysis elucidating palaeo-environmental effects captured in an in situ fossil forest – the last 80 years within an early Permian ecosystem. Palaeogeogr Palaeoclimatol Palaeoecol 487:278–295

Luthardt L, Hofmann M, Linnemann U, Gerdes A, Marko L, Rößler R (2018) A new U–Pb zircon age and a volcanogenic model for the early Permian Chemnitz Fossil Forest. Int J Earth Sci 107:2465. https://doi.org/10.1007/s00531-018-1608-8

Lyell C, Dawson JW (1853) On the remains of a reptile (*Dendrerpeton acadianum* Wyman and Owen), and of a land shell discovered in the interior of an erect fossil tree in the coal measures of Nova Scotia. Q J Geol Soc Lond 9:58–63

McLoughlin S (1993) Plant fossil distributions in some Australian Permian non-marine sediments. Sediment Geol 85:601–619

McLoughlin S (2011) *Glossopteris* – insights into the architecture and relationships of an iconic Permian Gondwanan plant. Journal of the Botanical Society of Bengal 65:93–106

McLoughlin S, Lindström S, Drinnan AN (1997) Gondwanan floristic and sedimentological trends during the Permian-Triassic transition: new evidence from the Amery Group, northern Prince Charles mountains, East Antarctica. Antarct Sci 9:281–298

Melo JHG, Loboziak S (2003) Devonian–Early Carboniferous miospore biostratigraphy of the Amazon Basin, northern Brazil. Rev Palaeobot Palynol 124:131–202

Milani EJ, Melo JHG, Souza PA, Fernandes LA, França AB (2007) Bacia do Paraná. Bol Geoc Petrobrás 15:265–287

Milner AC (1987) The Westphalian tetrapod fauna; some aspects of its geography and ecology. J Geol Soc Lond 144:495–506

Montañez IP, Tabor NJ, Niemeier D, DiMichele WA, Frank TD, Fielding CR, Isbell JL, Birgenheier LP, Rygel MC (2007) CO2-forced climate and vegetation instability during late Paleozoic deglaciation. Science 315:87–91

Montañez IP, McElwain JC, Poulsen CJ, White JD, DiMichele WA, Wilson JP, Griggs G, Hren MT (2016) Climate, pCO2 and terrestrial carbon cycle linkages during Late Paleozoic glacial-interglacial cycles. Nat Geosci 9:824–828

Oliveira E (1934) Ocorrência de plantas carboníferas da flora Cosmopolita no Estado do Piauhy. Anais da Academie Brasil Ciências 6:113–118

Opluštil S (2005) Evolution of the Middle Westphalian river valley drainage system in central Bohemia (Czech Republic) and it palaeogeographic implication. Palaeogeogr Palaeoclimatol Palaeoecol 222:223–258

Opluštil S (2010) Contribution to knowledge on ontogenetic developmental stages of *Lepidodendron mannebachense* Presl, 1838. Bull Geosci 85:303–316

Opluštil S, Pšenička J, Libertín M, Šimůnek Z (2007) Vegetation paterns of Westphalian and lower Stephanian mire assemblages preserved in tuff beds of the continental basins of Czech Republic. Rev Palaeobot Palynol 152:107–154

Opluštil S, Pšenička J, Libertín M, Bek J, Dašková J, Šimůnek Z, Drábková J (2009) Composition and structure of an *in situ* Middle

Pennsylvanian peat-forming plant assemblage in volcanic ash, Radnice Basin (Czech Republic). PALAIOS 24:726–746

Opluštil S, Pšenička J, Bek J, Wang J, Feng Z, Libertín M, Šimůnek Z, Bureš J, Drábková J (2014) T⁰ peat-forming plant assemblage preserved in growth position by volcanic ash-fall: a case study from the Middle Pennsylvanian of the Czech Republic. Bull Geosci 89:773–818

Opluštil S, Schmitz M, Cleal CJ, Martínek K (2016) A review of the Middle-Late Pennsylvanian west European regional substages and floral biozones, and their correlation to the Global Time Scale based on new U-Pb ages. Earth-Sci Rev 154:301–335

Pashin JC, Gastaldo RA (2009) Carboniferous of the black Warrior Basin. In: Greb SF, Chestnut DR Jr (eds) Carboniferous geology and biostratigraphy of the Appalachian and Black Warrior Basins. Kentucky Geological Survey, Lexington, pp 10–21. Special Publication 1, Series 12

Pérez Loinaze V (2007) A Mississippian miospore biozone for southern Gondwana. Palynology 31:101–118

Pfefferkorn HW, Gastaldo RA, DiMichele WA, Phillips TL (2008) Pennsylvanian tropical floras as a far-field record of changing climate. In: Fielding CR, Frank TD, Isbell JL (eds) Resolving the late Paleozoic gondwanan ice age in time and space, vol 441. Geological Society of America, Boulder, pp 305–316. Special Paper

Pfefferkorn HW, Alleman V, Iannuzzi R (2014) A greenhouse interval between icehouse times: climate change, long-distance plant dispersal, and plate motion in the Mississippian (late Visean-earliest Serpukhovian) of Gondwana. Gondwana Res 25:1338–1347

Phillips TL, DiMichele WA (1981) Paleoecology of Middle Pennsylvanian-age coal swamps in southern Illinois: Herrin Coal Member at the Sahara Mine No. 6. In: Niklas KJ (ed) Paleobotany, paleoecology and evolution, vol 1. Praeger Press, New York, pp 231–281

Phillips TL, Peppers RA (1984) Changing patterns of Pennsylvanian coal–swamp vegetation and implications of climatic control on coal occurrence. Int J Coal Geol 3:205–255

Phillips TL, Avcin MJ, Berggren D (1976) Fossil peat of the Illinois Basin. Ill State Geol Surv Educ Ser 11:1–39

Phillips TL, Kunz AB, Mickish DJ (1977) Paleobotany of permineralized peat (coal balls) from the Herrin (No. 6) Coal Member of the Illinois Basin. In: Given PH, Cohen AD (eds) Interdisciplinary studies of peat and coal origins, vol 7. Geological Society of America Microform Publication, Boulder, pp 18–49

Prevec R, Gastaldo RA, Neveling J, Reid SB, Looy CV (2010) An Autochthonous Glossopterid flora with Latest Permian palynomorphs and its depositional setting from the *Dicynodon* Assemblage Zone of the southern Karoo Basin South Africa. Palaeogeogr Palaeoclimatol Palaeoecol Palaeoecol 292:381–408

Pryor J (1988) Sampling methods for quantitative analysis of coal-ball plants. Palaeogeogr Palaeoclimatol Palaeoecol 63:313–326

Pšenička J, Opluštil S (2013) The epifytic plants in the fossil record and its example from in situ tuff from Pennsylvanian of Radnice Basin (Czech Republic). Bull Geosci 88:401–416

Retallack GJ, Veevers JJ, Morante R (1996) Global coal gap between Permian-Triassic extinction and Middle Triassic recovery of peat-forming plants. Geol Soc Am Bull 108:195–207

Roscher M, Schneider JW (2006) PermoCarboniferous climate: Early Pennsylvanian to Late Permian climate development of central Europe in a regional and global context. Geol Soc Lond Spec Publ 265:5–136

Rößler R, Zierold T, Feng Z, Kretzschmar R, Merbitz M, Annacker V, Schneider JW (2012) A snapshot of an Early Permian ecosystem preserved by explosive volcanism: new results from the petrified forest of Chemnitz, Germany. PALAIOS 27:814–834

Rößler R, Merbitz M, Annacker V, Luthardt L, Noll R, Neregato R, Rohn R (2014) The root systems of Permian arborescent sphenopsids: evidence from the Northern and Southern hemispheres. Palaeontographica Abt B 291:65–107

Rygel MC, Sheldon EP, Stimson MR, Calder JH, Ashley KT, Salg JL (2014) The Pennsylvanian Springhill Mines Formation: Sedimentological framework of a portion of the Joggins Fossil Cliffs UNESCO World Heritage Site. Atl Geol 50:249–289

Santos MECM, Carvalho MSS (2009) Paleontologia das bacias do Parnaíba, Grajaú e São Luís: reconstituições paleobiológicas. Serviço Geológico do Brasil – CPRM, Rio de Janeiro. 215 p

Shi GR, Waterhouse JB, McLoughlin S (2010) The Lopingian of Australasia: a review of biostratigraphy, correlations, palaeogeography and palaeobiogeography. Geol J 45:230–263

Šimůnek Z, Opluštil S, Drábková J (2009) Cordaites borassifolius (Sternberg) Unger (Cordaitales) from the Radnice Basin (Bolsovian, Czech Republic). Bull Geosci 84:301–336

Slater BJ, McLoughlin S, Hilton J (2015) A high-latitude Gondwanan Lagerstätte: the Permian permineralised peat biota of the Prince Charles Mountains, Antarctica. Gondwana Res 27:1446–1473

Spiekermann R, Uhl D, Benício JRW, Guerra-Sommer M, Jasper A (2018) A remarkable mass-assemblage of lycopsid remains from the Rio Bonito formation, lower Permian of the Paraná Basin, Rio Grande do Sul, Brazil. Palaeobiol Palaeoenviron 98:369–384

Spindler F, Werneburg R, Schneider JW, Luthardt L, Annacker V, Rößler R (2018) First arboreal 'pelycosaurs' (Synapsida: Varanopidae) from the early Permian Chemnitz Fossil Lagerstätte, SE-Germany, with a review of varanopid phylogeny. Paläontol Z 92:316–364

Sterzel JT (1875) Dic fossilen Pflanzen des Rothliegenden von Chemnitz in der Geschichte der Palaeontologie. Bericht der Naturwissenschaftlichen Gesellschaft zu Chemnitz 5:71–243

Suárez Soruco R, Lobo Boneta J (1983) La fase compresiva Eohercínica en el sector oriental de la Cuenca Cordillerana de Bolivia. Rev Téc Yacimien Petrol Fisc Boliv 9:189–202

Taylor EL, Ryberg PE (2007) Tree growth at polar latitudes based on fossil tree ring analysis. Palaeogeogr Palaeoclimatol Palaeoecol 255:246–264

Waldron JWF, Rygel MC, Gibling MR, Calder JH (2013) Evaporite tectonics and the late Paleozoic stratigraphic development of the Cumberland basin, Appalachians of Atlantic Canada. Geol Soc Am Bull 125:945–960

Wang J (2006) Tingia unita sp. nov. (Noeggerathiales) with strobilus from the Lower Permian of Wuda, Inner Mongolia, China. Chin Sci Bull 51:2624–2633

Wang J, Pfefferkorn HW, Bek J (2009) Paratingia wudensis sp. nov., a whole noeggerathialean plant preserved in an air fall tuff of earliest Permian age (Inner Mongolia, China). Am J Bot 96:1676–1689

Wang J, Pfefferkorn HW, Zhang Y, Feng Z (2012) Permian vegetational Pompeii from Inner Mongolia and its implications for landscape paleoecology and paleobiogeography of Cathaysia. Proc Natl Acad Sci U S A 109:4927–4932

Wang J, He XZ, Pfefferkorn HW, Wang JR (2013) Compaction rate of an early Permian volcanic tuff from Wuda coalfield, Inner Mongolia. Acta Geol Sin 87:1242–1249

Wilson JP, Montañcz IP, White JD, DiMichele WA, McElwain JC, Poulsen CJ, Hren MT (2017) Dynamic Carboniferous tropical forests: new views of plant function and potential for physiological forcing of climate. New Phytol 215:1333–1353

Winston RB (1986) Characteristic features and compaction of plant tissues traced from permineralized peat to coal in Pennsylvanian coals (Desmoinesian) from the Illinois Basin. Int J Coal Geol 6:21–41

Diving with Trilobites: Life in the Silurian–Devonian Seas

14

Oive Tinn, Tõnu Meidla, and Leho Ainsaar

Abstract

During the Silurian, shallow seas covered the margins of continental landmasses, the largest of these was the Gondwana supercontinent, which extended from the South Pole to the Equator. In addition, several smaller continents, including Laurussia in the equatorial region and Siberia in the Northern Hemisphere, occupied the globe. All oceans were inhabited by a rich and highly diverse marine biota in which all modern phyla occurred. The most iconic Paleozoic animals living in these oceans—trilobites—were the first invertebrates with complex eyes, and we could try to imagine seeing life as they may have experienced it. Other contemporary members of the arthropod phylum include eurypterids ("sea scorpions"), some of which developed immense body sizes and were among the first animals leaving their footprints on land. Large areas of sea floor were covered with brachiopods—bivalved lophophorates that superficially resemble clams—who were among the most common invertebrates in the Paleozoic oceans. Among the planktonic organisms, stunning colonial hemichordates, called graptolites, thrived alongside radiolarians (single-celled protozoa), early planktonic arthropods, and diverse phytoplankton*. We also encounter the largest reefs in the Earth's history, which were different from extant counterparts. These reef complexes were built by extinct tabulate and rugose corals, peculiar sponges (stromatoporoids), and a wide diversity of bryozoans (moss animals), together with calcified algae and bacteria. As time travellers, we will be impressed by the abundance of

fishes (mainly those we call agnathans—those without jaws) living alongside the first fishes with true jaw structures. And last, but not least, all these animals thrived in the most fascinating surroundings of algal "forests" of those times. Finally, we are going to witness one of the Big Five mass extinction events, which exterminated the great reef systems along with what was once a thriving fish fauna.

14.1 Introduction

We are going to leave land behind and take a tour in the oceans of the Silurian–Devonian periods, virtually diving into the seas that existed 444–359 million years ago. The time span of this excursion is immense—85 million years in total duration—but we will not age appreciably by the end of the chapter. Profound changes occurred during this time; the world was substantially different at the beginning compared to the end of these two geologic periods. We are going to start from the beginning of the Silurian when the world was struggling to recover from the end-Ordovician mass extinction event. We will see the relatively stable Silurian period with the appearance of immense reefs, the abundance of invertebrates and marine plankton, and the diversification of the fish fauna. We will move ahead in time and finish our trip at the end of the Devonian, in the midst of another devastating calamity.

Our tour will consist of two parts. First, we need to get acquainted with the paleogeography and climate of the early Paleozoic world. We will learn how the continents were distributed and how do they relate to the modern world. We will look at Silurian–Devonian outcrops and rocks from different paleocontinents. We will also see how climate and sea level changed during this interval of time (Box 14.1).

In the course of the second part of our trip, we will visit different marine habitats, including continental shelves, reefs, and lagoons. The assemblages we are going to see are part of the Paleozoic evolutionary fauna, which largely came

Electronic supplementary material A slide presentation and an explanation of each slide's content is freely available to everyone upon request via email to one of the editors: edoardo.martinetto@unito.it, ragastal@colby.edu, tschopp.e@gmail.com

*The asterisk designates terms explained in the Glossary.

O. Tinn (✉) · T. Meidla · L. Ainsaar
Department of Geology, University of Tartu, Tartu, Estonia
e-mail: oive.tinn@ut.ee; tonu.meidla@ut.ee; leho.ainsaar@ut.ee

Box 14.1: Climate and Sea Levels

Before starting our reconstruction of the Silurian–Devonian climate, we first need to step back a few million years and look at what Earth experienced near the end of the Ordovician Period [U1409]. Rocks of this age contain physical evidence for a short-lived global icehouse episode (perhaps not longer than about 0.5–1 My; Finnegan et al. 2011), marked by a major drop of sea level and continental glaciation in Africa. Named after the final stage of the Ordovician, this event is known as the Hirnantian glaciation. It coincided with a two-stage mass extinction, which has been regarded as the first of the Great Five mass extinction events of the Phanerozoic. It is recognized as the second most-severe extinction of marine life in Earth's history. The sub-events of this mass extinction were tied to the beginning of the glaciation and to its end, in response to major rearrangement and reorganization of the ecosystem under climatic pressure (Sheehan 2001). All aspects of the Ordovician cooling—beginning, duration, dynamics, and biotic effects—have initiated extensive discussions today and a multitude of alternative views exist in the literature.

Following the latest Ordovician deglaciation, global climate became milder, although there is evidence of a few glacial deposits in the polar region of South America that reveal the presence of several cooling events (Diaz-Martinez and Grahn 2007) or the existence of mountain glaciers. In general, the Silurian was a rather warm period, and geochemical data on stable carbon-and-oxygen isotopes from both invertebrate shells and limestone rocks [U1410] indicate that the ocean-atmosphere system underwent repeated episodes during which the stratified Silurian ocean overturned (Fig. 14.2), perhaps because of moderate cooling. Trends in the stable isotopes show four major positive stable carbon-isotope excursions in the Silurian (early Wenlock, late Wenlock, late Ludlow, Silurian–Devonian boundary), revealing that the comparatively short Silurian (25 My) period experienced recurrent changes in the global carbon cycle affecting climate.

Global climate continued to be warm for the first part of the Devonian and the period is considered a hothouse, non-glaciated epoch. Paleotemperatures calculated from the stable oxygen-isotope ($\delta^{18}O$) record of conodonts (Joachimski et al. 2009) indicate that the earliest Devonian, as well as the late Frasnian (373–372 Ma) and most of the Famennian (372–359 Ma)

were relatively warm periods. In contrast, the Middle Devonian represented a cooler epoch. There is some physical evidence for Southern Hemisphere glaciation immediately before the end of the Devonian in the south polar areas of Gondwana (South America, Central and South Africa; Isaacson et al. 2008) and evidence for contemporaneous mountain glaciers in tropical latitudes (in the Appalachians of eastern North America; Brezinski et al. 2008). However, other data have been used to suggest that atmospheric carbon dioxide concentrations were up to 12–17 times higher than the modern levels during much of the early Paleozoic era (Came et al. 2007). This condition would not have supported the development of extensive glaciation. Given that carbon dioxide is a greenhouse gas, global surface temperatures were considerably higher than modern ones during the early-middle Paleozoic.

The Silurian and Devonian periods were characterized by global sea-level highstand [U1411]. This means that former large lowland areas were widely flooded forming shallow shelf seas. Marine faunas were well adapted to widespread shallow shelf conditions and a rich fossil record is preserved in this setting. Interestingly, even during this general highstand we can distinguish a number of oscillations in its position—similar to the oscillating climatic conditions we visited in the late Paleozoic (Chaps. 12 and 13). To track sea-level changes, we should start from the Ordovician Period.

It has been widely accepted that the Middle Ordovician was the time with the highest sea levels recorded in the Phanerozoic (Munnecke et al. 2010), probably more than 200 m higher than at the present day (Haq and Schutter 2008). This global highstand interval was followed by an abrupt lowering during the latest Ordovician glaciation. Following the melting of continental ice, the Silurian period witnessed the beginning of another long-term rise in sea level that culminated in the mid-Silurian (Wenlock; 423–427 Ma). This was followed by a decline that lasted until the Early Devonian (Emsian; 408–393 Ma). The latest Early Devonian marked the beginning of another long-term rise that peaked in the early Late Devonian (Frasnian; 383–372 Ma). After a slight dip at the Frasnian/Famennian boundary and a recovery in the early Famennian, the long-term curve shows a gradual sea-level decline in the later Devonian towards the Devonian–Carboniferous transition (Johnson 1970; Simon et al. 2007).

into existence with the Great Ordovician Biodiversification Event (GOBE) and lasted until the end-Permian extinction. The biota suffered greatly at the end of the Ordovician, in the course of one of the five biggest mass extinctions during the Earth's history. Yet, biodiversity recovered during the early Silurian, especially with the radiation of nektonic (swimming) and demersal (free swimming close to the ocean floor) organisms in the Devonian.

We will meet most faunal groups that lived throughout the Silurian and Devonian periods. Some of these were benthic, meaning that they lived on or beneath the seafloor, like corals and sponges, brachiopods, gastropods, and echinoderms. Other animals were demersal or nektonic and were able to actively swim around, like many arthropods, cephalopods, and agnathans (jawless fishes). And finally, we will also see the large variety of early Paleozoic microscopic organisms that flourished in the water masses—swarming heterotrophic* zooplankton and photosynthesizing phytoplankton.

Our tour will end with harsh times at the end of the Devonian, when most of the Earth's biota, which had previously been so rich and diverse, started to decline, ecosystems fell apart, and large faunal groups disappeared forever. It is a time when we also witness the beginnings of Southern Hemisphere glaciation prior to the Carboniferous "coal forests" (Caputo et al. 2008). We will see how the largest reef systems that Earth has ever hosted, perished, the unique fish fauna died out, and oceans were depleted of planktonic organisms. We will analyze the faunal diversity trends together with geochemical data and sea level curves, and discuss what could have caused these disastrous extinction events in deep time.

14.2 Paleogeography

To understand the Silurian–Devonian world, we first have to become acquainted with its paleogeography, which was very different than the arrangements of continents on our present-day globe [U1401]. Over the years, Trond H. Torsvik and Robin M. Cocks (see Torsvik and Cocks 2013a, and the references therein), together with colleagues, have assembled a comprehensive geological data set and produced a series of detailed paleogeographic maps, which illustrate the early Paleozoic world. The positions and arrangements of the major continents have mainly been reconstructed on these maps with the help of paleomagnetic data, but the smaller continents and terranes have been placed using faunal and other data.

Viewing the Silurian–Devonian globe from space, we would see three dominant landmasses. These are the paleocontinents Gondwana (merging major southern continents of today), Laurussia (largely consisting of Laurentia, Baltica,

and Avalonia), and Siberia (Fig. 14.1). Additionally, there were a number of microcontinents, including North and South China, Tarim, and many minor terranes. At the same time, a large part of the planet was governed by oceans. Whereas the supercontinent of Gondwana was predominantly situated in the Southern Hemisphere, Baltica and Laurentia straddled the equator, and Siberia was the only larger isolated continent lying north of the equator (Cocks and Torsvik 2007). Most of the Northern Hemisphere was occupied by the immense Panthalassic Ocean (the name of which was taken from the Greek and means "all seas") (Nance et al. 2012). Before the collision of the continents of Laurentia and Baltica and the formation of the landmass known as Laurussia in the early to mid-Silurian, these two landmasses were separated by the Iapetus Ocean (named after one of the Greek Titans). The Rheic Ocean (named after the sister of Iapetus in Greek mythology) spread south of Laurussia and separated it from the continent of Gondwana. At the end of the Silurian, extensive rifting at the northern margin of Gondwana led to the opening of the Paleo-Tethys Ocean located between North Africa (which remained part of Gondwana) and the various terranes that now make up southern Europe (Iberia and Armorica). The Paleo-Tethys Ocean, especially its eastern part, widened considerably during the Early Devonian.

14.2.1 Gondwana

The Gondwana continent covered almost 100 million km², and consisted of a "core" and "peripheral" terranes. The "core" terrain was comprised of parts of today's South America, Antarctica, Africa, Australia, and India, whereas the "peripheral" areas included Avalonia, Armorica, Iberia, Perunica (Bohemia), Alpine and Balkan fragments, Turkey, Arabia, Himalayan fragments, South China, Sibumasu (Shan Thai), New Guinea, New Zealand, west South America, and Mexico. Today, if we travel to the Silurian deposits at the Jack Hills Gorge in Australia, and then decide to visit the Devonian strata of Crozon in Bretagne, France, we have to cross the Indian Ocean now separating Australia from Eurasia. However, in the Silurian–Devonian, these two places were situated along the shores of the largest paleocontinent of the early Paleozoic [U1402]. The tectonic assembly of Gondwana began in the Late Neoproterozoic–Cambrian (650–500 Ma), with the amalgamation of African and South American terranes to Antarctica, Australia, and India. Later, during the late Paleozoic, Gondwana merged with Laurussia and formed the supercontinent we know as Pangea. Gondwana straddled the equator 540 million years ago, and lay wholly in the Southern Hemisphere by 350 million years ago (Torsvik and Cocks 2013b).

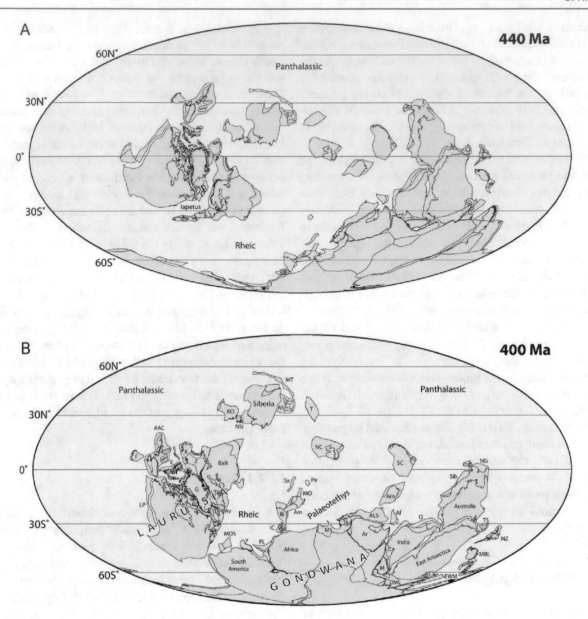

Fig. 14.1 Lower Paleozoic terranes at 440 Ma (early Silurian, Llandovery) and at 400 Ma (Early Devonian, Emsian) (Torsvik and Cocks 2013a, b)

14.2.2 Laurentia

The continent of Laurentia was named after the Laurentian Shield, a geologically stable area that today occupies most of Canada. If you visit the strata cropping out at the Gaspé Peninsula (see Chap. 15) or Anticosti Island [U1403], you will see the rocks formed as the result of Silurian–Devonian sedimentation along the margins of the Laurentia paleocontinent. The ancient landmass consisted of modern North America, northwestern Ireland, Scotland, Greenland, northern Alaska, and the Chukotka Peninsula of NE Russia. Laurentia had been an independent continent since late Neoproterozoic times. But, during the Silurian, it collided

with the paleocontinents of Baltica and Avalonia in the course of the Caledonian Orogeny, after which it formed the western sector of Laurussia (Cocks and Torsvik 2011).

14.2.3 Baltica

The landmass of Baltica roughly includes Scandinavia and eastern European lands around today's Baltic sea, most of Russia's west of the present-day Urals, parts of Belarus and Ukraine. Sediments that formed along the shores and off-shore of this paleocontinent can be observed in such places as Sweden, Estonia, and Latvia [U1404–1405]. Additionally, some regions in today's Ural Mountains and their northward extension into Novaya Zemlya were also part of the Baltica terrane and its associated peri-Baltica island arcs (Cocks and Torsvik 2005). The northwest margin of the ancient continent can be visited in the Scandinavian Caledonides. This mountain chain formed primarily during Baltica's collision with Laurentia after the Iapetus Ocean dwindled and closed, between the Late Ordovician and early Middle Devonian (Cocks and Torsvik 2007).

14.2.4 Avalonia

One of several small, early to middle Paleozoic microcontinents is Avalonia, which comprises a number of independent tectonic blocks that moved separately with respect to each other (Cocks et al. 1997). This short-lived microcontinent is named after the Avalonian Peninsula in Newfoundland, but its pieces can be recognized today on both sides of the Atlantic Ocean. At the western side of the ocean, it is composed of present-day eastern Newfoundland, New Brunswick, Nova Scotia, and the coastal areas of eastern USA (West Avalonia). The eastern segment comprises the Anglo-Welsh area, southern Ireland, and Belgium (together forming East Avalonia), and parts of northern Germany and northwestern Poland (Far Eastern Avalonia). Near the Ordovician–Silurian transition, Avalonia collided with Baltica, and fully amalgamated with it during the Silurian. Thereafter, the combined Avalonia–Baltica terrane was accreted to the even larger continent of Laurentia, forming the supercontinent Laurussia as a result of the Caledonian Orogeny.

One of the most characteristic rock types of the late Silurian, Devonian, and early Carboniferous is known as the Old Red Sandstone [U1405]. It is a distinctive, siliciclastic sedimentary deposit of continental origin, which was deposited across the subtropical to tropical areas of Laurussia (Kendall 2017). Consisting originally mostly of sand (visible today as sandstone) and sometimes gravel (conglomerate), its formation was the result of a number of major co-occurring global events. The first event was the Caledonian Orogeny, a period of major mountain building during which the Iapetus Ocean closed. Secondly, Earth's crust experienced increased rates of weathering promoted by the evolving continental vegetation. These earliest plants were small, with rather shallow horizontal roots (see Chap. 15). By the late Silurian, land plants had become larger and their root systems penetrated deeper and were more effective at physically and chemically weathering the regolith into soils. The vegetated areas expanded from coastal lowlands to higher elevations (upland territories), further away from wet areas and floodplains, to which early plants had been restricted. Increasing plant size resulted in higher rates of photosynthesis and the locking away of atmospheric carbon, influencing global climate by lowering carbon dioxide and increasing oxygen levels. Intensive plant-induced weathering of rocks enhanced production of fine-grained silicate sediments (mud) and a new array of soil types. The Old Red is dominated by terrestrial rocks, but non-marine, estuarine, and marine sediments are also found in many regions.

14.2.5 Siberia

The continental block of Siberia includes a large area within present-day eastern Russia, but it extends also into Mongolia, the northwestern part of China, and eastern Kazakhstan [U1406]. This continental plate moved northwards beginning in the late Neoproterozoic Era. The Silurian and Devonian rocks of this old terrane, very rich and fossiliferous, can be seen today as endless outcrops along the banks of many Siberian rivers.

14.2.6 Other Silurian–Devonian Microcontinents

Studies on sedimentary successions and fossils have indicated that several additional microcontinents were present at the time across the planet. Several of them were located north of Gondwana. One of these microcontinents covered most of today's south China and was separated from another microcontinent North China (Sinokorea), the coastal sediments of which crop out today in the Korean Peninsula and much of northern China [U1407–1408]. These microcontinents were probably situated not very far from each other (although they accreted only in the Mesozoic) and close to several of the Kazakh terranes. Annamia, which consists of today's Indochina Peninsula and some adjacent parts of China, remained close to Gondwana and possibly did not separate from it until the Paleo-Tethys Ocean opened in the Devonian. The nearby Tarim microterrane is located chiefly in western China today. It is considered the largest of the Kazakh terranes and was probably positioned near both the North China and Siberia blocks. It accreted to two other microterranes, the Junggar and Tien Shan microcontinents, in the Late Devonian.

The Arctic Alaska–Chukotka microcontinent consisted of Arctic southern Alaska, Seward and York terranes of Alaska, and the Wrangel Island terranes of northeastern Siberia. In addition, the microcontinent probably reaches the Chukchi, Northwind, and, perhaps, the Mendeleev areas that are submerged by the Arctic Ocean today (Cocks and Torsvik 2013). During the Silurian–Devonian, this microcontinent was situated at the edge of the Panthalassic Ocean, "drifting" towards Laurentia until accreting with its northern margin in the Late Devonian.

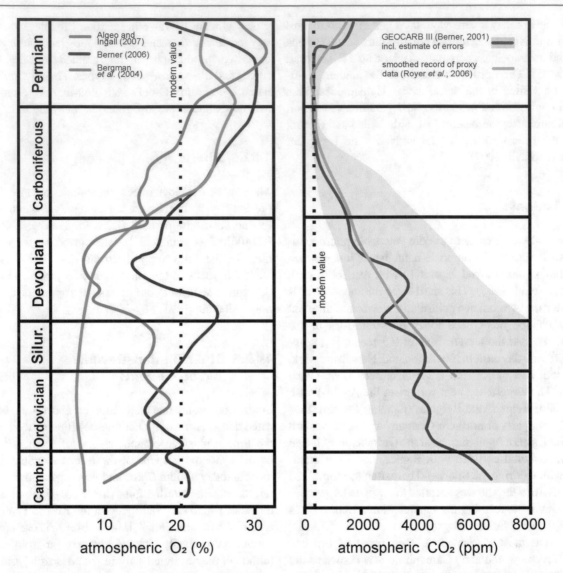

Fig. 14.2 Trends in atmospheric oxygen and carbon dioxide in the Paleozoic have been modeled differently by various authors. Note that, depending on the model, oxygen values in the Silurian and Devonian may have been more or less than present atmospheric levels (PAL). The concentration of modeled atmospheric CO_2 was much higher in the Silurian and Devonian than PAL, and decreased towards the end of the Devonian as a consequence of the spread of woody forests and sequestration of carbon (After Munnecke et al. 2010)

Much of southern Europe, including the Iberian Peninsula, was attached or close to Armorica. This microcontinent remained an integral part of Gondwana until the opening of the Paleo-Tethys Ocean near the Silurian–Devonian transition. Rocks that formed here can be found in France, Germany, and Italy. During our trip through the Mesozoic, we visited a more recent version of Armorica, where we explored the faunal changes during the end-Cretaceous mass extinction (Chap. 8).

14.3   Facies Zones and Ecologies

We are going to make our way through various marine environments of the Silurian–Devonian oceans, beginning in shallow shoreline waters and progressing to deeper neritic* and pelagic settings*. Similar to the modern seas, the distribution of sediments in the Silurian–Devonian marine basins was related to bathymetric*, climatic, and ecologic zonations. We demonstrate this relationship with the generalized model of a carbonate-dominated sedimentary basin, originally developed for the Silurian rocks in the Baltic paleobasin (Nestor 1997). Carbonate systems formed mostly in the tropical and subtropical latitudes, where land areas were distant or elevations low; these settings contain a rich fossil record, especially of diverse invertebrates and reefal structures. At higher latitudes and/or in tectonically active basins, where terrigenous clastic sediments were transported to the oceans from siliciclastic sourcelands, siliciclastic or mixed systems dominated and the sediment is typified by sandstones and siltstones in shallow areas, and shales in more offshore settings. The carbonate model [U1412] shows the

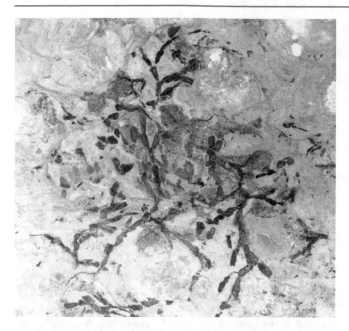

Fig. 14.3 Dasyclad alga *Palaeocymopolia silurica* (Courtesy of Oive Tinn)

lateral succession of sedimentary facies along a bathymetric profile, starting in the tidal flat/lagoonal settings and moving oceanward to shoal*, open shelf, transitional, depression, and further to abyssal zones. In the geological record, all these depositional settings can be differentiated from one another by characteristic sediments that also preserve fossil assemblages of variable diversity and composition.

It is beyond the scope of the current contribution to explore, in detail, all of the ecological associations found in an onshore to offshore transect. Hence, the chapter will present a synthesis of invertebrate macroecology. We are going to view organisms grouped according to their dominant mode of life into benthic (living close to, attached to, or living within the sea floor, e.g., Box 14.2), demersal (swimming animals who live close to the sea floor), planktonic (organisms drifting or floating in the water column where they can migrate vertically), and nektonic (organisms that can swim actively) strategies (Klug et al. 2010). It is important to keep in mind that there were also organisms that could adapt to different ecological conditions and modes of life at different stages of their life cycle.

Box 14.2: Algal Flora
When we stand on dry land and look around us, we can see plants nearly everywhere. When we go diving, we can view the seafloor covered with thickets of an infinite variety of algal forms. Where conditions are favorable, algae may cover seabeds all the way down to the deepest extent of the photic zone, as far as down in the

water column where there is still some amount of sunlight to fuel photosynthesis. The fact that algae were an important part of Paleozoic oceans is evidenced by features found in carbonate environments. These features are the bulk of carbonate sediments composed of algal debris proving that various groups of calcareous algae were common components of both reef and shallow marine communities in the Silurian and Devonian. However, as a whole, algae are not organisms that easily fossilize because most do not possess hard skeletal parts. Other factors that prevent their fossilization include the time during their life cycle that they spend as a macroscopic organism, their use as a food source by herbivores, and finally, the tendency to decompose rather rapidly after death. Most data about fossil algae come from fossil Lagerstätten, deposits of exceptionally preserved fossils. One of such is in Kalana, Estonia (Tinn et al. 2009), and exhibits limestone surfaces covered with exceptionally preserved non-calcified algal remains (Fig. 14.3). Thanks to this outcrop we know about early Silurian algal "forests" that created settings and habitats for marine invertebrates and vertebrates.

The Kalana Lagerstätte has revealed two main types of early Silurian algae. First, there are very distinctive unicellular, but macroscopic green algae from an order called Dasycladales [U1413] (Mastik and Tinn 2015). These extraordinary unicellular organisms are up to 6–10 cm tall, radially symmetrical, and have a specific construction, basically consisting of a system of interconnected tubes. Interestingly, this group has modern analogs, although they form just a negligible group of algae represented by a dozen or so living genera (Berger and Kaever 1992). Among modern dasycladales, there is a "living fossil" that is morphologically an almost exact copy of one of the algae found in Kalana, *Cymopolia barbata*. Nowadays, the genus inhabits supratidal (highest elevations affected by tides) and intertidal areas in tropical regions of the Caribbean, the Pacific, and the Atlantic oceans. The second type of algae is interpreted as a red alga based on its overall morphology, symmetry, and position of reproductive structures [U1414] (Mastik and Tinn 2017). These forms show an irregularly branching body (thallus) with lateral branches covered with tufts consisting of fine filaments; these algae were attached to the bottom with a strong holdfast. There is no analog in today's world, but similar algae found from the Late Ordovician of Laurentia paleocontinent (now Wisconsin, USA) demonstrate that these algal forms survived the severe mass extinction event at the end of the Ordovician.

14.3.1 Lagoonal/Tidal Flat Environment

Shallow water lagoons formed along coasts with gentle slopes, where reefs developed offshore and protected the lagoon from strong storms. Depending on the connections with the open ocean and the inflow of freshwater from rivers, salinity levels were rather variable. Saltwater concentrations in lagoons ranged from highly saline on one end and almost entirely fresh at the other end of the spectrum. Modern Estonia is one of the places where outcropping strata provide a record of such lagoons [U1415]. Here we encounter distinctive microbial structures called stromatolites. Stromatolites were widespread in the Precambrian and persisted in restricted environments throughout the Phanerozoic, reestablishing their geographical distribution in marine settings during major calamities (Kershaw et al. 2012). The principal organisms involved in stromatolite buildups are miscellaneous bacteria, microscopic algae, fungi, and protists, all of which are zoned along chemical and photic (light) gradients. Cyanobacteria, the driving forces in these structures, were able to actively trap and bind detrital sediments; many microbes in this community cycled nutrients and precipitated calcite or aragonite from dissolved calcium and carbonate ions in the water (Dupraz et al. 2009). The other kind of microbial structures in the lagoons are oncolites [U1416]. These are concentric spherical structures (think a child's toy marble), often forming around nuclei, such as pebbles or shell fragments. Besides these specific bacterial communities, shallow lagoonal areas also hosted distinct and rather monotonous arthropod faunas.

Variable lagoonal, estuarine, fluvial and lacustrine, and nearshore marine environments often preserve distinctive fossils of eurypterids, commonly also known as sea scorpions (Fig. 14.4) [U1417]. These arthropods were present from the Ordovician to the Permian, but their diversity peaked in the latest Silurian (Přidolian; Tetlie 2007). The majority of eurypterid fossils have been recorded from the paleocontinents of Baltica and Laurentia, and it has been assumed that most of these taxa were not able to cross vast oceans. There is always at least one exception. In eurypterids, that one exception is a group called the pterygotoids (which also involves the largest arthropods ever discovered; Braddy et al. 2008). These fossils have been found throughout the world intimating that this group was able to traverse ocean basins (Tetlie 2007). As many eurypterids were armed with well-developed grasping front appendages, they probably occupied a high trophic level as predators. And, predation may not have been restricted to aquatic environments. It is also assumed that eurypterids were capable of occasional visits to dry land (Braddy 2001). Frequently, eurypterid fossils are associated with another group of arthropods, ostracod-like animals called leperditiids [U1417]. Owing to their bivalved carapace, these distinctive arthropods superficially resemble ostracods, but their soft body inside the carapace was most likely anatomically different (Vannier et al. 2001). Leperditiids are often preserved as monospecific assemblages in very shallow marginal marine habitats, including tidal and reef flats, lagoons, and embayments that were often exposed to environmental stress. Hence, it is thought that leperditiids and eurypterids may have possessed specific adaptations, like resistance to desiccation and osmoregulation, to occupy these environments.

14.3.2 Reefs

As the Silurian and Devonian were periods of remarkable sea-level highstands, vast epicontinental oceans, and global greenhouse climates, these times are notable for their geo-

Fig. 14.4 Eurypterid *Eurypterus tetragonophthalmus* (Courtesy of Oive Tinn)

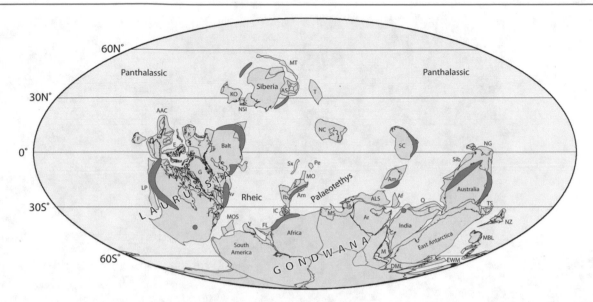

Fig. 14.5 Global distribution of Middle–Late Devonian reefs. Paleogeography after Torsvik and Cocks (2013a), reef distribution after Kiessling et al. (1999), Torsvik and Cocks (2013b), Cocks and Torsvik (2011)

graphically extensive reefs (Fig. 14.5). Physically immense Silurian–Devonian reef complexes have been documented in many parts of the world, and reef sediments are very apparent today in Canada [U1417] and Sweden [U1418]. In the Phanerozoic reef database compiled by Kiessling et al. (1999), the global number of reefs in the Wenlockian (433–427 Ma, Silurian) and Givetian/Frasnian (383–372 Ma, Devonian) per million years is outnumbered only by the abundance of Miocene (23–5 Ma) reef structures. It has been calculated that the major tropical barrier-reef belts stretched more than 2000 km in length, and these buildups are considered to be among the most extensive reef constructions in Earth history. Some estimates are that their surface area was up to ten times of those in the modern oceans (Copper and Scotese 2003). Only the 2100 km long Great Barrier Reef, off the northeastern coast of Australia, is comparable with the reefs formed during the Devonian reef optimum.

There are many definitions of reefs. Here, we regard reefs as biogenic, carbonate structures, raised above the surrounding sea floor (Stanley 2001). Sedimentologically, reefs are described by their massive appearance, an absence of clear bedding in the reef core, and a skeletal reef building framework consisting of corals, sponges, calcimicrobes, and algae. Silurian reefs have been documented from a wide range of environments, from lagoons with patch reefs and nearshore fringing structures to mid-shelf buildups (Fig. 14.6) [U1418–U1419] (Copper and Brunton 1991). Reef biotas varied from algae to poor patch-reef faunas in marginal marine settings, from rich open-marine coral–stromatoporoid–bryozoan communities [U1420–1424] to

sparsely fossiliferous lithistid (demosponges) or stromatactid mud-mounds in deeper settings [U1425] (Copper and Brunton 1991). Devonian metazoan reefs were not restricted to a narrow equatorial belt. These complexes reached as far as 40°–50°, or possibly even 60° of latitudes in both hemispheres [U1426] (Copper and Scotese 2003). By the Middle Devonian, reefs at low latitudes of 40° or less were dominated by rugose-and-tabulate corals and stromatoporoid sponges, generally with a significant calcimicrobial component. These calcium-producing microbes acted as binders and encrusters in the reef core. Reefs at higher latitudes (40°–50°) were generally mud-mounds of obscure origin, with rather sparse metazoan cover and those with coral–stromatoporoid frameworks were uncommon.

Reef complexes experienced a Late Devonian extinction at the Givetian–Frasnian boundary (383 Ma) as the result of a major decline in coral, stromatoporoid, and brachiopod diversity (Copper 2002). Subsequently, reefs experienced a second global expansion in the middle Frasnian (c. 377 Ma), but the faunal diversity of these buildups was low. Catastrophic reef decline began in the late Frasnian (c. 373 Ma) lasting probably about one million years. As a consequence of a worldwide elimination of the coral–stromatoporoid reef communities, the surviving patch reefs and reefal mud-mounds were constructed by calcimicrobes, lithistid sponges, green and red algae, and foraminiferans (Fig. 14.7) [U1427]. The surviving Famennian corals were primarily solitary, deep-water forms, which played only a minor role in reef composition. The last true stromatoporoid sponges died out at the Devonian–Carboniferous transition, changing the composition of all younger buildups.

Fig. 14.6 Silurian (Wenlock) limestones at Rönnklint, Gotland, Sweden. The upper part shows an imposing reef structure (Courtesy of Peep Männik)

Fig. 14.7 A "reef curve" showing the volume of Devonian reef structures. Changes in reef diversity during the Eifelian–Givetian and reef crises during the Late Devonian are seen as the expansion and subsequent loss of coral–stromatoporoid reef builders in the late Frasnian, and rise of episodic development of regional mud-mounds and calcimicrobial reefs in the Famennian (After Copper and Scotese 2003)

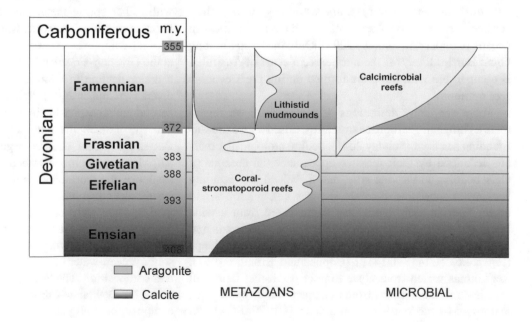

14.3.3 Open Shelf Area

We proceed on our voyage to the area known as the open-marine shelf. At first we are going to the depths of the ocean waters to visit benthic communities—bottom dwelling organisms—that lived on the sea floor as epifauna or below the sediment surface as infauna. Part of these animals were sessile, the others were vagile (free moving) organisms.

14.3.3.1 Bryozoans

Bryozoans were an important component of many reef complexes in addition to corals and calcimicrobes, sponges, and algae. These colonial animals were not just restricted to reefs, but they inhabited shallow shelf areas, as well. Bryozoans [U1428] are rather abundant today, living in both marine shelves and freshwater realms. Colonies of bryozoans grow on hard surfaces, as encrustations on sea weeds or sea shells, or form delicate meshworks. They are sessile, often combining etching and cementation as a means of their attachment to a substrate (Bromley and Heinberg 2006). Due to their calcareous skeleton, they also have a rich fossil record. And, like their relatives, the brachiopods (Taylor and Waeschenbach 2015), bryozoans feed on plankton by means of a specialized structure known as a lophophore.

14.3.3.2 Brachiopods

Brachiopods were a significant component of all Silurian and Devonian benthic communities [U1429]. In fact, these bivalved invertebrates inhabited a rather wide range of facies zones, ranging from continental shelves to basinal slopes; many taxa even lived in the deepest parts of the ocean at abyssal depths. Most often these suspension-feeding animals were attached to the sea floor, either by cementing themselves onto hard surfaces or by a special organ, called a pedicle, sticking out of a special opening in the posterior part of the shell (the foramen). The soft tissues of brachiopods are enclosed in a bivalved shell, which most often consists of the mineral calcite (for rhynchonelliformeans and craniiformeans). In one small group of brachiopods (linguliformeans), the shell is phosphatic instead. Extant brachiopods are rather uniform in their shape and are represented by fewer than 500 species, which comprise only about 5% of all named (more than 12,000 in total) brachiopod species since their appearance in the Paleozoic (Carlson 2016). Brachiopods are known from the early Cambrian (Terreneuvian, Epoch 2) to present. They reached their peak diversity in the Devonian, with the end-Permian extinction strongly reducing their taxonomic, morphological, and ecological diversity. Brachiopods were likely the most common macro-invertebrates in the Paleozoic coastal seas, and their very high population density may have supported formation of shell banks and coquinas [U1430]. One example is the earliest Silurian shell bank in Estonia that was deposited in the high-energy shoal environment of the Baltoscandian sea [U1412]. The zone of coquina beds, consisting of valves of the pentamerid brachiopod *Borealis borealis*, over which these beds are distributed is about 30 km wide, extending up to 200 km from east to west, and running parallel to the ancient coast line. The unit attains a thickness of up to 13.5 m (Nestor and Einasto 1997)!

14.3.3.3 Mollusks

The Mollusca is the second largest animal phylum after arthropods. Today, mollusks include some of the best known invertebrates including bivalves, gastropods, octopuses, squids, and their relatives.

Bivalves [U1431] appeared in basal Cambrian deposits, but remained rather rare and insignificant members of marine-invertebrate faunas. They lived as shallow burrowers or surface crawlers on microbially bound sediment surfaces (Cope and Kriz 2013). It was not until the latest Permian (Clapham and Bottjer 2007) when bivalves began to prosper and filled ecological niches that had been colonized previously by brachiopods. Although they existed as a rather small group in the early Paleozoic, they developed several anatomical innovations for success. These include, for instance, structures called siphons, which helped to burrow the animal deeper in the sediment while still feeding from the overlying water column. This innovation facilitated their diversification. Silurian bivalves are characterized by the development of numerous new forms of free-burrowing and epibyssate (anchored to a substrate with a specialized tissue) forms.

Today, gastropods [U1432] are a very common part of many invertebrate communities that thrive in marine and coastal environments. Snails also abound in freshwater habitats and taxa are adapted to do well on dry land. Recent mollusks come in a high variety of colors and forms. Their soft tissues can be enclosed with a calcareous (aragonitic) shell, whereas those taxa without a shell boast an infinite array of shapes and appearances. The earliest gastropods evolved in the Cambrian, were present in many Silurian–Devonian communities, and were probably herbivores or fed on detritus (detritivores) (Rohr 1999). The Silurian–Devonian gastropod communities are preserved in many depositional settings. They are found in open-marine high-energy clastic deposits, attached to hard substrates, and in soft bottom sediments characteristic of lagoonal areas with low energy and open circulation. Fossil snails also are preserved in brackish estuarine deposits as well as reefs. However, gastropods were only a minor component in marine benthic communities, rarely composing more than 1% of the fauna (Rohr 1999).

Cephalopods were an important part of the Silurian marine communities, and underwent a major radiation during the Devonian nekton revolution (Fig. 14.8). As cephalopods have rather variable anatomies and their fossils have been found in various settings, it is assumed that they had rather different

Fig. 14.8 Examples of common Silurian and Devonian cephalopods. (**a**) *Temperoceras ludense*. (**b**) *Anetoceras* (*Erbenoceras*) sp. (**c**) *Mimagoniatites fecundus* (Courtesy of Christian Klug)

modes of life. Taxa are interpreted to have been planktonic, demersal, or nektonic (Klug et al. 2010). For instance, orthocerids [U1433], a group distinguished for their straight (orthoconic) or slightly curved shells, often are preserved in black shales deposited in the deepest parts of ocean basins. Many of their anatomical characteristics indicate that these animals probably were not very efficient swimmers, and it is assumed that they migrated primarily vertically and/or drifted passively in the water column (Klug et al. 2010). Nautiloids with coiled or curved shells, which are preserved in shallow water deposits, have been considered as demersal organisms (Klug et al. 2010). Many cephalopods, including ammonoids and coiled nautiloids [U1434] with strong muscle-attachment structures and buoyancy devices (Klug et al. 2010), were probably nektonic. In addition to the ammonoids, the coleoids (squids and octopods), or at least their ancestors, evolved in the Devonian.

14.3.3.4 Echinoderms

Echinoderms [U1435–U1436] were an important and diverse part of early Paleozoic marine ecosystems. This phylum is characterized by radial (typically five-parted or pentamerous) symmetry, a distinctive mesodermal skeleton formed of calcitic plates, and a water-vascular system of reservoirs and canals, which act in the movement of tube feet. Considering large variations in their anatomy and morphology, echinoderms are usually divided into five major groups: stalked Crinozoa, globoid to discoid Echinozoa, star-shaped Asterozoa, stalked Blastozoa, and asymmetrical Homalozoa. Of these, the latter two existed in the Paleozoic only and now are extinct. The stalked echinoderms, crinoids, or "sea lilies" colonized the sea floors, were filter feeders, and dominated the higher-level benthos in the Paleozoic. Asteroids and ophiuroid echinoderms were also common, and it has been argued that unlike modern counterparts, many of them were deposit feeders. The members of this phylum are exclusively marine and mostly stenohaline (only able to tolerate a narrow salinity range); it seems they have never evolved a life strategy to inhabit freshwater environments. However, in the marine realm, they lived at all depths from the littoral zone to deep oceanic abyssal plains, comprising members of both vagile and sessile benthos, and even a few planktonic representatives are known.

14.3.3.5 Arthropods

Arthropods constitute a major and highly diverse group of invertebrates throughout the Phanerozoic (see also Box 14.3). Whereas insects comprise the most common and the largest group among arthropods, today, the largest group in the Paleozoic was, perhaps, the trilobites and their kin. These

Box 14.3: Ostracods [U1439]

These arthropods are tiny crustaceans that inhabit a wide range of marine, non-marine, and moist terrestrial habitats, including lakes. They are found from intertidal zones to abyssal depths. The ostracod body is enclosed in a small bivalved calcitic shell (carapace), which is readily preservable in sediments. As such, this feature has made the group the most abundant arthropod in the fossil record. They are known from the earliest Ordovician (Hou et al. 1996) and continue into the present, with about 33,000 extant and fossil species (Horne 2002). Today, the majority of ostracod species are marine and benthic. They crawl on the sediment or burrow in the upper sediment layer, but many species are pelagic. It is supposed that the majority of ostracods in the early Paleozoic, especially the group called paleocopes with their thick and rather unstreamlined carapaces, inhabited waters near the bottom of the ocean and were mostly benthic crawlers or swimmers (Siveter 1984).

Fig. 14.9 Examples of common Silurian and Devonian trilobites. (**a**) *Calymene blumenbachi* (photo: Geoscience collections of Estonia (http://geokogud.info/specimen_image/741) (**b**) *Cyphaspis ceratopthalma* (Courtesy of Brigitte Schoenemann)

forms make up the largest portion of the phylum. Trilobites were a highly successful group (Fig. 14.9) [U1437]. These arthropods appeared rather suddenly, early in the Cambrian. Their first fossil evidence has been dated to around 522 million years ago, and the last species went extinct during the End-Permian Mass Extinction, about 252 million years ago. Thus, they inhabited the seas and oceans of our planet for about 270 million years. Over 15,000 species and 5000 different genera of trilobites have been named (Fortey 2001). The majority of these animals were just a few centimeters long, the tiniest no more than a few millimeters, whereas the largest species reached a length of 72 cm in the Ordovician (Gutierrez-Marco et al. 2009). As trilobites lack close extant relatives, understanding their mode of life is not an easy task. We could assume, however, that variations in the construction of their exoskeleton provide hints that trilobites had rather diverse modes of life. There were tiny pelagic trilobites, probably feeding on phytoplankton and zooplankton, and there were much larger species, perhaps predators, feeding on other crustaceans. The majority of trilobites were bottom dwellers, living on or in the seafloor. They inhabited shallow shelf areas, reef habitats, and deep waters beyond the reach of light where some forms lack eyes. Others lived near the surface of the ocean. Many of the trilobites have left intricate behavioral traces on the surfaces of ancient ocean sediments, which have been of great help for understanding their life habits and ecologies. Many trilobites were equipped with a complex visual system (Schoenemann and Clarkson 2017), which does not have a true equivalent in the modern world [U1437–1438]. Just like the chemical composition of their entire exoskeleton, the lenses of their compound eyes are made of calcite, a simple mineral of calcium carbonate ($CaCO_3$). Their eye morphologies were diverse, much the same way as their body morphologies and lifestyles. Whereas many trilobite taxa had some sort of vision, viewing the world through eyes made up of a single to more than 700 lenses, a few species could boast eyes made up of more than 15,000 tiny, closely packed lenses.

14.3.3.6 Chordates

Conodonts are chordates (related to animals with a spinal cord) and are mostly known from apatitic, toothlike microfossils [U1440] preserved in marine rocks from the early Cambrian to the Triassic. To date, there have been fewer than a dozen whole-body fossils of conodont animals found in the fossil record. These specimens reveal that the organism was a small, slender eel-like animal with anatomical structures that have been interpreted as a notochord, chevron-shaped muscles, fins, and eyes (Fig. 14.10). The apatitic "teeth" were organized into a feeding apparatus consisting of 15 or 19 elements located in the head. Debate continues as to whether these animals are hemichordates or chordates (Donoghue et al. 2000), and their phylogenetic position

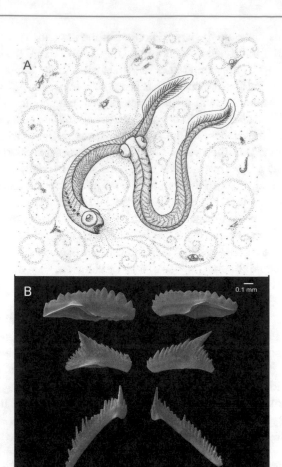

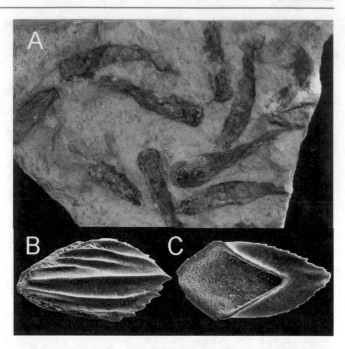

Fig. 14.11 Jawless fishes (agnathans) diversified in the Silurian. (**a**) Agnathan thelodont *Phlebolepis elegans*. (**b**) External view of a thelodont scale (http://geokogud.info/specimen_image/17476). (**c**) Internal view of a thelodont scale (http://geokogud.info/specimen_image/17474) (Courtesy of Tiiu Märss)

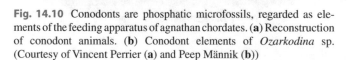

Fig. 14.10 Conodonts are phosphatic microfossils, regarded as elements of the feeding apparatus of agnathan chordates. (**a**) Reconstruction of conodont animals. (**b**) Conodont elements of *Ozarkodina* sp. (Courtesy of Vincent Perrier (**a**) and Peep Männik (**b**))

remains unclear (Turner et al. 2010). It is also not clear if conodonts were adapted to mostly a pelagic, free-swimming mode of life or a benthic to nektobenthic lifestyle. The group remains enigmatic.

All major Paleozoic vertebrate groups existed in the oceans of the Silurian and Devonian. These animals show morphological adaptations to a wide range of habitats, from freshwater and brackish environments to marginal and fully marine epicontinental shelves. One group appears to have been dominant.

Most vertebrate communities were dominated by a variety of jawless taxa, known as agnathans, during the Silurian. A peculiar group of fish-like animals called thelodonts [U1441] are mostly known based on their characteristic bony scales, which varied in size (0.5–2 mm) and morphology, depending on which part of the animal the scale developed. Although most of their fossil record consists of these scattered scale remains, a few complete fossils covered in thousands of preserved scales provide insight into their body form and let us speculate about their lifestyles (Fig. 14.11). Commonly, the size of these animals was about 15 cm, but a few species attained a length of 1 m (Long 2011). Many agnathans, as well as a similar group known as osteostracans [U1442], had distinctive armored headshields, each with a diagnostic shape and surface sculpture. The bony shield was perforated with two small eye holes and a nasal opening. The sides of the head shield had specific areas that may have served a sensory function; many osteostracans have a similar structure on top of the head shield. It is assumed that some osteostracans (e.g., *Tremataspis*) were largely bottom dwelling animals, moving only by the flips of their short tail. Other species were probably more mobile, capable of fast movement (Long 2011). Among the agnathans, the osteostracans are thought to be the closest to the evolutionary lineage that led to jawed vertebrates (Sansom 2009). The

Fig. 14.12 *Dunkleosteus terrelli*, an arthrodire placoderm from near Cleveland, Ohio, USA (Courtesy of Alex Ritchie)

other agnathan group, the heterostracans (which means "different shields") [U1443], can be identified based on the presence of several bony plates of different size and morphology, which typically covered the top and underside of their head.

The most diverse group of Devonian "fishes" (a paraphyletic group; see Box 7.3, Romano in Chap. 7) was the placoderms. Much like modern fishes, they were part of the global fauna and lived in all habitable freshwater and marine environments. Although they originated in the early Silurian, they had almost completely replaced the diverse fauna of Silurian jawless taxa by the Devonian. Their known evolutionary history spans around 70 million years. Their most impressive feature was their body armor, which consisted of strong and robust dermal bones. Many placoderms, like antiarchs [U1444], had a dorsoventrally compressed body and probably a benthic lifestyle. Placoderms also included the group called arthrodires, a member of which was the largest known predator of the time, *Dunkleosteus*, with fearsome blade-like jaws (Fig. 14.12) [U1445]. This group was shown to have had live birth of their young (vivipary), making them the oldest vertebrate with this reproductive strategy (Long et al. 2009). And finally, late Silurian and Devonian seas (Min and Schultze 1997) could boast a rich fauna of sarcopterygian or lobe-finned fishes [U1446]—the same strange-looking creatures, which today are represented by the iconic "living fossil," *Latimeria*. Anatomically, sarcopterygians share many common features with tetrapods (including humans; see also Romano in Chap. 7), the fleshy limbs being the most important among them.

14.3.4 Pelagic Zone

14.3.4.1 Acritarchs and radiolarians

Photosynthetic unicellular plants form an important component at the base of the marine trophic chain, serving as food for zooplankton, various groups of suspension feeders, and detritus feeding organisms. By their floating nature, phytoplankton groups had global distributions during most of the Paleozoic, forming diverse and rich assemblages. In the fossil record, phytoplankton are mostly represented by microscopic fossils called acritarchs (Fig. 14.13) [U1447], an informal group of organic-walled microfossils. The majority of acritarchs are interpreted as reproductive structures (cysts) of marine, mainly phytoplanktonic unicellular algae (Servais et al. 2016). A number of them show morphological similarities with organic-walled cysts of extant dinoflagellates, and they also display biogeographic distribution patterns that are comparable with that group. However, as many of the organic-walled fossils do not show the characteristic features of living dinoflagellates, their biological affinity remains uncertain. The diversity and abundance of acritarchs was not constant along the onshore to offshore gradient; the most diverse and abundant assemblages have been documented in mid-shelf environments, with a decline in diversity towards nearshore settings on one hand, and outer shelf/basinal settings on the other (Molyneux et al. 2013).

Farther distant from the continents, in the facies representing sedimentation in deeper water, we notice microscopic fossils of radiolarians [U1448]. Radiolarians are found as an important part of planktonic communities throughout the modern oceans. They are protozoa, unicellular eukaryotic organisms with an imposing silica-based skeleton, the diameter of which does usually not exceed 10–20 microns. Today, their mostly spherical or conical skeletons vary notably in construction and ornamentation. Fossil radiolarians are interpreted on the basis of their living counterparts. They bloomed under particularly favorable conditions, which resulted in the accumulations of cherts and siliceous limestones in the deep ocean basins.

14.3.4.2 Graptolites

Graptolites [U1449], which take their name from Greek, meaning "writings on the rock," were one of the most abundant pelagic animal groups during the Paleozoic. Interpreted as clonal, colonial hemichordates, they originated in the Cambrian, had their best days during the Ordovician, and suffered significant biodiversity losses in the end-Ordovician extinction. Almost all are extinct now, with a single genus, *Rhabdopleura*, regarded as the last living graptolite (Mitchell et al. 2013). This sessile animal in today's oceans could serve as a model for understanding the graptolite ani-

Fig. 14.13 Silurian and
Devonian plankton. (**a**)
Acritarch *Helios aranaides*.
(**b**) graptolite *Stomatograptus
grandis*. (**c**) Radiolarian
*Stigmosphaerostylus
vishnevskayae*. (**d**)
Chitinozoan *Ancyrochitina
longispina* (http://
geocollections.info/
file/106173). (Courtesy of
Claudia Rubinstein (**a**), Anna
Kozlowska (**b**), Olga Obut
(**c**))

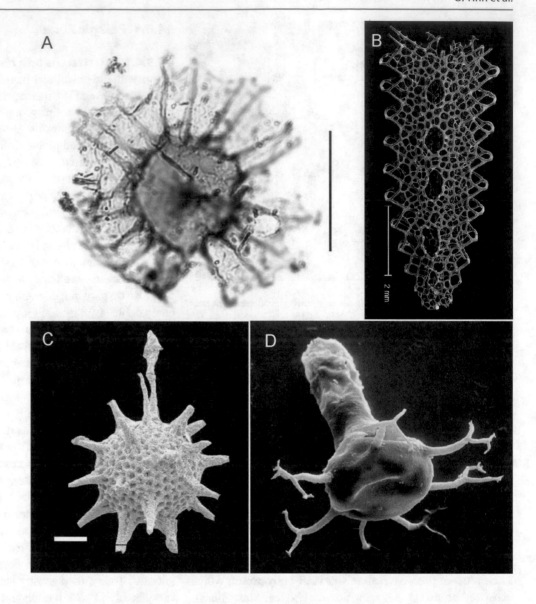

Fig. 14.13 Silurian and Devonian plankton. (**a**) Acritarch *Helios aranaides*. (**b**) graptolite *Stomatograptus grandis*. (**c**) Radiolarian *Stigmosphaerostylus vishnevskayae*. (**d**) Chitinozoan *Ancyrochitina longispina* (http://geocollections.info/file/106173). (Courtesy of Claudia Rubinstein (**a**), Anna Kozlowska (**b**), Olga Obut (**c**))

mal (zooid) and the biology of the extinct taxa, although it may not necessarily reflect their soft-body anatomy. Similar to the rhabdosome of *Rhabdopleura*, the colonies of graptolites were made up of long tubes (thecae) that housed microscopic animals (zooids). Each zooid was equipped with a distinctive feeding organ, the same two-armed structure, called a lophophore, found in brachiopods and bryozoans.

There were two main groups of graptolites: dendroids (bushy or fan-shaped forms) and graptoloids (superficially simpler forms with relatively few branches). Dendroids, including the living *Rhabdopleura*, are all interpreted to have been benthic organisms, living attached to some firm surface, like rocks, compacted or dewatered sediments, or skeletons of other organisms (Maletz 2015). Unfortunately, graptoloid graptolites do not have any modern ecological analogs. However, the morphology of these fossils and the

features of the sediments in which they are preserved indicate that graptoloids evolved a free-living planktonic lifestyle, filtering the water column to gather food. Some enigmatic fossils have been questionably interpreted as microscopic egg capsules (Box 14.4). Most graptoloid colonies were between 1 and 10 cm in length, but a few gigantic forms have also been reported, which reached more than a meter in diameter. A peculiar group of graptoloid graptolites, called retiolitids, show an extreme reduction of the skeleton, where the thecae are represented by a network of narrow struts and clusters. Thanks to their planktonic lifestyle, many graptoloid graptolites show a global distribution pattern, and their rapid evolution allows for correlation of Silurian–Devonian marine rocks over long distances. However, there were also a few graptolites that do show some provincialism and can be used only locally for the purposes of correlation.

Box 14.4: Chitinozoans

On our diving trip through the Paleozoic oceans, we also catch a glimpse of a number of minuscule enigmatic objects that do not reveal their biological affinity very easily [U1450]. Keen eyes notice tiny organic-walled structures, most of which are 50–250 microns in diameter, and occur in a range of elongate to spherical shapes. Some show an external ornamentation in the form of hairs or projections, some are hooked to algae or bottom-dwelling animals with these appendages. Most occur as isolated objects, but you could also find them joined to each other in long chains or even clusters. Many of these forms show a flat "base," some have a convex bottom. A few of these puzzling objects show an opening at one end, thus revealing their hollow nature. As their wall material seems to be similar to chitin, an organic molecule that makes up the cuticle of most arthropods, the fossils were named chitinozoans by the earliest researchers. However, as there is not much proof that these fossils could be the remains of animal carcasses, it is usually suggested that they represent egg capsules of unknown animals (Traverse 2008). These might have been either arthropods or even graptolites. It is also important to mention that these forms existed from the Cambrian to the end-Devonian. Ironically, nowadays, these enigmatic objects have proved to be valuable tools in biostratigraphy and for correlation of rocks across continents.

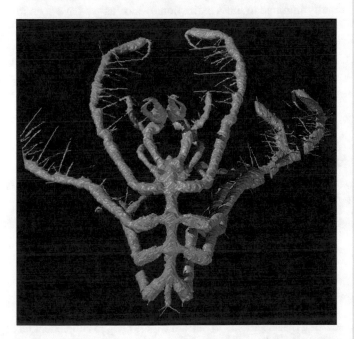

Fig. 14.14 A digital reconstruction of a Silurian sea spider, a pycnogonid *Haliestes dasos* from the Herefordshire Lagerstätte, UK. Courtesy of Derek J. Siveter, Mark D. Sutton, Derek E. G. Briggs and David J. Siveter

14.3.4.3 Pelagic Arthropods

Zooplankton of arthropods are common in today's oceans, but the fossil record of pelagic arthropods is sparse. The fossil record commonly consists of those taxa that possessed mineralized skeletons during their lifetime. Fossils of myodocopes [U1451], one of the two major groups of ostracods, often are preserved in environments that record dysoxic/anoxic conditions (e.g., black shale facies) where many benthic faunal elements are absent (Perrier et al. 2011). Other delicate arthropods are only preserved in Fossil Lagerstätten (Box 14.5). It is assumed (Perrier and Siveter 2013) that the early representatives of this group were largely nektobenthic, with most species living in benthic communities on well-oxygenated marine shelves in the

Box 14.5: Herefordshire Lagerstätte

There are very few Fossil Lagerstätten of Silurian age. The 425-million-year-old Herefordshire locality in the UK is one in which exceptionally preserved fossils from this period are found. The Herefordshire biota exhibits a diverse fauna [U1452], including a number of arthropods, polychaete worms, sponges, echinoderms, and many, as yet, unnamed taxa from a relatively deep-water outer shelf to slope environment. The preservation of the Herefordshire fossils is unique. They are preserved as three-dimensional calcitic void infills in carbonate concretions, which were deposited in fine-grained volcanic ash (Briggs and Siveter 1996). Most astounding about these fossils is their incredibly detailed preservation, allowing the study of their finest anatomical structures (Orr et al. 2000). Here, we present just a few fossils unearthed from the Herefordshire Lagerstätte to demonstrate how unique this window is to the Silurian marine biota.

Colymbosathon ecplecticos (the name can be translated as "the astounding swimmer with a large penis") is a myodocopid ostracod, classified in the extant family of Cylindroleberididae (Siveter et al. 2003). This fossil, morphologically very similar to its extant relatives, proves that the ostracod family, as far as its soft part anatomy goes, has remained incredibly static over the course of its evolutionary history.

Another extraordinary fossil is a sea spider, a pycnogonid *Haliestes dasos* (Siveter et al. 2004) (Fig. 14.14). This animal is a delicate marine arthropod without a biomineralized cuticle. It is thought that this species inhabited similar deep marine environments as do its extant relatives, and its mode of life was also relatively similar. The sea spiders have a really sparse fossil record, but phylogenetic analyses demonstrate that *Haliestes* belongs in, or near, the pycnogonid crown group, which may have originated by the Silurian period (Siveter et al. 2004).

The most controversial among the fossils presented here is *Enalikter aphson* (Siveter et al. 2015). This animal has been described as a megacheiran "great appendage" arthropod, but has also been interpreted as a polychaete arthropod (Struck et al. 2015).

early Silurian (Llandovery–Wenlock). However, they seem to have undergone an ecological shift at some point during late Wenlock or Ludlow, when they started to occupy pelagic niches.

14.4 Extinctions

The Silurian was a short and relatively stable period (25 My), without major crises, especially when compared with the younger Devonian period (60 My). During the Silurian, though, several minor extinction events are identified in the invertebrate-fossil record. They are called the Ireviken, Mulde, and Lau events, all named after localities found on Gotland Island, Sweden [U1453]. The oldest crisis is the Ireviken event at the Llandovery–Wenlock boundary (c. 433 Ma), which was a distressing time for many pelagic and hemipelagic organisms. Conodonts, graptolites, and trilobites experienced declines in biodiversity, while shallow water reefs were barely affected (Bond and Grasby 2017). The extinction event has been attributed to increasing anoxia, which first is recognized in deeper ocean sediments and spread, step-by-step, to shallow shelf areas. The younger Mulde and Lau events resulted in minor extinctions in the conodonts and graptolites, as well as in acritarchs, chitinozoans, and radiolarians. These extinctions occurred during the Ludlow epoch (427–423 Ma). Both events are associated with major positive excursions in the stable $\delta^{13}C$ isotope record, attributed to the widespread deposition of carbon-rich sediments. These deposits are associated with a eustatic fall in sea level. Currently, there are no convincing explanations available for these Silurian biotic crises.

The Devonian Period witnessed many marine extinctions of varying magnitudes [U1454]. Many of these are marked by the appearance of black shale beds in limestone sequences, including those in central Europe, North America, North Africa, Australia, and China. These widespread organic-rich black shales are thought to represent anoxic/hypoxic events during times of high sea level. Elevated extinction rates are found for many pelagic faunal groups, including conodonts and cephalopods, as well as many benthic taxa, including brachiopods and trilobites. These occur at the stratigraphic

position of every black shale horizon. But, there is one Devonian extinction event that eclipses all others.

To understand the importance of the greatest of the Devonian extinctions, we can look to the fossil record of lampreys and hagfishes. These modern jawless fishes (agnathans) are just a mere shadow of their former glory of diversity and abundance, because their kin were almost exterminated during the first and the greatest of the Devonian extinction events, the Kellwasser crisis. This event, which occurred at the Frasnian–Famennian transition (372 Ma), is considered one of the Big Five extinction events in the Phanerozoic. Many marine faunal groups suffered great losses in biodiversity, and the largest reef systems ever to have existed on our planet were severely impacted. The placoderms had become so diverse and abundant during the Devonian that the entire geologic period is sometimes called "the age of fishes." But this group first experienced a severe biodiversity decline at the Frasnian–Famennian boundary, and then underwent a rapid recovery that continued for several million years. Ultimately, all members of the group, both small and large, were unable to survive the Hangenberg event at the end of the Devonian.

The Devonian terminated with the Hangenberg anoxic event (359 Ma), which is associated with extinctions that mark the Devonian–Carboniferous transition (Fig. 14.15). Named after the black organic rich shale in Germany, the Hangenberg event can be found globally indicating that it was widespread across all ocean basins. An estimated 21% of marine genera and 16% of marine families may have been lost during this time (Bond and Grasby 2017). It has been argued that fishes, throughout their history, have proven virtually immune to mass extinction events, as we have seen in the Permian–Triassic and Cretaceous–Paleogene mass extinctions (see Romano in Chap. 7). The Late Devonian is the only major exception when fishes are significantly affected [U1455]. The Devonian extinctions have been attributed to a variety of causes including increased tectonic activity, the expulsion of a large volume of igneous magmatism, climate change, oceanic overturns, and eustatic fluctuations. Some researchers also have noted a possible relationship between the rise of woody forests and marine extinctions. In this coupled model, the increasing diversity and abundance of deeper rooted trees across coastal floodplains during the Late Devonian led to widespread soil formation, with enhanced physical and chemical weathering. A postulated increase in nutrient runoff, in the form of organic compounds, resulted in eutrophication in the Devonian seas and, ultimately, led to increased primary productivity. Increased marine primary productivity led to higher rates of organic decay stripping oxygen from the water column, promoting marine anoxia in bottom waters (Algeo and Scheckler 1998) which impacted benthic invertebrates.

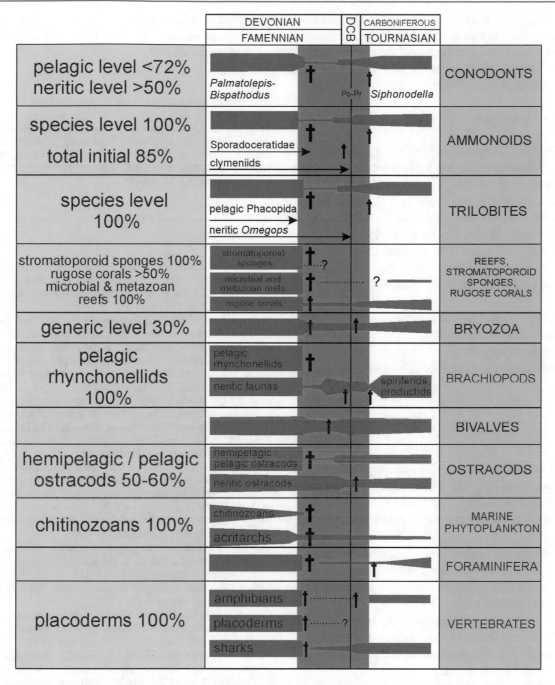

Fig. 14.15 Overview of fossil groups affected by the latest Devonian Hangenberg Crisis. Red bars denote radiations, extinctions, and diversity changes. Crosses denote extinctions during the event. The width of the bars shows the estimated relative abundance of taxa (not to scale). Note that the state of knowledge is different for the particular taxa (After Kaiser et al. 2016)

14.5 Conclusions

We have now finished a trip through 85 million years, during which time we have seen how our planet's seas and oceans appeared and changed from 445 to 359 million years ago. This is a very long interval of time, even in the terms of the geological time scale. A lot of things were altered from the Silurian to the Devonian. For instance, we began our trip from the near-equatorial paleocontinents of Baltica, Avalonia, and Laurentia, each of which had their own distinctive features and associated faunas. By the end of our trip, these land masses had assembled to form a large Laurussian paleocontinent. The marine fauna, especially vertebrates, went through substantial evolutionary change. At the beginning of our trip, we met small agnathans, or jawless fishes, including the thelodonts and osteostracans. By the time we finished our trip, we saw the rise in the diversity of lobe-finned fishes and the peak in the abundance of the placoderms. We have seen the rise and demise of the most extensive reef systems on Earth—neither seen before nor afterwards. And we have witnessed major calamities and the loss of the planet's biodiversity several times. All of these events have shaped the planet we now inhabit and know, today.

For Deeper Learning

- Cyanobacteria communities precipitating calcite or aragonite: Garcia-Pichel et al. 2004; Dupraz et al. 2009; Bosak et al. 2013.
- Dasycladales: Mastik and Tinn 2015; Tinn et al. 2015
- Geological data set and paleogeographic maps of the early Paleozoic world: Cocks et al. 1997; Cocks and Torsvik 2007, 2011, 2013; Torsvik and Cocks 2013b.
- Vertebrate groups and their morphological adaptations: Zigaite et al. 2013; Zigaite and Blieck 2013; Marss et al. 2014.
- Invertebrates: Brusca and Brusca 2003
- Radiolaria: Tetard et al. 2017
- Eurypterids: Braddy 2001
- Late Ordovician–early Silurian glaciation: Finnegan et al. 2011

Questions

1. Paleozoic marine ecosystems were very different from the ones we have, today. Name two major groups that went extinct since then.
2. What is GOBE and why is this event significant in Earth history?
3. How might the Hirnantian glaciation in the Late Ordovician impacted marine biodiversity?
4. Explain why plate tectonics and moving continents have an impact on marine life.
5. Explain how global climate change, beginning in the Ordovician into the Devonian, impacted marine life.
6. What are the four dominant modes of marine-invertebrate life strategies in the oceans?
7. Shallow, intertidal zones preserve a unique microbial structure in Paleozoic lagoons called stromatolites. How do these biosedimentary structures form?
8. How might the definition of a reef differ from a sedimentological and biological perspective?
9. Characterize the Silurian–Devonian benthic fauna of the open-marine shelves.
10. What makes trilobites fascinating?
11. Name three features that distinguish agnathan and osteostracan fishes from those in the modern oceans.
12. How can graptolites, chitinozoans, and radiolarian be used to correlate Paleozoic marine rocks?

13. What relationship(s) exists between extinction events and associated black shales in shaping marine biodiversity?

Acknowledgments Support from the Estonian Research Agency projects PUT 1484, IUT 20-34 and PRG 836 is deeply acknowledged. Many thanks are due to the colleagues who provided appropriate images: Andrei Dronov, Anna Kozłowska, Christian Klug, Peep Männik, Tiiu Märss, Viirika Mastik, Olga Obut, Vincent Perrier, Alex Ritchie, Zhan Renbin, Claudia Rubinstein, Brigitte Schoenemann, David J. Siveter, and Muriel Vidal.

References

Algeo TJ, Scheckler SE (1998) Terrestrial-marine teleconnections in the Devonian: links between the evolution of land plants, weathering processes, and marine anoxic events. Philos Trans R Soc B Biol Sci 353:113–128

Berger S, Kaever MJ (1992) Dasycladales. An illustrated monograph of a fascinating algal order. Thieme Verlag, Stuttgart, p 247

Bond DPG, Grasby SE (2017) On the causes of mass extinctions. Palaeogeogr Palaeoclimatol Palaeoecol 478:3–29

Bosak T, Knoll AH, Petroff AP (2013) The meaning of stromatolites. Annu Rev Earth Planet Sci 41:21–44

Braddy SJ (2001) Eurypterid palaeoecology: palaeobiological, ichnological and comparative evidence for a 'mass-moult-mate' hypothesis. Palaeogeogr Palaeoclimatol Palaeoecol 172:115–132

Braddy SJ, Poschmann M, Tetlie OE (2008) Giant claw reveals the largest ever arthropod. Biol Lett 4:106–109

Brezinski D, Cecil CB, Skema VW, Stamm R (2008) Late Devonian glacial deposits from the eastern United States signal an end of the mid-Paleozoic warm period. Palaeogeogr Palaeoclimatol Palaeoecol 268:143–151

Briggs DEG, Siveter DJ (1996) Soft-bodied fossils from a Silurian volcaniclastic deposit. Nature 382:248–250

Bromley RG, Heinberg C (2006) Attachment strategies of organisms on hard substrates: a palaeontological view. Palaeogeogr Palaeoclimatol Palaeoecol 232:429–453

Brusca CR, Brusca GJ (2003) Invertebrates. Sinauer Associates, Sunderland, p 936

Came RE, Eiler JM, Veizer J, Azmy K, Brand U, Weidman CR (2007) Coupling of surface temperatures and atmospheric CO_2 concentrations during the Palaeozoic era. Nature 449:198–1U3

Caputo MV, Melo JHG, Streel M, Isbell JL (2008) Late Devonian and early Carboniferous glacial records of South America. In: Fielding CR, Frank TD, Isbell JL (eds) Resolving the late paleozoic Gondwanan ice age in time and space, vol 441. Geological Society of America, pp 161–174

Carlson SJ (2016) The evolution of brachiopoda. Annu Rev Earth Planet Sci 44:409–438

Clapham ME, Bottjer DJ (2007) Permian marine paleoecology and its implications for large-scale decoupling of brachiopod and bivalve abundance and diversity during the Lopingian (late Permian). Palaeogeogr Palaeoclimatol Palaeoecol 249:283–301

Cocks LRM, Torsvik TH (2005) Baltica from the late Precambrian to mid-Palaeozoic times: the gain and loss of a terrane's identity. Earth Sci Rev 72:39–66

Cocks LRM, Torsvik TH (2007) Siberia, the wandering northern terrane, and its changing geography through the Palaeozoic. Earth Sci Rev 82:29–74

Cocks LRM, Torsvik TH (2011) The Palaeozoic geography of Laurentia and western Laurussia: a stable craton with mobile margins. Earth Sci Rev 106:1–51

Cocks LRM, Torsvik TH (2013) The dynamic evolution of the Palaeozoic geography of eastern Asia. Earth Sci Rev 117:40–79

Cocks LRM, McKerrow WS, vanStaal CR (1997) The margins of Avalonia. Geol Mag 134:627–636

Cope JCW, Kriz J (2013) The lower Palaeozoic palaeobiogeography of Bivalvia. In: Harper DAT, Servais T (eds) Early Palaeozoic biogeography and palaeogeography. Geological Society of London, London, pp 221–241

Copper P (2002) Reef development at the Frasnian/Famennian mass extinction boundary. Palaeogeogr Palaeoclimatol Palaeoecol 181:27–65

Copper P, Brunton F (1991) A global review of Silurian reefs. In: Bassett MG, Lane PD, Edwards D (eds) The murchison symposium: proceedings of an international conference on the Silurian System. Special Papers in Palaeontology, vol 44. pp 225–259

Copper P, Scotese CR (2003) Megareefs in middle Devonian super-greenhouse climates. Geological Society of America Special Paper, vol 370. pp 209–230

Diaz-Martinez E, Grahn Y (2007) Early Silurian glaciation along thewestern margin of Gondwana (Peru, Bolivia and northern Argentina): palaeogeographic and geodynamic setting. Palaeogeogr Palaeoclimatol Palaeoecol 245:62–81

Donoghue PCJ, Forey PL, Aldridge RJ (2000) Conodont affinity and chordate phylogeny. Biol Rev 75:191–249

Dupraz C, Reid RP, Braissant O, Decho AW, Norman RS, Visscher PT (2009) Processes of carbonate precipitation in modern microbial mats. Earth Sci Rev 96:141–162

Finnegan S, Bergmann K, Eiler JM, Jones DS, Fike DA, Eisenman I, Hughes NC, Tripati AK, Fischer WW (2011) The magnitude and duration of late Ordovician-early Silurian glaciation. Science 331:903–906

Fortey RA (2001) Trilobite systematics: the last 75 years. J Paleontol 75:1141–1151

Garcia-Pichel F, Al-Horani FA, Farmer JD, Ludwig R, Wade BD (2004) Balance between microbial calcification and metazoan bioerosion in modern stromatolitic oncolites. Geobiology 2:49–57

Gutierrez-Marco JC, Sa AA, Garcia-Bellido DC, Rabano I, Valerio M (2009) Giant trilobites and trilobite clusters from the Ordovician of Portugal. Geology 37:443–446

Haq BU, Schutter SR (2008) A chronology of Paleozoic Sea-level changes. Science 322:64–68

Horne DJ (2002) Ostracod biostratigraphy and palaeoecology of the Purbeck Limestone Group in southern England. Spec Pap Palaeontol 68:53–70

Hou X, Siveter DJ, Williams M, Walossek D, Bergstrom J, Smith MP, Thomas AT (1996) Appendages of the arthropod Kunmingella from the early Cambrian of China; its bearing on the systematic position of the Bradoriida and the fossil record of the Ostracoda. Philos Trans R Soc Lond Biol Sci 351:1131–1145

Isaacson PE, Diaz-Martinez E, Grader GW, Kalvoda J, Babek O, Devuyst FX (2008) Late Devonian-earliest Mississippian glaciation in Gondwanaland and its biogeographic consequences. Palaeogeogr Palaeoclimatol Palaeoecol 268:126–142

Joachimski MM, Breisig S, Buggisch W, Talent JA, Mawson R, Gereke M, Morrow JR, Day J, Weddige K (2009) Devonian climate and reef evolution: insights from oxygen isotopes in apatite. Earth Planet Sci Lett 284:599–609

Johnson JG (1970) Taghanic onlap and the end of North American Devonian provinciality. Geol Soc Am Bull 8:2077–2106

Kaiser SI, Aretz M, Becker RT (2016) The global Hangenberg crisis (Devonian–Carboniferous transition): review of a first-order mass extinction. Geol Soc Lond, Spec Publ 423(1):387–437

Kendall RS (2017) The old red sandstone of Britain and Ireland—a review. Proc Geol Assoc 128:409–421

Kershaw S, Crasquin S, Li Y, Collin PY, Forel MB, Mu X, Baud A, Wang Y, Xie S, Maurer F, Guo L (2012) Microbialites and global environmental change across the Permian-Triassic boundary: a synthesis. Geobiology 10:25–47

Kiessling W, Flugel E, Golonka J (1999) Paleoreef maps: evaluation of a comprehensive database on Phanerozoic reefs. AAPG Bull 83:1552–1587

Klug C, Kroger B, Kiessling W, Mullins GL, Servais T, Fryda J, Korn D, Turner S (2010) The Devonian nekton revolution. Lethaia 43:465–477

Long J (2011) The rise of fishes: 500 million years of evolution. The John Hopkins University Press, Baltimore, p 224

Long J, Trinajstic K, Johanson Z (2009) Devonian arthrodire embryos and the origin of internal fertilization in vertebrates. Nature 457:1124–1127

Maletz J (2015) Graptolite reconstructions and interpretations. Palaeontol Z 89:271–286

Marss T, Afanassieva O, Blom H (2014) Biodiversity of the Silurian osteostracans of the East Baltic. Earth Environ Sci Trans R Soc Edinb 105:73–148

Mastik V, Tinn O (2015) New dasycladalean algal species from the Kalana Lagerstätte (Silurian, Estonia). J Paleontol 89:262–268

Mastik V, Tinn O (2017) Leveilleites hartnageli Foerste, 1923 (Rhodophyta?) from the Ordovician of Laurentia and Silurian of Baltica: redescription and designation of a neotype. Palaeoworld 26:602–611

Min Z, Schultze HP (1997) The oldest sarcopterygian fish. Lethaia 30:293–304

Mitchell CE, Melchin MJ, Cameron CB, Maletz J (2013) Phylogenetic analysis reveals that Rhabdopleura is an extant graptolite. Lethaia 46:34–56

Molyneux SG, Delabroye A, Wicander R, Servais T (2013) Biogeography of early to mid Palaeozoic (Cambrian-Devonian) marine phytoplankton. In: Harper DAT, Servais T (eds) Early palaeozoic biogeography and palaeogeography. Geological Society Publishing House, Bath, pp 365–397

Munnecke A, Calner M, Harper DAT, Servais T (2010) Ordovician and Silurian Sea-water chemistry, sea level, and climate: a synopsis. Palaeogeogr Palaeoclimatol Palaeoecol 296:389–413

Nance RD, Gutierrez-Alonso G, Keppie JD, Linnemann U, Murphy JB, Quesada C, Strachan RA, Woodcock NH (2012) A brief history of the Rheic Ocean. Geosci Front 3:125–135

Nestor H (1997) Silurian. In: Raukas A, Teedumäe A (eds) Geology and mineral resources of Estonia. Estonian Academy Publishers, Tallinn, pp 89–196

Nestor H, Einasto R (1997) Ordovician and Silurian carbonate sedimentation basin. In: Raukas A, Teedumäe A (eds) Geology and mineral resources of Estonia. Estonian Academy Publishers, Tallinn, pp 192–204

Orr PJ, Briggs DEG, Siveter DJ, Siveter DJ, Cope JCW, Curtis CD (2000) Three-dimensional preservation of a non-biomineralized arthropod in concretions in Silurian volcaniclastic rocks from Herefordshire, England. J Geol Soc Lond 157:173–186

Perrier V, Siveter DJ (2013) Testing Silurian palaeogeography using 'European' ostracod faunas. In: Harper DAT, Servais T (eds) Early Palaeozoic biogeography and palaeogeography. Geological Society of London, London, pp 355–364

Perrier V, Vannier J, Siveter DJ (2011) Silurian bolbozoids and cypridinids (Myodocopa) from Europe: pioneer pelagic ostracods. Palaeontology 54:1361–1391

Rohr DM (1999) Lower Silurian to lower Devonian gastropod communities. In: Boucot AJ, Lawson JD (eds) Paleocommunities—a case study from the Silurian and lower Devonian. Cambridge University Press, Cambridge, pp 132–136

Sansom RS (2009) Phylogeny, classification and character polarity of the Osteostraci (vertebrata). J Syst Palaeontol 7:95–115

Schoenemann B, Clarkson ENK (2017) Vision in fossilised eyes. Earth Environ Sci Trans R Soc Edinb 106:209–220

Servais T, Perrier V, Danelian T, Klug C, Martin R, Munnecke A, Nowak H, Nutzel A, Vandenbroucke TRA, Williams M, Rasmussen CMO (2016) The onset of the 'Ordovician plankton revolution' in the late Cambrian. Palaeogeogr Palaeoclimatol Palaeoecol 458:12–28

Sheehan PM (2001) The late Ordovician mass extinction. Annu Rev Earth Planet Sci 29:331–364

Simon L, Godderis Y, Buggisch W, Strauss H, Joachimski MM (2007) Modeling the carbon and sulfur isotope compositions of marine sediments: climate evolution during the Devonian. Chem Geol 246:19–38

Siveter DJ (1984) Habitats and modes of life of Silurian ostracodes. Spec Pap Palaeontol 32:71–85

Siveter DJ, Sutton MD, Briggs DEG (2003) An ostracode crustacean with soft parts from the lower Silurian. Science 302:1749–1751

Siveter DJ, Sutton MD, Briggs DEG (2004) A Silurian sea spider. Nature 431:978–980

Siveter DJ, Briggs DEG, Siveter DJ, Sutton MD, Legg D, Joomun S (2015) *Enalikter aphson* is an arthropod: a reply to Struck et al. (2014). Proc R Soc B Biol Sci 282:20142663

Stanley G Jr (2001) The history and sedimentology of ancient reef systems. In: Stanley GD Jr (ed) Topics in geobiology, vol 17. Plenum Press, New York, p 488

Struck TH, Haug C, Haszprunar G, Prpic NM, Haug JT (2015) *Enalikter aphson* is more likely an annelid than an arthropod: a comment to Siveter et al. (2014). Proc R Soc B Biol Sci 282:20140946

Taylor PD, Waeschenbach A (2015) Phylogeny and diversification of bryozoans. Palaeontology 58:585–599

Tetard M, Monnet C, Noble PJ, Danelian T (2017) Biodiversity patterns of Silurian Radiolaria. Earth Sci Rev 173:77–83

Tetlie OE (2007) Distribution and dispersal history of Eurypterida (Chelicerata). Palaeogeogr Palaeoclimatol Palaeoecol 252:557–574

Tinn O, Meidla T, Ainsaar L, Pani T (2009) Thallophytic algal flora from a new Silurian Lagerstätte. Estonian J Earth Sci 58:38–42

Tinn O, Mastik V, Ainsaar L, Meidla T (2015) *Kalania pusilla*, an exceptionally preserved non-calcified alga from the lower Silurian (Aeronian, Llandovery) of Estonia. Palaeoworld 24:207–214

Torsvik TH, Cocks LRM (2013a) New global palaeogeographical reconstructions for the early Palaeozoic and their generation. Early Palaeozoic Biogeogr Palaeogeogr:5–24

Torsvik TH, Cocks LRM (2013b) Gondwana from top to base in space and time. Gondwana Res 24:999–1030

Traverse A (2008) Paleopalynology, 2nd edn. Springer, Berlin, p 813

Turner S, Burrow CJ, Schultze HP, Blieck A, Reif WE, Rexroad CB, Bultynck P, Nowlan GS (2010) False teeth: conodont-vertebrate phylogenetic relationships revisited. Geodiversitas 32:545–594

Vannier J, Wang S, Coen M (2001) Leperditicopid arthropods (Ordovician-late Devonian); functional morphology and ecological range. J Paleontol 75:75–95

Zigaite Z, Blieck A (2013) Palaeobiogeography of early Palaeozoic vertebrates. In: Harper DAT, Servais T (eds) Early Palaeozoic biogeography and palaeogeography, pp 449–460

Zigaite Z, Richter M, Karatajute-Talimaa V, Smith MM (2013) Tissue diversity and evolutionary trends of the dermal skeleton of Silurian thelodonts. Hist Biol 25:143–154

Back to the Beginnings: The Silurian-Devonian as a Time of Major Innovation in Plants and Their Communities

Patricia G. Gensel, Ian Glasspool, Robert A. Gastaldo,
Milan Libertin, and Jiří Kvaček

Abstract

Massive changes in terrestrial paleoecology occurred during the Devonian. This period saw the evolution of both seed plants (e.g., *Elkinsia* and *Moresnetia*), fully laminate∗ leaves and wood. Wood evolved independently in different plant groups during the Middle Devonian (arborescent lycopsids, cladoxylopsids, and progymnosperms) resulting in the evolution of the tree habit at this time (Givetian, Gilboa forest, USA) and of various growth and architectural configurations. By the end of the Devonian, 30-m-tall trees were distributed worldwide. Prior to the appearance of a tree canopy habit, other early plant groups (trimerophytes) that colonized the planet's landscapes were of smaller stature attaining heights of a few meters with a dense, three-dimensional array of thin lateral branches functioning as "leaves". Laminate leaves, as we know them today, appeared, independently, at different times in the Devonian. In the Lower Devonian, trees were not present and plants were shrubby (e.g., *Aglaophyton major*), preserved in a fossilized community at the Rhynic chert locality in Scotland and other places. Many of these stem-group plants (i.e., preceding the differentiation of most modern lineages) were leafless and rootless, anchored to the substrate by rhizoids. The earliest land plant macrofossil remains date back to the Silurian, with the early Silurian *Cooksonia barrandei* from central Europe representing the earliest vascular plant known, to date. This plant had minute bifurcating aerial axes terminating in expanded sporangia. Dispersed microfossils (spores and phytodebris) in continental and coastal marine sediments provide the earliest evidence for land plants (= Embryophytes), which are first reported from the Early Ordovician.

Electronic supplementary material A slide presentation and an explanation of each slide's content is freely available to everyone upon request via email to one of the editors: edoardo.martinetto@unito.it, ragastal@colby.edu, tschopp.e@gmail.com

∗The asterisk designates terms explained in the Glossary.

P. G. Gensel (✉)
Department of Biology, University of North Carolina,
Chapel Hill, NC, USA
e-mail: pgensel@bio.unc.edu

I. Glasspool · R. A. Gastaldo
Department of Geology, Colby College, Waterville, ME, USA
e-mail: ian.glasspool@colby.edu; robert.gastaldo@colby.edu

M. Libertin · J. Kvaček
National Museum, Prague, Czech Republic
e-mail: milan.libertin@nm.cz; jiri.kvacek@nm.cz

15.1 Introduction

Patricia G. Gensel and Milan Libertin

We are now approaching the end of our journey to vegetated landscapes that certainly are unfamiliar even to paleontologists. As we delve deeper and deeper in time, we will visit a succession of some of the earliest plant life to cover Earth. Until this point, our time has been spent in various woodland settings and, without a doubt, there were wooded topographies where we begin our "hike" [U1501]. But, unlike the past 390 million years, trees will not remain a part of the countryside as we step back further and further in time. Trees and tree-like architectures will disappear from the fossil record. We will see a significant decrease in the heights of the plants anchored to more primitive soils, and we will pass through the oldest shrub- or meadow-like areas where the tallest forms may have been only head high. As we continue further, plants become diminutive, barely brushing against our ankles but appearing, seemingly, like a carpet covering all available moist surfaces. And, finally, we reach a point where we no longer encounter evidence of the very earliest plants in sediments of a continental nature. Rather, minute and scattered remains attributed to land plants, including their microscopic parts, are now found preserved in ocean basins along

Fig. 15.1 *Archaeopteris* has been described as the earliest known "modern tree", having a woody trunk, growing to heights in excess of 30 m and bearing near-horizontal, helically arranged deciduous branches. (**a**) Artist reconstruction. Despite this, *Archaeopteris* has many features far removed from those of trees today. This progymno-sperm had a pteridophytic method of reproduction and bore some of the earliest planate leaves, though they are remarkably fern-like in appearance. (**b**) Lateral branch with alternately arranged ultimate branches bearing spirally arranged, laminate leaves. (**a**) © The Field Museum, GEO86500_125d and Karen Carr, Artist. With permission. (**b**) Image courtesy of Walter Cressler

with members of the Paleozoic fauna (see Chap. 14). Before we enter the unknown, let us begin with an overview of what is familiar and what is unfamiliar.

Traveling up or down any major Late Devonian river by "dugout" canoe, we first encounter the riparian forests lining the riverside in which the major tree, *Archaeopteris,* is known (Fig. 15.1). The name of this woody tree is a misnomer because it implies that the plant is an "early (archaeo) fern (pteris)". Rather, this canopy-forming tree possessed a weird combination, or mosaic, of gymnosperm and fern features, and is placed in a plant group distinctive from the others we've encountered thus far (see Sect. 15.1.1). When *Archaeopteris* colonized the landscape, the land's surface was covered by a well-established tiered community. Gallery forests grew adjacent to swampy areas populated by sprawling shrubs *(Rhacophyton)* and smaller groundcover plants, such as *Protobarinophyton*. Other low-lying environments were colonized by the earliest tree lycopsids, including *Lepidosigillaria* or *Cyclostigma*, similar to those that dominated wetland settings in the Carboniferous (see Chap. 13). By the latest Devonian, plants had evolved reproductive strategies to conquer seasonally dry regions outside of the wetlands. Here, several types of early seed plants, such as *Elkinsia*, were abundant. These forests that might seem familiar, in a general sense, become less familiar as we approach the Middle Devonian.

Middle Devonian forests can be separated into two types [U1502]. The forest structure that retains a familiar feeling is dominated by tree lycopsids such as *Protolepidodendropsis,*

which were persistent into the latest Paleozoic. The second forest is stranger. These puzzling forests were comprised of medium to very tall trees that bore no leaves. Each tree had a sort of crown made up of branches that branched multiple times, each branch terminated in a branch. These plants, the cladoxylopsids *Calamophyton* or *Eospermatopteris*, first appear in the Middle Devonian and are successful cohabitants until the early Late Devonian. Lianescent (vines) and herbaceous plants are known to occur in some of these early forests, as well as ancestors to the lycopsids, the zosterophylls (see Box 15.1), comparable to Early Devonian taxa. Plants get smaller the further we regress in time, with short-stature riparian and coastal marsh-like vegetation expanding in the late Early Devonian. These replaced open areas covered by a "green fuzz" of the earliest vegetation in moist settings adjacent to river or ocean margins (Silurian to earliest Devonian) [U1503]. Descendants of the early colonizers are still found in the mosses and liverworts. Yet, we have no evidence that any other earliest plant group remained relatively static and survived until today.

Fossils representing the earliest evidence of land plants, consisting of small to "large" macrofossils (large being a relative term of only several centimeters in length) and dispersed spores, take us into an even stranger world. While most of these fossils are found in Silurian and Devonian rocks, the earliest evidence of plants is known from the Ordovician (Katian or possibly earlier; Wellman 2010). Marine rocks preserve small sporangia with spore tetrads, an evolutionary feature ascribed to all higher plants (Steemans et al. 2009), as well as a myriad of dispersed spore assemblages containing similar types of spores from various parts of the globe. Before we machete our way through these unknown terranes, we need to gain an appreciation for the plant groups that occupied Silurian-Devonian landscapes.

15.1.1 Relationships

When fossil plants first were recognized as more than just a carbon smudge on a rock surface, all Early Devonian plants were collectively referred to as the "psilophyte flora" due to a similarity in growth architecture to a living plant (Arber 1921; Axelrod 1959). Living *Psilotum*, the whisk fern, grows in subtropical and tropical parts of the Americas, Africa, Asia, and Australasia. This plant does not have the appearance of any fern you may know. The plant body dichotomizes (evenly forks) as it grows from its flat-lying stem (rhizome), but it lacks both roots and leaves. The reproductive sporangia develop in the axils of a small "spine" called an enation, and this suite of characters was thought to be primitive and similar to all early land plants. Hence, the idea of a group of "psilophytes" first conquering land. We now recognize a number of unique plant groups during the Devonian-Silurian thanks, in large part, to the paleobotanist Harlan Banks. Today, *Psilotum* is no more recognized as a close relative of early land plants; it is rather considered a basal eusporangiate fern (see Chap. 11). With the addition of newly discovered plant fossils in the past 50 years and the advent of phylogenetic techniques since his classification, we now understand that early plant life was a bit more complex than initially proposed.

Banks (1968) presented a major reclassification of these plants in which he recognized at least four definable lineages, plus others of less well understood affinity [U1504]. His four categories of earliest vascular plants are the Rhyniophytina [U1505], Zosterophyllophytina [U1506], Trimerophytina [U1507], and Lycophytina [U1508]. At the same time, he also provided a more rigorous framework in which to consider the characteristics and affinities of these earliest land plants. With the discovery by Beck (1962) that *Archaeopteris* and other plants possessed a combination of woody stems and fern-like leaves with fern-like reproduction, Banks also recognized two more advanced groups, the aneurophytalean and archaeopteridalean progymnosperms [U1509]. Kenrick and Crane (1997) undertook a cladistic analysis focusing on Silurian-Early Devonian plants, but with inclusion of some of the younger groups, to better understand evolutionary relationships (Box 15.1). This resulted in several, sometimes major, changes in their classification (Fig. 15.2) [U1510]. For greater ease of description in our site visits to the Siluro-Devonian, however, the terms rhyniophytoid, rhyniophyte,

zosterophyll, trimerophyte, and progymnosperm, accompanied by diagnostic characters, will be used in the postcard descriptions to follow.

> **Box 15.1: Relationships of Siluro-Devonian Plants: Banks (1968) to Kenrick and Crane (1997) to Now**
>
> Banks (1968) subdivided the Early Devonian "psilophytes" into several distinct lineages, which were recognized as subdivisions. These were the Rhyniophytina, Zosterophyllophytina, and Trimerophytina, and he clarified lineages such as Lycophytina, the cladoxylopsids, and the progymnosperms. The cladistic study of Kenrick and Crane (KC; 1997) reorganized and clarified many of these groups and updated ideas about relationships of some Middle-Late Devonian plants.
>
> In the KC analysis, the Rhyniophytina of Banks is dismembered such that some taxa represent stem lineages possibly more related to lycopsids, whereas other taxa are now included in a redefined Rhyniopsida (e.g., *Rhynia, Stockmansella*). The Zosterophyllophytina are considered polyphyletic, with several stem* lineages and two more well-defined zosterophyllaceous clades, called Zosterophyllopsida by KC. They consist of basal (e.g., most *Zosterophyllum* spp., *Distichophytum*) and core (e.g., *Z. divaricatum, Oricilla, Barinophyton, Sawdonia, Serrulacaulis, Crenaticaulis*) groups. Some (or all?) of these plants are a possible sister group* or basal to Lycopsida. Lycophytina, according to KC, consists of plants ranging from stem taxa, including *Cooksonia* and *Renalia*, plus the Zosterophyllopsida and Lycopsida. Their Lycopsida include the "pre-lycophytes" (e.g., *Asteroxylon, Drepanophycus*) where sporangia originate from the stems, to true lycophytes including the Middle-Late Devonian Protolepidodendrales and (now) small tree lycophytes. It also is clear from several lines of evidence, but not shown in the cladogram (Fig. 15.2), that lycophytes diverged in the late Silurian and have been a separate lineage (consisting of several clades) since then (Gensel and Berry 2001). The Trimerophytina (e.g., *Psilophyton, Pertica, Trimerophyton*) are split into several lineages. In fact, the trimerophyte genus *Psilophyton* alone now is known to encompass several different taxa, each representing a distinct evolutionary line of plants. At least 12 species of *Psilophyton*, of varying degrees of preservation, are described, and these vary in size, branching pattern, and presence/absence of emergences. Trimerophytina are considered to be basal members of

Fig. 15.2 Phylogenetic relationships of Devonian land plant (= Embryophytes) groups modified from Kenrick and Crane (1997, The Origin and early Diversification of land plants, Smithsonian Press

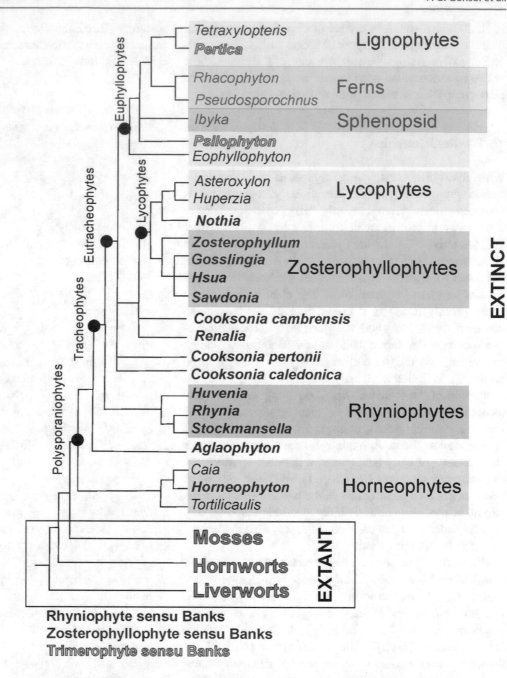

the euphyllophyte clade, or Euphyllophytina. This lineage includes ferns and their relatives and lignophytes (including seed plants). The progymnosperms, consisting of leafless, shrubby Aneurophytales and arborescent, leafy Archaeopteriales, form two clades basal to seed plants. They differ from the latter in being free-sporing, while sharing the presence of secondary xylem and phloem with seed plants. These, plus seed plants, are the lignophyte clade.

Some plants are difficult to place in any established lineage and are considered to be "of uncertain affinity" or some reference is made to possible affinities. For example, certain tiny plants with branched stems bearing sporangia, but in which vascular tissue is unknown, are termed rhyniophytoids (Gensel 2008). Other unplaced groups are the cryptospore-bearing plants (Edwards et al. 2014), and the nematophytes.

15.2 The Oldest Woodlands

Milan Libertin and Patricia G. Gensel

Wood evolved independently in several different plant groups during the Devonian (arborescent lycopsids, cladoxylopsids, some trimerophytes, and progymnosperms) resulting in the evolution of the tree habit first seen in the Middle Devonian. A virtual walk in tropical wet terrains stretching across the latest Devonian coastal plains would allow us to see groves and stands of *Archaeopteris*, one of the first woody trees to attain an impressive 30-m height (Fig. 15.1). These forests were distributed worldwide. From first glance, these plants would appear to be the analog of many Neogene landscapes [U1511]. These forests had a towering branched canopy on which individual leaves grew from twigs, providing the understory with a cool, moist shade. Yet, although the wood of any fallen tree might not appear exactly the same as modern conifer woods, what was even more different was the way in which they reproduced. *Archaeopteris* was free-sporing, like ferns, and upon closer inspection, even the leaves looked "fern-like" in a number of their features [U1512].

15.2.1 *Archaeopteris* Coastal Woodlands/Forests

Patricia G. Gensel

Coastal zones and floodplains that spread across Late Devonian continents hosted a vegetation that was ecologically partitioned into different paleoenvironments when compared to similar settings a bit deeper in time. Sediments in the area around Elkins, West Virginia, USA, and adjacent areas preserve Mid-Late Famennian (c. 368–359 Ma) plants [U1509]. These plants grew on soils developed on a southwestward prograding delta and floodplain complex, now called the Hampshire Formation. In addition to normal fluvial sandstone and mudstone deposits, these rocks contain at least two, about 1-m-thick coals (see Chap. 12). Detailed mapping of the coal and associated rocks and a paleoecological census of the plant remains show that the setting represents a low-lying deltaic shoreline, which was subjected to coastal storm events (Scheckler 1986). Plants preserved near the base of the lower coal include the enigmatic *Barinophyton sibiricum*, a core-zosterophyll. Overlying the peat swamp are planar laminated beds in which abundant *Sphenopteris* foliage, ovules, and other seed plant remains are often preserved as "leaf mats," which may have accumulated under tidal influence. These laminated beds terminate up section in a rooted zone indicating the development of an immature paleosol, most likely populated by *Rhacophyton* because considerable branching biomass of the plant occurs atop the rooting zone. These foliar axes are indicative of a scrambling growth habit for, what some authors consider to be, a "pre-fern" or an aneurophyte progymnosperm. There is some evidence to support the idea that tips of "fronds"/branch systems could root to propagate new plants such that vegetative production dominated the plant life cycle. If this is true, such a strategy could account for the high proportion of biomass in these coals.

The peat swamp, which now is a Late Devonian coal, was dominated by *Rhacophyton* biomass (Fig. 15.3). Although most of the biomass degraded to amorphous organic matter, pyrite concretions in the coal preserve their anatomy. The same proportion of *Rhacophyton* biomass is found in the second, or upper, coal. Sandwiched in between these peat swamps are deltaic sand-and-mud deposits in which abundant *Rhacophyton* [U1513] and *Archaeopteris* foliage and some stems, along with a tree lycopod similar to *Protolepidodendropsis pulchra*, are preserved. Other plant remains in deltaic deposits include cupules and seeds. The occurrence of this aneurophyte or "pre-fern" in wetland (peat) and better-drained soils supports the idea that *Rhacophyton* was broadly tolerant of soil differences (Scheckler 1986).

Non-peat-accumulating swamps, in general, are muddy because of a high water table in these immature soils of low relief. If we were to slog our way through these floodplains—we would not actually easily walk—we would want to keep our feet on top of the plant cover rather than stepping in the mud between them. Late Devonian swampy areas were dominated by the shrubby, scrambling *Rhacophyton*, making it easier for our trek, with possible rare occurrences of *Barinophyton*, the scrambling horsetail *Sphenophyllum*, and the plants that bore *Eviostachya* reproductive cones (strobili). We would encounter early seed plants, such as *Elkinsia* (Fig. 15.4) on slightly higher ground and growing on better-drained soils of the stream margins. Lining the rivers, growing a kilometer or so into the floodplain, or on slightly higher topographies, back of the coastal area, were stands of *Archaeopteris*. Tree lycopsids, the systematic affinities of which are not yet known, grew somewhere between the wettest peat and muddy *Rhacophyton* swamps and the gallery progymnosperm forests. These forms may be precursors to the giant scale trees of the Carboniferous (see Chaps. 12 and 13), but the most unique aspect of these landscapes is the appearance of the earliest seed-bearing plants, the gymnosperms. *Elkinsia* is the early seed plant that is most extensively known in the plant fossil record, to date (Rothwell et al. 1989; Serbet and Rothwell 1992). It has been reconstructed (Box 15.2) with a main stem with a rather unique anatomy [U1514]. In Late Devonian, seed-bearing plants diversified [U1515] and spread across the landscape, beginning to alter the planet's vegetation. We could walk through similar Late Devonian

Fig. 15.3 *Rhacophyton*, a fern-like plant. (**a**) Charcoalified remains of *Rhacophyton*, with a pinnate branching pattern (image courtesy of Walter Cressler). (**b**) Anatomy of central axis showing secondary, woody tissues (from Dittrich et al. (1983) Anatomy of *Rhacophyton ceratangium* from the Upper Devonian (Famennian) of West Virginia.

Rev. Palaeobot. Palynol. 40:127–147 with permission from Elsevier); (**c**) Reconstruction of vegetative and fertile parts of the plant (from Andrews and Phillips (1968), *Rhacophyton* from the Upper Devonian of West Virginia, Bot. J. Linn. Soc. 61 (284): 37–64, with permission from Oxford U. Press)

Box 15.2: The Early Seed Plant *Elkinsia*

The plant *Elkinsia* has a three-lobed vascular conducting strand usually consisting of only primary xylem. This is surrounded by a "sparganum" cortex, a feature seen in the lyginopterid seed ferns (see Chap. 12), characterized by a distinctive outer cortex of reinforced cells forming a pattern like Roman numerals on a clock face in cross section. When leaves emerged from the stem, a lobed leaf trace divided into two C-shaped bundles in the leaf and then divided up to four times more. Vegetative leaves are *Sphenopteris*-like in their leaf architecture. Cupulate organs (seeds) and synangia (pre-pollen organs) terminated fertile axes that divided in a cross-shaped organization and lacked leaves (Serbet and Rothwell 1992). Scheckler (1986) suggested that *Elkinsia* was a pioneering plant and Prestianni and Gerrienne (2010) concur.

landscapes in other parts of the world and witness similar vegetation and community organization. These include: Red Hill in Pennsylvania, USA; Taff's Well and Avon Gorge, Great Britain; Kerry Head, Ireland; the Condroz sandstones, Belgium; and Oese, Germany (Prestianni and Gerrienne 2010). Many of these sites contain Late Devonian plants that are preserved away from their site of growth, often in marginal marine or lagoonal sediments. Here, plants may be associated with other biotic components of the Late Devonian biosphere. We'll stop first at Red Hill near North Bend, less than a kilometer north of highway 120 in Gleasonton, Pennsylvania, USA.

15.2.2 Red Hill, Pennsylvania

Patricia G. Gensel

As the name implies, the Red Hill locality exposed a Late Devonian succession of red, primarily, mudrock

Fig. 15.4 The earliest known seed-bearing *Elkinsia*. (**a**) A reconstruction of the branching architecture on which both leaf-bearing and ovule-bearing axes occurred. (**b**) Laminate pinnules terminating axes. (**c**) Terminal cupules (ovule-bearing). (**d**) Thin section of ovule showing cupule (c) micropyle, and megaspore membrane (m). (modified from Serbet and Rothwell 1992)

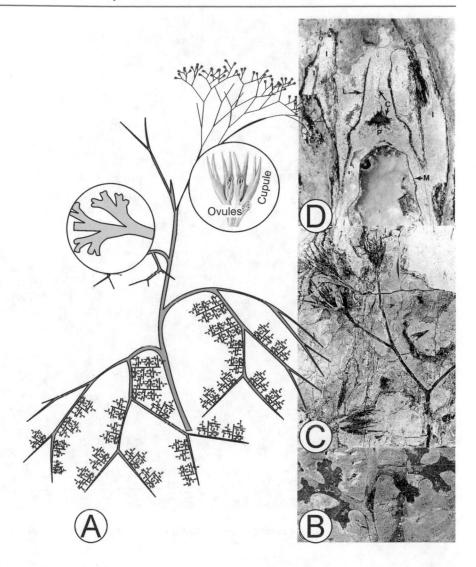

(Fig. 15.5). What is unique about this Famennian (372–359 Ma) sequence is the array of arthropods, fishes, and early tetrapods (vertebrates) with anatomical features that indicate these fishes and fish-like amphibians could survive out of the water. Fossil plants, the base of this food chain, are preserved in one thin interval but served as the habitat for terrestrial invertebrates including trigonotarbid arachnids and myriapods, both of which have been recovered from the site (Daeschler and Cressler III 2011). The succession consists of sandstone deposited in river channels, siltstone that filled abandoned freshwater channels and lakes, and floodplain paleosols adjacent to these ancient bodies of water.

The Red Hill paleoenvironment was an alluvial floodplain with meandering rivers often overflowing their banks and burying the vegetation associated with Vertisols* (soils with shrink-swell clays) and Calcisols* (calcium-rich soils). Meandering river channels produced an ever-changing environment, with abandoned channels becoming quiet-water lakes in which a teeming vertebrate community existed. Remains of both plants and terrestrial arthropods are preserved in oxbow lake or pond deposits [U1516]. Plants include the progymnosperm *Archaeopteris*, the pre-fern *Rhacophyton*, barinophytes, two kinds of tree size lycopsids, and seed plants (mostly represented by their cupulate ovules, and some foliage that may belong to this lineage; Cressler 2006; Cressler et al. 2010). Similar to other Late Devonian landscapes, *Archaeopteris* canopies were underlain by a *Rhacophyton* understory with various scrambling plants occupying the groundcover, most likely in light breaks. In more open sites, lycopsids grew adjacent to oxbow lakes and ponds, and seed plants colonized disturbed areas or those in which soil drainage was better. Channel deposits contain the vertebrate assemblage.

Fish and some of the earliest tetrapod fossils are preserved in freshwater lenses of silty mudrock in shallow channel margin, floodplain pond, and overbank pond deposits. Here, a stem tetrapod found outside of coeval rocks in Greenland was identified based on two shoulder

Fig. 15.5 The Famennian age, char-bearing outcrop at Red Hill, Pennsylvania, USA. (photo PG Gensel)

bones (i.e., cleithrum∗ and scapulacoracoid*). The shoulder-bone features in *Hynerpeton* are advanced, indicating that the animal was capable of both supporting its body and powering itself on land [U1517] (Daeschler et al. 1994). *Hynerpeton* and the more complete, three-dimensionally preserved *Tiktaalik* from Ellesmere Island, Arctic Canada (Daeschler et al. 2006), record the transition between fishes with fins and tetrapods with limbs and digits. In addition to the shoulder girdle, an unusual, isolated humerus and jaw fragments of two different tetrapods have been collected. A single bone from the snout of a tetrapod resembles that of an early Carboniferous tetrapod, *Pederpes*, from Scotland (Daeschler et al. 2009). These are our deep time ancestors. Without their transition from a fully aquatic to a terrestrial life strategy, all higher vertebrate groups we've seen would not have existed. But, because we did evolve from these stem tetrapod groups, it's now only a short 4-h drive north from Red Hill to the town of Gilboa, New York, where our next postcard is located.

15.2.3 Gilboa Quarry, New York, USA

Patricia G. Gensel

The localities we now visit in eastern New York State demonstrate that diverse forests, where large trees of different affinities grew in abundance, occurred very early in time. Here, those trees were not only leafless but also evolved a very different type of rooting structure to fix themselves in a poorly developed soil. We will have to make our way through at least two plant lineages living in the understory, each of which grew in a slightly different mode. Undoubtedly, various types of arthropods and spiders, as recognized from their fossilized exoskeletons (cuticles), were scuttling around in the undergrowth (e.g., Shear et al. 1987, 1989).

Late Middle Devonian fossil-tree stumps, preserved in life position and bearing radiating roots, were discovered in the 1870s at several horizons in the Riverside Quarry at Gilboa, New York [U1518]. These tree stumps were named *Eospermatopteris* by one of the first female paleontologists,

Fig. 15.6 The Middle Devonian cladoxylalean fern, *Eospermatopteris*. (**a**) Stump excavated from Schoharie Reservoir, Gilboa, New York. (photo: RA Gastaldo) (**b**) Tree cast and reconstruction of cladoxylalean fern (from Stein et al. 2007, Giant cladoxylopsid trees resolve the enigma of Earth's earliest forest stumps at Gilboa (Nature 446 (7138) with permission, Springer Nature)

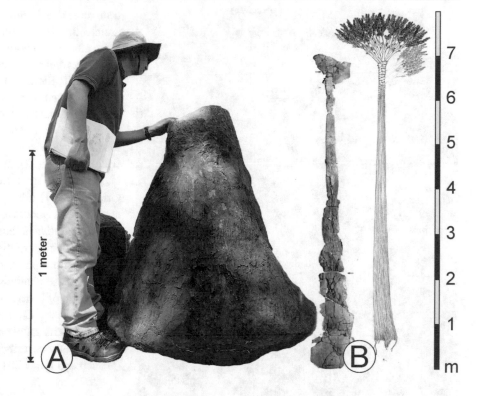

Winifred Goldring (1924; Fig. 15.6), and were widely cited as evidence of the Earth's "oldest forest" (e.g., Goldring 1927). These trees were not restricted to the quarried exposure because *Eospermatopteris* stumps also were found in nearby outcrops. In the early 2000s, impressions and flattened stem casts extending across a quarry floor not far from the Gilboa site, at South Mountain, showed a basal trunk comparable to *Eospermatopteris* and a crown region with attached, digitately divided, upward-extending branch systems (Fig. 15.6). Each branch division bore sterile and fertile appendages that allowed their identification with plants previously known from Belgium and Venezuela as the cladoxylopsid *Wattieza* (Pseudosporochnales) (Stein et al. 2007). *Wattieza* is a very strange plant [U1519]. It grew to a height of at least 8 m with a trunk bearing large branches in vertical ranks (Stein et al. 2007). The much-divided branches bore smaller divided, leafless terminal units that, in some cases, bore sporangia that shed spores. These trees belong to an extinct group, the pseudosporochnaleans, believed to be related to ferns—new data may revise that idea. It appears that the stems were fast-growing and "cheaply" constructed. The center of these trunks was hollow and encircled by many conducting (vascular) strands in the periphery (Stein et al. 2012). The trunks were surrounded by roots near the base, with a growth strategy similar to living palms or tree ferns. The sedimentology indicates a series of burial events affected these forests which were frequently disturbed, followed by their reestablishment in new soil horizons. Looking at the face of a rock exposure, if lucky enough, provides a view

about the spacing of trees along a single plane. To understand the forest structure, though, you have to be able to see it from above.

Stein and others were able to access the original Riverside Quarry site for a limited time when the area was cleaned of backfill and the quarry floor exposed. That exposure revealed remnants of a forest at soil level. New *Eospermatopteris* casts were retrieved from the backfill, and the cleaned forest floor showed numerous root mounds with radiating axes and a central depression (representing the base of the plant) with downward growing roots. A large portion of the quarry was plan-mapped and displayed the spatial distribution of the *Eospermatopteris/Wattieza* plants (the two named fossil-genera can be referred to the same reconstruction of an ancient "whole plant"). These trees often grew in clusters and were of different diameters at the time of preservation, suggesting an uneven aged stand (Stein et al. 2012). Roots extend from above the bases outward across the paleosol as well as downward from the basal region. Another important discovery was that at least two other plant types were found on the forest floor. One is a woody rhizome with adventitious roots and aerial branch systems. The branching pattern in aerial axes, along with the anatomy preserved in the rhizome, is consistent with these plants being related to the aneurophytalean progymnosperms, a group that is known from numerous other Middle Devonian localities. Like *Archaeopteris*, aneurophytes grew woody stems and reproduced by spores, but were smaller in stature (as presently known) and presumed to have been leafless and shrubby

Fig. 15.7 Trunk and tree base of *Archaeopteris*. (**a**) Silicified stump (photo: PG Gensel). (**b**) The tree base and rooting structure of *Archaeopteris* outlined on a paleosol exposed in the Cairo Quarry, Town of Cairo Public Works, New York, USA. (photo: RA Gastaldo)

[U1516]. The rhizomes curve around the bases of *Eospermatopteris* (*Wattieza*) tree bases, and sometimes approach them as if, in life, they may have been vines. The anatomy and branching system of these scrambling vines are similar to the aneurophytalean *Tetraxylopteris*. A few of these are preserved up to 4 m in length, associated with distal branches of an upright tree lycopsid [U1520], indicating the heterogeneous nature of this forest.

The forests in Gilboa are interpreted to have grown in a coastal wetland setting, with frequent marine incursions that buried successive forests. Originally, the paleoenvironment had been interpreted as a wetland swamp, but the underlying paleosol is well-developed, as are those from other Devonian sites (Morris et al. 2015), and may indicate that these plants grew in better drained conditions. Stein et al. (2012) indicated that this forest type may not be too different from coeval *Archaeopteris* dominated ones. It is, therefore, interesting that an exposed quarry floor in nearby Cairo, New York, provides evidence that both *Eospermatopteris* and *Archaeopteris* coexisted, the latter being more abundant (Fig. 15.7). *Archaeopteris* has a rooting system that is broadly spreading and branched and is considered to pene-

trate more deeply into the soil (Fig. 15.7). Hence, the rooting architecture of this plant is more modern-looking and, perhaps, altered soil composition and weathering patterns in the latest Devonian (Algeo and Scheckler 1998, 2010; Algeo et al. 2001; Stein 2018). There is evidence that these soils also supported other groundcover.

Smaller plants, of various systematic affinities, probably grew under or around these trees. The apparently rhizomatous lycopsid *Leclercqia* is abundant in the Gilboa region (Banks et al. 1972), and the zosterophylls *Serrulacaulis* and cf. *Sawdonia* sp. (Hueber and Banks 1979; Hueber and Grierson 1961) are preserved in nearby deposits of similar age. Several genera of aneurophytaleans also are known, including *Relimmia* and *Tetraxylopteris*. Most likely we only know of their more distal (terminal) branch systems, rather than the entire plant, itself. And, several authors think that some of these plants were shrubby besides their interpreted scrambling or nearly lianous growth strategies. The pseudo-sporochnalean *Calamophyton* is represented in North America by its distal branch systems, but nearly whole plants were recently described from quarries in Germany [U1521] (Giesen and Berry 2013).

15.3 Early Middle Devonian Coastal Marshes

Robert A. Gastaldo

Only about 900 km geographically from the Gilboa forests of New York State, but nearly eight million years older in time, we head to Devonian rocks exposed in the conifer forests of northern Maine, USA (Fig. 15.8a, d). Here, our postcard is of an idyllic trout stream in Baxter State Park, where the air is scented by balsam fir resins, and remnants of latest Early (Emsian) or earliest Middle (Eifelian) Devonian rocks [U1522]. The outcrops, exposed at the surface following the last glacial episode that scoured this landscape some 12,000 years ago, are not very impressive. All of these are of low relief, and many are covered in a carpet of recent mosses and club mosses, and we do not have to go back thousands of years in time to understand why the fossil plants preserved in the Trout Valley Formation became a turning point in our insight into early terrestrial communities. For that, we have to turn our attention to the second half of the twentieth century.

The United States Geological Survey (USGS) began a national mapping program following the end of the Second World War, targeting rural parts of the country that had, yet, to be detailed. Douglas Rankin, a USGS geologist, spent part of his early career with the Maine Geological Survey and began mapping northern Maine. Here, he discovered what looked to be compression-impression remains of rare plant

Fig. 15.8 Middle Devonian Trout Valley Formation, Maine, USA. (**a**) Outcrop localities along Trout Brook, Baxter State Park, Maine. (**b**) Low-angle trough cross-bedded siltstone of tidal origin in which the fossil flora is preserved. (**c**) Glacially exposed, fossiliferous bedrock along the margins of Trout Brook. P. = P Gensel and RA Gastaldo. (**d**) Bedding surface of tidal siltstone in which biotically oriented axes of *Psilophyton* are exposed. Scales in dm and cm. (photos: RA Gastaldo)

fossils then known as "psilophytes". These were examined by Erling Dorf who recognized their significance, and one of the first reports on these plants was published shortly thereafter (Dorf and Rankin 1962). With the recognition that early land plants are preserved in Maine and in coeval strata of New Brunswick, Canada, a series of studies over the past 60 years have added to our understanding of their evolutionary history and paleoecology (e.g., Kasper Jr et al. 1988; Allen and Gastaldo 2006). Both the environmental setting and the early land plants, themselves, are very different from the mosses and club mosses now covering the floor of Maine's northern balsam fir forests.

Rocks of the Trout Valley Formation were deposited as pebble conglomerate, fluvial and nearshore (marine) sandstone bodies, and muddy tidal flats. These sediments represent an estuarine coastal zone flanking an extinguished volcanic island, the remnants of which are now the Traveler Rhyolite against which the Trout Valley rocks lie (Allen and Gastaldo 2006). There is some evidence of in situ (autochthonous) preservation of the vegetation, in the form of very fine, vertically oriented rootlets, which colonized the mudflats [U1523]. But, the majority of fossil material is preserved on the bedding surfaces of tidal channels that traversed the mudflats (Fig. 15.8c, d). Here, aerial axes up to 50 cm in length, with lateral dichotomizing appendages, are aligned parallel to one another, (Fig. 15.8c) mimicking their original growth architectures in life. This "biological" orientation is the result of fibrous, longitudinal tissues, appearing as striations, which developed in the walls of these thin axes to assist in an erect growth habit. The two most conspicuous plants are *Psilophyton* and *Pertica* (Fig. 15.9), true vascular plants assigned to the early group called trimerophytes [U1520]. Intermixed or interbedded with dense *Psilophyton* and *Pertica* mats are other vascular plants belonging to various early clades [U1524]. These include *Sciadophyton* (embryophyte of unknown affinity); *Sporogonites* (a possible bryophyte); *Taeniocrada* (rhyniophyte); and *Drepanophycus*, *Kaulangiophyton*, and *Leclercqia* (lycopsids; Andrews et al. 1977; Kasper Jr et al. 1988; Allen and Gastaldo 2006).

Fig. 15.9 Early-Middle Devonian "trimerophytes". (**a**) *Psilophyton forbseii* showing pseudomonopodial main axes from which laterals branched dichotomously (3×; photo PG Gensel). (**b**) *Psilophyton crenulatum* recovered via maceration (Yale University image). (**c**) *Psilophyton coniculum* stem anatomy (from Trant and Gensel 1985, Branching in *Psilophyton*: a new species from the Lower Devonian of New Brunswick, Am. J. Bot. 72(8): 1256–1273, with permission from Wiley Press); (**d**) *Pertica quadrifaria*. (photo RA Gastaldo). (**e**) Axial anatomy of a new taxon reminiscent of *Pertica* (photo courtesy of PG Gensel)

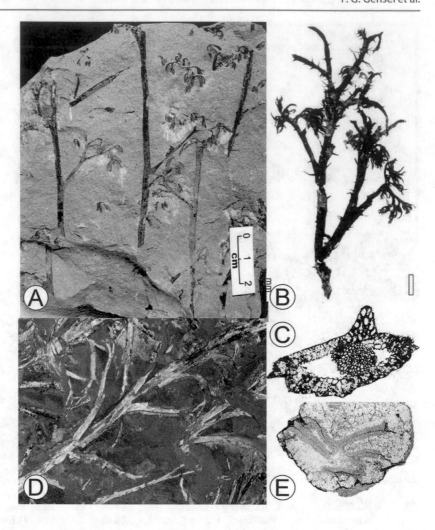

The biotically oriented assemblages in the Trout Valley Formation are unusual for several reasons. Trimerophyte, rhyniophyte, and lycophyte remains are preserved in association with molluscan bivalves of various affinities, eurypterids, and trace fossils (ichnofossils*), all of which are interpreted from brackish water (estuarine) settings [U1525] (Selover et al. 2005; Gastaldo 2016). Hence, it seems plausible that these early colonizing plants were tolerant of fluctuating coastal salinities. If true, this interpretation may also explain two facets of their preservation. Trimerophytes grew aerial axes from a surficial or shallowly buried axis or rhizome. Yet, the only evidence of rooting structures is vertically oriented rootlets that are unattached to a rhizome. Similarly, there is no evidence that the biotically ordered axes (Fig. 15.8d) are attached to any rhizomatous organ. The axis from which these plants developed is missing. Second, most of these aerial axes rotted and filled with mud before burial, resulting in internal casts. Saltwater incursion into these marshlands in response to coastal subsidence, then, would be responsible for their death, loss of any evidence of rhizomes through rotting, and the easy transport via tidal activity of the aerial axes into tidal channels where they are preserved. We have to dare to head farther northward into eastern Canada where these plant groups are best preserved to gain insights into coastal plains of the Devonian.

15.4 Late Early Devonian Floras of Gaspé and New Brunswick: Coastal Margins and Intermontane Rivers and Lakes

Patricia G. Gensel

Several localities in the Canadian provinces of New Brunswick and Quebec have been the source of information about Emsian (408–393 Ma) plants since the initial description of the genus *Psilophyton* by Dawson (1859) from the Gaspé Peninsula, Quebec, and adjacent regions of northern New Brunswick. Dawson (e.g., 1870, 1871) described other plant taxa, some included in *Psilophyton* and some not. Additional collections and studies by paleobotanists in both areas, especially in past decades, have clarified various taxa and produced a picture of a diverse late Early Devonian flora.

These plants are mostly smaller and simpler than those from the Middle Devonian.

15.4.1 Gaspé Battery Point Formation (Gaspé Sandstone Group), Emsian

The Battery Point Formation, outcropping on both the north and south shore of Gaspé Bay, Quebec, Canada, consists of a coarsening-up clastic wedge, located at 10–20° S latitude in Emsian times. This paleolatitude places it near the equator with a prevailing tropical climate. Unlike Red Hill (see Sect. 15.2.2), there are no calcretes or other sedimentary features indicative of seasonal dryness, and the plants probably grew under moderate rainfall. The Cap-aux-Os Member is the most plant-rich component of the Battery Point Formation and has been extensively studied sedimentologically (e.g., Griffing et al. 2000). It is in this depositional context that we understand these Early Devonian plants.

The sedimentary facies in the Cap-aux-Os Member are dominated by sandstones of various internal architectures, and three different fluvial associations are identified. River deposits at the base of the succession are large multistory sandstone bodies with interbedded thinner mudstone (Association 1). These rocks are overlain by gray mudstone with thin sandstone sheets or single-story sandstone bodies (Association 2). The uppermost interval (Association 3) consists of relatively coarse-grained, multistory sandstone bodies with uncommon thinner red mudstone intervals [U1526]. There is evidence of bedding with wave- or current-ripple marks, trace fossils, and disarticulated cephalaspid fish skeletons in some intervals. Desiccation-cracked mudstones preserve articulated lingulid brachiopods, and in dark gray shale and siltstone acritarch microfossils, small bivalves, and brachiopods can be found. The most current interpretation is that these rocks represent fluvial and delta-plain deposits close to the coastline, with some intervals interpreted as having been close to the tidal limit of tidally influenced lowlands (Griffing et al. 2000; Hotton et al. 2001). The vegetation that grew across these coastal zones commonly is found at or near their sites of growth.

Many fossil-plant assemblages, buried in situ, form monospecific stands, although attached rooting structures are not found. Rather, putative rhizomes and rooting structures are preserved in other intervals and may extend into underlying beds beneath some axes (Elick et al. 1998; Gensel and Berry 2001; Hotton et al. 2001). Plants preserved near their sites of growth include the majority of trimerophyte and rhyniophyte remains. Hotton et al. (2001) note that these occur in shaley (mudstone) drapes over channel-form sandstone bodies. They suggest that the plants were growing along channel margins or channel bar tops and probably detached and transported a short distance prior to burial.

Others, especially zosterophylls, were established in low, wet areas and buried by flood deposits. A few plants may have had specific growth conditions limiting their distribution. For example, the zosterophylls *Sawdonia ornata* and *Crenaticaulus* [U1527], commonly found associated with brackish and marine invertebrate fossils, are interpreted as being located near a coastline where washover occurred. *Spongiophyton* and *Prototaxites* were transported, sometimes in a more fragmentary form, from floodplains or a riparian habit, and deposited in channel-bar and channel-fill deposits of main river channels. But, one genus of early plant is cosmopolitan.

In our rambles, we will encounter large stands of *Psilophyton*, probably along the river channel margins and within stands of different zosterophylls in some of the marshes (see Sect. 15.3). *Psilophyton* plants grew to a height of several decimeters with slender (c. 4 mm), dichotomizing stems (Fig. 15.9b). Lateral branches divided in a similar pattern and are either the same or slightly smaller diameter than the central erect axes. They terminate in pointed (acuminate*) tips or pairs of spindle-shaped (fusiform) sporangia, forming loose clusters, and which dehisce (burst open) longitudinally [U1528]. The plant's anatomy consists of an unlobed core of vascular tissue in which the first developed conducting cells differentiate in the center. This anatomy is seen to persist in lateral branches. The anatomy of one of its species, *Psilophyton dawsonii* from the south shore of Gaspé, is the most completely known. As we've seen previously, *Psilophyton* is a very common element in Gaspé and New Brunswick and the Trout Valley Formation in Maine. It also is known from the Early Devonian of Wales, England, Belgium, possibly the Czech Republic, and Germany. Only one species is recorded in China and is of Pragian age (411–408 Ma). However, some early referrals of plant remains to *Psilophyton* have since been shown to be inaccurate. The devil lies in the details, also in plant identification, as we will see when studying some of Gaspé's plant fossils.

Some rooting structures, or "rhizomes," from Gaspé, attributed to *Psilophyton* by Dawson, are now known to represent another plant (Hotton et al. 2001). Their anatomy is very different from what we know from *Psilophyton dawsonii*. Smooth, wide axes, bifurcating at long intervals, bear rounded scars and exhibit a dark central strand. Unpublished specimens show similar axes with laterally attached ovoid sporangia with a thickened base (Gensel, pers. obs.), leaving a round scar when detached and, thus, are similar to *Stockmansella* (Fairon-Demaret 1985, 1986). That taxon, along with *Rhynia*, is currently viewed as part of the Rhyniopsida (Kenrick and Crane 1997). The Gaspé plant's water-conducting cells also resemble *Stockmansella* in exhibiting a unique wall pattern, with tiny holes perforating the walls and randomly oriented thickenings (Hueber 1983; Hotton et al. 2001; Gensel pers. obs.). A second taxon from this clade, *Huvenia*, may also be present in the Gaspé flora (Hotton et al. 2001). Rooting structures were not the only plant remains erroneously referred to as *Psilophyton* in the past.

Psilophyton princeps var. *ornatum* was described by Dawson based on vegetative remains, and its actual affinity was questioned for many years. Ultimately, the plant's suite of characters was recognized as similar to the zosterophyllophytes, and the plant renamed *Sawdonia ornata* (Hueber 1971). Stems are covered in tapered spine-like emergences and are similar to specimens from Abitibi River, Ontario, in which lateral sporangia occur (Hueber 1964; Hueber and Banks 1967). *Sawdonia* and other anatomically preserved zosterophylls exhibit an ovoid water-conducting central cylinder, which develops from the outside to the center (the opposite direction of what we know from *Psilophyton*). Recently studied fertile remains from Gaspé also reveal differences on the lateral sporangia of the plant, which are short-stalked and possess two valves of unequal size, with emergences covering the larger abaxial valve (Gensel and Berry 2016). Apart from rooting structures now referred to as rhyniopsids and vegetative remains reidentified as zosterophylls, stems initially attributed to *Psilophyton* were found to be different taxa as well.

Large stems up to 1 cm wide, from which regularly arranged lateral branches grew, terminating in tight clusters of fusiform∗ sporangia, are now attributed to the trimerophytes (Banks 1968). These plants also occur at Cap-aux-Os. One taxon was based on specimens originally labeled *Psilophyton robustius* (Dawson 1871) and was redescribed, more than 80 years later, as *Trimerophyton robustius* by Hopping (1956). These large axes exhibit a lateral branching pattern that divides into three branches (trichotomous), instead of two, with some branches terminating in tight clusters of sporangia. Other specimens, possibly from the same sequence, were described as *Pertica varia* by Granoff et al. (1976). The plant fossils consist of up to 0.75 m-long, incomplete main axes from which regularly arranged, clearly secondary lateral branches depart. These laterals may retain a central branch or continue to subdivide dichotomously. Some of the more dichotomous branches terminate in tight clusters of sporangia. These plants may represent the tallest of known late Early Devonian vascular plants and probably attained heights of a few meters. But, trimerophytes and zosterophylls were not the only Emsian coastal zone vegetation.

Lycopsids obtained from the Battery Point Formation include *Drepanophycus spinaeformis*, a plant with branching rhizomes, possible rooting structures and short to long, sometimes curved (falcate) leaves (Grierson and Hueber 1967). Fertile remains of these plants from New Brunswick show that stalked sporangia developed along the stem and occur among the leaves (Li and Edwards 1995). *Renalia hueberi*, probably an early member of the lycophyte lineage, occurs at

a different outcrop on the north shore of Gaspé Bay (Gensel 1976). The main stems are tiny, the axes of approximately 1 mm in width, and specimens are up to only a few centimeters in length. Lateral branches divide unequally and may be terminated in kidney-shaped (reniform) sporangia. A dark strand evident in stems indicates the presence of vascular tissue but cellular patterns are unknown. There are other plant types noted and described from this locality (Andrew and Gensel 1984; Hotton et al. 2001), but we end our postcard tour of the Gaspé with mention of a most enigmatic plant.

Prototaxites is the elephant in the Emsian room. It is a giant (Fig. 15.10). Dawson (1859) first described the taxon based on silicified trunk-like structures and, now, it is known to occur not only in the Gaspé Bay area but also at Pointe-a-la-Croix, Gaspé, in northern Maine, and several other Early to Late Devonian localities in the United States, Europe, and North Africa [U1529]. The best-preserved specimens consist of wide "stems" that attained nearly 1 m in diameter, whereas Moroccan specimens are up to 8 m in length (Boyce et al. 2007). Various ideas as to what these "logs" represent have been presented, ranging from a prototypical conifer (Dawson) to alga, rolled-up liverwort, and even fungus. Hueber (2001) suggested the preserved structures most closely resemble the fruiting body of a fungus. This, in turn, raises questions about a carbon source for such a large organism and has spurred additional research resulting in some controversial interpretations of this organism (Graham et al. 2010; Retallack and Landing 2014). However, we will not delve into this scientific discussion here and instead travel further to northern New Brunswick (Box 15.3).

Box 15.3: Evolution of Vascular Cambium Producing Secondary Xylem and Phloem

Gerrienne et al. (2011) reported the occurrence of plant stems appearing similar in size and morphology to *Psilophyton* from the Pragian of France and the Emsian of New Brunswick. These fossils possess aligned conducting cells (tracheids) in their xylem associated with features typical of secondary xylem as is produced by the activity of newly activated stem cells referred to as a vascular cambium (VC). In extant plants, and in other groups with a VC, both secondary xylem and secondary phloem (food-conducting cells) are normally produced. But, in these plants, preservation ends outside the secondary xylem with the exception of a few squashed thin-walled cells that may be remnants of the VC. This discovery pushes back in time the presence of secondary tissues, or wood, and suggests that perhaps attaining the type of wood prevalent in seed plants, characteristic of Devonian to recent plants, occurred in a stepwise manner.

Fig. 15.10 *Prototaxites,* a giant fungus. (**a**) Erect mold of *Prototaxites* in siltstone, Dalhousie Junction, New Brunswick, Canada. RA Gastaldo for scale; edges of mold marked by yellow dashed lines. (photo courtesy of M. Gibling). (**b**) Small silicified *Prototaxites* (photo courtesy of PG Gensel). (**c**) Transverse section of silicified specimen showing tubular nature of cells (from Retallack and Landing 2014. Affinities and architecture of Devonian trunks of *Prototaxites loganii.* Mycologia 106(6):1143–1158, with permission, Wiley Press)

Plants exhibiting secondary xylem include:

- *Armoricaphyton chateaupannense* (Strullu-Derrien et al. 2014); Pragian, France
- Unnamed plant (Gerrienne et al. 2011; Gensel 2018); Emsian, New Brunswick, Canada
- *Franhueberia gerriennii* (Hoffman and Tomescu 2013); Emsian of Gaspé, Canada
- As yet undescribed trimerophyte (Gensel 2018); Emsian, Gaspé, Canada

All these fossils exhibit a haplostelic primary xylem, with up to 24 rows of aligned tracheids. These tracheids also show signs of a doubling in places and

spaces where presumably less resistant cells of rays (typical of secondary xylem) occurred. The earliest occurrence of a bifacial VC, producing both secondary xylem and phloem, and also periderms, is Middle Devonian. This feature is found in aneurophytalean progymnosperms.

15.4.2 Campbellton Formation, Emsian, New Brunswick: Enlarging our Postcard View of an Early Devonian Landscape

Driving along the winding New Brunswick coast on Route 132 for about 4 h, we eventually arrive in Campbellton. We

Fig. 15.11 Lower Devonian assemblages from Campbellton, New Brunswick, Canada. (**a**) Typical outcrop along the shores of Chaleur Bay (photo courtesy of PG Gensel). (**b**) Transported assemblage of biologically ordered axes of lycopsid affinity (photo courtesy of RA Gastaldo)

will briefly visit another Early Devonian locality of northern New Brunswick contemporaneous, in part, with Gaspé. Here, plants are preserved in both coastal and inland settings. Outcrops of the Campbellton Formation extend, discontinuously, from Campbellton to Dalhousie, New Brunswick (not Nova Scotia). Detailed stratigraphic studies by Kennedy and Gibling (2011), along with a subsequent consideration of the depositional environments and plants preserved therein, were published by Kennedy et al. (2012). The western-most outcrops near Campbellton were deposited along a Devonian coastline, with one horizon burying remains of *Psilophyton crenulatum* in volcanic ashfall. Similar to other ashfall burial sites (see Chap. 13), preservation is exceptional (Fig. 15.11). Eastern outcrops are interpreted as intermontane fluvial or lacustrine, inland floodplain deposits, or mass flow deposits.

We've seen this vegetation before. Plants similar at the generic level to *Psilophyton*, and *Pertica* [U1530], *Drepanophycus*, and possibly *Sawdonia,* as well as new taxa of zosterophylls and lycopsids [U1531], occur. One of the

earliest occurrences of the lycophyte *Leclercqia* is here, as well as two strange taxa, *Chaleuria* and *Oocampsa*, which differ from each other and from the major Devonian plant lineages [U1532]. *Chaleuria* and *Oocampsa* both exhibit broad main stems, up to 1 cm in diameter, which may have been partly rhizomatous and partly upright. In contrast to other plants we've encountered, the lateral branches of these are dense and spirally arranged. The lateral branches in *Chaleuria* have dichotomizing, spirally arranged ultimate branches with fusiform, terminal sporangia in which two sizes and two kinds of spores were produced. This condition has been interpreted as incipient heterospory, a more derived mode of reproduction in spore-producing plants (Andrews et al. 1974). In *Oocampsa*, lateral branches are more dichotomously to pinnately divided, terminating in clusters of ovoid erect sporangia. Large, zonate spores are quite different from spores known from other Early Devonian plants (Wellman and Gensel 2004). Unfortunately, most plant assemblages were transported, to some degree, before burial and preservation, leaving us without any specific environmental context, although it is believed that these plants exhibited less niche partitioning than was suggested for Gaspé.

15.5 Rhynie, the Oldest Vegetated Hot Springs

Milan Libertín and Patricia G. Gensel

It is a quick trip across the Atlantic "pond" to the village of Rhynie in northeastern Scotland, approximately 50 km northwest of Aberdeen. It is hard to imagine that this area hosts one of the most important and famous localities of early Paleozoic plants when we walk through the countryside (Fig. 15.12a). There is no rock exposed at the surface. Rather, it lies beneath the green pasture lands studded with livestock as pictured on our postcard. The locality, known as the Rhynie chert, first was discovered by examining loose blocks turned up in the soil during plowing or as part of the stone walls edging the fields. Around 1912, during one of his collecting trips, Dr. William Mackie (for whom geology was a hobby) found fragments of fossiliferous chert in the dry stone walls, possibly while sitting on one and eating his lunch [U1533] (Andrew and Gensel 1984). He recognized that the chert contained plants entombed in the silicates and took them to Robert Kidston. Along with W. H. Lang, the chert was sectioned and studied, and they produced a series of papers (1917–1921) providing initial descriptions of the fossilized plants. In recent decades, trenches were dug in selected areas to collect additional plant material, and later excavations were undertaken to study the lateral and vertical extent of the deposits. Major

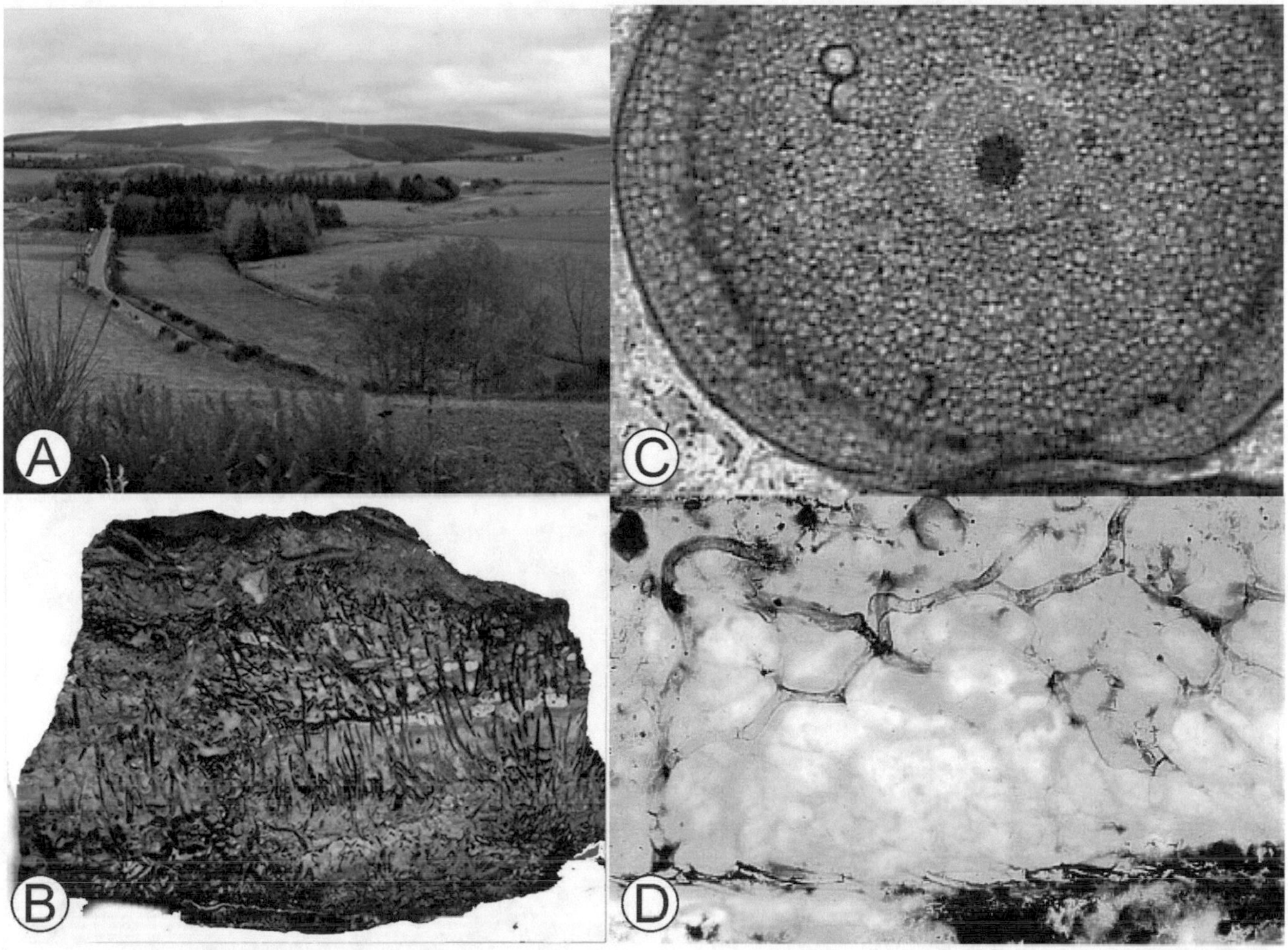

Fig. 15.12 The Early Devonian (Pragian) Rhynie chert, Aberdeenshire, preserved the oldest documented terrestrial ecosystem. (**a**) The Rhynie chert is not surficially exposed but, rather, is the bedrock beneath green pastures. (**b**) Polished specimen of the Rhynie chert showing vertically preserved stems of early land plants. (**c**) Transverse section of *Aglaophyton* stem showing anatomy. (With permission by the University of California-Berkeley Museum of Paleontology) (**d**) Fungal association preserved in cells

drilling and mapping led by geologists at the University of Aberdeen, with the assistance of many collaborators, has resulted in a much-improved understanding of the geology and of the environment in which this earliest ecosystem is preserved (Edwards et al. 2018a, https://www.abdn.ac.uk/rhynie).

The paleoenvironment in which the Rhynie chert formed is surrounded by sandstone and mudrock (shale) and is similar to today's hot spring-and-geyser landscape in Yellowstone National Park, Wyoming, USA (Rice et al. 2003). The fossiliferous cherts were deposited in a tectonic sedimentary basin in which volcanic activity occurred. Sandstone deposits reflect deposition in a braided river system with andesitic (igneous lava rock) flows associated with the fault systems along the margin of the basin. Surface water penetrated through porous sediment in the floodplain to a depth where these were heated by hydrothermal activity. Hydrothermally influenced sediments were intruded by dykes of andesitic

lava and accelerated hydrothermal reaction, pushing heated, silica-rich waters to the surface. Hot springs formed sinter layers that penetrated and enveloped everything living around the hot spring, preserving an intact biota in various stages of vegetative growth and reproduction (Fig. 15.12b).

Plants are preserved in growth position by the sinter, and their spatial distribution is locked into the deposit. The plant and invertebrate community lived around a shallow, temporary, freshwater lake, adjacent to active geysers. The permineralization was so thorough and rapid that even the finest anatomical details have been preserved, providing insight into the life history of several early plant genera. One of the most thoroughly investigated groups in the Rhynie chert is the rhyniophytes (Kerp 2018). In addition, several types of algae (Taylor et al. 1997), fungi (Remy et al. 1994; Taylor et al. 1999), and arthropods (proto-spiders, harvestmen, nematodes) are found (see summary by Dunlop and Garwood 2018).

One of the most common plants close to the hot springs is the genus *Aglaophyton*, currently considered a pro-tracheophyte [U1534]. It grew with an equally dividing (isotomous) branched rhizome, with hair-like extensions called rhizoids that acted to affix the plant to the ground surface and facilitate mineral uptake (Fig. 15.12b). Aerial axes developed from the rhizomes, some of which are preserved with terminal oval-shaped sporangia. Conducting cells in this plant exhibit a unique wall pattern, with anatomical affinities similar to some mosses (Fig. 15.12c). As with other early plants, *Aglaophyton* had no leaves or roots. Another taxon, *Rhynia*, initially considered similar to *Aglaophyton*, bore sporangia on terminating lateral branches and tracheid-like conducting cells (Box 15.4). It is considered part of the Rhyniopsida clade. Gametophytes [U1535],

Box 15.4: Evolution of Specialized Water-Conducting Cells in Silurian-Devonian Plants

Most bryophytes (mosses) lack specialized water-conducting cells, with the exception of certain taxa, where centrally located cells exhibit either smooth or finely pitted, or otherwise ornamented walls. All of these cells apparently lack lignin (a complex organic polymer providing structural support in cell walls). Extant vascular plants are defined, in part, by the presence of specialized, dead, lignified water-conducting cells in their xylem. These are called tracheids or, in flowering plants, tracheids and vessel elements. These cells exhibit particular patterns of lignified secondary walls—annular, helical, scalariform, or pitted—depending on whether cells differentiated early or later in stem or root or by plant group. The presence of lignin promotes preservation and has provided useful characters for distinguishing taxa or lineages and discriminating between stem and root structures.

We find several different developmental patterns when we examine the earliest plants and go back in time. For example, rhyniophytoids lack any evidence of such lignified cells, whereas others show centrally located cells that differ in wall thickness or pattern than those of the rest of the stem. These latter forms are more comparable to cells of some bryophytes. In other early plant lineages of the Silurian to mostly Early Devonian, the following unique types of water-conducting cells, considered lignified and closer to tracheids, have been recognized:

- A late Silurian–Early Devonian *Cooksonia*, *C. pertoni*, exhibits tracheids with two wall layers and annular secondary wall thickenings facing the cell center (lumen).

- S-type cells are typical of rhyniopsid tracheids. These have a two-layered wall that consists of a thin resistant layer facing the hollow center of the cell (lumen) and an outer, less resistant spongy layer. The resistant layer is covered in minute pores. The wall is helically thickened, with the gyre of the helix at different angles.

- G-type cells are found in zosterophylls, *Asteroxylon*, and probably *Baragwanathia* (Lycopsida). This cell type exhibits tracheids with two wall layers, the inner decay-resistant one with closely spaced annular thickenings and a non-resistant outer layer. The spaces in between the inner decay-resistant layer often show a number of small openings.

- P-type tracheids are typical of *Psilophyton* and *Pertica* (basal euphyllophytes or former trimerophytes). The two-layered wall exhibits an inner decay-resistant layer that is closely spaced, parallel, and interconnected. This layer developed overarching edges (scalariform-bordered thickenings), and the decay-resistant layer in between thickenings exhibits one or more rows of round openings (pits). Less resistant material formed the outer wall and the area within the scalariform thickenings.

Recent discoveries have shown some plants, similar to *Psilophyton*, but with secondary xylem, exhibit round-oval bordered pits in primary and P-type pitting in secondary xylem (*Franhueberia*). By end of Early Devonian, secondary wall thickenings (pitting patterns) were more similar to those of extant plants, first appearing in lycopsids.

which are multicellular, haploid, sexual structures in plants, are known in considerable detail, including the structures containing egg and sperm (Kerp et al. 2003; Taylor et al. 2005). Both of these taxa exhibit endophytic (within the plant cells) fungi that probably aided in water uptake (VA mycorrhizae). The Rhynie flora was more diverse than just rhyniophytes.

Nothia aphylla is another well-preserved plant more closely related to the zosterophyll clade (Kerp et al. 2001). It had spreading rhizomes, perhaps partly below ground, with rhizoids, and upright axes that branched dichotomously. The aerial axes were covered by elliptical emergences, many topped by a stoma (a specialized gas exchange structure). Sporangia producing only one type of spore developed near branch tops. To release the spores, each sporangium opened around its margins. *Nothia* was a geophyte in that it grew in sandy soils and reproduced clonally. Underground rhizomes survived from season to season, and elevated axes grew again annually. Another geophyte, the proto-lycopsids, also is part of the hot spring landscape.

A vascular plant, allied to the lycopsids, is the genus *Asteroxylon*. These plants also grew in sandy substrates more distant from the main sinter zone, but still were permineralized by silica (Kerp 2018). The rhizomes of *Asteroxylon* were positively geotropic, growing into the soil substrate [U1536]. Rooting structures branched equally (isotomous) whereas aerial stems, 1–2 cm wide and possibly up to 40 cm in height, branched unequally (anisotomous). These are covered with helically arranged, unvascularized leaf-like structures, and vascular strands extend into the cortex almost, but not quite, to the level of leaf-like attachment. Both aerial axes and the leaf-like structures possessed stomata. The internal anatomy exhibits several features not found in other groups at the time. The xylem in the aerial axes consists of lignified, simple conducting cells (tracheids) with closely spaced thickenings that encircle the cell (annular thickening). Their arrangement forms a star-shaped pattern. Similar to lycopsids, the sporangia of *Asteroxylon* are kidney-shaped (reniform) and developed on a short stalk (pedicel). Fertile zones are arranged spirally on axes interspersed among sterile ones (Kerp et al. 2013). This arrangement indicates the potential for periodic sexual reproduction promoted by changing environmental conditions. Dispersed spores described from these sediments indicate that the vegetation of the larger region was more diverse than the plant association preserved in the Rhynie chert (Wellman 2010).

Significant discoveries in this locality include another part of the Rhynie ecosystem, the fungi (Fig. 15.12d) [U1537]. Fungi serve several functions in an ecosystem, ranging from mutualistic to saprophytic. Rhynie fungi may be some of the best detailed forms, with mutualistic fungi allied to Glomales found inside plants (Taylor et al. 1992; Krings et al. 2017), as well as saprophytic forms degrading them (Taylor et al. 2003). The relationship between the water fungus *Sorodiscus*, which attacked the cells of the alga *Palaeonitella*, is one of the first examples of parasitism in the fossil record (Taylor et al. 1992).

The preservation of extremely minute details, which allows the investigation of vascular systems, reproductive organs, spores, generation of gametes, and even seasonal growth of plants, allows us to recreate a picture of the entire Rhynie hot spring ecosystem (Channing and Edwards 2009). These fossil Lagerstätten with complex preservation potential are very valuable [U1538]. They are windows, frozen in time, that enhance our understanding of early vascular plant evolution (Trewin and Kerp 2018). It was the exquisite preservation of the plants in the Rhynie chert that convinced earlier geologists and botanists that pre-Carboniferous terrestrial plants existed.

15.6 Bathurst Island, Canada: A Counterview to the Hot Springs

Patricia G. Gensel

We now travel from the modern conveniences found in one small village in Aberdeenshire, Scotland, to a very remote island setting in the high Arctic where we'll get a different perspective on late Silurian and Early Devonian vegetation. Back in time, this island was part of a large tectonic block located around the equator. In contrast to the mostly small and simple rhyniophytoids described from many Silurian localities in Laurussia, Baltica, and South American assemblages, a walk through these equatorial regions brings us to another worldly view. Late Silurian plants in eastern Bathurst Island, Nunavut, Canada, are somewhat familiar in their basic architecture and structure (Basinger et al. 1996; Kotyk et al. 2002). Plants attained several centimeters in length and stems were as wide as 4 mm, more closely resembling Early Devonian taxa. Unlike fossil-plant assemblages we've previously visited, these are preserved in offshore, deep marine, fly ash deposits ("Bathurst Island beds") securely dated as Silurian (late Ludlow or Ludfordian, 426–423 Ma) based on graptolites, conodonts, and brachiopods (see Chap. 14, for more information on these early animals). The sedimentological context of these assemblages indicates that the plants were deposited by mudflows in a marine basin, where they were quickly buried with little biological (bioturbation) activity to alter them. These mudrocks now are exposed mostly along stream margins on the island.

Transport of the plants to the marine realm resulted in their partial deterioration and most appear as incomplete portions of vegetative or fertile structures. Stem fragments, some with spines, are associated with fertile specimens that are referable to seven distinct taxa previously known *only* from the Early Devonian. They represent members of the zosterophyllopsids and plants bearing terminal sporangia more similar to typical rhyniophytoids, although larger in size.

The zosterophylls preserved here include taxa that vary mostly in the structure and organization of their sporangia. These reproductive structures can be organized in a helical or subopposite arrangement (different species of *Zosterophyllum*), or they can be borne in dense, two rowed spikes oriented toward one side of the stem (*Distichophytum*). A zosterophyll that had first been found on Bathurst Island, *Macivera gracilis*, exhibits sporangia that are longer than wide and located only in the distal regions of a branched stem. The Silurian species of these genera are smaller in size than their Devonian congenerics.

A brief visit to the Early Devonian (Pragian, 411–408 Ma) of Bathurst Island provides insight into the signifi-

Fig. 15.13 Late Silurian plants from Bathurst Island, Arctic Canada. (**a**) *Bathurstia* sp. (**b**, **c**) *Zosterophyllum* sp. (from Kotyk et al. 2002. Morphologically complex plant macrofossils from the late Silurian of Arctic Canada, Am. J. Bot. 80(6): 1004–1013, with permission, Wiley Press), color photos PG Gensel

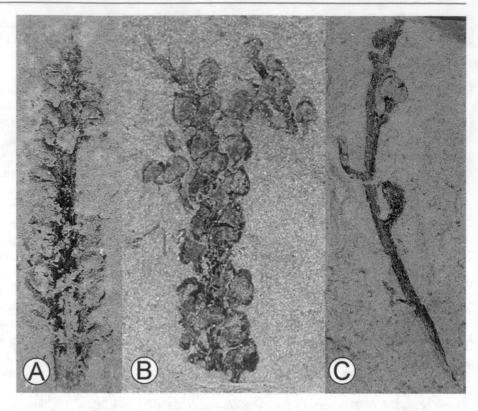

cance of these fossils [U1539]. The assemblage (Kotyk 1998) is dominated by zosterophylls, especially different types of the genera *Zosterophyllum* and *Distichophytum*. These differ mainly in size, being considerably larger than their Silurian counterparts. Here, we also find the (pre)-lycophyte *Drepanophycus*, some with putative rooting structures (Kotyk 1998). Kotyk and Basinger (2000) published a description of another zosterophyll, *Bathurstia denticulata*, where axes are covered with emergences and one specimen is exceptionally preserved attached to its rhizome [U1540]. Parts of the K-type branching pattern of this taxon appear to be rooting structures as well as aerial fertile axes (Fig. 15.13). Other taxa also occur on Bathurst Island and can be found in localities across different present continents.

Early Devonian and latest Silurian rhyniophyoids and zosterophllopsids are reported from China, Europe, and South America. For example, the Pragian Posongchong Formation in China is very rich in zosterophylls (Hao and Xue 2013). Fossils from localities in the Přídolí (423–419 Ma) of Podolia, Czech Republic (Bohemia), and Brazil are entities larger than the tiny rhyniophytoids noted above and below. Even without consideration of controversial plant assemblages in Australia (e.g., Tims and Chambers 1984), the Bathurst Island fossils, and possibly those from Podolia and Brazil, indicate that plants more complex than rhyniophytoids existed in the late Silurian. Additionally, the occurrence of very similar zosterophylls and lycopsids over more than 25 million years

tells us that these groups remained relatively static over that time span. Thus, the more complex and larger plants in these Silurian deposits indicate an earlier appearance of most vascular plant clades than the fossil record currently shows. This conclusion is supported by the dispersed spore record.

15.7 The Diminutive World of the Clee Hills of Shropshire

Ian Glasspool

In 1937, William H. Lang published "On the plant-remains from the Downtonian of England and Wales". This seminal paper focused on transitional Downtonian (uppermost Ludlow to Lochkovian; c. 425–411 Ma) successions from the Clee Hills of Shropshire in the Welsh Borderlands of the United Kingdom [U1541]. During the late Silurian, the locality was along the margin of the Laurussian continent. It's now time to head back across the pond to see what Lang first observed, and what has subsequently been much expanded upon by Dianne Edwards (e.g., Edwards et al. 2014).

Examining what had been thought to be unpromising fossil remains from terrestrial rocks, Lang was able to describe a highly diminutive (<10 cm tall; Edwards 1996), taxonomically simplistic, rootless, and leafless flora that included the first description of the now iconic genus *Cooksonia* [U1542]. Lang's localities, including Ludford Lane, Perton Quarry,

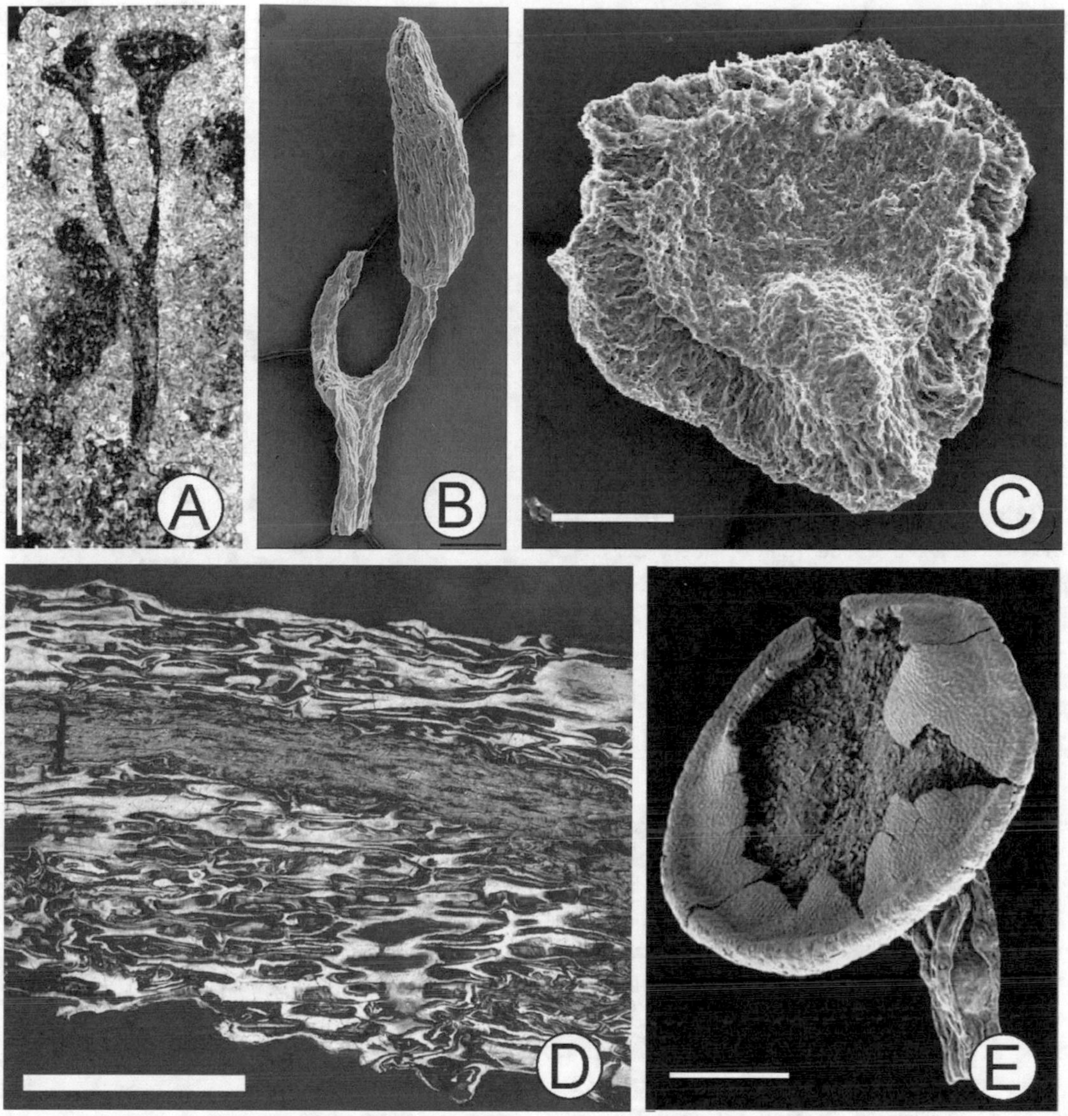

Fig. 15.14 Early Devonian plants from Shropshire, United Kingdom. (**a**) A compression specimen, the counterpart to the lectotype of *Cooksonia pertoni* from the Přídolí at Perton Lane, the surrounding black patches may be *Nematothallus*. NHM V58010 Scale bar = 2 mm. (from Edwards and Kenrick 2015. The early evolution of land plants, from fossils to genomics: a commentary on Lang (1937) 'On plant-remains from the basal Downtonian of England and Wales'. Phil. Trans. Roy. Soc. Lond., with permission). (**b**) A cryptospore-bearing plant with a simple, fusiform sporangium from the Lochkovian of North Brown Clee Hill assignable to *Fusiformitheca fanningiae*. Scale bar = 500 μm. (from Wellman et al. 1998. Permanent dyads in sporangia and spore masses from the Lower Devonian of the Welsh Borderland, Bot. J. Linn. Soc. 127(2): 117–147, with permission, Oxford Univ. Press.) (**c**) The lower surface of the thallus of *Nematothallus* sp., Ludlow, upper Silurian, Downton Castle Sandstone Formation, Ludford Lane. This specimen illustrates the tripartite organization, from left to center: cortex, palisade* tissue and fused basal layer. Scale bar = 200 μm. (Edwards et al. 2013. Contributions to the diversity in cryptogamic covers in the Mid-Palaeozoic: *Nematothallus*-revisited. Bot. J. Linn. Soc. 173:505–534, with permission, Oxford Univ. Press) (**d**) Differentially charred axis of *Hollandophyton colliculum* from the basal Přídolí of Ludford Lane. (from Glasspool et al. 2004. Charcoal in the Silurian as evidence of the earliest wildfires. Geology 32(5):381–383 with permission, Geol.Soc.America) (**e**) Charred, anatomically preserved *Cooksonia pertoni* subsp. *apiculispora* from the Lochkovian of Brown Clee Hill, Shropshire, containing *Aneurospora newportensis* spores. Scale bar = 500 μm. (from Edwards et al. 1992. A vascular conducting strand in the early land plant *Cooksonia*. Nature 357(6380): 683, with permission, Springer Nature)

Targrove, and Tin Mill Race, continue to be studied and are advancing our understanding of the evolution of early land plant body plans and their paleoecology, as well as offering insights into plant-arthropod interactions and latest Silurian–earliest Devonian food webs. Paleobotanically, these sites have demonstrated that *Cooksonia pertoni* had both tracheids and stomata, and that its spores varied over time (an example of cryptic evolution; Fanning et al. 1988). These localities have also yielded a far greater range of rhyniophytoids than had previously been appreciated, with *Cooksonia*-like plants such as *Hollandophyton colliculum*, *Tortilicaulis offaeus*, *Culullitheca richardsonii*, *Fusiformitheca fanningiae*, and others (Morris et al. 2011, 2018a). Much of this new evidence is not derived from the adpression fossils of *Cooksonia* that are so familiar in classic textbooks. Rather, our insights come from exquisite three-dimensional and anatomically preserved charred fossils that represent some of the earliest evidence of wildfire known on the planet (Fig. 15.14; Glasspool et al. 2006). Studied by SEM, these fossils exhibit incredible, even subcellular, anatomic details. These details reveal an early terrestrial flora characterized by "cryptogamic covers", a soil crust comprising a complex of bacteria, cyanobacteria, algae, fungi, lichens, nematophytes (an enigmatic group that may have fungal affinities; Edwards et al. 2018b), basal tracheophytes (e.g., *Cooksonia hemisphaerica*), and cryptospore-bearing plants (e.g., Edwards et al. 2014). Although these floras have been termed "Lilliputian" (Edwards 1996), their role in early terrestrialization and the evolving biogeochemical carbon cycle of the latest Silurian and earliest Devonian is anything but small.

15.8 Pre-Devonian Land Plants

Ian Glasspool, Jiří Kvaček, and Milan Libertín

There are a number of small-stature plants and enigmatic plant groups that appear in the pre-Devonian fossil record. Some of these forms look like plants, while others are more amorphous in their organization. We have seen the iconic oldest truly vascular plant (Lang 1937; Edwards et al. 1992), *Cooksonia*, in Shropshire, but species assigned to it are known from several localities in Europe, North America, northern Africa, South America, and China (Taylor et al. 2009). The oldest currently known species, and one of the larger plants, is *C. barrandei* from the Czech Republic (Fig. 15.15) [U1537]. Similar to many other localities from which it is described, the Czech rocks are not of continental origin. These fossils are described from the middle Silurian *Monograptus belophorus* marine Biozone of Wenlockian age (432 Ma; Libertín et al. 2018a, b). Like other members of the group, *C. barrandei* has twice-branched, relatively "robust" axes up to 1 mm in width, bearing terminal funnel-form sporangia [U1543]. Slightly younger examples of the genus *Cooksonia* (e.g., *C. pertoni, C. cambrensis,* and *C. hemisphaerica)* have been described from Wenlockian strata in County Tipperary, Ireland (Edwards et al. 1983). Due to many examples being exceptionally preserved as charcoal, the species *C. pertoni* is probably the most comprehensively studied of all *Cooksonia* species (see Morris et al. 2012). Whereas different examples of this species are morphologically and anatomically homologous, four subspecies are recognized based on differences in the spores found in situ in their sporangia (Fanning et al. 1988; Habgood et al. 2002; Morris et al. 2012).

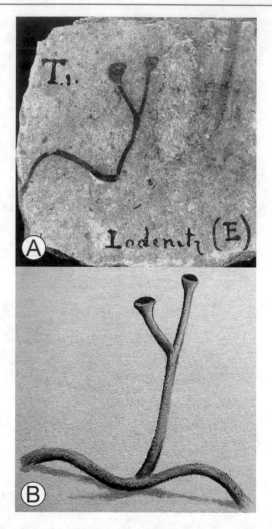

Fig. 15.15 *Cooksonia barrandei.* (**a**) Isotomously branched axis with sporangia, Loděnice, Špičatý vrch—Barrandovy Jámy, Czech Republic. Scale bar = 10 mm. (**b**) Reconstruction by Jiří Svoboda. (Both with permission Wiley Press)

Cooksonias are currently placed in the polysporangiophytes*, which are sporangia-bearing plants that may, or may not, contain vascular tissues. The small size of many *Cooksonia* and other rhyniophytoid taxa has led to the question of whether they were able to adequately photosynthesize, or if they may have remained attached to their gametophyte structure for that purpose (Boyce et al. 2007). There are a variety of *Cooksonia* forms, some of which have been assigned to more than one lineage. For example, the genus *Aberlemnia* is morphologically very similar to *Cooksonia*, but its sporangia are bilobate, opening with two flaps. Based on this character, Gonez and Gerrienne (2010) assigned it to the stem or basal lycopsids. Perhaps the best known, possible early lycopsid is *Baragwanathia* [U1544]. This plant gained notoriety due to its presumed late Silurian age, large size, and relative organizational complexity that often rendered it subject to debate (Hueber 1983; Garratt et al. 1984). First described by Lang and Cookson (1935) from the late Silurian of Yea, in Victoria, Australia, it now has been reported from other Early Devonian localities including Canada (Hueber 1983) and China (Hao and Xue 2013). *Baragwanathia* grew along the ground (procumbent) and had forking, branched axes that gave rise directly to occasional roots [U1545]. Similar to zosterophylls and other lycopsids, the central conducting cylinder formed from a star-shaped arrangement of tracheids. The sporangia grew in the axils of vascularized microphylls, which were helically arranged on the axes. Other stem group lycopsids of late Silurian age were discussed in Sect. 15.6.

Several other late Silurian plants had similar growth architectures to *Cooksonia*. *Steganotheca* (Edwards 1970) is somewhat more equally branched than *Cooksonia* and has elongate, striated, and flat-topped sporangia terminally arranged on gradually widening axes. This plant is known from the late Silurian to earliest Devonian. Although it is considered to be a vascular plant, there is no definitive proof, to date, about either the presence or character of tracheids in the axes. Recently, the genus *Tichavekia* was found in association with *Cooksonia* in the Prague Basin (Kraft et al. 2018). The plant axes attained lengths of at least 14 cm and branched up to five times equally (isotomously), each branch being no wider than 1 mm [U1539]. The terminal dichotomies of the plant were short and tipped with oval sporangia forming groups of four. In addition to the rhyniophytoids, other pre-Devonian organisms did not possess true stems, vascular tissue, or sporangia, such as the nematophytes.

The curious group of the nematophytes have long been considered neither alga nor vascular plant, leading some to consider them transitional forms [U1546]. New data indicate otherwise. One such genus is *Nematothallus* (Lang 1937) and often is preserved as either a resistant cuticle, with "cell outlines" or on which there are openings that may have functioned similarly to stomata. Some specimens consist of cuticle with underlying wefts or aligned tubes, and occasional banded tubes may occur intermixed. It now seems that *Nematothallus* is either a fungus or lichen, and some other layered tubes with cuticle might represent lichens (Fig. 15.16; Honegger et al. 2012; Edwards et al. 2013). Some of the ornamented (banded) tubes may represent epibionts (microbes living on or within these organisms). Another intriguing type of fossil is *Parka* (Fleming 1831), a flattened oval (thallus) with rounded bodies on it, present in the late Silurian and continuing into the Early Devonian. Its similarity to a charophycean green alga, *Coleochaete* (where zygotes appear as round to oval bodies on the algal thallus) is interesting because molecular phylogenies identify charophyceans as the sister group of the land plants. An organism like *Parka* can help us figure out the aspect of their possible common ancestor.

Other enigmatic plants include flattened axial structures, up to 20 cm in length, that most probably represent cuticles

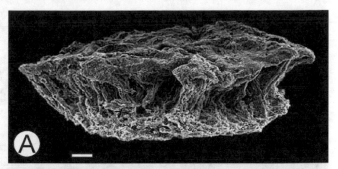

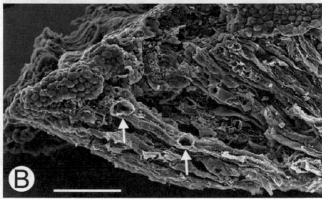

Fig. 15.16 (**a, b**): Fragment of *Nematothallus williamii*, Lochkovian, Shropshire in two magnifications of the same specimen showing a pattern typical for *Nematothallus* cuticle. Arrows indicate positions of lateral branches or areas. Scale bar 100 μm. (from Edwards et al. 2013. Contributions to the diversity in cryptogamic covers in the Mid-Palaeozoic: *Nematothallus*-revisited. Bot. J. Linn. Soc. 173:505–534, with permission, Oxford Univ. Press)

of stems. One example is *Orestovia* (Ergolskaya 1936) occurring in the Early Devonian of the Kuznetsk basin of Siberia. Primitive stomata, conducting cells, and spores have been interpreted in the taxon (Kräusel and Venkatachala 1966). Details of their sunken stomata, shown in thin sections, are known from specimens that lacked any associated spores or conducting cells (Gensel and Johnson 1994). *Orestovia* and some related forms have extremely thick and resistant cuticles and form thick deposits of so-called paper coal in the Lower Devonian of Russia, which have been used as a fuel source.

15.9 The Oldest Evidence for the Colonization of Land

Milan Libertín, Jiří Kvaček, and Ian Glasspool

The oldest evidence of land being colonized by plants comes from the dispersed spore record (Gensel 2008; Rubenstein et al. 2010). Derived plants can be distinguished from algal precursors by their spores, which are developed into tetrads via meiosis, encased in a sporopollenin wall, and, subse-

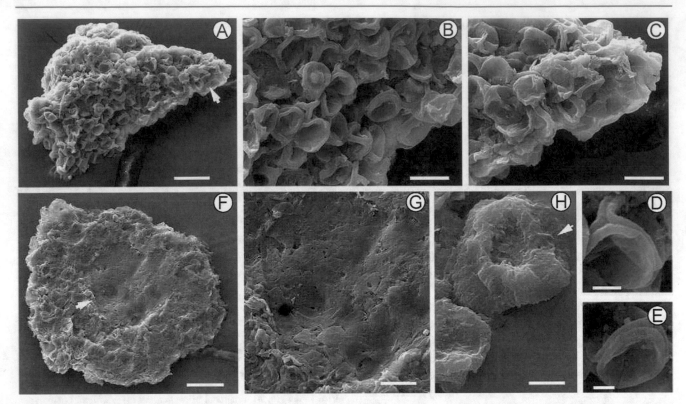

Fig. 15.17 Fossil plant fragments from the Ordovician (Llanvirn, 475 million years ago) of Oman. (**A**) SEM of fragment of sporangium containing naked permanent tetrads. Note the presence of sporangium covering in the bottom right-hand corner (arrow). Scale bar 50 μm. (**B**) Close-up of a illustrating the spore contents. Scale bar 20 μm. (**C**) Close-up of **A** illustrating spores overlying the sporangium covering. Scale bar 20 μm. (**D, E**) Close-up of **A** illustrating individual spore tet- rads. Scale bar 5 μm. (**F**) Specimen CW47f. SEM of relatively complete sporangium, with a large patch of sporangium covering preserved (arrow). Scale bar 75 μm. (**G**) Close-up of **F** illustrating the nature of the sporangium covering. Scale bar 30 μm. (**H**) Specimen CW47i. SEM of an envelope-enclosed permanent tetrad that is preserved in a frag- mentary sporangium. Note the muri ornamenting the envelope (arrow). Scale bar 10 μm. (Wellman et al. 2003. Nature 425(6955):248–9 © Springer Nature with permission)

Box 15.5: Cryptospores Differ from Acritarchs and Trilete Spores

Permanent tetrahedral tetrads in which trilete marks, typical of vascular plant spores, first are detected were reported by Gray and Boucot (1971) from early-to-mid-Llandovery (444–433 Ma) rocks of New York State. They regarded permanent tetrads and permanent monads lacking a haptotypic* mark as being derived from land plants, not algae. In fact, Gray (1985) argued they exhibited features closer to liverworts. Richardson (1985) coined a term for these, plus dyads and monads lacking a haptotypic mark, namely cryptospores.

Cryptospores occur as monads (single spore), permanent dyads (two fused spores), or tetrads (four fused spores) (Figs. 15.17, 15.18, and 15.19) [U1542]. Some tetrads are enclosed in an outer envelope, whereas other examples are not. Spore assemblages containing cryptospores are found in Cambrian (but see below) to Devonian sediments, mainly from the paleotropics, and in marine to terrestrial deposits. They are different from acritarchs in that they exhibit a more robust wall and/or occur in tetrads.

The oldest fragments of a sporangium in which cryptospores, in the form of permanent tetrads, are found, comes from the Llanvirn (Ordovician; 475 Ma) of Oman. Analysis of their wall ultrastructure supports a possible liverwort affinity (Wellman et al. 2003). Other ultrastructural studies of cryptospore walls from Darriwilian-aged (467–458 Ma) material show the presence of homogeneous wall structure. This is a characteristic of living embryophytes where the layer is secreted by an active tapetum. The presence of a tapetum has been used to imply that these Ordovician cryptospores developed inside a sporangium, although fossilized sporangia of this antiquity are not known (Taylor et al. 2017). Other cryptospores exhibit a multi-laminated wall as found in many liverwort spores. The extant liverwort *Haplomitrium gibbsiae* has also been shown to regularly produce cryptosporic permanent dyad pairs (Renzaglia et al. 2015).

Fig. 15.18 Cryptospores. (**A**) Permanent spore tetrad in tightly attached tetrahedral configuration. *Tetrahedraletes* sp. of Hirnantian age. (**B**) Cryptospore spore tetrad in planar configuration, *Tetraplanisporites* of Hirnantian age. (**C**) Broken tetrahedral cryptospore tetrad of *Imperfectotrileites vavrdovii*; spores have broken away from a permanent tetrad. (**D**) Permanent cryptospore dyad, *Dyadospora murusdensa*. (**E**) Permanent cryptospore dyad, *Dyadospora* cf. *murusattenuata*. (**F**) Permanent cryptospore tetrad enclosed in a reticulate synoecosporal wall, *Velatitetras* cf. *retimembrana*. All originate from the Power Glen Formation, Hirnantian age. Balls Falls Provincial Park, Ontario, Canada. (Images courtesy of P Strother)

The parent plant fossils from which cryptospores have been obtained were recently placed into a basal group of early land plants described by Edwards et al. (2014). Cambrian forms have been referred to "algae" aligned with the embryophyte clade (probably extinct forms), rather than to the sister clade of the green algae (Strother 2016). Cryptospores (Fig. 15.18) dominated spore assemblages until the Late Ordovician when a few trilete forms, single spores resulting from disassociated tetrads, appeared in small numbers. These became more diverse and abundant in the Silurian, particularly the Wenlock (Steemans et al. 2009, 2010; Wellman et al. 2013).

The oldest trilete spores are known from the mid- to Late Ordovician of Saudi Arabia based on chitinozoan and acritarch biostratigraphy (Steemans et al. 2009). These forms range from Katian (453–445 Ma) to Hirnantian (see Chap. 14) and may represent the earliest evidence of vascular plants [U1548]. However, trilete spores occur in some mosses, although many are alete (without a lete mark). It remains probable that plants producing trilete monads may have come from a broader morphological group of basal embryophyta.

quently, separated (Strother and Taylor 2018). The majority of the early spores (Box 15.5) are cryptospores, occurring in obligate tetrads, dyads (pairs), or singularly as monads. Ultrastructural data and in situ cryptospores suggest a basal embryophyte affinity. Trilete spores*, ones with a Y-shaped scar delimiting site of opening for spore germination, are typical of vascular plants (tracheophytes) and only a few bryophytes. Early records of trilete spores consistently came from the Llandovery (basal Silurian, 444–433 Ma) until a recent report by Steemans et al. (2009), in which several types of trilete spores were reported from the Upper Ordovician (Katian, 543–445 Ma) of Saudi Arabia. In many Silurian samples, cryptospores and/or trilete spores may co-occur with isolated cuticles or tubes of uncertain affinity, which could be remnants of nematophytes.

To gain an impression of how the earliest land plant vegetation may have looked, we must return and visit two places in the Welsh Basin, both in Shropshire. The first locality is from the latest Silurian (Přídolí, c. 419 Ma); a second locality, a profile of the Brown Clee Hill, is about four million years younger. Very small but remarkably well-preserved mesofossils containing cryptospores occur in both sites (e.g., Morris et al. 2018a). Based on studies

Fig. 15.19 Cryptospores. (**A**). Permanent cryptospore tetrad. *Tetrahedraletes medinensis*. (**B**). Permanent cryptospore tetrad enclosed in a reticulate synoecosporal wall, *Velatitetras retimembrana*. (**C**). Two specimens of permanent cryptospore tetrads, one in tetrahedral configuration, the other in a cross configuration. These are considered to be taphonomic variants of the same taxon, *Tetrahedraletes medinensis*. All cryptospores originate from the Tuscarora Formation, Llandovery (Aeronian) age. Mill Hall, PA, USA. (Images courtesy of P Strother)

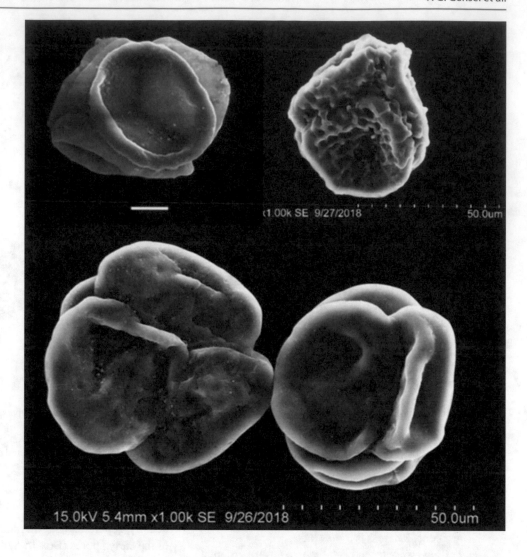

from these localities, it is possible to say that producers of cryptospores grew to only a few millimeters in size and probably had a stature similar to small mosses (bryophytes) (Steemans et al. 2009; Kenrick et al. 2012). Even at these small sizes, we can discriminate several different cryptospore-bearing groups. *Partitatheca* includes plants with dichotomously branched axes terminated by sporangia with stomata, which conform to the appearance of *Cooksonia*. But, these plants produced permanent dyads with a laminated spore-wall structure that are assigned to the dispersed spore genus *Cymbohilates* (Edwards et al. 2012). *Lenticulatheca,* recognized in the same assemblage, has discoid sporangia, containing similar dyads of the same group (*Cymbohilates*). Axes of *Grisellatheca* were terminated by short, dichotomously branched axes bearing slightly elongate sporangia, and these contain permanent tetrads of the *Tetrahedraletes*-type (Edwards et al. 2014). Dispersed forms of cryptospore occur earlier in southern

Gondwana, and apparently radiated into Avalonia, then Euroamerica, and Baltica (Wellman et al. 2013).

15.10 Discussion

Patricia G. Gensel and Milan Libertin

Colonization of land by photosynthesizing plants is one of the most important evolutionary events in the history of the biosphere and appears to have occurred in several steps, beginning in the Late Ordovician, or possibly the Cambrian, and continuing into the Devonian (Strother and Taylor 2018). There are several requirements for an organism to survive and propagate on land. One critical abiotic parameter was the presence of an ozone layer, at least a weak one, to block cosmic (radioactive) and ultraviolet radiation, which damages cellular DNA. Only after an ozone layer was generated, could organisms colonize land! Algae were the first plants to abandon fully marine chemistries and move, initially, into the intertidal brackish zone and, later, freshwater environments. Algal groups diversified during the early Paleozoic and included single-celled flagellated organisms to highly diversified thalli, measuring up to several meters in length. Some green algae, primarily brittleworts, Zygnematophyceae, and related charophycean algae achieved significant diversity during the Ordovician, and it seems likely that they inhabited shallow water settings and perpetually damp nearshore environments. A move to conquer moist to dry land followed with the advent of evolutionary innovations [U1549]. Although the unfiltered sunlight on land facilitated photosynthesis, heat affected internal cellular water and nutrient relationships. To prevent desiccation, a sheathing in a resistant compound evolved. Several lines of evidence are used to clarify relationships of chlorophytes (green algae) and their descendants, including photosynthetic pigments (chlorophyll A and B), a common storage product (starch), and molecular features. Chlorophytes, though, are not closely related to terrestrial green plants. Rather, a separate branch of the green algae, the charophytes are considered to be their sister taxa (Fig. 15.20; McCourt et al. 2004). The emergence of terrestrial plants is still not well understood, because the soft tissue of these colonizers had a very low preservation potential, which is also true for the earliest vascular plants.

Two primary hypotheses are put forth for the emergence of vascular plants. The first is that vascular plants evolved from mosses and their ancestors, plants that produced one type of sporangium (monosporangiate* plants; Graham et al. 2000). The alternative view is that they evolved from the pre-

decessors of hornworts, from plants in which more than one type of sporangia developed (polysporangiate; Puttick et al. 2018). On the basis of the most recent phylogenetic analysis (Morris et al. 2018b), it seems that vascular plants have a common ancestor with hornworts.

Beginning some 432 million years ago until the end of the Devonian, approximately 75 million years in duration, we have seen in this chapter a considerable change in plant type, size, diversity, and complexity. We have witnessed major innovations in plant organs such as the first appearance of leaves and roots, changing reproductive modes, and the evolution of wood (secondary xylem).

The end of our adventurous journey is the most important event in the history of plant evolution. This is their adaptation from a fully aquatic to a fully terrestrial environment. In addition to features discussed earlier, this phenomenon is associated with the development of a two-parted life strategy, involving evolution of a longer-lived, complex sporophyte generation, along with adaptations necessary to sustain life on land. In plants other than bryophytes, an independent diploid sporophyte generation, namely the diploid roots, stems, and leaves represent an evolutionary novelty. Hypotheses as to the evolution of a dominant sporophyte-based plant include the homologous and antithetic theories.

The homologous origin of alternation of land plant generations that was originally introduced by Čelakovský (1874) supposes that land plants arose from ancestors of green algae with isomorphic (equal morphologies) haploid and diploid phases. On the other hand, the antithetic (or interpolation) hypothesis supposes a heteromorphic (two different morphologies) haploid and diploid phase, where

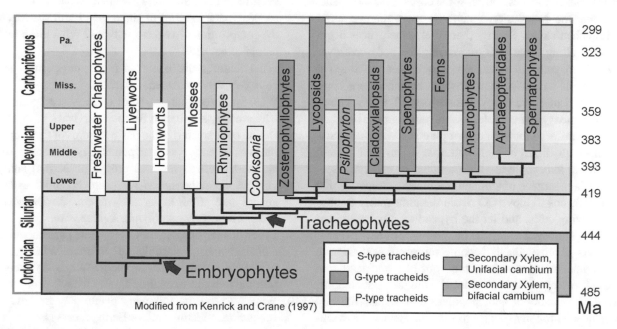

Fig. 15.20 Phylogenetic relationships between the major Paleozoic plant groups. Land plants correspond to embryophytes, and vascular plants to tracheophytes (modified from Kenrick and Crane 1997)

the haploid gametophyte phase was gradually reduced. The diploid sporophyte became more complex as mitotic cell division in the zygote formed a multicellular structure as a result of delayed meiosis (Bower 1908). Ideas as to how this dominant sporophyte generation evolved over the gametophyte are still under discussion, with ideas about Early Devonian gametophytes recently presented (Kerp et al. 2003; Kenrick 2018).

Land plant diversification [U1550] significantly influenced the development of the lithosphere and, in turn, other Earth systems in many different chemical and physical ways [U1551]. Probably, the most significant of these was that of soil development and the stabilization of continental sediments. Terrestrial plants affected weathering and erosion and, as a consequence, river systems (Gibling and Davies 2012; Gibling et al. 2014). The expansion and diversification of land plants impacted climate, especially in incorporation of CO_2 into organic carbon that either was held by plants, incorporated into soil, or transported to the oceans (Berner and Kothavala 2001). The fossil record of this time period demonstrates that nearly every important phase of plant evolution happened in the Devonian (and according to H. P. Banks, "the rest is icing on the cake").

15.11 Conclusions

As we have walked through these landscapes, it is clear that the earliest land plants, now extinct, were very different from those that we see around us at present and some interpretation is needed [U1552]. The earliest land plants of the latest Silurian and Early Devonian generally lacked recognizable roots and leaves and, in some cases, vascular tissue, giving one the impression that these were nothing more than branched sticks. Shortly thereafter, emergences begin to develop along these axes, increasing the body area over which photosynthesis could possibly occur. As internal water-and-gas exchange relationships became more complex, we find that two innovations evolve. The first is evidence of some type of vascular tissue, ranging from lignified tracheids similar to those in extant plants to water-conducting cells with different wall patterns, or no wall pattern more similar to bryophyte-grade conducting cells. The second is the appearance of stomata, regulatory structures that provide a means to move CO_2 from the atmosphere to photosynthesizing cells, and let the byproduct, O_2, be emitted back to the atmosphere despite the presence of a waterproof cuticle. As photosynthesis became more efficient and the need to uptake water increased, root-like and true root structures, many with mycorrhizal (fungal) associations, evolved. Propagation and population sustainability are always needed for any species to survive. Hence, the evolution of the sporangium, the structure in which meiosis occurs to produce haploid spores. Sporangia were borne first terminally or laterally on stems, singly or in groups. Sporangia vary widely in shape, presence, or absence and, if present, location on the stem, type of dehiscence structure, and spore type. But, releasing spores into a hostile environment also required protection from desiccation, a problem solved by terrestrial plant ancestors that had evolved a resistant and robust spore-wall chemical, sporopollenin. These innovations set the stage for the conquest of all continental environments, the establishment of a myriad of ecosystems, and an ever-changing planetary surface, with the comings and goings of plant groups over the course of the Phanerozoic.

Questions

1. What defines a forest? When do the earliest forests occur? How are they different from extant ones? Be able to describe two types of plants that form the canopy of an early forest. What forms of preservation have enabled us to recognize the existence of forests? What limitations do we have in terms of characterizing early forests?

2. Where are plants preserved, and inferred to have grown, during the Late, Middle, and Early Devonian, respectively? What type of vegetation was present?

3. Which plant groups are embryophytes?

4. Some plant stems, and/or sporangia, are covered with emergences, and these frequently are used to define taxa. What are these structures?

5. Name several major innovations in plant size, anatomical organization, architecture, or reproduction that occurred in the Silurian-Devonian. What is the significance of each in terms of changes to Earth systems or to the composition of vegetation types? What is the earliest record of each?

6. What features distinguish a progymnosperm, such as *Tetraxylopteris* or *Archaeopteris*?

7. What are three characteristics of early seed plants, and why are they significant for their survival?

8. Define or characterize the basic features of the four main lineages of early vascular plants (rhyniaceans, zosterophylls, trimerophytes, and progymnosperms) and provide an exemplar genus for each. Potential exemplar genera are *Cooksonia, Sawdonia, Zosterophyllum, Psilophyton, Pertica, Rhynia, Aglaophyton*

9. Lycopsids (zosterophylls + lycophytes) are known to be a distinct lineage since the late Silurian. What defines a lycophyte versus a zosterophyll? When do lycopsids become tree-like? Are they similar today?

10. What role did the following plants play in structuring vegetation, affecting soils or Earth processes, or in evolutionary changes in lineages? *Eospermatopteris/Wattieza,*

Rhacophyton, Archaeopteris, Protolepidodendropsis, Leclercqia, Elkinsia, and its relatives.

11. What is a Fossil Lagerstätte? Why might the Rhynie chert be considered a Fossil Lagerstätte?

12. Some paleobotanists suggest that lichens, which today are pioneer plants in establishing soils, existed during the Devonian. What taxa may represent lichens? Fungi? How might nematophytes address this question?

Acknowledgments The authors want to thank the following colleagues, in alphabetical order, for images used as figures and in the accompanying lecture slides: Dr. Christopher Berry, Cardiff University, UK; Dr. Walter Cressler, West Chester University of Pennsylvania, USA; Dr. Jeffrey Doran, Canada; Prof. Dr. Hans Kerp, Universität Münster, Germany; Dr. Douglas Jensen, Converse College, Spartanburg, SC, USA; Prof. Stephen Scheckler, Virginia Polytechnic Institute and University, Blacksburg, USA; Dr. William Stein, State University of New York-Binghamton, USA; Dr. Paul Strother, Boston College, USA.

References

Algeo TJ, Scheckler SE (1998) Terrestrial-marine teleconnections in the Devonian: links between the evolution of land plants, weathering processes, and marine anoxic events. Philos Trans R Soc Lond B 353:113–130

Algeo TJ, Scheckler SE (2010) Land plant evolution and weathering rates changes in the Devonian. J Earth Sci 21(Supp. 1):75–78

Algeo TJ, Scheckler SE, Maynard JB (2001) Effects of early vascular land plants on weathering processes and global chemical fluxes during the Middle and Late Devonian. In: Gensel PG, Edwards D (eds) Plants invade the land: evolutionary and environmental perspectives. Columbia Univ Press, New York, pp 213–236

Allen JP, Gastaldo RA (2006) Sedimentology and taphonomy of the Early to Middle Devonian plant-bearing beds of the Trout Valley Formation, Maine. In: DiMichele WA, Greb S (eds) Wetlands through time. Special publication, vol 399. Geological Society of America, Boulder, pp 57–78

Andrews HN, Gensel PG, Forbes WH (1974) An apparently heterosporous plant from the Middle Devonian of New Brunswick. Palaeontology 17:387–408

Andrews HN, Kasper EA, Forbes WH, Gensel PG, Chaloner WG (1977) Early Devonian flora of the Trout Valley Formation of northern Maine. Rev Palaeobot Palynol 23:255–285

Arber EAN (1921) Devonian floras. Cambridge University Press, Cambridge, 109 p

Axelrod DI (1959) Evolution of the Psilophyte Paleoflora. Evolution 13:264–275

Banks HP (1968) The early history of land plants. In: Drake E (ed) Evolution and environment: a symposium presented on the one hundredth anniversary of the foundation of the Peabody Museum of Natural History at Yale University. Yale University Press, New Haven and London, pp 73–107

Banks HP, Bonamo PM, Grierson JD (1972) *Leclercqia complexa* gen. et sp. nov., a new lycopod from the late Middle Devonian of eastern New York. Rev Palaeobot Palynol 14:19–40

Basinger JF, Kotyk ME, Gensel PG (1996) Early land plants from the Late Silurian-Early Devonian of Bathurst Island, Canadian Arctic Archepelago. Geol Surv Can Curr Res 1996B:51–60

Beck CB (1962) Reconstructions of *Archaeopteris*, and further consideration of its phylogenetic position. Am J Bot 49:373–382

Berner RA, Kothavala Z (2001) Geocarb III: a revised model of atmospheric CO_2 over Phanerozic time. Am J Sci 301:182–204

Bower FO (1908) The origin of a land flora, a theory based on the facts of alternation. Macmillan, London, 727 p

Boyce KC, Hotton CL, Fogel ML, Cody GD, Hazen RM, Knoll AH, Hueber FM (2007) Devonian landscape heterogeneity recorded by a giant fungus. Geology 35:399–402

Čelakovský J (1874) Über die verschiedenen Formen und Bedetutung des Generationswechsels der Pflanzen. Jahresbericht der Königl Böhmischen Gesellschaft der Wissenschaften Prague 1874:22–61

Channing A, Edwards D (2009) Yellowstone hot spring environments and the palaeo-ecophysiology of Rhynie chert plants: towards a synthesis. Plant Evol Divers 2:111–143

Cressler WL (2006) Plant paleoecology of the Late Devonian Red Hill locality, north-central Pennsylvania, an *Archaeopteris*-dominated wetland plant community and early tetrapod site. In: Greb SF, DiMichele WA (eds) Wetlands through time. Special paper. Geological Society of America, Boulder, pp 79–102

Cressler WL, Prestianni C, LePage BA (2010) Late Devonian spermatophyte diversity and paleoecology at Red Hill, north-Central Pennsylvania, USA. Int J Coal Geol 83:91–102

Daeschler EB, Clack JA, Shubin NH (2009) Late Devonian tetrapod remains from Red Hill, Pennsylvania, USA: how much diversity? Acta Zool 90:306–317

Daeschler EB, Cressler WL III (2011) Late Devonian paleontology and paleoenvironments at Red Hill and other fossil sites in the Catskill Formation of north-central Pennsylvania. In: Ruffolo RM, Ciampaglio CN (eds) Geological Society of America field guide, vol 20. GSA, Boulder, pp 1–16

Daeschler EB, Shubin NH, Jenkins FA Jr (2006) A Devonian tetrapod-like fish and the evolution of the tetrapod body plan. Nature 440:757–763

Daeschler EB, Shubin NH, Thomson KS, Amaral WW (1994) A Devonian tetrapod from North America. Science 265:639–642

Dawson JW (1859) On fossil plants from the Devonian rocks of Canada. Q J Geol Soc Lond 15:477–488

Dawson JW (1870) The primitive vegetation of the Earth. Nature 2:85–88

Dawson JW (1871) The fossil plants of the Devonian and Upper Silurian formations of Canada. Geol Surv Can 1:1–19

Dorf E, Rankin D (1962) Early Devonian plants from the Traveler Mountain area, Maine. J Paleontol 36:999–1004

Dunlop JA, Garwood RJ (2018) Terrestrial invertebrates in the Rhynie chert ecosystem. Philos Trans R Soc Lond Ser B 373(1739):20160493

Edwards D (1970) Fertile Rhyniophytina from the Lower Devonian of Britain. Palaeontology 13:451–461

Edwards D (1996) New insights into early land ecosystems: a glimpse of a Lilliputian world. Rev Palaeobot Palynol 90:159–174

Edwards D, Axe L, Honegger R (2013) Contributions to the diversity in cryptogamic covers in the mid-Palaeozoic: *Nematothallus* revisited. Bot J Linn Soc 173:505–534

Edwards D, Davies KL, Axe L (1992) A vascular conducting strand in the early land plant *Cooksonia*. Nature 357:683–685

Edwards D, Feehan J, Smith DG (1983) A late Wenlock flora from Co. Tipperary, Ireland. Bot J Linn Soc 86:19–36

Edwards D, Honegger R, Axe L, Morris JL (2018b) Anatomically preserved Silurian 'nematophytes' from the Welsh Borderland (UK). Bot J Linn Soc 187:272–291

Edwards D, Kenrick P, Dolan L (2018a) History and contemporary significance of the Rhynie cherts—our earliest preserved terrestrial ecosystem. Philos Trans R Soc Lond B Biol Sci 373:20160489

Edwards D, Morris JL, Richardson JB, Kenrick P (2014) Cryptospores and cryptophytes reveal hidden diversity in early land floras. New Phytol 202:50–78

Edwards D, Richardson JB, Axe L, Davies KL (2012) A new group of Early Devonian plants with valvate sporangia containing sculptured permanent dyads. Bot J Linn Soc 168:229–257

Elick JM, Driese SG, Mora CI (1998) Very large plant and root traces from the Early to Middle Devonian: implications for early terrestrial ecosystems and atmospheric $p(CO_2)$. Geology 26:143–146

Ergolskaya ZV (1936) Petrographical examination of the Barzas coals. Trudy Centralnogo Nauchno-Issledovatelskogo Geologo-Razvedocnogo Instituta. Trans Cent Geol Prospect Inst 70:5–54

Fairon-Demaret M (1985) The fossil plants of the Emsian of Sart Tilman Belgium 1. *Stockmansia langii* new genus new combination. Review of Palaeobotany. Palynology 44:243–260

Fairon-Demaret M (1986) *Stockmansella*, a new name for *Stockmansia* Fairon-Demaret (fossil). Taxon 35:334

Fanning U, Richardson JB, Edwards D (1988) Cryptic evolution in an early land plant. Evol Trends Plants 2:13–24

Fleming J (1831) On the occurrence of scales of vertebrate animals in the old red sandstone of Fifeshire Edinburgh. J Nat Geogr Sci 3:81–86

Garratt MJ, Tims JD, Rickards RB, Chambers TC, Douglas JG (1984) The appearance of *Baragwanathia* (Lycophytina) in the Silurian. Bot J Linn Soc 89:355–358

Gastaldo RA (2016) New paleontological insights into the Emsian–Eifelian Trout Valley Formation, Baxter State Park's scientific forest management area, Aroostook County, Maine. PALAIOS 31:339–346

Gensel PG (1976) *Renalia hueberi*, a new plant from the Lower Devonian of Gaspé. Rev Palaeobot Palynol 22:19–37

Gensel PG, Andrew HN (1984) Plant life in the Devonian. Praeger Press, Westport, 380 p

Gensel PG (2008) The earliest land plants. Annu Rev Ecol Evol Syst 39:459–477

Gensel PG (2018) Early Devonian woody plants and implications for the early evolution of vascular cambia. In: Krings M, Harper CM, Cuneo NR, Rothwell GW (eds) Transformative paleobotany. Elsevier, Amsterdam, pp 21–33

Gensel PG, Berry CM (2001) Early lycophyte evolution. Am Fern J 91:74–98

Gensel PG, Berry CM (2016) Sporangial morphology of the early Devonian zosterophyll *Sawdonia ornata* from the type locality (Gaspé). Int J Plant Sci 177:618–632

Gensel PG, Johnson NG (1994) The cuticular structure and stomatal organization of *Orestovia* sp. cf. *O. petzii* from the Kuznetsk basin, western Siberia. Palaeontogr Abt B 233:1–10

Gerrienne P, Gensel PG, Strullu-Derrien C, Lardeux H, Steemans P, Prestianni C (2011) A simple type of wood in two early Devonian plants. Science 333:837

Gibling MR, Davies NS (2012) Paleozoic landscapes shaped by plant evolution. Nat Geosci 5:99–105

Gibling MR, Davies NS, Falcon-Lang HJ, Bashforth AR, DiMichele WA, Rygel MC, Ielpi A (2014) Palaeozoic co-evolution of rivers and vegetation: a synthesis of current knowledge. Proc Geol Assoc 125:524–533

Giesen P, Berry CM (2013) Reconstruction and growth of the early tree *Calamophyton* (Pseudosporochnales, Cladoxylopsida) based on exceptionally complete specimens from Lindlar, Germany (Mid-Devonian): organic connection of *Calamophyton* branches and *Duisbergia* trunks. Int J Plant Sci 174(4):665–686

Glasspool IJ, Edwards D, Axe L (2006) Charcoal in the Early Devonian: a wildfire-derived Konservat–Lagerstätte. Rev Palaeobot Palynol 142:131–136

Goldring W (1924) The Upper Devonian forest of seed ferns in eastern New York. N Y State Mus Bull 251:50–72

Goldring W (1927) The oldest known petrified forest. Science 24:514–529

Gonez P, Gerrienne P (2010) *Aberlemnia caledonica* gen. et comb. nov., a new name for *Cooksonia caledonica* Edwards 1970. Rev Palaeobot Palynol 163:64–72

Graham LE, Cook ME, Busse JS (2000) The origin of plants: body plan changes contributing to a major evolutionary radiation. Proc Natl Acad Sci USA 25:4535–4540

Graham LE, Cook ME, Hanson DT, Pigg KB, Graham JM (2010) Structural, physiological, and stable carbon isotopic evidence that the enigmatic Paleozoic fossil *Prototaxites* formed from rolled liverwort mats. Am J Bot 97:268–275

Granoff JA, Gensel PG, Andrews HN (1976) A new species of *Pertica* from the Devonian of eastern Canada. Palaeontogr Abt B 155:119–128

Gray J (1985) The microfossil record of early land plants; advances in understanding of early terrestrialization, 1970–1984. Philos Trans R Soc Lond B 309:167–195

Gray J, Boucot AJ (1971) Early Silurian spore tetrads from New York: earliest new world evidence for vascular plants? Science 173:918–921

Grierson JD, Hueber FM (1967) Devonian lycopods from northern New Brunswick. Alta Soc Pet Geol Calg 2:823–836

Griffing DH, Bridge JS, Hotton CL (2000) Coastal-fluvial palaeoenvironments and plant palaeoecology of the Lower Devonian (Emsian), Gaspé Bay, Québec, Canada. In: Friend PF, Williams BPJ (eds) New perspectives on the Old Red Sandstone. Geological Society, London, pp 61–84

Habgood KS, Edwards D, Axe L (2002) New perspectives on *Cooksonia* from the Lower Devonian of the Welsh Borderland. Bot J Linn Soc 139:339–359

Hao S-G, Xue JA (2013) The Early Devonian Posongchong flora of Yunnan, a contribution to an understanding of the evolution and early diversification of vascular plants. Science Press, Beijing, 366 p

Hoffman C, Tomescu AMF (2013) An early origin of secondary growth: *Franhueberia gerriennei* gen. et sp. nov. from the Lower Devonian of Gaspé (Quebec, Canada). Am J Bot 100:754–763

Honegger R, Edwards D, Axe L (2012) The earliest records of internally stratified cyanobacterial and algal lichens from the Lower Devonian of the Welsh Borderland. New Phytol 197:264–275

Hopping C (1956) On a specimen of *Psilophyton robustius* Dawson, from the Lower Devonian of Canada. Proc R Soc Edinb B 66:10–28

Hotton CL, Hueber FM, Griffing DH, Bridge JS (2001) Early terrestrial plant environments: an example from the Emsian of Gaspé, Canada. In: Gensel PG, Edwards DH (eds) Plants invade the land: evolutionary and environmental perspectives. Columbia University Press, New York, pp 179–212

Hueber FM (1964) The psilophytes and their relationship to the origin of ferns. Torrey Bot Club Mem 21:5–9

Hueber FM (1971) *Sawdonia ornata*: a new name for *Psilophyton princeps* var. *ornatum*. Taxon 20:641–642

Hueber FM (1983) A new species of *Baragwanathia* from the Sextant Formation (Emsian), northern Ontario, Canada. Bot J Linn Soc 86:57–79

Hueber FM (2001) Rotted wood-alga fungus: the history and life of *Prototaxites* Dawson 1859. Rev Palaeobot Palynol 116:123–159

Hueber FM, Banks HP (1967) *Psilophyton princeps*: the search for organic connection. Taxon 16:81–85

Hueber FM, Banks HP (1979) *Serrulacaulis furcatus* gen. sp. nov., a new zosterophyll from the Lower Upper Devonian of New York state. Rev Palaeobot Palynol 28:169–189

Hueber FM, Grierson JD (1961) On the occurrence of *Psilophyton princeps* in the early Upper Devonian of New York. Am J Bot 48:473–479

Kasper AE Jr, Gensel PG, Forbes WH, Andrews HN Jr (1988) Plant Paleontology in the state of Maine - a review: Maine geological survey studies in Maine. Geology 1:109–128

Kennedy K, Gensel PG, Gibling MR (2012) Paleoenvironmental inferences from the classic Early Devonian plant-bearing locality of the Campbellton Formation, New Brunswick, Canada. PALAIOS 27:424–438

Kennedy K, Gibling MR (2011) The Campbellton Formation, New Brunswick, Canada: paleoenvironments in an important Early Devonian terrestrial locality. Can J Earth Sci 48:1561–1580

Kenrick P (2018) Changing expressions: a hypothesis for the origin of the vascular plant life cycle. Philos Trans R Soc Lond B Biol Sci 373(1739):20170149. https://doi.org/10.1098/rstb.2017.0149

Kenrick P, Crane PR (1997) The origin and early diversification of land plants – a cladistic study: Smithsonian series in comparative evolutionay biology. Smithsonian Institution Press, Washington, 441 p

Kenrick P, Wellman CH, Schneider H, Edgecombe GD (2012) A timeline for terrestrialization: consequences for the carbon cycle in the Palaeozoic. Philos Trans R Soc Lond B 367:519–536

Kerp H (2018) Organs and tissues of Rhynie chert plants. Philos Trans R Soc Lond Ser B 373(1739):20160495. https://doi.org/10.1098/rstb.2016.0495

Kerp H, Hass MH, Mosbrugger V (2001) New data on Nothia aphylla Lyon 1964 ex El-Saadawy et Lacey 1979, a -poorly known plant from the Lower Devonian Rhynie Chert. In: Gensel PG, Edwards D (eds) Plants invade the land: evolutionary & environmental perspectives. Columbia University Press, New York, pp 52–82

Kerp H, Trewin NH, Hass H (2003) New gametophytes from the Early Devonian Rhynie Chert. Trans R Soc Edinb 94:411–428

Kerp H, Wellman CH, Krings M, Kearney P, Hass H (2013) Reproductive organs and in situ spores of Asteroxylon mackiei Kidston & Lang, the most complex plant from the lower Devonian Rhynie Chert. Int J Plant Sci 174:293–308

Kotyk ME (1998) Late silurian and early devonian fossil plants of Bathurst Island, Arctic Canada. Master's thesis, Department of Geological Sciences, University of Saskatchewan, Saskatoon, Canada

Kotyk ME, Basinger JF (2000) The Early Devonian (Pragian) zosterophyll Bathurstia denticulata Hueber. Can J Bot 78:193–207

Kotyk ME, Basinger JF, Gensel PG, DeFreitas TA (2002) Morphologically complex plant macrofossils from the late Silurian of Arctic Canada. Am J Bot 89:1004–1013

Kraft P, Pšenička J, Sakala J, Frýda J (2018) Initial plant diversification and dispersal event in upper Silurian of the Prague Basin. Palaeogeogr Palaeoclimatol Palaeoecol 514:144–155

Kräusel R, Venkatachala BS (1966) Devonische Spongiophytaceen aus Ost-und West-Asien. Senkenbergia Lethaia 47:215–251

Krings M, Harper CJ, Taylor EL (2017) Fungi and fungal interactions in the Rhynie Chert: a review of the evidence, with the description of Perexiflasca tayloriana gen. et sp. nov. Philos Trans R Soc Lond B 373:20160500

Lang WH (1937) On the plant-remains from the Downtonian of England and Wales. Philos Trans R Soc Lond B 227:245–291

Lang WH, Cookson IC (1935) On a flora, including vascular land plants, associated with Monograptus, in rocks of Silurian age, from Victoria, Australia. Philos Trans R Soc Lond B 224:421–449

Li CS, Edwards D (1995) A re-investigation of Halle's Drepanophycus spinaeformis Göpp. from the Lower Devonian of Yunnan Province, southern China. Bot J Linn Soc 118:163–192

Libertín M, Kvaček J, Bek J, Štorch P (2018b) Plant diversity of the mid Silurian (Lower Wenlock, Sheinwoodian) terrestrial vegetation preserved in marine sediments from the Barrandian area, The Czech Republic. Foss Impr 74:327–333

Libertín M, Kvaček J, Bek J, Žárský V, Štorch P (2018a) Sporophytes of polysporangiate land plants from the early Silurian period may have been photosynthetically autonomous. Nat Plants 4:269–271

McCourt RM, Delwiche CF, Karol KG (2004) Charophyte algae and land plant origins. Trends Evol Ecol 19:661–666

Morris JL, Edwards D, Richardson JB (2018a) The advantages and frustrations of a plant Lagerstätte as illustrated by a new taxon from the Lower Devonian of the Welsh Borderland, UK. In: Krings M, Harper CM, Cuneo NR, Rothwell GW (eds) Transformative paleobotany. Elsevier, Amsterdam, pp 49–67

Morris JL, Edwards D, Richardson JB, Axe L, Davies KL (2012) Further insights into trilete spore producers from the Early Devonian (Lochkovian) of the Welsh Borderland, UK. Rev Palaeobot Palynol 185:35–36

Morris JL, Leake JR, Stein WE, Berry CM, Marshal JEA, Wellman CH, Milton JA, Hillier S, Mannolini F, Quirk J, Beerling DJ (2015) Investigating Devonian trees as geo-engineers of past climates: linking palaeosols to palaeobotany and experimental geobiology. Palaeontology 58:787–801

Morris JL, Puttick MN, Clark JW, Edwards D, Kenrick P, Pressel S, Wellman C, Yang Z, Schneider H, Donoghue PCJ (2018b) The timescale of early land plant evolution. Proc Natl Acad Sci USA 115:E2274–E2283

Morris JL, Richardson JB, Edwards D (2011) Lower Devonian plant and spore assemblages from Lower Old Red Sandstone strata of Tredomen Quarry, South Wales. Rev Palaeobot Palynol 165:183–208

Prestianni C, Gerrienne P (2010) Early Seed plant radiation: an ecological hypothesis. In: Vecoli M, Clement G, Meyer-Berthaud B (eds) The terrestrialization process: modelling complex interactions at the biosphere-geosphere interface, vol 339. Geological Society Special Publication, London, pp 71–80

Puttick MN, Morris JL, Williams TA, Cox CJ, Edwards D, Kenrick P, Pressel S, Wellman C, Schneider H, Pisani D, Donoghue PCJ (2018) The interrelationships of land plants and the nature of the ancestral embryophyte. Curr Biol 28:733–745

Remy W, Taylor TN, Hass H, Kerp H (1994) 400 million year old vesicular arbuscular mycorrhizae (VAM). Proc Natl Acad Sci USA 91:11841–11843

Renzaglia KS, Crandall-Stotler B, Duckett J, Schuette S, Strother PK (2015) Permanent spore dyads are not 'a thing of the past': on their occurrence in the liverwort Haplomitrium (Haplomitriopsida). Bot J Linn Soc 179:658–669

Retallack GJ, Landing E (2014) Affinities and architecture of Devonian trunks of Prototaxites loganii. Mycologia 106:1143–1156

Rice CM, Trewin NH, Anderson LI (2003) Geological setting of the early Devonian Rhynie cherts, Aberdeenshire, Scotland: an early terrestrial hot spring system. J Geol Soc Lond 159:203–214

Richardson JB (1985) Lower Palaeozoic sporomorphs: their stratigraphical distribution and possible affinities. Philos Trans R Soc Lond B 309:201–205

Rothwell GW, Scheckler SE, Gillespie WH (1989) Elkinsia gen. nov., a Late Devonian gymosperm with cupulate ovules. Bot Gaz 150:170–189

Rubenstein CV, Gerrienne P, De la Puente GS, Astini RA, Steemans P (2010) Early Middle Ordovician evidence for land plants in Argentina (eastern Gondwana). New Phytol 188:365–369

Scheckler SE (1986) Geology, floristics and paleoecology of Late Devonian coal swamps from Appalachian Laurentia (USA). Ann Soc Geol Belg 109:209–222

Selover RW, Gastaldo RA, Nelson RE (2005) An estuarine assemblage from the Middle Devonian Trout Valley Formation of northern Maine. PALAIOS 20:192–197

Serbet R, Rothwell GW (1992) Characterizing the most primitive seed ferns. A reconstruction of Elkinsia polymorpha. Int J Plant Sci 153:602–621

Shear WA, Palmer JM, Coddington JA, Bonamo PM (1989) A Devonian spinneret: early evidence of spiders and silk use. Science 246:479–481

Shear WA, Selden PA, Rolfe WDI, Bonamo PM, Grierson JD (1987) A spider and other arachnids from the Devonian of Gilboa, New York (Arachnida, Trigonotarbida). Am Mus Novit 2901:1–74

Steemans P, Le Hérissé A, Melvin J, Miller MA, Paris F, Verniers J, Wellman CH (2009) Origin and radiation of the earliest vascular land plants. Science 324:353

Steemans P, Wellman CH, Gerrienne P (2010) Paleogeographic and paleoclimatic considerations based on Ordovician to Lochkovian vegetation. Geol Soc Lond Spec Publ 339:49–58

Stein WE (2018) Middle Devonian terrestrial ecosystem in the Catskill region: evidence from tree-size rooting trace fossils. Paper no 27-6. NE GSA. Geol Soc Am Abstr Programs 50(2). https://doi.org/10.1130/abs/2018NE-310545.

Stein WE, Berry CM, VanAller Hernick L, Mannolini F (2012) Surprisingly complex community discovered in the mid-Devonian fossil forest at Gilboa. Nature 483:78–81

Stein WE, Mannolini F, VanAller Hernick L, Landing E, Berry CM (2007) Giant cladoxylopsid trees resolve the enigma of the Earth's earliest forest stumps at Gilboa. Nature 446:904–907

Strother PK (2016) Systematics and evolutionary significance of some new cryoptospores from the Cambrian of eastern Tennessee, USA. Rev Palaeobot Palynol 227:28–41

Strother PK, Taylor WA (2018) The evolutionary origin of the plant spore in relation to the antithetic origin of the plant sporophyte. In: Krings M, Harper CJ, Cuneo NR, Rothwell GW (eds) Transformative paleobotany: papers to commemorate the life and legacy of Thomas N. Taylor. Academic Press, Cambridge, pp 3–20

Strullu-Derrien C, Kenrick P, Taffereau P, Cochard H, Bonnemain J-L, LeHerisse A, Lardeux H, Radel E (2014) The earliest wood and its hydraulic properties documented in ca 407-million-year-old fossils using synchrotron microtomography. Bot J Linn Soc 174:423–437

Taylor TN, Hass H, Kerp H (1997) A cyanolichen from the Lower Devonian Rhynie Chert. Am J Bot 84:992–1004

Taylor TN, Hass H, Kerp H (1999) The oldest fossil ascomycetes. Nature 399:648

Taylor TN, Kerp H, Hass H (2005) Life history biology of early land plants: deciphering the gametophyte phase. Proc Natl Acad Sci USA 102:5892–5897

Taylor TN, Klavins SD, Krings M, Taylor EL, Kerp H, Hass H (2003) Fungi from the Rhynie chert: a view from the dark side. Earth Environ Sci Trans R Soc Edinb 94:457–473

Taylor TN, Remy W, Hass H (1992) Parasitism in a 400-million-year-old green alga. Nature 357:493–494

Taylor TN, Taylor EL, Krings M (2009) Palaeobotany, the biology and evolution of fossil plants, 2nd edn. Elsevier, Amsterdam, 1230 p

Taylor WA, Strother PK, Vecoli M, Al-Hajri S (2017) Wall ultrastructure of the oldest embryophytic spores: implications for early land plant evolution. Rev Micropaleontol 60(3):281–288. https://doi.org/10.1016/j.revmic.2016.12.002

Tims JD, Chambers TC (1984) Rhyniophytina and Trimerophytina from the early land flora of Victoria, Australia. Palaeontology 27:265–279

Trewin N, Kerp H (2018) The Rhynie and Windyfield cherts, Early Devonian, Rhynie, Scotland. In: Fraser NC, Suess H-D (eds) Terrestrial conservation Lagerstätten: windows into the evolution of life on land. Dunedin Academic Press, Edinburgh, pp 1–38

Wellman C (2010) The invasion of the land by plants: when and where? New Phytol 188:306–309

Wellman CH, Gensel PG (2004) Morphology and wall ultrastructure of the spores of the Lower Devonian plant Oocampsa catheta Andrews et al.. Review of Palaeobotany. Palynology 130:269–295

Wellman CH, Osterloff PL, Mohiuddin U (2003) Fragments of the earliest land plants. Nature 425:282–285

Wellman CH, Steemans P, Miller MA (2013) Spore assemblages from Upper Ordovician and lowermost Silurian sediments recovered from the Qusaiba-1 shallow core hole, Qasim region, central Saudi Arabia. Rev Palaeobot Palynol 212:111–126

Glossary

Abaxial The lower surface, generally, of a leaf; or the side furthest away from/facing away from a plant axis.

Acuminate Pointed, refers to the apex.

Adaxial The upper surface, generally, of a leaf; or the side nearest to/facing the plant axis.

Adventitious roots Root structures that develop from any non-root tissue, commonly originating from stems above the ground's surface; may act to prop up or stabilize the trunk.

Allochthonous A fossil assemblage preserved at some distance away from its original environment or habitat, having been transported and, possibly, disarticulated.

Alluvial Deposits produced by the transport of sediment adjacent to rivers in floodplains during major floods.

Altiplano A high plateau in Latin America, the second largest area after the Qinghai-Tibet Plateau.

Amphistomatic The presence of stomata on both sides (abaxial and adaxial) of a leaf.

Anastomose Irregularly branching and rejoining to form a mesh-like pattern.

Anoxic A condition in the water column that is devoid of oxygen.

Apatite Phosphatic mineral $[Ca_{10}(PO_4)_6(OH)_2]$ in vertebrate skeletons also found in some invertebrate groups (e.g., annelids, brachiopods).

Arborescent Tree or tree-like habit in growth, structure, or appearance of a plant, generally >4 m in height.

Archaic Characteristics found in ancestral lineages.

Autochthonous A fossil assemblage buried and preserved in place; burial in its original growth position, habit, or environment.

Autotrophic An organism that is able to produce its own food supply (e.g., photosynthesis) to enlarge or maintain its mass without the exploitation of other organisms.

Basal Close to the origin of a particular clade; basal organisms often have a generalized morphology.

Bathymetric Relating to the measurement of water depth at various places in a body of water and to the information derived from such measurements.

Bifurcate Divided into two branches, dichotomized.

Biome An assemblage of organisms that have common characteristics in an environment established over a range of continents, controlled by a shared physical climate.

Brachydont Dentition with low-crowned teeth.

Bunodont Dentition with large grinding teeth.

Calcaneus A heal bone (sometimes also called calcaneum).

Calcisols Calcium-rich soils in which calcite-cemented nodules or horizons form in the soil profile.

Calcite A common mineral ($CaCO_3$) of calcium, carbon, and oxygen found in invertebrate skeletons and forming limestone rocks.

Carpel A folded, sheet-like plant structure with its margins fused together to enclose an ovule; at maturity becomes part of a fruit; generally believed to be an evolutionary feature derived from a modified leaf.

Cenogram analysis The analysis of body-size distribution in mammal communities.

Chlamydospermous Seeds having a thin, single integument prolonged into a long apical beak, the entire structure surrounded by a thick layer derived from bracts.

Clade A group of organisms that includes a common ancestor and all its descendants; clades are usually identified through phylogenetic analysis.

Clastic sediment Rock fragments, varying in size from boulders to clay, that originated from the physical weathering of preexisting rock.

Cleithrum A bone in the skull of fishes, which is detached in tetrapods and is an element of the shoulder girdle; it is lost in higher vertebrates.

Coal ball Peat that is permineralized by the introduction of mineral-charged waters, generally of a marine origin, resulting in the precipitation of calcite or silica preserving plant-cell structures. In some instances, the interaction of sulfur-reducing bacteria with decaying plant matter may result in the crystallization of pyrite, also forming a coal ball.

Compression A fossil in which volatiles (water and gasses) are lost from cells and more resistant organic matter is flattened under the pressure of burial, resulting in a carbon film of the original organic material.

© Springer Nature Switzerland AG 2020
E. Martinetto et al. (eds.), *Nature through Time*, Springer Textbooks in Earth Sciences, Geography and Environment,
https://doi.org/10.1007/978-3-030-35058-1

Coquina Sedimentary rock that is mostly composed of sorted fossil-shell debris cemented by calcite.

Core The part of a clade that excludes its basal members, which often are a sister group in a cladogram. The term, here, is applied if some of the clades in the core group have no extant members (see below).

Crown Part of a clade that includes all the subclades with currently existing species. More basal, completely extinct forms in this same clade are referred to as "stem groups" (see below).

Cupule A structure, usually cup-like, which partly to totally envelops a seed or a fruit from its base.

Cursorial, cursoriality Limbs adapted to running.

Cuticle In plants, the cuticle is a composite structure of extracellular water-resistant layer (the cuticular layer) composed of cutin and organic solvent-soluble lipids, which covers the aboveground epidermis of land plants and that is only broken by stomata and lenticels. The cuticle provides some physical protection, but its main purpose is to prevent water loss. Several groups of invertebrate animals possess an exoskeleton that is also called cuticle (e.g., arthropods), composed of chitin.

Cyclothem Alternating stratigraphic successions of marine and continental sediments, often interbedded with coal seams, that are a consequence of sea-level dynamics in response to glacial and deglacial processes.

Dendrochronology The interpretation of age of an individual tree derived from tree ring analysis; tree rings reflect the environmental conditions under which the plant grew, providing proxy data on climate, precipitation, and other factors that may have affected growth during the life of the plant.

Diamictite A poorly or non-sorted clastic sedimentary rock of continental glacial origin in which variously sized clasts, ranging from boulders to sand, occur in a muddy matrix.

Dichotomous A branching pattern in which the axis is divided into two equal (isotomous) or unequal (anisotomous) axes (branches). See also entry for bifurcate.

Dispersal The extension of a population into new areas or regions, not previously occupied by the parents, by successive generations of organisms or their propagules.

Distichous Parts along the plant axis (stem or trunk) that are arranged alternately in two opposite, vertical rows.

Dropstones Large rocks (cobble to boulder) first transported by ice to a standing body of water and then to the sediment–water interface in response to melting; often buried by sediment derived from continued melting of floating ice.

Durophagy A feeding strategy specialized on hard resources, including invertebrate exoskeletons and plant reproductive structures (e.g., shells or nuts).

Dysoxic Low concentration of oxygen in a water column that lies between fully oxygenated (8.0–2.0 mL/1 O_2) and suboxic (0.2–0.0 mL/1 O_2) and anoxic (0.0 mL/l O_2); dysoxic (0.2–2.0 mL/l O_2).

Ecoregions According to Olson et al. (2001), these are relatively large units of land containing a distinct assemblage of natural communities and species, with boundaries that approximate the original extent of natural communities prior to major anthropogenic land-use change. Olson et al. (2001) identify 867 ecoregions in only 14 biomes.

Ejecta Rocks that have been ejected into the atmosphere during volcanic eruptions, generally of igneous origin, or bolide (asteroids, meteorites) impacts.

Ektexine The outer wall of a pollen grain, the exine. Includes up to four layers and the external sculpture elements.

Endemic, endemism A clade or a species is confined or restricted to a specific area or region, only occurring there and in no other place on Earth.

Endexine The inner part of the wall of a pollen grain, the exine.

End moraine systems The accumulation of glacial-derived debris transported by a moving ice sheet at its distal margin, where ice melts and sediment is dropped.

Ephemeral A feature that appears and disappears from time to time, such as a lake, stream, or river which exists during the rainy season but dries up in the dry season; or plants with a short flowering time.

Epicontinental A feature that overlies continental crust, normally used for a seaway.

Epirelief A positive topographic feature, often referred to a structure (e.g., trace fossil in a geological context) that protrudes above a horizontal (bedding) plane.

Erratics Large rocks transported by a mass of ice.

Estuarine Pertinent to a partly enclosed, coastal area of brackish water into which a river discharges, or joins the ocean.

Eustatic Relating to, or characterized by, a global change in sea level.

Eutrophication Increased nutrient input to a water body stimulating photosynthesis and algal growth, with subsequent oxygen depletion in the water column.

Extirpation The condition when a taxon (at any hierarchical level) no longer exists in a geographical area where it previously occurred, but continues to exist elsewhere.

Extrabasinal element Fossil occurrences of terrestrial organisms not common to the depositional environment, or those considered to have grown in an area away from the depositional setting.

Fascicle A small cluster or bundle of plant parts (leaves, flowers, etc.) lacking an elongate axis.

Floristic regions Geographic areas with a relatively uniform plant composition, established in response to climatic and/or historical reasons.

Fluvial A depositional system pertaining to rivers and the sediments deposited, therein.

Fluvial-eolian Sediment derived from river and wind transport.

Formation A succession of sedimentary rocks that be consistently recognized by a set of characteristics and physically mapped throughout an area; a formal unit that can be subdivided into members and or organized into larger groups.

Fusiform Spindle-shaped.

Gametophyte A haploid (N) plant body.

Geothermal activity Heat flux to the surface originating from deep in Earth's crust.

Glacial An interval of time (>10 to 20 ka or longer) during which colder climates prevail between two warm intervals when expansive ice is present on the planet.

Glacigenic Sediments or morphologies generated by the movement of an ice mass.

Glaciofluvial Sediments or morphologies generated by the flux of the meltwater originating from an ice mass.

Glaciolacustrine Sediments or morphologies typical of a standing water body associated (below or lateral) with an ice mass.

Guard cells Pairs of specialized crescent-shaped epidermal cells that immediately border the stomatal pore and, together with it, form the stoma. Through changes in their turgor, the guard cells open or close the stoma, controlling gas exchange and limiting water loss in a plant.

Gyttja Dark mud-rich sediment with a high decayed organic content accumulated in a lake or marsh.

Hadley Center Named in honor of George Hadley, is one of the UK's leading centers for the study of scientific issues associated with climate change.

Haptotypic Site or scar of contact left where spores or pollen grains were in contact immediately after meiosis; often the site where a grain splits open to release the gametophyte or pollen tube.

Heterosporous A sporophyte (2N) plant in which two types of spores develop, differing in size and sex (male microspore and female megaspore); each germinates and grows into separate male and female (N) gametophytes.

Heterotroph An organism that does not produce its own food, but relies on other sources of organic carbon, mainly plant or animal matter.

Hiatus A time gap in the sedimentary record.

Highstand A time during which sea levels are at their highest, generally when base level was above the edge of a continental shelf.

Histological A feature dealing of cellular organization defined as a tissue (e.g., epithelium); histological thin sections are prepared cuts of tissues where their structure and organization can be observed.

Holoarctic Concerning the assemblage of the Nearctic and Palearctic biogeographic regions (see Olson et al. 2001).

Holotype A single specimen of a living or fossil organism that is selected as the reference of a taxon to which all others are compared (e.g., a species).

Homologous A tissue, skeletal element, or organ similar in position, structure, characteristics, and evolutionary origin but not necessarily in function.

Hothouse An interval of geologic time during which continental glaciers are small in aerial extent and geographically restricted, or not present, CO_2 and other greenhouse gas concentrations are elevated, and sea-surface temperatures may be high when compared to prior states.

Humerus Upper arm bone.

Hypostomatic The presence of stomata only on the abaxial (lower) leaf surface.

Hypsodonty Teeth with an increased tooth-crown height.

Icehouse An interval of geologic time during which continental glaciers are prevalent and extensive, CO_2 and other greenhouse gas concentrations are low, and sea-surface temperatures may be low when compared to prior states.

Ichnofossils Trace fossils that record biological activity of an organism.

Ignimbrite A pumice-dominated rock of volcanic origin, formed from the cooling of pyroclastic material ejected from an explosive eruption; layers can be welded into rock if the temperature of the pyroclastic material is high (>535 °C).

Interbedded A sedimentary succession comprised of beds with different characters, often sand and mud, which are repeated in deposits.

Interglacial A time slice (c. 10–20 ka or longer) of definitely warmer climate between two glacial intervals.

Interstadials A short time or duration when warmer climate prevails in a cold glacial interval.

Intertropical Convergence Zone The equatorial belt of low atmospheric pressures in which the winds from both hemispheres meet and lose speed, forcing moist air upward resulting in heavy rainfall.

Intracratonic basin Broad, shallow, saucer-shape sediment-accumulating area within the confines of a craton.

Intramontane basin An alluvium-filled valley between mountain ranges, often formed over a graben, in which sediment accumulates.

Iron curtain A strongly controlled political boundary, almost impossible to cross for eastern Europeans, which divided Europe into a western and eastern block during the second half of the twentieth century.

Kame terraces Elevated plains, often appearing as a hanging landscape, formed between the lateral margin of a glacier and its enclosing valley wall.

K-selectors   Animals or plants exhibiting a reproductive strategy, where adults produce few offspring; in animals, the strategy includes care of and protection of the young for an extended time with sexual maturity often delayed in life.

Lacustrine   Sediments accumulated or formed in lakes, or environments associated with lakes.

Lagerstätten   Localities in which extremely well-preserved fossils, often also with soft-tissue preservation, occur; term originates from the German mining industry for "mother lode."

Lamina, laminated   A thin layer; in an animal refers to a thin sheet or strut of bone; in a plant refers to the photosynthetic organ (e.g., broad, expanded leaf blade); in rock refers to any sedimentary bed with a thickness less than one centimeter.

Lateritic   Reddish clay-derived, brick-hard material forming a distinctive soil.

Ligule   A scale-like flap of tissue found at the base of the upper surface of a microphyllous leaf.

Littoral   Semi-emergent life habit.

Longirostrine   Animals with elongated skulls, due to the presence of a long snout (or rostrum).

Loph count, lophedness   The number of elongated ridges between tooth cusps.

Lowstand   A time during which sea levels are at their lowest, generally when base level was at some distance below the edge of a continental shelf.

Mangrove   Salt-tolerant plants, commonly trees and shrubs, which live in muddy coastal intertidal zones, often developing adventitious rooting systems for stabilization.

Megafauna   Animals of large or giant size that are part of the general fauna.

Megasites   Used in this book for large trampled surfaces bearing abundant trace fossils.

Megaspore   Sexually produced, large (mm-size) vascular plant spore (haploid; N) from which a female gametophyte generation develops.

Megasporophyll   A modified leaf on which a female sporangium (megasporangium) is born and in which megaspores develop via meiosis.

Mesodont   Dentition where teeth are neither low- nor high-crowned.

Mesophyll   Ground tissue of a leaf, located between the upper and lower epidermis, composed of parenchyma bearing chloroplasts for photosynthesis; this tissue is differentiated into the palisade parenchyma (photosynthetic) and the spongy mesophyll (site of gas exchange).

Microphyll   A type of simple, generally small, plant leaf with one single, unbranched midvein.

Micropyle   A small opening in the integuments of the ovule of seed-bearing plants through which sperm are able to access the egg.

Microspore   Sexually produced, small vascular plant (haploid; N) spores from which a male gametophyte generation develops.

Microsporophyll   A modified leaf on which a male sporangium (microsporangium) is born and in which microspores develop via meiosis.

Mixed mesophytic forest   A vegetation type with many tree species, dominated by deciduous taxa in association with some evergreens; developed both in North America and East Asia in the cooler part of the subtropical climate area (= warm temperate, Köppen's Cfa), with high rainfall during the growing season.

Monocarpic   A plant-reproductive strategy where an individual undergoes a single reproductive cycle, becomes non-functional, and dies.

Monolete spore   Meiosis of a spore-mother cell wherein the four resultant spores are organized along the vertical axis; when split apart, there is a single line on each spore through which germination occurs.

Monosporangiate   Describing a plant bearing only one type of spore (either megaspores or microspores) in the reproductive organ.

Morphogenus/morphospecies   A genus or species that is recognized based upon morphological characteristics alone; different fossil parts of a single biological species can be assigned to different morphospecies. In plants, the Code of Nomenclature established strict definitions in the late twentieth century that were replaced, more recently, by the terms fossil-genus/fossil-species.

Multi-cusped   Teeth with several cusps, or projections; our molars are multi-cusped teeth.

Munsell color chart   The system that specifies colors based on three properties: hue, value (lightness), and chroma (color purity); created by Albert H. Munsell in the first decade of the twentieth century and adopted for geoscience research.

Mutant   An organism with a modified genome compared to its ancestors.

Neritic zone   Relatively shallow marine environment, between the low tide mark and the depth to which sunlight penetrates ocean waters; characterized by abundant nutrients, sunlight, and a rich biota.

Nitrophilous   An organism favoring abundant nitrogen content in the substrate.

Ombrotrophic   A soil or vegetation that receives all water and nutrients from rainfall rather than from groundwater, often resulting in organic matter accumulation and the formation of a peat (histosol).

Opportunistic taxon   Species without a high degree of specialization capable of either adapting to perturbed or stressed environmental conditions, or one in which reproduction is prioritized over survival; often referred to as

R-strategists exhibiting high growth rates and fecundity, with low survival odds for many of the offspring.

Overbank   Areas adjacent to drainage systems in a river-dominated landscape where sediment is transported and deposited on the surrounding alluvial plain.

Paleo-   A prefix used, normally without hyphen, before another word to refer to the past; 172 see the following definitions.

Paleobiome   A biome of the past.

Paleochannel   A sandstone body in the rock record that represents deposits of a once active river and may include overlying mudstone, representing stream channel filled following abandonment.

Paleopedological   Regarding ancient soil structure.

Paleosol   A volume of solid substrate exhibiting weathering profiles and features resulting from soil formation (pedogenesis) in deep time, which was successively buried and preserved by overlying sediment.

Palisade parenchyma   A layer or layers of elongate photosynthetic cells, containing chloroplasts, occurring just below the leaf cuticle.

Paludal   Sediments that were deposited in swamps.

Palynology   The study of cryptospores, spores, pollen, and certain microscopic planktonic organisms.

Papilla   A short projection arising from a cell, usually found on the epidermis.

Paratropical   Features associated with tropical environments that are close to the transition (ecotone) to non-tropical environments; particularly used for vegetation characterized by plants that are not present in the sub-tropical (= warm temperate) ecoregions.

Parautochthonous   A fossil assemblage buried and preserved in the general environment in which the organism lived, but not in growth position or in place.

Pedogenesis   Soil formation associated with weathering and the development of a soil profile.

Pedogroup   A group of geologic substrates characterized by the development of a same type of soil.

Pelagic zone   Water column of the open ocean; often referred to as habitat above the sea bottom.

Periglacial   Settings close to a glacier.

Phyllotaxis   The arrangement of leaves on an axis or stem conforming to one of several different mathematical relationships.

Phylogenetic diversification   A process during which a clade radiates and evolves many novel taxa.

Physiography   Physical organization of an environment or territory (e.g., mountain and coast plain).

Phytoliths   Microscopic silica inclusions in plants.

Phytoplankton   Populations of freely floating photosynthetic organisms, mainly protists.

Pinna   The first-order subdivision of the rachis of a frond in a pinnately compound leaf.

Pinnate   A feather-like arrangement of leaflets borne on either side of a central rachis or petiole; also used to describe a similar pattern of leaf vein arrangement.

Pinnate leaf   A compound leaf in which leaflets (pinnules) are attached along the rachis, an extension of the petiole, and originate, or are born from lateral axes (pinnae).

Pinnatifid   Resembling pinnate segmentation, but with pinnae that remain confluent (merged together) along their base.

Pinnules   The subdivisions of the rachis of a frond above the first order in a pinnately compound leaf.

Pith cast   Sediment (lithified or unlithified) that filled the continuous central cavity, or central area where stem spongy tissue decayed to leave a void, of a plant, replicating (casting) the features of this interior space.

Pollen   Micrometer-scale male reproductive cell (microgametophytes) of seed plants, which produces male gametes (sperm).

Pollen chamber   A cavity near the top of an ovule into which pollen grains fall and are stored, prior to germination, followed by the development of a pollen tube, and male gamete production that fertilizes the egg.

Polycarpic   A plant-reproductive strategy where an individual undergoes multiple reproductive cycles during its lifespan.

Polysporangiophyte   An individual plant bearing more than one sporangium.

Prepollen   The male gametophyte generation in some groups of fossil gymnosperms (seed ferns [pteridosperms], cordaites, and conifers) characterized by proximal openings of the grain and the absence of a pollen-tube development upon germination.

Productive meadows   Grasslands with a high plant biomass production.

Pronation   Inward rotation of the fore- or hind feet, which can happen during a locomotor cycle, or as a consequence of evolution.

Proto-carpel   A folded, sheet-like structure bearing an ovule in extinct plants, the margins of which are not fused together; generally believed to be a modified leaf.

Rachis   The main axis of a frond or compound leaf.

Radiation   A process during which a clade diversifies, and many novel species evolve in a short time interval.

Radiometric dating methods   Techniques that exploit the instability of some isotopes, whose proportion in a modern sample, compared to that in the original material, can provide a numerical age based on the known decay rate of the isotope.

Ramp   Associated with limestone-forming environments in which there is a low gradient from a shallow-water shoreline, or lagoon, to the ocean-basin floor bottom, and may include reef fragments transported basinward during storm activity.

Rheotrophic A soil or vegetation that receives nearly all of its water and nutrients from groundwater.

Rhizomatous Plant-growth strategy wherein stems grow parallel to the ground, generally below or at the soil surface, from which roots develop downward and aerial axes or leaves upward.

Rhizome A flat-lying stem; see rhizomatous.

Riparian Plants that are growing either along the banks of a waterway or in wetlands adjacent to rivers and streams.

Saprotroph Organisms feeding on or deriving nutrients from decaying organic matter.

Scapulacoracoid An element of the shoulder girdle, formed through the fusion of two separate bones, the scapula and the coracoid.

Sclerophyll Having hard, thick, and leathery leaves.

Section A formal taxonomic subdivision of plants below the rank of genus and subgenus; sometimes abbreviated as "sect." after the genus name.

Seed The propagating structure of spermatophytes where the fertilized mature ovule contains the plant embryo which, under appropriate growth conditions, will germinate and grow into a new sporophyte (2N) plant.

Septum A wall or tissue dividing a cavity or structure into smaller ones.

Serrated An edge with little protrusions or teeth, like a blade of a saw or a steak knife.

Shoal A shallow, submarine area in which sediment or unconsolidated rock rises from the sediment–water interface to near the water's surface.

Shoreline The boundary between a water body and dry land.

Siliciclastic Terrigenous sediments derived from the weathering of bedrock, consisting mostly of quartz or silicate minerals.

Sister clade, or sister group, or sister species, or sister taxon Two clades or species that are so closely related that they share a common ancestor, and do not share a common ancestor with another clade or species.

Size selectivity When environmental conditions favor organisms of a specific size over others, which can result in the extinction of particular size classes, and/or the radiation of others.

Speleothems Secondary mineral deposits formed in a cave, most commonly made up by thin-layered calcium carbonate resulting from the dissolution and reprecipitation of limestone.

Spongy mesophyll Leaf tissue, comprising variously shaped, loosely arranged chlorophyllous parenchyma cells with large intercellular spaces.

Sporangium A structure composed of a sterile wall and sporocytes, in which spores develop via meiosis.

Spore A resistant cell resulting from either sexual or asexual reproduction, found in many plants, algae, and fungi, which may be adapted for dispersal and for survival during periods of stressed environmental conditions.

Sporophyll A developmentally modified leaf on which sporangia form.

Stadial A time interval during which colder climate prevails within a warm interglacial.

Stem group or lineage The part of a clade with the extinct basal forms, as opposed to crown, or core (see above).

Steppe Treeless (or almost so) vegetation developed in regions with low rainfall where plants grow as a discontinuous vegetation cover and dominated by grasses, herbs, and shrubs with short lifespans.

Stoma/pl. stomata A pore in the epidermis of the plant, bordered by a pair of guard cells, which provides a means for gaseous exchange between its internal tissues and the atmosphere.

Stomatal complex Refers to the stoma, its guard cells, and any associated subsidiary cells (epidermal cells associated with the guard cells, but which are morphologically distinct from other epidermal cells).

Strut A rod or bar that is part of a framework, or a structure.

Subclade A clade within a clade. For example, Aves (birds) is a subclade of Dinosauria, which is a subclade of Archosauria.

Subgenus Used in taxonomy to group species into clades, the relationships of which are considered to be somewhere between species and genera.

Subtropical humid forest A vegetation type developed in the area of subtropical (= warm temperate, Köppen's Cfa) climate, characterized by high rainfall during the growing season; the name can also be applied to the biome characteristic of this climate type.

Supergroup A highly comprehensive unit in the hierarchy of sedimentary rock bodies (lithostratigraphy), comprising at least two groups (of large rock bodies), and many more formations.

Syncline A fold in a rock sequence where younger rock layers are centrally positioned along the axis on a horizontal flat surface.

Synthem A rock body separated by beneath and above by unconformities, often representing a sedimentary cycle.

Taiga Tree-dominated vegetation of the high latitudes, where mostly conifer species of *Larix*, *Picea*, and *Pinus* are able to exploit the short growing season.

Taphonomy A subdiscipline of paleontology studying the processes that led to the specific arrangement of fossils in a particular locality; processes can include, but are not limited to, disturbance by other organisms including predation, transport of the material by rivers, or partial decomposition.

Tephra A sedimentary deposit mainly composed of material produced by a volcanic eruption.

Thermophilous Organisms or communities that favor growth under warm conditions.

Tiered vegetation Stratified organization of a community wherein plants grow close to the soil surface as ground-cover, overtopped by shrubby or short-statured taxa (understory) which, in turn, is beneath one or more upper-most levels of trees in a canopy.

Till Mass of glacially transported boulders, pebbles, and sand, unsorted and unstratified, in a finer matrix.

Tillite Lithified sedimentary rock consisting of glacially transported boulders, pebbles. and sand, unsorted and unstratified, in a finer matrix.

Tracheid A cell type with lignified walls, but without per-foration plates, found in vascular plants; these cells are elongated and function in the internal transport of water and dissolved minerals.

Trilete spore Meiosis of a spore-mother cell wherein the four resultant spores are organized in a tetrahedral con-figuration; when split apart, each proximal (interior) face is marked by a "Y" through which germination occurs.

Tundra Treeless vegetation of the high latitudes, where tree growth is hindered by low temperatures and short growing seasons, such that lichens, mosses, herbs, and dwarf shrubs dominate.

Tundra steppe A subdivision of tundra developed in regions with lower rainfall resulting in a discontinuous vegetation cover dominated by grasses with short growth cycles.

Varves Lake sediments composed of pairs or couplets of thin (1 mm or less) layers of contrasting aspect (color and/or texture) deposited within a single year.

Vascular bundles Strands of primary vascular tissue (xylem and phloem) constituting the vascular system of the plant.

Vertisols Clay-rich soils with a high content of shrink-swell clays formed in regions with a dry season.

Viviparous Animals that give live birth, as opposed to lay-ing eggs.

Xeric Region experiencing low or scarce rainfall, at least seasonally; mostly applied to characterize vegetation.

Zoidiogamy The fertilization by means of motile anthero-zoids developed in prepollen grains without the produc-tion of a pollen tube.

Taxonomic Index

See below for a synthetic chart of the terrestrial plant phylogeny and nomenclature adopted in this work. Compiled by E. Tschopp, R.A. Gastaldo, P. Gensel, E. Kustatscher, E. Martinetto.

Scientific name	Vernacular name, description	Higher taxonomy	Figures, Boxes	Slides
Abelisauroidea	A clade of carnivorous dinosaurs	Theropoda, Ceratosauria		
Abies	Extant genus, firs	Conifers, Pinaceae	1.26; 2.16	U0207
Abrictosaurus	A small, bipedal herbivorous dinosaur	Ornithischia, Heterodontosauridae	7.7	
A. consors			7.7	
Acacia	Extant genus, woody legumes	Eudicots, Fabaceae		
Acalypha	Extant genus, copperleaves	Eudicots, Euphorbiaceae		
Acanthomorpha	Spiny-finned teleost fish	Osteichthyes, Actinopterygii		
Acer	Extant genus, maples	Angiosperms, eudicots, Sapindaceae	1.21	U0147, U1110
A. gr. cappadocicum	An extant group of taxa from western Eurasia		1.21	U0147
A. lobelii	An extant species from southern Europe			U0147
A. opulifolium	An extant species from southern Europe			
A. pseudoplatanus	An extant species from Europe			
A. sect. Saccharina	An extant group of taxa			
Aceratheriinae	Family of hornless rhinoceros	Perissodactyla, Rhinocerotidae	3.5	
Acinonyx	Cheetah	Carnivora, Felidae		
Acipenseridae	Sturgeons	Osteichthyes, Actinopterygii		
Acritarcha	Organic microfossils	Eukaryota		
Acrocanthosaurus	A large carnivorous dinosaur	Theropoda, Carcharodontosauridae		U0707
Actinidia	Extant genus, kiwi fruits	Angiosperms, eudicots, Actinidiaceae	2.12	U0251
A. arguta	An extant species from eastern Eurasia		2.12	
A. faveolata	Fossil-species		2.12	
Actinidiaceae	Kiwi fruit family	Angiosperms, eudicots, Ericales		U0251
Actinistia	Coelacanths	Osteichthyes, Sarcopterygii		
Actinocalyx	Fossil-genus	Angiosperms, eudicots, Ericales		
Actinopterygii	Ray-fin fishes	Vertebrata, Osteichthyes	7.12; 7.13	
Aculea	Fossil-genus of ferns from Antarctica	Pteridophytes		
Adoxaceae	Moschatel family	Angiosperms, eudicots, Dipsacales, Adoxaceae, considered in most recent papers as synonym of Viburnaceae		

(continued)

© Springer Nature Switzerland AG 2020
E. Martinetto et al. (eds.), *Nature through Time*, Springer Textbooks in Earth Sciences, Geography and Environment,
https://doi.org/10.1007/978-3-030-35058-1

Scientific name	Vernacular name, description	Higher taxonomy	Figures, Boxes	Slides
Aesculus	Extant genus, buckeyes	Angiosperms, eudicots, Sapindaceae		
Aethophyllum	Herbaceous fossil conifer	Coniferopsida, Voltziales		
Aetosauria	Terrestrial, armored relatives of crocodiles	Diapsida, Archosauria		
Afrocasia	An araceous (aroid) leaf fossil-genus	Angiosperms, monocots, Araceae	5.27	U0518
A. kahlertiana	Fossil-species		5.27	U0518
Agathoxylon	Fossil wood anatomy resembling a living austral conifer (*Agathis*)	Conifers, Araucariaceae		U0640
A. arizonicum	Fossil-species			U0640
Agilodocodon	An early, tree-living mammal	Synapsida, Therapsida	7.9	
A. scansorius			7.9	
Aglaophyton	Fossil-genus representing a sporophyte phase; groundcover plant	Plantae, "early land plants"		U1535
A. major	Fossil-species			
Agnatha	Jawless fish	Animalia, Vertebrata		
Alamatus	Fossil-genus of ferns from Antarctica	Pteridophytes		U0616
A. bifida	Fossil-species			U0616
Alangium	Extant genus, relative to dogwoods	Angiosperms, eudicots, Cornaceae	2.2	
A. deutschmannii	Fossil-species from Europe		2.2	
Albinykus	An alvarezsaur; small, long-legged dinosaur	Theropoda, Alvarezsauridae		
Alces	Moose	Artiodactyla, Cervidae	3.2	
A. alces			3.2	
Alchornea	Extant genus of tropical trees	Angiosperms, eudicots, Eurphorbiaceae	1.11	
Alethopteris	Fossil-genus of seed fern foliage	"Pteridosperms", Alethopteridaceae		
Alismataceae		Angiosperms, monocots, Alismatales		
Allantodiopsis	Fossil-genus of ferns	Pteridophytes, Dryopteridaceae		
Allericarpus	Fossil-genus based on fruits from Europe	Angiosperms, eudicots, Ericales		
Alligatoridae	Alligators	Archosauria, Crocodyliformes		
Alloiopteris	Fern with bipinnate fronds	Pteridophytes, Zygopteridaceae		U1124
Allosaurus	Carnivorous bipedal dinosaur	Theropoda, Allosauridae		U0825
Alnus	Extant genus, alders	Angiosperms, eudicots, Betulaceae	1.7; 2.6	U0213
A. alnobetula subsp. *fruticosa*	Alder shrub		1.7	
A. incana	Extant species			
A. julianiformis	Fossil-species		2.6	U0213
Alvarezsauroidea	Possible insect-eaters with shortened arms with a single claw	Dinosauria, Theropoda	7.4	U0708
Alvinia	Conifer fossil-genus	Conifers, Cheirolepidiaceae		
Amanda	Fossil-fenus of fern-like fronds with fertile spike	Pteridophytes		
Amaranthaceae	Amaranth	Angiosperms, eudicots, Caryophyllales		
Amborella	Extant genus	Angiosperms, eumagnoliids, Amborellales		
Amborellales		Angiosperms, eumagnoliids		
Amentotaxus	Extant genus	Conifers, Cephalotaxaceae		
Ammonoidea	Ammonites	Mollusca, Cephalopoda		
Amphibia	Amphibians	Vertebrata, Tetrapoda		

(continued)

Scientific name	Vernacular name, description	Higher taxonomy	Figures, Boxes	Slides
Amyelon	Fossil-genus based on anatomically preserved roots	Gymnosperms, Cordaitales	12.12	
Anacardiaceae	Cashew or sumac family	Angiosperms, eudicots, Sapindales		
Anachoropterids	Group of anatomically preserved late Paleozoic ferns	Pteridophytes		
Anchitherium	Browsing, deer-like horse	Perissodactyla, Equidae	3.5	
A. aurelianense			3.5	
Ancylotherium	Large browsing chalicothere (odd-toed ungulate)	Perissodactyla, Chalicotheriidae	3.5	
A. pentelicum			3.5	
Ancyrochitina		Animalia, Chitinozoa	14.12	U1450
A. longispina			14.12	U1450
Andropogoneae	Sorghum	Angiosperms, monocots, Poaceae		
Aneimites	Fossil-genus of fern-like fronds	"Pteridosperms"		
Anetoceras	Devonian ammonite	Ammonoidea, Anarcestida	14.3	U1434
Aneurophytales	Leafless, shrubby progymnosperms	Tracheophytes, "progymnosperms"		U1509, U1516
Angiosperms	Flowering plants	Plantae, spermatophytes		
Anguilliformes	Eels	Osteichthyes, Actinopterygii		
Anguimorpha	Worm-lizards	Diapsida, Squamata		
Anisodon		Perissodactyla, Chalicotheriidae	3.2; 3.5	
A. grande			3.2; 3.5	
Anisoptera	Dragonflies	Insecta, Odonata		
Ankylosauria	Quadrupedal dinosaurs with extensive armor	Ornithischia, Thyreophora		U0710
Ankyropteris	Fossil-genus of ferns	Pteridophytes, Zygopteridaceae		
A. brongniartii	Fossil-species of vine or climbing ferns			
Annelida	Segmented worms	Animalia, Lophotrochozoa		
Annularia	Leaf taxon of horsetails	Sphenophytes, Calamitaceae	12.4	
Anomochlooideae	Basal grasses	Angiosperms, monocots, Poaceae		
Anomozamites	Fossil-genus	Gymnosperms, Bennettitales	6.9	
Anseriformes	Geese, ducks	Aves, Anserimorphae		
Anthophila	Bees	Insecta, Hymenoptera		
Anthracotheriidae	Hippopotamus-like ungulates	Mammalia, Artiodactyla	3.4	
Anthracotherium	Early relatives of hippopotamus	Artiodactyla, Anthracotheriidae	3.4	
A. magnum			3.4	
Antiarchi	Heavily armored placoderm fishes	Vertebrata, Placodermi		
Antilopinae	Antelope family	Artiodactyla, Bovidae		
Anura	Frogs	Tetrapoda, Amphibia		
Apatosaurus	A long-necked dinosaur	Sauropoda, Diplodocidae		
Aphlebia	Fossil-genus of ferns, but the common name applies also to the basal elements of fern pinnae, with a form that differs from the other pinnae	Pteridophytes		
A. hvistendahliae	Fossil-species of ferns, based on the basal elements of fern pinnae			
Apocrita	Wasps	Insecta, Hymenoptera		
Apodemus	Field mice	Rodentia, Muridae		
Aquatifolia	Cretaceous leaf fossil-genus	Angiosperms, Nymphaeaceae		
Aquifoliaceae	Holly family	Angiosperms, eudicots	1.11	U0126
Aquilapollenites	Cretaceous pollen fossil-genus	Angiosperms		
Araceae	Angiosperm family	Angiosperms, monocots, Alismatales		

(continued)

Scientific name	Vernacular name, description	Higher taxonomy	Figures, Boxes	Slides
Araliaceae	Angiosperm family, including ivy	Angiosperms, eudicots, Apiales	1.11	
Araliaephyllum	European Cretaceous angiosperm	Angiosperms, Chloranthaceae?	5.8	U0506
A. kowalewskianum	Fossil-species		5.8	U0506
Araneae	Spiders	Arthropoda, Arachnida		
Araucaria	Extant genus of conifers	Conifers, Araucariaceae	6.3	U0623, U0628, U1222
A. araucana	Monkey puzzle tree, extant species			
A. angustifolia	Extant species			U0628
A. bidwillii	Extant species			U0623
A. columnaris	Extant species			U0623
A. heterophylla	Norfolk Island pine, extant species			U0623
A. lanceolata	Fossil-species		6.3	
Araucariaceae	Conifer family	Conifers, Pinales		U0600, U0612, U0614, U0624, U0628, U0654
Araucarites	Fossil-genus of conifers	Conifers, Araucariaceae	6.3	U0628
A. phillipsii	Fossil-species			U0628
Arberiaceae	Fossil-genus of male reproductive structures	Gymnosperms, Glossopteridales		U1216
Archaeanthus	One of the oldest angiosperm fossil-genera	Angiosperms		
Archaefructus	Small aquatic angiosperm from the Cretaceous of Asia	Angiosperms, Archaefructaceae	5.16	U0516
A. liaoningensis	Fossil-species		5.16	U0516
A. sinensis	Fossil-species		5.16	U0516
Archaeocalamites	Fossil-genus	Sphenophytes, Calamitaceae		U1304
Archaeohyracidae	Small-sized notoungulates	Mammalia, Notoungulata	3.3	
Archaeonoeggerathia	Fossil-genus; Beck suggested a progymnosperm connection; it is not assured	Noeggerathiales, higher affinity uncertain		
Archaeopteridales	Woody arborescent, leafy progymnosperms	Plantae, "progymnosperms"		U1509, U1512
Archaeopteris	Devonian tree	"Progymnosperms", Archaeopteridaceae	15.1; 15.7	U1511, U1512, U1513
Archaeopteryx	One of the first birds	Theropoda, Archaeopterygidae		U0715, U0716
Archeampelos	Fossil-genus	Angiosperms, eudicots, Cercidiphyllaceae?		
A. acerifolia	Fossil-species			
Archosauria	Most recent common ancestor of birds and crocodiles, and all its descendants	Eureptilia, Archosauriformes		
Archosauriformes	Most recent common ancestor of crocodiles and the extinct *Proterosuchus*, and all its descendents	Tetrapoda, Eureptilia		
Arccaccac	Palm family	Angiosperms, monocots, Arecales	1.11	
Arecales	Palms	Angiosperms, monocots		
Aristidoideae	Grasses	Angiosperms, monocots, Poaceae		
Aristonectes	Filter-feeding plesiosaur	Diapsida, Sauropterygia		
Arsinoitherium	Horned mammal	Embrithopoda, Arsinoitheriidae	3.4	
Artemisia	Extant genus, ragweeds	Angiosperms, eudicots, Asteraceae		U0174
Arthrodira	Group of armored placoderms	Vertebrata, Placodermi		
Arthropitys	Fossil-genus	Sphenophytes, Calamitaceae		
Artiodactyla	Even-toed ungulates	Mammalia, Ungulata		

(continued)

Scientific name	Vernacular name, description	Higher taxonomy	Figures, Boxes	Slides
Arundinoideae	Grasses	Angiosperms, monocots, Poaceae		
Arvicola	Water vole	Rodentia, Cricetidae		
Arvicolinae	Voles, lemmings, and muskrats family	Rodentia, Cricetidae		
Aspidistes	Fossil-genus of ferns	Pteridophytes, Aspidiaceae	6.11	
A. thomasii	Fossil-species		6.11	
Asplenium	Extant genus of ferns	Pteridophytes, Aspleniaceae	11.3	U1106, U1107, U1110, U1111
A. scolopendrium	Extant fern species		11.3	U1110
Asteraceae	Daisy family	Angiosperms, eudicots, Asterales		U0219
Asterochlaena	Fern fossil-genus, possibly of zygopterid affinity	Pteridophytes, Zygopteridaceae?		U1122
A. laxa	Fossil-species			U1122
A. ramosa	Fossil-species			
Asteroideae	Asters	Angiosperms, eudicots, Asteraceae		
Asterophyllites	Sphenopsid leaf fossil-genus	Sphenophytes, Calamitaceae	12.4	U1328
Asteroxylon	Early rhizomatous lycopsid	Lycopsids, Asteroxylaceae		U1536
A. mackiei	Fossil-species			U1536
Asterozoa	Invertebrate group including starfish and brittle stars	Animalia, Echinodermata		
Astrapotherium	Tapir-to-hippo-sized browsers	Mammalia, Astrapotheriidae	3.3	
Atlascopcosaurus		Ornithischia, Ornithopoda		U0814
A. loadsi				U0814
Atreipus	Ichnogenus (trace fossil) attributed to early Ornithischian dinosaurs	Dinosauria, Ornithischia		
Ausktribosphenidae	Early mammal clade	Synapsida, Mammalia		
Ausktribosphenos		Mammalia, Ausktribosphenidae		U0812
A. nyktos				U0812
Aussiedraco	Pterosaur	Archosauria, Ornithocheiroidea		
Australovenator	Megaraptorid, carnivorous dinosaur from the Cretaceous of Australia	Theropoda, Megaraptoridae		U0811
A. wintonensis				U0811
Austrobaileyales		Angiosperms, eumagnoliids		
Austrosaurus	Long-necked dinosaur	Sauropoda, Somphospondyli		
Averhoites	Leaf fossil-genus	Angiosperms		
Aves	Birds	Theropoda, Avialae		
Avialae	Subclade of Aves	Dinosauria, Theropoda	7.4	
Azolla	Extant genus, aquatic floating ferns	Pteridophytes, Salviniaceae		U1111
A. aff. nana	Miocene species in Europe			U1111
A. aff. rossica	Miocene species of Europe			U1111
A. aff. ventricosa	Miocene species of Europe			U1111
Azonotriletes	Fossil-genus based on spores	Pteridophyta	9.4	
Bachitherium	Early ruminant	Artiodactyla, Bachitheriidae	3.4	
Bacteria	Prokaryotic microorganisms	–		U0918, U1533
Baiera	Fossil-genus	Ginkgophytes, Ginkgoaceae	6.10	U0631
B. furcata	Jurassic species of Europe			U0631
Bambusoideae	Bamboos	Angiosperms, monocots, Poaceae		
Banksia	Extant genus of Australian wildflowers	Angiosperms, eudicots, Proteaceae		U0602
Barapasaurus	A long-necked dinosaur	Sauropoda, Gravisauria		
Barinophyton	Devonian groundcover plant	"Early land plants". zosterophyllophytes, Barinophytaceae		U1516
B. sibiricum	Fossil-species			
Barosaurus	A long-necked dinosaur	Sauropoda, Diplodocidae		

(continued)

Scientific name	Vernacular name, description	Higher taxonomy	Figures, Boxes	Slides
Barrandeina	Understory fern in Devonian forests	Pteridophytes, Barrandeinaceae	11.17	U1130
B. dusliana	Fossil-species			U1130
Batoidea	Cartilaginous fishes called rays	Chondrichthyes, Elasmobranchii		
Belemnitida	Belemnites	Mollusca, Cephalopoda		
Bellarinea	Cretaceous conifer from Australia	Pinophytes, Conifers, Podocarpaceae	6.3	U0612
Bennettitales	Extinct group of cycad-like plants	Spermatophytes		U0504, U0629, U0652, U0907
Bergiopteris	Late Paleozoic progymnosperm from Gondwana	"Progymnosperms"	13.11	
B. reidsdomae	Fossil-species	"Progymnosperms"	13.11	
Bernaultia	Seed fern fossil-genus based on cm-sized pollen-bearing structures	"Pteridosperms", Medullosaceae		
Besanosaurus		Ichthyopterygia, Besanosauridae		
Betula	Extant genus, birches	Angiosperms, eudicots, Betulaceae	1.7; 1.26	
B. divaricata	Birch shrub			
B. fruticosa	Birch shrub			
B. nana	Birch shrub			
B. pendula	Birch		1.7	
B. pubescens	Birch			
Betulaceae	Birch family	Angiosperms, eudicots, Fagales		
Bevhalstia	Herbaceous, aquatic Cretaceous plant	Angiosperms?		
Bignoniaceae	Bignonias	Angiosperms, eudicots, Lamiales		
Bilsdalea	Jurassic conifer fossil-genus	Gymnosperms, pinophytes, Conifers	6.9	
B. dura	Fossil-species		6.9	
Birgeria	Fast-swimming fish	Actinopterygii, Birgeriidae	7.13	
B. americana			7.13	
Bishops	Cretaceous mammal from Australia	Mammalia, Ausktribosphenidae		U0812
Bison	Bison	Artiodactyla, Bovidae		
B. priscus	Steppe bison			
Bivalvia	Clams, oysters, mussels	Animalia, Mollusca		
Blastoidea	Stalked echinoderms	Echinodermata, Blastozoa		
Blastozoa		Animalia, Echinodermata		
Blattodea	Cockroaches	Arthropoda, Insecta		
Blechnum	Deer or hard fern	Pteridophytes, Blechnaceae		U1111
B. dentatum	Extant species			U1111
Bobasatrania	Unhurried deep-bodied fish	Actinopterygii, Bobasatraniidae		
Bobosaurus		Sauropterygia, Pistosauria		
Borealis	Brachiopod from shallow shelf areas of the Silurian of Europe	Brachiopoda, Rhynchonelliformea		U1430
Borreria		Angiosperms, eudicots, Rubiaceae	1.11	
Bos	Wild and domestic cattle	Artiodactyla, Bovidae	3.2	
B. primigenius	European auroch		3.2	
B. taurus	Cows			
Bothrodendron	Fossil-genus	Lycopsids, Bothrodendraceae		
Botrychiopsis	Permian seed fern	Pteridophytes, Botryopteridaceae	13.11	U1336, U1339
B. plantiana	Fossil-species		13.11	
Botryopteridaceae	Primitive fern family	Pteridophytes		
Bouliachelys	A turtle	Cryptodira, Dermochelyoidea		
Brachaucheninae	Subclade of Pliosauridae	Sauropterygia, Pliosauridae		
Brachiopoda	Lophophorate bivalves commonly called lamp shells	Animalia, Lophophorata		

(continued)

Scientific name	Vernacular name, description	Higher taxonomy	Figures, Boxes	Slides
Brachiosaurus	A giraffe-like, long-necked dinosaur	Sauropoda, Brachiosauridae	7.2	
Brachychirotherium	Ichnogenus often associated with aetosaurs	Diapsida, Archosauria		
Brachylophosaurus	A duck-billed dinosaur	Ornithischia, Hadrosauridae	7.7	
B. canadensis			7.7	
Brasenites	Leaf fossil-genus	Angiosperms		
Brasilodendron	A sub-arborescent cormose lycopsid	Lycopsids, Lycopodiopsidaceae	13.10	U1336
B. pedroanum	Fossil-species		13.10	U1336
Brontosaurus	A long-necked Jurassic dinosaur from North America	Sauropoda, Diplodocidae	7.2	U0823, U0827
Bryophytes	An informal plant group including the liverworts, hornworts, and mosses	Plantae, Embryophyta		
Bryozoa	Moss animals	Animalia, Lophophorata		U0813, U1428
Bubalus	Water buffalo	Artiodactyla, Bovidae		
B. murrensis	European water buffalo			
Budvaricarpus	Cretaceous fruit fossil-genus	Angiosperms		
Buteo	Common buzzard	Aves, Accipitridae	7.1	
Buteoxylonales	A poorly understood late Paleozoic seed fern group	"Pteridosperms"		
Byrsonima	Extant genus, locust berries	Angiosperms, eudicots, Malpighiaceae		
Caimaninae	Caimans (crocodilian alligatorid)	Crocodyliformes, Alligatoridae		
Calamitea	Paleozoic horsetail fossil-genus	Sphenophytes, Calamitaceae		
Calamitean sphenopsids	Tree-sized horsetails	Sphenophytes, Calamitaceae		
Calamites	Paleozoic horsetail fossil-genus	Sphenophytes, Calamitaceae	11.14; 12.3; 12.4	U0642, U1316, U1317, U1325
C. striata	Fossil-species			
Calamophyton	Devonian forest-forming tree	Pteridophytes. Cladoxylopsida, Pseudosporochnales		U1502, U1521
Calamopityales	Extinct group of seed ferns	Plantae, "Pteridosperms"		
Calamostachys	Carboniferous cone fossil-genus	Sphenophytes, Calamitaceae		
Calamus	Extant genus of rattan palms	Angiosperms, monocots, Arecaceae		
Calcimicrobes	Calcareous colonial microfossils	–		
Calligonum	Extant genus, shrubs	Angiosperms, eudicots, Polygonaceae	1.31	
C. cf. *polygonoides*	Extant species		1.31	
Callistophytales	Carboniferous vine or climber	"Pteridosperms"		
Callistophyton	Fossil-genus	"Pteridosperms", Callistophytaceae		
Callospermarion	Fossil-genus	"Pteridosperms", Callistophytaceae		
Calycanthaceae	Sweet shrubs or spice bushes	Angiosperms, eumagnoliids		
Calymene	A Silurian trilobite	Trilobita, Calymenidae	14.4	U1437
C. blumenbachi			14.4	U1437
Calymmatotheca	Fossil seed-fern genus	"Pteridosperms"	13.2	U1306
Camarasaurus	A long-necked dinosaur	Sauropoda, Camarasauridae		U0825
Camptosaurus	A relatively large, bipedal ornithischian	Ornithopoda, Camptosauridae		
Canis	Dogs	Carnivora, Canidae	1.9	
C. dirus	Dire wolves		1.9	
C. lupus	Gray wolf			
Cannabaceae	Hemp, hops, hackberry	Angiosperms, eudicots, Rosales		
Canrightia	Fossil-genus based on fruits	Angiosperms		
Canrightiopsis	One of the earliest angiosperms, from the Cretaceous of Europe	Angiosperms, Chloranthaceae	5.5	U0505

(continued)

Scientific name	Vernacular name, description	Higher taxonomy	Figures, Boxes	Slides
C. dinisii	Fossil-species		5.5	U0505
Capra	Goat	Artiodactyla, Bovidae		
C. ibex	Ibex			
Cardiopteris	Extant genus of vines	Angiosperms, eudicots, Cardiopteridaceae		
Carex	Extant genus of sedges	Angiosperms, monocots, Cyperaceae		
Carici rupestris-Kobresietea	Not a species, but an association of plants			
Carmichaelia	New Zealand broom	Angiosperms, eudicots, Fabaceae		
Carpinus	Extant genus, hornbeams	Angiosperms, eudicots, Betulaceae	1.26	U0231
C. betulus	Hornbeam			
Carya	Extant genus, hickories	Angiosperms, eudicots, Juglandaceae	1.37	U0219, U0230
Caryanthus	Fossil-genus, infructescence of a member of the *Normapolles* complex	Angiosperms, eudicots		
Caryophyllaceae	Carnation family	Angiosperms, eudicots, Caryophyllales		
Caspiocarpus	Fossil-genus	Early angiosperm		
Cassia	Extant genus, tree legumes	Angiosperms, eudicots, Fabaceae		
Castanopsis	Extant genus, chinquapins	Angiosperms, eudicots, Fagaceae	2.16	
Castor	Beaver	Rodentia, Castoridae		
C. fiber plicidens	Beaver			
Castorocauda lutrasimilis	A stem mammal	Synapsida, Mammalia	7.9	
Castorocauda	A stem mammal	Therapsida, Docodontidae	7.9	
Casuarina	Extant genus, evergreen shrubs and trees	Angiosperms, eudicots, Casuarinaceae		
Casuarinaceae		Angiosperms, eudicots, Fagales		
Cathaya	Extant genus of conifers	Pinophytes, conifers, Pinaceae	1.37	
C. argyrophylla	Extant conifer only growing in China		1.37	
C. vanderburghii	European fossil-species		2.2	
Cathaysiopteris	Permian leaf fossil-genus	Pteridophytes, Gigantopteridales	12.8	
Cathiaria	Cretaceous fossil-genus of fruits	Angiosperms		
Caulopteris	Tree fern fossil-genus associated with *Psaronius*	Pteridophytes, Marattiaceae		U1117
Caytonanthus	Pollen-bearing fossil-genus of the Caytoniales	"Pteridosperms", Caytoniales	6.10	
C. oncodes			6.10	
Caytonia	Fossil-genus based on leaves of a seed fern	"Pteridosperms", Caytoniales	5.1; 6.10	U0504, U0630
C. nathorstii	Jurassic fossil-species		6.10	
Caytoniales		Plantae, "Pteridosperms"		U0504, U0630
Cedrus	Extant genus, cedars	Pinophytes, conifers, Pinaceae	1.22	U0207
Celtis	Extant genus, hackberries	Angiosperms, eudicots, Cannabaceae	1.31	U0163
C. cf. caucasica	Extant species		1.31	U0163
Cephalanthus	Extant genus, buttonbushes	Angiosperms, eudicots, Rubiaceae		
Cephalaspidida	Extinct group of jawless fish	Vertebrata, Osteostraci		
Cephalopoda	Mollusc group including squids, octopus, and nautilus	Animalia, Mollusca		
Cephalotaxus	Extant genus, plum yews	Conifers, Cephalotaxaceae	2.2	
C. cf. rhenana	Fossil-species		2.2	
Ceratopetalum	Extant genus	Angiosperms, eudicots, Cunoniaceae		
Ceratophyllaceae	Hornwort Family	Angiosperms, Ceratophyllales		
Ceratophyllales	Aquatic "hornworts"	Angiosperms		

(continued)

Scientific name	Vernacular name, description	Higher taxonomy	Figures, Boxes	Slides
Ceratophyllum	Extant genus	Angiosperms, Ceratophyllales, Ceratophyllaceae		U0163, U0507
C. cf. *demersum*	Extant species		1.31	U0163
Ceratopsia	Horned and frilled dinosaurs	Ornithischia, Marginocephalia		U0710
Ceratosaurus	A large carnivorous dinosaur	Theropoda, Ceratosauridae		
Ceratotherium	A genus of rhinoceros	Perissodactyla, Rhinocerotidae	3.2; 3.5	
C. neumayri	Mixed-feeding rhinoceros		3.5	
C. simum	White rhinoceros		3.2	
Cercidiphyllum	Extant genus, katsuras	Angiosperms, eudicots, Cercidiphyllaceae		U0418
Cervus	Deer	Artiodactyla, Cervidae		U0148
C. elaphus	Red deer			U0148
Cetiosaurus	A long-necked dinosaur	Sauropoda, Eusauropoda	7.1; 7.2	
Chalicotheriidae	Clawed perissodactyls	Mammalia, Perissodactyla	3.5	
Chamaecyparis	Extant genus, false cypresses	Conifers, Cupressaceae		
C. obtusa	Extant species			
Chansitheca	Fossil leptosporangiate-fern fossil-genus	Pteridophytes	11.11	U1118
C. wudaensis	Fossil-species		11.11	U1118
Chasmatosuchus	A 2 m long archosauriform similar to a crocodile (but not closely related)	Diapsida, Archosauriformes	10.9	U1024
Cheirolepidiaceae	Extinct conifer family	Conifers, Pinales		U0614
Chelycarapookus	A turtle from the Cretaceous of Australia	Tetrapoda, Testudines		U0814
C. arcuatus				U0814
Chenopodioideae	Chenopods	Angiosperms, eudicots, Amaranthaceae		
Chimaeriformes	Ghost sharks	Vertebrata, Chondrichthyes		
Chiroptera	Bats	Synapsida, Mammalia		
Chirotherium	A trace fossil generally attributed to archosaurs	Diapsida, Archosauria		
Chitinozoa	Flask-shaped, organic walled marine microfossils	Animalia		
Chloranthaceae	An early-diverging family of flowering plants	Angiosperms, eumagnoliids, Chloranthales	1.11	U0126, U0507
Chloranthales		Angiosperms, eumagnoliids		
Chloranthistemon	Fossil-genus	Angiosperms, eumagnoliids, Chloranthaceae		
C. crossmanensis	Fossil-species			
Chloranthoids	Informal group of early diverging angiosperm including Chloranthales and allies	Angiosperms		
Chloridoideae	Grasses of the PACMAC clade	Angiosperms, monocots, Poaceae		
Choerolophodon	Extinct Miocene proboscid	Mammalia, Proboscidea		
Chondrichthyes	Cartilaginous fish	Vertebrata, Gnathostomata	7.12	U0747
Chondrostei	Clade including sturgeons and paddlefishes	Osteichthyes, Actinopterygii		
Chordata	Animals with a notochord	Animalia, Deuterostomia		
Cinnamomophyllum	Fossil angiosperm-leaf fossil-genus	Angiosperms, eumagnoliids, ?Lauraceae	5.29	
Cinnamomum	Extant genus, cinnamon trees	Angiosperms, eumagnoliids, Lauraceae		U0205, U0206
Circumpolles	A gymnosperm pollen type	Pinophytes, Cheirolepidiaceae	9.4	
Citipati	A beaked, feathered, bipedal dinosaur	Theropoda, Oviraptorosauria	8.4; 8.5	U0723, U0816, U0819
Cladium	Extant genus of sedges	Angiosperms, monocots, Cyperaceae		

(continued)

Scientific name	Vernacular name, description	Higher taxonomy	Figures, Boxes	Slides
Cladophlebis	Extant genus, member of the royal fern family	Pteridophytes, Osmundaceae	6.3; 6.11; 11.9	U0612, U0615, U0632, U1113, U1115, U1328
C. australis	Cretaceous fossil-species from Australia		6.3	U0612
C. denticulata	Jurassic European fossil-species		6.11; 11.9	U0632, U1115
C. harrisii	Fossil-species		6.11	
Cladoxylopsida	Cladoxylopsids, potential ancestors of ferns and horsetails	Plantae, pteridophytes?		U1130
Claraia	Scallop-like bivalve molluscs	Bivalvia, Pterinopectinidae	10.8	U1008, U1023
Clathropteris	Fern fossil-genus	Pteridophytes, Dipteridaceae		
C. meniscoides	Fossil-species			
Clavatipollenites	Fossil-genus based on pollen grains, probably one of the earliest angiosperms, from the Cretaceous of Europe	Angiosperms, chloranthoids	5.4	U0505
C. hughesii	Fossil-species		5.4	U0505
Cleithrolepis	Unhurried deep-bodied fish	Actinopterygii, Cleithrolepididae		
Clevosaurus	Tuatara relative from the Triassic of Europe	Rhynchocephalia, Clevosauridae		U0838, U0839
Cnidaria	Aquatic, mostly marine invertebrates animals including corals, sea anemones, and jellyfish	Animalia, Parahoxozoa		
Cochlearites	Triassic European bivalve			U0906
Coelodonta		Perissodactyla, Rhinocerotidae		
C. antiquitatis	Woolly rhinoceros			
Coelophysis	A small carnivorous dinosaur	Theropoda, Coelophysidae	7.1	
Coelurosauria	Clade of mostly carnivorous dinosaurs	Theropoda, Avetheropoda	7.4	
Coelurus	A relatively small, carnivorous dinosaur	Theropoda, Tyrannosauroidea		
Colaniella		Eukaryota, Foraminifera		
Coleoidea	Squids and octopods	Mollusca, Cephalopoda		
Coleoptera	Beetles	Arthropoda, Insecta		U0653
Colpoxylon	Medullosan seed-fern fossil-genus	"Pteridosperms"		
Colymbosathon		Ostracoda, Myodocopida		U1452
Combretaceae	White mangrove family	Angiosperms, eudicots, Myrtales		
Compsognathidae	Clade of small, carnivorous dinosaurs	Theropoda, Tyrannoraptora	7.4	
Conchostracans	Paraphyletic group commonly called clam shrimps	Arthropoda, Crustacea		
Coniferales	Taxonomic order for conifers, a group that may be paraphyletic	Gymnosperms, Conifers		
Coniferophytes	A probably paraphyletic plant group including the extinct Cordaitales, the living conifers and allies; Gnetales are excluded. Synonym of pinophytes	Gymnosperms, pinophytes	12.2	U1221, U1328
Coniopteris	Cosmopolitan Mesozoic fern fossil-genus	Pteridophytes, Dicksoniaceae	8.7	U0627, U0634, U1113
C. hymenophylloides	Fossil-species			U0634
C. murrayana	Fossil-species			
C. simplex	Fossil-species			U0634
Conodonta	Extinct agnathan chordates resembling eels	Animalia, Vertebrata		
Conohyus	Omnivorous tetraconodontine pig	Artiodactyla, Suidae	3.5	

(continued)

Scientific name	Vernacular name, description	Higher taxonomy	Figures, Boxes	Slides
Cooksonia	An early land plant from the Silurian and Devonian	Plantae, "early land plants"		U1504, U1510, U1542, U1543
C. barrandei	Earliest land plant from the Silurian of Europe			U1543
C. pertoni	Fossil-species			U1504, U1542
Copaifera	Extant genus of copal-bearing legumes	Angiosperms, eudicots, Fabaceae		
Coprosma	Extant genus of Austral-Pacific evergreens	Angiosperms, eudicots, Rubiaceae		
Cordaitales	Extinct, possible early conifers	Gymnosperms, pinophytes	12.12	U1221
Cordaites	Early confer-like trees	Gymnosperms, Cordaitales, pinophytes	12.12	U1116, U1328, U1329, U1330
C. borassifolius	Fossil-species		12.12	
Coriaria	Extant genus of Mediterranean and Asian shrubs	Angiosperms, eudicots, Coriariaceae		
Coricladus	Fossil-genus of coniferophytes	Pinophytes, Voltziales		
C. quiteriensis	Fossil-species			
Corispermum	Extant genus including bugseeds, tickseeds, and tumbleweeds	Angiosperms, eudicots, Amaranthaceae		
C. crassifolium	Extant species		1.7	
Cornaceae	Dogwood family	Angiosperms, eudicots, Cornales		
Cornales	Dogwood order	Angiosperms, angiosperms, eudicots		
Cornus	Extant genus, dogwoods	Angiosperms, eudicots, Cornaceae		
Corriebaatar	A multituberculate mammal	Multituberculata, Corriebaataridae		U0812
C. marywaltersae				U0812
Corylites	Fossil-genus of hazel relatives	Angiosperms, eudicots, Betulaceae	5.12	
Corylus	Extant genus, hazels	Angiosperms, eudicots, Betulaceae	1.26	
C. ferox	Extant species of hazel		1.26	
Corynella	Extant genus, legumes	Angiosperms, eudicots, Fabaceae		
Corynepteris	Fossil-genus	Pteridophytes, Zygopteridaceae	11.14	U1126
C. angustissima	Fossil-species, interpreted as an understory fern		11.15	U1126
Corystospermales	Order of Triassic seed ferns	"Pteridosperms"		
Cotinus	Extant genus, smoke trees	Angiosperms, eudicots, Anacardiaceae		U0163
C. cf. coggygria	Extant species		1.31	U0163
Craigia	Extant genus	Angiosperms, eudicots, Malvaceae	1.37	U0219
C. yunnanensis	Extant species		1.37	
Craniata	Craniiformean brachiopods	Lophophorata, Brachiopoda		
Cratochelone	A turtle	Cryptodira, Protostegidae		
Cremohipparion	Hipparionine horses	Mammalia, Perissodactyla		
Crenaticaulis	Fossil-genus of slender, dichotomously branching, leafless early land plants, dominating Devonian marshlands	Plantae, zosterophyllophytes		U1527
Crinoidea	Sea lilies	Echinodermata, Crinozoa		
Crinozoa	Sessile echinoderms	Animalia, Echinodermata		
Crocodylia	Crocodiles	Archosauriformes, Eusuchia		
Crocodyliformes	Crocodile-relatives	Archosauria, Pseudosuchia		
Crocodylomorpha	Crocodiles and their relatives	Archosauria, Pseudosuchia		
Crossotheca	Fossil-genus based on cross-shaped pollen organs	"Pteridosperms", Lyginopteridales	12.6	

(continued)

Scientific name	Vernacular name, description	Higher taxonomy	Figures, Boxes	Slides
Crossozamia	Fossil-genus based on cycad megasporophylls	Cycadophyta, Cycadales		
Crurotarsi		Diapsida, Archosauriformes		
Crustacea	Crustaceans	Animalia, Arthropoda		U0813, U1439
Cryptoclididae	Subclade of Plesiosauroidea	Diapsida, Sauropterygia	Box 7.2	
Cryptodira	Hidden-neck turtles	Tetrapoda, Testudines		
Cryptomeria	Extant conifer genus native to East Asia	Conifers, Cupressaceae		U0230, U0231
Cteniogenys	A small Jurassic reptile from North America	Archosauromorpha, Choristodera		U0828, U0830
Ctenoluciidae	Pikes	Actinopterygii, Characiformes		
Cunoniaceae	Southern hemisphere tropical and wet-temperate family	Angiosperms, eudicots, Oxalidales		
Cupressaceae	Cypress family	Conifers, Pinales		U0614
Cyanobacteria	A phylum of bacteria obtaining their energy through photosynthesis	Bacteria, Terrabacteria		
Cyathea	Extant genus of tree ferns	Pteridophytes, Cyatheaceae		U1106, U1107
C. deckenii	Extant species			U1107
Cyatheaceae	Scaly tree-fern family	Pteridophytes, Cyatheales		U1107
Cycadales	Cycads	Cycadophyta		U0652
Cycadophyta	Cycads	Plantae, tracheophytes		
Cyclea	Extant genus of climbers	Angiosperms, eudicots, Menispermaceae	2.2	U0205, U0206
C. palatinati-bavariae	Fossil-species, European woody climber	Angiosperms, eudicots, Menispermaceae	2.2	
Cyclobalanopsis	Extant genus, ring-cupped oaks	Angiosperms, eudicots, Fagaceae	2.16	
Cyclostigma	Early tree lycopsid colonizing Devonian wetlands	Lycopsids, other affinity uncertain		U1304, U1511
Cyclostomata	Jawless fishes	Animalia, Vertebrata	7.12	
Cygnus	Swan	Aves, Anatidae		
Cylindroleberididae		Ostracoda, Myodocopida		
Cymopolia	Green algal genus	Chlorophyta, Dasycladaceae		
C. barbata	Fossil-species			
Cynepteris	Triassic fern with unusual, net-like leaf venation from North America	Pteridophytes, Cynepteridaceae		U0643
C. lasiophora	Fossil-species			
Cyperaceae	Sedges	Angiosperms, monocots, Poales		
Cyphaspis	A small Devonian trilobite	Trilobita, Proetida	14.4	U1438
C. ceratopthalma			14.4	U1438
Czekanowskia	Fossil-genus of Czekanowskiales	Gymnosperms, Czekanowskiales		
C. blackii	Fossil-species		6.10	
C. furcula	Fossil-species		6.10	
C. microphylla	Fossil-species		6.10	
C. thomasii	Fossil-species		6.10	
Czekanowskiales	Extinct gymnosperm group related to *Ginkgo;* also known as Leptostrobales	Gymnosperms		
Dacrycarpus	Extant conifer genus	Conifers, Podocarpaceae		
Dacrydium	Extant genus, rimus	Conifers, Podocarpaceae		
Dakotanthus	Fossil-genus, a pentamerous (5-parted) Cretaceous flower from North America	Angiosperms, eudicots, Quillajaceae		U0512
Danthonioideae	PACMAD clade grasses	Angiosperms, monocots, Poaceae		
Daonella	Oyster-like saltwater clams	Bivalvia, Posidoniidae		
Dapedium	Unhurried deep-bodied fish	Actinopterygii, Dapediidae		

(continued)

Scientific name	Vernacular name, description	Higher taxonomy	Figures, Boxes	Slides
Daptocephalus	Permian mammal-like vertebrate of Africa	Therapsida, Dicynodontia		U1010
Dasycladales	Unicellular green algae	Chlorophyta, Ulvophyceae		
Debeya	Cretaceous leaf fossil-genus	Angiosperms		
Decapoda	Crayfish	Crustacea, Malacostraca		
Decapodiformes	Squids	Mollusca, Coleoidea		
Deinocheirus	Beaked, feathered dinosaur	Theropoda, Ornithomimosauria	7.5; 8.4	U0816
Deinonychus	Small, feathered, carnivorous dinosaur	Theropoda, Dromaeosauridae		
Deinotherium	Miocene to Early Pleistocene elephant relative	Proboscidea, Deinotheriidae		
D. bozasi			3.2	
Delnortea	Fossil-genus of possible seed ferns, late Paleozoic	"Pteridosperms"?		
Delphinidae	Oceanic dolphins	Artiodactyla, Cetacea		
Dendraena	Fossil-genus of late Paleozoic true ferns	Pteridophytes		
D. pinnatilobata	Fossil-species			
Graptolithina	Dendroids, colonial hemichordates	Hemichordata, Pterobranchia		
Dendrolagus	Tree kangaroo	Marsupialia, Macropodidae		
Desmopteris	Fossil-genus of late Paleozoic true ferns	Pteridophytes		
D. alethopteroides	Fossil-species			
D. longifolia	Fossil-species			
Desmosponges	Sponges with silica spicules, spongin fibers, or both in their skeleton	Porifera, Demospongiae		
Dewalquea	Fossil-genus	Angiosperms, eudicots, Platanaceae		
Diamantinasaurus	A long-necked dinosaur from the Cretaceous of Australia	Sauropoda, Titanosauria		U0810, U0811
D. matildae				U0810, U0811
Diaphorodendron	Lycopsid similar to *Lepidodendron*	Lycopsids, Diaphorodendraceae	12.1	U1322
D. scleroticum	Fossil-species interpreted as an arborescent lycopsid	Lycopsids		U1322
Diaphorodenraceae	Family of arborescent lycopsids	Lycopsids		
Diapsida	Clade including araeoscelidians, lizards, and archosaurs	Vertebrata, Tetrapoda		
Diceros	Black rhinoceros	Perissodactyla, Rhinocerotidae		
D. bicornis			3.2	
Dichopteris	Jurassic seed-fern fossil-genus	"Pteridosperms"	9.3	
D. visianica	Fossil-species			
Dicksonia	Extant genus, tree ferns	Pteridophytes, Dicksoniaceae	6.11	U0634
D. mariopteris	Jurassic fossil-species from Europe		6.11	U0634
D. sellowiana	Extant South American species			U0634
Dicksoniaceae	Tree fern family	Pteridophytes, Cyatheales		U0627, U0634
Dicotyledons	Dicots, Dicotyledones, recognized as a paraphyletic group	Spermatophytes, angiosperms		
Dicranophyllum	Permian fossil-genus of leafy shoots with uncertain systematic position, possibly conifers	Gymnosperms	12.13	
D. hallei	Fossil-species			
Dicrocerus	Small, brachydont deer	Artiodactyla, Cervidae		
D. elegans			3.5	
Dicroidium	Arboreal seed fern	"Pteridosperms", Corystospermaceae	6.17; 6.18	U0649, U0650, U0651

(continued)

Scientific name	Vernacular name, description	Higher taxonomy	Figures, Boxes	Slides
D. odontopteroides	Fossil-species		6.18	
Dictyophyllum	Fossil-genus of leptosporangiate ferns	Pteridophytes, Dipteridaceae		U1115
D. irregularis	Fossil-species			U1115
D. nervulosum	Fossil-species			
D. nilssonii	Fossil-species			
Dictyopteridiaceae	Clade of seed ferns with winged seeds on one surface of a flattened shield-shaped organ	"Pteridosperms", Dictyopteridiopsida		U1216
Dicynodon	Mammal-like vertebrate from the Permian of Africa	Therapsida, Dicynodontidae	10.3	U1010, U1012
Dicynodontidae		Synapsida, Therapsida		U1012
Didelphodon	Cretaceous megatherian	Mammalia, Stagodontidae		
Dilcherocarpon	Cretaceous fossil-genus of fruits	Angiosperms, eudicots, Combretaceae		
Diluvicursor	A small Cretaceous ornithischian dinosaur from Australia	Ornithischia, Ornithopoda		U0814
D. pickeringi				U0814
Dinoflagellata	Flagellate microorganisms	Eukaryota, Alveolata		
Dinophyton	Triassic gymnosperm from North America	Plantae, tracheophytes		U0646
D. spinosus	Fossil-species			U0646
Dinosauria	Dinosaurs (including birds)	Diapsida, Archosauria		U0721, U0823, U0910
Dioonitocarpidium	Fossil-genus based on Mesozoic seeds	Gymnosperms, Cycadales		
Diplodocidae	Subclade of Eusauropoda	Sauropoda, Neosauropoda	7.2	
Diplodocus	Long-necked Jurassic dinosaur from North America	Sauropoda, Diplodocidae	7.3	U0823, U0825, U0827
Diplothmema	Y-forked seed-fern leaf	"Pteridosperms", Lyginopteridales		U1305
D. gothanica	Fossil-species			
Dipnoi	Lungfishes	Sarcopterygii, Dipnomorpha		
Diprotodon	Extinct, giant wombat-like animals from Australia	Marsupialia, Diprotodontidae		U0133
D. optatum			1.13	
Diprotodontidae	Oligocene to Pleistocene Australian endemic marsupial	Mammalia, Marsupialia		
Dipteridaceae	Umbrella ferns	Pteridophytes, Gleicheniales		U1114, U1115
Dipterocarpaceae	Family of tropical rain forest trees	Angiosperms, eudicots, Malvales		U1231
Dipterocarpus	Extant genus	Angiosperms, eudicots, Dipterocarpaceae	2.16	
Distichophytum	An early vascular plant	Plantae, zosterophyllophytes		
Docofossor	A digging, mole-like mammal	Mammalia, Docodontidae	7.9	
D. brachydactylus			7.9	
Donlesia	Mesozoic fossil-genus based on fruits	Angiosperms, Ceratophyllaceae		
Dorcatherium	Mouse deer	Artiodactyla, Tragulidae	3.5	
Dreadnoughtus	A long-necked dinosaur	Sauropoda, Titanosauria		
Drepanophycus	Pre-lycopsid rhizomatous plant	Lycopodiopsida, Drepanophycaceae		U1508, U1524
Drimys	Extant genus	Angiosperms, eumagnoliids, Winteraceae	2.10	
Drimys winteri	Extant South American plant species	Angiosperms, Winteraceae	2.10	
Drinker	A small ornithischian dinosaur	Ornithischia, Neornithischia		
Dromaeosauridae	Small, feathered carnivorous dinosaurs	Theropoda, Maniraptora	7.4; 8.2	

(continued)

Scientific name	Vernacular name, description	Higher taxonomy	Figures, Boxes	Slides
Dryas	Extant genus, mountain avens	Angiosperms, eudicots, Rosaceae		
D. octopetala	Extant plant species			
Dryosaurus	A bipedal ornithischian	Ornithischia, Iguanodontia		
Dunkleosteus	A gigantic Devonian predatory fish from North America	Placodermi, Dunkleosteidae	14.7	U1445
D. terrelli			14.7	
Dystrophaeus	A long-necked dinosaur	Sauropoda, Eusauropoda		
Echinodermata		Animalia, Bilateria	.	
Echinozoa	Free-living echinoderms	Bilateria, Echinodermata		
Ehretia	Extant genus, trees related to the borage family	Angiosperms, eudicots, Ehretiaceae, formerly assigned to Boraginaceae	2.2	
E. europaea	European fossil-species		2.2	
Ehrhartoideae	Grasses of the BOP clade	Angiosperms, monocots, Poaceae		
Elaeocarpaceae	Mostly tropical and subtropical flowering plants	Angiosperms, eudicots, Oxalidales		
Elasmosauridae	Long-necked xenopsarians	Sauropterygia, Plesiosauria		
Elatocladus	Mesozoic fossil-genus of leafy shoots	Gymnosperms, Coniferophyta		
E. setosus	Fossil-species		6.9	
E. zamioides	Fossil-species		6.9	
Elephas	Elephant	Proboscidea, Elephantidae		
E. maximus	Asian elephant		3.2	
Eleutherophyllum	Lower Carboniferous taxon showing both horsetail and club moss characteristics	Affinity uncertain		
Elkinsia	One of the earliest seed plants, Devonian	Gymnosperms, Elkinsiaceae	15.4	U1511, U1514, U1515
E. polymorpha	Fossil-species			U1514
Elsemaria	Angiosperm fossil-genus from Japan	Angiosperm	5.15	U0510
E. kokubunii	Fossil-species		5.15	
Embothrium	Chilean firebush	Angiosperms, eudicots, Proteaceae		
E. coccineum	Extant species		2.10	
Embrithopoda	Extinct clade of rhino-like ungulates	Mammalia, Paenungulata	3.4	
Emporia	A genus of snout moths	Lepidoptera, Pyralidae		
Enalikter		Arthropoda, Megacheira?		U1452
E. aphson				U1452
Enantiornithes	An early subclade of birds: enantiornithine birds	Theropoda, Aves		
Engelhardia	Extant genus fo trees native to southeast Asia	Angiosperms, eudicots, Juglandaceae		U0219
Entelodontidae	Large, pig-like animals	Mammalia, Artiodactyla	3.4	
Eoangiopteris	Fossil-genus of ferns	Pteridophytes, pteridophytes, Marattiaceae		
Eocercidiphyllites	Shrubby angiosperms, fossil-genus	Angiosperms, eudicots		
Eocycas	Fossil-genus of cycads	Gymnosperms, Cycadales	11.10	
Eomaia	An early, tree-living therian mammal	Mammalia, Theria		U0728
Eoplatanus	A Cretaceous platanoid from North America	Angiosperms, eudicots, Platanaceae	5.20	U0512
E. serrata	Fossil-species		5.20	
Eospermatopteris	Devonian tree dominating the first forests	Pteridophytes, cladoxylopsids	15.6	U1502, U1518, U1519
Ephedra	Extant genus, evergreen shrub-like plant	Gnetophytes, Ephedraceae		
Ephedraceae	Gymnospermous shrubs	Gnetophytes, Ephedrales		

(continued)

Scientific name	Vernacular name, description	Higher taxonomy	Figures, Boxes	Slides
Ephemeroptera	Mayflies	Insecta, Palaeoptera		
Equicalastrobus	Triassic member of the horsetail family from North America	Pteridophytes, Equisetaceae		U0642
E. chinleana	Fossil-species			
Equidae	Horse family	Mammalia, Perissodactyla		U0308
Equisetales	Horsetails	Plantae, pteridophytes		U0907
Equisetum	Extant genus, horsetails	Pteridophytes, sphenophytes, Equisetaceae		U0632
E. columnare	Jurassic fossil-species from Europe			U0632
Equus	Genus including horses, donkeys, and zebras	Perissodactyla, Equidae		
E. ferus	Wild horse		3.2	
Erbenoceras	Devonian ammonite	Cephalopoda, Ammonoidea	14.3	U1434
Erdmanithecales	Extinct seed plant order	Gymnosperms related to angiosperms		U0504
Eremopteris	Fossil-genus based on glossopterid reproductive organs	Glossopteridales		
Eremotherium	A giant Pleistocene ground sloth from South America	Xenarthra, Megatheriidae		U0127
Eretmophyllum	Fossil-genus	Ginkgophytes, Ginkgoales	6.10	
Ericaceae	Heather family	Angiosperms, eudicots, Ericales		
Ericales	Heathers	Angiosperms, eudicots, Asterids		
Erketu	A long-necked dinosaur	Sauropoda, Titanosauriformes		
Erlikosaurus	A Cretaceous therizinosaur dinosaur from Asia	Theropoda, Therizinosauridae		U0821
Erpetoichthys	Reedfishes	Actinopterygii, Polypteridae		
Erythrosuchidae		Diapsida, Archosauriformes		U0923
Erythrosuchus	A Triassic archosauriform	Archosauriformes, Erythrosuchidae	10.9	U1024
Ettingshausenia	Large-leaved Cretaceous plane trees from Eurasia	Angiosperms, eudicots, Platanaceae		U0508, U0509
Eubrontes	Ichnogenus for tracks of carnivorous dinosaurs	Dinosauria, Theropoda		
Eucalyptus	Extant genus, blue gum trees	Angiosperms, eudicots, Myrtaceae		
Eucommia	Extant genus, small trees of China	Angiosperms, eudicots, Eucommiaceae		
Eucommiidites	Angiosperm-like pollen fossil-genus	Erdtmanithecales		
Eudicotyledones, eudicots	Flowering plant group	Plantae, angiosperms		U0501, U0512
Eukaryota	Organisms in which cells have a membrane-bound nucleus			
Eumagnoliids		Plantae, angiosperms		
Eumorphotis	Early Triassic bivalve that thrived after the End-Permian Mass Extinction	Mollusca, Bivalvia	10.8	U1023
Euphyllophytina	Early fern-like plants	Plantae, tracheophytes,		
Euphorbiaceae	Spurge family	Plantae, angiosperms	1.11	
Eurya	Extant genus	Angiosperms, eudicots, Pentaphylacaceae	1.26; 2.2	
E. stigmosa	Fossil-species		2.2	
Eurypteridae	Sea scorpions	Arthropoda, Chelicerata		
Eurypterus	A sea scorpion genus	Arthropoda, Eurypteridae	14.11	U1417
E. tetragonophthalmus			14.11	U1417
Eusauropoda	Sauropoda subclade	Dinosauria, Sauropodomorpha	7.2	
Eusphenopteris	Late Paleozoic fern-like pinnate leaf	"Pteridosperms", Lyginopteridales		
Eutheria		Mammalia, Theria		
Evazoum	Triassic footprint genus			

(continued)

Scientific name	Vernacular name, description	Higher taxonomy	Figures, Boxes	Slides
Eviostachya	Reproductive cones	Sphenophytes, Calamitaceae		
Evolsonia	Sterile leaf, Permian	affinity uncertain	12.8	
Fabaceae	Legumes	Angiosperms, eudicots, Fabales		
Fagaceae	Oak family	Angiosperms, eudicots, Fagales		
Fagales	Oak order	Angiosperms, eudicots, Rosids		U0232
Fagus	Extant genus, beeches	Angiosperms, eudicots, Fagaceae	2.16	U0233, U0234
F. sylvatica	European beech			
Fairlingtonia	Early Cretaceous eudicot	Angiosperms, angiosperms, eudicots		
Fedekurtzia	A fossil-genus of seed ferns, late Paleozoic	"Pteridosperms"	13.2	U1305
F. cf. argentina	Fossil-species		13.2	
Felidae	Cat family	Mammalia, Carnivora		
Felis	Small- and medium-sized cats	Carnivora, Felidae		
F. silvestris	Wildcat			
Festucetalia	Phytosociological name			
F. lenensis	Phytosociological name used for a plant association			
Ficus	Extant genus, figs	Angiosperms, eudicots, Moraceae	1.26	
Filicales	Ferns	Plantae, pteridophytes		
Filograna	Marine polychaete-worm genus from the Permian of Italy	Polychaeta, Serpulidae	9.16	U0927
Foraminifera	Single-celled planktonic animals with chitin, calcite, agglutinated shells	Eukaryota, Rhizaria		U0108, U0906, U1023, U1029
Fortuna	Fossil-genus	Plantae, incertae sedis		
F. marsiliodes	Fossil-species, small aquatic herb			
Fraxinopsiales		Plantae, Gnetopsida	6.17	
Fraxinopsis	Fossil-genus	Gnetopsida, Fraxinopsiales	6.17	
Fraxinus	Extant genus, ashes	Angiosperms, eudicots, Oleaceae	1.31	U0163
F. cf. oxycarpa	Extant species		1.31	U0163
Frenelopsis	Fossil-genus of conifers	Conifers, Cheirolepidiaceae		
Fruitadens	A small-bodied, likely omnivorous ornithischian dinosaur	Ornithischia, Heterodontosauridae		
Fryopsis	Fossil-genus, lower Carboniferous sterile leaf	Affinity uncertain, possible pteridosperm		
Fungi	The clade including yeasts, rusts, smuts, mildews, molds, and mushrooms	Eukaryota, Zoosporia		U0605, U0614, U0918, U1343, U1503, U1504, U1524, U1533, U1541, U1546
Fusulinida	Extinct group of foraminifera	Eukaryota, Foraminifera		
Gadiformes	Cod fishes	Osteichthyes, Actinopterygii		
Galeamopus	A long-necked dinosaur	Sauropoda, Diplodocidae		
G. pabsti			7.3	
Galleonosaurus	Cretaceous ornithopod dinosaur from Australia	Ornithischia, Ornithopoda		U0812
G. dorisae				U0812
Gangamopteris	Permian glossopterid leaf fossil-genus	Gymnosperms, Glossopteridales	13.11	U1339
G. angustifolia	Fossil-species		13.11	
G. buriadica	Fossil-species			
G. spatulata	Fossil-species		13.11	
Gargoyleosaurus	An ankylosaur	Ornithischia, Nodosauridae		
Garudimimus	An ornithomimosaur from the Cretaceous of Asia	Theropoda, Ornithomimosauria		U0821
G. brevipes				U0821
Gastropoda	Snails	Animalia, Mollusca		

(continued)

Scientific name	Vernacular name, description	Higher taxonomy	Figures, Boxes	Slides
Gavialidae	Gavials	Archosauria, Crocodyliformes		
Gazella	Gazelle	Artiodactyla, Bovidae	3.5	
Geinitzia	Fossil-genus of conifers	Gymnosperms, Conifers		
G. rigida	Fossil-species		6.9	
Gekkota	Gekko family	Diapsida, Squamata		
Genyornis	Extinct, giant flightless bird from Sahul	Aves, Dromornithidae		U0133
G. newtoni			1.13	
Geranium	Extant genus, cranesbills	Angiosperms, eudicots, Geraniaceae		
Gigantonoclea	Permian leaf with unusual frond architecture from North China	"Pteridosperms", Gigantopteridales		
Gigantopteridales	Permian seed-ferns	Plantae, "pteridosperms"	12.8	
Gigantopteris	Large-leafed Permian fossil-genus with reticulate venation	"Pteridosperms", Gigantopteridales	12.8	
G. whitei	Fossil-species			
Ginkgo	Extant genus	Ginkgophytes, Ginkgoales	6.10	U0401, U0412, U0631, U1209
G. biloba	Maidenhair tree		12.11	U1220
G. cranei	Fossil-species			
G. huttonii	A species from the Jurassic of Europe			U0631
Ginkgoales		Gymnosperms, ginkgophytes	12.11	U0652, U1220, U1235
Ginkgoites	Triassic ginkgo-like tree from North America	Ginkgophytes, Ginkgoales		U0646, U0652
G. watsoniae	Fossil-species			U0646
Ginkgophyllum	Fossil-genus based upon ginkgo-like leaves	Ginkgophytes, Ginkgoales	12.11	
G. grassetii	Fossil-species			
Ginkgophytes		Plantae, gymnosperms		U0631, U0646, U0652
Giraffidae	Family of ruminants	Mammalia, Artiodactyla		
Gleicheniaceae	Forked true ferns	Pteridophytes, Gleicheniales		U1113
Glomerula	Polychaete worm genus	Polychaeta, Sabellidae	9.16	
Glossopteridales	Extinct gymnosperm group, mainly from Gondwana	Gymnosperms, formerly comprised in the "pteridosperms"	12.9; 12.10; 13.12	U1339
Glossopteris	Principal fossil-genus of glossopterid leaves, extended to a "Whole-Plant" concept for a late Paleozoic tree	Gymnosperms, Glossopteridales	12.9	U1335, U1336, U1338
G. browniana	Fossil-species			
Glossotheca	Fossil-genus of pollen organs	Gymnosperms, Glossopteridales	12.9	
Glyptostroboxylon	Fossil-genus	Conifers, Cupressaceaeiaceae	1.22	
Glyptostrobus	Extant genus, chinese swamp cypresses	Conifers, Cupressaceae	2.5; 2.7	U0174, U0176, U0219, U0220, U0221, U0222, U0223, U0225
G. europaeus	Common European fossil-species		2.5	U0176
G. pensilis	The single extant species of the genus		2.5	U0223
Gnetales	Extant order including only three genera: *Ephedra*, *Gnetum*, and *Welwitschia*	Plantae, gnetophytes		U0504
Gnetophytes	Small, diverse gymnosperm phylum with plants having some features similar to angiosperms	Plantae		
Gnetopsida		Plantae, gnetophytes		
Gnetum	Extant genus, tropical evergreen trees, shrubs and lianas	Gnetophytes, Gnetaceae		

(continued)

Scientific name	Vernacular name, description	Higher taxonomy	Figures, Boxes	Slides
Gondwanatheria	A group of cynodonts	Therapsida, Mammaliaformes		
Gorgonopsidae	Extinct therapsid carnivorous order	Synapsida, Therapsida		
Grallator	An ichnogenus produced by carnivorous dinosaurs	Dinosauria, Theropoda		
Gunneraceae		Angiosperms, eudicots, Gunnerales		
Gymnostoma	Extant tree genus in the western Pacific and Malesia	Angiosperms, eudicots, Casuarinaceae		
Hadrosauridae	Duck-billed dinosaurs	Ornithischia, Ornithopoda	8.2	
Hadrosauroidea	Duck-billed dinosaurs	Ornithischia, Ornithopoda		
Haliestes	Sea spider	Arthropoda, Pycnogonida	14.13	U1452
H. dasos			14.13	U1452
Halletheca	Fossil-genus of pollen organs	"pteridosperms", Medullosales		
Hamamelidaceae	Witch-hazel family	Angiosperms, eudicots, Saxifragales		
Hamamelis	Extant genus, witch hazels	Angiosperms, eudicots, Hamamelidaceae		
Haplocanthosaurus	A long-necked dinosaur	Sauropoda, Neosauropoda		
Hauffiosaurus	A longirostrine pliosaurid	Sauropterygia, Plesiosauria		
Hausmannia	Fossil-genus of cosmopolitan Mesozoic fern leaves	Pteridophytes, Dipteridaceae		
Haya	A small, bipedal Cretaceous ornithischian from Asia	Ornithischia, Parksosauridae		U0820
H. griva				U0820
Hebe	Extant genus, shrubs and trees	Angiosperms, eudicots, Plantaginaceae		
Hederago	Extant genus of climbers	Angiosperms		
Hedyosmum	Extant genus of woody angiosperms	Angiosperms, Chloranthaceae	1.11	U0126
Heidiphyllum	Triassic woody conifer of Africa	Conifers, Voltziales	6.17	U0649, U0652, U0653
Helios	A microorganism	Eukaryota, Acritarcha	12.12	U1447
H. aranaides			14.12	U1447
Hemichordata	Marine deuterostome animals; during development, the first opening (blastopore) develops as the anus	Animalia, Ambulacraria		
Hemipteroids	Insect group with features considered reduced or simplified from the typical primitive body-plan	Animalia, Arthropoda, Insecta		
Heterangium	Fossil-genus of vine stems or axes	"Pteridosperms", Lyginopteridaceae		
Heterobalanus	Extant subgenus of oaks, commonly called mountain oaks	Angiosperms, eudicots, Fagaceae		
Heterobrachiocrinus	Permian crinoid genus	Animalia, Echinodermata, Crinoidea		
Heterostraci	Extinct group of jawless vertebrates	Vertebrata, Agnatha		
Hindeodus	A conodont genus used in biostratigraphy	Conodonta, Anchignathodontidae		U1006, U1007
H. parvus	Species used to identify earliest Triassic marine deposits			U1006, U1007
Hipparion	An ancient horse	Perissodactyla, Equidae	3.5	
Hippocampus	Seahorses	Actinopterygii, Syngnathidae		
Hippotherium	An extinct genus of horses	Perissodactyla, Equidae		
Hispanotherium	Rhinoceros	Perissodactyla, Rhinocerotidae		
Hizemodendron	A creeping lepidodendrid	Lycopsids, Lepidodendrales	12.2	
Holostei	Clade including bowfin and gars	Actinopterygii, Neopterygii		

(continued)

Scientific name	Vernacular name, description	Higher taxonomy	Figures, Boxes	Slides
Holothuroidea	Sea slugs	Echinodermata, Echinozoa		
Homalodotheriidae	High-browsing notoungulates	Mammalia, Notoungulata		
Homalodotherium	Extinct South American genus of hoofed mammals	Notoungulata, Homalodotheriidae	3.3	
Homalozoa	Obsolete term used for Paleozoic subphylum of echinoderms (= carpoids)	Animalia, Echinodermata		
Hominini	Clade including all the species more closely related to *Homo* than to *Pan* (chimpanzees)	Primates, Hominidae		
Homo	Hominin	Primates, Hominini		U0134, U0135, U0161, U0165
H. erectus				U0165
H. habilis				U0165
Hoploaceratherium		Perissodactyla, Rhinocerotidae		
H. tetradactylum	Browsing, hornless (aceratherine) rhinoceros		3.5	
Huayangosaurus	A relative of *Stegosaurus*	Ornithischia, Huayangosauridae	7.7	
Humiria	Extant genus, bastard bulletwoods	Angiosperms, eudicots, Humiriaceae		
Huvenia	Early Devonian vascular plant	"Early land plants", Rhyniaceae		U1505
Hyaenidae	Hyenas	Mammalia, Carnivora	1.35	
Hybodontidae	Extinct clade of sharks	Vertebrata, Chondrichthyes		
Hylonomus	The earliest reptile	Tetrapoda, Romeriida		
Hymenotheca	Fossil-genus	Angiosperms, monocots, Hydrocharitaceae		
Hynerpeton	Stem tetrapod	Vertebrata, Tetrapoda		U1517
Hyracodon	Lightly built, pony-like mammal	Perissodactyla, Hyracodontidae	3.4	
Hyracoidea	Hyraxes	Placentalia, Afrotheria		
Hyrcantha	Fossil-genus	Angiosperms		
Iberomys	A subgenus of voles	Rodentia, Cricetidae		
Icacinaceae	Tropical family of trees, shrubs, and lianas	Angiosperms, eudicots, Icacinales		
Icacinoxylon	Wood Fossil-genus of Icacinaceae			
Ichthyosauria	Marine reptiles	Tetrapoda, Diapsida		
Iguania	Iguanas	Diapsida, Squamata		
Iguanodontia	A clade of early duck-billed dinosaurs	Ornithischia, Ornithopoda		
Ilex	Extant genus, hollies	Angiosperms, eudicots, Aquifoliaceae	1.11	U0126
Illicium	Extant genus, star anises	Angiosperms, Schisandraceae		
Indovitis	Late Cretaceous genus of grapes	Angiosperms, eudicots, Vitaceae	5.17	
I. chitaleyae	Fossil-species, earliest member of the grape family		5.17	
Indricotherium	Giant relative of rhinoceroses	Perissodactyla, Hyracodontidae	3.4	
Insecta	Insects	Animalia, Arthropoda		
Interatheriidae	Small-sized notoungulates	Mammalia, Notoungulata		
Isisfordia	A crocodylomorph	Archosauria, Eusuchia		
Isochirotherium	A Triassic ichnogenus from Europe	Diapsida, Archosauriformes	9.15	U0922
Isoetales	Quillworts and relatives	Lycopsids, Isoetaceae		U1202
Isoetes	Extant genus, quillworts	Lycopsids, Isoetaceae		
Isoetites	Fossil-genus	Lycopsids, Isoetaceae		
Jaguariba	Early Cretaceous fossil-genus of crown water lilies	Angiosperms, Nymphaeaceae		
J. wiersemana	Fossil-species		5.22	
Joffrea	Fossil-genus	Angiosperms, eudicots, Cercidiphyllaceae		U0418
J. speirsii	Fossil-species			U0418

(continued)

Scientific name	Vernacular name, description	Higher taxonomy	Figures, Boxes	Slides
Juglandaceae	Hickories, walnuts, and relatives	Angiosperms, eudicots, Fagales		
Juglans	Extant genus, walnuts	Angiosperms, eudicots, Juglandaceae		U0149
J. sect. *Cardiocaryon*	A group of walnuts also called butternuts in America			
J. bergomensis	Fossil-species			U0149
Juniperus	Extant genus, junipers	Conifers, Cupressaceae	6.8	U0163
J. cf. *polycarpos*	Extant species			U0163
Kaatedocus	A long-necked dinosaur	Sauropoda, Diplodocidae		
Kajanthus	Early Cretaceous angiosperm fossil-genus	Angiosperms, eudicots, Lardizabalaceae	5.5	U0505
K. lusitanicus	Cretaceous eudicot flower from Europe		5.5	U0505
Kannaskoppia	Cupulate seed fern structure	"Pteridosperms", Petriellales	6.17	
Karinopteris	Carboniferous bifurcate leaf fossil-genus	"Pteridosperms", Lyginopteridales		
Kaulangiophyton	Devonian pre-lycopod	Lycopsids		U1524
Kayentapus	An ichnogenus of a theropod dinosaur	Dinosauria, Theropoda	9.2	
Kegelidium	A lower Carboniferous fossil-genus of sporangia	"Pteridosperms"	13.2	U1306
K. lamegoi	Fossil-species			U1306
Kentrosaurus	A relative of *Stegosaurus*	Ornithischia, Stegosauridae	7.7	
Kidstonia	Carboniferous foliage	?fern		
K. heracleensis	Fossil-species			
Klitzschophyllites	Early aquatic angiosperms	Angiosperms, monocots		
Kobresietea	A vegetation type named after the sedge genus *Kobresia*, recently included in *Carex*			
Komlopteris	An Eocene seed fern	"Pteridosperms"		
Koolasuchus	A Cretaceous temnospondyl from Australia	Tetrapoda, Chigutisauridae		U0812
K. cleelandi				U0812
Koompassia	Extant genus, tall tropical rainforest legume trees	Angiosperms, eudicots, Fabaceae	2.17	U0256, U0257
K. malaccensis	Kempas or Tualang		2.17	
Koparion	A troodontid	Theropoda, Troodontidae		
Kraeuselisporites	Fossil-genus based on Permian–Triassic spores	Lycopsids	10.5	U1016
K. apiculatus	A spore from the Permian–Triassic boundary of Europe		10.5	U1016
Kryoryctes	A monotreme from the Cretaceous of Australia	Mammalia, ?Monotremata		U0814
K. cadburyi				U0814
Kuehneotherium	A stem mammaliaform	Mammaliaformes, Kuehneotheriidae		
Kunbarrasaurus	An ankylosaur with a long tail	Ornithischia, Ankylosauria		
Kylikipteris	Mesozoic fern foliage fossil-genus	Pteridophytes, Dipteridaceae		
K. arguta	Fossil-species			
Kyphosichthys	An unhurried, deep-bodied fish	Actinopterygii, Neopterygii		
Lacertilia	Non-monophyletic group; limbed lizards	Diapsida, Squamata		
Lagarostrobos	Extant genus, huon pines	Conifers, Podocarpaceae		
L. franklinii	Extant species			
Lagenostoma	Carboniferous fossil-genus of fossil seeds	"Pteridosperms", Lyginopteridaceae		
Lagomorpha	Clade including hares, rabbits, and pikas	Synapsida, Mammalia		
Lambeosaurinae	A clade of duck-billed dinosaurs	Ornithischia, Hadrosauridae	8.2	

(continued)

Scientific name	Vernacular name, description	Higher taxonomy	Figures, Boxes	Slides
Laria	Fossil-genus	Angiosperms, Malvaceae		U0213
L. rueminiana	Fossil-species		2.6	U0213
Larix	Extant genus, larches	Conifers, Pinaceae		
L. gmelinii	Extant species		1.7	
Latimeria	West Indian Ocean extant coelacanth	Actinistia, Latimeriidae		
Lauraceae	Laurel family	Angiosperms, eumagnoliids, Laurales		
Laurales	Laurel order	Angiosperms, eumagnoliids		
Laveineopteris	Vegetative leaves of a medullosalean pteridosperm	"Pteridosperms", Cyclopteridaceae		
Leaellynasaura	A small ornithopod dinosaur from the Cretaceous of Australia	Ornithischia, Ornithopoda		U0814, U0815
Leclercqia	Rhizomatous Devonian lycopsid	Lycopsids, Protolepidodendraceae		U1524, U1531
Ledum	Labrador tea; a subsection of *Rhododendron* in the last classifications	Angiosperms, eudicots, Ericaceae		
L. palustre	Extant species		1.7	
Leitneria	Extant genus, corkwoods	Angiosperms, eudicots, Simaroubaceae		
Leontiniidae	Large notoungulates	Mammalia, Notoungulata		
Leperditiids	Bivalve enclosed aquatic crustacean	Arthropoda, Crustacea, Ostracoda		
Lepidodendraceae		Lycopsids, Lepidodendrales	12.1; 12.2	U1204
Lepidodendron	Dominant forest-building and peat-accumulating tree during late Paleozoic	Lycopsids, Lepidodendraceae	11.14; 12.1; 12.2; 13.3	U1317
L. cf. wortheni	Fossil-species		13.3	
L. lycopodioides	Fossil-species			
L. mannabachense	Fossil-species			
L. simile	Fossil-species			
Lepidodendropsis	Lower Carboniferous forest-forming tree	Lycopsids, Lepidodendraceae		
Lepidophloios	Upper Carboniferous forest tree	Lycopsids, Lepidodendraceae	11.14; 12.2	U1317, U1322
L. acerosus	Fossil-species			
L. hallii	Fossil-species			U1322
Lepidopteris	Late Permian to Triassic seed fern leaf	"Pteridosperms", Peltaspermaceae	6.17; 6.18	
L. stormbergensis	Fossil-species		6.18	
Lepidosauromorpha		Tetrapoda, Diapsida		
Lepidosigillaria	Early Late Devonian tree lycopsid	Lycopsids, Isoetales		U1511
Lepidostrobophyllum	Lepidodendrid megasporophyll	Lycopsids, Lepidodendraceae		
Lepidostrobus	Lepidodendrid reproductive cone	Lycopsids, Lepidodendraceae		
Leptoceratops	A small, quadrupedal ceratopsian	Ornithischia, Leptoceratopsidae		
Leptocleidia	Short-necked xenopsarians	Sauropterygia, Plesiosauria		
Leptostrobaleans	A group of seed ferns commonly called Czekanowskiales	Gymnosperms, "ptcridosperms", Czekanowskiales		
Lesothosaurus	A bipedal ornithischian	Ornithischia, Genasauria	7.1; 7.7	
L. diagnosticus			7.7	
Lesqueria	Early Cretaceous fruiting axis referred to angiosperms	Angiosperms, eumagnoliids, Magnoliales		
Lidgettoniaceae	A family of Permian seed ferns	Gymnosperms, Glossopteridales		U1216
Limnobiophyllum	Fossil-genus, a relative of duckweeds	Angiosperms, monocots, Araceae		
Limulidae	Horseshoe crabs	Arthropoda, Chelicerata		

(continued)

Scientific name	Vernacular name, description	Higher taxonomy	Figures, Boxes	Slides
Lindleycladus	Conifer fossil-genus	Gymnosperms, Coniferales		
L. lanceolatus	Fossil-species		6.9	
Lingula	Extant inarticulate brachiopod genus	Brachiopoda, Lingulidae	10.8	U1023
Lingulidae	Family of brachiopods	Brachiopoda, Lingulida		
Linguliformea	Clade of brachiopods	Lophophorata, Brachiopoda		
Linopteris	Seed-fern fossil-genus	"Pteridosperms", Medullosales	13.3	
L. lepida	Fossil-species		13.3	
Liopleurodon	A very large plesiosaur	Sauropterygia, Pliosauridae		
Liquidambar	Extant genus, sweetgums	Angiosperms, eudicots, Altingiaceae		
Lirainosaurus	A small-sized, long-necked dinosaur	Sauropoda, Titanosauria		
Liriodendron	Extant genus, tulip trees	Angiosperms, eumagnoliids, Magnoliaceae		U0230, U0231, U0604
Liriodendropsis	Fossil-genus	Angiosperms, eumagnoliids, Magnoliaceae		
Lissamphibia	Clade including all modern amphibians	Tetrapoda, Amphibia		
Lithioperna	Lagoonal to subtidal bivalve	Mollusca, Bivalvia		U0906
Lithiotis	Early Jurassic rudist-like bivalve	Mollusca, Bivalvia		U0906
Lithistid	Sponges	Porifera, Demospongiae		U1427
Lithocarpus	Extant genus, stone oaks	Angiosperms, eudicots, Fagaceae	2.16	
Litopterna	Extinct order of hoofed mammals	Mammalia, Ungulata	3.3	
Litsea	Extant genus, may chang	Angiosperms, eumagnoliids, Lauraceae		
Lobatopteris	Pinnules of the *Psaronius* tree fern frond	Pteridophytes, Marattiales		U1121, U1126
L. aspidioides	Fossil-species			U1126
Lonicera	Extant genus, honeysuckles	Angiosperms, eudicots, Caprifoliaceae		
Lophiiformes	Anglerfishes	Osteichthyes, Actinopterygii		
Lophophorata		Animalia, Lophotrochozoa		
Loranthaceae	Showy mistletoe family	Angiosperms, eudicots, Santalales		
Lusicarpus	Cretaceous fossil-genus based on fruits	Angiosperms		
Lusistemon	Early Cretaceous fossil-genus of staminate flower	Angiosperms, Eudicot		
Lycophytina	Early vascular plants	Plantae, tracheophytes		U1508
Lycopodiacidites	Cretaceous spore resembling *Lycopodium*	Incertae sedis		U0905
L. regulatus	Fossil-species			U0905
Lycopodiales	Club mosses	Plantae, lycopsids		U1202
Lycopodites	Early lycopod-like fossil resembling extant plants	Lycopsids, Lycopodiaceae		
L. riograndensis	Fossil-species			
Lycopsida	Club mosses	Plantae, lycopsids		U1309
Lyginodendron	Fossil-genus	Plantae, "pteridosperms"		
Lyginopteridales	Seed ferns	Plantae, "pteridosperms"		
Lyginopteris	Fossil-genus	"Pteridosperms", Lyginopteridaceae	121.6	
L. oldhamia	Fossil-species		12.6	
Lygodium	Climbing fern	Pteridophytes, Schizaeales		U1106, U1107, U1111
L. kaulfusii	Fossil-species			U1111
Lystrosaurus	Mammal-like tetrapod from the Permian and Triassic of Africa	Therapsida, Lystrosauridae		U1010, U1012
Macaca	Macaques	Primates, Cercopithecidae		

(continued)

Scientific name	Vernacular name, description	Higher taxonomy	Figures, Boxes	Slides
M. sylvanus	Barbary macaque			
Macginitiea	Sycamore or plane tree fossil-genus	Angiosperms, eudicots, Platanaceae		
Macrauchenia	Camel-sized litoptern	Mammalia, Macraucheniidae	3.3	
Macraucheniidae	High-browsing notoungulates	Mammalia, Litopterna		
Macropodidae	Kangaroo family	Mammalia, Marsupialia		
Magnolia	Deciduous or evergreen trees or shrubs, mostly thermophilous	Angiosperms, eumagnoliids, Magnoliaceae		U0230, U0231, U0513
Magnoliaceae		Angiosperms, eumagnoliids, Magnoliales		
Magnoliales		Angiosperms, eumagnoliids		
Eumagnoliids		Plantae, angiosperms		
Maiopatagium	An early gliding mammal	Mammalia, Eleutherodontidae	7.9	
M. furculiferum			7.9	
Majonica	Voltzialean reproductive organ	Conifers, Majonicaeae		
Mallotus	Extant genus in spurge family	Angiosperms, eudicots, Euphorbiaceae		
M. maii	European Cenozoic fossil-species		2.2	
Malpighiales	One of the largest orders of flowering plants	Angiosperms, eudicots, Rosids		
Malvaceae	Mallow family	Angiosperms, eudicots, Malvales		
Mammalia	Mammals	Synapsida, Mammaliaformes		
Mammaliaformes	Early stem mammal-like animals	Synapsida, Therapsida		
Mammut	Mastodon	Proboscidea, Mammutidae		
Mammuthus	Mammoth	Proboscidea, Elephantidae		U0172
M. columbi	Columbian mammoth			
M. meridionalis	Large southern elephants		1.35	U0172
M. primigenius	Woolly mammoth			
Mangifera	Extant genus, mangos	Angiosperms, eudicots, Anacardiaceae		
Maniraptora	Raptor-like dinosaurs	Theropoda, Maniraptoriformes	7.4	U0721, U0723, U0804, U0812
Maniraptoriformes	Raptor-like dinosaurs	Theropoda, Coelurosauria	7.4	
Marattia	Extant genus of primitive, large, eusporangiate fern with synangia formed from two rows of sporangia	Pteridophytes, Marattiaceae		
M. intermedia	Fossil-species			
Marattiaceae	Wetland ferns	Pteridophytes, Marattiales		U1113, U1115
Marattiopsis	Mesozoic fern fossil-genus	Pteridophytes, Marattiaceae		
Marchantiaceae	Liverwort family	Marchantiophyta, Marchantiales		
Marchantiophyta	Liverworts	Plantae, embryophytes		
Marshosaurus	A relatively small, carnivorous dinosaur	Theropoda, Piatnitzkysauridae		
Marskea	Jurassic conifer	Conifers, Taxaceae	6.9	
M. jurassica	Fossil-species			
Mastixia	Extant genus of resinous evergreen trees	Angiosperms, eudicots, Cornales		U0211, U0212
Matonia	Extant leptosporangiate fern genus	Pteridophytes, Matoniaceae		U0633
M. braunii	Fossil-species			
M. pectinata	Fossil-species			U0633
Matoniaceae	Leptosporangiate fern family	Pteridophytes, Gleicheniales		U0633
Mauldinia	Fossil-genus, member of the laurel family	Angiosperms, eumagnoliids, Lauraceae		U0506
M. bohemica	Cretaceous fossil-species from Europe		5.7	U0506

(continued)

Scientific name	Vernacular name, description	Higher taxonomy	Figures, Boxes	Slides
Maurita	Extant genus of fan palms	Angiosperms, monocots, Arecaceae	1.11	
M. flexuosa	Wetland tree known as moriche palm			
Mayoa	Fossil-genus	Angiosperms, monocots, Araceae		
Medullosa	Fossil-genus of seed-ferns	"Pteridosperms", Medullosales	12.7	
M. leuckartii	Fossil-species		12.7	
Medullosales	Seed-fern order	Plantae, "pteridosperms"	12.7	
Megacheira	Extinct class of predatory arthropods	Arthropoda, Arachnomorpha		
Megalodontidae	Extinct family of bivalve molluscs	Bivalvia, Heterodonta	9.11	
Megaphyton	Impressions of tree-fern trunks	Pteridophytes, Marattiales		
Megaraptoridae	Carnivorous dinosaur clade	Theropoda, Avetheropoda		
Melastomataceae	Annual or perennial herbs, shrubs, or small trees	Angiosperms, eudicots, Myrtales		
Meles	Badger	Carnivora, Mustelidae		
M. meles	European badger			
Meliaceae	Mahogany family	Angiosperms, eudicots, Sapindales		
Meliosma	Extant deciduous tree in tropical to warm temperate regions	Angiosperms, eudicots, Sabiaceae	2.2	
M. canavesana	European fossil-species		2.2	
Menispermaceae	Moonseed family	Angiosperms, eudicots, Ranunculales		
Merycoidodon	Extinct genus of herbivore	Artiodactyla, Merycoidodontidae	3.4	
Merycoidodontidae	Oreodonts	Mammalia, Artiodactyla	3.4	
Mesohippus	Muntjac-sized, early three-toed horse	Perissodactyla, Equidae	3.4	
Mesotherium	Superficially rodent-like, burrowing notoungulates	Notoungulata, Mesotheriidae	3.3	
Metasequoia	Extant genus, dawn redwood	Pinophytes, Conifers, Cupressaceae		U0225
M. foxi	Dawn redwood fossil-species			
M. occidentalis	Only extant species			
Metatheria	All mammals more closely related to marsupials than to placentals	Mammalia, Theria		
Metazoa	Animals	Eukaryota, Holozoa		U1425, U1427
Metriorhynchoidea	Fully marine crocodyliforms	Crocodyliformes, Thalattosuchia		
Micrairoideae	PACMAD clade grasses	Angiosperms, monocots, Poaceae		
Microcachrys	Extant genus, creeping pines	Conifers, Podocarpaceae		
Microcleididae	A subclade of Plesiosauroidea	Sauropterygia, Plesiosauria		
Microconchida	Group of small, spirally-coiled, encrusting fossil "worm" tubes	Animalia, Tentaculita		
Microphyllopteris	A fossil-genus of pioneer ferns	Pteridophytes, Gleicheniaceae		U1113
Microstonyx	Large, omnivorous pig	Artiodactyla, Suidae	3.5	
Microtus	Voles	Rodentia, Cricetidae		
M. (Iberomys) breccensis	Murid			
Microvictoria	Aquatic members of the water lilies	Angiosperms, Nymphaeaceae		
Mimagoniatites	Devonian ammonoid genus	Cephalopoda, Agoniatitidae	14.3	U1434
M. fecundus			14.3	U1434
Mimosoideae	Tropical to subtropical trees, herbs, lianas, and shrubs	Angiosperms, eudicots, Fabaceae		
Myriapoda	Myriapods	Animalia, Arthropoda		
Mollusca	Mollusks	Animalia, Lophotrochozoa		
Monetianthus	Cretaceous water-lily	Angiosperm, Nymphaeaceae		

(continued)

Scientific name	Vernacular name, description	Higher taxonomy	Figures, Boxes	Slides
Monocots	Flowering plants, the seeds of which typically contain only one embryonic leaf	Plantae, angiosperms		U0401
Monoporites	Pollen of grass-like plant	Angiosperms, monocots		
Monotremata	A basal clade of mammals	Mammalia, Australosphenida		
Pteridophytes	A large informal group of spore-bearing plants comprising horstails, all of the fern, and extinct fern-like clades			
Montsechia	Small aquatic angiosperm	Plantae, angiosperms		
Moraceae	Mulberry family	Angiosperms, eudicots, Rosales		
Moresnetia	One of the earliest Devonian seed plants	Gymnosperms, "pteridosperms"		U1511, U1515
Morganucodon	A stem mammaliaform	Therapsida, Morganucodontidae		
Morus	Extant genus, mulberries	Angiosperms, eudicots, Moraceae	1.26	
Mosasauria	Marine reptiles	Diapsida, Squamata		
Multituberculata	Early mammal group	Mammalia, Theriiformes		
Mus	Mouse	Rodentia, Muridae		
Muttaburrasaurus	A large ornithopod	Ornithischia, Ornithopoda		
Mymoorapelta	An ankylosaur	Ornithischia, Nodosauridae		
Myodocopida	One of two ostracod groups	Crustacea, Ostracoda		
Myrica	Extant genus, bayberries	Angiosperms, eudicots, Myricaceae		
Myriophyllum	Extant genus, water milfoils	Angiosperms, eudicots, Haloragaceae		U0163
M. cf. *spicatum*	Extant species		1.31	U0163
Myrsine	Extant genus, colic woods	Angiosperms, eudicots, Primulaceae		
Myrtaceae	Myrtle family	Angiosperms, eudicots, Myrtales		
Myrtoidea	Fossil-genus based on entire-margined lauroid leaves	Angiosperms, eumagnoliids, Lauraceae		U0506, U0508
M. geinitzii	Fossil-species			U0506
Myrtus	Extant genus, myrtles	Angiosperms, eudicots, Myrtaceae		
Mythunga	A pterosaur genus	Pterosauria, Pterodactyloidea		
Myxini	Hagfishes	Vertebrata, Cyclostomata		
Nanantius	A bird genus	Avialae, Enantiornithes		
Nataligma	Stalked ovulate cones from the Triassic of Africa	Gnetopsida, Nataligmales	6.17	U0652
Nataligmales		Plantae, Gnetopsida		
Nautilidae	Coiled cephalopod nautilus family	Cephalopoda, Nautiloidea		
Nautiloidea		Mollusca, Cephalopoda		
Nehvizdyella	Mesozoic ginkgoalean reproductive structure	Ginkgophytes, Ginkgoales		
Nelumbo	Extant genus, sacred lotus	Angiosperms, eudicots, Nelumbonaceae		
N. puertae	Fossil-species		5.24	
Nelumbonaceae	The sacred lotus family	Angiosperms, eudicots, Proteales		
Nemegtosaurus	A long-necked dinosaur	Sauropoda, Lithostrotia		
Nemejcopteris	Fossil-genus, a groundcover true fern	Pteridophytes, Zygopteridaceae	11.10	U1118, U1328
N. haiwangii	Fossil-species			U1118
Neocoelurosaurs, Neocoelurosauria	Subclade of Coelurosauria	Dinosauria, Theropoda	7.4	
Neopterygii	Fishes including gars, bowfins, and derived teleosts	Osteichthyes, Actinopterygii		U0748, U0749
Neoschwagerina	A genus of foraminifera	Eukaryota, Rhizaria		
Nesodon	Late Oligocene to Miocene South American mammal	Notoungulata, Toxodontidae	3.3	

(continued)

Scientific name	Vernacular name, description	Higher taxonomy	Figures, Boxes	Slides
Neuralethopteris	Carboniferous leaf seed-fern fossil-genus	"Pteridosperms", Alethopteridaceae	13.3	
N. cf. *schlehanii*	Fossil-species		13.3	
Neuropteris	Fossil-genus of seed ferns	"Pteridosperms", Neurodontopteridaceae	12.7; 13.6	
N. flexuosa	Fossil-species		13.6	
Nipa	Extant genus of palms	Angiosperms, monocots		
Nitraria	Extant genus of arid-adapted shrubs	Angiosperms, eudicots, Nitrariaceae		
Nodosauridae	Subclade of Ankylosauria	Ornithischia, Ankylosauria		
Noeggerathia	Late Carboniferous and early Permian leaf	Noeggerathiales, affinity uncertain		
Noeggerathiales	Extinct order of vascular plants	Tracheophytes, affinity uncertain		U1309, U1328, U1330
Nothofagaceae	Southern beech family	Angiosperms, eudicots, Fagales		
Nothofagidites	Pollen similar to today's Southern beech	Angiosperms, eudicots, Fagales		U0602
Nothofagus	Extant genus, southern beeches	Angiosperms, eudicots, Nothofagaceae	2.10	U0244, U0245
N. antarctica	Antarctic beech			
N. obliqua	Patagonian oak			
N. pumilio	Lenga beech		2.10	
Nothorhacopteris	South American foliage of Carboniferous age associated in cupulate ovules	"Pteridosperms"		U1308
N. kellaybelenesis	Fossil-species			U1308
Nothosauria	Marine reptiles	Diapsida, Sauropterygia		
Notochelone	An extinct sea turtle	Testudines, Protostegidae		
Notoungulata	Endemic South American ungulates	Mammalia, Ungulata	3.3	
Nymphaea	Extant genus, water lilies	Angiosperms, Nymphaeaceae		U0611
N. alba	Extant species			
Nymphaeaceae	Water lily family	Angiosperms, Nymphaeales		
Nymphaeales	Water lilies	Plantae, angiosperms		
Nyssa	Extant genus, tupelos	Angiosperms, eudicots, Nyssaceae	1.37	
N. sylvatica	Black gum		1.37	
Nyssaceae	Tupelo family	Angiosperms, eudicots, Cornales		
Ochotona	Pika	Lagomorpha, Ochotonidae		
Ocotea	Extant genus, member of the laurel family	Angiosperms, eumagnoliids, Lauraceae		
Octopoda	Octopuses	Cephalopoda, Coleoidea		
Odontoceti	Toothed whales	Artiodactyla, Cetacea		
Odontopteris	Fronds of medullosalean seed ferns	"Pteridosperms", Medullosales	12.7	
Oligocarpia	Carboniferous filicalean fern	Pteridophytes, Gleicheniaceae		
O. lindsaeoides	Fossil-species			
Onoclea	Extant genus of deciduous ferns in eastern Asia and eastern North America	Pteridophytes, Onocleaceae		U0419, U0420
O. sensibiis	"Sensitive fern"			U0419, U0420
Ophiuroidea	Brittle stars	Echinodermata, Asterozoa		
Opisthocoelicaudia	A long-necked Cretaceous dinosaur from Asia	Sauropoda, Lithostrotia	8.4	U0818
O. skarzynskii				U0818
Oraristrix	Rancho La Brea owl genus	Aves, Strigidae		
Orcinus	Marine dolphin genus	Artiodactyla, Delphinidae		
O. orca	Killer whales	Synapsida, Mammalia		
Oricilla	Core zosterophyllopsid	Plantae, zosterophyllophytes		

(continued)

Scientific name	Vernacular name, description	Higher taxonomy	Figures, Boxes	Slides
Orites	Extant genus	Angiosperms, eudicots, Proteaceae		
Ornithischia	Bird-hipped dinosaurs	Archosauria, Dinosauria		U0710, U0711, U0712, U0713, U0714
Ornitholestes	A relatively small, carnivorous dinosaur	Theropoda, Maniraptoromorpha		
Ornithomimosauria	Beaked, feathered dinosaurs	Theropoda, Maniraptoriformes	7.4	U0708
Ornithopoda	Duck-billed dinosaurs	Ornithischia, Cerapoda		
Orthocerida	Cephalopods with long, straight or gently curved exoskeletons	Mollusca, Cephalopoda		
Orthopteroidea	Winged insects	Arthropoda, Insecta		
Ortiseia	Fossil-genus of conifer-like trees	Pinophytes, Utrechtiaceae		
Oryctodromeus	A small, burrowing dinosaur	Ornithischia, Parksosauridae	8.4	U0816
Osmunda	Extant genus comprising the royal ferns and othe temperate zone ferns with divided fronds often attaining 1.5 m in height	Pteridophytes, Osmundaceae		U0632, U1111
O. japonica	Extant species			
O. palustris	Extant species			
O. parschlugiana	European fossil-species of royal ferns			U1111
O. regalis	Extant royal fern			U1111
Osmundaceae	Royal fern family	Pteridophytes, Osmundales		U0615, U0632, U1114
Osmundales	Royal fern order	Plantae, pteridophytes		U1120, U1122
Osmundastrum	Extant genus, cinnamon ferns	Pteridophytes, Osmundaceae		
O. cinnamomeum	Extant cinnamon fern			
Osmundopsis	A fossil-genus of royal ferns	Pteridophytes, Osmundaceae		
O. sturii	Fossil-species			
Osteichthyes	Bony fishes	Animalia, Vertebrata	7.12	
Osteostraci	A group of bony-armored jawless fish	Animalia, Vertebrata		
Ostracoda	Seed shrimps	Arthropoda, Crustacea		
Othnielosaurus	A small, bipedal herbivorous dinosaur	Ornithischia, Neornithischia		
O. consors			7.7	
Otovicia	Fossil-genus of conifer-like trees	Pinophytes, Utrechtiaceae		
Otozamites	Fossil-genus of leaves assigned to Bennettitales	Gymnosperms, Bennettitales	6.9	
Ottokaria	Ovulate structures of glossopterid seed ferns	"Pteridosperms", Glossopteridaceae	12.9	
Otwayemys	An extinct, basal turtle from the Cretaceous of Australia	Tetrapoda, Testudinata		U0814
O. cunicularius				U0814
Oviraptorosauria	Beaked, feathered dinosaurs	Theropoda, Pennaraptora	7.4	U0708
Ozarkodina	A conodont genus ranging from Ordovician to Devonian	Conodonta, Spathognathodontidae	14.5	U1440
Pabiania	Cretaceous angiosperm leaf fossil-genus	Angiosperms, eumagnoliids, Lauraceae		
Pachycardia	Triassic unionoid clam	Bivalvia, Unionida		
Pachycephalosauria	Dome-headed dinosaurs	Ornithischia, Marginocephalia		
Pachypleurosauria	Aquatic, lizard-like reptiles	Sauropterygia, Nothosauroidea		
Pachytesta	Carboniferous fossil-genus of seeds	"Pteridosperms", Medullosales		
Paedotherium	Extinct, quadrupedal mammal	Notoungulata, Hegetotheriidae	3.3	
Pagiophyllum	Carboniferous to Cretaceous coniferous foliage	Conifers		U0616, U0628, U0644, U0905

(continued)

Scientific name	Vernacular name, description	Higher taxonomy	Figures, Boxes	Slides
P. insigne	Fossil-species		6.9	
P. maculosum	Jurassic European species			U0628
P. navajoensis	Triassic species from North America			U0644
P. rotzoanum	Jurassic European species			U0905
Palaeocarpinus	Fossil-genus, Paleogene fruit aggregate with nutlets	Angiosperms, eudicots, Betulaceae		
Palaeocarya	Neogene winged fruit fossil-genus	Angiosperms, eudicots, Juglandaceae		
Palaeoloxodon	Extinct genus of straight-tusked elephants	Proboscidea, Elephantidae		
Palaeomastodon	Extinct genus with short upper and lower tusks	Proboscidea, Palaeomastodontidae	3.4	
Palaeopterygii	Large, predatory ray-fin fishes	Vertebrata, Osteichthyes		U0749
Palaeoryx	Mesodont, mixed-feeding antelope	Artiodactyla, Bovidae	3.5	
Palaeocymopolia	Ordovician green algae with segmented dichotomously branching noncalcified thallus	Plantae, Chlorophyta, Dasycladales	14.14	
P. silurica	Fossil-species		14.14	
Paleohypsodontus	Hypsodont, bovid-like ruminant	Ruminantia, Bovidae		
Paleoochna	Fossil-genus, Paleocene fruits	Angiosperms, eudicots, Ochnaceae		
P. tiffneyi	Fossil-species			
Paleotubus	A tube worm from the Permian of Europe	?Serpulid?	9.16	U0927
P. sosiensis			9.16	
Palermocrinus	Paleozoic stalked crinoid	Echinodermata, Crinoidea		
Paliurus	Extant genus, shrubs or small trees growing in dry regions	Angiosperms, eudicots, Rhamnaceae		U0213
P. tiliifolius	Fossil-species		2.6	U0213
Pandanaceae	Pandan; screwpine	Angiosperms, monocots, Pandanales		
Pandemophyllum	Cretaceous entire-margined lauroids from North America	Angiosperms, eumagnoliids, Lauraceae		U0512
Panicoideae	Grass clade including sugarcane, maize, sorghum, and switchgrass	Angiosperms, monocots, Poaceae		
Papuacedrus	Extant genus of medium-sized to large evergreen trees	Conifers, Cupressaceae		
Paradinandra	Cretaceous charcoalified fossil flowers from Sweden	Angiosperms, eudicots, Ericales		
Paralligator	Extinct Cretaceous crocodile-like genus from Mongolia	Crocodyliformes, Paralligatoridae		
Paralycopodites	Fossil-genus of Carboniferous arborescent lycopsids	Lycopsids, Lepidodendrales		
P. simile	Fossil-species			
Paranthropus	An extinct genus of hominins	Primates, Hominidae		
Paraprotophyllum	Cretaceous, deciduous large-leaved platanoids	Angiosperms, eudicots, Platanaceae	5.12	U0509
Parasaurolophus	A duck-billed dinosaur with a long head crest	Ornithischia, Hadrosauridae	7.1; 7.6	
Parataxodium	A taxodiod cupressaceous tree	Conifers, Cupressaceae		
Paratingia	A Permian cycad-like plant	Tracheophytes, Noeggerathiales	11.10	U1116, U1119, U1328
P. wudensis	Fossil-species			
Paripteris	Carboniferous fossil-genus of seed fern foliage	"Pteridosperms", Medullosales		
Parthenocissus	Extant genus, Virginia creepers	Angiosperms, eudicots, Vitaceae		

(continued)

Scientific name	Vernacular name, description	Higher taxonomy	Figures, Boxes	Slides
Patagosaurus	A long-necked dinosaur from the Jurassic of Argentina	Sauropoda, Eusauropoda	8.8	U0832, U0833
Patagotitan	A long-necked dinosaur from the Cretaceous of Argentina	Sauropoda, Lognkosauria		
Paulophyton	A lower Carboniferous basal, leafless plant with terminal sporangia	Affinity uncertain	13.2	U1304
P. sommeri	Fossil-species		13.2	
Pecinovia	Late Cretaceous fossil-genus of inflorescence axes, flowers, and isolated stamens	Angiosperms, eudicots		
Pecopteris	A fossil-genus of leaf fronds, probably belonging to *Psaronius* tree ferns	Pteridophytes		U1117, U1121, U1328
P. lativenosa	Fossil-species			
P. norinii	Fossil-species			
Pederpes	An early tetrapod from the Carboniferous of Scotland	Tetrapoda, Whatcheeriidae		
Peltaspermales	Extinct clade of seed ferns	Plantae, "pteridosperms"		
Pelycosauria	Informal grouping of basal synapsids	Tetrapoda, Synapsida		
Pentaphylacaceae		Angiosperms, eudicots, Ericales		
Percidae	Perches	Osteichthyes, Actinopterygii		
Perissodactyla	Odd-toed ungulates	Mammalia, Ungulata		
Pertica	Devonian fossil-genus of vascular plants from North America	Plantae, Trimerophyta or basal fern-like plants	15.9	U1520, U1524, U1528
Petromyzontiformes	Lampreys	Vertebrata, Agnatha		
Pharoideae	Basal grasses	Angiosperms, monocots, Poaceae		
Phlebolepis	Extinct Silurian genus of jawless fish	Agnatha, Phlebolepididae	14.6	U1441
P. elegans			14.6	U1441
Phlebopteris	Mesozoic fern fossil-genus	Pteridophytes, Matoniaceae		U0633, U0643
P. angustiloba	Fossil-species			
P. formosa	Fossil-species			
P. polypodioides	European Jurassic fossil-species			U0633
P. smithii	Triassic fossil-species from North America			U0643
Phoebe	Extant genus of the Lauraceae	Angiosperms, eumagnoliids, Lauraceae		
Phyllitis	Extant herbaceous fern	Pteridophytes, Aspleniaceae	11.3	U1110
P. scolopendrium	Hart's-tongue or hart's-tongue fern		11.3	U1110
Phyllopteroides	Fossil-genus, Cretaceous ferns	Pteridophytes, Osmundaceae Pteridophytes		U0612
P. laevis	Lower Cretaceous fossil-species from Australia			U0612
Picea	Extant genus, spruces	Conifers, Pinaceae	2.16	U0146, U0207
P. smithiana	Extant species		1.26	
Pinaceae	Pine family	Conifers, Pinales		U0624
Pinacosaurus	An ankylosaur from the Cretaceous of China	Ornithischia, Ankylosauridae		U0819
P. grangeri				U0819
Pinus	Extant genus, pines	Conifers, Pinaceae	1.26	U0146, U0219
P. cembra	Swiss pine			
P. roxburghii	Chir pine		1.26	
P. subgen. Haploxylon	Soft pine			
P. wallichiana	Blue pine		1.26	
Pistosauroidea	Triassic marine-reptile clade from the Northern Hemisphere	Diapsida, Sauropterygia		

(continued)

Scientific name	Vernacular name, description	Higher taxonomy	Figures, Boxes	Slides
Pittosporum	Extant genus, cheesewoods	Angiosperms, eudicots, Pittosporaceae		
Placentalia	Mammals that retain the fetus in the uterus until a relatively late stage of development	Mammalia, Eutheria		U0728
Placodermi	Clade of extinct, armored fishes	Vertebrata, Gnathostomata		
Placodontia	Clade of Triassic marine reptiles with crushing teeth	Diapsida, Sauropterygia		
Plantago	Extant genus, plantains	Angiosperms, eudicots, Plantaginaceae		
Platanaceae	Plane-tree family	Angiosperms, eudicots, Proteales		U0418
Platananthus	Fossil-genus of plane-tree inflorescences	Angiosperms, eudicots, Platanaceae		
Platanus	Extant genus, American sycamore or plane trees	Angiosperms, eudicots, Platanaceae		U0230, U0231
Plateosaurus	A large, Triassic bipedal, sauropodomorph of Europe and Greenland	Sauropodomorpha, Plateosauridae		U0835, U0836, U0837, U0839
P. quenstedti			8.10	
Platypterygius	Cretaceous marine reptile	Ichthyosauria, Ophthalmosauridae		U0809
P. australis				U0809
Plecoptera	Stoneflies	Arthropoda, Insecta		
Plesiohadros	An ornithopod from the Cretaceous of Mongolia	Ornithischia, Hadrosauroidea		
Plesiosauria	Mesozoic marine reptiles with four paddle-shaped limbs	Diapsida, Sauropterygia		
Plesiosauroidea	A subclade of Plesiosauria from the Jurassic and Cretaceus	Sauropterygia, Plesiosauria		
Plesiosauromorphs	An informal group of short-headed, long-necked plesiosaurs	Sauropterygia, Plesiosauria		
Pleurodira	Side-neck turtles	Tetrapoda, Testudines		
Pleuromeia	A Triassic fossil-genus of spore-bearing plants	Lycopsids, Isoetaceae	10.4	U1016
P. sternbergii	Dominant fossil-species in the Early Triassic vegetation across Eurasia		10.4	U1016
Pleuronectiformes	Flatfishes	Osteichthyes, Actinopterygii		
Pliomys	Extinct genus of forest voles	Rodentia, Cricetidae		
Plionarctos	Extinct Miocene to Pleistocene genus of North American and European bear	Carnivora, Ursidae		U0238
Pliosauridae	A subclade of Plesiosauria from the Jurassic and Cretaceous	Sauropterygia, Plesiosauria		
Pliosauromorphs	Informal group of long-headed, short-necked plesiosaurs	Sauropterygia, Plesiosauria		
Plocostoma	A sea snail	Mollusca, Gastropoda		
Poaceae	Grass family	Angiosperms, monocots, Poales		U0219, U0309
Poales	Order including grasses and sedges	Angiosperms, monocots		
Podocarpaceae	Plum pines	Conifers	1.11	U0126, U0243, U0614
Podocarpales	Order often included in Pinales	Conifers, Podocarpales or Pinales		
Podocarpus	Extant genus, yew pines; evergreen shrubs or trees	Conifers, Podocarpaceae	1.11	U0126
Poebrotherium	An early Eocene-to-Miocene camel from North America	Artiodactyla, Camelidae		
Polycalyx	Fossil-genus, ovulate organ	Gymnosperms, "pteridosperms"		
Polychaeta	Bristle worms	Animalia, Annelida		
Polycotylidae	Subclade of Cretaceous Plesiosauria	Sauropterygia, Plesiosauria		

(continued)

Scientific name	Vernacular name, description	Higher taxonomy	Figures, Boxes	Slides
Polygalaceae	Milkwort family	Angiosperms, eudicots, Fabales		
Polygonaceae	Knotweed family	Angiosperms, eudicots, Caryophyllales		
Polylepis	Extant genus of shrubs and trees from mid- to high-elevation in the tropical Andes	Angiosperms, eudicots, Rosaceae	1.10	U0123, U0126
Polyodontidae	Paddlefishes	Osteichthyes, Actinopterygii		
Polypodiaceae	A family of epiphytic and terrestrial ferns	Pteridophytes, Polypodiales		U1113
Polypteridae	Bichirs	Osteichthyes, Actinopterygii		
Polypteriformes	Clade including bichirs and reedfishes	Osteichthyes, Actinopterygii		
Pooideae	Subfamily of C3-photosynthesizing grasses	Angiosperms, monocots, Poaceae		
Populus	Extant genus, poplars and aspens	Angiosperms, eudicots, Salicaceae	1.31	U0163
P. cf. *nigra*	Black poplar		1.31	U0163
Porifera	Sponges	Eukaryota, Animalia		
Posidonia	Seagrass	Angiosperms, monocots, Posidoniaceae		
Potamogeton	Extant genus of mostly freshwater, aquatic plants	Angiosperms, monocots, Potamogetonaceae		U0163
P. cf. *perfoliatus*	Extant species		1.31	U0163
Potoniea	Fossil-genus of medullosalean pollen organs	"Pteridosperms", Potonieaceae		
Praedodromeus	Extinct genus of false ground beetles	Coleoptera, Trachypachidae		
Praemegaceros	Extinct Pleistocene-to-Holocene deer relatives from Eurasia	Artiodactyla, Cervidae		
Pragocladus	Fossil-genus	Angiosperms		
Prisca	Cretaceous spiked inflorescence (flower)	Angiosperms, eumagnoliids, Lauraceae		
Proboscidea	Elephant relatives	Mammalia, Tethytheria		
Proganochelys	Triassic stem-turtle	Tetrapoda, Testudinata		U0838, U0839
Programinitis	Grass-like monocots	Angiosperms, monocots		
P. burmitis	Fossil-species			
"Progymnosperms"	Group of woody, spore-bearing plants, now extinct, ancestral to the gymnosperms	Tracheophytes		
Promyalina	A mussel genus	Mollusca, Bivalvia, Myalinidae	10.8	U1023
Pronephrium	Extant tropical rainforest fern	Pteridophytes, Thelypteridaceae		U1111
P. stiriacum	Fossil-species			U1111
Propomatoceros	A Permian tube worm from Europe	Lophotrochozoa, Annelida	9.16	U0927
P. permianus			9.16	
Proteaceae	*Banksia* family	Angiosperms, eudicots, Proteales		
Proteacidites	Fossil pollen of the *Banksia* family	Angiosperms, eudicots, Proteaceae		U0602
Protemnodon	Extinct genus similar to a very large wallaby; Pliocene and Pleistocene of Australia	Marsupialia, Macropodidae		
Proterosuchidae	Extinct clade of long-snouted Permian-and-Triassic reptiles	Diapsida, Archosauriformes		
Protists	An informal group of eukaryotes that are neither plants, animals, or fungi	Eukaryota		
Protobarinophyton	Early groundcover plant from the Devonian	Plantae, zosterophyllophytes, Barinophytaceae		U1511
Protoceratops	A sheep-sized ceratopsian	Ornithischia, Protoceratopsidae	8.4; 8.5	U0816, U0819
P. andrewsi				U0819

(continued)

Scientific name	Vernacular name, description	Higher taxonomy	Figures, Boxes	Slides
Protolepidodendrales	Extinct clade of Devonian-to-Carboniferous lycopsids	Plantae, lycopsids		
Protolepidodendron	Fossil-genus	Lycopsids, Protolepidodendrales	11.17	U1128, U1130
P. scharianum	Fossil-species			U1130
Protolepidodendropsis	Fossil-genus of tree lycopsids	Lycopsids, Protolepidodendrales		U1502
P. pulchra	Fossil-species			
Protomonimia	Fossil-genus, angiosperm reproductive structures from Asia	Angiosperms, Magnoliales	5.15	U0510
P. kasai-nakajhongii	Fossil-species		5.15	
Protopteridium	Middle Devonian plant with a pinnate two-dimensional branching pattern	"Progymnosperms", Aneurophytales	11.17	U1128, U1130
P. hostimense	Fossil-species			U1130
Protorosauria		Diapsida, Archosauromorpha		
Prototaxites	Enigmatic large organism, possibly a fungus, from the Silurian to Devonian	Fungi, Prototaxitaceae	15.10	U1524, U1529
Protozoa	Informal group for unicellular eukaryotes	Eukaryota		U0918
Protungulatum	Early Paleocene placental mammal from North America	Mammalia, Eutheria		
Protypotherium	An extinct Miocene mammal genus from South America	Notoungulata, Interatheriidae	3.3	
Prunus	Extant genus including plums, cherries, peaches, nectarines, apricots, and almonds	Angiosperms, eudicots, Rosaceae		U0163, U0434
P. cathybrownae	Eocene fossil-species from North America			U0434
P. cf. padus	European extant bird cherry		1.31	U0163
Psalixochlaenaceae	Extinct late Paleozoic fern family	Pteridophytes		
Psaronius	Carboniferous and Permian tree fern fossil-genus	Pteridophytes, Psaroniaccae	11.10; 11.12; 11.13; 11.14 12.5	U1116, U1117, U1119, U1120, U1121, U1126, U1330
Pseudoasterophyllites	Herbaceous halophytic early angiosperm	Plantae, angiosperms	5.10	U0507
P. cretaceous	Cretaceous fossil-species from Europe		5.10	U0507
Pseudodama	Deer relatives from the Miocene in Europe	Artiodactyla, Cervidae		
Pseudolarix	Extant genus, golden larch	Conifers, Pinaceae	2.2	
P. schmidtgenii	European fossil-species			
Pseudopanax	Extant genus, lancewoods	Angiosperms, eudicots, Araliaceae		
Pseudosporochnales	Extinct Devonian-to-Carboniferous order of ferns or fern-like plants	Pteridophytes?, Cladoxylopsids		U1519, U1521
Pseudosporochnus	Fossil-genus of fern-like plants	Pteridophytes?, Cladoxylopsids, Pseudosporochnales	11.16; 11.17	U1128, U1129, U1130
P. krejcii	Devonian fossil-species			U1129, U1130
Pseudovoltzia	Late Paleozoic conifer	Pinophytes, Majonicaceae		
Psilophyton	Devonian fossil-genus of vascular plants	Plantae, "early land plants", Trimerophyte or similar	15.9	U1503, U1507, U1524, U1528, U1530
P. crenulatum	Fossil-species		15.11	
P. dawsonii	Fossil-species			

(continued)

Scientific name	Vernacular name, description	Higher taxonomy	Figures, Boxes	Slides
Psilotum	Extant genus, whisk ferns	Pteridophytes, Psilotaceae		
Psittacosaurus	A small, bipedal ceratopsian	Ornithischia, Psittacosauridae	8.4	U0726, U0816, U0822
P. mongoliensis	Cretaceous species from Asia			U0822
Pteridophytes	Vascular plants reproducing by spores	Eukaryota, Plantae		
"Pteridosperms"	Extinct group of seed plants whose foliage superficially resembles that of ferns	Plantae, tracheophytes		
Pterocarya	Extant genus, wingnuts	Angiosperms, eudicots, Juglandaceae		
Pterophyllum	Mesozoic leaf fossil-genus	Gymnosperms, Bennettitales	6.9	U1328
Pterosauria	Flying archosaurs	Archosauria, Ornithodira		
Pteruchus	Arboreal seed fern pollen organ	"Pteridosperms", Corystospermaceae		U0651
Pterygotoids	Eurypterids	Arthropoda, Merostomata		
Ptilophyllum	Mesozoic leaf fossil-genus	Gymnosperms, Bennettitales	6.9	U0629
P. pecten	Jurassic fossil-species from Europe			U0629
Puccinellia	Extant genus, alkali grasses	Angiosperms, monocots, Poaceae	1.7	
Pueliodeae	Basal grasses	Angiosperms, monocots, Poaceae		
Pycnogonida	Sea spiders	Arthropoda, Chelicerata		
Pyracantha	Extant genus, firethorns	Angiosperms, eudicots, Rosaceae	1.26	
Pyrgopolon	Permian tube worm from Europe	Annelida, Serpulidae	9.16	U0927
P. gaiae			9.16	
Pyrotherium	Oligocene hippopotamus-sized browser from South America	Ungulata, Pyrotheriidae	3.3	
Qantassaurus	A Cretaceous bipedal, herbivorous dinosaur from Australia	Dinosauria, Ornithischia		U0812
Q. intrepidus				U0812
Quercus	Extant genus, oaks	Angiosperms, eudicots, Fagaceae	1.26; 2.16	U0163, U0219, U0237
Q. cf. *castaneifolia*	Chestnut-leaved oak			U0163
Q. cf. *cerris*	Turkey oak		1.31	U0163
Q. incana	Bluejack oak			
Q. lanuginosa	Pubescent oak			
Q. sect. *Heterobalanus*	Mountain oak			
Q. semecarpifolia	Kharsu Oak			
Q. subgen. *Cyclobalanopsis*	Ring-cupped oaks		1.26	
Q. subgen. *Lepidobalanus*	White oaks		1.26	
Questora	Late Paleozoic woody stem	Gymnosperms, early Medullosales		
Radiolaria	Protozoans with intricate mineral skeletons	Eukaryota, Rhizaria		U1029, U1448
Rajidae	Skates	Vertebrata, Chondrichthyes		
Ranunculaceae	Crowfoot family	Angiosperms, eudicots, Ranunculales		
Rebellatrix	Fast-swimming fish from the Triassic	Osteichthyes, Actinopterygii		U0751
Rhodophyta	Red algae	Eukaryota, Archaeplastida		
Regalecidae	Oarfish	Osteichthyes, Actinopterygii		
Rehderodendron	Extant small deciduous southeast Asian trees	Angiosperms, eudicots, Styracaceae		
Relimmia	Devonian fossil-genus	"Progymnosperms", Aneurophytales		
Renalia	Devonian vascular plant fossil-genus	"Early land plants"		U1510, U1526

(continued)

Scientific name	Vernacular name, description	Higher taxonomy	Figures, Boxes	Slides
Repenomamus	A cat-sized Cretaceous mammal from China	Mammalia, Gobiconodontidae		U0726
R. giganticus				U0726
Retimonocolpites	Triassic-to-Paleogene monosulcate-pollen fossil-genus	Angiosperms	5.14	U0510
Retiolitids	Silurian graptolites	Hemichordata, Graptolithina		
Rhabdodon	A basal iguanodontian from the Cretaceous of Europe	Ornithischia, Rhabdodontidae		
Rhabdopleura	Small, worm-shaped animals	Hemichordata, Graptolithina		
Rhacophyton	Late Devonian sprawling, fern-like shrub	Pteridophytes, Rhacophytales	15.3	U1511, U1513, U1516
Rhacopteris	Fossil-genus, probably of seed ferns	Pteridophytes		
Rhamnaceae	Buckthorn family	Angiosperms, eudicots, Rosales		
Rhodeopteridium	Probable pteridosperm fossil-genus	Gymnosperms		U1336
Rhododendron	Extant genus, rhododendrons	Angiosperms, eudicots, Ericaceae		
R. ponticum	Common rhododendron			
R. ponticum var. *sebinense*	Fossil-subspecies			
Rhomaleosauridae	Basal clade of Jurassic Plesiosauria	Sauropterygia, Plesiosauria		
Rhomaleosaurus	A short-necked plesiosaur	Plesiosauria, Rhomaleosauridae	7.10	
Rhynchosauria	Extinct Triassic group of diapsids	Diapsida, Archosauromorpha		
Rhynchippus	Extinct, hippo-like mammal from the Oligocene of South America	Notoungulata, Notohippidae	3.3	U1513
Rhynchocephalia	Clade including extant tuatara and extinct sphenodonts	Diapsida, Lepidosauria		
Rhynchonelliformea	Brachiopod clade	Animalia, Brachiopoda		
Rhynchosauroides	Late Triassic footprint (ichnofossil)	Animalia, Reptilia		
Rhynia	Devonian fossil-genus for one of the earliest land plants	Plantae, "early land plants", Rhyniopsida		U1505
Rhyniophytina	Early vascular plants	Eukaryota, Plantae, "early land plants",		U1505, U1510
Rhyniopsida	Early vascular plants	Plantae, "early land plants", Rhyniophytina		U1505
Ribes	Extant genus, including the edible currants	Angiosperms, eudicots, Saxifragales		U0163
R. cf. *orientale*	Extant species		1.31	U0163
Rigbya	Cupulate glossopterid reproductive structures	"Pteridosperms", Rigbyaceae	12.9	
Rigbyaceae	Family of glossopterid reproductive structures	Gymnosperms, Glossopteridales		
Rissikia	Mesozoic coniferous tree	Conifers, Podocarpaceae	6.17	
Rodentia	Rodents	Mammalia, Glires		
Rogersia	Fossil-leaf genus	Angiosperms		
R. angustifolia	Fossil-species	Angiosperms	5.18	
Rosa	Extant genus, roses	Angiosperms, eudicots, Rosaceae		
R. acicularis	Wild rose			
Rosaceae	Rose family	Angiosperms, eudicots, Rosales		
Rubiaceae	Coffee family	Angiosperms, eudicots, Gentianales	1.11	
Rubidgea	A large-bodied Permian apex-predator	Synapsida, Gorgonopsidae		
Rubus	Extant genus, raspberries	Angiosperms, eudicots, Rosaceae		
R. idaeus	European red raspberry		1.7	

(continued)

Scientific name	Vernacular name, description	Higher taxonomy	Figures, Boxes	Slides
Rugosa	Extinct Ordovician-to-Permian coral group	Cnidaria, Hexacorallia		U1421
Rutaceae	Lemon family	Angiosperms, eudicots, Sapindales		
Sabalites	Fossil-genus based on palm leaves	Angiosperms, monocots		
Sabellidae	Feather duster worms	Annelida, Polychaeta		
Sabiaceae	Flowering plant family from tropical to warm temperate regions of southern Asia and the Americas	Angiosperms, eudicots, Proteales		
Sagenopteris	Triassic-to-Cretaceous seed fern fossil-genus	"Pteridosperms", Caytoniaceae		U0630
S. phillipsii	Fossil-species			U0630
Salix	Extant genus, willows	Angiosperms, eudicots, Salicaceae	1.31	U0114, U0163
S. cf. *alba*	White willow		1.31	U0163
S. cf. *pseudomedemii*	Extant species		1.31	U0163
Salmonidae	Salmon family	Osteichthyes, Actinopterygii		
Salvinia	Floating water fern	Pteridophytes, Salviniaceae	11.5	U1111, U1112
S. cerebrata	Fossil-species of megaspores			U1111
S. reussii	Fossil-species based on whole plant remains		11.5	U1111, U1112
Salviniales	Floating water ferns	Pteridophytes		U1105
Sambucus	Extant genus, elder	Angiosperms, eudicots, Adoxaceae or Viburnaceae		U1110
Samotherium	An extinct Miocene-and-Pliocene giraffe relative from Eurasia and Africa	Artiodactyla, Giraffidae	3.5	
S. major	Mesodont, mixed-feeding giraffe		3.5	
Sansevieria	Extant genus, devil's tongues	Angiosperms, monocots, Asparagaceae		
S. ehrenbergii	East African wild sisal			
Sapindaceae	Soapberry family	Angiosperms, eudicots, Sapindales		
Sapindopsis	Cretaceous leaf fossil-genus	Angiosperms, eudicots, Platanaceae	5.19; 5.25; 5.26	U0511, U0518
S. anhouryi	Fossil-species		5.26	U0518
S. magnifolia	Fossil-species		5.19	
Saportanthus	Cretaceous fossil-genus of flowering plant from Portugal	Angiosperms, eumagnoliids, Laurales		U0505
S. parvus	Fossil-species		5.5	U0505
Sarcopterygii	Lobe-fin fishes	Vertebrata, Osteichthyes	7.12	U0749
Sargentodoxa	Extant genus, deciduous climbing shrub	Angiosperms, eudicots, Ranunculales	2.2	U0205, U0206
S. gossmannii	Fossil-species based on seeds		2.2	
Sargodon	Unhurried deep-bodied fish	Actinopterygii, Semionotiformes		
Sassafras	Extant genus, deciduous trees from North America and Asia	Angiosperms, eumagnoliids, Lauraceae		U0230
Saurichthys	Elongate ambush predators	Actinopterygii, Saurichthyiformes	7.13	U0751
Saurolophus	A large Cretaceous hadrosaurid from North America and Asia	Ornithischia, Hadrosauridae		U0818
S. angustirostris				U0818
Saurophaganax	A large Jurassic carnivorous dinosaur from North America	Theropoda, Allosauridae		
Sauropoda	Mesozoic long-necked dinosaurs	Dinosauria, Sauropodomorpha	6.8	U0833
Sauropodomorpha	Omnivorous or herbivorous, lizard-hipped dinosaurs	Archosauria, Dinosauria		U0835

(continued)

Scientific name	Vernacular name, description	Higher taxonomy	Figures, Boxes	Slides
Sauropterygia	Aquatic reptiles	Vertebrata, Diapsida		U0740
Savannasaurus	A long-necked Cretaceous dinosaur from Australia	Sauropoda, Titanosauria		U0810
S. elliottorum				U0810
Sawdonia	Fossil-genus, zosterophyll, ranging from Early Devonian to late Middle/early Late Devonian	"Early land plants", Zosterophyllophytes, Sawdoniales		U1527
S. ornata	Fossil-species			
Scandianthus	A fossil-genus of flowers	Angiosperms, Saxifragales		
Scarrittia	Extinct Oligocene genus of hoofed mammal	Notoungulata, Leontiniidae	3.3	
Schefflera	Trees, shrubs or lianas with woody stems and palmately compound leaves	Angiosperms, eudicots, Araliaceae	1.11	
Schima	Extant genus, evergreen members of the tea family	Angiosperms, eudicots, Theaceae		
S. wallichii	Extant species			
Schizacaceae	Curly grass fern family	Pteridophytes, Schizaceae		U1107, U1113
Schoenoplectus	Extant genus, club-rush	Angiosperms, monocots, Cyperaceae		
Sciadophyton	Devonian embryophyte of unknown affinity, possible gametophyte phase	"Early land plants", Rhyniophytina, Rhyniaceae		U1524
Sciadopityaceae	Umbrella-pine family	Conifers, Pinaceae		
Sciadopitys	Extant genus, Japanese umbrella-pine	Conifers, Sciadopityaceae		U0207
Scincomorpha	Skink family	Diapsida, Squamata		
Scleractinia	Stony corals	Cnidaria, Hexacorallia		
Scolecopteris	Late Paleozoic fern fossil-genus	Pteridophytes, Marattiaceae		
Scolosaurus	Cretaceous quadrupedal, armored ankylosaur from North America	Ornithischia, Ankylosauridae	7.6	
Scorpiones	Scorpions	Arthropoda, Arachnida	13.9	
Segnosaurus	Cretaceous bipedal, herbivorous dinosaur from Mongolia	Theropoda, Therizinosauroidea	8.4	U0816
Selachimorpha	Sharks	Vertebrata, Chondrichthyes		
Selaginella	Extant genus, spikemosses	Lycopsids, Selaginellaceae		
S. remotifolia	Extant species			
Selaginellales		Plantae, lycopsids		U1202
Semecarpus	Extant genus, varnish tree	Angiosperms, eudicots, Anacardiaceae		
Senftenbergia	Paleozoic fossil-genus of seed-fern fructifications	"Pteridosperms", Tedeleaceae		U1107
S. plumosa	Carboniferous fossil-species			U1107
Sequoia	Extant genus, redwood	Conifers, Cupressaceae		U0165
Serpentes	Snakes	Diapsida, Squamata		
Serpula	Calcareous tubeworm	Polychaeta, Serpulidae	9.16	U0927
S. distefanoi	Extinct Permian species		9.16	
Serpulidae	Tube worms	Annelida, Polychaeta		U0928
Serrulacaulis	Devonian fossil-genus of land plants with branching axes and sporangia	Plantae, "early land plants", Zosterophyllophytes		
Sespia	Extinct Oligocene genus of oreodont from North America	Artiodactyla, Merycoidodontidae		
Shamosuchus	Cretaceous crocodyliform from Mongolia	Crocodyliformes, Paralligatoridae		
Shunosaurus	A long-necked Jurassic dinosaur from Asia	Sauropoda, Eusauropoda		

(continued)

Scientific name	Vernacular name, description	Higher taxonomy	Figures, Boxes	Slides
Sigillaria	Carboniferous-and-Permian fossil-genus of spore-bearing, arborescent lycopsids	Lycopsids, Sigillariaceae	11.10	U1116, U1119, U1316, U1317, U1328, U1330
S. ichthyolepis	Fossil-species			
S. sternbergii	Fossil-species			
Sigillariaceae	Spore-bearing, arborescent lycopsids	Lycopodiopsida, Lepidodendrales		U1204
Silesauridae	Clade of Triassic quadrupedal animals closely related to dinosaurs	Diapsida, Archosauria		
Siluriformes	Catfishes	Osteichthyes, Actinopterygii		
Sinomenium	Extant genus, moonseed	Angiosperms, eudicots, Menispermaceae	2.2	
S. cantalense	Fossil-species based on fruit remains		2.2	
Smilodon	Pleistocene-and-Holocene saber-toothed cats	Carnivora, Felidae		
Solenites	A fossil-genus of seed plants	Gymnosperms, Czekanowskiales	6.10	
S. vimineus	Fossil-species			
Sollasia	Carboniferous-to-Triassic extinct desmosponge genus	Animalia, Porifera		
Sonapteris	Carboniferous fern fossil-genus	Pteridophytes, Botryopteridaceae		
S. bekii	Fossil-species			
Sonchus	Extant genus, sow thistles	Angiosperms, eudicots, Asteraceae	1.7	
S. arvensis	Perennial sow thistle			
Sorbus	Extant genus, rowans	Angiosperms, eudicots, Rosaceae		
S. aucuparia	European rowan			
Spencerites	Herbaceous Carboniferous lycopsid fossil-genus	Lycophyta	11.14	U1325
S. leismanii	Fossil-species			
Sphaerostoma	Parasitic flatworm	Trematoda, Opecoelidae	12.6	
Sphenobaiera	Permian-to-Cretaceous leaf fossil-genus	Ginkgophytes, Ginkgoales	6.10; 12.11	U0649, U0652, U1220
Sphenophyllales	Climbing horsetails	Pteridophytes, sphenophytes		
Sphenophyllum	Late Paleozoic fossil-genus	Pteridophytes, sphenophytes, Sphenophyllaceae	11.14	U1328
Sphenophytes	Horsetails	Pteridophytes, sphenophytes		U0642, U0907, U1304, U1345, U1347
Sphenopteridium	Fossil-genus of late Paleozoic seed-fern foliage	Gymnosperms, "pteridosperms"		
Sphenopteris	Fossil-genus of late Paleozoic seed-fern foliage	"Pteridosperms"	11.7; 11.14	U0616, U1113, U1328
S. cirrhifolia	Fossil-species			
S. gruenbachiana	Fossil-species		11.7	U1113
S. mixta	Fossil-species		11.14	
S. pottsvillea	Fossil-species			
Spiraea	Extant genus, member of the rose family	Angiosperms, eudicots, Rosaceae	1.31	U0163
S. cf. *crenata*	Extant species			U0163
Spongiophyton	Devonian thallose fossil, affinity unknown	Eukaryota, Plantae		
Sporogonites	Possible Devonian bryophyte	Plantae, bryophytes		U1524
Squamata	Lizards, snakes, and worm lizards	Diapsida, Lepidosauria		
Stamnostoma	Carboniferous seed-fern fossil-genus	"Pteridosperms", Moresnetiaceae		
Staphylea	Extant genus, bladdernuts	Angiosperms, eudicots, Staphyleaceae		

(continued)

Scientific name	Vernacular name, description	Higher taxonomy	Figures, Boxes	Slides
Stegosaurus	A quadrupedal dinosaur with plates along its back	Ornithischia, Stegosauridae	7.1; 7.6; 7.7	U0823, U0825
S. stenops	Jurassic species of *Stegosaurus*		7.7	
Steinmannia		Angiosperms, monocots, Asparagales		
Stellaria	Extant genus, starworts	Angiosperms, eudicots, Caryophyllaceae	1.7	
S. jacutica	Extant species			
Stephania	Extant genus, herbaceous perennial vines	Angiosperms, eudicots, Menispermaceae	1.26	
Stephanorhinus	Extinct Pleistocene genus of Eurasian rhino	Perissodactyla, Rhinocerotidae		
S. cf. hemitoechus	Narrow-nosed rhinoceros			
Stephanospermum	Fossil-genus of seed-fern ovules	"Pteridosperms", Medullosales		
Stewartia	Extant genus, mountain camellia	Angiosperms, eudicots, Theaceae		
S. monadelpha	Extant species			
Sthenurus	Extinct Pliocene-to-Pleistocene genus of short-faced giant wallaby	Marsupialia, Macropodidae	1.13	
Stigmaria	Fossil- genus for underground rooting structures of Carboniferous lycopsid trees	Lycopsids, Lepidodendraceae		U1203
Stigmosphaerostylus	A radiolarian	Protista, Sarcomastigophora, Radiolaria	14.12	U1448
S. vishnevskayae			14.12	U1448
Stockmansella	Devonian fossil-genus of plants	Plantae, Rhyniopsida		U1505
Stokesosaurus	Jurassic early relative of *Tyrannosaurus* from North America	Theropoda, Tyrannosauroidea		
Stomatograptus	A graptolite	Hemichordata, Graptolithina	14.12	U1449
S. grandis			14.14	U1449
Stopesia	Fossil-genus based on seeds	Trimeniaceae		
Stromatoporoidea	Clade of aquatic, reef-building Ordovician-to-Devonian sponges	Animalia, Porifera		
Struthio	Ostrich	Aves, Struthionidae		
Struthiosaurus	A Cretaceous nodosaurid ankylosaur from Europe	Ornithischia, Nodosauridae		
Styracaceae	Snowbell family	Angiosperms, eudicots, Ericales		
Styrax	Extant genus, snowbells	Angiosperms, eudicots, Styracaceae		
Subholostei	Small, predatory ray-fin fishes	Vertebrata, Osteichthyes		U0749
Supersaurus	A long-necked Jurassic dinosaur from Europe and North America	Sauropoda, Diplodocidae		
Sus	Pigs	Artiodactyla, Suidae		
S. barbatus	Bearded pig		3.2	
S. scrofa	Wild boar			
Sutcliffia	Carboniferous fossil of seed-fern stem genus	"Pteridosperms", Medullosales		
Suuwassea	A long-necked Jurassic dinosaur from North America	Sauropoda, Flagellicaudata		
Symplocos	Extant genus, shrubs and trees with white or yellow flowers	Angiosperms, eudicots, Ericales	1.37; 2.2	U0205, U0206
S. cocincinnensis	Extant species		1.37	
S. schereri	Fossil-species		2.2	
Synapsida	Clade including mammals and their ancestors	Tetrapoda, Amniota		

(continued)

Scientific name	Vernacular name, description	Higher taxonomy	Figures, Boxes	Slides
Synchysidendron	Fossil-genus of arboreal club mosses	Lycopsids, Isoetales (lepidodendrids)		U1322
S. resinosum	Fossil-species			U1322
Tabulata	Exticnt group of tabulate Ordovician-to-Permian corals	Cnidaria, Anthozoa		
Taeniocrada	Devonian fossil-genus of plants with leafless, flattened stems	Plantae, Rhyniophyta		
Taeniopteris	Late Paleozoic to Mesozoic genus of fossil ferns or cycads	Incertae sedis but probably leaf of seed plant		U1328
Talarurus	Cretaceous ankylosaur from Mongolia	Ornithischia, Ankylosauridae		U0821
T. plicatospineus				U0821
Tanystropheidae	Extinct group of Triassic reptiles with long necks	Diapsida, Archosauromorpha		
Tanystropheus	An extinct Triassic reptile with an extremely long neck from Eurasia	Archosauromorpha, Tanystropheidae	9.6; 10.9	U0912, U1024
T. longobardicus			9.6	U0912
Tapiridae	Tapir family	Mammalia, Perissodactyla		
Tapirus	Tapir	Perissodactyla, Tapiridae		
Tarbosaurus	A close Asian relative of Tyrannosaurus from the Cretaceous	Theropoda, Tyrannosauroidea		U0818
Tarchia	An armored Cretaceous ankylosaur from Mongolia	Ornithischia, Ankylosauridae	7.5	
Taxodium	Extant genus, bald cypresses	Conifers, Cupressaceaeiaceae		U0174, U0237
Taxus	Extant genus, yews	Conifers, Taxaceae		
T. wallichiana	Wallich yew		1.26	
Tayassu		Artiodactyla, Tayassuidae	3.2	
T. pecari	White-lipped peccary		3.2	
Tazoudasaurus	A long-necked Jurassic dinosaur from North Africa	Dinosauria, Sauropoda		
Teinolophos	An extinct Cretaceous egg-laying mammal	Mammalia, Monotremata		U0812
T. trusleri				U0812
Teixeiraea	Early Cretaceous fossil-genus of flowers	Angiosperms, angiosperms, eudicots		
Telemachus	Fossil-genus, Seed cone of a woody conifer whose mother "whole plant" has been reconstructed	Conifers, Voltziaceae		U0652
Teleoceras	Extinct Miocene-to-Pliocene genus of grazing rhinoceros	Perissodactyla, Rhinocerotidae		U0238
Teleostei	Subclade of ray-fin fishes	Osteichthyes, Neopterygii		U0747
Teleostomorpha	Early members of teleost fishes	Osteichthyes, Neopterygii		
Temnospondyli	Extinct clade of early, salamander-like tetrapods	Vertebrata, Tetrapoda		
Temperoceras	A genus of Ordovician-to-Devonian orthoconic nautiloid	Nautiloidea, Geisonoceratidae	14.3	U1433
T. ludense	Silurian species		14.3	U1433
Testudines	Turtles	Tetrapoda, Amniota	7.11	
Tetracentron	Extant tree growing at mid- to high-latitudes in Asia	Angiosperms, eudicots, Trochodendraceae		
Tetraclinis	Extant genus, sictus tree	Conifers, Cupressaceae		
Tetrapoda	Four-limbed animals	Vertebrata, Osteichthyes		
Tetrastigma	Extant genus, lianas	Angiosperms, eudicots, Vitaceae	2.2	U0205, U0206
T. chandlerae	Fossil-species			
Tetraxylopteris	Devonian fossil-genus of vascular plants	"Progymnosperms", Aneurophytales		U1520
Thalattosauria	Triassic marine reptiles	Tetrapoda, Diapsida		

(continued)

Scientific name	Vernacular name, description	Higher taxonomy	Figures, Boxes	Slides
Thaumatopteris	Mesozoic fern fossil-genus	Pteridophytes, Dipteridaceae		
T. brauniana	Fossil-species			
Theaceae	Tea family	Angiosperms, eudicots, Ericales		
Thecodontosaurus	Triassic bipedal, herbivorous dinosaur	Dinosauria, Sauropodomorpha	7.1	
Thelodonti	Ordovician-to-Devonian clade of extinct jawless fishes with distinctive scales	Animalia, Vertebrata		
Therapsida	Synapsic clade including modern mammals	Amniota, Synapsida		
Theria	Subclade of mammals	Synapsida, Mammalia		
Therizinosaurus	Cretaceous herbivore theropod with enormous claws from Asia	Theropoda, Therizinosauroidea	7.1	
Theropoda	Clade including all carnivorous dinosaurs and modern birds	Archosauria, Dinosauria		
Thoatherium	An Miocene extinct, camel/llama-like mammals from Argentina	Ungulata, Proterotheriidae	3.3	
Thomashuxleyia	Extinct Eocene genus of notoungulate from South America	Notoungulata, Isotemnidae	3.3	
Thoracopterus	Triassic large-finned over-water gliding fishes	Actinopterygii, Perleidiformes		U0751
Thrombolites	Clotted, cyanobacteria mats			
Thucydia	Late Paleozoic fossil-genus of woody gymnosperms related to conifers	Pinophytes, Thucydiaceae		
Thunnini	Tuna tribe	Actinopterygii, Scombridae		
Thylacoleo	Marsupial lion	Marsupialia, Thylacoleonidae		
T. carnifex				
Thymus	Extant genus, thymes	Angiosperms, eudicots, Lamiaceae		U0163
T. cf. kotschyanus	Extant species		1.31	U0163
Thyreophora	Dinosaur clade including stegosaurs and ankylosaurus	Dinosauria, Ornithischia		
Ticinosuchus	Extinct Triassic genus of terrestrial, crocodile-like archosaur from Europe	Archosauromorpha, Pseudosuchia		
Tiktaalik	Devonian transitional vertebrate between classic "fish" and tetrapods	Vertebrata, Sarcopterygii		
Tilia	Extant genus, lime trees	Angiosperms, eudicots, Malvaceae		U0163
T. cf. begoniifolia	Extant species		1.31	U0163, U1110
Tingia	Fossil-genus of late Paleozoic trees	"Early land plants", "Noeggerathiales"		U1328, U1330
T. unita	Fossil-species			
Titanosauria	Subclade of Titanosauriformes	Sauropoda, Neosauropoda	8.2	
Titanosauriformes	Jurassic and Cretaceous long-necked dinosaurs	Sauropoda, Neosauropoda		
Todites	Mesozoic fern fossil-genus	Pteridophytes, Osmundaceae		U0632
T. princeps	Fossil-species		6.11	
T. thomasii	Fossil-species		6.11	
T. williamsonii	Jurassic fossil-species of Europe		6.11	U0632
Tomcatia	Cretaceous fossil-genus of seeds with features found in three gymnosperm orders: Bennettitales-Erdtmanithecales-Gnetales	Plantae, spermatophytes, gymnosperms		
Tomiodendron	Fossil-genus of arboreal club mosses	Lycopsids, Isoetaceae		

(continued)

Scientific name	Vernacular name, description	Higher taxonomy	Figures, Boxes	Slides
Torvosaurus	A large Jurassic carnivorous dinosaur from Europe and North America	Theropoda, Megalosauridae		
Toxodon	Miocene-to-Holocene extinct genus of South American mammals	Notoungulata, Toxodontidae	3.3	
Tragoportax	Extinct Miocene antelope	Artiodactyla, Bovidae	3.5	
T. amalthea	Brachydont-mesodont, mixed-feeding antelope		3.5	
Tragulidae	Mouse deer family	Mammalia, Artiodactyla	3.5	
Tremataspis	Silurian extinct genus of jawless fish from Estonia	Vertebrata, Osteostraci		U1442
Tribosphenida	An early mammal clade	Synapsida, Mammalia		
Triceratops	Cretaceous quadrupedal dinosaur with horns and a frill from North America	Ornithischia, Ceratopsidae	7.1; 7.6	U0601
Trigonocarpus	Carboniferous fossil-genus of seed-fern ovules	"Pteridosperms", Medullosales	12.7	
T. noeggerathii	Fossil-species		12.7	
Trigonotarbida	Silurian-to-Permian extinct clade of arachnids	Arthropoda, Arachnida		
Trilobita	Trilobites	Animalia, Arthropoda		
Trimenia	Extant small tree genus in southeast Asia and Australasia	Angiosperms		
Trimerophytina	Early vascular plant group	Plantae, "early land plants", tracheophytes		U1507, U1510, U1520
Trimerophyton	Devonian vascular plant with trifurcate lateral branches	Plantae, Trimerophytina		
T. robustius	Fossil-species			
Trionychidae	Softshell turtles	Testudines, Cryptodira		
Triphyllopteris	Fossil-genus of possible seed-fern foliage	Gymnosperms?		U1308
T. boliviana	Fossil-species			U1308
Trochodendraceae	Flowering plants with secondary xylem without vessel elements	Angiosperms, eudicots, Trochodendrales		
Trochodendrales	Flowering plants with secondary xylem without vessel elements	Angiosperms, angiosperms, eudicots		
Trochodendroides	Fossil-genus	Angiosperms, eudicots, Cercidiphyllaceae	5.12	U0509
Trochodendroids	Informal group of plants related to extant *Trochodendron*	Angiosperms		U0509
Troodon	Small, feathered Cretaceous dinosaur from North America	Theropoda, Troodontidae		
Troodontidae	Small, Jurassic-to-Cretaceous feathered dinosaurs	Theropoda, Eumaniraptora	7.4	U0708
Tropidogyne	A pentamerous flower	Angiosperms, eudicots, Cunoniaceae		
Tsagandelta	Extinct, Cretaceous carnivorous mammal from Asia	Mammalia, Deltatheridiidae		
Tsuga	Extant genus, hemlocks	Conifers, Pinaceae	1.26; 1.37	U0207, U0219
T. dumosa	Himalayan hemlock		1.26	
Tubicaulis	Fossil-genus of small climbing Pennsylvanian-age ferns	Pteridophytes		U1123
T. bertheri	Fossil-species			
T. grandeuryi	Fossil-species			
T. solenites	Fossil-species			U1123
Tucanopollis	Fossil-genus based on pollen, probably from an extinct plant associated with both *Ceratophyllum* and Chloranthaceae			U0507

(continued)

Scientific name	Vernacular name, description	Higher taxonomy	Figures, Boxes	Slides
Turpinia	Muttonwoods, an extant genus of trees and shrubs	Angiosperms, eudicots, Staphyleaceae		
T. ettingshausenii	Fossil-species		2.2	
Tyrannosauroidea	Clade including *Tyrannosaurus*	Theropoda, Coelurosauria	7.4	
Tyrannosaurus	Cretaceous carnivorous dinosaur	Theropoda, Tyrannosauroidea	8.4	U0706, U0816, U0818
Ullmannia	Fossil-genus of Permian conifers	Conifers		
Ulmaceae	Elm family	Angiosperms, eudicots, Rosales		U1106
Ulmus	Extant genus, elms	Angiosperms, eudicots, Ulmaceae		U0219, U1110
Ulodendraceae	Extinct late Paleozoic club moss family	Tracheophytes, lycopsids		U1204
Umkomasia	Triassic arboreal seed fern	"Pteridosperms", Corystospermaceae		U0651, U0652
Ungulata	Hoofed mammals	Synapsida, Mammalia	3.3	
Unionidae	Freshwater mussels	Mollusca, Bivalvia		
Unionites	Permian-to-Triassic bivalve	Mollusca, Bivalvia	10.8	U1023
Urodela	Salamanders	Vertebrata, Tetrapoda		
Ursus	Bear	Carnivora, Ursidae		
U. etruscus	Extinct Pliocene-to- Pleistocene Etruscan bear			
Urtica	Extant genus, nettles	Angiosperms, eudicots, Urticaceae	1.7	
U. dioica	Common nettle; a nitrophilous pioneer plant			
Urticaceae	Nettle family	Angiosperms, eudicots, Rosales		
Utrechtia	Late Paleozoic fossil-genus of conifer-like trees	Pinophytes, Voltziaceae		
Vaccinium	Extant genus, shrubs or dwarf shrubs in the heath family	Angiosperms, eudicots, Ericaceae	1.7	
V. vitis-idaea	Partridgeberry			
Valecarpus	Fossil-genus based on fruits	Angiosperms		
Valeriana	Extant genus, valerians	Angiosperms, eudicots, Caprifoliaceae		
Varanus	Monitor lizards	Squamata, Varanidae		
V. komodoensis	Komodo dragon			
V. priscus	Extinct Pleistocene species, the largest true (monitor) lizard			
Velociraptor	Cretaceous small carnivorous dinosaur	Theropoda, Dromaeosauridae	8.5	U0706, U0819
V. mongoliensis				U0819
Veronica	Extant genus, speedwells	Angiosperms, eudicots, Plantaginaceae		
Vertebraria	Rooting structures of a Permian seed fern	Gymnosperms, Glossopteridales		
Vesicaspora	Fossil-genus based on pollen of seed ferns	"Pteridosperms", Callistophytaceae		
Viburnum	Extant genus, cranberrybushes	Angiosperms, eudicots, Adoxaceae or Viburnaceae		
Vitaceae	Grape family	Angiosperms, eudicots, Vitales		
Vitis	Extant genus, grapes	Angiosperms, eudicots, Vitaceae		
Volkheimeria	Jurassic long-necked dinosaur from South America	Dinosauria, Sauropoda	8.8	U0832, U0833
Voltzia	Late Paleozoic fossil-genus of woody gymnosperms related to conifers	Plantae, pinophytes		U0913
V. heterophylla	Fossil-species of woody gymnosperms related to conifers	Plantae, pinophytes		U0913

(continued)

Scientific name	Vernacular name, description	Higher taxonomy	Figures, Boxes	Slides
Voltziales	Late Paleozoic extinct order of woody gymnosperms related to conifers	Plantae, pinophytes		U0652
Vulpes	Fox	Carnivora, Canidae		
V. vulpes	Red fox			
Walchia	Carboniferous-to-Permian conifer fossil-genus from North America and Europe	Conifers, Voltziaceae	12.13	
W. piniformis	Fossil-species		12.13	
Watsonulus	Triassic early neopterygian	Osteichthyes, Actinopterygii	7.13	
W. eugnathoides				
Wattieza	Fossil-genus of Devonian tree related to modern ferns and horsetails	Pteridophytes, Pseudosporochnales		U1518, U1519
Weichselia	Fossil-genus of Jurassic and Cretaceous salt-marsh ferns	Pteridophytes, Matoniaceae		
Weinmannia	Woody angiosperm	Angiosperms, eudicots, Cunoniaceae		
Welwitschiaceae	Family with only one southwestern African genus	Gnetophytes, Welwitschiales		
Winteraceae	Tropical tree-and-shrub family	Angiosperms, eumagnoliids, Canellales		
Wintonotitan	Cretaceous long-necked dinosaur from Australia	Sauropoda, Titanosauriformes		U0810
W. wattsi				U0810
Woodwardia	Extant genus, chain ferns	Pteridophytes, Blechnaceae		U1111
W. muensteriana	Fossil-species			U1111
W. virginica	Virginia chain fern			
Xenarthra	Anteaters, tree sloths, and armadillos	Mammalia, Placentalia		
Xenopsaria	Jurassic-to-Cretaceous Plesiosauria subclade	Sauropterygia, Plesiosauroidea		
Yamaceratops	Cretaceous early ceratopsian from Mongolia	Ornithischia, Auroraceratopsidae		U0820
Y. dorngobiensis				U0820
Zamites	Triassic-to-Eocene fossil-genus of trees	Gymnosperms, Bennettitales	6.9	U0629, U0645
Z. gigas	Jurassic fossil-species from Europe			U0629
Z. powellii	Triassic fossil-species from North America			U0645
Zanthoxylum	Deciduous and evergreen trees and shrubs	Angiosperms, eudicots, Rutaceae	2.2	
Z. cf. ailanthiforme	Fossil-species			
Zeilleropteris	Permian seed fern fossil-genus	"Pteridosperms", Gigantopteridales		
Zelkova	Extant genus, European and southwest and eastern Asian deciduous tree	Angiosperms, eudicots, Ulmaceae		
Zelleropteris	Permian leaf genus	Gymnosperms, Gigantopteridales	12.8	
Zingiberales	Gingers and allies	Angiosperms, monocots		
Zingiberopsis	Fossil-genus of ginger-like plants	Angiosperms, monocots		
Zizyphoides	Paleocene leaf fossil-genus	Angiosperms, eudicots, Trochodendraceae		
Zizyphus	Extant genus, member of the buckthorn family	Angiosperms, eudicots, Rhamnaceae	1.26	
Zlatkocarpus	Cretaceous fossil-genus of fruits	Angiosperms, Chloranthoids	5.9	U0506
Z. brnikensis	Fossil-species		5.9	
Zlivifructus	Cretaceous fossil-genus of fruits	Angiosperms, Fagales		

(continued)

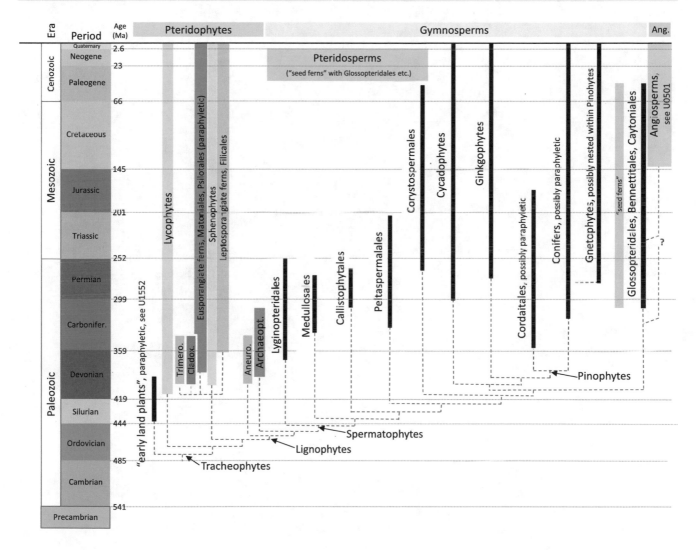

Synthetic chart of the terrestrial plant phylogeny adopted in this work, showing the position of the main groups named in the index. Compiled by R.A. Gastaldo, P. Gensel, E. Kustatscher, E. Martinetto. See slide U1552 in Chap. 15 for details on "early land plants" and Silurian-Devonian phylogeny. See slide U0501 in Chap. 5 for phylogeny of the main angiosperm clades.

Index

Printed in the United States
by Baker & Taylor Publisher Services